AF581902

Impaired Mitochondrial Bioenergetics under Pathological Conditions

Impaired Mitochondrial Bioenergetics under Pathological Conditions

Editors

Salvatore Nesci
Giorgio Lenaz

MDPI • Basel • Beijing • Wuhan • Barcelona • Belgrade • Manchester • Tokyo • Cluj • Tianjin

Editors
Salvatore Nesci
Veterinary Medical Sciences
University of Bologna
Ozzano Emilia
Italy

Giorgio Lenaz
Biomedical and Neuromotor Sciences
University of Bologna
Bologna
Italy

Editorial Office
MDPI
St. Alban-Anlage 66
4052 Basel, Switzerland

This is a reprint of articles from the Special Issue published online in the open access journal *Life* (ISSN 2075-1729) (available at: www.mdpi.com/journal/life/special_issues/Impaired_mitochondrial).

For citation purposes, cite each article independently as indicated on the article page online and as indicated below:

LastName, A.A.; LastName, B.B.; LastName, C.C. Article Title. *Journal Name* **Year**, *Volume Number*, Page Range.

ISBN 978-3-0365-4648-3 (Hbk)
ISBN 978-3-0365-4647-6 (PDF)

Contents

 life

MDPI

Editorial

Impaired Mitochondrial Bioenergetics under Pathological Conditions

Salvatore Nesci [1,*] and Giorgio Lenaz [2,*]

[1] Department of Veterinary Medical Sciences, Alma Mater Studiorum University of Bologna, Via Tolara di Sopra, 50, 40064 Ozzano Emilia, BO, Italy
[2] Department of Biomedical and Neuromotor Sciences, Alma Mater Studiorum University of Bologna, Via Massarenti 9, Pad 11, 40138 Bologna, BO, Italy
* Correspondence: salvatore.nesci@unibo.it (S.N.); giorgio.lenaz@unibo.it (G.L.)

Citation: Nesci, S.; Lenaz, G. Impaired Mitochondrial Bioenergetics under Pathological Conditions. *Life* **2022**, *12*, 205. https://doi.org/10.3390/life12020205

Received: 18 January 2022
Accepted: 27 January 2022
Published: 29 January 2022

Mitochondria are the powerhouses of cells; however, mitochondrial dysfunction causes energy depletion and cell death in various diseases. The altered oxidative phosphorylation and ion homeostasis are associated with reactive oxygen species (ROS) production, resulting from the disassembly of respiratory supercomplexes and electron transfer chain disruption [1]. In pathological conditions, the dysregulation of mitochondrial homeostasis promotes Ca^{2+} overload in the matrix and ROS accumulation, which induce the mitochondrial permeability transition and pore formation responsible for morphological mitochondrial changes [2] linked to membrane dynamics, and ultimately, cell death.

The Special Issue, entitled "Impaired Mitochondrial Bioenergetics under Pathological Conditions", includes 19 contributions, among which 5 are research articles and 14 are reviews. This research topic aimed to shed further light on the role of altered bioenergetics in diseases, a paramount argument in both basic and medical research. The energy transduction system has a supramolecular structure in order to better meet the bioenergetic demand of cellular activities.

Herein, the biochemical machinery alterations of oxidative phosphorylation (OXPHOS), hosted by the inner mitochondrial membrane, responsible for mitochondrial dysfunctions, and involved in several neurodegenerative and age-related diseases, have been discussed by considering: (*i*) the physio(patho)logical role of respiratory supercomplexes and (*ii*) the bifunctional features of ATP synthase as a life and death enzyme [3]. The structures of the mitochondrial electron transport chain in different supercomplexes (SCs) are all characterized by the presence of Complex III. Rugolo et al. highlight the genetic alterations that hinder the assembly of Complex III, which cause noticeable perturbation of the SCs architecture associated with several significant metabolic disturbances [4]. In mammalian mitochondria, Complex I is the largest respiratory complex, with 31 additional supernumerary subunits. Even if the accessory subunits are not necessary for enzyme activity, Zickermann's group and Vik's group discuss the mutations in these subunits which modify the Complex I assembly and lead to detrimental enzyme activity associated with many mitochondrial disease states [5,6]. A novel mutation of human mtDNA in the ATP6 gene has detrimental consequences on ATP synthase on yeast, with a pathogenicity resembling that which compromises human health [7]. However, altered expression of nuclear and mitochondrial genes of the ATP synthase is associated with neurodegenerative diseases and other age-related diseases in humans [8]. Indeed, Pamplona's group report that ageing and the alteration of mitochondrial oxidative stress and lipid metabolism represent a positive loop in the progression of Alzheimer's diseases [9]. Neurodegeneration and primary mitochondrial diseases, triggered by a defective respiratory chain, comprise impaired respiratory complex assembly and function. However, Needs et al. [10] consider that mitochondrial protein import is important for normal organelle physiology and observed an interlink with the regulation of respiratory complex assembly and function in human diseases. Moreover, in the paper by Franco et al., the human mitochondrial disorders that affect the OXPHOS activity have been suggested to be studied by using the *Saccharomyces cerevisiae* as a model to gain an understanding of the underlying molecular consequences of pathogenic mutations in mitochondria [11].

Ageing-related diseases are supported by mitochondrial-mediated regulated and programmed cell death in the myocardial structure, which promotes the development of heart failure [12]. Pinton's group examines the disequilibrium in mitochondrial bioenergetics that leads to unbalance of the energy consumption/generation with increasing age and the imbalance in mitochondrial dynamics as impaired conditions in the failing heart [13]. Structural integrity and the coordination of mitochondrial enzyme activity in energy metabolism and redox homeostasis is also ensured by mitochondrial phospholipid cardiolipin (CL). However, diabetic cardiomyopathy and heart failure are cardiomyopathies described in Barth syndrome, a dysfunctional CL disease [14].

In metabolic diseases arising with age, mitochondrial ROS production might have a key role in inflammaging. Exercise induces improvements in mitochondrial function limiting ROS production by reducing inflammation and ensuring an adaptation of immune cell metabolism to fight against virus infection [15]. Indeed, Elesela and Lukacs consider that viruses exploit mitochondrial dynamics to induce negative impacts on the cell metabolism and promote viral life processes during infection [16]. The mitochondrial shaping and mitochondrial ROS production and release provide an intricate vicious circle, subject to maintaining physiological states or contributing to pathological conditions [17].

The accumulated evidence of cell bioenergetics of different tissues highlights a correlation with the pathogenesis of type 2 diabetes. In particular, mitochondrial dysfunction is related to the coupling mechanism of metabolism and the exocytosis of insulin and glucagon in diabete [18]. An increased risk of developing diabetes mellitus is found in patients with the m.3243A>G mutation, and excessive palmitate showed a negative effect on respiratory rates, promoting insulin resistance [19]. Therefore, measurement of mitochondrial bioenergetics by a new frozen respirometry protocol utilizing non-invasive analysis might facilitate the clinical monitoring of mitochondrial function in samples collected at remote sites [20]. In addition, mitochondrial oxidation and substrate oxidation measurement by in vitro myopathy assays was also considered a prognostic method for use on equine muscle biopsies [21].

In summary, studies on impaired mitochondrial bioenergetics in pathology could provide molecular tools to counteract diseases associated with mitochondrial dysfunctions.

Author Contributions: Writing—original draft preparation, S.N.; supervision, writing—review and editing, G.L. All authors have read and agreed to the published version of the manuscript.

Funding: This research received no external funding.

Data Availability Statement: Not applicable.

Conflicts of Interest: The authors declare no conflict of interest.

References

1. Maranzana, E.; Barbero, G.; Falasca, A.I.; Lenaz, G.; Genova, M.L. mitochondrial respiratory supercomplex association limits production of reactive oxygen species from complex I. *Antioxid. Redox Signal.* **2013**, *19*, 1469–1480. [CrossRef] [PubMed]
2. Nesci, S. Mitochondrial permeability transition, F1FO-ATPase and calcium: An enigmatic triangle. *EMBO Rep.* **2017**, *18*, 1265–1267. [CrossRef] [PubMed]
3. Nesci, S.; Trombetti, F.; Pagliarani, A.; Ventrella, V.; Algieri, C.; Tioli, G.; Lenaz, G. Molecular and supramolecular structure of the mitochondrial oxidative phosphorylation system: Implications for pathology. *Life* **2021**, *11*, 242. [CrossRef]
4. Rugolo, M.; Zanna, C.; Ghelli, A.M. Organization of the respiratory supercomplexes in cells with defective complex III: Structural features and metabolic consequences. *Life* **2021**, *11*, 351. [CrossRef] [PubMed]
5. Kahlhöfer, F.; Gansen, M.; Zickermann, V. Accessory subunits of the matrix arm of mitochondrial complex I with a focus on subunit NDUFS4 and its role in complex I function and assembly. *Life* **2021**, *11*, 455. [CrossRef]
6. Dang, Q.-C.L.; Phan, D.H.; Johnson, A.N.; Pasapuleti, M.; Alkhaldi, H.A.; Zhang, F.; Vik, S.B. Analysis of human mutations in the supernumerary subunits of complex i. *Life* **2020**, *10*, 296. [CrossRef]

7. Ding, Q.; Kucharczyk, R.; Zhao, W.; Dautant, A.; Xu, S.; Niedzwiecka, K.; Su, X.; Giraud, M.-F.; Gombeau, K.; Zhang, M.; et al. Case report: Identification of a novel variant (m.8909T>C) of human mitochondrial ATP6 gene and its functional consequences on yeast ATP synthase. *Life* **2020**, *10*, 215. [CrossRef]
8. Galber, C.; Carissimi, S.; Baracca, A.; Giorgio, V. The ATP synthase deficiency in human diseases. *Life* **2021**, *11*, 325. [CrossRef]
9. Jové, M.; Mota-Martorell, N.; Torres, P.; Ayala, V.; Portero-Otin, M.; Ferrer, I.; Pamplona, R. The causal role of lipoxidative damage in mitochondrial bioenergetic dysfunction linked to Alzheimer's disease pathology. *Life* **2021**, *11*, 388. [CrossRef]
10. Needs, H.I.; Protasoni, M.; Henley, J.M.; Prudent, J.; Collinson, I.; Pereira, G.C. Interplay between mitochondrial protein import and respiratory complexes assembly in neuronal health and degeneration. *Life* **2021**, *11*, 432. [CrossRef]
11. Franco, L.V.R.; Bremner, L.; Barros, M.H. Human mitochondrial pathologies of the respiratory chain and ATP synthase: Contributions from studies of *Saccharomyces cerevisiae*. *Life* **2020**, *10*, 304. [CrossRef] [PubMed]
12. No, M.-H.; Choi, Y.; Cho, J.; Heo, J.-W.; Cho, E.-J.; Park, D.-H.; Kang, J.-H.; Kim, C.-J.; Seo, D.Y.; Han, J.; et al. Aging promotes mitochondria-mediated apoptosis in rat hearts. *Life* **2020**, *10*, 178. [CrossRef]
13. Morciano, G.; Vitto, V.A.M.; Bouhamida, E.; Giorgi, C.; Pinton, P. Mitochondrial bioenergetics and dynamism in the failing heart. *Life* **2021**, *11*, 436. [CrossRef] [PubMed]
14. Wasmus, C.; Dudek, J. Metabolic alterations caused by defective cardiolipin remodeling in inherited cardiomyopathies. *Life* **2020**, *10*, 277. [CrossRef] [PubMed]
15. Casuso, R.A.; Huertas, J.R. Mitochondrial functionality in inflammatory pathology-modulatory role of physical activity. *Life* **2021**, *11*, 61. [CrossRef]
16. Elesela, S.; Lukacs, N.W. Role of mitochondria in viral infections. *Life* **2021**, *11*, 232. [CrossRef]
17. Brillo, V.; Chieregato, L.; Leanza, L.; Muccioli, S.; Costa, R. Mitochondrial dynamics, ROS, and cell signaling: A blended overview. *Life* **2021**, *11*, 332. [CrossRef]
18. Grubelnik, V.; Zmazek, J.; Markovič, R.; Gosak, M.; Marhl, M. Mitochondrial dysfunction in pancreatic alpha and beta cells associated with type 2 diabetes mellitus. *Life* **2020**, *10*, 348. [CrossRef]
19. Motlagh Scholle, L.; Schieffers, H.; Al-Robaiy, S.; Thaele, A.; Lehmann Urban, D.; Zierz, S. Palmitate but not oleate exerts a negative effect on oxygen utilization in myoblasts of patients with the m.3243A>G mutation: A pilot study. *Life* **2020**, *10*, 204. [CrossRef]
20. Acin-Perez, R.; Benincá, C.; Shabane, B.; Shirihai, O.S.; Stiles, L. Utilization of human samples for assessment of mitochondrial bioenergetics: Gold standards, limitations, and future perspectives. *Life* **2021**, *11*, 949. [CrossRef]
21. Kruse, C.-J.; Stern, D.; Mouithys-Mickalad, A.; Niesten, A.; Art, T.; Lemieux, H.; Votion, D.-M. In vitro assays for the assessment of impaired mitochondrial bioenergetics in equine atypical myopathy. *Life* **2021**, *11*, 719. [CrossRef] [PubMed]

life

MDPI

Review

Utilization of Human Samples for Assessment of Mitochondrial Bioenergetics: Gold Standards, Limitations, and Future Perspectives

Rebeca Acin-Perez [1,2,*], Cristiane Benincá [1,2], Byourak Shabane [1,2], Orian S. Shirihai [1,2,3,4] and Linsey Stiles [1,2,3,*]

1 Department of Medicine, Endocrinology, David Geffen School of Medicine, University of California, Los Angeles, CA 90095, USA; CBeninca@mednet.ucla.edu (C.B.); BShabane@mednet.ucla.edu (B.S.); OShirihai@mednet.ucla.edu (O.S.S.)
2 Metabolism Theme, David Geffen School of Medicine, University of California, Los Angeles, CA 90095, USA
3 Department of Molecular and Medical Pharmacology, University of California, Los Angeles, CA 90095, USA
4 Molecular Biology Institute, University of California, Los Angeles, CA 90095, USA
* Correspondence: RAcinPerez@mednet.ucla.edu (R.A.-P.); LStiles@mednet.ucla.edu (L.S.)

Abstract: Mitochondrial bioenergetic function is a central component of cellular metabolism in health and disease. Mitochondrial oxidative phosphorylation is critical for maintaining energetic homeostasis, and impairment of mitochondrial function underlies the development and progression of metabolic diseases and aging. However, measurement of mitochondrial bioenergetic function can be challenging in human samples due to limitations in the size of the collected sample. Furthermore, the collection of samples from human cohorts is often spread over multiple days and locations, which makes immediate sample processing and bioenergetics analysis challenging. Therefore, sample selection and choice of tests should be carefully considered. Basic research, clinical trials, and mitochondrial disease diagnosis rely primarily on skeletal muscle samples. However, obtaining skeletal muscle biopsies requires an appropriate clinical setting and specialized personnel, making skeletal muscle a less suitable tissue for certain research studies. Circulating white blood cells and platelets offer a promising primary tissue alternative to biopsies for the study of mitochondrial bioenergetics. Recent advances in frozen respirometry protocols combined with the utilization of minimally invasive and non-invasive samples may provide promise for future mitochondrial research studies in humans. Here we review the human samples commonly used for the measurement of mitochondrial bioenergetics with a focus on the advantages and limitations of each sample.

Keywords: bioenergetics; fibroblasts; frozen tissue; leukocytes; mitochondria; oxygen consumption; platelets; respirometry; skeletal muscle

Citation: Acin-Perez, R.; Benincá, C.; Shabane, B.; Shirihai, O.S.; Stiles, L. Utilization of Human Samples for Assessment of Mitochondrial Bioenergetics: Gold Standards, Limitations, and Future Perspectives. *Life* **2021**, *11*, 949. https://doi.org/10.3390/life11090949

Academic Editors: Giorgio Lenaz and Salvatore Nesci

Received: 28 July 2021
Accepted: 23 August 2021
Published: 10 September 2021

1. Introduction

Mitochondrial function is essential to meet energy demand and coordinate cellular function. Mitochondria are often referred to as the powerhouse of the cell. However, in addition to converting nutrients into energy that can be used by the cell, mitochondria also play an important role in intercellular signaling, calcium buffering, biosynthesis, and apoptosis [1]. Consumed nutrients get broken down into small molecules that can be used either to enter catabolic or anabolic processes. In the catabolic processes, nutrients break down into small molecules that are used to fuel the mitochondrial electron transport chain (ETC) to either produce adenosine triphosphate (ATP) through oxidative phosphorylation (OXPHOS) or generate heat. On the other hand, the anabolic processes involve building up molecules that can be used by the cells as storage or for structural purposes. The balance between catabolism and anabolism defines the metabolic signature in every tissue [2–5]. Therefore, mitochondrial metabolism is critical for maintaining energetic homeostasis and mitochondrial dysfunction promotes the development and progression of metabolism- and aging-related disorders.

Mitochondrial diseases are genetically inherited, clinically heterogeneous disorders mainly affecting post-mitotic tissues with high energy demand, such as the brain, heart, and skeletal muscle. Since symptoms are varied, affecting multiple organs in the body, diagnosing mitochondrial disease can be challenging. Genetic testing (Figure 1) is the most reliable way to diagnose and categorize a mitochondrial disorder but requires the right choice of starting material and sequencing technique [6]. Specifically, for mitochondrial deoxyribonucleic acid (mtDNA) encoded diseases, the variability observed by mutational heteroplasmy and copy number in different tissues needs to be considered. A negative genetic result does not exclude a mtDNA-related disorder and sampling of tissues such as the urinary tract and buccal epithelial cells or skeletal muscle is often needed. Moreover, some mtDNA variants, specifically deletions and depletion-causing mutations, are detectable only in the high-energy demand affected tissue [7]. Therefore, even with genetic testing available, molecular diagnosis often requires functional characterization. For that, muscle biopsy is the gold standard sample to perform enzymatic activities and/or histology in clinical settings. In research, besides the use of muscle biopsies, skin biopsy-derived fibroblasts are also used for functional characterization of mitochondrial disorders with newly discovered mutations and genes. In this case, assessing the ETC function is required for analysis of mitochondrial-dependent cell metabolism with measuring oxygen consumption, namely respirometry, as the gold standard measurement of mitochondrial function. Respirometry provides an integrated measurement of oxidative mitochondrial metabolism. Platforms and applications for measuring respiration in fresh samples are well established and have been extensively described and reviewed [8–24].

However, measuring respiration in primary tissue samples from humans can be challenging given that a muscle biopsy may not be accessible, sample amount may be limiting, and, when measuring respiration using the traditionally established protocols, there is the limitation that fresh tissue and extensive processing is required. This is due to the fact that freeze-thawing samples damages membranes resulting in cytochrome c release from the mitochondrial intermembrane space, which makes the electron transport chain inefficient as cytochrome c is an essential electron carrier. In addition, damage to mitochondrial membranes causes respiration to uncouple energy production. For these reasons, respirometry must be performed quickly and is done primarily on-site, which limits the use of respirometry as a diagnostic approach, for use in clinical trials, and in previously frozen samples stored in biobanks. These limitations make translational mitochondrial research challenging and largely unfeasible on a large scale. To address these challenges, new approaches are emerging to make use of existing samples for respirometry and use less invasive samples, which will be discussed later in this review.

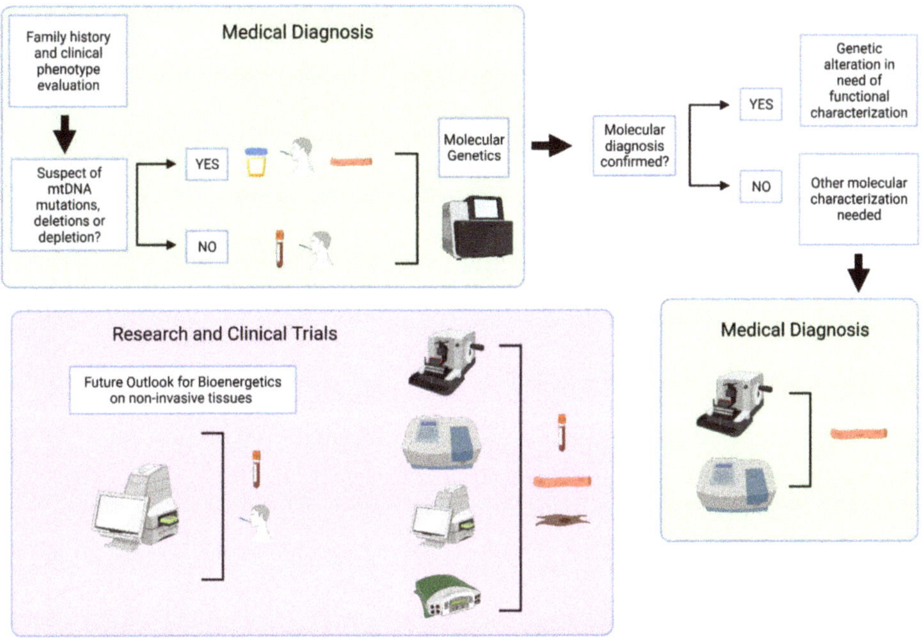

Figure 1. Overview of human samples and bioenergetic assessment used for clinical diagnosis and research. The primary approach for medical diagnosis of mitochondrial disorders (green boxes) is genetic testing of non-invasive or minimally invasive samples. Follow-up testing may be required in skeletal muscle biopsies when additional molecular characterizations are needed for diagnosis or to determine the extent of mitochondrial dysfunction. These include histology and biochemical analysis. For mitochondrial function evaluation in research (purple box), skeletal muscle, primary fibroblasts, and circulating blood samples are routinely used for molecular and biochemical analysis, histology, and bioenergetic testing. Further optimization and development of non-invasive samples may provide a future path for expanding bioenergetics research in human subjects to include large-scale population studies and clinical trials. Created with BioRender.com on 10 August 2021.

2. Materials and Methods

2.1. Cell Culture

Lymphoblastoid cell lines were cultured according to the manufacturer's instructions in Roswell Park Memorial Institute (RPMI)-1640 medium supplemented with 2 mM glutamine and 15% fetal bovine serum (FBS). Frozen LCLs went through three freeze-thaw cycles with liquid nitrogen before use in enzymatic assays or respirometry. Lymphocytes were cryopreserved and did not require cell culture. Thawed lymphocytes were treated with 0, 0.1, 0.5, or 1 μM rotenone for 10 min and then frozen at −80 °C.

2.2. Material and Reagents

Table 1 presents all the reagents with the associated supplier and catalog number that were used to generate data for this review.

Table 1. Vendor and catalog number information for reagents used in the proof of concept assays completed for this review.

Reagent	Supplier	Catalog Number
Acetyl-CoA	Sigma	A2181
Adenosine 5′Diphosphate (ADP)	Sigma	A5285
ATP	Sigma	A26209
Alamethicin	Sigma	A4665
Antimycin A (AA)	Sigma	A8674
Ascorbic Acid (Asc)	Fisher	A61-100
Azide	Sigma	S8032
Coenzyme Q1	Sigma	C7956
Cytochrome c	Sigma	C2506
Dulbecco's Modified Eagle Medium (DMEM)	Sigma	D5030
5,5-dithiobis(2-nitrobenzoic acid) (DTNB)	Sigma	D218200
Ethylene glycol-bis(2-aminoethylether)- N,N,N′,N′-tetraacetic acid (EGTA)	Sigma	E4378
FBS	Corning	35010CV
Carbonyl cyanide p-trifluoromethoxyphenylhydrazone (FCCP)	Enzo	BML-CM120
Glucose	Sigma	G5767
Glutamine	Gibco	25030081
4-(2-hydroxyethyl)-1-piperazineethanesulfonic acid (HEPES)	Gibco	15630080
Reduced nicotinamide adenine dinucleotide (NADH)	Sigma	N8129
Magnesium Chloride	Sigma	M0250
Mannitol	Sigma	M9546
MitoTracker Deep Red (MTDR)	Invitrogen	M22426
Oligomycin (Oligo)	Sigma	495455
Oxaloacetic acid	Sigma	O4126
Phosphate Buffered Saline (PBS)	Gibco	14190250
Poly-D-Lysine	Sigma	P6407
Potassium cyanide	Sigma	60178
Potassium phosphate	Sigma	P5655
RPMI-1640	Gibco	11875119
Rotenone (Rot)	Sigma	R8875
Sodium Pyruvate	Gibco	11360070
Succinic Acid (Succ)	Sigma	S9512
Sucrose	Sigma	S0389
N, N, N′, N′-tetramethyl-p-phenylenediamine (TMPD)	Sigma	87890
XF plasma membrane permeabilizer (XF PMP)	Agilent	102504-100

2.3. *XF96 Extracellular Flux Analysis in Fresh and Frozen Cells*

In all respirometry assays, lymphocytes and LCLs were spun down on poly-d-lysine-coated plates on the day of the assay. LCLs were plated at 200,000 cells per well and lymphocytes were plated at 400,000 cells/well. Cells were added to the plate at a total volume of 175 µL per well and centrifuged at 400× *g* for 5 min with no break. Table 2 presents the mitochondrial-targeted compounds and substrates used and the associated mechanism of action.

2.3.1. Intact Cell Respirometry

Intact lymphocytes respirometry was measured in a DMEM assay medium containing 5 mM glucose, 2 mM glutamine, 1 mM pyruvate, and 5 mM HEPES. Injections resulted in a final concentration of 2 µM oligomycin from Port A, 3 µM FCCP from Ports B and C, and 2 µM antimycin A and rotenone from Port D. Measurements included three basal measurements, three measurements after injection of oligomycin, four FCCP measurements, and three antimycin A and rotenone measurements. Respiration measurements were conducted with four technical replicates.

Table 2. The mitochondrial compounds/substrates used to investigate mitochondrial function in fresh and frozen samples. The compounds used to measure mitochondrial respiratory function and their corresponding functions are listed. Abbreviations: ADP: Adenosine diphosphate; ATP: Adenosine triphosphate; CI: Complex I; CII: Complex II; CIII: Complex III; CIV: Complex IV; CV: Complex V; ETC: Electron transport chain; NADH: Reduced nicotinamide adenine dinucleotide; OXPHOS: Oxidative phosphorylation; TMPD: N, N, N′, N′-tetramethyl-p-phenylenediamine.

Mitochondrial Compound/Substrate	Function
ADP	Induce State 3, ADP-stimulated respiration
ATP	Induce Complex V (CV) hydrolysis in uncoupled mitochondria
Alamethicin	Permeabilize the mitochondrial inner membrane
Antimycin A	Complex III (CIII) inhibitor
Ascorbate	Maintains TMPD in the reduced state
Azide	Complex IV (CIV) inhibitor
Cytochrome c	Soluble component of ETC; added to replace the cytochrome c that is lost with freezing
FCCP	Chemical uncoupler of OXPHOS to measure maximal respiration
NADH	Complex I (CI) substrate
MitoTracker Deep Red	Fluorescent dye to measure mitochondrial content
Oligomycin	CV inhibitor
Rotenone	CI inhibitor
recombinant perfringolysin O (rPFO) or XF PMP	Permeabilize the plasma membrane
Succinate	Complex II (CII) substrate
TMPD	Electron donor to cytochrome c/CIV

2.3.2. Permeabilized Cell Respirometry

Permeabilized lymphocytes respirometry was measured in Mitochondrial Assay Solution (MAS) (70 mM Sucrose, 220 mM Mannitol, 5 mM KH_2PO_4, 5mM $MgCl_2$, 1 mM EGTA, 2 mM HEPES; pH 7.2) containing 4 mM ADP, 2 μM rotenone, 5 mM succinate, and 5 nM XF PMP. Cells were spun down in Seahorse assay medium and then washed twice with MAS before adding MAS supplemented with ADP, succinate, rotenone, and XF PMP. Cells were incubated at 37 °C for 10 min before initiating the respirometry assay with the permeabilized cells starting in State 3 respiration. Injections resulted in a final concentration of 2 μM oligomycin from Port A, 1 μM FCCP from Port B, and 2 μM antimycin A and rotenone from Port C, and 0.5 mM TMPD and 1mM Asc from Port D. Measurements included three State 3 measurements, three measurements after injection of oligomycin, two FCCP measurements, two antimycin A and rotenone measurements, and three measurements after injection of TMPD/Asc. Respiration measurements were conducted with four technical replicates.

2.3.3. RIFS

Respirometry in previously frozen lymphocytes was measured in MAS. Cells were resuspended and spun down in 20 μL of unsupplemented MAS. The samples were centrifuged at 2100× *g* for 10 min and stopped without a break. The samples were brought to a final volume of 150 μL with MAS supplemented with 10 μg/mL cytochrome c and 2.5 μg/mL alamethicin. Injections resulted in a final concentration of either 2 μM rotenone and 5 mM succinate or 1 mM NADH from Port A, 2 μM antimycin A and rotenone from Port B, 0.5 mM TMPD and 1mM Asc from Port C, and 50 mM sodium azide from Port D. Measurements included one measurement before injection of substrates (NADH or succinate and rotenone), two maximal respiration measurements after injection of substrates, two measurements after injection of antimycin A and rotenone, two TMPD/Asc measurements, and two measurements after inhibition of CIV with azide. Respiration was measured in triplicate.

2.3.4. Complex V ATP Hydrolysis Assay

ATP hydrolysis was measured in frozen samples in MAS. Cells were resuspended and spun down in unsupplemented MAS (20 μL of cell suspension). After spinning the cells in the XF96 plate 130 μL MAS containing 5 mM succinate plus 2 μM rotenone was added per well. Injections resulted in a final concentration of 2 μM antimycin A from Port A, 10 mM ATP plus 1 μM FCCP from Port B and C. Port B and C are injected consecutively to assess maximal ATP concentration. Measurements included two measurements before injection of Port A, two measurements after injection of Port A and three maximal ATP hydrolytic activity (ECAR) after injection of Port B/C. Respiration was measured in triplicate.

2.4. *MitoTracker Deep Red in Frozen Cells*

Cells were stained with 500 nM MTDR in PBS for 10 min in a black 96-well plate imaging plate. Cells were then washed with PBS to remove the MTDR. Fluorescence intensity was measured from the bottom of the plate in a Tecan Spark 10 M Multimode Plate Reader using an excitation wavelength of 633 nm and an emission wavelength of 678 nm.

2.5. *Enzymatic Assays in Frozen Cells*

2.5.1. Complex I Activity

CI activity was measured in cell lysates following NADH oxidation at 340 nm as described previously [25].

2.5.2. Citrate Synthase Activity

Citrate synthase activity was performed in cell lysates using DTNB, as described in [5].

2.6. *Statistical Analysis*

Assays were run as proof of concept that the RIFS protocol is comparable to traditional CI spectrophotometric assays and only run in one independent experiment. Therefore, statistics were not performed on the data.

3. Gold Standards: Human Samples for Assessment of Mitochondrial Bioenergetic Function

3.1. *Bioenergetics Testing in Mitochondrial Disorders*

Impaired oxidative phosphorylation is a hallmark of mitochondrial disorders with progressive clinical manifestations that can vary from single organ to multisystem disorders, often affecting organs with high energetic demands [26]. Mitochondrial diseases have an estimated prevalence of one in 4300 adults, making them the most common inherited metabolic disorder [27–30]. Defects in CI are the most commonly observed respiratory chain disorder [31–33], making the complex a good candidate for bioenergetic testing.

In patients suspected to have mitochondrial disorders, genetic testing and metabolic screening are recommended before biochemical assays to evaluate mitochondrial function due to the precision in categorizing and less-invasive nature of these measurements. Analyte testing of blood, urine, or even cerebrospinal fluid (CSF) in specific cases, for pyruvate, lactate, amino acids, creatine kinase, and carnitine, can provide information about alterations in metabolism in these patients and provide a biochemical metabolic signature of mitochondrial disease [34,35]. Additionally, pathogenic variants in over 400 genes, both mitochondrial and nuclear, have been identified as contributors to mitochondrial disorders [26]. Pathogenic mutations in mtDNA and nuclear DNA (nDNA) can be used to confirm primary mitochondrial disease due to abnormal OXPHOS function. A genetics-first approach has been facilitated by the availability of exome and whole genome sequencing that has improved the identification of mitochondrial disease genes diagnosis [36–38]. Identification of nuclear mitochondrial disease genes are also used to diagnose mitochondrial dysfunctions [39–42]. Secondary mitochondrial dysfunction, which can be influenced by environmental factors in addition to genetic mutations, adds a further level of complexity and can lead to metabolic and neurodegenerative diseases [43]. While advances in sequencing have made rapid diagnosis possible in a large number of patients, mtDNA or nDNA mutations cannot always be identified even in the presence of a clinical mitochondrial phenotype. Therefore, tissue testing for functional validation may still be necessary [42,44].

Functional, biochemical assays to detect alterations in enzymatic activity, respiration capacity, or ATP synthesis in primary tissue can be used to confirm mitochondrial defects as well as histopathology and mtDNA copy number measurement. However, it is important to note that these assays cannot distinguish between primary defects in OXPHOS enzymes and secondary mitochondrial dysfunctions. Measurement of respirometry and ATP synthesis requires fresh tissue since freezing will disrupt the inner mitochondrial membrane and uncouple oxygen consumption from ATP synthesis. Given the need for same day processing and assaying, the expertise required to run these samples, and limitations in the availability of tissue, bioenergetic alternatives to respirometry and ATP synthesis measurements can be run in frozen samples, such as enzymatic assays measuring ETC complex activities.

Although the electron transport chain and ATP synthesis uncouple when samples are frozen, the mitochondrial electron transport complexes remain structurally intact and functional [45,46]. For that reason, a common method used to assess mitochondrial function in frozen samples has been to measure spectrophotometric enzymatic assays that provide information on the activity of individual ETC complexes or the combination of CI + CIII or CII + CIII. The protocols to measure enzymatic activities often use supraphysiological concentrations of some reagents and non-physiological electron donors and acceptors. Table 3 provides an overview of assays to measure enzymatic activities in mitochondria. Although these measurements were successfully used in a relatively high-throughput manner to diagnose primary mitochondrial diseases, namely diseases caused by a primary defect in electron transport chain function [47–49], they cannot provide a single measurement of the coordinated function of the electron transport chain function working at more physiological rates. Taken together, these observations suggest that spectrophotometric assays might be less sensitive to detect defects in mitochondrial architecture and complex ultrastructure, such as the ones associated with secondary mitochondrial dysfunction. However, since the integrated measurement of mitochondrial respiration or ATP synthesis is often unfeasible in these samples these assays provide a useful alternative.

Table 3. Spectrophotometric approaches to measure mitochondrial enzymatic activities for assessment of mitochondrial bioenergetic function. The table presents the metabolic enzyme of interest, the metabolic pathway the measurement would evaluate, and methodology references. Abbreviations: ATP: Adenosine triphosphate; CI: Complex I; CII: Complex II; CIII: Complex III; CIV: Complex IV; DH: Dehydrogenase; ETC: Electron transport chain; NADH: Reduced nicotinamide adenine dinucleotide; OXPHOS: Oxidative phosphorylation; TCA: Tricarboxylic acid cycle.

Enzyme	Metabolic Pathway	References
CI: NADH/CoQ1 Oxidoreductase	ETC, OXPHOS	[50,51]
CII: Succinate DH	ETC, OXPHOS, TCA	[25,50,51]
CIII: Coenzyme Q: Cytochrome c—oxidoreductase	ETC, OXPHOS	[50,51]
CIV: Cytochrome c oxidoreductase	ETC, OXPHOS	[50,51]
Combined CI + CIII	ETC, OXPHOS	[51]
Combined CII + CIII	ETC, OXPHOS	[51]
Combined CI + CIII + CIV	ETC, OXPHOS	[52]
CV: ATP hydrolysis	ETC, OXPHOS	[53,54]
Creatine kinase	ATP homeostasis	[55]
Adenylate kinase	ATP homeostasis	[55]
Citrate synthase	TCA	[50,51]
α-ketoglutarate DH	TCA	[56,57]
Isocitrate DH	TCA	[58]
Malate DH	TCA	[55–57]
Aconitase	TCA, Redox balance	[50,56,59]
Manganese Superoxide dismutase (MnSOD)	Redox balance	[60]
Nicotinamide nucleotide transhydrogenase	Redox balance	[61–63]
Catalase	Redox balance	[56]
Glycerol-3 phosphate DH	Glycerophosphate shuttle	[25,51]
Pyruvate DH	Glucose and Fatty acid oxidation	[64]
β-Hydroxyacyl CoA DH	Fatty acid oxidation	[50,55]
Short-chain hydroxyl-acyl-CoA DH	Fatty acid oxidation	[58,65,66]
β-hydroxybutyrate DH	Ketone body	[67–69]
Hexokinase	Glucose metabolism	[55,70]
Arginase	Urea cycle	[71]

3.2. Skeletal Muscle Biopsies

Muscle biopsies have long been the gold standard for investigating changes in mitochondrial function via respirometry in human samples [9,18]. Measurements of mitochondrial function can be performed in intact muscle fibers, permeabilized muscle fibers, or isolated mitochondria from muscle biopsies [13,18]. Skeletal muscle biopsies provide a terminally differentiated, post-mitotic sample that requires less mtDNA replication and may maintain stable levels of mtDNA heteroplasmy compared with mitotic cells [72–74]. Skeletal muscles have both high energy demand and mitochondrial content making skeletal muscle a valuable primary tissue sample to measure mitochondrial respiration and ATP synthesis. Measuring respiration in muscle fibers offers the ability to probe specific complexes of the ETC, providing information about underlying mechanisms of altered mitochondrial function. As mentioned previously, mitochondrial bioenergetics can also be assessed in these samples through spectrophotometric assays to measure different metabolic pathways (Table 3) and, in the case of citrate synthase, provide a measure of mitochondrial content. Histology can provide information about mitochondrial structure and bioenergetics through commonly used stains such as modified Gomori trichrome (muscle structure, mitochondrial accumulation, and ragged red fiber), cytochrome oxidase (CIV), succinate dehydrogenase (CII, nDNA encoded), and NADH dehydrogenase (CI), as well as stains to assess intramuscular lipid accumulation [75].

While a wealth of mitochondrial information can be gained from skeletal muscle, this technique can be invasive and not available outside clinical settings or the amount of material obtained can be limiting depending on the technique used [75,76]. For example, open biopsies that require a clinical setting are invasive but provided plenty of material while needle biopsies are less invasive, but the amount of sample obtained may be limiting. While a needle biopsy does not require an operative setting, it does require local anesthesia, sterile conditions, and some specialization to obtain the sample [75,77]. Therefore, there has been a push towards the use of more accessible samples with the development of new techniques. To that point, in vivo, non-invasive applications to measure mitochondrial respiratory capacity are available, such as near-infrared spectroscopy (NIRS), which corresponds with the high-resolution respirometry in muscle biopsies [78]. While the use of this and other in vivo techniques is suitable to study human mitochondrial function, there is still a need for more specific, higher throughput techniques that require less expertise to run for use in translational research studies and clinical applications.

While testing of muscle biopsies was considered the gold standard, before the genomics era, to measure mitochondrial bioenergetic function and diagnosis of primary mitochondrial disease [79], alternative methods would be preferentially useful for diagnosis and research when genetic screening is not enough for categorization of the deficiency [80]. Identifying suitable alternatives to muscle biopsies and systemic biomarkers of mitochondrial function has become a major research focus. This is due to the fact the evaluation of mitochondrial bioenergetic function remains an important parameter for basic research and, in some cases, diagnosis, as a way to determine the extent of mitochondrial dysfunction. Functional readouts of mitochondrial bioenergetics also have applications for translation research and for clinical trials where genetic monitoring may not provide a suitable alternative.

3.3. Human Fibroblasts

Fibroblasts from human patients can be generated from minimally invasive samples (1 mm punch skin biopsies) and maintain DNA mutations, although heteroplasmy is not necessarily maintained with passages, and cumulative cellular damage [81]. Respirometry measurements are straightforward in fibroblasts across multiple platforms [82–84]. Table 4 provides a general overview of the advantages and limitations of measuring mitochondrial bioenergetics in intact cells, permeabilized cells, and isolated mitochondria, all of which can be utilized in fibroblasts.

Table 4. Advantages and disadvantages of respirometry measurements in intact and permeabilized cells and isolated mitochondria. Adapted from [12]. Abbreviations: ETC: Electron transport chain; OCR: Oxygen consumption rate.

	Intact	Permeabilized	Isolated Mitochondria
Advantages	• Greater physiological context • Cellular control of metabolism/nutrient preference • Intact mitochondrial architecture • Information for oxygen consumption rate (OCR) and glycolysis can be obtained simultaneously • Limited material needed	• Experimental control of substrates • Mechanistic analysis of ETC and mitochondrial metabolism • Native intracellular environment • Intact mitochondrial architecture • Limited material needed	• Experimental control of substrates • Extra-mitochondrial metabolism does not limit O_2 consumption • Mechanistic analysis of ETC and mitochondrial metabolism • Reproducible, robust OCRs with high dynamic range • Normalizing to total protein should normalize for mitochondrial content
Drawbacks	• Cell culture conditions, confluence, and experimental media influence outcomes • Less mechanistic insight • May be more difficult to interpret since so many processes can alter OCR • Changes in OCR could reflect changes in mitochondrial content • Normalization is required	• No cellular control of metabolism/nutrient preference • Careful titration of permeabilizing agent required • Changes in OCR could reflect changes in mitochondrial content • Normalization is required	• No cellular control of metabolism/nutrient preference • Loss of intracellular environment • Loss of mitochondrial morphology • Isolation procedure can damage mitochondria • Selection bias • More starting material required

Fibroblast cell respirometry has been reported to be a faster and more sensitive measure of electron transport chain defects than traditional spectrophotometry enzymatic assays [85]. In a study of Leigh and Leigh-Like Syndrome patients, respirometry was able to detect impairment in the mitochondrial respiratory chain in 50% of patients that could not be identified with measuring individual ETC complex activities alone [84]. This study demonstrates that detection of causative mutations, biochemical, and respiratory defects vary between individuals and that using combinations of these approaches has the best diagnostic rates. Additionally, enzymatic impairment is not always consistent between muscle biopsies and fibroblasts, suggesting that testing multiple samples, when possible, is the best practice.

Patient fibroblasts provide an important research tool to measure mitochondrial bioenergetics as well as a full characterization of mitochondria including, but not limited to, mitochondrial morphology, turnover, and in-depth analysis of mitochondrial metabolism [86,87]. For research purposes, commercially available fibroblasts can be obtained for different mitochondrial disorders and diseases of aging. This makes it possible to better understand the full range of mitochondrial impairment caused by certain mutations or in particular diseases. It also opens up the ability for pre-clinical assessment of mitochondrial function in these cells in response to interventions, such as compound treatments. Fibroblasts can present the metabolic signatures observed in mitochondrial disease patients, such as a metabolic shift from oxidative phosphorylation to increased glycolysis for ATP synthesis [88,89]. Therefore, these cells can be a particularly important tool to test compounds that are not expected to improve mitochondrial function in healthy, control cells but only under mitochondrial dysfunction. Compound testing in fibroblasts can facilitate determining compound effects on metabolically stressed cells without needing to use exogenous stressors such as calcium, reactive oxygen species (ROS), or mitochondrial inhibitors that can be difficult to titrate and often have narrow exposure windows between impaired function and apoptosis.

Considerations for Utilizing Fibroblasts for Respirometry Studies

Given that fibroblasts proliferate in culture, the amount of sample is not limited to the same extent as with tissue biopsies. However, some considerations arise from measuring metabolism in cultured fibroblasts. In vitro culture conditions, such as the amount of glucose in the culture medium, have been shown to change cellular metabolism across multiple cell lines including fibroblasts [90–94]. Costa et al. demonstrate that culturing fibroblasts in low glucose conditions, a more physiological condition than standard high glucose conditions, results in remodeling of the mitochondrial network and a push towards a more oxidative phenotype without causing an extensive metabolic reconfiguration of substrate preference compared with high glucose cultured cells [95]. On the other hand, growing cells in galactose cause both mitochondrial and metabolic remodeling resulting in increased oxygen consumption rates, ATP levels, and mitochondrial biogenesis. Galactose culturing is a useful approach for screening for drug-induced mitochondrial toxicity as it increases sensitivity to mitochondrial respiratory chain inhibitors [96]. Galactose culturing can also improve the testing of interventions that increase mitochondrial bioenergetic function by preventing compensatory ATP synthesis through glycolysis [97]. There are additional considerations besides media composition, for example, modifying the culturing oxygen concentration has been shown to increase the in vitro lifespan of fibroblasts [98]. Therefore, it is important to consider the conditions under which fibroblasts are grown and how these conditions could affect mitochondrial function and metabolism when investigating mitochondrial phenotypes in vitro.

Furthermore, when utilizing primary fibroblasts it is important to consider population doubling time as these cells have limited proliferative capacity and will become senescent over time [99]. Indeed, human skin fibroblasts display hallmarks of aging senescence and are used as a model of aging in vitro [100–102]. However, changes in fibroblast metabolism with increasing passage numbers have technical implications for the experimental use of these cells for the investigation of mitochondrial function. Glucose uptake and lactate production increase with fibroblast population doublings, which is consistent with a switch towards glycolysis as cells become senescent [103,104]. Senescent fibroblasts also display changes in mitochondrial bioenergetics with increased mitochondrial membrane potential heterogeneity, a decreased respiratory control ratio, and uncoupling of mitochondrial respiration from ATP synthesis without changes in total respiratory capacity [105]. There is also evidence that mitochondrial content increases in senescent fibroblasts [100]. However, this may represent compensation for increased cell volume that occurs during senescence and not necessarily a change in mitochondrial density [105]. Subculturing of fibroblasts can also induce shifts in mitochondrial heteroplasmy in cells with mtDNA mutations [106]. Taken together, these data highlight the importance of considering passage doubling when using fibroblasts as a model for studying mitochondrial function and comparing across similar passage numbers.

Primary fibroblasts represent the biological and chronological age of the subjects from which they are derived [81,107]. Fibroblasts derived from old subjects display slower proliferation rates and alterations in mitochondrial function and metabolism compared with fibroblasts from young subjects [107–109]. Therefore, a final consideration for using fibroblasts for investigating mitochondrial function is what control should be used for patient cells including considerations of age, sex, and genetics. Age and sex-matched fibroblasts should be used as controls and, whenever possible, those cells should be derived from a relative, such as an unaffected sibling, to better control for genetic differences between the patient and control fibroblasts. Another option is to generate isogenic controls using the corrected disease-causing mutation, for example by clustered regularly interspaced short palindromic repeats (CRISPR)—associated protein 9 (CRISPR-CAS9) correction or overexpressing the wild-type complementary DNA (cDNA), making the comparison more precise by keeping the same genetic background [110].

3.4. *Summary of Benchmark Human Samples to Study Mitochondrial Bioenergetics*

There is growing evidence that impaired mitochondrial function contributes to aging and age-related diseases. Age-related impairments in mitochondrial function include decreased OXPHOS capacity and ATP synthesis and well as increased ROS generation and accumulation of mtDNA mutations and deletions [111–113]. Therefore, monitoring mitochondrial function could provide a useful clinical tool for both diagnosis and monitoring the effectiveness of treatments. To accomplish this, there is a need for a less invasive way to monitor mitochondrial function than testing primary tissues. While fibroblasts provide an important research model that can be used for disease diagnosis and pre-clinical research, this approach is not suitable for clinical testing and monitoring due to poor scalability. Muscle biopsies are optimal for bioenergetics measurements, nevertheless collecting these samples is often considered as a last resort due to availability and invasiveness [42]. Therefore, other minimally invasive samples are needed to provide a translational approach to monitoring mitochondrial respiratory function. There is growing evidence that respirometry measurements in circulating blood cells may be a systemic biomarker of mitochondrial bioenergetic function. Blood-based bioenergetics present a model that is minimally invasive to obtain and more feasible to test in clinical settings and on a larger scale in clinical trials.

4. Blood-Based Bioenergetics—Respirometry as a Systemic Biomarker of Mitochondrial Function

Mitochondrial content, substrate preference, protein composition, and mitochondrial morphology vary between tissues based on energetic and metabolic demands. Therefore, there are clear advantages to measuring mitochondrial function in the tissue of interest, which is the approach commonly used in basic research. However, primary tissue samples can be difficult to obtain in human subjects and the amount of sample can be limiting for downstream bioenergetic applications. Therefore, functional measurements of mitochondria cannot always be used in translational research or to identify impairment in mitochondrial function clinically, even if it would be a useful readout. While there is differential mitochondrial function between tissues, there is also evidence that circulating cells can act as a biomarker for overall mitochondrial function. Using platelets and leukocytes as a less invasive biomarker of mitochondrial function has emerged as a potential approach for situations when testing primary tissue samples is not feasible. These cells represent an attractive systemic biomarker candidate and an alternative to invasive tissue biopsies in human studies, since blood can be easily obtained in sufficient quantities for respirometry assays and other bioenergetic measurements. Additionally, isolated blood cells can be cryopreserved for later testing if the cells need to be isolated at an off-site location and shipped for respirometry analysis. Cryopreservation allows for samples that are collected at different times to be stored and subsequently run together in the same experiment, which may cut down on variability. However, long-term cryopreservation can also decrease cell viability and impair mitochondrial bioenergetic parameters and should be considered when using this approach [114].

Protocols for measuring respirometry in circulating blood cells have been well established for both intact and permeabilized cells and previously described [115–117]. Oxygen consumption measurements in circulating blood cells are compatible with any respirometry platform. Given that platelets and leukocytes are non-adherent, they are well suited for measurements in oxygraph closed-chamber respirometers, but they can also be tested in plate-based respirometers by coating plates to promote cell adhesion. The primary advantage of using a plate-based respirometer, such as the Seahorse Extracellular Flux Analyzer, is minimizing the amount of sample required for measurements, which is about ten times less in the Seahorse XF96 compared with an oxygraph [118].

Tyrrell et al. have compared respirometry in permeabilized muscle fibers to intact platelets and monocytes in healthy vervet/African green monkeys [119]. Their data demonstrate that the maximal respiratory capacity of circulating blood cells correlates with the maximal OXPHOS capacity of permeabilized skeletal muscle from the same animal. Both platelet and monocyte maximal oxygen consumption displayed a significant correlation with (1) the combined CI and CII-linked OXPHOS (State 3 respiration) in permeabilized skeletal muscle fibers and (2) the respiratory control ratio in isolated skeletal muscle mitochondria. Comparisons of skeletal muscle bioenergetics and circulating blood cells have also been completed in human subjects [120]. Braganza et al. show significant correlations between bioenergetic parameters in platelets and permeabilized skeletal muscle fibers from the same subject. Maximal respiration in platelets significantly correlated with maximal respiration in muscle and platelet proton leak correlated with State 4 (leak) respiration in muscle. Furthermore, basal and ATP-linked respiration was decreased in older subjects compared with a younger cohort. However, not all studies demonstrate correlations between circulating blood cells and muscle fibers. In a study of 32 women, Rose et al. did not observe a correlation in intact respirometry in peripheral blood mononuclear cells (PBMCs) or platelets with permeabilized muscle fibers, though there were some correlations with permeabilized platelets and permeabilized muscle [118]. Taken together, these results suggest that bioenergetic profiling of platelets and leukocytes may be used as a biomarker of mitochondrial function and, in some cases, reflect the bioenergetic capacity of muscle tissues and potentially other highly metabolic primary tissues.

Pecina et al. could identify defects in CI in permeabilized lymphocytes isolated from pediatric patients, some with known mitochondrial disorders, using high-resolution respirometry [121]. The authors used a combined approach testing functional methods and protein analysis and were able to detect several types of isolated OXPHOS disorders. They suggest that lymphocytes are a suitable sample to detect isolated defects of CI, CIV, and ATP synthase and that repeated analysis of lymphocytes could allow for monitoring changes in mitochondrial function during disease progression. Additionally, changes in platelet and leukocyte mitochondrial respiratory function have been reported in several diseases including type 2 diabetes, human immunodeficiency virus (HIV), neurodegenerative diseases, as well as in aging [120,122–132]. Interestingly, there does not seem to be a single respiratory parameter that demonstrates consistent differences in healthy controls compared with disease patients. Changes in respiration, both increased and decreased, have been found in basal, ATP-linked, and maximal respiration [133], can be variable, and sex-associated differences in oxygen consumption rates have been reported [130,134]. The wide range of mitochondrial changes observed in these studies could be due to different bioenergetic impairments and compensations between diseases. Additionally, there could be differences arising from the subpopulation of cells tested or potential confounding factors such as the presence of other, contaminating cell types, or inadvertent cell activation during isolation.

Considerations for Utilizing Circulating Blood Cells for Respirometry Studies

An important consideration for blood-based bioenergetics is selecting an appropriate cell type to use for respirometry experiments and whether mixed cell populations, such as PBMCs, are suitable samples for respirometry. Rausser et al. used a high-throughput mitochondrial phenotyping platform to compare PBMCs to leukocyte subtypes from the same individual [135]. They show large functional differences in both mitochondrial content and respiratory chain enzymatic activity between immune cell subtypes. Furthermore, PBMC cell type composition can vary between individuals and within the same individual over time. Therefore, they caution that using PBMCs for mitochondrial bioenergetics measurements can mask age- and sex-related changes. Furthermore, Darley-Usmar et al. have shown that respiration rates, the ratio of OXPHOS to glycolysis, as well as mitochondrial protein composition vary between circulating blood cell subpopulations [116]. For example, platelets have the highest basal and ATP-linked OCR compared with leukocyte cell types but lower maximal respiration and reserve capacity [116,136], which could alter bioenergetic results and interpretation if platelet contamination is present in the test samples.

Platelets are a common contaminant in human PBMC samples and vary depending on blood collection and PBMC isolation strategies [137]. Urata et al. have demonstrated that platelet contamination is common and can vary to a large extent amongst PBMC preparations [138]. They measured mtDNA copy number before and after platelet depletion and showed that the amount of platelet contamination varied between individuals and that PBMC mtDNA content is overestimated almost two-fold due to platelet contamination. This work was framed in the context of using PBMC mtDNA copy number as a potential biomarker to monitor disease but has implications for blood-based bioenergetics, as well. Additionally, cellular and mitochondrial metabolism including mitochondrial bioenergetic profiles changes in resting cells compared with activated cells [122,139]. Thrombin-activated platelets increase basal and ATP-linked oxygen consumption and glycolysis compared with control cells [140]. Therefore, it is important to consider cell activation, optimal collection, and isolation conditions when isolating platelets and leukocytes from whole blood [141,142].

For further information on respirometry in circulating blood cells, Braganza et al. have published a comprehensive review of blood-based bioenergetics outlining the research published in platelets and leukocytes as well as considerations for the use of blood-based bioenergetics as a translational and clinical tool [133]. While more research is needed to understand what cell populations are most representative of systemic mitochondrial function and specific diseases, blood-based bioenergetics offers a minimally invasive alternative to muscle biopsies and could provide a systemic biomarker of mitochondrial function and key readout for clinical trials. However, one major consideration for the utilization of this approach is that samples generally need to be collected and immediately processed to isolate circulating platelets and leukocytes from whole blood. Further isolating subpopulations of cells requires more starting material and processing. Therefore, there is still a need for a non-invasive sample that requires minimal on-site processing.

5. Mitochondrial Function in Previously Frozen Specimens: Overview and Commentary on an Updated Approach to Respirometry

Several attempts have been made to carry out respirometry on cryopreserved tissue samples. In these approaches, samples were collected and frozen in different solutions trying to preserve the structure and integrity of the mitochondrial membrane. However, the outcome of these attempts was variable and not standardized [143–147]. Mitochondrial complexes and supercomplexes can be isolated from frozen samples and separated in native gels. These isolated complexes and supercomplexes preserve their enzymatic activity and more interestingly, mitochondrial supercomplexes respire [45,46]. These works demonstrate that the electron transport chain is not destroyed by freeze-thawing and that previously frozen sample respirometry is feasible [46]. The challenge is to convert the assay performed in isolated mitochondrial supercomplexes to previously frozen samples, without the need to isolate the ETC components.

We have recently published a new approach to measure respirometry in frozen samples (RIFS) that reconstitutes maximal mitochondrial respiration in previously frozen samples [148,149]. Using RIFS, maximal oxygen consumption of the ETC is measured by providing physiological electron donors and acceptors with oxygen consumption as an integrated readout. It is important to consider that in previously frozen samples, the natural provider of electron equivalents, the tricarboxylic acid (TCA) cycle, is not directly feeding the ETC. For that reason, we need to bypass the components that are lost during the freeze-thaw step by providing electron donors directly to the ETC and correct for variable permeabilization of mitochondrial membranes, when necessary. For example, NADH is used rather than pyruvate + malate since NADH, a reducing equivalent that donates electrons to CI of the ETC, does not have a permeability barrier in frozen tissue. The protocol is straightforward in almost all cases except for tissues that contain a high proportion of fibers and collagen such as skeletal muscle or tissues that are rich in membranes such as the brain. In these tissues, an additional protocol step is required for either enzymatic digestion (skeletal muscle) or additional membrane permeabilization (brain) for the reagents and electron donors, particularly NADH, to be accessible to mitochondria.

The RIFS protocol was optimized and validated using the Seahorse XF96 Extracellular Flux Analyzer, but should be suitable for any respirometry platform. The RIFS assay includes injection of NADH to measure maximal CI respiratory capacity, succinate to measure maximal CII respiratory capacity, as well as TMPD to measure CIV activity. TMPD provides electrons to CIV by directly reducing cytochrome c. Inhibitors of CI, CIII, and CIV are used as controls in each assay to determine sensitivity to respiratory chain complexes. RIFS preserves 90–95% of the maximal respiratory capacity in frozen samples and can be applied to isolated mitochondria, cells, and tissue homogenates with high sensitivity. As an example in cells, Figure 2 compares fresh and frozen respiration in human lymphocytes from two donors. In fresh intact and permeabilized cells, these lymphocyte samples demonstrated low inter-individual variability and the expected response to injected compounds in all assays (Figure 2A,B). However, oxygen consumption rates were lower than expected in permeabilized cells possibly due to incomplete permeabilization. RIFS formulation allows for simultaneous measurements of respiration driven by electron entry through CI and CII while determining individual CIV activity. Additionally, CV, or ATP synthase, activity can be measured via ATP hydrolysis. In lymphocyte cell lysates, respiration driven through CI and CII, as well as CIV and CV activity can be measured with low inter-individual variability. Furthermore, CV is sensitive to dose-dependent inhibition with oligomycin (Figure 2C–E).

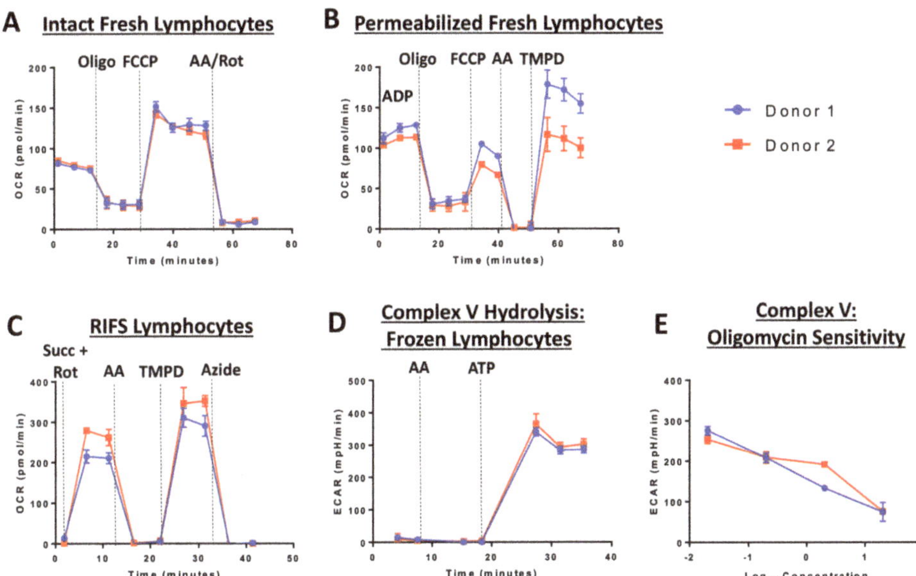

Figure 2. Extracellular flux analysis in fresh and frozen lymphocytes from two donors. Respirometry was measured in fresh (**A**,**B**) and frozen lymphocytes (**C**–**E**). (**A**) Oxygen consumption was measured in intact cells before and after injection of oligomycin (CV inhibitor), FCCP (chemical uncoupler), and antimycin A (CIII inhibitor) and rotenone (CI inhibitor). (**B**) Respiration in permeabilized cells with CII substrates. A final injection of TMPD was included to measure CIV. (**C**) RIFS protocol with succinate and rotenone to drive respiration through CII. (**D**) CV activity measured by media acidification as a result of ATP hydrolysis. (**E**) Oligomycin-sensitivity in the CV measurement. Abbreviations: AA: Antimycin A; ATP: Adenosine triphosphate; FCCP: Carbonyl cyanide p-trifluoromethoxyphenylhydrazone; oligo: Oligomycin; Rot: Rotenone; Succ: Succinate; TMPD: N, N, N′, N′-tetramethyl-p-phenylenediamine.

To measure CV activity, the extracellular acidification (ECAR) is measured to monitor the acidification of the experimental medium in response to hydrolysis of injected, saturating levels of ATP, as previously described in Divakaruni et al. [150]. CV activity can be measured in previously isolated, frozen mitochondria as well as tissue and cell homogenates. In this application, the addition of the CV inhibitor oligomycin acts as an important control to determine that the increased ECAR after injection of ATP is a result of ATP hydrolysis and not non-specific acidification.

This ECAR approach will provide a measure of maximal hydrolytic capacity as a readout of CV enzymatic activity. However, it is important to note that ATP hydrolysis, both through ECAR measurements or with spectrophotometric assays, may underestimate CV deficiencies depending on the specific mutated subunit. For example, mutations in the mtDNA-encoded ATP synthase F_0 subunit 6 that cause severe impairment in ATP synthesis may show little or no defects in ATP hydrolysis [151]. Therefore, while CV activity is a useful readout, it cannot replace measurement of ATP synthesis or respirometry in fresh samples or provide information on physiological levels of ATP hydrolysis.

The simplified sample preparation with RIFS, particularly with homogenates, and the 96-well format of the Seahorse XF96 Analyzer allows using significantly less biological material. RIFS does not require isolation of mitochondria or the use of detergents or plasma membrane permeabilizers, which simplifies the methodology and minimizes changes induced by partial or over permeabilization due to the use of detergents. Acin-Perez et al. provide validation in mouse, zebrafish, and human samples [148]. Another advantage of using total tissue lysates is that it accounts for tissue-specific mitochondrial function, where, for example, soleus showed significantly higher respiration than quadriceps, as previously described [152]. In homogenates and cell lysates, respirometry can also be normalized to mitochondrial content to provide a measure of respiration per functional unit. In addition to traditional methods to measure mitochondrial mass, we have recently described a high throughput MitoTracker Deep Red protocol that requires minimal biological samples that can be run in parallel to RIFS and used to normalize respiration [148,149]. Since differences in mitochondrial content can account for differences in respiration rates in homogenates, the normalization step can differentiate between changes in mitochondrial content and function when homogenate samples or cell lysates are loaded per total protein. Normalization to mitochondrial content is not necessary with isolated mitochondria since respiration in these samples is already normalized to mitochondrial content when loaded per total protein.

The RIFS technique demonstrated that highly oxidative tissues such as heart, brown adipose tissue, and brain showed higher mitochondrial respiration rates per milligram of tissue whereas white adipose tissue, known for its low mitochondrial content, showed the lowest respiration rates. These results suggest that RIFS reveals physiological differences depending on the energy and metabolic demand of the tissue. When comparing RIFS to traditional methods used in frozen samples such as spectrophotometric enzymatic assays, the latter provides information about the maximal activity of the individual complexes or some of them combined, whereas RIFS allows for an integrative measure of the ETC from CI or CII to the natural electron acceptor that is oxygen. Additionally, RIFS often requires less sample to perform the assay compared to enzymatic assays and can show increased sensitivity to specific inhibition of the ETC complexes, as displayed in Figures 3 and 4 where results were obtained with half as much starting material compared with enzymatic assays. In addition, the assay is straightforward where the number of reagents needed is minimal in comparison to spectrophotometric assays. The simple, standardized protocol of RIFS could limit the inter-laboratory variability that has been reported with enzymatic spectrophotometric assays [153].

Snap freezing samples in liquid nitrogen is the best approach for RIFS. However, due to possible delays between sample collection, freezing, and storage for clinical samples, an initial concern was how quickly samples need to be frozen to preserve mitochondrial function. For that reason, different sampling procedures were tested to determine which approach best preserved maximal respiratory rates when samples could not be immediately flash-frozen in liquid nitrogen. Samples placed on ice for up to 3 h after collection as well as samples immediately stored at −20 °C preserved the integrity and function of mitochondria ETC complexes suggesting that this approach could be suitable for samples that require a delay in processing [148,149].

5.1. Applications for Respirometry in Previously Frozen Samples

There are numerous applications where RIFS can be used in preexisting, stored frozen samples. In adipose samples from pheochromocytoma patients, differences in maximal respiratory capacity that were observed in fresh clinical samples were maintained after long-term storage at −80 °C [154]. RIFS generated comparable data to those initially obtained in freshly isolated mitochondria including increased respiration through CI, CII, and CIV and increased mitochondrial content. Furthermore, RIFS made it possible to assay all the samples together compared to fresh samples that had to be run one at a time. This application could be particularly important in age-related studies since it enables a longitudinal design in which bioenergetic profiling can be monitored and compared over years across the same patients. RIFS could also be used as a functional test for patients with mitochondrial disease using less invasive samples. We tested the RIFS protocol in lymphoblastoid cell lines (LCLs), an immortalized B lymphocyte cell line, from patients with the mitochondrial disorder Leber optic atrophy (LHON). These cells were derived from patients with a mtDNA mutation in the NADH dehydrogenase subunit 4 gene. Both RIFS and spectrophotometric assays were able to detect impaired CI activity compared with control cells and revealed similar differences between the LHON lines (Figure 3A). Moreover, we were able to detect impaired NADH-driven respiration with half the amount of sample or the same amount of sample while simultaneously getting information about CII-driven respiration and CIV activity, both of which were, for the most part, similar to the control cells with the exception of LHON LCL line 1 that showed some decrease in maximal CII respiration (Figure 3B). Comparing a citrate synthase assay to the MTDR assay to measure mitochondrial content, the results from both assays were comparable, but again RIFS required less starting material, fewer reagents, and was less time-sensitive than the citrate synthase assay (Figure 3C).

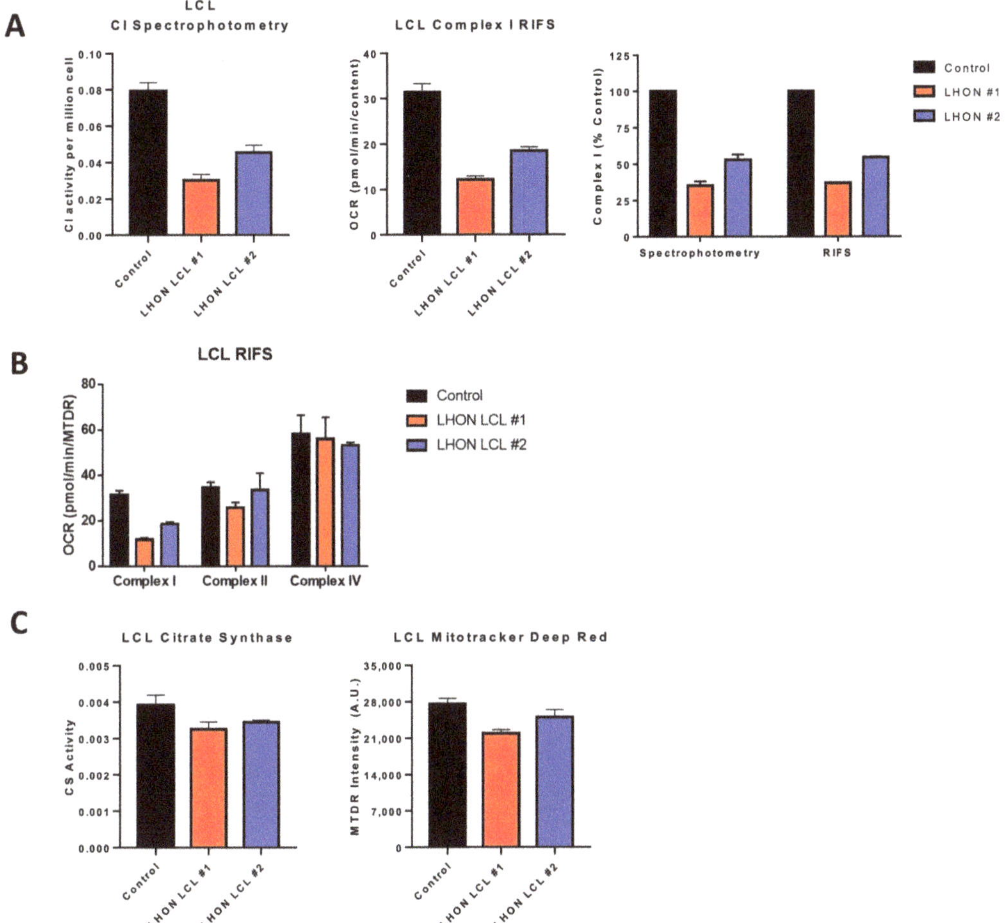

Figure 3. Comparison of RIFS and spectrophotometric approaches in Leber optic atrophy lymphoblastoid cell lines. (**A**) CI activity compared with the RIFS protocol. The spectrophotometric CI assay required 0.5×10^6 cells/well compared with 0.2×10^6 cells/well with RIFS. The data were also normalized to control cells as a more direct comparison of the two techniques. (**B**) CI, CII, and CIV results were obtained using RIFS. (**C**) Mitochondrial content measurements with citrate synthase activity and the MitoTracker Deep Red plate reader assay. Abbreviations: LCL: Lymphoblastoid cell line; LHON: Leber optic atrophy; RIFS: Respirometry in frozen samples.

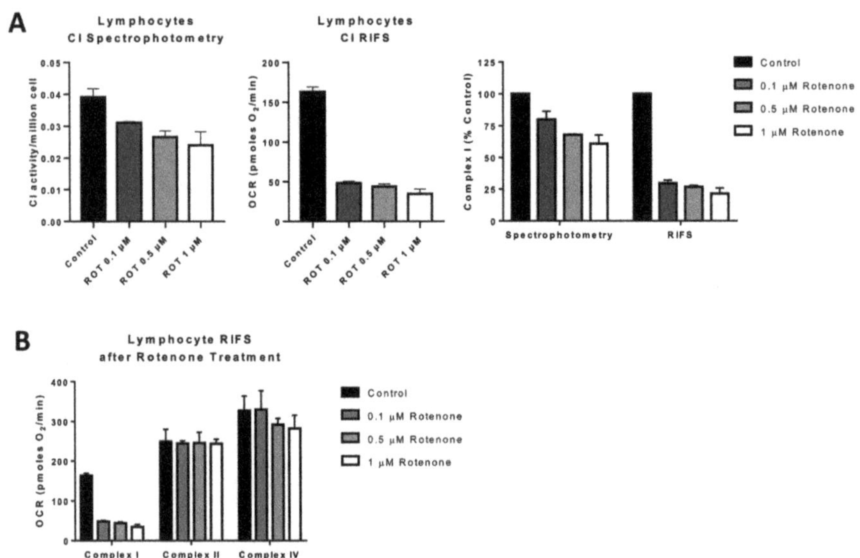

Figure 4. Comparison of RIFS and spectrophotometric approaches to measure Complex I in human lymphocytes treated with rotenone prior to freezing. (**A**) CI activity compared with the RIFS protocol. The spectrophotometric CI assay required 1×10^6 cells/well compared with 0.4×10^6 cells/well with RIFS. The data were also normalized to control cells as a more direct comparison of the two techniques. (**B**) Compiled CI, CII, and CIV results from two donors were obtained using RIFS. Abbreviations: CI: Complex I; RIFS: Respirometry in frozen samples; ROT: Rotenone.

An additional potential application for RIFS is in monitoring environmental toxicants to which mitochondria are a key target [155]. For that reason, applications targeted to direct screening of compounds leading to mitochondrial toxicity can benefit from RIFS by testing the functional impact of environmental exposures within a large population [155,156]. We tested rotenone exposure in lymphocytes treated with different concentrations of rotenone before freezing and compared the RIFS protocol to a CI activity spectrophotometric assay. Both approaches were able to reveal impairment in CI, but the RIFS assay was more than twice as sensitive and required half as much starting material (Figure 4A). Additionally, given the high oxygen consumption rates with the RIFS protocol, it is likely that less material is required than was used in these proof-of-concept assays (0.4×10^6 cells/well). The requirement of less starting material could be particularly important when the amount of sample is limited. In addition to CI, RIFS also provide information on CII and CIV demonstrating that the rotenone-induced defect was limited to CI (Figure 4B).

5.2. Limitations of RIFS

While the RIFS protocol was not developed to replace traditional respirometry approaches, it is an alternative method when using fresh samples is not feasible due to sample size and availability, or immediate access to instrumentation is not possible, a problem particularly relevant for testing human samples. Table 5 provides a comparison of the advantages and disadvantages of measuring respiration in fresh and frozen tissue samples. Respirometry in fresh samples will undoubtedly provide the most complete information about mitochondrial function and the bioenergetics of a sample given the coupled nature of fresh samples, that is, respiration is coupled to ATP synthesis. However, RIFS can provide a convenient alternative when respiration in fresh samples is not feasible or as a complementary method to determine maximal respiratory capacity in an integrated, yet uncoupled, system.

Table 5. Advantages and disadvantages of respirometry measurements in fresh and frozen tissue samples. Abbreviations: ATP: Adenosine triphosphate; TCA: Tricarboxylic acid cycle.

	Fresh	Frozen
Advantages	• Greater physiological context • Coupled mitochondria that synthesize ATP • Intact mitochondria that maintain morphology with some techniques • Metabolic pathways and transporters can be investigated • Respiration can be measured in intact and permeabilized cells and isolated mitochondria	• Integrated oxygen consumption measurement • Maximal respiratory capacity often correlates with disease phenotypes • Minimal sample requirement and processing • Samples can be collected off-site • Samples can be stored and run together in one assay • Maximal respiration can be measured in homogenates and normalized to mitochondrial content
Drawbacks	• Samples have to be processed and run on the same day • Mitochondrial isolation requires sufficient starting material • Control and patient samples cannot always be run simultaneously increasing variability	• Samples are not coupled and ATP synthesis capacity cannot be assessed • Broken mitochondria that do not maintain mitochondrial morphology or membrane integrity • Metabolic pathways cannot be assessed • TCA cycle and transporters are not active

Since freeze-thaw cycles will disrupt the integrity of the mitochondrial membranes and results in the loss of soluble matrix components, an important limitation is that coupled respiration or ATP synthesis cannot be measured with RIFS. These limitations should be considered when using this approach as it will not be sensitive to changes in substrate transport, substrate oxidation, ATP turnover, or TCA cycle flux. However, it has been demonstrated that disease-related phenotype often correlates with a decrease in maximal respiration [119,157–159] and that defects in coupled respiration are mostly related to some mechanism of uncommon drug toxicity [160]. Therefore, there are multiple applications for RIFS when testing fresh samples is logistically impracticable. While the data presented in this review are preliminary as proof of concepts for use of RIFS in circulating blood cells, the data suggest it could complement the current blood-based bioenergetics approaches. Furthermore, as the field continues to search for a biomarker of mitochondrial function, the RIFS approach provides an additional method for measuring maximal mitochondrial respiration and may be useful for non-invasive samples.

6. Future Outlook: Respirometry in Non-Invasive Samples

In the last decade, there have been substantial improvements in approaches to measure mitochondrial function in human samples. In particular, measurement of mitochondrial function in circulating platelets and leukocytes has elevated translational mitochondrial research and improved the prospects of clinical monitoring of mitochondrial bioenergetics. However, blood collection still needs to be completed in a medical setting and requires fresh processing. This processing can be intensive, especially when isolating sub-populations of PBMCs, which can present challenges in the ability to scale up for clinical trials. Additionally, blood-based bioenergetics do not always demonstrate changes in mitochondrial function observed in primary tissues [161], which suggests that subpopulation testing is an important consideration as subpopulation bioenergetic correlations may differ between tissues and diseases. More research is needed to understand the bioenergetic differences between subpopulations, population variability, and selection of the appropriate cell type so that this approach can be standardized. With the growing number of mitochondrial drug targets being developed and tested in clinical trials, there is a need for further minimally/non-invasive samples that can be scaled up for clinical testing and large-scale population studies. The RIFS protocol has the potential to be used for these applications if it can be optimized from non-invasive samples such as cells from urine or saliva, or buccal mucosa cells collected from cheek swabs, which can be easily obtained and immediately frozen without processing for later testing. Since RIFS requires minimal starting material, it could be modified for use in non-invasive samples that require minimal processing. While it is still unknown whether RIFS will be a suitable approach for non-invasive samples, there is some evidence that buccal swabs may be an interesting foundation for the development of non-invasive samples.

Buccal mucosa cells represent a potential non-invasive sample that can be used for mitochondrial bioenergetics profiling. Buccal swab testing of electron transport chain activity is possible [162], but has been reported to show poor sensitivity and specificity for mitochondrial disease [41]. Frederiksen et al. showed that there is a significant correlation in mutation load of the common A3243G mtDNA point mutation in four different tissues: blood leucocytes, buccal cells, skeletal muscle cells, and urine epithelial cells [163]. They suggest that genetic testing of buccal cells and urine epithelial cells may provide an important supplement to the diagnosis of mitochondrial diseases, but also note that skeletal muscle was still the most informative tissue with the highest mutation load [163]. Buccal swabs from a 6-year old patient with suspected myoclonic epilepsy and ragged red fibers (MERRF) presented common mtDNA deletions and decreased ETC activities compared with controls [164]. In a study of 40 patients with suspected mitochondrial disease, when comparing enzymatic activities in buccal mucosa cells to muscle biopsies, there was a 77% correlation for an isolated CI deficiency, 86% for combined CI and CIV defects, and 100% in CIV. Overall, detection of mitochondrial impairment was 82% in buccal swabs compared to muscle biopsies [165]. Alterations in buccal mitochondrial DNA and function have also been reported in lung cancer patients and autism spectrum disorder patients (ASD) [166,167]. Additionally, there is evidence that buccal swab samples can be used to monitor mitochondrial enzymatic activity in response to treatment. In a study of ASD children with or without mitochondrial disease, CI, CIV, and citrate synthase activity in response to fatty acid and folate supplementation was measured in buccal extracts [168]. The results suggest that changes in CI and citrate synthase activities can be observed in buccal swabs in response to common interventions and that increased activity in these enzymes was more discernable in the mitochondrial disease subgroup.

While studies measuring mitochondrial function in buccal swabs are still limited, enough data demonstrate mitochondrial functional changes in buccal swabs to warrant further investigation. It is possible that an integrated readout, such as RIFS, would provide both better sensitivity and specificity for buccal samples. However, a potential limitation to buccal swab bioenergetics measurements could be bacterial contamination that decreases measurement specificity to the human samples [169,170]. Removal of bacteria from these samples will likely be an important step to improve sensitivity in the respirometry assay and ensure that outcomes are accurate. While this approach in buccal mucosa is still hypothetical and would require optimization and validation, it could provide a method to rapidly screen maximal mitochondrial respiratory capacity and provide a complementary approach to blood-based bioenergetic measurements and measurements in primary tissues.

7. Summary

Impaired mitochondrial function has been shown to contribute to the development and progression of various diseases. There is an unmet need to clinically monitor mitochondrial function. However, until now this has been limited due to the labor-intensive immediate processing of the samples. Blood-based bioenergetics have met some of this need but do not exclude the necessity of immediate processing. With the development of RIFS as a complementary technique to measure mitochondrial function in previously frozen samples, the possibility of clinical monitoring of mitochondrial function in samples collected at remote sites, or retrospectively in samples residing in tissue biobanks becomes more apparent. While these emerging approaches require further optimization, in some cases, and validation of population variability and sensitivity to detect differences between healthy and disease patients, they represent a potential step forward in translation bioenergetic monitoring in human samples.

Author Contributions: Conceptualization, L.S. and R.A.-P.; investigation, B.S., L.S. and R.A.-P.; writing—original draft, L.S. and R.A.-P.; writing—review and editing, B.S., C.B., L.S., R.A.-P. and O.S.S.; writing—revised draft, C.B., L.S. and R.A.-P. All authors have read and agreed to the published version of the manuscript.

Funding: O.S.S. is funded by American Diabetes Association 1-19-IBS-049.

Institutional Review Board Statement: Ethical review and approval were waived for this study due to the use of de-identified/commercially available samples.

Informed Consent Statement: Patient consent was waived due to the use of de-identified/commercially available samples.

Data Availability Statement: Data available upon request.

Acknowledgments: The authors would like to acknowledge Tom Grammatopoulos for providing materials. The authors would like to thank Corey Osto and Shai Rambod, as well as Ilan Benador, Daniel Dagan, Ajit Divakaruni, Marc Liesa-Roig, Marcus Oliveira, Martin Picard, and Amy Wang for scientific discussions and valuable advice.

Conflicts of Interest: The following competing interests did not influence the writing of this review: O.S.S. is a cofounder of Enspire Bio; O.S.S. and L.S. are cofounders of MyChon Health.

Abbreviations

Frequently Used Abbreviations: AA: Antimycin A; ADP: Adenosine diphosphate; Asc: Ascorbate; ATP: Adenosine triphosphate; CI: Complex I; CII: Complex II; CIII: Complex III; CIV: Complex IV; CV: Complex V; DH: Dehydrogenase; DNA: Deoxyribonucleic acid; ECAR: Extracellular acidification rate; ETC: Electron transport chain; FCCP: Carbonyl cyanide p-trifluoromethoxyphenylhydrazone; LCL: Lymphoblastoid cell line; LHON: Leber optic atrophy; NADH: Reduced nicotinamide adenine dinucleotide; nDNA: Nuclear DNA; mtDNA: Mitochondrial DNA; MTDR: MitoTracker Deep Red; OCR: Oxygen consumption rate; Oligo: Oligomycin; OXPHOS: Oxidative phosphorylation; PBMC: Peripheral blood mononuclear cell; RIFS: Respirometry in frozen samples; ROS: Reactive oxygen species; Rot: Rotenone; Succ: Succinate; TCA: Tricarboxylic acid cycle; TMPD: N, N, N′,N′-tetramethyl-p-phenylenediamine.

References

1. Herst, P.M.; Rowe, M.R.; Carson, G.M.; Berridge, M.V. Functional mitochondria in health and disease. *Front. Endocrinol.* **2017**, *8*, 296. [CrossRef]
2. Fernandez-Vizarra, E.; Enriquez, J.A.; Perez-Martos, A.; Montoya, J.; Fernandez-Silva, P. Tissue-specific differences in mitochondrial activity and biogenesis. *Mitochondrion* **2011**, *11*, 207–213. [CrossRef]
3. Spinelli, J.B.; Haigis, M.C. The multifaceted contributions of mitochondria to cellular metabolism. *Nat. Cell Biol.* **2018**, *20*, 745–754. [CrossRef]
4. Eisner, V.; Picard, M.; Hajnoczky, G. Mitochondrial dynamics in adaptive and maladaptive cellular stress responses. *Nat. Cell Biol.* **2018**, *20*, 755–765. [CrossRef]
5. Benador, I.Y.; Veliova, M.; Liesa, M.; Shirihai, O.S. Mitochondria bound to lipid droplets: Where mitochondrial dynamics regulate lipid storage and utilization. *Cell Metab.* **2019**, *29*, 827–835. [CrossRef] [PubMed]
6. Stenton, S.L.; Prokisch, H. Genetics of mitochondrial diseases: Identifying mutations to help diagnosis. *EBioMedicine* **2020**, *56*, 102784. [CrossRef]
7. Herbers, E.; Kekalainen, N.J.; Hangas, A.; Pohjoismaki, J.L.; Goffart, S. Tissue specific differences in mitochondrial DNA maintenance and expression. *Mitochondrion* **2019**, *44*, 85–92. [CrossRef] [PubMed]
8. Lanza, I.R.; Nair, K.S. Mitochondrial metabolic function assessed in vivo and in vitro. *Curr. Opin. Clin. Nutr. Metab. Care* **2010**, *13*, 511–517. [CrossRef] [PubMed]
9. Gnaiger, E. Capacity of oxidative phosphorylation in human skeletal muscle: New perspectives of mitochondrial physiology. *Int. J. Biochem. Cell Biol.* **2009**, *41*, 1837–1845. [CrossRef]
10. Divakaruni, A.S.; Paradyse, A.; Ferrick, D.A.; Murphy, A.N.; Jastroch, M. Analysis and interpretation of microplate-based oxygen consumption and pH data. *Methods Enzymol.* **2014**, *547*, 309–354. [CrossRef] [PubMed]
11. Divakaruni, A.S.; Rogers, G.W.; Murphy, A.N. Measuring mitochondrial function in permeabilized cells using the seahorse XF analyzer or a clark-type oxygen electrode. *Curr. Protoc. Toxicol.* **2014**, *60*, 25.2.1–25.2.16. [CrossRef]
12. Brand, M.D.; Nicholls, D.G. Assessing mitochondrial dysfunction in cells. *Biochem. J.* **2011**, *435*, 297–312. [CrossRef]

13. Ost, M.; Doerrier, C.; Gama-Perez, P.; Moreno-Gomez, S. Analysis of mitochondrial respiratory function in tissue biopsies and blood cells. *Curr. Opin. Clin. Nutr. Metab. Care* **2018**, *21*, 336–342. [CrossRef]
14. Horan, M.P.; Pichaud, N.; Ballard, J.W. Review: Quantifying mitochondrial dysfunction in complex diseases of aging. *J. Gerontol. A Biol. Sci. Med. Sci.* **2012**, *67*, 1022–1035. [CrossRef] [PubMed]
15. Djafarzadeh, S.; Jakob, S.M. High-resolution respirometry to assess mitochondrial function in permeabilized and intact cells. *J. Vis. Exp.* **2017**, *120*, e54985. [CrossRef] [PubMed]
16. Zhang, J.; Nuebel, E.; Wisidagama, D.R.; Setoguchi, K.; Hong, J.S.; Van Horn, C.M.; Imam, S.S.; Vergnes, L.; Malone, C.S.; Koehler, C.M.; et al. Measuring energy metabolism in cultured cells, including human pluripotent stem cells and differentiated cells. *Nat. Protoc.* **2012**, *7*, 1068–1085. [CrossRef] [PubMed]
17. Doerrier, C.; Garcia-Souza, L.F.; Krumschnabel, G.; Wohlfarter, Y.; Meszaros, A.T.; Gnaiger, E. High-resolution FluoRespirometry and OXPHOS protocols for human cells, permeabilized fibers from small biopsies of muscle, and isolated mitochondria. *Methods Mol. Biol.* **2018**, *1782*, 31–70. [CrossRef] [PubMed]
18. Pesta, D.; Gnaiger, E. High-resolution respirometry: OXPHOS protocols for human cells and permeabilized fibers from small biopsies of human muscle. *Methods Mol. Biol.* **2012**, *810*, 25–58. [CrossRef]
19. Jones, A.E.; Sheng, L.; Acevedo, A.; Veliova, M.; Shirihai, O.S.; Stiles, L.; Divakaruni, A.S. Forces, fluxes, and fuels: Tracking mitochondrial metabolism by integrating measurements of membrane potential, respiration, and metabolites. *Am. J. Physiol. Cell Physiol.* **2021**, *320*, C80–C91. [CrossRef]
20. Lanza, I.R.; Nair, K.S. Functional assessment of isolated mitochondria in vitro. *Methods Enzymol.* **2009**, *457*, 349–372. [CrossRef] [PubMed]
21. Nolfi-Donegan, D.; Braganza, A.; Shiva, S. Mitochondrial electron transport chain: Oxidative phosphorylation, oxidant production, and methods of measurement. *Redox Biol.* **2020**, *37*, 101674. [CrossRef] [PubMed]
22. Hill, B.G.; Shiva, S.; Ballinger, S.; Zhang, J.; Darley-Usmar, V.M. Bioenergetics and translational metabolism: Implications for genetics, physiology and precision medicine. *Biol. Chem.* **2019**, *401*, 3–29. [CrossRef] [PubMed]
23. Gnaiger, E. Polarographic oxygen sensors, the oxygraph, and high-resolution respirometry to assess mitochondrial function. In *Drug-Induced Mitochondrial Dysfunction*; Dykens, J.A., Will, Y., Eds.; Wiley: Hoboken, NJ, USA, 2008; pp. 327–348. [CrossRef]
24. Salabei, J.K.; Gibb, A.A.; Hill, B.G. Comprehensive measurement of respiratory activity in permeabilized cells using extracellular flux analysis. *Nat. Protoc.* **2014**, *9*, 421–438. [CrossRef] [PubMed]
25. Garaude, J.; Acin-Perez, R.; Martinez-Cano, S.; Enamorado, M.; Ugolini, M.; Nistal-Villan, E.; Hervas-Stubbs, S.; Pelegrin, P.; Sander, L.E.; Enriquez, J.A.; et al. Mitochondrial respiratory-chain adaptations in macrophages contribute to antibacterial host defense. *Nat. Immunol.* **2016**, *17*, 1037–1045. [CrossRef]
26. Schlieben, L.D.; Prokisch, H. The dimensions of primary mitochondrial disorders. *Front. Cell Dev. Biol.* **2020**, *8*, 600079. [CrossRef]
27. Gorman, G.S.; Schaefer, A.M.; Ng, Y.; Gomez, N.; Blakely, E.L.; Alston, C.L.; Feeney, C.; Horvath, R.; Yu-Wai-Man, P.; Chinnery, P.F.; et al. Prevalence of nuclear and mitochondrial DNA mutations related to adult mitochondrial disease. *Ann. Neurol.* **2015**, *77*, 753–759. [CrossRef] [PubMed]
28. Gorman, G.S.; Chinnery, P.F.; DiMauro, S.; Hirano, M.; Koga, Y.; McFarland, R.; Suomalainen, A.; Thorburn, D.R.; Zeviani, M.; Turnbull, D.M. Mitochondrial diseases. *Nat. Rev. Dis. Primers* **2016**, *2*, 16080. [CrossRef]
29. Schaefer, A.M.; Taylor, R.W.; Turnbull, D.M.; Chinnery, P.F. The epidemiology of mitochondrial disorders—Past, present and future. *Biochim. Biophys. Acta* **2004**, *1659*, 115–120. [CrossRef]
30. Schaefer, A.; Lim, A.; Gorman, G. *Epidemiology of Mitochondrial Disease*; Mancuso, M., Klopstock, T., Eds.; Springer: Cham, Switzerland, 2019. [CrossRef]
31. Loeffen, J.L.; Smeitink, J.A.; Trijbels, J.M.; Janssen, A.J.; Triepels, R.H.; Sengers, R.C.; van den Heuvel, L.P. Isolated complex I deficiency in children: Clinical, biochemical and genetic aspects. *Hum. Mutat.* **2000**, *15*, 123–134. [CrossRef]
32. Hoppel, C.L.; Kerr, D.S.; Dahms, B.; Roessmann, U. Deficiency of the reduced nicotinamide adenine dinucleotide dehydrogenase component of complex I of mitochondrial electron transport. Fatal infantile lactic acidosis and hypermetabolism with skeletal-cardiac myopathy and encephalopathy. *J. Clin. Investig.* **1987**, *80*, 71–77. [CrossRef]
33. Swalwell, H.; Kirby, D.M.; Blakely, E.L.; Mitchell, A.; Salemi, R.; Sugiana, C.; Compton, A.G.; Tucker, E.J.; Ke, B.X.; Lamont, P.J.; et al. Respiratory chain complex I deficiency caused by mitochondrial DNA mutations. *Eur. J. Hum. Genet.* **2011**, *19*, 769–775. [CrossRef] [PubMed]
34. Clarke, C.; Xiao, R.; Place, E.; Zhang, Z.; Sondheimer, N.; Bennett, M.; Yudkoff, M.; Falk, M.J. Mitochondrial respiratory chain disease discrimination by retrospective cohort analysis of blood metabolites. *Mol. Genet. Metab.* **2013**, *110*, 145–152. [CrossRef] [PubMed]
35. Thompson Legault, J.; Strittmatter, L.; Tardif, J.; Sharma, R.; Tremblay-Vaillancourt, V.; Aubut, C.; Boucher, G.; Clish, C.B.; Cyr, D.; Daneault, C.; et al. A metabolic signature of mitochondrial dysfunction revealed through a monogenic form of leigh syndrome. *Cell Rep.* **2015**, *13*, 981–989. [CrossRef] [PubMed]
36. Schon, K.R.; Ratnaike, T.; van den Ameele, J.; Horvath, R.; Chinnery, P.F. Mitochondrial diseases: A diagnostic revolution. *Trends Genet.* **2020**, *36*, 702–717. [CrossRef] [PubMed]
37. Riley, L.G.; Cowley, M.J.; Gayevskiy, V.; Minoche, A.E.; Puttick, C.; Thorburn, D.R.; Rius, R.; Compton, A.G.; Menezes, M.J.; Bhattacharya, K.; et al. The diagnostic utility of genome sequencing in a pediatric cohort with suspected mitochondrial disease. *Genet. Med.* **2020**, *22*, 1254–1261. [CrossRef]

38. Watson, E.; Davis, R.; Sue, C.M. New diagnostic pathways for mitochondrial disease. *J. Transl. Genet. Genom.* **2020**, *4*, 188–202. [CrossRef]
39. Chinnery, P.F. Mitochondrial disease in adults: What's old and what's new? *EMBO Mol. Med.* **2015**, *7*, 1503–1512. [CrossRef]
40. Pfeffer, G.; Pyle, A.; Griffin, H.; Miller, J.; Wilson, V.; Turnbull, L.; Fawcett, K.; Sims, D.; Eglon, G.; Hadjivassiliou, M.; et al. SPG7 mutations are a common cause of undiagnosed ataxia. *Neurology* **2015**, *84*, 1174–1176. [CrossRef]
41. Muraresku, C.C.; McCormick, E.M.; Falk, M.J. Mitochondrial Disease: Advances in clinical diagnosis, management, therapeutic development, and preventative strategies. *Curr. Genet. Med. Rep.* **2018**, *6*, 62–72. [CrossRef]
42. Niyazov, D.M.; Kahler, S.G.; Frye, R.E. Primary mitochondrial disease and secondary mitochondrial dysfunction: Importance of distinction for diagnosis and treatment. *Mol. Syndromol.* **2016**, *7*, 122–137. [CrossRef]
43. Sulaiman, S.A.; Rani, Z.Z.M.; Radin, F.Z.M.; Murad, N.A. Advancement in the diagnosis of mitochondrial diseases. *J. Transl. Genet. Genom.* **2020**, *4*, 159–187. [CrossRef]
44. Joyce, N.C.; Oskarsson, B.; Jin, L.W. Muscle biopsy evaluation in neuromuscular disorders. *Phys. Med. Rehabil. Clin. N. Am.* **2012**, *23*, 609–631. [CrossRef]
45. Acin-Perez, R.; Enriquez, J.A. The function of the respiratory supercomplexes: The plasticity model. *Biochim. Biophys. Acta* **2014**, *1837*, 444–450. [CrossRef]
46. Acin-Perez, R.; Fernandez-Silva, P.; Peleato, M.L.; Perez-Martos, A.; Enriquez, J.A. Respiratory active mitochondrial supercomplexes. *Mol. Cell* **2008**, *32*, 529–539. [CrossRef]
47. Barrientos, A. In vivo and in organello assessment of OXPHOS activities. *Methods* **2002**, *26*, 307–316. [CrossRef]
48. Barrientos, A.; Fontanesi, F.; Diaz, F. Evaluation of the mitochondrial respiratory chain and oxidative phosphorylation system using polarography and spectrophotometric enzyme assays. *Curr. Protoc. Hum. Genet.* **2009**, *63*, 19–23. [CrossRef] [PubMed]
49. Birch-Machin, M.A.; Turnbull, D.M. Assaying mitochondrial respiratory complex activity in mitochondria isolated from human cells and tissues. *Methods Cell Biol.* **2001**, *65*, 97–117. [CrossRef] [PubMed]
50. Janssen, R.C.; Boyle, K.E. Microplate assays for spectrophotometric measurement of mitochondrial enzyme activity. *Methods Mol. Biol.* **2019**, *1978*, 355–368. [CrossRef] [PubMed]
51. Lapuente-Brun, E.; Moreno-Loshuertos, R.; Acin-Perez, R.; Latorre-Pellicer, A.; Colas, C.; Balsa, E.; Perales-Clemente, E.; Quiros, P.M.; Calvo, E.; Rodriguez-Hernandez, M.A.; et al. Supercomplex assembly determines electron flux in the mitochondrial electron transport chain. *Science* **2013**, *340*, 1567–1570. [CrossRef] [PubMed]
52. Acin-Perez, R.; Hernansanz-Agustin, P.; Enriquez, J.A. Analyzing electron transport chain supercomplexes. *Methods Cell Biol.* **2020**, *155*, 181–197. [CrossRef]
53. Haraux, F.; Lombes, A. Kinetic analysis of ATP hydrolysis by complex V in four murine tissues: Towards an assay suitable for clinical diagnosis. *PLoS ONE* **2019**, *14*, e0221886. [CrossRef]
54. Morava, E.; Rodenburg, R.J.; Hol, F.; de Vries, M.; Janssen, A.; van den Heuvel, L.; Nijtmans, L.; Smeitink, J. Clinical and biochemical characteristics in patients with a high mutant load of the mitochondrial T8993G/C mutations. *Am. J. Med. Genet. A* **2006**, *140*, 863–868. [CrossRef]
55. Chi, M.M.; Hintz, C.S.; Coyle, E.F.; Martin, W.H., 3rd; Ivy, J.L.; Nemeth, P.M.; Holloszy, J.O.; Lowry, O.H. Effects of detraining on enzymes of energy metabolism in individual human muscle fibers. *Am. J. Physiol.* **1983**, *244*, C276–C287. [CrossRef] [PubMed]
56. Moreno-Loshuertos, R.; Acin-Perez, R.; Fernandez-Silva, P.; Movilla, N.; Perez-Martos, A.; Rodriguez de Cordoba, S.; Gallardo, M.E.; Enriquez, J.A. Differences in reactive oxygen species production explain the phenotypes associated with common mouse mitochondrial DNA variants. *Nat. Genet.* **2006**, *38*, 1261–1268. [CrossRef] [PubMed]
57. Tretter, L.; Adam-Vizi, V. Inhibition of Krebs cycle enzymes by hydrogen peroxide: A key role of [alpha]-ketoglutarate dehydrogenase in limiting NADH production under oxidative stress. *J. Neurosci.* **2000**, *20*, 8972–8979. [CrossRef] [PubMed]
58. Acin-Perez, R.; Iborra, S.; Marti-Mateos, Y.; Cook, E.C.L.; Conde-Garrosa, R.; Petcherski, A.; Munoz, M.D.M.; Martinez de Mena, R.; Krishnan, K.C.; Jimenez, C.; et al. Fgr kinase is required for proinflammatory macrophage activation during diet-induced obesity. *Nat. Metab.* **2020**, *2*, 974–988. [CrossRef]
59. Gardner, P.R.; Nguyen, D.D.; White, C.W. Aconitase is a sensitive and critical target of oxygen poisoning in cultured mammalian cells and in rat lungs. *Proc. Natl. Acad. Sci. USA* **1994**, *91*, 12248–12252. [CrossRef]
60. Oyanagui, Y. Reevaluation of assay methods and establishment of kit for superoxide dismutase activity. *Anal. Biochem.* **1984**, *142*, 290–296. [CrossRef]
61. Francisco, A.; Ronchi, J.A.; Navarro, C.D.C.; Figueira, T.R.; Castilho, R.F. Nicotinamide nucleotide transhydrogenase is required for brain mitochondrial redox balance under hampered energy substrate metabolism and high-fat diet. *J. Neurochem.* **2018**, *147*, 663–677. [CrossRef]
62. Ronchi, J.A.; Figueira, T.R.; Ravagnani, F.G.; Oliveira, H.C.; Vercesi, A.E.; Castilho, R.F. A spontaneous mutation in the nicotinamide nucleotide transhydrogenase gene of C57BL/6J mice results in mitochondrial redox abnormalities. *Free Radic. Biol. Med.* **2013**, *63*, 446–456. [CrossRef] [PubMed]
63. Shimomura, K.; Galvanovskis, J.; Goldsworthy, M.; Hugill, A.; Kaizak, S.; Lee, A.; Meadows, N.; Quwailid, M.M.; Rydstrom, J.; Teboul, L.; et al. Insulin secretion from beta-cells is affected by deletion of nicotinamide nucleotide transhydrogenase. *Methods Enzymol.* **2009**, *457*, 451–480. [CrossRef]

64. Acin-Perez, R.; Hoyos, B.; Zhao, F.; Vinogradov, V.; Fischman, D.A.; Harris, R.A.; Leitges, M.; Wongsiriroj, N.; Blaner, W.S.; Manfredi, G.; et al. Control of oxidative phosphorylation by vitamin A illuminates a fundamental role in mitochondrial energy homoeostasis. *FASEB J.* **2010**, *24*, 627–636. [CrossRef]
65. Hale, D.E.; Cornell, J.E.; Bennett, M.J. Stability of long-chain and short-chain 3-hydroxyacyl-CoA dehydrogenase activity in postmortem liver. *Clin. Chem.* **1997**, *43*, 273–278. [CrossRef] [PubMed]
66. Binstock, J.F.; Schulz, H. Fatty acid oxidation complex from Escherichia coli. *Methods Enzymol.* **1981**, *71*, 403–411. [CrossRef] [PubMed]
67. Grinblat, L.; Pacheco Bolanos, L.F.; Stoppani, A.O. Decreased rate of ketone-body oxidation and decreased activity of D-3-hydroxybutyrate dehydrogenase and succinyl-CoA:3-oxo-acid CoA-transferase in heart mitochondria of diabetic rats. *Biochem. J.* **1986**, *240*, 49–56. [CrossRef] [PubMed]
68. Lehninger, A.L.; Sudduth, H.C.; Wise, J.B. D-beta-Hydroxybutyric dehydrogenase of muitochondria. *J. Biol. Chem.* **1960**, *235*, 2450–2455. [CrossRef]
69. DeBalsi, K.L.; Wong, K.E.; Koves, T.R.; Slentz, D.H.; Seiler, S.E.; Wittmann, A.H.; Ilkayeva, O.R.; Stevens, R.D.; Perry, C.G.; Lark, D.S.; et al. Targeted metabolomics connects thioredoxin-interacting protein (TXNIP) to mitochondrial fuel selection and regulation of specific oxidoreductase enzymes in skeletal muscle. *J. Biol. Chem.* **2014**, *289*, 8106–8120. [CrossRef]
70. Scheer, W.D.; Lehmann, H.P.; Beeler, M.F. An improved assay for hexokinase activity in human tissue homogenates. *Anal. Biochem.* **1978**, *91*, 451–463. [CrossRef]
71. Nuzum, C.T.; Snodgrass, P.J. Multiple assays of the five urea-cycle enzymes in human liver homogenates. In *The Urea Cycle*; Wiley: New York, NY, USA, 1976; pp. 325–349.
72. Lawless, C.; Greaves, L.; Reeve, A.K.; Turnbull, D.M.; Vincent, A.E. The rise and rise of mitochondrial DNA mutations. *Open Biol.* **2020**, *10*, 200061. [CrossRef]
73. Menzies, R.A.; Gold, P.H. The turnover of mitochondria in a variety of tissues of young adult and aged rats. *J. Biol. Chem.* **1971**, *246*, 2425–2429. [CrossRef]
74. Korr, H.; Kurz, C.; Seidler, T.O.; Sommer, D.; Schmitz, C. Mitochondrial DNA synthesis studied autoradiographically in various cell types in vivo. *Braz. J. Med. Biol. Res.* **1998**, *31*, 289–298. [CrossRef] [PubMed]
75. Bourgeois, J.M.; Tarnopolsky, M.A. Pathology of skeletal muscle in mitochondrial disorders. *Mitochondrion* **2004**, *4*, 441–452. [CrossRef]
76. Shanely, R.A.; Zwetsloot, K.A.; Triplett, N.T.; Meaney, M.P.; Farris, G.E.; Nieman, D.C. Human skeletal muscle biopsy procedures using the modified Bergstrom technique. *J. Vis. Exp.* **2014**, 51812. [CrossRef]
77. Hayot, M.; Michaud, A.; Koechlin, C.; Caron, M.A.; Leblanc, P.; Prefaut, C.; Maltais, F. Skeletal muscle microbiopsy: A validation study of a minimally invasive technique. *Eur. Respir. J.* **2005**, *25*, 431–440. [CrossRef] [PubMed]
78. Ryan, T.E.; Brophy, P.; Lin, C.T.; Hickner, R.C.; Neufer, P.D. Assessment of in vivo skeletal muscle mitochondrial respiratory capacity in humans by near-infrared spectroscopy: A comparison with in situ measurements. *J. Physiol.* **2014**, *592*, 3231–3241. [CrossRef] [PubMed]
79. Parikh, S.; Goldstein, A.; Koenig, M.K.; Scaglia, F.; Enns, G.M.; Saneto, R.; Anselm, I.; Cohen, B.H.; Falk, M.J.; Greene, C.; et al. Diagnosis and management of mitochondrial disease: A consensus statement from the Mitochondrial Medicine Society. *Genet. Med.* **2015**, *17*, 689–701. [CrossRef] [PubMed]
80. Saneto, R.P. Mitochondrial diseases: Expanding the diagnosis in the era of genetic testing. *J. Transl. Genet. Genom.* **2020**, *4*, 384–428. [CrossRef]
81. Auburger, G.; Klinkenberg, M.; Drost, J.; Marcus, K.; Morales-Gordo, B.; Kunz, W.S.; Brandt, U.; Broccoli, V.; Reichmann, H.; Gispert, S.; et al. Primary skin fibroblasts as a model of ParkinsoN's disease. *Mol. Neurobiol.* **2012**, *46*, 20–27. [CrossRef]
82. Jonckheere, A.I.; Huigsloot, M.; Janssen, A.J.; Kappen, A.J.; Smeitink, J.A.; Rodenburg, R.J. High-throughput assay to measure oxygen consumption in digitonin-permeabilized cells of patients with mitochondrial disorders. *Clin. Chem.* **2010**, *56*, 424–431. [CrossRef]
83. Iyer, S.; Bergquist, K.; Young, K.; Gnaiger, E.; Rao, R.R.; Bennett, J.P., Jr. Mitochondrial gene therapy improves respiration, biogenesis, and transcription in G11778A Leber's hereditary optic neuropathy and T8993G Leigh's syndrome cells. *Hum. Gene Ther.* **2012**, *23*, 647–657. [CrossRef]
84. Ogawa, E.; Shimura, M.; Fushimi, T.; Tajika, M.; Ichimoto, K.; Matsunaga, A.; Tsuruoka, T.; Ishige, M.; Fuchigami, T.; Yamazaki, T.; et al. Clinical validity of biochemical and molecular analysis in diagnosing Leigh syndrome: A study of 106 Japanese patients. *J. Inherit. Metab. Dis.* **2017**, *40*, 685–693. [CrossRef] [PubMed]
85. Invernizzi, F.; D'Amato, I.; Jensen, P.B.; Ravaglia, S.; Zeviani, M.; Tiranti, V. Microscale oxygraphy reveals OXPHOS impairment in MRC mutant cells. *Mitochondrion* **2012**, *12*, 328–335. [CrossRef]
86. Lahuerta, M.; Aguado, C.; Sanchez-Martin, P.; Sanz, P.; Knecht, E. Degradation of altered mitochondria by autophagy is impaired in Lafora disease. *FEBS J.* **2018**, *285*, 2071–2090. [CrossRef]
87. Hannibal, L.; Theimer, J.; Wingert, V.; Klotz, K.; Bierschenk, I.; Nitschke, R.; Spiekerkoetter, U.; Grunert, S.C. Metabolic profiling in human fibroblasts enables subtype clustering in glycogen storage disease. *Front. Endocrinol.* **2020**, *11*, 579981. [CrossRef]
88. Tokuyama, T.; Hirai, A.; Shiiba, I.; Ito, N.; Matsuno, K.; Takeda, K.; Saito, K.; Mii, K.; Matsushita, N.; Fukuda, T.; et al. Mitochondrial dynamics regulation in skin fibroblasts from mitochondrial disease patients. *Biomolecules* **2020**, *10*, 450. [CrossRef] [PubMed]

89. Lin, D.S.; Kao, S.H.; Ho, C.S.; Wei, Y.H.; Hung, P.L.; Hsu, M.H.; Wu, T.Y.; Wang, T.J.; Jian, Y.R.; Lee, T.H.; et al. Inflexibility of AMPK-mediated metabolic reprogramming in mitochondrial disease. *Oncotarget* **2017**, *8*, 73627–73639. [CrossRef] [PubMed]
90. Gstraunthaler, G.; Seppi, T.; Pfaller, W. Impact of culture conditions, culture media volumes, and glucose content on metabolic properties of renal epithelial cell cultures. Are renal cells in tissue culture hypoxic? *Cell Physiol. Biochem.* **1999**, *9*, 150–172. [CrossRef] [PubMed]
91. Van den Bogert, C.; Spelbrink, J.N.; Dekker, H.L. Relationship between culture conditions and the dependency on mitochondrial function of mammalian cell proliferation. *J. Cell Physiol.* **1992**, *152*, 632–638. [CrossRef] [PubMed]
92. Atkuri, K.R.; Herzenberg, L.A.; Niemi, A.K.; Cowan, T.; Herzenberg, L.A. Importance of culturing primary lymphocytes at physiological oxygen levels. *Proc. Natl. Acad. Sci. USA* **2007**, *104*, 4547–4552. [CrossRef]
93. Golpour, M.; Akhavan Niaki, H.; Khorasani, H.R.; Hajian, A.; Mehrasa, R.; Mostafazadeh, A. Human fibroblast switches to anaerobic metabolic pathway in response to serum starvation: A mimic of warburg effect. *Int. J. Mol. Cell Med.* **2014**, *3*, 74–80.
94. Balin, A.K.; Fisher, A.J.; Anzelone, M.; Leong, I.; Allen, R.G. Effects of establishing cell cultures and cell culture conditions on the proliferative life span of human fibroblasts isolated from different tissues and donors of different ages. *Exp. Cell Res.* **2002**, *274*, 275–287. [CrossRef]
95. Costa, C.F.; Pinho, S.A.; Pinho, S.L.C.; Miranda-Santos, I.; Bagshaw, O.; Stuart, J.; Oliveira, P.J.; Cunha-Oliveira, T. Mitochondrial and metabolic remodeling in human skin fibroblasts in response to glucose availability. *bioRxiv* **2021**. [CrossRef]
96. Pereira, S.P.; Deus, C.M.; Serafim, T.L.; Cunha-Oliveira, T.; Oliveira, P.J. Metabolic and phenotypic characterization of human skin fibroblasts after forcing oxidative capacity. *Toxicol. Sci.* **2018**, *164*, 191–204. [CrossRef]
97. Iannetti, E.F.; Smeitink, J.A.M.; Willems, P.; Beyrath, J.; Koopman, W.J.H. Rescue from galactose-induced death of Leigh Syndrome patient cells by pyruvate and NAD(.). *Cell Death Dis.* **2018**, *9*, 1135. [CrossRef] [PubMed]
98. Chen, Q.; Fischer, A.; Reagan, J.D.; Yan, L.J.; Ames, B.N. Oxidative DNA damage and senescence of human diploid fibroblast cells. *Proc. Natl. Acad. Sci. USA* **1995**, *92*, 4337–4341. [CrossRef] [PubMed]
99. Hayflick, L. The limited in vitro lifetime of human diploid cell strains. *Exp. Cell Res.* **1965**, *37*, 614–636. [CrossRef]
100. Tigges, J.; Krutmann, J.; Fritsche, E.; Haendeler, J.; Schaal, H.; Fischer, J.W.; Kalfalah, F.; Reinke, H.; Reifenberger, G.; Stuhler, K.; et al. The hallmarks of fibroblast ageing. *Mech. Ageing Dev.* **2014**, *138*, 26–44. [CrossRef]
101. Tollefsbol, T.O. Techniques for analysis of biological aging. *Methods Mol. Biol.* **2007**, *371*, 1–7. [CrossRef]
102. Cristofalo, V.J.; Volker, C.; Allen, R.G. Use of the fibroblast model in the study of cellular senescence. In *Aging Methods and Protocols (Methods in Molecular Medicine)*; Barnett, Y.A., Barnett, C.R., Eds.; Humana Press: Totowa, NJ, USA, 2000; Volume 38.
103. Bittles, A.H.; Harper, N. Increased glycolysis in ageing cultured human diploid fibroblasts. *Biosci. Rep.* **1984**, *4*, 751–756. [CrossRef]
104. James, E.L.; Michalek, R.D.; Pitiyage, G.N.; de Castro, A.M.; Vignola, K.S.; Jones, J.; Mohney, R.P.; Karoly, E.D.; Prime, S.S.; Parkinson, E.K. Senescent human fibroblasts show increased glycolysis and redox homeostasis with extracellular metabolomes that overlap with those of irreparable DNA damage, aging, and disease. *J. Proteome Res.* **2015**, *14*, 1854–1871. [CrossRef]
105. Hutter, E.; Renner, K.; Pfister, G.; Stockl, P.; Jansen-Durr, P.; Gnaiger, E. Senescence-associated changes in respiration and oxidative phosphorylation in primary human fibroblasts. *Biochem. J.* **2004**, *380*, 919–928. [CrossRef]
106. Latorre-Pellicer, A.; Lechuga-Vieco, A.V.; Johnston, I.G.; Hamalainen, R.H.; Pellico, J.; Justo-Mendez, R.; Fernandez-Toro, J.M.; Claveria, C.; Guaras, A.; Sierra, R.; et al. Regulation of mother-to-offspring transmission of mtDNA heteroplasmy. *Cell Metab.* **2019**, *30*, 1120–1130.e1125. [CrossRef]
107. Schneider, E.L. Aging and cultured human skin fibroblasts. *J. Investig. Dermatol.* **1979**, *73*, 15–18. [CrossRef] [PubMed]
108. Greco, M.; Villani, G.; Mazzucchelli, F.; Bresolin, N.; Papa, S.; Attardi, G. Marked aging-related decline in efficiency of oxidative phosphorylation in human skin fibroblasts. *FASEB J.* **2003**, *17*, 1706–1708. [CrossRef] [PubMed]
109. Goldstein, S.; Moerman, E.J.; Porter, K. High-voltage electron microscopy of human diploid fibroblasts during ageing in vitro. Morphometric analysis of mitochondria. *Exp. Cell Res.* **1984**, *154*, 101–111. [CrossRef]
110. Wong, R.C.B.; Lim, S.Y.; Hung, S.S.C.; Jackson, S.; Khan, S.; Van Bergen, N.J.; De Smit, E.; Liang, H.H.; Kearns, L.S.; Clarke, L.; et al. Mitochondrial replacement in an iPSC model of Leber's hereditary optic neuropathy. *Aging* **2017**, *9*, 1341–1350. [CrossRef]
111. Herbst, A.; Lee, C.C.; Vandiver, A.R.; Aiken, J.M.; McKenzie, D.; Hoang, A.; Allison, D.; Liu, N.; Wanagat, J. Mitochondrial DNA deletion mutations increase exponentially with age in human skeletal muscle. *Aging Clin. Exp. Res.* **2020**, *33*, 1811–1820. [CrossRef] [PubMed]
112. Short, K.R.; Bigelow, M.L.; Kahl, J.; Singh, R.; Coenen-Schimke, J.; Raghavakaimal, S.; Nair, K.S. Decline in skeletal muscle mitochondrial function with aging in humans. *Proc. Natl. Acad. Sci. USA* **2005**, *102*, 5618–5623. [CrossRef] [PubMed]
113. Chistiakov, D.A.; Sobenin, I.A.; Revin, V.V.; Orekhov, A.N.; Bobryshev, Y.V. Mitochondrial aging and age-related dysfunction of mitochondria. *Biomed. Res. Int.* **2014**, *2014*, 238463. [CrossRef]
114. Keane, K.N.; Calton, E.K.; Cruzat, V.F.; Soares, M.J.; Newsholme, P. The impact of cryopreservation on human peripheral blood leucocyte bioenergetics. *Clin. Sci.* **2015**, *128*, 723–733. [CrossRef]
115. Traba, J.; Miozzo, P.; Akkaya, B.; Pierce, S.K.; Akkaya, M. An optimized protocol to analyze glycolysis and mitochondrial respiration in lymphocytes. *J. Vis. Exp.* **2016**, *117*, 54918. [CrossRef] [PubMed]
116. Kramer, P.A.; Chacko, B.K.; Ravi, S.; Johnson, M.S.; Mitchell, T.; Darley-Usmar, V.M. Bioenergetics and the oxidative burst: Protocols for the isolation and evaluation of human leukocytes and platelets. *J. Vis. Exp.* **2014**, *85*, 51301. [CrossRef] [PubMed]

117. Sjovall, F.; Ehinger, J.K.; Marelsson, S.E.; Morota, S.; Frostner, E.A.; Uchino, H.; Lundgren, J.; Arnbjornsson, E.; Hansson, M.J.; Fellman, V.; et al. Mitochondrial respiration in human viable platelets—Methodology and influence of gender, age and storage. *Mitochondrion* **2013**, *13*, 7–14. [CrossRef]
118. Rose, S.; Carvalho, E.; Diaz, E.C.; Cotter, M.; Bennuri, S.C.; Azhar, G.; Frye, R.E.; Adams, S.H.; Borsheim, E. A comparative study of mitochondrial respiration in circulating blood cells and skeletal muscle fibers in women. *Am. J. Physiol. Endocrinol. Metab.* **2019**, *317*, E503–E512. [CrossRef] [PubMed]
119. Tyrrell, D.J.; Bharadwaj, M.S.; Jorgensen, M.J.; Register, T.C.; Molina, A.J. Blood cell respirometry is associated with skeletal and cardiac muscle bioenergetics: Implications for a minimally invasive biomarker of mitochondrial health. *Redox Biol.* **2016**, *10*, 65–77. [CrossRef] [PubMed]
120. Braganza, A.; Corey, C.G.; Santanasto, A.J.; Distefano, G.; Coen, P.M.; Glynn, N.W.; Nouraie, S.M.; Goodpaster, B.H.; Newman, A.B.; Shiva, S. Platelet bioenergetics correlate with muscle energetics and are altered in older adults. *JCI Insight* **2019**, *5*, e128248. [CrossRef]
121. Pecina, P.; Houstkova, H.; Mracek, T.; Pecinova, A.; Nuskova, H.; Tesarova, M.; Hansikova, H.; Janota, J.; Zeman, J.; Houstek, J. Noninvasive diagnostics of mitochondrial disorders in isolated lymphocytes with high resolution respirometry. *BBA Clin.* **2014**, *2*, 62–71. [CrossRef]
122. Nicholas, D.; Proctor, E.A.; Raval, F.M.; Ip, B.C.; Habib, C.; Ritou, E.; Grammatopoulos, T.N.; Steenkamp, D.; Dooms, H.; Apovian, C.M.; et al. Advances in the quantification of mitochondrial function in primary human immune cells through extracellular flux analysis. *PLoS ONE* **2017**, *12*, e0170975. [CrossRef]
123. Avila, C.; Huang, R.J.; Stevens, M.V.; Aponte, A.M.; Tripodi, D.; Kim, K.Y.; Sack, M.N. Platelet mitochondrial dysfunction is evident in type 2 diabetes in association with modifications of mitochondrial anti-oxidant stress proteins. *Exp. Clin. Endocrinol. Diabetes* **2012**, *120*, 248–251. [CrossRef]
124. Hartman, M.L.; Shirihai, O.S.; Holbrook, M.; Xu, G.; Kocherla, M.; Shah, A.; Fetterman, J.L.; Kluge, M.A.; Frame, A.A.; Hamburg, N.M.; et al. Relation of mitochondrial oxygen consumption in peripheral blood mononuclear cells to vascular function in type 2 diabetes mellitus. *Vasc. Med.* **2014**, *19*, 67–74. [CrossRef]
125. Willig, A.L.; Kramer, P.A.; Chacko, B.K.; Darley-Usmar, V.M.; Heath, S.L.; Overton, E.T. Monocyte bioenergetic function is associated with body composition in virologically suppressed HIV-infected women. *Redox Biol.* **2017**, *12*, 648–656. [CrossRef] [PubMed]
126. Einsiedel, L.; Cherry, C.L.; Sheeran, F.L.; Friedhuber, A.; Wesselingh, S.L.; Pepe, S. Mitochondrial dysfunction in CD4+ lymphocytes from stavudine-treated HIV patients. *Mitochondrion* **2010**, *10*, 534–539. [CrossRef] [PubMed]
127. Annesley, S.J.; Lay, S.T.; De Piazza, S.W.; Sanislav, O.; Hammersley, E.; Allan, C.Y.; Francione, L.M.; Bui, M.Q.; Chen, Z.P.; Ngoei, K.R.; et al. Immortalized ParkinsoN's disease lymphocytes have enhanced mitochondrial respiratory activity. *Dis. Model. Mech.* **2016**, *9*, 1295–1305. [CrossRef]
128. Smith, A.M.; Depp, C.; Ryan, B.J.; Johnston, G.I.; Alegre-Abarrategui, J.; Evetts, S.; Rolinski, M.; Baig, F.; Ruffmann, C.; Simon, A.K.; et al. Mitochondrial dysfunction and increased glycolysis in prodromal and early ParkinsoN's blood cells. *Mov. Disord.* **2018**, *33*, 1580–1590. [CrossRef] [PubMed]
129. Michalak, S.; Florczak-Wyspianska, J.; Rybacka-Mossakowska, J.; Ambrosius, W.; Osztynowicz, K.; Baszczuk, A.; Kozubski, W.; Wysocka, E. Mitochondrial respiration in intact peripheral blood mononuclear cells and sirtuin 3 activity in patients with movement disorders. *Oxid. Med. Cell Longev.* **2017**, *2017*, 9703574. [CrossRef] [PubMed]
130. Maynard, S.; Hejl, A.M.; Dinh, T.S.; Keijzers, G.; Hansen, A.M.; Desler, C.; Moreno-Villanueva, M.; Burkle, A.; Rasmussen, L.J.; Waldemar, G.; et al. Defective mitochondrial respiration, altered dNTP pools and reduced AP endonuclease 1 activity in peripheral blood mononuclear cells of Alzheimer's disease patients. *Aging* **2015**, *7*, 793–815. [CrossRef] [PubMed]
131. Ehinger, J.K.; Morota, S.; Hansson, M.J.; Paul, G.; Elmer, E. Mitochondrial respiratory function in peripheral blood cells from HuntingtoN's disease patients. *Mov. Disord. Clin. Pract.* **2016**, *3*, 472–482. [CrossRef]
132. Tyrrell, D.J.; Bharadwaj, M.S.; Van Horn, C.G.; Marsh, A.P.; Nicklas, B.J.; Molina, A.J. Blood-cell bioenergetics are associated with physical function and inflammation in overweight/obese older adults. *Exp. Gerontol.* **2015**, *70*, 84–91. [CrossRef]
133. Braganza, A.; Annarapu, G.K.; Shiva, S. Blood-based bioenergetics: An emerging translational and clinical tool. *Mol. Aspects Med.* **2020**, *71*, 100835. [CrossRef]
134. Silaidos, C.; Pilatus, U.; Grewal, R.; Matura, S.; Lienerth, B.; Pantel, J.; Eckert, G.P. Sex-associated differences in mitochondrial function in human peripheral blood mononuclear cells (PBMCs) and brain. *Biol. Sex Differ.* **2018**, *9*, 34. [CrossRef]
135. Rausser, S.; Trumpff, C.; McGill, M.A.; Junker, A.; Wang, W.; Ho, S.H.; Mitchell, A.; Karan, K.R.; Monk, C.; Segerstrom, S.C.; et al. Mitochondrial phenotypes in purified human immune cell subtypes and cell mixtures. *bioRxiv* **2021**. [CrossRef]
136. Chacko, B.K.; Kramer, P.A.; Ravi, S.; Johnson, M.S.; Hardy, R.W.; Ballinger, S.W.; Darley-Usmar, V.M. Methods for defining distinct bioenergetic profiles in platelets, lymphocytes, monocytes, and neutrophils, and the oxidative burst from human blood. *Lab. Investig.* **2013**, *93*, 690–700. [CrossRef] [PubMed]
137. Banas, B.; Kost, B.P.; Goebel, F.D. Platelets, a typical source of error in real-time PCR quantification of mitochondrial DNA content in human peripheral blood cells. *Eur. J. Med. Res.* **2004**, *9*, 371–377. [PubMed]
138. Urata, M.; Koga-Wada, Y.; Kayamori, Y.; Kang, D. Platelet contamination causes large variation as well as overestimation of mitochondrial DNA content of peripheral blood mononuclear cells. *Ann. Clin. Biochem.* **2008**, *45*, 513–514. [CrossRef] [PubMed]

139. Jones, N.; Cronin, J.G.; Dolton, G.; Panetti, S.; Schauenburg, A.J.; Galloway, S.A.E.; Sewell, A.K.; Cole, D.K.; Thornton, C.A.; Francis, N.J. Metabolic adaptation of human CD4(+) and CD8(+) T-cells to T-cell receptor-mediated stimulation. *Front. Immunol.* **2017**, *8*, 1516. [CrossRef]
140. Ravi, S.; Chacko, B.; Sawada, H.; Kramer, P.A.; Johnson, M.S.; Benavides, G.A.; O'Donnell, V.; Marques, M.B.; Darley-Usmar, V.M. Metabolic plasticity in resting and thrombin activated platelets. *PLoS ONE* **2015**, *10*, e0123597. [CrossRef] [PubMed]
141. Maurer-Spurej, E.; Pfeiler, G.; Maurer, N.; Lindner, H.; Glatter, O.; Devine, D.V. Room temperature activates human blood platelets. *Lab. Investig.* **2001**, *81*, 581–592. [CrossRef]
142. Vollmar, B.; Slotta, J.E.; Nickels, R.M.; Wenzel, E.; Menger, M.D. Comparative analysis of platelet isolation techniques for the in vivo study of the microcirculation. *Microcirculation* **2003**, *10*, 143–152. [CrossRef]
143. Garcia-Roche, M.; Casal, A.; Carriquiry, M.; Radi, R.; Quijano, C.; Cassina, A. Respiratory analysis of coupled mitochondria in cryopreserved liver biopsies. *Redox Biol.* **2018**, *17*, 207–212. [CrossRef]
144. Kuznetsov, A.V.; Kunz, W.S.; Saks, V.; Usson, Y.; Mazat, J.P.; Letellier, T.; Gellerich, F.N.; Margreiter, R. Cryopreservation of mitochondria and mitochondrial function in cardiac and skeletal muscle fibers. *Anal. Biochem.* **2003**, *319*, 296–303. [CrossRef]
145. Larsen, S.; Wright-Paradis, C.; Gnaiger, E.; Helge, J.W.; Boushel, R. Cryopreservation of human skeletal muscle impairs mitochondrial function. *Cryo Lett.* **2012**, *33*, 170–176.
146. Nukala, V.N.; Singh, I.N.; Davis, L.M.; Sullivan, P.G. Cryopreservation of brain mitochondria: A novel methodology for functional studies. *J. Neurosci. Methods* **2006**, *152*, 48–54. [CrossRef]
147. Yamaguchi, R.; Andreyev, A.; Murphy, A.N.; Perkins, G.A.; Ellisman, M.H.; Newmeyer, D.D. Mitochondria frozen with trehalose retain a number of biological functions and preserve outer membrane integrity. *Cell Death Differ.* **2007**, *14*, 616–624. [CrossRef]
148. Acin-Perez, R.; Benador, I.Y.; Petcherski, A.; Veliova, M.; Benavides, G.A.; Lagarrigue, S.; Caudal, A.; Vergnes, L.; Murphy, A.N.; Karamanlidis, G.; et al. A novel approach to measure mitochondrial respiration in frozen biological samples. *EMBO J.* **2020**, *39*, e104073. [CrossRef] [PubMed]
149. Osto, C.; Benador, I.Y.; Ngo, J.; Liesa, M.; Stiles, L.; Acin-Perez, R.; Shirihai, O.S. Measuring mitochondrial respiration in previously frozen biological samples. *Curr. Protoc. Cell Biol.* **2020**, *89*, e116. [CrossRef]
150. Divakaruni, A.S.; Andreyev, A.Y.; Rogers, G.W.; Murphy, A.N. In situ measurements of mitochondrial matrix enzyme activities using plasma and mitochondrial membrane permeabilization agents. *Anal. Biochem.* **2018**, *552*, 60–65. [CrossRef] [PubMed]
151. Baracca, A.; Barogi, S.; Carelli, V.; Lenaz, G.; Solaini, G. Catalytic activities of mitochondrial ATP synthase in patients with mitochondrial DNA T8993G mutation in the ATPase 6 gene encoding subunit a. *J. Biol. Chem.* **2000**, *275*, 4177–4182. [CrossRef] [PubMed]
152. Jacobs, R.A.; Diaz, V.; Meinild, A.K.; Gassmann, M.; Lundby, C. The C57Bl/6 mouse serves as a suitable model of human skeletal muscle mitochondrial function. *Exp. Physiol.* **2013**, *98*, 908–921. [CrossRef] [PubMed]
153. Gellerich, F.N.; Mayr, J.A.; Reuter, S.; Sperl, W.; Zierz, S. The problem of interlab variation in methods for mitochondrial disease diagnosis: Enzymatic measurement of respiratory chain complexes. *Mitochondrion* **2004**, *4*, 427–439. [CrossRef]
154. Vergnes, L.; Davies, G.R.; Lin, J.Y.; Yeh, M.W.; Livhits, M.J.; Harari, A.; Symonds, M.E.; Sacks, H.S.; Reue, K. Adipocyte browning and higher mitochondrial function in periadrenal but not SC fat in pheochromocytoma. *J. Clin. Endocrinol. Metab.* **2016**, *101*, 4440–4448. [CrossRef]
155. Meyer, J.N.; Leung, M.C.; Rooney, J.P.; Sendoel, A.; Hengartner, M.O.; Kisby, G.E.; Bess, A.S. Mitochondria as a target of environmental toxicants. *Toxicol. Sci.* **2013**, *134*, 1–17. [CrossRef] [PubMed]
156. Zolkipli-Cunningham, Z.; Falk, M.J. Clinical effects of chemical exposures on mitochondrial function. *Toxicology* **2017**, *391*, 90–99. [CrossRef] [PubMed]
157. Tyrrell, D.J.; Bharadwaj, M.S.; Van Horn, C.G.; Kritchevsky, S.B.; Nicklas, B.J.; Molina, A.J. Respirometric profiling of muscle mitochondria and blood cells are associated with differences in gait speed among community-dwelling older adults. *J. Gerontol. A Biol. Sci. Med. Sci.* **2015**, *70*, 1394–1399. [CrossRef]
158. Tyrrell, D.J.; Bharadwaj, M.S.; Jorgensen, M.J.; Register, T.C.; Shively, C.; Andrews, R.N.; Neth, B.; Keene, C.D.; Mintz, A.; Craft, S.; et al. Blood-based bioenergetic profiling reflects differences in brain bioenergetics and metabolism. *Oxid. Med. Cell Longev.* **2017**, *2017*, 7317251. [CrossRef] [PubMed]
159. Winnica, D.; Corey, C.; Mullett, S.; Reynolds, M.; Hill, G.; Wendell, S.; Que, L.; Holguin, F.; Shiva, S. Bioenergetic differences in the airway epithelium of lean versus obese asthmatics are driven by nitric oxide and reflected in circulating platelets. *Antioxid. Redox Signal.* **2019**, *31*, 673–686. [CrossRef] [PubMed]
160. Reily, C.; Mitchell, T.; Chacko, B.K.; Benavides, G.; Murphy, M.P.; Darley-Usmar, V. Mitochondrially targeted compounds and their impact on cellular bioenergetics. *Redox Biol.* **2013**, *1*, 86–93. [CrossRef]
161. Hedges, C.P.; Woodhead, J.S.T.; Wang, H.W.; Mitchell, C.J.; Cameron-Smith, D.; Hickey, A.J.R.; Merry, T.L. Peripheral blood mononuclear cells do not reflect skeletal muscle mitochondrial function or adaptation to high-intensity interval training in healthy young men. *J. Appl. Physiol.* **2019**, *126*, 454–461. [CrossRef] [PubMed]
162. Rodinova, M.; Trefilova, E.; Honzik, T.; Tesarova, M.; Zeman, J.; Hansikova, H. Non-invasive screening of cytochrome c oxidase deficiency in children using a dipstick immunocapture assay. *Folia Biol.* **2014**, *60*, 268–274.
163. Frederiksen, A.L.; Andersen, P.H.; Kyvik, K.O.; Jeppesen, T.D.; Vissing, J.; Schwartz, M. Tissue specific distribution of the 3243A->G mtDNA mutation. *J. Med. Genet.* **2006**, *43*, 671–677. [CrossRef]

164. Yorns, W.R., Jr.; Valencia, I.; Jayaraman, A.; Sheth, S.; Legido, A.; Goldenthal, M.J. Buccal swab analysis of mitochondrial enzyme deficiency and DNA defects in a child with suspected myoclonic epilepsy and ragged red fibers (MERRF). *J. Child. Neurol.* **2012**, *27*, 398–401. [CrossRef]
165. Goldenthal, M.J.; Kuruvilla, T.; Damle, S.; Salganicoff, L.; Sheth, S.; Shah, N.; Marks, H.; Khurana, D.; Valencia, I.; Legido, A. Non-invasive evaluation of buccal respiratory chain enzyme dysfunction in mitochondrial disease: Comparison with studies in muscle biopsy. *Mol. Genet. Metab.* **2012**, *105*, 457–462. [CrossRef] [PubMed]
166. Goldenthal, M.J.; Damle, S.; Sheth, S.; Shah, N.; Melvin, J.; Jethva, R.; Hardison, H.; Marks, H.; Legido, A. Mitochondrial enzyme dysfunction in autism spectrum disorders: A novel biomarker revealed from buccal swab analysis. *Biomark. Med.* **2015**, *9*, 957–965. [CrossRef] [PubMed]
167. Kolbasina, N.A.; Gureev, A.P.; Serzhantova, O.V.; Mikhailov, A.A.; Moshurov, I.P.; Starkov, A.A.; Popov, V.N. Lung cancer increases H_2O_2 concentration in the exhaled breath condensate, extent of mtDNA damage, and mtDNA copy number in buccal mucosa. *Heliyon* **2020**, *6*, e04303. [CrossRef]
168. Delhey, L.M.; Nur Kilinc, E.; Yin, L.; Slattery, J.C.; Tippett, M.L.; Rose, S.; Bennuri, S.C.; Kahler, S.G.; Damle, S.; Legido, A.; et al. The effect of mitochondrial supplements on mitochondrial activity in children with autism spectrum disorder. *J. Clin. Med.* **2017**, *6*, 18. [CrossRef]
169. Garcia-Closas, M.; Egan, K.M.; Abruzzo, J.; Newcomb, P.A.; Titus-Ernstoff, L.; Franklin, T.; Bender, P.K.; Beck, J.C.; Le Marchand, L.; Lum, A.; et al. Collection of genomic DNA from adults in epidemiological studies by buccal cytobrush and mouthwash. *Cancer Epidemiol. Biomark. Prev.* **2001**, *10*, 687–696.
170. Mahfuz, I.; Cheng, W.; White, S.J. Identification of Streptococcus parasanguinis DNA contamination in human buccal DNA samples. *BMC Res. Notes* **2013**, *6*, 481. [CrossRef]

 life

MDPI

Article

In Vitro Assays for the Assessment of Impaired Mitochondrial Bioenergetics in Equine Atypical Myopathy

Caroline-J. Kruse [1,*], David Stern [2], Ange Mouithys-Mickalad [3], Ariane Niesten [4], Tatiana Art [1], Hélène Lemieux [5] and Dominique-M. Votion [2]

[1] Department of Functional Sciences, Faculty of Veterinary Medicine, Physiology and Sport Medicine, Fundamental and Applied Research for Animals & Health (FARAH), University of Liège, 4000 Liège, Belgium; Tatiana.Art@Uliege.be

[2] Equine Pole, Fundamental and Applied Research for Animals & Health (FARAH), Faculty of Veterinary Medicine, University of Liège, 4000 Liège, Belgium; d.stern@Uliege.be (D.S.); dominique.votion@Uliege.be (D.-M.V.)

[3] Center for Oxygen, Research & Development (CORD), Center for Interdisciplinary Research on Medicines (CIRM), Institute of Chemistry, B6a, University of Liège, Allée du Six Août, 11, 4000 Liège, Belgium; amouithys@Uliege.be

[4] Center for Oxygen, Research & Development (CORD), Fundamental and Applied Research for Animals & Health (FARAH), Institute of Chemistry, B6a, University of Liège, Allée du Six Août, 11, 4000 Liège, Belgium; Ariane.Niesten@Uliege.be

[5] Faculty Saint-Jean and Department of Medicine, University of Alberta, 8406-91 Street, Edmonton, AB T6C 4G9, Canada; helene.lemieux@ualberta.ca

* Correspondence: caroline.kruse@Uliege.be

Abstract: Equine atypical myopathy is a seasonal intoxication of grazing equids. In Europe, this poisoning is associated with the ingestion of toxins contained in the seeds and seedlings of the sycamore maple (*Acer pseudoplatanus*). The toxins involved in atypical myopathy are known to inhibit ß-oxidation of fatty acids and induce a general decrease in mitochondrial respiration, as determined by high-resolution respirometry applied to muscle samples taken from cases of atypical myopathy. The severe impairment of mitochondrial bioenergetics induced by the toxins may explain the high rate of mortality observed: about 74% of horses with atypical myopathy die, most within the first two days of signs of poisoning. The mechanism of toxicity is not completely elucidated yet. To improve our understanding of the pathological process and to assess therapeutic candidates, we designed in vitro assays using equine skeletal myoblasts cultured from muscle biopsies and subjected to toxins involved in atypical myopathy. We established that equine primary myoblasts do respond to one of the toxins incriminated in the disease.

Keywords: atypical myopathy; high-resolution respirometry; toxicity assays; cell culture; equine primary myoblasts

Citation: Kruse, C.-J.; Stern, D.; Mouithys-Mickalad, A.; Niesten, A.; Art, T.; Lemieux, H.; Votion, D.-M. In Vitro Assays for the Assessment of Impaired Mitochondrial Bioenergetics in Equine Atypical Myopathy. *Life* **2021**, *11*, 719. https://doi.org/10.3390/life11070719

Academic Editors: Giorgio Lenaz, Salvatore Nesci and Gopal J. Babu

Received: 4 May 2021
Accepted: 15 July 2021
Published: 20 July 2021

1. Introduction

Equine atypical myopathy (AM) is a severe environmental intoxication linked to the ingestion of certain maple (*Acer*) seeds and seedlings. In Europe, the incriminated tree is the sycamore maple (*Acer pseudoplatanus*) [1], whereas in the United States, the box elder (*Acer negundo*) has been linked to the poisoning [2]. Two toxins, hypoglycin A (HGA) [3–5] and methylenecyclopropylglycine (MCPrG) [6] are involved in the poisoning [7]. These molecules are not toxic per se but once in the body are transformed into their active metabolites, methylenecyclopropylacetyl-CoA (MCPA-CoA) and methylenecyclopropylformyl-CoA (MCPF-CoA), respectively [8–10]. Both toxins are known inhibitors of fatty-acid ß-oxidation, which results in an impaired capacity of energy production using oxidative metabolism [10–13]. The MCPA-CoA inhibits also dehydrogenases involved in the degradation of branched-chain amino acids [10]. The ingestion of maple toxins led to the detection of toxins, conjugated toxic metabolites, and fatty esters in blood [1,2,5,7,14–17].

Clinical signs in intoxicated horses include muscle weakness and stiffness, eventual recumbency, and, in 74% of cases, death [18]. Macroscopic, histologic, and histochemical analyses confirm multifocal degeneration and necrosis with variable severity between cases and muscles [19]. Indeed, muscular lesions seem to be more constant and severe in the myocardium and respiratory and postural muscles [19,20], therefore, oxidative muscle fibers appear to be particularly affected by the toxin [19]. Transmission electron microscopy revealed several ultrastructural changes affecting especially mitochondria, as matrix loss and cristae fragmentation [19].

Previous studies performed on skeletal muscle show that structural alterations are associated with mitochondrial functional consequences [13]. Using high-resolution respirometry (HRR), a severe depression of mitochondrial oxidative phosphorylation (OXPHOS) and electron transfer system capacities (ET capacity) in AM affected horses was found [13] using standardized substrate-uncoupler-inhibitor titration (SUIT) protocols validated for respirometric assessments of equine muscle cells [21].

Since AM outbreaks are seasonal (i.e., autumnal and spring outbreaks resulting from the consumption of seeds and seedlings, respectively) and do not occur to the same extend every year [22], in vivo sampling is naturally limited. Also, because of the acute nature of AM and the rapid progression of the condition with a mean survival time of 38 h [23], complementary examinations and sampling might be difficult to perform. Additionally, the use of surrogate animals to study AM may not be valid, since both rodents and rabbits display damage of different organs than horses after HGA intoxication and do not show signs of rhabdomyolysis [24,25]. Because of these obstacles, an in vitro model was designed attempting to reproduce mitochondrial dysfunction by adding methylenecyclopropylacetyl (MCPA) (i.e., the sole toxin commercially available) during HRR experiments. Ultimately, our final goal will be to define a treatment for AM based on the ability of the drug to restore an adequate mitochondrial function. The aim of the present study was: (1) to reproduce specific changes in OXPHOS capacity and respiratory control patterns observed in skeletal muscle of AM affected horses using conventional SUIT protocols [13]; (2) to measure the effect of MCPA on fatty acids utilization, and (3) to determine the cytotoxicity and viability of equine primary muscle cells subjected to HGA and MCPA.

2. Materials and Methods

2.1. Cell Culture of Equine Primary Myoblasts

Equine primary skeletal myoblast cultures were purchased from RevaTis® (RevaTis, Aye, Belgium). A vial containing 2.5 million cells was thawed at 37 °C in a water bath. The cells were then cultivated in an 80 cm^2 culture Tflask and multiplied from passage 5 to passage 8. All manipulations were performed under streamline flow hood.

Cells were cultured in the maintenance media Dulbecco's modified Eagles's low-glucose DMEM 1 g/L (Lonza, Verviers, Belgium) supplemented with 20% fetal bovine serum, 1% L-glutamine, 1% penicillin-streptomycin, and 0.5% amphotericin B (Thermo Fisher Scientific, Karlsruhe, Germany). When 80% confluence was reached, culture medium was removed, and the flasks were washed using Dulbecco's phosphate-buffered saline (DPBS) without Ca^{2+} and Mg^{2+}. Cells were subsequently trypsinized (TrypLE™ Express, Thermo Fisher Scientific, Karlsruhe, Germany), centrifuged, and counted according to Ceusters et al. [26]. The equivalent of one 175 cm^2 culture T-flask was used to continue cell culture to further passages.

At each passage, part of the cells was immediately processed for HRR. Ten thousand cells per well were seeded in a 96 clear bottom white plate (Cellstar® Greiner Bio-One, Vilvoorde, Belgium) and supplemented with DMEM medium until further analysis. Plates and flasks were incubated at 37 °C and 5% CO_2 in a humidified incubator.

2.2. Toxicity Assays

The cell toxicity of MCPA was assessed using the CellTox^{TM} Green assay (Promega Benelux, Leiden, The Netherlands). This assay was developed to determine toxic effects by binding DNA of cells with impaired membrane integrity.

Plated cells were kept in a humidified incubator at 37 °C and 5% CO_2 until 80% of confluence was reached. Confluence was assessed by light microscopy after 24 to 36 h. Once confluent, the cells were washed with DPBS without Ca^{2+} and Mg^{2+}. The cells were then exposed to different concentrations of MCPA (Merck, Darmstadt, Germany) with DMSO as solvent and HGA (Toronto Research Chemicals, Ontario M3J2K8, Canada) suspended in DMEM medium without phenol red. CellTox^{TM} Green dye was added to each well. The bottom of the plate was covered with a BrightMax^{TM} seal (Greiner Bio-One, Vilvoorde, Belgium) and subsequently placed in the EnSpire® Multimode Plate Reader (PerkinElmer, Waltham, MA, USA) for reading. Data was recorded for 24 h after toxin exposure in order to determine the evolution of cytotoxicity.

Additionally, a real-time viability assay through reducing potential measurement was performed. The nonlytic dye contained in the RealTime-Glo^{TM} MT Cell Viability Assay (Promega Benelux, Leiden, The Netherlands) allows for continuous reading and is based on the continuous reduction of the viability substrate by the viable cells contained in each well. The DMSO concentrations in each well were of 0.5% in order to exclude DMSO induced toxicity to the cells.

2.3. High-Resolution Respirometry

After centrifugation, the cells were counted and then placed in MiR05 mitochondrial respiration medium (0.5 mM EGTA, 3 mM $MgCl_2 \cdot 6 H_2O$, 20 mM taurine, 10 mM KH_2PO_4, 20 mM HEPES, 1 g/L BSA, 60 mM potassium-lactobionate, 110 mM sucrose, pH 7.1) at 37 °C. In total, 2.5 to 3 million cells were added to each 2 mL Oxygraph-2k chamber (Oroboros Instruments, Innsbruck, Austria). Oxygen concentration (μM) and oxygen flux per million cells (pmol $O_2 \cdot s^{-1} \cdot 10^6$ cells^{-1}) were recorded online using DatLab software (Oroboros Instruments, Austria). Experiments were conducted using specific SUIT protocols and oxygen levels were maintained between 200 and 500 μM O_2 to avoid any oxygen-related limitations of respiration and to align with previous studies [13,21,27]. Oxygen flux was expressed as respiration per million cells (pmol $O_2 \cdot s^{-1} \cdot 10^6$ cells^{-1}), or as control ratios, namely flux control ratios (*FCR*). These ratios have the advantage of being independent of cell count, mitochondrial content, and density, indicating qualitative changes of mitochondrial respiratory control [28].

Several SUIT protocols were applied, using both fatty acids and reduced nicotinamide adenine dinucleotide (NADH)-linked substrates in order to assess mitochondrial pathways at different levels of integration. In order to have a global overview, the electron transfer (ET) pathway was fed by the NADH junction (N-junction) substrates pyruvate, glutamate in the presence of malate. Fatty acids are catabolized via ß-oxidation and support ET pathways through both reduced flavin adenine dinucleotide ($FADH_2$) junction (F-junction) and N-junction (see level 5; Figure 1). Therefore, ß-oxidation relies on the combination of F-junction and N-junction pathways.

Muscle cells suspended in MiR05 were added to the Oxygraph-2k chambers containing a final volume of 2 mL per chamber maintained at 37.0 °C. In SUIT1, electron flow was sustained by the NADH-linked substrate glutamate and co-substrate malate (GM; 10 and 2 mM, respectively) followed by a saturating concentration of ADP (D; 2.5 mM). In SUIT2, the initial substrate was pyruvate with malate as co-substrate (PM; 5 and 2 mM, respectively) with subsequent addition of digitonin (Dig; 10 μg$\cdot 10^6$ cells^{-1}) and ADP (D; 2.5 mM) followed by addition of glutamate (G; 10 mM). Three fatty acids' protocols were performed and started with the addition of acetylcarnitine (Act; 5 mM; SUIT3), octanoylcarnitine (Oct; 0.5 mM; SUIT4), and palmitoylcarnitine (Pal; 0.04 mM; SUIT5), and malate (2 mM; SUIT3, SUIT4, SUIT5) as co-substrate. In SUIT3 to 5, electrons from both the electron-transferring flavoprotein complex (CETF) and the Complex I (CI) entered the Q-junction. In all protocols, digitonin (Dig; 0.01 mM) was added before ADP. Optimal digitonin concentration was determined by careful titration experiments as previously described [29]. In all SUIT protocols, ADP-stimulated respiration represents OXPHOS capacity, *P* whereas ET capacity, *E* was obtained by addition of the uncoupler FCCP, rendering this state as not limited by the capacity of the phosphorylation system [28].

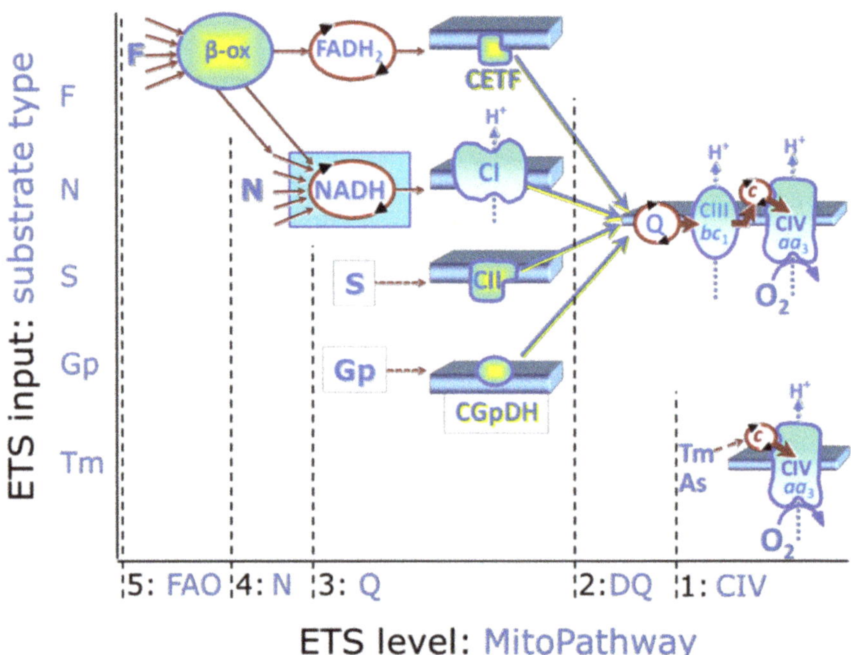

Figure 1. ET-substrate types are linked to ET-pathway types in substrate-uncoupler-inhibitor titration (SUIT) protocols. Electrons from multiple upstream origins feed into the Q junction. These origins include (5) fatty acid ß-oxidation (FAO) providing electrons from $FADH_2$ to the electron-transferring flavoprotein complex (CETF; F pathway), (4) dehydrogenases and enzymes converging at the N-junction and providing electrons from NADH to complex I (N pathway), (3) succinate (S) providing electrons from $FADH_2$ to Complex II (CII; S pathway). From the Q junction, electrons are then transferred to Complex III, cytochrome c and complex IV, before their transfer to O_2 to form H_2O. Figure from Gnaiger (2020) with permission [28]. Abbreviations: $FADH_2$ = flavin adenine dinucleotide; NADH = Nicotinamide adenine dinucleotide.

In every protocol, succinate (S; 10 mM) was subsequently added for electron flow from Complex II (CII) into the Q-junction to give the flux through the NS-pathway (N and S pathway combined) or the F- and S-pathway combined. By stepwise addition of the non-physiological uncoupler FCCP (U; 0.05 µM, followed by 0.025 µM steps until maximal oxygen flux was reached), ET-capacity, *E*, was obtained. Electron input into the Q-junction through CII only was subsequently measured by inhibition of CI with rotenone (Rot; 0.5 µM). Finally, residual oxygen consumption state was obtained by addition of antimycin A (Ama; 2.5 µM) to inhibit Complex III (CIII). For each protocol, MCPA [9 mM] was added to a parallel chamber after ADP addition. The concentration was defined beforehand by toxicity assays. All protocols are summarized in Table 1.

For all protocols, a second oxygraphic run was performed to establish if cytochrome *c* (Cyt *c*; 10 µM) addition induced an increase in O_2 flux. Cytochrome *c* release is considered as an essential quality control because of the possible limitation of active respiration when the outer mitochondrial membrane has been damaged by the laboratory procedures, allowing the loss of cytochrome *c* located in the intermembrane space [28].

Table 1. Substrate-uncoupler-inhibitor titration (SUIT) protocols performed.

	SUIT Protocols
SUIT1 [1]	1GM; 2D; 3S; 4U; 5Rot; 6Ama
SUIT2 [1]	1PM; 2D; 3G; 4S; 5U; 6Rot; 7Ama
SUIT3 [2]	1ActM; 2D; 3S; 4U; 5Rot; 6Ama
SUIT4 [2]	1OctM; 2D; 3S; 4U; 5Rot; 6Ama
SUIT5 [2]	1PalM; 2D; 3S; 4U; 5Rot; 6Ama

[1] NS-pathway. [2] F-pathway and S-pathway. Abbreviations: GM = Glutamate & malate; PM = Pyruvate & malate; ActM = Acetylcarnitine & malate; OctM = Octanoylcarnitine & malate; PalM = Palmitoylcarnitine & malate; D = ADP; G = Glutamate; S = Succinate; U = Uncoupler (FCCP); Rot = Rotenone; Ama = Antimycine A.

2.4. *Total Protein Content*

Total protein content was measured using the Pierce™ 660 nm Protein Assay. Absorbance was recorded using a spectrophotometer and MPM6 analysis software. Measurements were performed for each passage of one vial in duplicate (N = 1, n = 2).

2.5. *Data Analysis*

Raw respirometric data was normalized by *FCR* to allow direct comparison with the results presented in Lemieux et al. [13]. The pathway control ratios of oxygen flux with substrate types provided separately for the N- and S-pathways were normalized for flux through the combined NS-pathway (i.e., N/NS and S/NS). All respiratory states were corrected for residual oxygen consumption. For treated cells, MCPA was added after ADP and raw data were collected after MCPA addition. The different nominators and denominators were calculated by subtracting residual oxygen consumption from every calculated raw data. Regarding our ratios, CI_P and CI + CII_P are relative to OXPHOS capacity whereas CI + CII_E and CII_E are relative to ET capacity. Also, it is assumed that S-pathway linked respiration is not influenced by uncoupling [13,21] and therefore $CII_E = CII_P$. The ratios obtained for SUIT1 and SUIT2 in MCPA-treated cells were compared to ratios calculated in MCPA affected horses. Two vials from the same horse were used for respirometric analysis and runs were performed in duplicate (N = 2, n = 2), when not specified otherwise. Additionally, for each raw parameter in MCPA treated cells, a percentage from the control cells was calculated. Similarly, the effect of MCPA was evaluated with SUIT3, SUIT4, and SUIT5.

For a same respirometric parameter, a two-tailed *t*-test was performed between control and treated cells. Statistical significance was set at $p < 0.05$. In the tables, means are represented ± standard error of the mean (SEM). Additionally, ratios of MCPA treated cells were compared to previously published ratios of AM affected horses in Lemieux et al. via unpaired two-tailed *t*-test. For this comparison, ratios of different passages were pooled.

Regarding toxicity assays, data were analyzed by GraphPad Prism and a non-linear regression (sigmoidal, 4PL, least squares fit) was performed. For cell reduction potential, a non-linear regression (sigmoidal, 4PL, least squares fit) was also performed for each tested MCPA concentration at different times.

3. Results

3.1. *Toxicity Assays*

When analyzed independently, the kinetics of recorded fluorescence show a steep augmentation reaching a plateau at approximatively 10 h after toxin exposure. Plateau values were grouped for each concentration. After a 24-h exposure, wells submitted to 15 mM MCPA displayed between 78 and 105% of maximal response compared to the lysis solution (Promega Benelux, Leiden, The Netherlands). An important increase of cytotoxicity was observed between 10 and 15 mM (Figure 2). Best-fit values for total data, regardless of the passage indicated an IC50 of 15.7 mM. This value should however be interpreted with caution since a complete confidence interval could not be calculated. Also, it is notable that the differences between individual passages do result in very different responses to MCPA. Indeed, cytotoxicity ranged widely from 25 to 67% for 8 mM and from 29 to 76% for 10 mM. Lowest toxicity was measured at passage 8 with 25 and 28% for 8 mM and 29 and 31% for 10 mM.

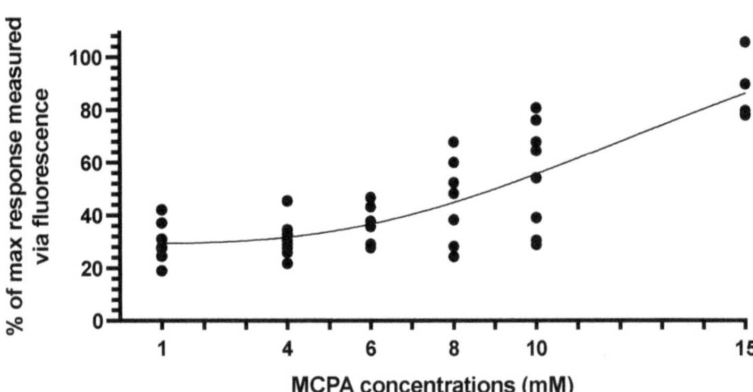

Figure 2. Methylenecyclopropylacetyl (MCPA) induced cytotoxicity measured by CellTox™ Green.

Regarding cell reducing potential, a non-linear regression (sigmoidal, 4PL, least squares fit) was also performed for each tested MCPA concentration at different times. Best-fit values indicated an IC50 of 8.09 ± 0.55 mM depending on the time recorded (Figure 3). In neither toxicity assay, HGA addition to cell cultures at concentrations ranging from 0.25 to 1 mM resulted in changes compared to control cells (data not shown). According to best-fit values, MCPA addition for subsequent experiments (i.e., HRR) was set at 9 mM.

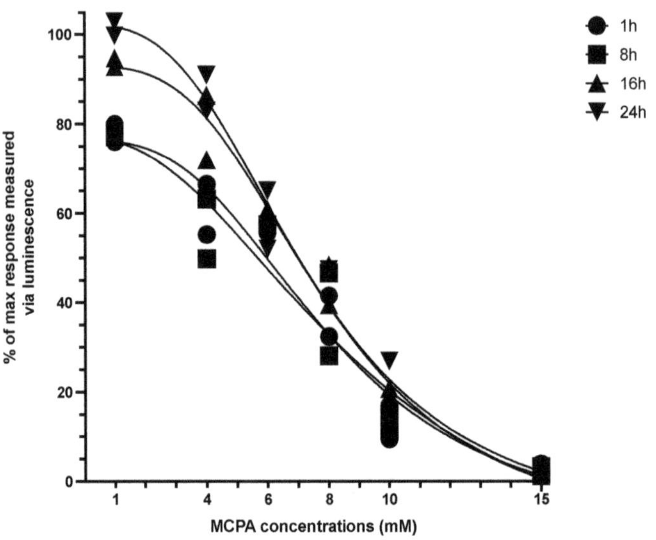

Figure 3. Cell reduction capacity depending on methylenecyclopropylacetyl (MCPA) concentrations at different time points.

3.2. High-Resolution Respirometry

At any passage, the addition of cytochrome *c* resulted in a slight (i.e., $\leq$10%) increase in O_2 flux thus confirming the preservation of the outer mitochondrial membrane integrity. When comparing obtained *FCR* of control vs. MCPA treated cells in the SUIT1 protocol (with glutamate and malate), a significant difference was observed for the ratio CI_P/CII_E, which was increased with 9 mM MCPA (Figure 4). For the three ratios with CI + CII at the denominator, there were no significant changes with MCPA treatment.

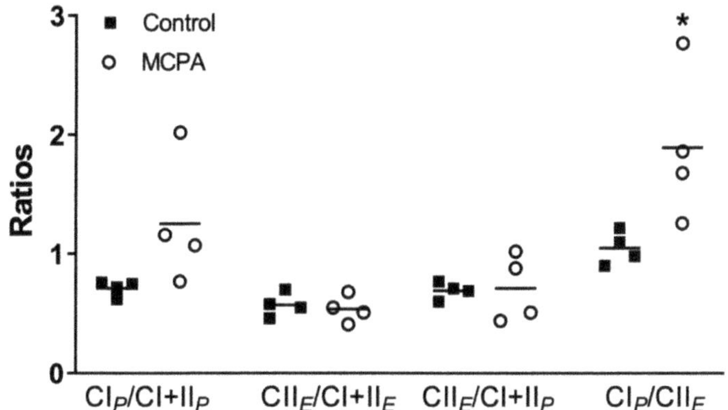

Figure 4. Ratios in control cells and cells treated with 9 mM methylenecyclopropylacetyl (MCPA) for the substrate-uncoupler-inhibitor titration protocol 1 (SUIT1). Each dot represents one passage, and the same passage are represented with or without MCPA. The bar represents the mean. * Significantly different from controls ($p < 0.05$) with a t-test were indicated with a *. Abbreviations: CI_P = Complex I linked OXPHOS capacity; CII_E = Complex II linked ET capacity; CI + CII_P = Complex I&II linked OXPHOS capacity; CI + CII_E = Complex I&II linked ET capacity.

In SUIT2, significant differences between groups were also detected. Compared to the control cells, the MCPA treated cells showed an increase in the ratios of $CI_P{}^a$/CI + II_P and $CI_P{}^a/CII_E$ and a decrease in the ratio CII_E/CI + II_P (Figure 5). Interestingly, the ratio CI_P/CI + II_P was significantly different between groups when pyruvate and malate were used as substrates, but not when glutamate was added (Figure 5). The ratio of CII/CI + II was significant only when the denominator (CI + II) was taken under the *P* state, but not when it was taken under the *E* state (Figure 5).

Without considering cell passages, a decrease in oxygen flux of 46% in average (SEM $\pm$ 0.08) was recorded after MCPA addition to the chamber in SUIT1. This decrease remained relatively constant for the rest of the protocol: CI + II_P = 43% of control (SEM $\pm$ 0.15), CI + II_E = 40% of control (SEM $\pm$ 0.2), CII_E = 32% (SEM $\pm$ 0.13) (Table 2). Similarly, a decrease in oxygen flux to 53% of control cells (SEM $\pm$ 0.11) was recorded after MCPA addition to the chamber (CI_P). A constant decrease was noted for the rest of the protocol: CI + II_P = 43% of control (SEM $\pm$ 0.10), CI + II_E = 40% of control (SEM $\pm$ 0.17), CII_E = 29% (SEM $\pm$ 0.09).

Table 2. Respirometric value percentage of MCPA treated cells compared to control cells.

Protocol	CI_P	F_P	CI + II_P	CI + II_E	CII_E
SUIT 1	46%	-	43%	40%	32%
SUIT 2	53%	-	43%	40%	29%
SUIT 3	-	67%	61%	72%	51%
SUIT 4	-	74%	62%	56%	49%
SUIT 5	-	67%	60%	58%	37%

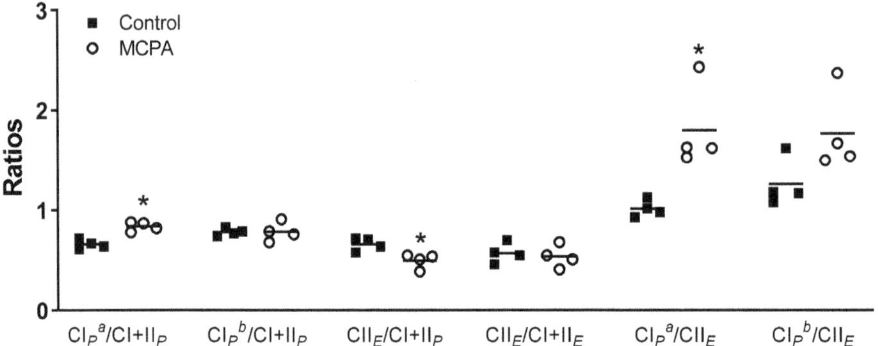

Figure 5. Ratios in control cells and cells treated with 9 mM methylenecyclopropylacetyl (MCPA) for the substrate-uncoupler-inhibitor titration protocol 2 (SUIT2). Each dot represents one passage, and the same passage are represented with or without MCPA. The bar represents the mean. * Significantly different from controls ($p < 0.05$) with a t-test were indicated with a *. [a] using glutamate and malate as N pathway substrates. [b] using pyruvate, malate and glutamate as N pathway substrates.

Similarly, for fatty acid protocols using acetylcarnitine as substrate, oxygen flux immediately after MCPA addition was on average 67% of the control value (SEM ± 0.2). When Succinate was added, 61% of control (SEM ± 0.24) and after FCCP addition 72% of control (SEM ± 0.08). Finally, after rotenone addition, 51% (SEM ± 0.12) of control were attained. When the fatty acid substrate was octanoylcarnitine, oxygen flux immediately after MCPA addition was on average 74% of the control value (SEM ± 0.24). For the rest of the protocol: CI + II_P = 62% of control (SEM ± 0.14), CI + II_E = 56% of control (SEM ± 0.24), CII_E = 49% (SEM ± 0.15). Fatty acid protocols using palmitoylcarnitine as substrate, oxygen flux immediately after MCPA addition was on average 67% of the control value (SEM ± 0.07). For the rest of the protocol: CI + II_P = 60% of control (SEM ± 0.11), CI + II_E = 58% of control (SEM ± 0.31), CII_E = 37% (SEM ± 0.25). Hence, a decrease in respiration was recorded in all SUIT protocols used. Also, whatever the protocol, initial LEAK respirometric parameters were similar between control and treated cells.

Regardless of the substrate sustaining electron flow used, the addition of MCPA depressed respiration in treated cells. This effect was noted at each cell passage.

3.3. Total Protein Content

In order to compare macroscopic cell count to an internal measure, total protein content was analyzed (Table 3). Despite a similar cell count, the total protein content seemed to vary between runs (*no statistical analysis performed*). Since total protein content was not measured for each vial used, internal normalization by ratios was preferred. These ratios are independent of cell count or tissue mass [28].

Table 3. Total protein content (±SEM) at each passage performed in duplicate (n = 2). Protein concentration is expressed in μg/μL).

P5	P6	P7	P8
4.17 ± 0.44	4.89 ± 0.55	3.79 ± 0.32	3.85 ± 0.12

4. Discussion

Regardless of the protocol, an immediate effect of MCPA on mitochondrial electron transfer system (ETS) complexes was observed. This is interesting as it corroborates the in vivo observations obtained in horses with AM [11]. Similar to findings in Lemieux et al. [13], mitochondrial respiration seemed to be more depressed with glutamate-malate sustained respiration (SUIT1) compared to pyruvate-glutamate-malate (SUIT2).

Indeed, a severe depression of both OXPHOS and ET capacity could be reproduced in vitro. In all five SUIT protocols, the addition of MCPA resulted in an immediate effect on the N- and S-pathway but also on the F-pathway, sustained by fatty acid ß-oxidation. When analyzing SUIT2, the contribution of the S-pathway was similar to the N-pathway in affected and control horses [13]. However, in our study, the S-pathway seemed to be more affected and therefore resulted in a strong diminution of OXPHOS and ET capacity, no matter which protocol was used. This finding is probably linked to a longer exposure time to the toxin since CII-linked activity is measured after CI and CETF sustained O_2 flux.

Regarding fatty acid substrates, it is also important to note that ß-oxidation supplies electron transfer through the N-junction as well as the rate-limiting F-junction pathway branches. Even though a progressive decrease in respiration was also recorded with SUIT3, 4 and 5, F-pathway combined with S-pathway sustained respiration resulted in better supported respiration than the NS-pathway.

Overall, MCPA addition to the oxygraphy chamber resulted in a generalized inhibitory effect, acting either on all ETS complexes, or having a specific target in downstream ETS components as Q, CIII or CIV. In order to address this question, enzymatic assays testing either the individual complexes' response to the toxin or related enzymes upstream (e.g., PDH) and other SUIT protocols targeting specific downstream complexes will need to be applied in future studies. Also, it cannot be excluded that the IC50 calculated on the basis of toxicity and viability assays was too high to identify the first target and resulted in a general decrease of O_2 flux.

It is also noteworthy that despite the recording of a depression in mitochondrial respiration, without considering the passage and the cell batch, the differences between the different passages should be further explored in order to define the cause of variability. The use of undifferentiated equine primary myoblasts implies that metabolism and mitochondrial function may significantly differ compared with differentiated myotubes [30]. Therefore, it seems plausible to suspect that the oxidative phenotype, which depends on oxidative capacity and fiber type composition [30], would be impacted by myogenic differentiation and therefore is not completely reflected in situ metabolism. While the toxic effect of MCPA and MCPrG on ß-oxidation is well documented [9,10] it is worth noting that in this study as well as in the study performed by Lemieux et al. [13], the pathologic pathway is not restricted to an inhibition of ß-oxidation since SUIT1 and SUIT2 both result in a severe depression of the OXPHOS and ETS homeostasis.

The concentrations used in the cytotoxicity/viability study ranged from 1 to 15 mM MCPA and from 0.25 to 1 mM HGA. Regarding the latter, no effect was observed at the aforementioned concentrations. This might indicate a lack of metabolization by the cell culture used within the time of the experiment. When analyzing both cytotoxicity and cell reduction capacity, it appears that a wider range of concentration must be tested in future experiments. Indeed, at 8 mM MCPA, cell reduction capacity, which is cumulative, increases. This may be imputed to the immediate effect of MCPA on the cells. It is, however, unclear if only some cells are less impacted by low concentrations of MCPA. In any event, at the lowest concentration after several hours the reductive capacity of the cells is restored and similar to control cells. However, it is clear that even at 1 mM, there is an immediate reaction to the toxin indicating cellular impairment. Also, best-fit value calculated for IC50 in our cytotoxicity experiment was 15.7 mM. This concentration would indicate that almost two times more MCPA is necessary to induce a cytotoxic effect compared to an effect on cell reductive capacity. When analyzed in detail, the calculated cytotoxicity IC50 value for passages 6 and 7 is 9.02 mM ($\pm$0.58 mM), similar to the IC50 obtained for the reducing potential. This raises the question of the fluctuation of cell types and maturity between passages as well as the sensitivity of these assays for these types of cells. Concentrations of MCPA (i.e., 9 mM) induced a severe decrease in mitochondrial respiration by HRR, compatible with a 50% decrease depending on the evaluated mitochondrial complex. However, the range of concentrations needs to be larger in future experiments in order to determine the minimal concentration at which a toxic effect of MCPA can be recorded.

Also, despite a similar cell count, total protein content varied between runs. The differences between passages may also be related to a variable response on the cellular level to the toxin.

In any case, our study manages to replicate mitochondrial alterations in response to MCPA intoxication. However, current mitochondrial activity assessment still relies on end-point assays, which yield limited kinetic and therefore prognostic information [31]. Indeed, our assays are based on oxygen flux recordings depending on substrates, pathways, and oxidative phosphorylation. Since in this case the clinical picture is a myopathy secondary to poisoning, it is essential to determine if the mitochondrial damage is consequential to the ß-oxidation defect and therefore toxic lipid accumulation. To determine if altered oxidative phosphorylation/mitochondrial respiration is causal or consequential to the clinical symptoms observed in AM affected horses, tissue-tissue interactions might need to be monitored to detect early if onsets of mitochondrial stress precede acute rhabdomyolysis. Even though cellular adaptations might be far-fetched and unrealistic in such an acute disease, defining the onset of stress in the first affected tissue will enhance chances of therapy. Since the mainly affected muscle fibers are oxidative, a mitochondrial dysfunction leading to a shift from oxidative phosphorylation to glycolysis, these cells may be less equipped to assume their role because of their limited ability to generate ATP by alternative means or because of the ultrastructural mitochondrial changes [19].

So far, many factors have been cited as potential contributors in the pathophysiology of mitochondrial dysfunction involved in a wide variety of disorders as decreased mitochondrial content, altered substrate delivery, muscle inflammation, morphological distortion of mitochondria due to glycogen cytoplasmic accumulation, oxidative damage and mitochondrial damage induced by gluco- and lipotoxicity secondary to intracellular substrate accumulation [32–34]. Since the direct effect of MCPA can be replicated on a cellular model, a down-regulation of nuclear and mitochondrial genes in AM does not seem plausible. However, the larger scale consequences on organs and organelles of HGA and MCPrG metabolization are to date unknown and may also constitute a therapeutic target strategy. Indeed, if the toxins have an impact on mitochondrial proteostasis, the damage may occur at different scales; the horse's whole metabolism can be impacted, the mitochondrion's interaction with the cell and the mitochondrion itself may be damaged, which will activate pathways to counteract the damage [35].

In the same line, MCPA-carnitine concentrations quantified in serum of AM-affected horses went up to several thousand nmol/L [7,15]. It is therefore imperative to compare our results to a direct dosage of MCPA-carnitine in muscle of AM affected horses as well as to more sensitive techniques, able to detect event slight augmentations of low concentrations. Additionally, purchased cells originated from one donor horse, and the reaction of these cells are therefore not to be extrapolated to all animals susceptible to HGA and MCPA intoxication. Through an immortal cell line, an easy-to-use and alternative can be found [36,37], providing a pure population of cells to reproduce results obtained in this preliminary study. A standardized cell culture with an immortal cell line will also minimize horse-associated reactions to the toxin as well as passage-dependent responses.

In conclusion, our cellular in vitro model reproduced MCPA linked toxicity to a certain extend. For result reproduction, cytotoxicity assessment and in fine high throughput screening of therapeutic molecules, the use of an immortalized cell line is the next step.

Author Contributions: C.-J.K. and D.-M.V.: Conceptualization and methodology; C.-J.K. and A.N.: investigation; C.-J.K. and D.S.: data curation; C.-J.K. and D.-M.V.: writing—original draft preparation; H.L., D.-M.V. and C.-J.K.: review and editing; A.M.-M., A.N. and D.-M.V.: proofreading; C.-J.K. and H.L.: visualization; D.-M.V. and T.A.: supervision and project administration; D.-M.V. and T.A.: funding acquisition. All authors have read and agreed to the published version of the manuscript.

Funding: This research was funded by "Fonds spéciaux à la recherche" of the University of Liège. The first author is the recipient of a "Fonds De La Recherche Scientifique—FNRS" grant.

Institutional Review Board Statement: Not applicable.

Informed Consent Statement: Not applicable.

Data Availability Statement: Equine primary myoblasts were obtained from RevaTis and are commercially available.

Acknowledgments: The authors gratefully acknowledge the valuable help received from T. Arnould for protein extraction at the Laboratory of Biochemistry and Cell Biology (URBC) from the Namur Research Institute for Life Sciences (NARILIS) of the University of Namur (Belgium). Our acknowledgements extend to L. Gillet and the Laboratory of Immunology and Vaccinology of the University of Liège (Belgium) for the use of the EnSpire® Multimode plate reader and Clovis Wouters financed by a fellowship from Pommier Nutrition and LABÉO (UniCaen and ULiège) for proofreading.

Conflicts of Interest: The authors declare no conflict of interest.

References

1. Votion, D.-M.; van Galen, G.; Sweetman, L.; Boemer, F.; de Tullio, P.; Dopagne, C.; Lefère, L.; Mouithys-Mickalad, A.; Patarin, F.; Rouxhet, S.; et al. Identification of methylenecyclopropyl acetic acid in serum of European horses with atypical myopathy. *Equine Vet. J.* **2013**, *46*, 146–149. [CrossRef]
2. Valberg, S.J.; Sponseller, B.T.; Hegeman, A.D.; Earing, J.; Bender, J.B.; Martinson, K.L.; Patterson, S.E.; Sweetman, L. Seasonal pasture myopathy/atypical myopathy in North America associated with ingestion of hypoglycin A within seeds of the box elder tree. *Equine Vet. J.* **2013**, *45*, 419–426. [CrossRef] [PubMed]
3. Unger, L.; Nicholson, A.; Jewitt, E.M.; Gerber, V.; Hegeman, A.; Sweetman, L.; Valberg, S. Hypoglycin A concentrations in seeds of acer pseudoplatanus trees growing on atypical myopathy-affected and control pastures. *J. Vet. Intern. Med.* **2014**, *28*, 1289–129. [CrossRef] [PubMed]
4. Westermann, C.M.; van Leeuwen, R.; van Raamsdonk, L.W.D.; Mol, H.G.J. Hypoglycin A concentrations in maple tree species in the netherlands and the occurrence of atypical myopathy in horses. *J. Vet. Intern. Med.* **2016**, *30*, 880–884. [CrossRef] [PubMed]
5. Baise, E.; Habyarimana, J.A.; Amory, H.; Boemer, F.; Douny, C.; Gustin, P.; Marcillaud-Pitel, C.; Patarin, F.; Weber, M.; Votion, D.M. Samaras and seedlings of Acer pseudoplatanus are potential sources of hypoglycin A intoxication in atypical myopathy without necessarily inducing clinical signs. *Equine Vet. J.* **2016**, *48*, 414–417. [CrossRef] [PubMed]
6. Fowden, L.; Pratt, H.M. Cyclopropyl amino acids of the genus Acer: Distribution and biosynthesis. *Phytochemistry* **1973**, *12*, 1677–1681. [CrossRef]
7. Bochnia, M.; Sander, J.; Ziegler, J.; Terhardt, M.; Sander, S.; Janzen, N.; Cavalleri, J.V.; Zuraw, A.; Wensch-Dorendorf, M.; Zeyner, A. Detection of MCPG metabolites in horses with atypical myopathy. *PLoS ONE* **2019**, *14*, e0211698. [CrossRef] [PubMed]
8. Von Holt, C. Methylenecyclopropaneacetic acid, a metabolite of hypoglycin. *Biochim. Biophys. Acta* **1966**, *125*, 1–10. [CrossRef]
9. Melde, K.; Buettner, H.; Boschert, W.; Wolf, H.P.; Ghisla, S. Mechanism of hypoglycaemic action of methylenecyclopropylglycine. *Biochem. J.* **1989**, *259*, 921–924. [CrossRef]
10. Melde, K.; Jackson, S.; Bartlett, K.; Stanley, H.; Sherrattt, A.; Ghisla, S. Metabolic consequences of methylenecyclopropylglycine poisoning in rats. *Biochem. J.* **1991**, *274*, 395–400. [CrossRef]
11. Von Holt, C.; Chang, J.; Von Holt, M.; Böhm, H. Metabolism and metabolic effects of hypoglycin. *Biochim. Biophys. Acta* **1964**, *90*, 611–613. [CrossRef]
12. Westermann, C.M.; Dorland, D.; van Diggelen, O.P.; Schoonderwoerd, K.; Bierau, J.; Waterham, H.R.; van der Kolk, J.H. Decreased oxidative phosphorylation and PGAM deficiency in horses suffering from atypical myopathy associated with acquired MADD. *Mol. Gen. Metab.* **2011**, *104*, 273–278. [CrossRef] [PubMed]
13. Lemieux, H.; Boemer, F.; van Galen, G.; Serteyn, D.; Amory, H.; Baise, E.; Cassart, D.; van Loon, G.; Marcillaud-Pitel, C.; Votion, D.M. Mitochondrial function is altered in horse atypical myopathy. *Mitochondrion* **2016**, *30*, 35–41. [CrossRef] [PubMed]
14. Westermann, C.M.; Dorland, L.; Votion, D.M.; de Sain-van der Velden, M.G.; Wijnberg, I.D.; Wanders, R.J.; Spliet, W.G.; Testerink, N.; Berger, R.; Ruiter, J.P.; et al. Acquired multiple Acyl-CoA dehydrogenase deficiency in 10 horses with atypical myopathy. *Neuromuscul. Disord.* **2008**, *18*, 355–364. [CrossRef] [PubMed]
15. Bochnia, M.; Ziegler, J.; Sander, J.; Uhlig, A.; Schaefer, S.; Vollstedt, S.; Glatter, M.; Abel, S.; Recknagel, S.; Schusser, G.F.; et al. Hypoglycin A content in blood and urine discriminates horses with atypical myopathy from clinically normal horses grazing on the same pasture. *PLoS ONE* **2015**, *10*, e0136785. [CrossRef]
16. Karlíková, R.; Široká, J.; Jahn, P.; Friedecký, D.; Gardlo, A.; Janečková, H.; Hrdinová, F.; Drábková, Z.; Adam, T. Equine atypical myopathy: A metabolic study. *Vet. J.* **2016**, *216*, 125–132. [CrossRef]
17. Boemer, F.; Detilleux, J.; Cello, C.; Amory, H.; Marcillaud-Pitel, C.; Richard, E.; van Galen, G.; van Loon, G.; Lefère, L.; Votion, D.M. Acylcarnitines profile best predicts survival in horses with atypical myopathy. *PLoS ONE* **2017**, *12*, e0182761. [CrossRef]
18. van Galen, G.; Marcillaud Pitel, C.; Saegerman, C.; Patarin, F.; Amory, H.; Baily, J.D.; Cassart, D.; Gerber, V.; Hahn, C.; Harris, P.; et al. European outbreaks of atypical myopathy in grazing equids (2006–2009): Spatiotemporal distribution, history and clinical features. *Equine Vet. J.* **2012**, *44*, 614–620. [CrossRef]
19. Cassart, D.; Baise, E.; Cherel, Y.; Delguste, C.; Antoine, N.; Votion, D.; Amory, H.; Rollin, F.; Linden, A.; Coignoul, F.; et al. Morphological alterations in oxidative muscles and mitochondrial structure associated with equine atypical myopathy. *Equine Vet. J.* **2007**, *39*, 26–32. [CrossRef]
20. Palencia, P.; Rivero, J.L.L. Atypical myopathy in two grazing equids in northern Spain. *Vet. Rec.* **2007**, *161*, 346–348. [CrossRef]

21. Votion, D.-M.; Gnaiger, E.; Lemieux, H.; Mouithys-Mickalad, A.; Serteyn, S. Physical fitness and mitochondrial respiratory capacity in horse skeletal muscle. *PLoS ONE* **2012**, *7*, e34890. [CrossRef]
22. Votion, D.M.; François, A.C.; Kruse, C.; Renaud, B.; Farinelle, A.; Bouquieaux, M.C.; Marcillaud-Pitel, C.; Gustin, P. Answers to the frequently asked questions regarding horse feeding and management practices to reduce the risk of atypical myopathy. *Animals* **2020**, *24*, 365. [CrossRef] [PubMed]
23. van Galen, G.; Saegerman, C.; Marcillaud Pitel, C.; Patarin, F.; Amory, H.; Baily, J.D.; Cassart, D.; Gerber, V.; Hahn, C.; Harris, P.; et al. European outbreaks of atypical myopathy in grazing horses (2006–2009): Determination of indicators for risk and prognostic factors. *Equine Vet. J.* **2012**, *44*, 621–625. [CrossRef] [PubMed]
24. Chen, K.K.; Anderson, R.C.; McCowen, M.C.; Harris, P.N. Pharmacologic action of hypoglycin A and B. *J. Pharmacol. Exp. Ther.* **1957**, *121*, 272–285. [PubMed]
25. Brooks, S.E.; Audretsch, J.J. Studies on hypoglycin toxicity in rats. I. Changes in hepatic ultrastructure. *Am. J. Pathol.* **1970**, *59*, 161–180.
26. Ceusters, J.D.; Mouithys-Mickalad, A.A.; de la Rebière de Pouyade, G.; Franck, T.J.; Votion, D.M.; Deby-Dupont, G.P.; Serteyn, D.A. Assessment of reactive oxygen species production in cultured equine skeletal myoblasts in response to conditions of anoxia followed by reoxygenation with or without exposure to peroxidases. *Am. J. Vet. Res.* **2012**, *73*, 426–434. [CrossRef] [PubMed]
27. Pesta, D.; Gnaiger, E. High-Resolution Respirometry: OXPHOS protocols for human cells and permeabilized fibers from small biopsies of human muscle. *Methods Mol. Biol.* **2012**, *810*, 25–58. [CrossRef]
28. Gnaiger, E. *Mitochondrial Pathways and Respiratory Control. An Introduction to OXPHOS Analysis*, 5th ed.; Bioenergetics Communications: Axams, Austria, 2020. [CrossRef]
29. Lemieux, H.; Subarsky, P.; Doblander, C.; Wurn, M.; Troppmair, J.; Gnaiger, E. Impairment of mitochondrial respiratory function as an early biomarker of apoptosis induced by growth factor removal. *BioRxiv* **2017**. [CrossRef]
30. Remels, A.H.V.; Langen, R.C.J.; Schrauwen, P.; Schaart, G.; Schols, A.M.W.J.; Gosker, H.R. Regulation of mitochondrial biogenesis during myogenesis. *Mol. Cell. Endocrinol.* **2010**, *315*, 113–120. [CrossRef] [PubMed]
31. Bavli, D.; Prill, S.; Ezra, E.; Levy, G.; Cohen, M.; Vinken, M.; Vanfleteren, J.; Jaeger, M.; Nahmias, Y. Real-time monitoring of metabolic function in liver-onchip microdevices tracks the dynamics of Mitochondrial dysfunction. *Proc. Natl. Acad. Sci. USA* **2016**, *113*, E2231–E2240. [CrossRef]
32. Schrauwen, P.; Schrauwen-Hinderling, V.; Hoeks, J.; Hesselink, M.K.C. Mitochondrial dysfunction and lipotoxicity. *BBA Mol. Cell. Biol. Lipids* **2010**, *1801*, 266–271. [CrossRef] [PubMed]
33. Cui, H.; Kong, Y.; Zhang, H. Oxidative Stress, Mitochondrial Dysfunction, and Aging. *J. Signal Trans.* **2012**, *2012*, 1–13. [CrossRef] [PubMed]
34. Farah, B.L.; Sinha, R.A.; Wu, Y.; Singh, B.K.; Lim, A.; Hirayama, M.; Landau, D.J.; Bay, B.H.; Koeberl, D.; Yen, P.M. Hepatic mitochondrial dysfunction is a feature of Glycogen Storage Disease Type Ia (GSDIa). *Sci. Rep.* **2017**, *7*, 1–12. [CrossRef]
35. Moehle, E.A.; Shen, K.; Dillin, A. Mitochondrial proteostasis in the context of cellular and organismal health and aging. *J. Biol. Chem.* **2019**, *14*, 5396–5407. [CrossRef] [PubMed]
36. Baydoun, A.R. Cell culture techniques. In *Principles and Techniques of Biochemistry and Molecular Biology*; Baydoun, A.R., Wilson, K., Eds.; Cambridge University Press: Cambridge, UK, 2005; pp. 71–102. [CrossRef]
37. Kaur, G.; Dufour, J.M. Cell lines. *Spermatogenesis* **2012**, *2*, 1–5. [CrossRef]

 life

MDPI

Review

Accessory Subunits of the Matrix Arm of Mitochondrial Complex I with a Focus on Subunit NDUFS4 and Its Role in Complex I Function and Assembly

Flora Kahlhöfer [1,2,3], Max Gansen [1,2] and Volker Zickermann [1,2,*]

1 Structural Bioenergetics Group, Institute of Biochemistry II, Medical School, Goethe University Frankfurt, Max-von-Laue Str. 9, D-60438 Frankfurt, Germany; flora.kahlhoefer@gmail.com (F.K.); s0915628@stud.uni-frankfurt.de (M.G.)
2 Centre for Biomolecular Magnetic Resonance, Institute for Biophysical Chemistry, Goethe University Frankfurt, Max-von-Laue-Str. 9, D-60438 Frankfurt, Germany
3 Department of Cardiology, Internal Medicine, Goethe University Frankfurt, Theodor-Stern-Kai 7, D-60590 Frankfurt, Germany
* Correspondence: Zickermann@med.uni-frankfurt.de

Abstract: NADH:ubiquinone-oxidoreductase (complex I) is the largest membrane protein complex of the respiratory chain. Complex I couples electron transfer to vectorial proton translocation across the inner mitochondrial membrane. The L shaped structure of complex I is divided into a membrane arm and a matrix arm. Fourteen central subunits are conserved throughout species, while some 30 accessory subunits are typically found in eukaryotes. Complex I dysfunction is associated with mutations in the nuclear and mitochondrial genome, resulting in a broad spectrum of neuromuscular and neurodegenerative diseases. Accessory subunit NDUFS4 in the matrix arm is a hot spot for mutations causing Leigh or Leigh-like syndrome. In this review, we focus on accessory subunits of the matrix arm and discuss recent reports on the function of accessory subunit NDUFS4 and its interplay with NDUFS6, NDUFA12, and assembly factor NDUFAF2 in complex I assembly.

Keywords: mitochondrial disease; Leigh syndrome; NADH dehydrogenase; respiratory chain; oxidative phosphorylation; assembly factor

Citation: Kahlhöfer, F.; Gansen, M.; Zickermann, V. Accessory Subunits of the Matrix Arm of Mitochondrial Complex I with a Focus on Subunit NDUFS4 and Its Role in Complex I Function and Assembly. *Life* **2021**, *11*, 455. https://doi.org/10.3390/life11050455

Academic Editors: Giorgio Lenaz and Salvatore Nesci

Received: 11 April 2021
Accepted: 14 May 2021
Published: 19 May 2021

1. Introduction

Mitochondrial complex I (proton pumping NADH:ubiquinone oxidoreductase) is the largest and most intricate membrane protein complex of the respiratory chain [1–4]. It is a redox-driven proton pump that couples electron transfer from NADH to ubiquinone (Q) with vectorial proton translocation across the inner mitochondrial membrane. With a proton pump stoichiometry of 4 H^+ per NADH consumed, complex I contributes about 40% of the proton motive force that drives ATP synthase. Mitochondrial complex I from a broad range of species can reversibly switch from an active A form into an inactive D form [5,6]. The A/D transition is thought to protect against excessive formation of reactive oxygen species [7,8]. The structure of complex I has been determined by X-ray crystallography [9,10] and cryo-EM [1,11–16] and is now well described.

Mammalian complex I comprises 45 subunits [17]. We have established *Yarrowia lipolytica* as a yeast genetic model organism to study eukaryotic complex I [18]. *Y. lipolytica* complex I comprises 43 subunits of which 40 are orthologues of mammalian complex I [19]. In this review, we use the nomenclature for human complex I also for orthologous proteins from other organisms (Table 1). The large number of polypeptides is divided into central subunits and accessory subunits [20,21]. The 14 central subunits are conserved from bacteria to man and are assigned to three functional modules [21]. The N module (central subunits NDUFS1, NDUFV1, NDUFV2) for NADH oxidation and the Q module (central subunits NDUFS2, NDUFS3, NDUFS7, NDUFS8) for Q reduction are located in the matrix arm of complex I (Figure 1, Table 1). In the membrane arm, the P module (central subunits ND1 to ND6 and ND4L) for proton translocation is subdivided in a proximal P_P and a distal P_D module [22]. The genes for the seven central subunits of the membrane arm represent a substantial part of the mitochondrial genome. All other complex I subunits and all assembly factors are encoded by nuclear DNA. The N module harbors the NADH oxidation site with the initial electron acceptor FMN. The N and Q modules together comprise eight FeS clusters [23,24]. Cluster N1a is thought to have a function for transient storage of electrons to prevent excessive ROS formation and/or to control NADH binding in the active site [25]. The other seven FeS clusters are arranged in an electron transfer chain connecting the NADH oxidation site with the Q reduction site [26]. Cluster N2 is the immediate electron donor for Q. In contrast to other Q reactive enzymes, the Q reduction site of complex I is buried in the protein structure and is located remotely from the membrane phase [27]. The hydrophobic Q has to transit through a tunnel into the Q module to receive electrons from N2 [9,28,29]. It is generally accepted that the energy driving the proton pumps is released in the Q module. However, the coupling mechanism of complex I has remained controversial.

The large majority of accessory subunits is only found in eukaryotic complex I. A notable exception are subunits NDUFS4, NDUFS6, and NDUFA12 that are already present in complex I from α-proteobacteria [30]. The accessory subunits are arranged around the core of central subunits [10,16,31,32]. In general, their function is less clear, but in many cases, severe complex I assembly defects were found after knock out (KO) of individual genes coding for accessory subunits in human cell lines [33].

Mutations causing a broad spectrum of pathological conditions were reported for central and accessory subunits and assembly factors, and they have been reviewed recently [32,34–36].

Here, we focus on accessory subunit NDUFS4 in the matrix arm of complex I and the interplay of accessory subunits with assembly factor NDUFAF2 during the attachment of the N module.

Table 1. Subunits of the peripheral arm of respiratory complex I.

Homo sapiens	*Yarrowia lipolytica*	*Bos taurus*	**Comment**
central subunits peripheral arm			
NDUFS1	NUAM	75-kDa	2x Fe_4S_4; 1x Fe_2S_2
NDUFV1	NUBM	51-kDa	FMN; NADH; 1x Fe_4S_4
NDUFS2	NUCM	49-kDa	Q-binding
NDUFS3	NUGM	30-kDa	
NDUFV2	NUHM	24-kDa	1x Fe_2S_2
NDUFS8	NUIM	TYKY	2x Fe_4S_4
NDUFS7	NUKM	PSST	Q-binding; 1x Fe_4S_4
accessory subunits peripheral arm			
NDUFA2	NI8M	B8	
NDUFS4	NUYM	18-kDa/AQDQ	
NDUFS6	NUMM	13-kDa	Zn^{2+}
NDUFA12	N7BM	B17.2	paralog of assembly factor NDUFAF2
NDUFA7	NUZM	B14.5a	
NDUFA5	NUFM	B13	
NDUFA9	NUEM	39-kDa	NADPH
NDUFA6	NB4M	B14	LYRM6
NDUFAB1	ACPM1	SDAP	ACPM
NDUFV3		9-kDa	
	ST1		sulfur transferase

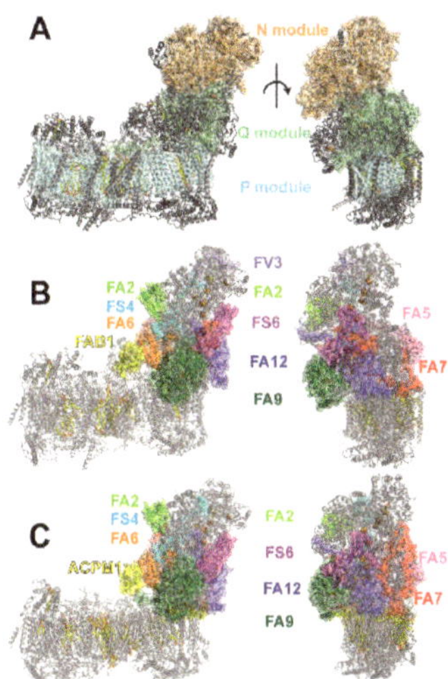

Figure 1. Functional modules of complex I and accessory subunits of the matrix arm. (**A**) The central subunits of complex I are assigned to functional modules for NADH oxidation (N module, orange), ubiquinone reduction (Q module, green), and proton pumping (P module, cyan). (**B**) Accessory subunits of the matrix arm of human complex I (PDB ID: 5xtd) are shown in color, all other subunits are shown in gray. (**C**) Accessory subunits of the matrix arm of *Y. lipolytica* complex I (PDB ID: 6rfr; color code as in (**B**)); the sulfur transferase subunit ST1 is not part of the model; note that FV3 is not present in *Y. lipolytica* complex I.

2. Accessory Subunits of the Matrix Arm in Yeast and Mammalian Complex I

The matrix arm of mammalian and yeast complex I comprises 10 accessory subunits. Overall, the same set of subunits is found, but subunit NDUFV3 of mammalian complex I is not present in the yeast enzyme complex. NDUFV3 is the only subunit for which tissue specific isoforms have been reported [37–39]. On the other hand, only *Y. lipolytica* complex I is associated with the sulfur transferase subunit ST1 [40,41]. Binding of ST1 is substoichiometric and the deletion of the ST1 gene has no impact on complex I function or biogenesis.

NDUFA9 is the largest accessory subunit of the matrix arm. It has the fold of a short chain dehydrogenase [42] and binds NADPH [43]. The cofactor is present in all structures of the eukaryotic complex with sufficient resolution and is therefore a tightly bound component of the subunit [1,11]. The NADPH molecule is too far away from the nearest FeS cluster to allow electron transfer and its function remains unknown. It has been shown recently that NDUFA9 binds the head groups of several phospholipid molecules, which is remarkable for a subunit of the peripheral arm [19]. The subunit is thought to undergo a conformational change in the A/D transition [44] and the relaxation of the protein structure in the C-terminal domain of the subunit has been reported for the D form of mammalian complex I [45].

Mammalian complex I binds two copies of the mitochondrial acyl carrier protein (ACPM) subunit NDUFAB1. In contrast, the yeast enzyme comprises two different but closely related ACPM variants [46]. In all cases, a fatty acid is appended to the phosphopantethein group of the ACPM [47,48]. This fatty acid is inserted into the interior of a mitochondrial LYR (Lys-Tyr-Arg motif) protein that forms a heterodimer with an ACPM [32,49,50]. The mitochondrial LYR proteins were initially implicated in FeS cluster biogenesis [51], but are now recognized to be associated with different macromolecular complexes in the mitochondrion [49]. ACPM/LYRM heterodimers are bound to the Q module of complex I (NDUFAB1α/NDUFA6) [50] and to the tip of the membrane arm (NDUFABβ/NDUFB9) [52]. It is interesting to note that free NDUFAB1 has an essential function in mitochondrial fatty acid synthesis to generate the octanoic acid precursor for lipoic acid [53]. The ACPMs associated with complex I carry longer chain fatty acids and a regulatory function is debated [54]. We have shown that binding of the LYRM protein NDUFA6 to the matrix arm is essential for the Q reductase activity [50]. More recently, we determined the structures of NDUFA6 mutants and showed that single exchanges at the contact site with the functionally important ND3 loop have a strong impact on the interface region of the matrix and membrane arms [55].

NDUFA5 has been noticed in connection with the A/D transition, because the interface of this subunit and accessory subunit NDUFA10 must rearrange during deactivation [45]. Since NDUFA10 is lacking in complex I from *Y. lipolytica*, the longer lifetime of the A form in mammalian complex I might be connected with this specific structural feature [13].

NDUFA2 has a thioredoxin-like fold, but its function has remained unclear. In mammals, the subunit has two cysteine residues, but in *Y. lipoytica*, only one cysteine is conserved.

The three subunits NDUFS4, NDUFS6, and NDUFA12 are distinguished by the fact that they are already found in complex I from α proteobacteria [30]. NDUFS4 has attracted a lot of attention, because it is a hot spot for pathogenic mutations. Knock-out mouse models (*Ndufs4* KO) are widely used to study Leigh syndrome (LS) [56,57]. Moreover, NDUFS4 can be singled out because, in mammalian species, it harbors a canonical serine phosphorylation site [58]. However, analysis of bovine complex I by mass spectrometry did not provide evidence for phosphorylation of the subunit [59,60]. Phosphorylation is thought to play an important role during import and/or maturation of the precursor protein [61,62]. NDUFS6 has a zinc binding site [63]. It is interesting to note that NDUFA12 is a paralog of assembly factor NDUFAF2 [64]. Several lines of evidence have indicated that the interplay of subunits NDUFS4, NDUFS6, and NDUFA12 with assembly factor NDUFAF2 is critical for the attachment of the N module to nascent complex I.

3. Leigh Syndrome and the *Ndufs4* KO Mouse Model

In humans, inactivation of the NDUFS4 gene on chromosome 5 is known to cause severe neurologic disorders [65–68]. In most cases, LS or Leigh-like syndrome is diagnosed (Table 2) [69–75]. LS is a rare disease with a prevalence of roughly 1:40.000 live births and a generally poor prognosis [76–78]. A recent meta-analysis showed that 35% of LS cases are associated with defects in respiratory complex I [79]. In 2016, a ratio of 22 cases of NDUFS4-linked LS for a group of 198 patients with complex I-linked LS was reported [73]. Genotyping of microsatellite DNA markers and array-comparative genomic hybridization has been utilized for diagnosis and might be used for patients with a high pre-test probability in the future [80,81]. Blue native (BN) PAGE consistently revealed abnormal assembly profiles in skin fibroblasts from affected patients and was proposed as a reliable and specific screening method [82]. *Ndufs4* KO mouse models as well as human and murine cell lines have been used extensively to study LS and to explore strategies to counteract the pathophysiological consequences of complex I deficiency [56,57]. Attempts to alleviate disease progression such as expression of plant NDH-2 [83], administration of redox-modulators [84], or targeting of NAD^+-metabolism [85,86] have been reported. Inhibition of mTOR by rapamycin was shown to dramatically improve survival and health in *Ndufs4* KO mice [87], probably by rescuing a dysfunctional α-ketoglutarate/glutamate/glutamine metabolic axis [88]. The metabolite α-ketoglutarate is thought to sustain sufficient OXPHOS capacity and substrate level phosphorylation even when complex I activity is compromised [89]. In addition, there is evidence that glutamatergic neurons, in particular, drive disease development [88]. The link between mTOR inhibition and the neuron-specific neurotransmitter metabolism opens up a further possible explanation for the positive effect of rapamycin. mTOR is present in two distinct complexes, mTORC1 and mTORC2. mTORC2 was initially described as rapamycin insensitive; however, chronic rapamycin treatment is thought to decrease the formation of new functional mTORC2 [90], resulting in a decrease of PKC-β-dependent pro-inflammatory signaling [91]. Rapamycin treatment thus exerts its positive effect via the inhibition of both mTORC complexes, resulting in changes in metabolism and a decreased tendency to inflammation. In another promising approach at the preclinical stage, it was shown that hypoxia treatment with 11% O_2 not only ameliorated symptoms but, in fact, led to the reversal of neurological impairment in the *Ndufs4* KO mouse model [92–94]. It was recently demonstrated that hypoxic breathing normalizes a detrimental hyperoxia in brain tissue, while activation of the hypoxia-inducible factor (HIF) is not a crucial factor [95]. A new perspective on LS has recently been opened by the observation that switching from glycolytic metabolism to OXPHOS is critical for early neuronal morphogenesis [96]. Defective metabolic reprogramming due to mutations in OXPHOS complexes is thought to be incompatible with normal brain development and might lead to early termination of pregnancy in more cases than previously known.

Table 2. Summary of pathogenic mutations in NDUFS4, NDUFS6, NDUFA12, and NDUFAF2.

Subunit	Mutation DNA	Mutation Protein	Disease	Reference
NDUFS4	c.44 G > A	p.Trp15*	Leigh like syndrome	[65]
	c.44 G > A	no complex I assembly	Leigh like syndrome	[68]
	c.99-1 G > A c.462delA	p.Ser34Ilefs*4 p.Lys154Asnfs*34	Leigh syndrome	[97]
	c.221delC	p.Thr74Ilefs*17	Complex I deficiency	[97]
	c.289delG	p.Tyr97*	Leigh like syndrome	[68]
	c.291delG	p.Trp97*	Leigh syndrome	[73]
	c.316 C > T	p.Arg106*	Leigh like syndrome	[98]
	c.340 T > C	p.Trp114Arg	Leigh syndrome	[99]
	c.355 G > C c.462delA	p.Asp119His p.Lys154Asnfs*34	Leigh syndrome	[72]
	c.393dupA	p.Glu132Argfs*15	Leigh syndrome	[70]
	c.462delA	p.Lys154Asnfs*34	Leigh syndrome	[69]
	c.466-470 AAGTC duplication	frameshift, elongation of the carboxyl terminus by 14 residues	Leigh like syndrome	[68,75]
NDUFS6	c.186+2 T > A	splicing abnormality, deletion	Complex I deficiency	[100]
	c.313_315delAAAG	p.104Lys_106Thrfs	Complex I deficiency	[101]
	c.343 C > A c.309 + 5 G > A	p.Cys115Arg	Leigh syndrome	[102]
	c.344 G > A	p.Cys115Tyr	lactic acidemia	[103]
NDUFA12	c.86G > A	p.Arg29Lys	Leigh syndrome	[104]
	c.178C > T	p.Arg60*	Leigh syndrome	[105]
	c.178C > T	p.Arg60*	Mucolipidosis Type II, Leigh syndrome	[106]
	c.224G > A	p.Trp75*	Leigh syndrome	[104]
	c.253G > T	p.Glu85*	Leigh syndrome	[104]
	c.395delA	p.Lys132Argfs*50	Leigh syndrome	[104]
NDUFA2	c.1A > T	p.M1L	hypotonia, nystagmus, ataxia, acute episodes of encephalopathy	[107]
	c.9G > A	p.Trp3*	Leigh syndrome	[108]
	c.103delA	p.Ile35Serfs*	Leigh syndrome	[97]
	c.114C > G	p.Y38*	Leigh syndrome	[109]
	c.182C > T	p.R45*	progressive encephalopathy	[110]
	c.221G > A	p.Trp74*	Leigh syndrome	[97]

Gene therapy approaches in the *Ndufs4* KO mouse model were also pursued as an alternative to pharmacological therapy options [111,112]. Adeno-associated viral vector (AAV)-based gene replacement showed promising results in *Ndufs4* KO mice, but differences in the blood brain barrier between mouse and human are still an obstacle for future clinical applications [112].

4. NDUFS4-Linked Complex I Dysfunction at the Molecular Level

Several lines of evidence indicate that NDUFS4 plays a role in the late stage of complex I assembly [66,68,113,114]. In animal models and patient cell lines, quantitative mass spectrometry showed that deletion of NDUFS4 caused an increase of assembly factor NDUFAF2 and induced a near complete loss of accessory complex I subunit NDUFA12 [115]. In BN PAGE, an 830 kDa subcomplex harbouring NDUFAF2, but lacking the N module, has been observed. However, in intact tissue, substantial rotenone sensitive Q reductase activity was found, which argues against the complete loss of the N module under in vivo conditions [116]. It is interesting to note that integration of complex I into supercomplexes appears to have a stabilizing function for complex I lacking NDUFS4 [117].

We have studied the impact of a NDUFS4 gene deletion on complex I function and assembly in the aerobic yeast *Y. lipolytica* [118]. We found that in the yeast KO strain, complex I levels were decreased and ubiquinone (Q) reductase activity in membranes was reduced. Complexome profiling of intact mitochondria showed that assembly factor NDUFAF2 was bound to complex I, but in clear contrast to the situation observed for mammalian species [115], we did not find a substantial decrease of NDUFA12. In the yeast system, large scale purification of complex I is straightforward. We found that in purified complex I from the NDUFS4 deletion strain, all subunits except NDUFS4 were present and the amount to NDUFAF2 was clearly substoichometric. This suggests that NDUFAF2 was only loosely attached to complex I before solubilization and was easily removed during protein purification. The purified complex showed reduced ubiquinone reductase activity while the formation of ROS under turnover conditions was increased. The EPR spectrum of mutant complex I showed a marked change in the N1b and N3 signals. The cryo-EM structure of the mutant (Figure 2) [19] offered a straightforward explanation for the biochemical and spectroscopic data. We found that the absence of NDUFS4 exposes clusters N1b and N3 to solvent. Thus, the change in EPR spectra is caused by the loss of the shielding function of the accessory subunit. Interestingly, in *T. thermophilus* complex I, the NDUFS1 subunit has an extra loop that partially matches the position of the NDUFS4 subunit in mitochondrial complex I [118]. The increased ROS formation of the mutant might be linked with the greater solvent accessibility of FeS clusters or a longer dwell-time of electrons on FMN, which is known to be critical for the generation of superoxide [119].

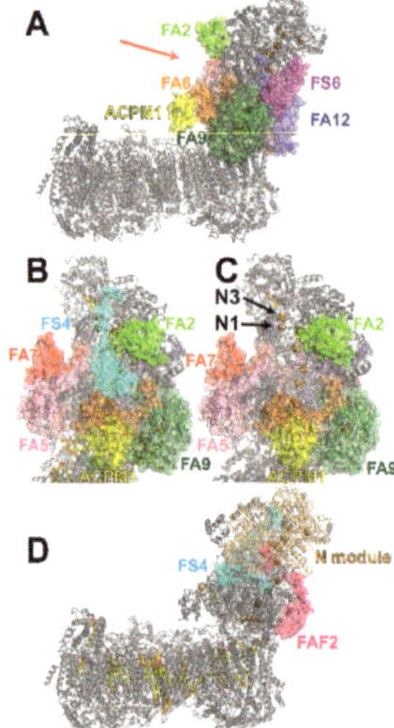

Figure 2. Structure of complex I lacking NDUFS4 and of an assembly intermediate harboring assembly factor NDUFAF2. (**A**) Cryo-EM structure of *Y. lipolytica* complex I purified from *ndufs4Δ* strain (PDB ID: 6rfs); the red arrow indicates direction of view for (**B**) and (**C**). (**B**) Detail view on NDUFS4 in wild type *Y. lipolytica* complex I (PDB ID: 6rfr), direction of view (see (**A**)). (**C**) Same as (**B**) for *ndufs4Δ* mutant (PDB ID: 6rfs); FeS clusters N1b and N3 are solvent exposed. (**D**) Structure of complex I assembly intermediate purified from *Y. lipolytica ndufs6Δ* strain (PDB ID: 6rfq). The assembly intermediate harbors assembly factor NDUFAF2 and all subunits, except NDUFS6 and NDUFA12. For clarity, only NDUFAF2 and NDUFS4 are shown in color. The position of NDUFAF2 matches the position of NDUFA12 in wild type complex I.

5. The Role of NDUFS4, NDUFS6, and NDUFA12 in Complex I Assembly

The intricate assembly pathway of mammalian complex I has been studied in detail [120–122]. Five submodules are initially formed and then combined in a stepwise process to yield complete complex I. At least 15 assembly factors are known to associate with submodules and play an indispensable role in the assembly process [120]. Assembly factor NDUFAF2 was originally identified as a c-Myc controlled mitochondrial protein (Mimitin) with similarity to complex I subunit NDUFA12 [123]. Whole genome subtraction of fermentative and non-fermentative yeasts gave strong indications that NDUFAF2 is a complex I assembly factor and a null mutation of the associated gene was shown to cause progressive encephalopathy [110]. Analysis of mutants in the fungus *Neurospora crassa* [114] and complementation assays using human mitochondria derived from patients [113] showed that NDUFAF2 function is tightly associated with the attachment of the N module and that NDUFS4, NDUFS6, and NDUFA12 must work together to release the assembly factor in the final step of complex I biogenesis. Since NDUFAF2 and NDUFA12 are paralogs, it had been proposed that both polypeptides occupy the same position in the mature enzyme complex and in the preceding assembly intermediate [64]. We have shown that deletion of the gene encoding NDUFS6 in *Y. lipolytica* caused accumulation of an assembly intermediate that lacked NDUFA12, while NDUFAF2 remained firmly bound [63]. The Q reductase activity of the NDUFS6 KO mutant was reduced to 44%. Mutations in the zinc binding site stalled complex I assembly to varying degrees. Pathogenic mutations in NDUFS6 have been reported (Table 2) and exchange of a cysteine ligand of the metal binding site was shown to cause fatal neonatal lactic acidosis [103]. Taking advantage of straightforward His-tag affinity purification of *Y. lipolytica* complex I, we obtained a preparation of the assembly intermediate of sufficient quality for high-resolution structure determination by cryo-EM (Figure 2) [19]. The structure shows that in the assembly intermediate NDUFAF2 in fact matches the position of NDUFA12 in mature complex I. The NDUFAF2 structure also clashes with the position of NDUFS6. The structure thus offers a straightforward explanation for why NDUFS6 and NDUFA12 are required for the release of NDUFAF2. At first sight, the role of NDUFS4 was less clear, because the subunit appeared to be separated from the assembly factor binding site. Interestingly, no cryo-EM density was observed for a sequence stretch of about 100 amino acids in the C-terminal part of the assembly factor indicating disorder. A finger-like protrusion of NDUFS4 penetrates a narrow cleft between the N and Q modules and comes close to the site where the NDUFAF2 structure is unresolved. We have proposed that in the assembly intermediate NDUFS4 has already pushed out a domain of the assembly factor which becomes flexible after detachment from the complex. The major part of the assembly factor remains bound because NDUFS6 is lacking and the NDUFS6/NDUFA12 tandem cannot be formed for complete removal of NDUFAF2. In the NDUFS4 KO, a weak association of complex I with NDUFAF2 is possible because the protein surface occupied by this accessory subunit in the wild type is still available for the assembly factor in the mutant. The C-terminal end of the assembly factor binds to NDUFS1 and anchors the N module. We propose that before the binding of NDUFS4, the un-modelled sequence stretch of the assembly factor is bound to the assembly intermediate and forms a platform for the docking of the N module. Thus, the C-terminal part of the assembly factor guides the N module to its attachment site, while the N terminal domain is responsible for a stable connection with the nascent complex. These results give a consistent picture for *Y. lipolytica*, but cannot explain why NDUFAF2 remains firmly bound in mammalian NDUFS4 KO cells [115]. We propose that there is no fundamental difference in the N module assembly but that only the relative contribution of NDUFS4, NDUFS6, and NDUFA12 for the detachment of NDUFAF2 is different. In *Y. lipolytica*, NDUFS4 plays a minor part, while, in mammals, the lack of NDUFS4 precludes NDUFAF2 detachment, which in turn blocks the association with NDUFA12. This may also explain a weaker binding of the N module in the mutant complex I.

A recent report showed that the N module is turned over faster than the rest of complex I [124]. The N module is thought to be more exposed to oxidative damage because of superoxide formation at the FMN cofactor [119,125]. Selective exchange of dysfunctional N module is advantageous, because it has a lower energetic cost than *de novo* synthesis of the complete enzyme complex. Interestingly, the three subunits discussed here are among the group of subunits with the highest exchange rate in agreement with their role in the attachment of the N module.

6. Conclusions

The central subunits of complex I harbor all bioenergetic core functions. Nevertheless, there is increasing evidence that mutations in accessory complex I subunits can have dramatic consequences and cause fatal disease. The pathophysiology of NDUFS4-linked Leigh syndrome is increasingly well understood. However, therapeutic approaches are still at an experimental stage. Loss of NDUFS4 affects complex I assembly and causes detrimental structural changes in assembled complex I. While animal models and mammalian cell lines are indispensable to study LS and possible therapeutic approaches, the yeast *Y. lipolytica* offers the advantage of straightforward gene manipulation and large-scale purification of complex I variants for biochemical, spectroscopic, and structural analysis of complex I and complex I variants.

Author Contributions: F.K., M.G. and V.Z. wrote the manuscript and prepared the figures. All authors have read and agreed to the published version of the manuscript.

Funding: This work was supported by the Deutsche Forschungsgemeinschaft (DFG grant ZI552/4-2 to V.Z.).

Institutional Review Board Statement: Not applicable.

Informed Consent Statement: Not applicable.

Data Availability Statement: Not applicable.

Conflicts of Interest: The authors declare no conflict of interest.

References

1. Parey, K.; Wirth, C.; Vonck, J.; Zickermann, V. Respiratory Complex I—Structure, Mechanism and Evolution. *Curr. Opin. Struct. Biol.* **2020**, *63*, 1–9. [CrossRef] [PubMed]
2. Yoga, E.G.; Angerer, H.; Parey, K.; Zickermann, V. Respiratory Complex I—Mechanistic Insights and Advances in Structure Determination. *Biochim. et Biophys. Acta (BBA) Bioenerg.* **2020**, *1861*, 148153. [CrossRef]
3. Hirst, J. Mitochondrial Complex I. *Annu. Rev. Biochem.* **2013**, *82*, 551–575. [CrossRef]
4. Sazanov, L.A. A Giant Molecular Proton Pump: Structure and Mechanism of Respiratory Complex I. *Nat. Rev. Mol. Cell Biol.* **2015**, *16*, 375–388. [CrossRef] [PubMed]
5. Kotlyar, A.B.; Vinogradov, A.D. Slow Active/Inactive Transition of the Mitochondrial NADH-Ubiquinone Reductase. *Biochim. et Biophys. Acta (BBA) Bioenerg.* **1990**, *1019*, 151–158. [CrossRef]
6. Maklashina, E.; Kotlyar, A.B.; Cecchini, G. Active/De-Active Transition of Respiratory Complex I in Bacteria, Fungi, and Animals. *Biochim. et Biophys. Acta (BBA) Bioenerg.* **2003**, *1606*, 95–103. [CrossRef]
7. Chouchani, E.T.; Methner, C.; Nadtochiy, S.M.; Logan, A.; Pell, V.R.; Ding, S.; James, A.M.; Cochemé, H.M.; Reinhold, J.; Lilley, K.S.; et al. Cardioprotection by S-Nitrosation of a Cysteine Switch on Mitochondrial Complex I. *Nat. Med.* **2013**, *19*, 753–759. [CrossRef] [PubMed]
8. Dröse, S.; Stepanova, A.; Galkin, A. Ischemic A/D Transition of Mitochondrial Complex I and its Role in ROS Generation. *Biochim. et Biophys. Acta (BBA) Bioenerg.* **2016**, *1857*, 946–957. [CrossRef] [PubMed]
9. Baradaran, R.; Berrisford, J.M.; Minhas, G.S.; Sazanov, L.A. Crystal Structure of the Entire Respiratory Complex I. *Nat. Cell Biol.* **2013**, *494*, 443–448. [CrossRef] [PubMed]
10. Zickermann, V.; Wirth, C.; Nasiri, H.; Siegmund, K.; Schwalbe, H.; Hunte, C.; Brandt, U. Mechanistic Insight from the Crystal Structure of Mitochondrial Complex I. *Science* **2015**, *347*, 44–49. [CrossRef]
11. Agip, A.-N.A.; Blaza, J.N.; Fedor, J.G.; Hirst, J. Mammalian Respiratory Complex I through the Lens of Cryo-EM. *Annu. Rev. Biophys.* **2019**, *48*, 165–184. [CrossRef] [PubMed]
12. Kampjut, D.; Sazanov, L.A. The Coupling Mechanism of Mammalian Respiratory Complex I. *Science* **2020**, *370*, eabc4209. [CrossRef]
13. Grba, D.N.; Hirst, J. Mitochondrial Complex I Structure Reveals Ordered Water Molecules for Catalysis and Proton Translocation. *Nat. Struct. Mol. Biol.* **2020**, *27*, 1–9. [CrossRef]

14. Klusch, N.; Senkler, J.; Yildiz, Ö.; Kühlbrandt, W.; Braun, H.-P. A Ferredoxin Bridge Connects the Two Arms of Plant Mitochondrial Complex I. *Plant Cell* **2021**. [CrossRef] [PubMed]
15. Soufari, H.; Parrot, C.; Kuhn, L.; Waltz, F.; Hashem, Y. Specific Features and Assembly of the Plant Mitochondrial Complex I Revealed by Cryo-EM. *Nat. Commun.* **2020**, *11*, 1–7. [CrossRef] [PubMed]
16. Guo, R.; Zong, S.; Wu, M.; Gu, J.; Yang, M. Architecture of Human Mitochondrial Respiratory Megacomplex I2III2IV2. *Cell* **2017**, *170*, 1247–1257.e12. [CrossRef] [PubMed]
17. Carroll, J.; Fearnley, I.M.; Skehel, J.M.; Shannon, R.J.; Hirst, J.; Walker, J.E. Bovine Complex I Is a Complex of 45 Different Subunits. *J. Biol. Chem.* **2006**, *281*, 32724–32727. [CrossRef] [PubMed]
18. Kerscher, S.; Drose, S.; Zwicker, K.; Zickermann, V.; Brandt, U. Yarrowia Lipolytica, a Yeast Genetic System to Study Mito-chondrial Complex I. *Biochim. Biophys. Acta* **2002**, *1555*, 83–91. [CrossRef]
19. Parey, K.; Haapanen, O.; Sharma, V.; Köfeler, H.; Züllig, T.; Prinz, S.; Siegmund, K.; Wittig, I.; Mills, D.J.; Vonck, J.; et al. High-Resolution cryo-EM Structures of Respiratory Complex I: Mechanism, Assembly, and Disease. *Sci. Adv.* **2019**, *5*, eaax9484. [CrossRef]
20. Kmita, K.; Zickermann, V. Accessory Subunits of Mitochondrial Complex I. *Biochem. Soc. Trans.* **2013**, *41*, 1272–1279. [CrossRef]
21. Brandt, U. Energy Converting NADH: Quinone Oxidoreductase (Complex I). *Annu. Rev. Biochem.* **2006**, *75*, 69–92. [CrossRef] [PubMed]
22. Hunte, C.; Zickermann, V.; Brandt, U. Functional Modules and Structural Basis of Conformational Coupling in Mitochondrial Complex I. *Science* **2010**, *329*, 448–451. [CrossRef] [PubMed]
23. Ohnishi, T. Iron–Sulfur Clusters/Semiquinones in Complex I. *Biochim. et Biophys. Acta (BBA) Bioenerg.* **1998**, *1364*, 186–206. [CrossRef]
24. Roessler, M.M.; King, M.S.; Robinson, A.J.; Armstrong, F.A.; Harmer, J.; Hirst, J. Direct Assignment of EPR Spectra to Structurally Defined Iron-Sulfur Clusters in Complex I by Double Electron–Electron Resonance. *Proc. Natl. Acad. Sci. USA* **2010**, *107*, 1930–1935. [CrossRef]
25. Schulte, M.; Frick, K.; Gnandt, E.; Jurkovic, S.; Burschel, S.; Labatzke, R.; Aierstock, K.; Fiegen, D.; Wohlwend, D.; Gerhardt, S.; et al. A Mechanism to Prevent Production of Reactive Oxygen Species by Escherichia Coli Respiratory Complex I. *Nat. Commun.* **2019**, *10*, 1–9. [CrossRef]
26. Sazanov, L.A. Structure of the Hydrophilic Domain of Respiratory Complex I from Thermus Thermophilus. *Science* **2006**, *311*, 1430–1436. [CrossRef]
27. Zickermann, V.; Bostina, M.; Hunte, C.; Ruiz, T.; Radermacher, M.; Brandt, U. Functional Implications from an Unexpected Position of the 49-kDa Subunit of NADH:Ubiquinone Oxidoreductase. *J. Biol. Chem.* **2003**, *278*, 29072–29078. [CrossRef] [PubMed]
28. Warnau, J.; Sharma, V.; Gamiz-Hernandez, A.P.; Di Luca, A.; Haapanen, O.; Vattulainen, I.; Wikström, M.; Hummer, G.; Kaila, V.R.I. Redox-Coupled Quinone Dynamics in the Respiratory Complex I. *Proc. Natl. Acad. Sci. USA* **2018**, *115*, E8413–E8420. [CrossRef]
29. Fedor, J.G.; Jones, A.J.Y.; Di Luca, A.; Kaila, V.R.I.; Hirst, J. Correlating kinetic and structural data on ubiquinone binding and reduction by respiratory complex I. *Proc. Natl. Acad. Sci. USA* **2017**, *114*, 12737–12742. [CrossRef]
30. Yip, C.-Y.; Harbour, M.E.; Jayawardena, K.; Fearnley, I.M.; Sazanov, L.A. Evolution of Respiratory Complex I. *J. Biol. Chem.* **2011**, *286*, 5023–5033. [CrossRef]
31. Zhu, J.; Vinothkumar, K.R.; Hirst, J.Z.J. Structure of Mammalian Respiratory Complex I. *Nat. Cell Biol.* **2016**, *536*, 354–358. [CrossRef]
32. Fiedorczuk, K.; Letts, J.A.; Degliesposti, G.; Kaszuba, K.; Skehel, G.D.M.; Sazanov, L.A. Atomic Structure of the Entire Mammalian Mitochondrial Complex I. *Nat. Cell Biol.* **2016**, *538*, 406–410. [CrossRef]
33. Stroud, D.A.; Surgenor, E.E.; Formosa, L.E.; Reljic, B.; Frazier, A.E.; Dibley, M.; Osellame, L.D.; Stait, T.; Beilharz, T.H.; Thorburn, D.R.; et al. Accessory Subunits are Integral for Assembly and Function of Human Mitochondrial Complex I. *Nat. Cell Biol.* **2016**, *538*, 123–126. [CrossRef] [PubMed]
34. Dang, Q.-C.L.; Phan, D.H.; Johnson, A.N.; Pasapuleti, M.; AlKhaldi, H.A.; Zhang, F.; Vik, S.B. Analysis of Human Mutations in the Supernumerary Subunits of Complex I. *Life* **2020**, *10*, 296. [CrossRef] [PubMed]
35. Rodenburg, R.J. Mitochondrial Complex I-Linked Disease. *Biochim. et Biophys. Acta (BBA) Bioenerg.* **2016**, *1857*, 938–945. [CrossRef] [PubMed]
36. Ghezzi, D.; Zeviani, M. Human Diseases Associated with Defects in Assembly of OXPHOS Complexes. *Essays Biochem.* **2018**, *62*, 271–286. [CrossRef]
37. Bridges, H.R.; Mohammed, K.; Harbour, M.E.; Hirst, J. Subunit NDUFV3 is Present in Two Distinct Isoforms in Mammalian Complex I. *Biochim. et Biophys. Acta (BBA) Bioenerg.* **2017**, *1858*, 197–207. [CrossRef]
38. Dibley, M.G.; Ryan, M.T.; Stroud, D.A. A Novel Isoform of the Human Mitochondrial Complex I Subunit NDUFV3. *FEBS Lett.* **2016**, *591*, 109–117. [CrossRef]
39. Guerrero-Castillo, S.; Cabrera-Orefice, A.; Huynen, M.A.; Arnold, S. Identification and Evolutionary Analysis of Tissue-Specific Isoforms of Mitochondrial Complex I Subunit NDUFV3. *Biochim. et Biophys. Acta (BBA) Bioenerg.* **2017**, *1858*, 208–217. [CrossRef] [PubMed]
40. Abdrakhmanova, A.; Dobrynin, K.; Zwicker, K.; Kerscher, S.; Brandt, U. Functional Sulfurtransferase is Associated with Mitochondrial Complex I fromYarrowia Lipolytica, but is Not Required for Assembly of its Iron-Sulfur Clusters. *FEBS Lett.* **2005**, *579*, 6781–6785. [CrossRef]

41. Parey, K.; Brandt, U.; Xie, H.; Mills, D.J.; Siegmund, K.; Vonck, J.; Kühlbrandt, W.; Zickermann, V. Cryo-EM Structure of Respiratory Complex I at Work. *eLife* **2018**, *7*. [CrossRef] [PubMed]
42. Jörnvall, H.; Persson, B.; Krook, M.; Atrian, S.; Gonzalez-Duarte, R.; Jeffery, J.; Ghosh, D. Short-Chain Dehydrogenases/Reductases (SDR). *Biochemistry* **1995**, *34*, 6003–6013. [CrossRef]
43. Abdrakhmanova, A.; Zwicker, K.; Kerscher, S.; Zickermann, V.; Brandt, U. Tight Binding of NADPH to the 39-kDa Subunit of Complex I is Not Required for Catalytic Activity but Stabilizes the Multiprotein Complex. *Biochim. et Biophys. Acta (BBA) Bioenerg.* **2006**, *1757*, 1676–1682. [CrossRef] [PubMed]
44. Babot, M.; Labarbuta, P.; Birch, A.; Kee, S.; Fuszard, M.; Botting, C.H.; Wittig, I.; Heide, H.; Galkin, A. ND3, ND1 and 39kDa Subunits are more Exposed in the De-Active form of Bovine Mitochondrial Complex I. *Biochim. et Biophys. Acta (BBA) Bioenerg.* **2014**, *1837*, 929–939. [CrossRef]
45. Agip, A.-N.A.; Blaza, J.N.; Bridges, H.R.; Viscomi, C.; Rawson, S.; Muench, S.P.; Hirst, J. Cryo-EM Structures of Complex I from Mouse Heart Mitochondria in Two Biochemically Defined States. *Nat. Struct. Mol. Biol.* **2018**, *25*, 548–556. [CrossRef] [PubMed]
46. Dobrynin, K.; Abdrakhmanova, A.; Richers, S.; Hunte, C.; Kerscher, S.; Brandt, U. Characterization of Two Different Acyl Carrier Proteins in Complex I from Yarrowia Lipolytica. *Biochim. et Biophys. Acta (BBA) Bioenerg.* **2010**, *1797*, 152–159. [CrossRef] [PubMed]
47. Angerer, H.; Schönborn, S.; Gorka, J.; Bahr, U.; Karas, M.; Wittig, I.; Heidler, J.; Hoffmann, J.; Morgner, N.; Zickermann, V. Acyl Modification and Binding of Mitochondrial ACP to Multiprotein Complexes. *Biochim. et Biophys. Acta (BBA) Bioenerg.* **2017**, *1864*, 1913–1920. [CrossRef]
48. Carroll, J.; Fearnley, I.M.; Shannon, R.J.; Hirst, J.; Walker, J.E. Analysis of the Subunit Composition of Complex I from Bovine Heart Mitochondria*S. *Mol. Cell. Proteom.* **2003**, *2*, 117–126. [CrossRef]
49. Angerer, H. Eukaryotic LYR Proteins Interact with Mitochondrial Protein Complexes. *Biology* **2015**, *4*, 133–150. [CrossRef]
50. Angerer, H.; Radermacher, M.; Mańkowska, M.; Steger, M.; Zwicker, K.; Heide, H.; Wittig, I.; Brandt, U.; Zickermann, V. The LYR Protein Subunit NB4M/NDUFA6 of Mitochondrial Complex I Anchors an Acyl Carrier Protein and is Essential for Catalytic Activity. *Proc. Natl. Acad. Sci. USA* **2014**, *111*, 5207–5212. [CrossRef] [PubMed]
51. Adam, A.C.; Bornhövd, C.; Prokisch, H.; Neupert, W.; Hell, K. The Nfs1 Interacting Protein Isd11 has an Essential Role in Fe/S Cluster Biogenesis in Mitochondria. *EMBO J.* **2006**, *25*, 174–183. [CrossRef]
52. Zhu, J.; King, M.S.; Yu, M.; Klipcan, L.; Leslie, A.G.W.; Hirst, J. Structure of Subcomplex Iβ of Mammalian Respiratory Complex I Leads to New Supernumerary Subunit Assignments. *Proc. Natl. Acad. Sci. USA* **2015**, *112*, 12087–12092. [CrossRef] [PubMed]
53. Kastaniotis, A.J.; Autio, K.J.; Kerätär, J.M.; Monteuuis, G.; Mäkelä, A.M.; Nair, R.R.; Pietikäinen, L.P.; Shvetsova, A.; Chen, Z.; Hiltunen, J.K. Mitochondrial Fatty Acid Synthesis, Fatty Acids and Mitochondrial Physiology. *Biochim. et Biophys. Acta (BBA) Mol. Cell Biol. Lipids* **2017**, *1862*, 39–48. [CrossRef]
54. Nowinski, S.M.; Van Vranken, J.G.; Dove, K.K.; Rutter, J. Impact of Mitochondrial Fatty Acid Synthesis on Mitochondrial Biogenesis. *Curr. Biol.* **2018**, *28*, R1212–R1219. [CrossRef] [PubMed]
55. Yoga, E.G.; Parey, K.; Djurabekova, A.; Haapanen, O.; Siegmund, K.; Zwicker, K.; Sharma, V.; Zickermann, V.; Angerer, H. Essential Role of Accessory Subunit LYRM6 in the Mechanism of Mitochondrial Complex I. *Nat. Commun.* **2020**, *11*, 1–8. [CrossRef]
56. Ingraham, C.A.; Burwell, L.S.; Skalska, J.; Brookes, P.S.; Howell, R.L.; Sheu, S.-S.; Pinkert, C.A. NDUFS4: Creation of a Mouse Model Mimicking a Complex I Disorder. *Mitochondrion* **2009**, *9*, 204–210. [CrossRef]
57. Breuer, M.E.; Willems, P.H.; Smeitink, J.A.; Koopman, W.J.; Nooteboom, M. Cellular and Animal Models for Mitochondrial Complex I Deficiency: A Focus on the NDUFS4 Subunit. *IUBMB Life* **2013**, *65*, 202–208. [CrossRef]
58. Papa, S.; Sardanelli, A.M.; Cocco, T.; Speranza, F.; Scacco, S.C.; Technikova-Dobrova, Z. The Nuclear-Encoded 18 kDa (IP) AQDQ Subunit of Bovine Heart Complex I is Phosphorylated by the Mitochondrial cAMP-Dependent Protein Kinase. *FEBS Lett.* **1996**, *379*, 299–301. [CrossRef]
59. De Rasmo, D.; Palmisano, G.; Scacco, S.; Technikova-Dobrova, Z.; Panelli, D.; Cocco, T.M.; Sardanelli, A.M.; Gnoni, A.; Micelli, L.; Trani, A.; et al. Phosphorylation Pattern of the NDUFS4 Subunit of Complex I of the Mammalian Respiratory Chain. *Mitochondrion* **2010**, *10*, 464–471. [CrossRef] [PubMed]
60. Chen, R.; Fearnley, I.M.; Peak-Chew, S.Y.; Walker, J.E. The Phosphorylation of Subunits of Complex I from Bovine Heart Mitochondria. *J. Biol. Chem.* **2004**, *279*, 26036–26045. [CrossRef]
61. De Rasmo, D.; Panelli, D.; Sardanelli, A.M.; Papa, S. cAMP-Dependent Protein Kinase Regulates the Mitochondrial Import of the Nuclear Encoded NDUFS4 Subunit of Complex I. *Cell. Signal.* **2008**, *20*, 989–997. [CrossRef]
62. Papa, S.; De Rasmo, D.; Scacco, S.; Signorile, A.; Technikova-Dobrova, Z.; Palmisano, G.; Sardanelli, A.M.; Papa, F.; Panelli, D.; Scaringi, R.; et al. Mammalian Complex I: A Regulable and Vulnerable Pacemaker in Mitochondrial Respiratory Function. *Biochim. et Biophys. Acta (BBA) Bioenerg.* **2008**, *1777*, 719–728. [CrossRef]
63. Kmita, K.; Wirth, C.; Warnau, J.; Guerrero-Castillo, S.; Hunte, C.; Hummer, G.; Kaila, V.R.I.; Zwicker, K.; Brandt, U.; Zickermann, V. Accessory NUMM (NDUFS6) Subunit Harbors a Zn-Binding Site and is Essential for Biogenesis of Mitochondrial Complex I. *Proc. Natl. Acad. Sci. USA* **2015**, *112*, 5685–5690. [CrossRef] [PubMed]
64. Kensche, P.R.; Duarte, I.; Huynen, A.M. A Three-Dimensional Topology of Complex I Inferred from Evolutionary Correlations. *BMC Struct. Biol.* **2012**, *12*, 19. [CrossRef]

65. Petruzzella, V.; Vergari, R.; Puzziferri, I.; Boffoli, D.; Lamantea, E.; Zeviani, M.; Papa, S. A Nonsense Mutation in the NDUFS4 gene Encoding the 18 kDa (AQDQ) Subunit of Complex I Abolishes Assembly and Activity of the Complex in a Patient with Leigh-Like Syndrome. *Hum. Mol. Genet.* **2001**, *10*, 529–535. [CrossRef]
66. Petruzzella, V.; Papa, S. Mutations in Human Nuclear Genes Encoding for Subunits of Mitochondrial Respiratory Complex I: The NDUFS4 Gene. *Gene* **2002**, *286*, 149–154. [CrossRef]
67. Petruzzella, V.; Panelli, D.; Torraco, A.; Stella, A.; Papa, S. Mutations in the NDUFS4 Gene of Mitochondrial Complex I Alter Stability of the Splice Variants. *FEBS Lett.* **2005**, *579*, 3770–3776. [CrossRef]
68. Scacco, S.; Petruzzella, V.; Budde, S.; Vergari, R.; Tamborra, R.; Panelli, D.; Heuvel, L.P.V.D.; Smeitink, J.A.; Papa, S. Pathological Mutations of the Human NDUFS4 Gene of the 18-kDa (AQDQ) Subunit of Complex I Affect the Expression of the Protein and the Assembly and Function of the Complex. *J. Biol. Chem.* **2003**, *278*, 44161–44167. [CrossRef]
69. Anderson, S.L.; Chung, W.K.; Frezzo, J.; Papp, J.C.; Ekstein, J.; DiMauro, S.; Rubin, B.Y. A Novel Mutation in NDUFS4 Causes Leigh Syndrome in an Ashkenazi Jewish Family. *J. Inherit. Metab. Dis.* **2008**, *31*, 461–467. [CrossRef] [PubMed]
70. Lamont, R.E.; Beaulieu, C.L.; Bernier, F.P.; Sparkes, R.; Innes, A.M.; Jackel-Cram, C.; Ober, C.; Parboosingh, J.S.; Lemire, E.G. A Novel NDUFS4 Frameshift Mutation Causes Leigh Disease in the Hutterite Population. *Am. J. Med. Genet. Part A* **2016**, *173*, 596–600. [CrossRef] [PubMed]
71. Budde, S.M.S.; Heuvel, L.P.W.J.V.D.; Smeets, R.J.P.; Skladal, D.; Mayr, J.A.; Boelen, C.; Petruzzella, V.; Papa, S.; Smeitink, J.A.M. Clinical Heterogeneity in Patients with Mutations in the NDUFS4 Gene of Mitochondrial Complex I. *J. Inherit. Metab. Dis.* **2003**, *26*, 813–815. [CrossRef]
72. Leshinsky-Silver, E.; Lebre, A.-S.; Minai, L.; Saada, A.; Steffann, J.; Cohen, S.; Rötig, A.; Munnich, A.; Lev, D.; Lerman-Sagie, T. NDUFS4 Mutations cause Leigh Syndrome with Predominant Brainstem Involvement. *Mol. Genet. Metab.* **2009**, *97*, 185–189. [CrossRef] [PubMed]
73. Ortigoza-Escobar, J.D.; Oyarzabal, A.; Montero, R.; Artuch, R.; Jou, C.; Jiménez, C.; Gort, L.; Briones, P.; Muchart, J.; López-Gallardo, E.; et al. Ndufs4 Related Leigh Syndrome: A Case Report and Review of the Literature. *Mitochondrion* **2016**, *28*, 73–78. [CrossRef] [PubMed]
74. Finsterer, J.; Zarrouk-Mahjoub, S. NDUFS4-Related Leigh Syndrome in Hutterites. *Am. J. Med. Genet. Part A* **2017**, *173*, 1450–1451. [CrossRef]
75. Heuvel, L.V.D.; Ruitenbeek, W.; Smeets, R.; Gelman-Kohan, Z.; Elpeleg, O.; Loeffen, J.; Trijbels, F.; Mariman, E.; de Bruijn, D.; Smeitink, J. Demonstration of a New Pathogenic Mutation in Human Complex I Deficiency: A 5-bp Duplication in the Nuclear Gene Encoding the 18-kD (AQDQ) Subunit. *Am. J. Hum. Genet.* **1998**, *62*, 262–268. [CrossRef]
76. Rahman, S.; Blok, R.B.; Dahl, H.-H.M.; Danks, D.M.; Kirby, D.M.; Chow, C.W.; Christodoulou, J.; Thorburn, D.R. Leigh syndrome: Clinical Features and Biochemical and DNA Abnormalities. *Ann. Neurol.* **1996**, *39*, 343–351. [CrossRef] [PubMed]
77. Leigh, D. Subacute Necrotizing Encephalomyelopathy in an Infant. *J. Neurol. Neurosurg. Psychiatry* **1951**, *14*, 216–221. [CrossRef] [PubMed]
78. Gerards, M.; Sallevelt, S.C.; Smeets, H.J. Leigh Syndrome: Resolving the Clinical and Genetic Heterogeneity Paves the Way for Treatment Options. *Mol. Genet. Metab.* **2016**, *117*, 300–312. [CrossRef] [PubMed]
79. Chang, X.; Wu, Y.; Zhou, J.; Meng, H.; Zhang, W.; Guo, J. A Meta-Analysis and Systematic Review of Leigh Syndrome: Clinical Manifestations, Respiratory Chain Enzyme Complex Deficiency, and Gene Mutations. *Medicine* **2020**, *99*, e18634. [CrossRef]
80. Bénit, P.; Steffann, J.; Lebon, S.; Chretien, D.; Kadhom, N.; De Lonlay, P.; Goldenberg, A.; Dumez, Y.; Dommergues, M.; Rustin, P.; et al. Genotyping Microsatellite DNA Markers at Putative Disease Loci in Inbred/Multiplex Families with Respiratory Chain Complex I Deficiency Allows Rapid Identification of a Novel Nonsense Mutation (IVS1nt −1) in the NDUFS4 Gene in Leigh Syndrome. *Qual. Life Res.* **2003**, *112*, 563–566. [CrossRef]
81. Lombardo, B.; Ceglia, C.; Tarsitano, M.; Pierucci, I.; Salvatore, F.; Pastore, L. Identification of a Deletion in the NDUFS4 Gene Using Array-Comparative Genomic Hybridization in a Patient with Suspected Mitochondrial Respiratory Disease. *Gene* **2014**, *535*, 376–379. [CrossRef] [PubMed]
82. Assouline, Z.; Jambou, M.; Rio, M.; Bole-Feysot, C.; De Lonlay, P.; Barnerias, C.; Desguerre, I.; Bonnemains, C.; Guillermet, C.; Steffann, J.; et al. A Constant and Similar Assembly Defect of Mitochondrial Respiratory Chain Complex I Allows Rapid Identification of NDUFS4 Mutations in Patients with Leigh Syndrome. *Biochim. et Biophys. Acta (BBA) Mol. Basis Dis.* **2012**, *1822*, 1062–1069. [CrossRef]
83. Catania, A.; Iuso, A.; Bouchereau, J.; Kremer, L.S.; Paviolo, M.; Terrile, C.; Bénit, P.; Rasmusson, A.G.; Schwarzmayr, T.; Tiranti, V.; et al. Arabidopsis Thaliana Alternative Dehydrogenases: A Potential Therapy for Mitochondrial Complex I Deficiency? Perspectives and Pitfalls. *Orphanet J. Rare Dis.* **2019**, *14*, 236. [CrossRef] [PubMed]
84. De Haas, R.; Das, D.; Garanto, A.; Renkema, H.G.; Greupink, R.; Broek, P.V.D.; Pertijs, J.; Collin, R.W.J.; Willems, P.; Beyrath, J.; et al. Therapeutic Effects of the Mitochondrial ROS-Redox Modulator KH176 in a Mammalian Model of Leigh Disease. *Sci. Rep.* **2017**, *7*, 1–11. [CrossRef] [PubMed]
85. Karamanlidis, G.; Lee, C.F.; Garcia-Menendez, L.; Kolwicz, S.C.; Suthammarak, W.; Gong, G.; Sedensky, M.M.; Morgan, P.G.; Wang, W.; Tian, R. Mitochondrial Complex I Deficiency Increases Protein Acetylation and Accelerates Heart Failure. *Cell Metab.* **2013**, *18*, 239–250. [CrossRef]
86. Lee, C.F.; Caudal, A.; Abell, L.; Gowda, G.A.N.; Tian, R. Targeting NAD+ Metabolism as Interventions for Mitochondrial Disease. *Sci. Rep.* **2019**, *9*, 1–10. [CrossRef]

87. Johnson, S.C.; Yanos, M.E.; Kayser, E.-B.; Quintana, A.; Sangesland, M.; Castanza, A.; Uhde, L.; Hui, J.; Wall, V.Z.; Gagnidze, A.; et al. mTOR Inhibition Alleviates Mitochondrial Disease in a Mouse Model of Leigh Syndrome. *Science* **2013**, *342*, 1524–1528. [CrossRef] [PubMed]
88. Johnson, S.C.; Kayser, E.-B.; Bornstein, R.; Stokes, J.; Bitto, A.; Park, K.Y.; Pan, A.; Sun, G.; Raftery, D.; Kaeberlein, M.; et al. Regional Metabolic Signatures in the Ndufs4(KO) Mouse Brain Implicate Defective Glutamate/α-Ketoglutarate Metabolism in Mitochondrial Disease. *Mol. Genet. Metab.* **2020**, *130*, 118–132. [CrossRef]
89. Kayser, E.-B.; Sedensky, M.M.; Morgan, P.G. Region-Specific Defects of Respiratory Capacities in the Ndufs4(KO) Mouse Brain. *PLoS ONE* **2016**, *11*, e0148219. [CrossRef] [PubMed]
90. Sarbassov, D.D.; Ali, S.M.; Sengupta, S.; Sheen, J.-H.; Hsu, P.P.; Bagley, A.F.; Markhard, A.L.; Sabatini, D.M. Prolonged Rapamycin Treatment Inhibits mTORC2 Assembly and Akt/PKB. *Mol. Cell* **2006**, *22*, 159–168. [CrossRef]
91. Martin-Perez, M.; Grillo, A.S.; Ito, T.K.; Valente, A.S.; Han, J.; Entwisle, S.W.; Huang, H.Z.; Kim, D.; Yajima, M.; Kaeberlein, M.; et al. PKC Downregulation upon Rapamycin Treatment Attenuates Mitochondrial Disease. *Nat. Metab.* **2020**, *2*, 1472–1481. [CrossRef]
92. Ferrari, M.; Jain, I.H.; Goldberger, O.; Rezoagli, E.; Thoonen, R.; Cheng, K.-H.; Sosnovik, D.E.; Scherrer-Crosbie, M.; Mootha, V.K.; Zapol, W.M. Hypoxia Treatment Reverses Neurodegenerative Disease in a Mouse Model of Leigh Syndrome. *Proc. Natl. Acad. Sci. USA* **2017**, *114*, E4241–E4250. [CrossRef] [PubMed]
93. Jain, I.H.; Zazzeron, L.; Goli, R.; Alexa, K.; Schatzman-Bone, S.; Dhillon, H.; Goldberger, O.; Peng, J.; Shalem, O.; Sanjana, N.E.; et al. Hypoxia as a Therapy for Mitochondrial Disease. *Science* **2016**, *352*, 54–61. [CrossRef] [PubMed]
94. Grange, R.M.; Sharma, R.; Shah, H.; Reinstadler, B.; Goldberger, O.; Cooper, M.K.; Nakagawa, A.; Miyazaki, Y.; Hindle, A.G.; Batten, A.J.; et al. Hypoxia Ameliorates Brain Hyperoxia and NAD+ Deficiency in a Murine Model of Leigh Syndrome. *Mol. Genet. Metab.* **2021**, *133*, 83–93. [CrossRef]
95. Jain, I.H.; Zazzeron, L.; Goldberger, O.; Marutani, E.; Wojtkiewicz, G.R.; Ast, T.; Wang, H.; Schleifer, G.; Stepanova, A.; Brepoels, K.; et al. Leigh Syndrome Mouse Model Can Be Rescued by Interventions that Normalize Brain Hyperoxia, but Not HIF Activation. *Cell Metab.* **2019**, *30*, 824–832.e3. [CrossRef] [PubMed]
96. Inak, G.; Rybak-Wolf, A.; Lisowski, P.; Pentimalli, T.M.; Jüttner, R.; Glažar, P.; Uppal, K.; Bottani, E.; Brunetti, D.; Secker, C.; et al. Defective Metabolic Programming Impairs Early Neuronal Morphogenesis in Neural Cultures and an Organoid Model of Leigh Syndrome. *Nat. Commun.* **2021**, *12*, 1–22. [CrossRef]
97. Calvo, E.S.; Tucker, E.J.; Compton, A.; Kirby, D.M.; Crawford, G.; Burtt, N.P.; Rivas, M.; Guiducci, C.; Bruno, D.L.; Goldberger, A.O.; et al. High-Throughput, Pooled Sequencing Identifies Mutations in NUBPL and FOXRED1 in Human Complex I Deficiency. *Nat. Genet.* **2010**, *42*, 851–858. [CrossRef] [PubMed]
98. Budde, S.; Heuvel, L.V.D.; Janssen, A.; Smeets, R.; Buskens, C.; DeMeirleir, L.; Van Coster, R.; Baethmann, M.; Voit, T.; Trijbels, J.; et al. Combined Enzymatic Complex I and III Deficiency Associated with Mutations in the Nuclear Encoded NDUFS4 Gene. *Biochem. Biophys. Res. Commun.* **2000**, *275*, 63–68. [CrossRef] [PubMed]
99. Kohda, M.; Tokuzawa, Y.; Kishita, Y.; Nyuzuki, H.; Moriyama, Y.; Mizuno, Y.; Hirata, T.; Yatsuka, Y.; Yamashita-Sugahara, Y.; Nakachi, Y.; et al. A Comprehensive Genomic Analysis Reveals the Genetic Landscape of Mitochondrial Respiratory Chain Complex Deficiencies. *PLoS Genet.* **2016**, *12*, e1005679. [CrossRef] [PubMed]
100. Kirby, D.M.; McFarland, R.; Ohtake, A.; Dunning, C.; Ryan, M.T.; Wilson, C.; Ketteridge, D.; Turnbull, D.M.; Thorburn, D.R.; Taylor, R.W. Mutations of the Mitochondrial ND1 Gene as a Cause of MELAS. *J. Med. Genet.* **2004**, *41*, 784–789. [CrossRef]
101. Pronicka, E.; Piekutowska-Abramczuk, D.; Ciara, E.; Trubicka, J.; Rokicki, D.; Karkucińska-Więckowska, A.; Pajdowska, M.; Jurkiewicz, E.; Halat, P.; Kosińska, J.; et al. New Perspective in Diagnostics of Mitochondrial Disorders: Two Years' Experience with Whole-Exome Sequencing at a National Paediatric Centre. *J. Transl. Med.* **2016**, *14*, 1–19. [CrossRef] [PubMed]
102. Ogawa, E.; Shimura, M.; Fushimi, T.; Tajika, M.; Ichimoto, K.; Matsunaga, A.; Tsuruoka, T.; Ishige, M.; Fuchigami, T.; Yamazaki, T.; et al. Clinical Validity of Biochemical and Molecular Analysis in Diagnosing Leigh Syndrome: A Study of 106 Japanese Patients. *J. Inherit. Metab. Dis.* **2017**, *40*, 685–693. [CrossRef] [PubMed]
103. Spiegel, R.; Shaag, A.; Mandel, H.; Reich, D.S.; Penyakov, M.; Hujeirat, Y.; Saada, A.; Elpeleg, O.; Shalev, A.S. Mutated NDUFS6 is the Cause of Fatal Neonatal Lactic Acidemia in Caucasus Jews. *Eur. J. Hum. Genet.* **2009**, *17*, 1200–1203. [CrossRef] [PubMed]
104. Torraco, A.; Nasca, A.; Verrigni, D.; Pennisi, A.; Zaki, M.S.; Olivieri, G.; Assouline, Z.; Martinelli, D.; Maroofian, R.; Rizza, T.; et al. Novel NDUFA12 Variants are Associated with Isolated Complex I Defect and Variable Clinical Manifestation. *Hum. Mutat.* **2021**. [CrossRef] [PubMed]
105. Ostergaard, E.; Rodenburg, R.J.; Brand, M.V.D.; Thomsen, L.L.; Duno, M.; Batbayli, M.; Wibrand, F.; Nijtmans, L. Respiratory Chain Complex I Deficiency due to NDUFA12 Mutations as a New Cause of Leigh Syndrome. *J. Med. Genet.* **2011**, *48*, 737–740. [CrossRef]
106. Speer, R.R.; Ezeanya, U.C.; Beaudoin, S.J.; Glass, K.M.; Oji-Mmuo, C.N. Term Neonate Presenting with the Combined Occurrence of Mucolipidosis Type II and Leigh Syndrome. *J. Pediatr. Genet.* **2019**, *9*, 137–141. [CrossRef] [PubMed]
107. Barghuti, F.; Elian, K.; Gomori, J.M.; Shaag, A.; Edvardson, S.; Saada, A.; Elpeleg, O. The Unique Neuroradiology of Complex I Deficiency Due to NDUFA12L Defect. *Mol. Genet. Metab.* **2008**, *94*, 78–82. [CrossRef] [PubMed]
108. Herzer, M.; Koch, J.; Prokisch, H.; Rodenburg, R.; Rauscher, C.; Radauer, W.; Forstner, R.; Pilz, P.; Rolinski, B.; Freisinger, P.; et al. Leigh Disease with Brainstem Involvement in Complex I Deficiency due to Assembly Factor NDUFAF2 Defect. *Neuropediatrics* **2010**, *41*, 30–34. [CrossRef]

109. Hoefs, S.J.; Dieteren, C.E.; Rodenburg, R.J.; Naess, K.; Bruhn, H.; Wibom, R.; Wagena, E.; Willems, P.H.; Smeitink, J.A.; Nijtmans, L.G.; et al. Baculovirus Complementation Restores a Novel NDUFAF2 Mutation Causing Complex I Deficiency. *Hum. Mutat.* **2009**, *30*, E728–E736. [CrossRef] [PubMed]
110. Ogilvie, I.; Kennaway, N.G.; Shoubridge, E.A. A Molecular Chaperone for Mitochondrial Complex I Assembly is Mutated in a Progressive Encephalopathy. *J. Clin. Investig.* **2005**, *115*, 2784–2792. [CrossRef]
111. Reynaud-Dulaurier, R.; Benegiamo, G.; Marrocco, E.; Al-Tannir, R.; Surace, E.M.; Auwerx, J.; Decressac, M. Gene Replacement Therapy Provides Benefit in an Adult Mouse Model of Leigh Syndrome. *Brain* **2020**, *143*, 1686–1696. [CrossRef] [PubMed]
112. Silva-Pinheiro, P.; Cerutti, R.; Luna-Sanchez, M.; Zeviani, M.; Viscomi, C. A Single Intravenous Injection of AAV-PHP.B-hNDUFS4 Ameliorates the Phenotype of Ndufs4 Mice. *Mol. Ther. Methods Clin. Dev.* **2020**, *17*, 1071–1078. [CrossRef]
113. Lazarou, M.; McKenzie, M.; Ohtake, A.; Thorburn, D.R.; Ryan, M.T. Analysis of the Assembly Profiles for Mitochondrial- and Nuclear-DNA-Encoded Subunits into Complex I. *Mol. Cell. Biol.* **2007**, *27*, 4228–4237. [CrossRef]
114. Pereira, B.; Videira, A.; Duarte, M. Novel Insights into the Role of Neurospora crassa NDUFAF2, an Evolutionarily Conserved Mitochondrial Complex I Assembly Factor. *Mol. Cell. Biol.* **2013**, *33*, 2623–2634. [CrossRef]
115. Adjobo-Hermans, M.J.; de Haas, R.; Willems, P.H.; Wojtala, A.; Vries, S.E.V.E.-D.; Wagenaars, J.A.; Brand, M.V.D.; Rodenburg, R.J.; Smeitink, J.A.; Nijtmans, L.G.; et al. NDUFS4 Deletion Triggers Loss of NDUFA12 in Ndufs4 Mice and Leigh Syndrome Patients: A Stabilizing Role for NDUFAF2. *Biochim. et Biophys. Acta (BBA) Bioenerg.* **2020**, *1861*, 148213. [CrossRef] [PubMed]
116. Kruse, S.E.; Watt, W.C.; Marcinek, D.J.; Kapur, R.P.; Schenkman, K.A.; Palmiter, R.D. Mice with Mitochondrial Complex I Deficiency Develop a Fatal Encephalomyopathy. *Cell Metab.* **2008**, *7*, 312–320. [CrossRef] [PubMed]
117. Calvaruso, M.A.; Willems, P.; Brand, M.V.D.; Valsecchi, F.; Kruse, S.; Palmiter, R.; Smeitink, J.; Nijtmans, L. Mitochondrial Complex III Stabilizes Complex I in the Absence of NDUFS4 to Provide Partial Activity. *Hum. Mol. Genet.* **2011**, *21*, 115–120. [CrossRef]
118. Kahlhöfer, F.; Kmita, K.; Wittig, I.; Zwicker, K.; Zickermann, V. Accessory Subunit NUYM (NDUFS4) is Required for Stability of the Electron Input Module and Activity of Mitochondrial Complex I. *Biochim. et Biophys. Acta (BBA) Bioenerg.* **2017**, *1858*, 175–181. [CrossRef]
119. Galkin, A.; Brandt, U. Superoxide Radical Formation by Pure Complex I (NADH:Ubiquinone Oxidoreductase) from Yarrowia lipolytica. *J. Biol. Chem.* **2005**, *280*, 30129–30135. [CrossRef]
120. Formosa, L.E.; Dibley, M.; Stroud, D.A.; Ryan, M.T. Building a complex complex: Assembly of Mitochondrial Respiratory Chain Complex I. *Semin. Cell Dev. Biol.* **2018**, *76*, 154–162. [CrossRef] [PubMed]
121. Formosa, L.E.; Muellner-Wong, L.; Reljic, B.; Sharpe, A.J.; Jackson, T.D.; Beilharz, T.H.; Stojanovski, D.; Lazarou, M.; Stroud, D.A.; Ryan, M.T. Dissecting the Roles of Mitochondrial Complex I Intermediate Assembly Complex Factors in the Biogenesis of Complex I. *Cell Rep.* **2020**, *31*, 107541. [CrossRef]
122. Guerrero-Castillo, S.; Baertling, F.; Kownatzki, D.; Wessels, H.J.; Arnold, S.; Brandt, U.; Nijtmans, L. The Assembly Pathway of Mitochondrial Respiratory Chain Complex I. *Cell Metab.* **2017**, *25*, 128–139. [CrossRef]
123. Tsuneoka, M.; Teye, K.; Arima, N.; Soejima, M.; Otera, H.; Ohashi, K.; Koga, Y.; Fujita, H.; Shirouzu, K.; Kimura, H.; et al. A Novel Myc-Target Gene, mimitin, That Is Involved in Cell Proliferation of Esophageal Squamous Cell Carcinoma*. *J. Biol. Chem.* **2005**, *280*, 19977–19985. [CrossRef] [PubMed]
124. Stepanova, A.; Sosunov, S.; Niatsetskaya, Z.; Konrad, C.; Starkov, A.A.; Manfredi, G.; Wittig, I.; Ten, V.; Galkin, A.; Sosunov, S. Redox-Dependent Loss of Flavin by Mitochondrial Complex I in Brain Ischemia/Reperfusion Injury. *Antioxid. Redox Signal.* **2019**, *31*, 608–622. [CrossRef] [PubMed]
125. Kussmaul, L.; Hirst, J. The Mechanism of Superoxide Production by NADH:Ubiquinone Oxidoreductase (complex I) from Bovine Heart Mitochondria. *Proc. Natl. Acad. Sci. USA* **2006**, *103*, 7607–7612. [CrossRef] [PubMed]

 life

Review

Mitochondrial Bioenergetics and Dynamism in the Failing Heart

Giampaolo Morciano [1,2,*], Veronica Angela Maria Vitto [2], Esmaa Bouhamida [2], Carlotta Giorgi [2] and Paolo Pinton [1,2,*]

1 Maria Cecilia Hospital, GVM Care&Research, 48033 Cotignola, Italy
2 Laboratory for Technologies of Advanced Therapies (LTTA), Section of Experimental Medicine, Department of Medical Sciences, University of Ferrara, 44121 Ferrara, Italy; vttvnc@unife.it (V.A.M.V.); bhmsme@unife.it (E.B.); carlotta.giorgi@unife.it (C.G.)
* Correspondence: mrcgpl@unife.it (G.M.); paolo.pinton@unife.it (P.P.)

Abstract: The heart is responsible for pumping blood, nutrients, and oxygen from its cavities to the whole body through rhythmic and vigorous contractions. Heart function relies on a delicate balance between continuous energy consumption and generation that changes from birth to adulthood and depends on a very efficient oxidative metabolism and the ability to adapt to different conditions. In recent years, mitochondrial dysfunctions were recognized as the hallmark of the onset and development of manifold heart diseases (HDs), including heart failure (HF). HF is a severe condition for which there is currently no cure. In this condition, the failing heart is characterized by a disequilibrium in mitochondrial bioenergetics, which compromises the basal functions and includes the loss of oxygen and substrate availability, an altered metabolism, and inefficient energy production and utilization. This review concisely summarizes the bioenergetics and some other mitochondrial features in the heart with a focus on the features that become impaired in the failing heart.

Keywords: mitochondria; heart failure; bioenergetics; mitochondrial dynamics

Citation: Morciano, G.; Vitto, V.A.M.; Bouhamida, E.; Giorgi, C.; Pinton, P. Mitochondrial Bioenergetics and Dynamism in the Failing Heart. *Life* **2021**, *11*, 436. https://doi.org/10.3390/life11050436

Academic Editors: Giorgio Lenaz and Salvatore Nesci

Received: 24 March 2021
Accepted: 7 May 2021
Published: 12 May 2021

1. An overview of Mitochondrial Bioenergetics and Function in the Healthy Heart

Over recent years, mitochondrial dysfunction was recognized as the hallmark of manifold heart diseases (HDs) in their onset and development [1,2]. Despite the advancement of specific prevention guidelines and relevant therapeutic strategies, HD remains the main cause of death in Western countries [3]. Mitochondria occupy 30% of the total volume of cardiomyocytes and are localized in three different areas among cardiac fibers [4]; these structures are highly dynamic during all phases of heart development and exhibit continuous changes in terms of their bioenergetics and biology during cardiomyocyte differentiation [5]. This mitochondrial abundance is justified by the high-energy supply required by the heart that is provided by mitochondria, defined as "the powerhouse of the cells". In fact, via oxidative phosphorylation (OXPHOS), mitochondria generate 95% of the adenosine triphosphate (ATP) needed to maintain cardiac activities [6]. In cardiac tissue, the principal source of energy is generated by fatty acid oxidation (FAO), which occurs in the mitochondrial matrix. The products obtained by beta-oxidation enter the Krebs cycle and are ultimately utilized by the electron transport chain (ETC), which takes place along the inner mitochondrial membrane (IMM), and this phenomenon creates a proton electrochemical gradient and consequently generates the mitochondrial transmembrane potential ($\Delta\Psi_m$). Under physiological conditions, the $\Delta\Psi_m$ is -180 mV across the IMM and thus acts as a driving force for cations, such as calcium (Ca^{2+}), which is considered the most important second messenger in cells. Mitochondrial Ca^{2+} plays a crucial role in controlling cell physiopathology and cell fate [7]. In the heart, its importance is highlighted by the fact that it is fundamental for the contractile function of cardiomyocytes [8]; indeed, excitation and contraction are paired to cardiomyocyte depolarization and Ca^{2+} fluxes.

Concomitant with an action potential, the sarcolemma depolarizes, and Ca^{2+} enters a dedicated area across the sarcolemma through L-type voltage-operated calcium channels (VOCCs). Here, the Ca^{2+} concentration increases from very low levels (nanomolar ranges) to approximately 10 μM to give rise to the so-called Ca^{2+} sparklet. These Ca^{2+} sparklets induce ryanodine receptor 2 (RyR2) opening and Ca^{2+} release from the main store inside cells, the sarcoplasmic reticulum (SR), which might generate higher Ca^{2+} sparks into the cytosol and thus allow cell shortening and blood pumping in the heart.

Mitochondrial Ca^{2+} also leads to the generation of ATP via the activation of dehydrogenases [9] and increasing the activity of the ATP synthase [10] when muscles are stimulated. ATP cycling, as well as Ca^{2+}, constitutes a key point for the control of both muscle contraction and relaxation. Indeed, ATP is used by the sarcoendoplasmic reticulum Ca^{2+}-ATPase cardiac isoform 2a (SERCA2a) to remove Ca^{2+} from the cytosol in the diastolic phase to ensure muscle relaxation and the correct Ca^{2+} reuptake into the SR for the next muscle contraction. Thousands of these events occur with each action potential. Although other systems of Ca^{2+} removal exist in the cells, such as Plasma membrane Ca^{2+} ATPase (PMCA) and Na^{+}/Ca^{2+} exchanger (NCX), SERCA2a handles about 75% of cytosolic Ca^{2+}.

Ca^{2+} is also responsible for activating cell death pathways. Indeed, it is widely reported that persistent accumulation of Ca^{2+} into the cytosol and mitochondrial Ca^{2+} overload trigger the opening of the mitochondrial permeability transition pore complex (PTPC), a biological event that, if sustained, leads to cell death [11,12].

An important consequence of mitochondrial function is reactive oxygen species (ROS) production [13]. Indeed, ROS are physiologically generated by the ETC via the reduction of oxygen to superoxide and by the Krebs cycle in the mitochondrial matrix. Although a balance exists between ROS production and scavenging systems and that ROS regulate some physiological processes, a significant dysregulation may occur in or as a consequence of diseases; indeed, ROS bursts can damage proteins, lipids, and mitochondrial DNA (mtDNA), which trigger inflammation and ultimately cell death as well as PTPC opening [14].

Heart failure (HF) is a severe condition for which there is currently no cure, and only medications are administered to maintain a patient's life as normally as possible. Approximately 30 million people worldwide are suffering from HF, and these patients need new therapies that address the disease rather than only providing symptom relief. The clinical scenario is represented by two main conditions: either a reduction of left ventricular (LV) ejection fraction, named as heart failure with reduced ejection fraction (HFrEF) [15], which results in an enlarged LV that cannot contract as it should, or a preserved LV ejection fraction (HFpEF), where the LV preserves its ability to contract but, being less than the necessary volume of blood which enters in the left chamber during an improper diastole, it does not meet the body's requirement. Mitochondria, intracellular Ca^{2+} and ATP cycling not by chance are associated with significant alterations during HF, and understanding these changes might have a potential impact on the search for new pharmacological treatments. Two of the main intracellular alterations in HF are the elevated cytosolic Ca^{2+} levels in diastole and its decrease in availability in the SR during systole. This leads to an inefficient cycle of excitation contraction coupling (ECC). Several studies in human and animal models have associated these defects in Ca^{2+} cycling to SERCA2a dysfunction [16,17] in terms of reduced expression and impaired regulation [18].

HF develops through stages of cardiac adaptation in terms of macroscopic (e.g., hypertrophy) and microscopic (e.g., metabolism) changes. During cardiac hypertrophy, a metabolic shift from FAO to glycolysis occurs [19]. This finding is supported by a seminal paper that, using the carbon isotope labelling techniques [20], showed the importance of the glycolytic pathway in producing ATP in hypertrophic hearts to compensate for the significant decrease in OXPHOS [21]. Despite enhanced glycolysis, various studies showed either decreased or normal glucose oxidation, leading to an uncoupling between the uptake of glucose and its oxidation [19]. If the energy for low work load becomes inefficient, which means deeply impaired contractility, the advanced stage of HF begins which continues to be characterized by repression of FAO and high glycolysis. Here, the use of ketone bodies and branched-chain amino acids (BCAAs) as alternative metabolic substrates are prevailing. These can be further converted to acetyl-CoA to enter the tricarboxylic acid cycle and ETC to produce ATP.

This review aims to describe mitochondrial dynamics and bioenergetics in the heart with references to HF.

2. Mitochondria-Related Metabolic Abnormalities in the Failing Heart

The mammalian heart is considered a metabolic omnivore with the capacity to oxidize fatty acids, ketone bodies, carbohydrates (glucose and lactate), and BCAAs to meet its high energy demand after birth [22]. Cardiac metabolism maintains a dynamic state of equilibrium for efficient energy transfer and is a highly concerted plethora of chemical reactions leading to the conversion of ATP both to sustain cell function and allow contraction, growth, repair, and regeneration. The metabolic alterations in the failing myocardium have been explored; from many points of view [23], HF is considered a return to fetal stages due to the shift from FAO-based metabolism (which mainly manifests as a 35% decrease in the ATP concentration and changes in substrate utilization) to glycolysis (the main active pathway during the fetal period) accompanied by progressive degeneration of the myocardium [24,25] (Figure 1). Accordingly, cardiac metabolism in HF was recognized as a field of active research for well over a century [26] and will be discussed in the first part of this review.

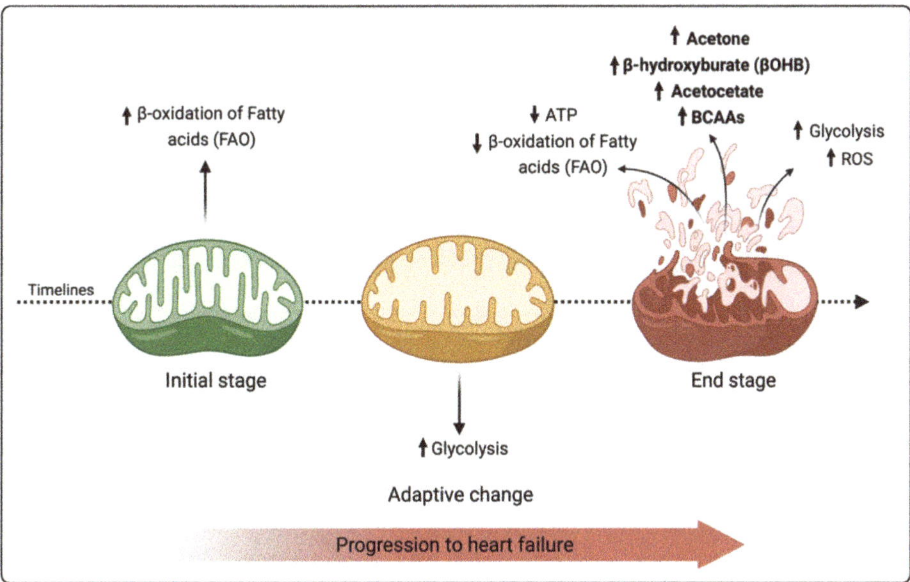

Figure 1. Metabolic and mitochondrial ultrastructural changes in HF. The figure summarizes the first part of the review which reports all metabolic changes and main mitochondrial ultrastructural abnormalities accompanying HF development in humans.

2.1. *Substrate Changes*

2.1.1. Fatty Acids

The beta-oxidation of fatty acids was first demonstrated by Knoop in 1904. FAO represents a predominant fuel source for the myocardium and typically exhibits high flexibility following changes in substrate availability [27]. FAs pass the plasma membrane via tissue-specific transporters and are processed to enter mitochondria for oxidation. Each cycle of FAO produces acetyl-CoA, NADH, and $FADH_2$ for ATP production, and this production is estimated to equal 50% to 70% of the ATP consumed during contraction [28]. In HF, a reduction in the expression of genes encoding FAO enzymes was observed in patients and animal models [29], and this decrease is linked to a reduction in mitochondrial respiration; both of these effects are directly correlated with the stage (worsening from early to advanced) and the cause of HF. Accordingly, one of the most consistent metabolic changes in HF is the marked downregulation of fatty acid utilization, which is mainly described in end-stage HF studies [24,30,31] (Figure 1). Cardiac lipid accumulation is also observed in failing hearts from patients who are affected by diabetes and obesity-related metabolic complications, triggering a phenomenon known as "lipotoxicity" [32]. Reduced levels of FAO pathways and the accumulation of incompletely oxidized fatty acids trigger a mismatch between the supply and oxidation of FAs [33]; moreover, the oxidation of glucose and lactates becomes impaired, which leads to an uncoupling of yet increased glycolysis and less cardiac efficiency and function.

In addition to respiratory inhibition, lipotoxicity leads to other mitochondrial dysfunctions caused by a free FAs-dependent drastic permeability of mitochondrial membranes increasing proton conductance [34] and, being long-chain acyl-CoA esters, potent inhibitors of adenine nucleotide translocator (ANT) [35], with serious implications in ROS generation. In the first scenario, free FAs lead to the dissipation of the membrane potential by continuously passing across the IMM from the intermembrane space (IMS) to the matrix and there, releasing protons. Indeed, the matrix has higher pH than IMS. In a second step, FAs come back to the IMS as anions taking advantage of ANT; here, for each of these cycles, the mitochondrial matrix is enriched by one proton. This action would affect membrane permeability and bioenergetics, and trigger cell death [36,37]. Recently, it was demonstrated that introducing omega-3 FA into the diet modifies the composition of free FA, shifting from the precursors of inflammatory states (arachidonic acid) to those involved in their resolution (eicosapentaenoic acid), thus ameliorating cardiac dysfunction [38].

In addition to their impact as energy-providing substrates, FAs act as mediators of signal transduction and as ligands for nuclear receptor peroxisome proliferator-activated receptors (PPARs)-α. The PPAR-α/PXRα complex (retinoid X receptor pathway) and its transcriptional partner Peroxisome proliferator-activated receptor gamma coactivator 1-alpha (PGC1-α) are considered master regulators of mitochondrial biogenesis [39] and FAO by sensing dietary needs and pathological states. Interestingly, in almost all studies in the field, these two compounds were found to be significantly downregulated and thus become the culprits of the metabolic shift toward glycolysis in HF [40–42]. In support of this finding, preclinical studies observed repression of OXPHOS, increased oxidative stress, and accelerated HF following pressure overload in cardiac PGC1-α-knockout (KO) transgenic animals.

2.1.2. Glucose

In an effort to counteract the decrease in ATP generation due to OXPHOS impairment, glycolysis rates become elevated in HF [43].

In 1907, Locke and Rosenheim were the first to study myocardial glucose uptake in an isolated rabbit heart model [44]: Glucose and its metabolites play several roles in cardiac myocytes, and their uptake is regulated by the membrane translocation of glucose transporter (GLUT) 1 and GLUT4 followed by phosphorylation into 6-phosphoglucose (G6P). Of note, glycolysis and glucose oxidation are differentially regulated in the heart. Therefore, in failing hearts, the elevated rate of glycolysis is not needed to translate into enhanced glucose oxidation [45]. A plethora of studies support this notion: Patients with HF present enhanced levels of cardiac glycolysis without increases in glucose oxidation, lactate, and pyruvate accumulation [46]. Instead, significant abolition of glucose oxidation was detected in many animal models of HF and in a pool of patients with congestive HF [47] and is currently considered a metabolic marker that proceeds with the development of cardiac deterioration in HF [48]. This feature is due to deregulation of the overall mitochondrial oxidative capacity and alterations in pyruvate dehydrogenase (PHD) activity, which lead to a reduction in the conversion of pyruvate into acetyl-CoA [49–51].

Nevertheless, the literature reports apparently conflicting evidence for glucose utilization in HF over time. Additional studies showed that the glucose oxidation rates either remained unchanged in a compensated HF rat model [21] or were elevated in canine cardiac-pacing experiments [41]. The reason for this contradictory evidence is unclear. The most accredited hypothesis is that glucose oxidation varies according to the severity of HF (and to a small extent to the cause); it starts to increase at initial stages and then remains unchanged until it decreases at advanced stages [52].

2.1.3. Ketones

New metabolic changes, including ketone bodies and branched-chain amino acids (BCAAs), were recently identified as alternative substrates in end-stage HF (Figure 1).

Ketone bodies are small molecular energy substrates that are rapidly mobilized and produced during fasting and starvation through hepatic ketogenesis and provide three types of ketones: acetone, β-hydroxyburate (βOHB), and acetoacetate. Ketones can enter cardiomyocytes and be translocated to the mitochondrial matrix, where βOHB is oxidized into acetoacetate by the key enzyme βOHB dehydrogenase (BDH1); thus, acetoacetate is stimulated by succinyl-CoA to acetoacetyl-CoA and then converted to acetyl-CoA by acetyl-CoA acetyltransferase. Consequently, acetyl-CoA is able to enter the tricarboxylic acid cycle and ETC to produce ATP [53,54].

Recent studies by Aubert and associates found that BDH1 was significantly upregulated at least two- to three-fold in an HF mouse model, and this increase is accompanied by significant stimulation of ketone metabolism, which suggests the role of ketone bodies as alternative substrates when glucose oxidation is downregulated in failing hearts [55]. Consistent with previous findings, enhanced myocardial ketone body oxidation rates were found in an ex-vivo-isolated murine heart [56], and significantly higher ketone body levels were detected in the peripheral blood of patients affected by HF than in normal subjects [57]. Although metabolomics studies confirmed that HF utilizes ketones as an alternative substrate in an effort to supply energy at the end stage of human HF, this change does not apport beneficial effects to the evolution of the pathology. Indeed, the plasma ketone concentration and acetone are reportedly linked with malignant prognosis in patients with chronic HF [58,59]. These findings might provide new insights for therapeutic approaches toward HF [60].

2.1.4. BCAAs

Similarly, BCAAs are an important group of essential amino acids and an integral part of myocardial energy metabolism and thus play a crucial role in the pathophysiology of end-stage failing hearts. BCAAs act as key building blocks for peptide synthesis and are effective sources for the biosynthesis of sterol, keto bodies, and glucose. In addition, BCAAs are essential for normal growth and function at the cellular and organism levels [61].

Cardiac BCAA absolute levels were found to be significantly enhanced in a mouse model of HF that underwent transverse aortic constriction [62]. This evidence was correlated with a significant reduction in BCAA catabolic enzymes and in the % of EF, and was associated with insulin resistance [62–64]. Moreover, recent studies reported that branched chain alpha-keto acids (BCKAs), which are produced after the initial step of BCAA metabolism, are also increased in the myocardium of HF patients [63]. Enhancing BCAA oxidation in the myocardium can ameliorate the progression of HF by improving cardiac function (increasing the % EF and reverting insulin sensitivity) in a mouse model, which suggests that the imbalance between BCAA availability and use in failing hearts might be correlated with contractile dysfunction [63]. The improved BCCA catabolism was obtained by using 3,6-dichlorobenzothiophene-2-carboxylic acid (BT2), an inhibitor of the branched-chain α-keto acid dehydrogenase kinase (BCKDK) which phosphorylates and inactivates the key enzyme (branched-chain α-keto acid dehydrogenase) responsible for providing substrates used by the TCA cycle [64].

Cardiac BCAA catabolism is described to be deregulated also in myocardial infarction (MI)-operated mice, and this effect contributes to post-MI HF and remodeling by enhancing the mammalian target of rapamycin (mTOR) signaling, a modulator of various anabolic (e.g., protein synthesis) and catabolic (e.g., autophagy) pathways [65]. Alterations in mTOR signaling contribute to the progression of failing hearts [66,67]. Furthermore, the amino acids aspartate and glutamate play a key role in the transfer of reducing equivalents across the mitochondrial membrane for the oxidation of cytosolic NADH through the mitochondrial ETC.

3. Mitochondrial Dynamism in the Failing Heart

The development of real-time methods involving the use of chemical dyes and fluorescent proteins or the use of transmission electron microscopy (TEM) has allowed scientists to see mitochondria and study their real morphology, which is highly dynamic and impacts cell function, physiology, and disease [68]. It is interesting to note how several lines of evidence show key roles for mitochondrial fitness in a broad range of cardiac and metabolic disorders, including the development and progression of HF and hypertrophy [69]. In cardiomyocytes, mitochondria are clustered among the myofibrils and the plasma membrane (subsarcolemmal mitochondria or SSM), across the sarcomere myofibrils (intermyofibrillar mitochondria or IMF), and tightly adjacent to the nucleus [70]. Together, these mitochondria are associated with the sarcoplasmic reticulum (SR), a subcellular compartment that participates in cardiac contraction by controlling the processes of Ca^{2+} storage, and release and reuptake in skeletal muscles [71,72].

Overall, the mitochondrial morphology is controlled by known phenomena of balanced fusion and fission that interconnect with each other and to events involving an increased mitochondrial mass (biogenesis) and "in excess" organelle removal (mitophagy) [73]. Whether changes in these pathways are a cause or consequence, or simply accompany the pathology of heart diseases, remains unclear [74].

3.1. *Ultrastructural Abnormalities*

Important ultrastructural abnormalities were detected from an analysis of mitochondria in HF tissues, and these abnormalities, which include the loss of mitochondrial granules, swelling with disorganization of the cristae, vacuolar degeneration, mitochondrial fragmentation, deep structural lesions (disruption of the OMM and IMM and loss of the electrodense matrix), and a strong reduction in the mitochondrial volume [52,75–78], become increasingly prominent from the initial to end stages of the disease. All of these abnormalities can often contribute to cytochrome c release, which ultimately leads to apoptosis, and were found with both IFM and nuclear mitochondria but not SSM mitochondria [79–82]. Moreover, the surrounding environment undergoes corresponding changes, including disorganization of sarcomeres and T-tubules.

For correct maintenance of the morphology and structure of the cristae, some proteins located in the IMM that result in being impaired in HF are important; these include mitofilin and cardiolipin [83,84]. Mitofilin is a structural protein, whereas cardiolipin is a phospholipid that is needed to support energy production and regulation of the mitochondrial structure, biogenesis, and dynamics [56–58]. Recent studies demonstrated that the downregulation of mitofilin in living cells induces the formation of mitochondria with disorganized inner membranes, loss of mitochondrial function, increased ROS production, and apoptosis [85]. Notably, low levels of mitofilin were also found in tissues of human hearts with HF [86]. However, the downregulation of cardiolipin in HF induces an increase in the production of ROS and mitochondrial dysfunction in cardiomyocytes until death [87]. The maintenance of mitochondrial function through the prevention of mitofilin and cardiolipin is thus important to limit the development of HF.

3.2. Biogenesis

As previously mentioned, several studies using animal models of HF showed that the expression of some key regulators of energy metabolism, such as PGC-1α and β, is strongly reduced [88–90]. Normally, PGC-1α is a protein with a key role in mitochondrial biogenesis and improves OXPHOS and FAO [91]. Its overexpression, which usually occurs in response to external stimuli (i.e., physical exercise and exposure to cold temperatures), triggers a significant increase in the volume of the whole mitochondrial network. Moreover, PGC-1α is able to bind several transcription factors, including nuclear respiratory factor (NRF) 1 and 2, to activate transcription factor A (TFAM), which is responsible for initiating the transcription of nuclear-encoded proteins with structural and mtDNA replication properties [92]. In response to PGC-1α deregulation, abnormal mitochondrial biogenesis rates were observed in animals with HF and in patients with different etiologies [86]. As opined by Karamanlidis and colleagues, the main mitochondrial dysfunctions (including their reduced turnover) in failing hearts are due to defects in the mtDNA amount and integrity instead of alterations in the gene expression profile [93]. Indeed, in a significant number of failing human hearts, a reduction in mtDNA-encoded proteins derived from a loss of mtDNA was also observed [93,94].

Nevertheless, several kinds of cardiovascular diseases including HF in advanced stages were observed accompanying the course of mitochondrial-based pathologies caused by mtDNA mutations [95,96]. Among those, MELAS (mitochondrial encephalomyopathy, lactic acidosis, and stroke-like episodes) syndrome was observed in concomitance to 3243 A>G mtDNA substitution [97,98].

In addition to its role in mitochondrial biogenesis and its relationship with energy metabolism, PGC-1α can also improve contractility and endothelial function in models of HF [99–102]. Recent studies performed using PGC-1α-KO mice showed an accelerated development of HF, which was linked to downregulation of OXPHOS genes [90] and deficiencies in the function and reserve of cardiac energy [103,104].

3.3. Fusion and Fission Machinery

The process of mitochondrial biogenesis is tightly interconnected with two additional pathways named fusion and fission, and these pathways allow the origin of an interconnected and fragmented mitochondrial network, respectively [105]. It was shown that proteins contributing to mitochondrial dynamics are greatly expressed in the mammalian myocardium, and their ablation is deadly [72,106–109]. This finding suggests that in cardiomyocytes, mitochondrial dynamism is essential for noncanonical functions governing mitochondrial quality control, Ca^{2+} signaling, cardiac development, and cell death [110].

Although an imbalance in mitochondrial dynamics usually accompanies several cardiovascular diseases, such as ischemia-reperfusion (I/R) and diabetic cardiomyopathy [111], very few studies have addressed these pathways in HF [112]. Mitochondrial dynamics represent continuous events of fission and fusion occurring at all developmental stages of a cell and contribute to cell division, cell death, mtDNA, and nutrient exchange to sustain metabolism in the heart properly [79–81,113]. Proteins involved in mitochondrial fusion and fission were highly preserved during evolution [114]; they utilize GTP energy to guide conformational changes and exhibit distinct mitochondrial sublocalization [105]. Mitochondrial fusion is a complex mechanism involving three steps: tethering, OMM fusion, and IMM fusion. Tethering occurs through one or two homologous proteins that belong to the superfamily of mitochondrial transmembrane GTPases. Proteins that are located on the OMM and involved in fusion are mitofusin 1 (Mfn1) and 2 (Mfn2) [115,116]. These mediators can form three different molecular complexes: (i) Mfn1 homotypic oligomers, (ii) Mfn1-Mfn2 heterotypic oligomers, and (iii) Mfn2 homotypic oligomers [117,118]. Fusion of the OMM is normally followed immediately by IMM fusion directed by another dynamin superfamily GTPase, optic atrophy 1 (Opa1), which leads to the final formation of filamentous organelles [119]. This protein is located on the IMM and is redistributed to the IMS. In addition to controlling IMM fusion, Opa1, according to certain protein modifications, can also exert specific effects on the cristae structure [120].

Previous studies showed the essential role of mitofusins during adulthood, which involve conferring protection against long-term cardiac dysfunction, and in early embryonic cardiac development [121]. Unexpectedly, their absence induces mitochondrial fragmentation, and this effect is accompanied by respiratory dysfunction and progressive HF [108]. Additionally, a reduction in Opa1 is representative of mitochondrial variations affected by HF in humans. All these protein changes reflect a population of smaller mitochondria with a truly fragmented network controlled by posttranscriptional events and always associated with apoptosis and consequent loss of functional cardiomyocytes [122,123]. Among all established models of HF, Opa1 is the only protein that always shows a decrease in expression during HF [123]. Indeed, perhaps in an effort to compensate for this loss in OPA1, mitofusins were found to increase following HF.

A report study conducted by Menezes T. and colleagues found that Mfn1 is a substrate target for PKCβII (a Ca^{2+}-dependent protein kinase). PKCβII accumulates in the OMM during HF and produces extensive mitochondrial fragmentation following the phosphorylation of Mfn1 at serine 86, which results in a loss of GTPase activity, as detected in both animal and human samples.

Contrary to what is known, Mfn1-dependent fusion in cardiomyocytes fails in response to dysregulation of Ca^{2+} cycling and inefficient cardiac contractility [124]. Being all features that characterize HF, further studies should be performed in in vivo samples to confirm the possibility that targeting Mfn1 expression can be fully considered as therapeutic opportunity.

Mitochondrial fission in mammalian cells is led by the cytoplasmic protein dynamin-related protein-1 (Drp-1), which forms complexes at fission sites on the OMM [125], and fission protein-1 (Fis-1), which encircles the outer mitochondria and promotes the assembly of protein complexes on the OMM [126,127]. Indeed, Fis-1, mitochondrial fission factor (Mff), and mitochondrial dynamics proteins 49 and 51 (MiD49 and MiD51) act as receptors for the recruitment of Drp1 to the mitochondrial surface [128]. Drp1 performs multiple functions, often at the intersection between mitochondrial fission and mitophagy [129]. Drp1 cytoplasm-mitochondria shuttling is crucial for its roles and this is finely regulated by many post-translational modifications including phosphorylation, SUMOylation, palmitoylation, and ubiquitination [129].

In HF models, Fis-1 appears to play a minor role because its levels usually remain unchanged; otherwise, the overall amount of Drp1 tends to be higher in human samples of HF while these data are not confirmed in animal models [123]. In light of this finding, on one hand, Drp1 might also play minor roles in HF, including its crucial direct role in cell death when it is upregulated; on the other hand, post-translational modifications (PTM) rather than changes in the overall amount of the protein may occur in HF. For example, it was observed that Sentrin/SUMO-specific protease 5 (SENP5) is able to deSUMOylate and repress Drp1-dependent mitochondrial fission. SENP5 is upregulated in HF and induces a phenotype of apoptotic cardiomyocytes [130]. However, the most well described PTM for Drp1 is phosphorylation, especially at Serine 616 (S616), which ensures Drp1 translocation at mitochondria and fragmentation. In a mouse model subjected to pressure overload, mitophagy is transiently upregulated in the hypertrophic heart in the first week, in a manner dependent from the phosphorylation of Drp1 at S616. This pathway is not further maintained in later stages of hypertrophy where this repression determines irreversible dysfunctions and HF [131].

3.4. *Mitophagy*

Both under baseline conditions and in response to stress, cells activate a highly regulated mechanism, called mitophagy, to digest senescent and damaged mitochondria [74]. The organelles are engulfed by autophagosomes and are subsequently delivered to lysosomes for degradation. This mechanism is crucial for the maintenance of cellular homeostasis; defects in mitophagy trigger and amplify mitochondrial dysfunction (due to the accumulation of aberrant mitochondria), and this effect is accompanied by the development of cardiomyopathies and ultimately HF with severe contractile dysfunction [132]. This process could be regulated by mitochondrial PTEN-induced kinase 1 (PINK1) and the cytosolic ubiquitin ligase Parkin. Normally, when mitochondria are healthy, PINK1 is imported into the mitochondrial matrix through the translocase of the outer membrane (TOM) complex. In contrast, when a damaged mitochondrion exhibits a loss of mitochondrial membrane potential, PINK1 accumulates on the OMM [133], which results in the recruitment of Parkin from the cytosol to the mitochondrial membrane [134]. However, to initiate Parkin-mediated mitophagy, two events induced by PINK1 are needed. The first event is the phosphorylation of Mfn2, which causes attraction on the mitochondrial surface of Parkin, whereas the second event consists of the phosphorylation of Parkin at Ser65 (ubiquitin-like domain), which increases its E3 ligase activity [135,136]. These steps introduce the ubiquitination of mitochondrial proteins to promote phagosome recruitment and the successive degradation of mitochondrial proteins by the lysosome. Studies suggest that for degradation by the autophagosome, mitochondrial protein ubiquitination via the Lys63 linkage plays a signaling role in the recognition of damaged organelles. In this mechanism, some adaptor proteins, such as NBR1 and p62, bind to ubiquitinated mitochondrial proteins and interact directly with LC3 on the autophagosome, creating a link between mitochondria and autophagosomes [137,138]. In addition, the ubiquitination of mitofusins via Lys48 linkage induces their degradation by the ubiquitin/proteasome system [139] to prevent the fusion of both healthy and damaged mitochondria, which aids the final aim of mitophagy.

From these findings, it is clear that mitophagy is a critical mitochondrial quality control mechanism in myocytes [140–142]. Defects in mitophagy associated with the proteins PINK1 and Parkin have negative consequences for cardiomyocytes. Although cardiac function is normal in young Parkin-KO mice, as they age, they accumulate abnormal mitochondria, develop irreversible HF, and show increased cell death following myocardial infarction [141]. Similarly, the absence of PINK1 leads to cardiac mitochondrial dysfunction with a great burst of ROS and irreversible cardiac hypertrophy [140]. The loss of PINK1 in mice also increases susceptibility to pressure overload-mediated HF and I/R injury [140,143]. Similar findings were also documented in humans affected by advanced HF, and these can be characterized by inefficient mitophagy as a consequence of the reduction in the overall PINK1 levels. Although the knowledge of mitophagy as a cause or consequence of HF remains controversial, a recent report suggested that a shift between two isoforms of AMP-activated protein kinase (AMPK) alpha (from 2 to 1) occurs in failing hearts and is responsible for its pathogenesis [144]. Restoring AMPKα2 and thus increasing the phosphorylation of PINK1 at serine 495 increases mitophagy to efficient levels to prevent the progression of HF [144].

Increasing evidence in the last decade highlighted a role also for Parkin- and PINK1-independent pathways in the activation of mitophagy under conditions of stress, as those occurring under I/R [132,145]. These molecular routes would include additional players such as cardiolipin, Bcl2/adenovirus E1B 19 kDa protein-interacting protein 3 (Bnip3), a receptor for LC3-II binding at the OMM and FUNDC1, a protein that under hypoxia becomes phosphorylated at S17 by Unc-51 similar to autophagy activating kinase (Ulk1), and dephosphorylated by phosphoglycerate mutase family member 5 (PGAM5) at S13, triggering mitophagy [145,146]. Moreover, and of great interest, an in vivo model of cardiac ischemia showed the activation of a phosphorylations cascade involving S555 of Ulk1, S179 of Rab9, and S616 of Drp1. These actively participate in the formation of a multiprotein complex in which Ulk1 is phosphorylated by AMPK, indirectly inducing the phosphorylation of Drp1 (and thus mitochondrial fission and mitophagy) via assembling with the Rab9-Rip1-Drp1 axis [147].

The fine balance among all these pathways (biogenesis, fusion, fission, and mitophagy) is crucial either to prevent or overcome injuries following stressful conditions and can thus be considered a promising and feasible therapeutic target [1,25,94]. To understand this fine balance better, increased mitophagy might cause excessive mitochondrial clearance, which would leave the myocytes with too few mitochondria to produce sufficient ATP. During acute cardiac injury, such as MI or I/R, a limited increase in mitophagy could be beneficial to clear damaged mitochondria, but in chronic cardiac diseases, such as HF, sustained upregulation of mitophagy might be harmful [132].

These studies clearly show that the dysregulation of mitophagy has the potential to lead to the accumulation of abnormal mitochondria, contractile dysfunction, and ultimately the progressive loss of myocytes.

3.5. Mitochondrial Permeability Transition Pore Complex (PTPC)

Dissipation of the mitochondrial membrane potential, which is the signal for PINK1- and Parkin-dependent mitophagy, is usually caused by permeabilization of the IMM in a phenomenon called mitochondrial permeability transition (MPT). MPT occurs when proteinaceous channels, by a still partially unknown entity, open across the IMM and OMM, which causes uncontrolled fluxes of solutes and thereby osmotic stress, mitochondrial swelling, and pro-apoptotic factor release into the cytosol [148–150]. Years of intense research found strong similarities between ATP synthase and PTPC, and, in fact, the latter is currently considered a molecular rearrangement of the dimeric form of ATP synthase into monomers with contributions from a plethora of additional proteins with either structural [151] or modulatory properties [152–154], as observed following exposure to stressful conditions [150]. The alteration of the ATP synthase structure (in the putative PTPC conformation) exerts major effects on the disturbance of mitochondrial energy and respiration and hence on cardiac performance. Currently, only little information is available regarding the involvement of the PTPC in HF, and, thus, knowledge on its cause/effect consequentiality remains elusive. Of note, as mentioned above, metabolic changes in HF (especially lipids accumulation) might first contribute to membranes' permeability and membrane potential dissipation, triggering the opening of the PTPC [155].

The activation of mitochondrial phospholipases in the failing heart reportedly prompts the formation of toxic metabolites in a Ca^{2+}-dependent manner. These hydroxyeicosatetraenoic acids induce PTPC opening by Ca^{2+} overload, which further worsens the HF scenario and increases the percentage of nonfunctional mitochondria and cell death [156]. This final endpoint was also suggested by independent groups: The final stages of HF are characterized by increased oxidative stress, increased diastolic Ca^{2+} overload, and episodes of ischemia, all of which prompt abrupt PTPC opening [157]. The pharmacological inhibition of the PTPC, such as by the use of cyclosporin A (CsA), reduced PTPC opening, the dissipation of mitochondrial potential and respiratory deficits, ameliorate the adverse conditions surrounding HF [158].

Another crucial determinant in PTPC opening is ROS. Major sources of ROS in the heart are caused by ETC function, nicotinamide adenine dinucleotide phosphate (NADPH) oxidases (Noxs), and uncoupled NO synthases (NOS). Evidence about the importance in limiting oxidative stress in the heart comes from in vivo studies using transgenic mice with impaired levels of intracellular antioxidants, such as SOD2 [159,160], peroxiredoxins [161], glutathione peroxidase [162], and thioredoxin reductase [163]. In all cases, the alteration of the antioxidant expression correlated with the extent of oxidative stress, with important implications in cardiovascular diseases [164]. In the same way, PTPC opening is also dependent on the action of scavenger enzymes in the cell, especially on the glutathione content and function [165]. This phenomenon was in principle analyzed in single cardiomyocytes and described as a vicious cycle of ROS-induced ROS release (RIRR) [165,166] in which the PTPC opening (triggered by ROS) is essential to generate a ROS burst in contiguous mitochondria, and is associated with a significant dissipation of their membrane potential. The use of selective inhibitors, such as rotenone for complex I of ETC and bongkrekic acid for PTPC, inhibited RIRR, providing the proof of evidence that the ROS burst derives from mitochondria and is generated solely by PTPC opening consequences [165].

Further research on the modulation of PTPC in the failing heart might lead to the discovery of new therapeutic approaches for the treatment of symptoms and regression.

3.6. Calcium Cycling and Handling

Ca^{2+} plays an important role in the pathophysiology of the heart [1,167,168]. Its homeostasis is considered an essential mediator in regulating ECC and modulating systolic and diastolic function. Here, Ca^{2+} transduces action potentials in mechanical force. In healthy hearts, the action potential depolarizes the sarcolemma, allowing the passage of low amounts of Ca^{2+} from the extracellular space to the cytosol via L-type calcium channels (LTCCs). As consequence, the magnitude of Ca^{2+} action increases, due to the induced and highly synchronized Ca^{2+} sparks from the RyRs of SR in the whole cell [169,170]. After systole, Ca^{2+} should be removed from the cytosol to guarantee the correct muscle relaxation and, thus, the diastolic phase. SERCA2a, PMCA, and NCX are deputed to do this [171] in an ATP-dependent manner in the first two cases, evidencing further roles and the importance ascribed to ATP in muscle relaxation before a new EC cycle. All these proteinaceous channels are highly regulated by cytosolic proteins which often contribute to the end of the EC coupling but also to pathological states.

SR membranes are closely juxtaposed to mitochondria giving rise to a subcellular compartment named as mitochondria-associated membranes (MAMs) which, in lieu of material exchange, signal transduction and metabolic regulation [172,173]. During ECC it was demonstrated that Ca^{2+} fluctuations occur in mitochondria after SR Ca^{2+} release [174] but they remain controversial if they take part in the regulation of cytosolic Ca^{2+} levels, as it was estimated that mitochondria can handle up to 15% of cytosolic Ca^{2+}, compared to other extrusion systems [175]. However, mitochondrial Ca^{2+} uptake has great repercussions on cardiomyocytes fate; in healthy bodies, by ensuring bioenergetics and cardiac contractility [9,10,176], and in disease bodies, where an overload can induce the PTPC opening upon I/R and a vicious cycling of ROS generation, oxidation of RyR2, and SR Ca^{2+} leak in HF, associated with mitochondrial fragmentation and dysfunction [177]. Mitochondrial Ca^{2+} dysregulation was reported also as defects in MAMs' morphology where the distance among contact sites between SR and mitochondria is increased in stages immediately preceding HF such as hypertrophy, chronic noradrenaline stimulation, and aging [178,179].

Myocytes from failing hearts are characterized by reduced and slower cytosolic Ca^{2+} transients ($[Ca^{2+}]_i$); indeed, SR Ca^{2+} storage and release are impaired (Figure 2). Although this feature is mainly associated with a reduction in the SERCA2a protein amount, which becomes significantly downregulated in failing hearts compared with nonfailing hearts, its regulation might also play a crucial role because serine 16 phospholamban phosphorylation is reportedly reduced accordingly [16]. Moreover, increased Ca^{2+} leakage from the SR was explored as a secondary cause due to the dissociation of FKBP12.6 from RyR2 following protein kinase A (PKA)-dependent hyperphosphorylation at serine 2808 [180,181] (Figure 2). RyR2 expression does not change in HF but assumes aberrant gating based on the previously mentioned model.

Confirming the undoubtedly essential function of SERCA2a, the restoration of its expression with adeno-associated (AAV) vectors was classified as successful gene therapy in preclinical models of HF [17,182]. Recently, multiple ongoing clinical trials (as reviewed in [183]) are studying the use of AAV to improve SERCA2a expression and function. In addition, pharmacological treatments, such as the use of Istaroxime [184], stimulate SERCA2a and completely restore Ca^{2+} cycling.

The recorded elevation in cytosolic Ca^{2+} load is also caused by an altered influx from the extracellular milieu. Indeed, pathological SR Ca^{2+} depression activates the function of LTCCs and store-operated calcium entry (SOCE), which rewire a notable amount of extracellular Ca^{2+} into the cytosol [185] (Figure 2).

Concomitant with this finding, an additional molecular mechanism was observed in some models of HF, including humans: An increase in NCX function attempts to remove excess cytosolic Ca^{2+} to compensate for inefficient SERCA2a function and also further depresses SR calcium [186,187]. In contrast, other studies found that NCX is impaired by working in a reverse mode due to high [Na$^+$] conditions [188]. These combinatorial effects lead to a complete change in the spatiotemporal fluxes of intracellular Ca^{2+}, which reflect a markedly defective systolic contraction and diastolic relaxation of the heart [189].

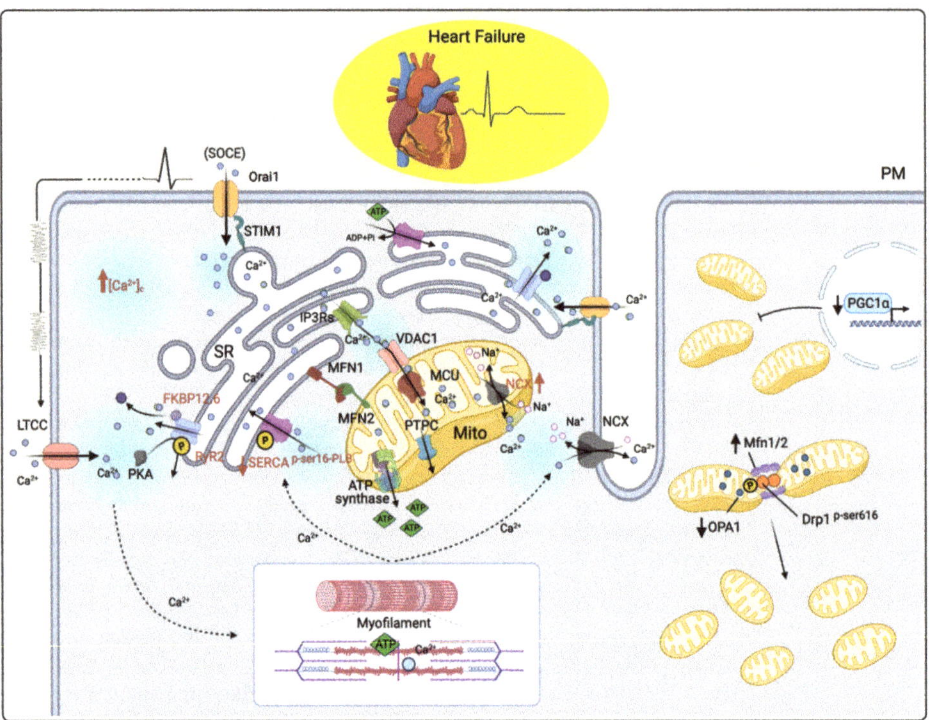

Figure 2. Calcium cycling in the failing heart. The figure summarizes calcium cycling in the normal heart and how it changes in the failing heart. Proteins responsible for these alterations are highlighted in red (e.g., SERCA2a expression and molecular regulation; RyR2 aberrant gating), and the pathways to which they refer are described in the text. The importance of ATP in muscle contraction and relaxion is also depicted. On the right, a graphical abstract highlighting biogenesis and fusion-fission machinery in HF.

4. Epigenetics in HF

In the last decade, and this is still an evolving field of research, evidence showed the involvement of epigenetics signature in HF. Although it is known that many factors affect epigenetics patterns (environmental, diet, lifestyle, pollutants) [190], and, in turn, epigenetics may be the cause of given pathological states (including HF) [191], we would summarize how mitochondrial function intersects with epigenetics in HF. In detail, the consequences of mitochondrial alterations or metabolites produced during the development of HF are emphasized accordingly to the topic of this review.

Epigenetics programs include DNA methylation, histone acetylation/methylation, and noncoding RNAs [192]. Chromatin acetylation involves acetyltransferases and deacetylases which make it more or less accessible to DNA binding elements, respectively. The role of this couple of enzymes was strongly matched to irreversible cardiac remodeling due to either the activation (when acetylated) or depletion (when deacetylated) of GATA Binding Protein 4 (GATA4) and Myocyte Enhancer binding Factor 2 (MEF2), transcription factors responsible for the gene expression rewiring during cardiac hypertrophy [193–195]. Chromatin methylations are reversible and can be related to activation and inhibition of the transcriptional activity; Kaneda et al. demonstrated the significant relationship between a triple addition of the methyl group to lysine 4 or 9 of histone H3 and HF [196]. Interestingly, three genes located close to these modifications are RYR2, CACNB2 (encoding for the subunit beta 2 of the voltage-dependent calcium channel), and CACNA2D1 (encoding for the alpha-2 and delta subunits of the voltage-dependent calcium channel complex), responsible for intracellular Ca^{2+} cycling [196]. Instead, DNA methylation is usually synonymous with gene expression "shutdown". In animal models of pressure overload and cardiac hypertrophy, increased DNA methylation led to significant repression of the transcriptional activity in the left ventricle and allowing for cardiac contractility [197,198]. Otherwise, noncoding RNAs alter gene expression in a wide variety of modes which are not the focus of this review and are reviewed elsewhere [199].

It is clear that changes in metabolism play a crucial role in HF. Additionally, metabolites' availability influences epigenetics patterns, because most of them act as cofactors for enzymes inducing chromatin modifications and control the transcriptional program as a consequence [200]. Of interest, some of these metabolites are part of those intermediates becoming altered and concurring in HF development; these are ketones (such as βOHB, α-ketoglutarate), Acetyl-CoA, NAD+, FAD+, and mitochondrial ROS.

Acetyl-CoA is one of the main regulators of protein acetylation due to the possibility to supply acetyl groups to enzymes in charge of modifying histones [201]. Thus, the rate of FAO (and its changes during HF) would be associated to either higher or lower gene expression on the basis of the Acetyl-CoA produced [202]. This indeed constitutes a critical factor in cycles of acetylation/deacetylation. βOHB also increases acetylation of proteins being a potent inhibitor of histone deacetylases [203]; this action can lead to beneficial effects in HF due to the activation of neighboring genes such as Forkhead Box O3 (FOXO3), which confers resistance to oxidative stress and sirtuins, already known to have strong cardioprotective roles [204]. Moreover, NAD+, an essential cofactor for many enzymes involved in energy production in cardiomyocytes [205,206] and found to be decreased during HF stages, is also essential as a cofactor for sirtuins function [207]. Sirtuins are a family of proteins able to deacetylate chromatin with crucial roles. In this context, the absence of NAD+ has negative effects on the expression of sirtuins which determine no longer protective roles to counteract HF [208]. NAD+ levels are further finely regulated by Nicotinamide Nucleotide Adenylyltransferase 2 (NMNAT2), whose upregulation protect from hypertrophy [205]. During HF, a significant amount of mitochondrial ROS are produced; as well as the role already addressed, they alter both DNA and histone methylation [200], the first one by oxidizing guanosine in 8-oxo-20-deoxyguanosine and producing as a consequence a hypomethylation state of the DNA; this triggers the onset of inflammation and oxidative stress due to the activation of gene expression involved in those pathways [209]; the second, by reducing S-adenosil metionina (SAM), which is responsible for methylation of histones [210].

5. Conclusions and Future Perspectives

HF is caused by a multitude of risk factors and is involved in many cardiovascular pathologies such as coronary artery disease, hypertension, faulty heart valves, and congenital defects. For this reason, it is of great interest and involves a considerable percentage of patients. Cardiology has made great strides worldwide in prevention and patient care, but many steps of basic research are still poorly understood, and translational approaches should be improved. If on the one hand, the metabolic switch and related consequences during the adaptative stages and advanced HF are fairly known, further research about Ca^{2+} cycling, handling, signaling, and interconnections with mitochondrial fusion-fission machinery deserves attention. In particular, a focus on the Ca^{2+} spatio-temporal route and which proteins are involved in it, from the extracellular milieu to the cytosol, would be noteworthy. Then, the contribution of mitochondria as organelles closely related to SR and able to buffer Ca^{2+} to orchestrate a plethora of downstream events would also be noteworthy. Although animal models for the in vivo study of MAMs are lacking, and investigations on patients continue to be limited, efforts in this way may improve the creation of targeted therapeutic strategies to recover mitochondrial and SR function in HF.

Author Contributions: Conceptualization, G.M. and P.P.; software, V.A.M.V.; validation, G.M., P.P. and C.G.; investigation, G.M., V.A.M.V. and E.B.; resources, G.M., V.A.M.V. and E.B.; writing—original draft preparation, V.A.M.V. and E.B.; writing—review and editing, G.M.; visualization, G.M.; supervision, G.M.; funding acquisition, P.P., C.G. and G.M. All authors have read and agreed to the published version of the manuscript.

Funding: This research was funded by the Italian Association for Cancer Research (IG-23670 to P.P. and IG-19803 to C.G.), Progetti di Rilevante Interesse Nazionale (PRIN2017E5L5P3 to P.P. and PRIN20177E9EPY to C.G.), Italian Ministry of Health (GR-2013-02356747 to C.G. and GR-2019-12369862 to G.M.), local funds from the University of Ferrara, the European Research Council Grant (853057-InflaPML to C.G.).

Institutional Review Board Statement: Not applicable.

Informed Consent Statement: Not applicable.

Acknowledgments: P.P. is grateful to Camilla degli Scrovegni for the continuous support.

Conflicts of Interest: The authors have no conflicts of interests.

References

1. Bonora, M.; Wieckowski, M.R.; Sinclair, D.A.; Kroemer, G.; Pinton, P.; Galluzzi, L. Targeting Mitochondria for Cardiovascular Disorders: Therapeutic Potential and Obstacles. *Nat. Rev. Cardiol.* **2019**, *16*, 33–55. [CrossRef] [PubMed]
2. Siasos, G.; Tsigkou, V.; Kosmopoulos, M.; Theodosiadis, D.; Simantiris, S.; Tagkou, N.M.; Tsimpiktsioglou, A.; Stampouloglou, P.K.; Oikonomou, E.; Mourouzis, K.; et al. Mitochondria and Cardiovascular Diseases—From Pathophysiology to Treatment. *Ann. Transl. Med.* **2018**, *6*, 256. [CrossRef]
3. Brown, D.A.; Perry, J.B.; Allen, M.E.; Sabbah, H.N.; Stauffer, B.L.; Shaikh, S.R.; Cleland, J.G.F.; Colucci, W.S.; Butler, J.; Voors, A.A.; et al. Expert Consensus Document: Mitochondrial Function as a Therapeutic Target in Heart Failure. *Nat. Rev. Cardiol.* **2017**, *14*, 238–250. [CrossRef]
4. Piquereau, J.; Caffin, F.; Novotova, M.; Lemaire, C.; Veksler, V.; Garnier, A.; Ventura-Clapier, R.; Joubert, F. Mitochondrial Dynamics in the Adult Cardiomyocytes: Which Roles for a Highly Specialized Cell? *Front. Physiol.* **2013**, *4*, 102. [CrossRef] [PubMed]
5. Chung, S.; Dzeja, P.P.; Faustino, R.S.; Perez-Terzic, C.; Behfar, A.; Terzic, A. Mitochondrial Oxidative Metabolism Is Required for the Cardiac Differentiation of Stem Cells. *Nat. Clin. Pract. Cardiovasc. Med.* **2007**, *4* (Suppl. 1), S60–S67. [CrossRef] [PubMed]
6. Murphy, E.; Ardehali, H.; Balaban, R.S.; DiLisa, F.; Dorn, G.W.; Kitsis, R.N.; Otsu, K.; Ping, P.; Rizzuto, R.; Sack, M.N.; et al. Mitochondrial Function, Biology, and Role in Disease: A Scientific Statement from the American Heart Association. *Circ. Res.* **2016**, *118*, 1960–1991. [CrossRef] [PubMed]
7. Lenaz, G.; Baracca, A.; Fato, R.; Genova, M.L.; Solaini, G. New Insights into Structure and Function of Mitochondria and Their Role in Aging and Disease. *Antioxid. Redox Signal.* **2006**, *8*, 417–437. [CrossRef] [PubMed]
8. Cao, J.L.; Adaniya, S.M.; Cypress, M.W.; Suzuki, Y.; Kusakari, Y.; Jhun, B.S.; O-Uchi, J. Role of Mitochondrial Ca2+ Homeostasis in Cardiac Muscles. *Arch Biochem. Biophys.* **2019**, *663*, 276–287. [CrossRef]
9. Traaseth, N.; Elfering, S.; Solien, J.; Haynes, V.; Giulivi, C. Role of Calcium Signaling in the Activation of Mitochondrial Nitric Oxide Synthase and Citric Acid Cycle. *Biochim. Biophys. Acta* **2004**, *1658*, 64–71. [CrossRef]
10. Jouaville, L.S.; Pinton, P.; Bastianutto, C.; Rutter, G.A.; Rizzuto, R. Regulation of Mitochondrial ATP Synthesis by Calcium: Evidence for a Long-Term Metabolic Priming. *Proc. Natl. Acad. Sci. USA* **1999**, *96*, 13807–13812. [CrossRef]
11. Halestrap, A.P. Calcium-Dependent Opening of a Non-Specific Pore in the Mitochondrial Inner Membrane Is Inhibited at PH Values below 7. Implications for the Protective Effect of Low PH against Chemical and Hypoxic Cell Damage. *Biochem. J.* **1991**, *278 Pt 3*, 715–719. [CrossRef]
12. Crompton, M.; Costi, A.; Hayat, L. Evidence for the Presence of a Reversible Ca2+-Dependent Pore Activated by Oxidative Stress in Heart Mitochondria. *Biochem. J.* **1987**, *245*, 915–918. [CrossRef]
13. Genova, M.L.; Lenaz, G. The Interplay Between Respiratory Supercomplexes and ROS in Aging. *Antioxid. Redox Signal.* **2015**, *23*, 208–238. [CrossRef] [PubMed]
14. Giorgi, C.; Marchi, S.; Simoes, I.C.M.; Ren, Z.; Morciano, G.; Perrone, M.; Patalas-Krawczyk, P.; Borchard, S.; Jędrak, P.; Pierzynowska, K.; et al. Mitochondria and Reactive Oxygen Species in Aging and Age-Related Diseases. *Int. Rev. Cell Mol. Biol.* **2018**, *340*, 209–344. [CrossRef]
15. Simmonds, S.J.; Cuijpers, I.; Heymans, S.; Jones, E.A.V. Cellular and Molecular Differences between HFpEF and HFrEF: A Step Ahead in an Improved Pathological Understanding. *Cells* **2020**, *9*, 242. [CrossRef] [PubMed]

16. Schwinger, R.H.; Münch, G.; Bölck, B.; Karczewski, P.; Krause, E.G.; Erdmann, E. Reduced Ca(2+)-Sensitivity of SERCA 2a in Failing Human Myocardium Due to Reduced Serin-16 Phospholamban Phosphorylation. *J. Mol. Cell Cardiol.* **1999**, *31*, 479–491. [CrossRef] [PubMed]
17. Del Monte, F.; Harding, S.E.; Schmidt, U.; Matsui, T.; Kang, Z.B.; Dec, G.W.; Gwathmey, J.K.; Rosenzweig, A.; Hajjar, R.J. Restoration of Contractile Function in Isolated Cardiomyocytes from Failing Human Hearts by Gene Transfer of SERCA2a. *Circulation* **1999**, *100*, 2308–2311. [CrossRef] [PubMed]
18. Zhihao, L.; Jingyu, N.; Lan, L.; Michael, S.; Rui, G.; Xiyun, B.; Xiaozhi, L.; Guanwei, F. SERCA2a: A Key Protein in the Ca2+ Cycle of the Heart Failure. *Heart Fail. Rev.* **2020**, *25*, 523–535. [CrossRef]
19. Leong, H.S.; Brownsey, R.W.; Kulpa, J.E.; Allard, M.F. Glycolysis and Pyruvate Oxidation in Cardiac Hypertrophy—Why so Unbalanced? *Comp. Biochem. Physiol. A Mol. Integr. Physiol.* **2003**, *135*, 499–513. [CrossRef]
20. Allard, M.F.; Schönekess, B.O.; Henning, S.L.; English, D.R.; Lopaschuk, G.D. Contribution of Oxidative Metabolism and Glycolysis to ATP Production in Hypertrophied Hearts. *Am. J. Physiol.* **1994**, *267*, H742–H750. [CrossRef]
21. Doenst, T.; Pytel, G.; Schrepper, A.; Amorim, P.; Färber, G.; Shingu, Y.; Mohr, F.W.; Schwarzer, M. Decreased Rates of Substrate Oxidation Ex Vivo Predict the Onset of Heart Failure and Contractile Dysfunction in Rats with Pressure Overload. *Cardiovasc. Res.* **2010**, *86*, 461–470. [CrossRef]
22. Sambandam, N.; Lopaschuk, G.D.; Brownsey, R.W.; Allard, M.F. Energy Metabolism in the Hypertrophied Heart. *Heart Fail. Rev.* **2002**, *7*, 161–173. [CrossRef]
23. Tomin, T.; Schittmayer, M.; Sedej, S.; Bugger, H.; Gollmer, J.; Honeder, S.; Darnhofer, B.; Liesinger, L.; Zuckermann, A.; Rainer, P.P.; et al. Mass Spectrometry-Based Redox and Protein Profiling of Failing Human Hearts. *Int. J. Mol. Sci.* **2021**, *22*, 1787. [CrossRef]
24. Stanley, W.C.; Recchia, F.A.; Lopaschuk, G.D. Myocardial Substrate Metabolism in the Normal and Failing Heart. *Physiol. Rev.* **2005**, *85*, 1093–1129. [CrossRef]
25. Rosano, G.M.; Vitale, C. Metabolic Modulation of Cardiac Metabolism in Heart Failure. *Card Fail. Rev.* **2018**, *4*, 99–103. [CrossRef]
26. Taegtmeyer, H. Energy Metabolism of the Heart: From Basic Concepts to Clinical Applications. *Curr. Probl. Cardiol.* **1994**, *19*, 59–113. [CrossRef]
27. Wisneski, J.A.; Gertz, E.W.; Neese, R.A.; Mayr, M. Myocardial Metabolism of Free Fatty Acids. Studies with 14C-Labeled Substrates in Humans. *J. Clin. Investig.* **1987**, *79*, 359–366. [CrossRef] [PubMed]
28. Lopaschuk, G.D.; Ussher, J.R.; Folmes, C.D.L.; Jaswal, J.S.; Stanley, W.C. Myocardial Fatty Acid Metabolism in Health and Disease. *Physiol. Rev.* **2010**, *90*, 207–258. [CrossRef] [PubMed]
29. Sack, M.N.; Rader, T.A.; Park, S.; Bastin, J.; McCune, S.A.; Kelly, D.P. Fatty Acid Oxidation Enzyme Gene Expression Is Downregulated in the Failing Heart. *Circulation* **1996**, *94*, 2837–2842. [CrossRef] [PubMed]
30. Heather, L.C.; Cole, M.A.; Lygate, C.A.; Evans, R.D.; Stuckey, D.J.; Murray, A.J.; Neubauer, S.; Clarke, K. Fatty Acid Transporter Levels and Palmitate Oxidation Rate Correlate with Ejection Fraction in the Infarcted Rat Heart. *Cardiovasc. Res.* **2006**, *72*, 430–437. [CrossRef] [PubMed]
31. Kato, T.; Niizuma, S.; Inuzuka, Y.; Kawashima, T.; Okuda, J.; Tamaki, Y.; Iwanaga, Y.; Narazaki, M.; Matsuda, T.; Soga, T.; et al. Analysis of Metabolic Remodeling in Compensated Left Ventricular Hypertrophy and Heart Failure. *Circ. Heart Fail.* **2010**, *3*, 420–430. [CrossRef]
32. Sharma, S.; Adrogue, J.V.; Golfman, L.; Uray, I.; Lemm, J.; Youker, K.; Noon, G.P.; Frazier, O.H.; Taegtmeyer, H. Intramyocardial Lipid Accumulation in the Failing Human Heart Resembles the Lipotoxic Rat Heart. *FASEB J.* **2004**, *18*, 1692–1700. [CrossRef]
33. Lai, L.; Leone, T.C.; Keller, M.P.; Martin, O.J.; Broman, A.T.; Nigro, J.; Kapoor, K.; Koves, T.R.; Stevens, R.; Ilkayeva, O.R.; et al. Energy Metabolic Reprogramming in the Hypertrophied and Early Stage Failing Heart: A Multisystems Approach. *Circ. Heart Fail.* **2014**, *7*, 1022–1031. [CrossRef]
34. Wojtczak, L.; Wieckowski, M.R. The Mechanisms of Fatty Acid-Induced Proton Permeability of the Inner Mitochondrial Membrane. *J. Bioenerg. Biomembr.* **1999**, *31*, 447–455. [CrossRef] [PubMed]
35. Pande, S.V.; Blanchaer, M.C. Reversible Inhibition of Mitochondrial Adenosine Diphosphate Phosphorylation by Long Chain Acyl Coenzyme A Esters. *J. Biol. Chem.* **1971**, *246*, 402–411. [CrossRef]
36. Aon, M.A.; Bhatt, N.; Cortassa, S.C. Mitochondrial and Cellular Mechanisms for Managing Lipid Excess. *Front. Physiol.* **2014**, *5*, 282. [CrossRef]
37. Lemaitre, R.N.; McKnight, B.; Sotoodehnia, N.; Fretts, A.M.; Qureshi, W.T.; Song, X.; King, I.B.; Sitlani, C.M.; Siscovick, D.S.; Psaty, B.M.; et al. Circulating Very Long-Chain Saturated Fatty Acids and Heart Failure: The Cardiovascular Health Study. *J. Am. Heart Assoc.* **2018**, *7*, e010019. [CrossRef] [PubMed]
38. Toko, H.; Morita, H.; Katakura, M.; Hashimoto, M.; Ko, T.; Bujo, S.; Adachi, Y.; Ueda, K.; Murakami, H.; Ishizuka, M.; et al. Omega-3 Fatty Acid Prevents the Development of Heart Failure by Changing Fatty Acid Composition in the Heart. *Sci. Rep.* **2020**, *10*, 15553. [CrossRef] [PubMed]
39. Di, W.; Lv, J.; Jiang, S.; Lu, C.; Yang, Z.; Ma, Z.; Hu, W.; Yang, Y.; Xu, B. PGC-1: The Energetic Regulator in Cardiac Metabolism. *Curr. Issues Mol. Biol.* **2018**, *28*, 29–46. [CrossRef]
40. Lei, B.; Lionetti, V.; Young, M.E.; Chandler, M.P.; d'Agostino, C.; Kang, E.; Altarejos, M.; Matsuo, K.; Hintze, T.H.; Stanley, W.C.; et al. Paradoxical Downregulation of the Glucose Oxidation Pathway despite Enhanced Flux in Severe Heart Failure. *J. Mol. Cell Cardiol.* **2004**, *36*, 567–576. [CrossRef]

41. Osorio, J.C.; Stanley, W.C.; Linke, A.; Castellari, M.; Diep, Q.N.; Panchal, A.R.; Hintze, T.H.; Lopaschuk, G.D.; Recchia, F.A. Impaired Myocardial Fatty Acid Oxidation and Reduced Protein Expression of Retinoid X Receptor-Alpha in Pacing-Induced Heart Failure. *Circulation* **2002**, *106*, 606–612. [CrossRef] [PubMed]
42. Van Bilsen, M. "Energenetics" of Heart Failure. *Ann. N. Y. Acad. Sci.* **2004**, *1015*, 238–249. [CrossRef] [PubMed]
43. Tran, D.H.; Wang, Z.V. Glucose Metabolism in Cardiac Hypertrophy and Heart Failure. *J. Am. Heart Assoc.* **2019**, *8*, e012673. [CrossRef]
44. Locke, F.S.; Rosenheim, O. Contributions to the Physiology of the Isolated Heart: The Consumption of Dextrose by Mammalian Cardiac Muscle. *J. Physiol.* **1907**, *36*, 205–220. [CrossRef] [PubMed]
45. Lopaschuk, G.D.; Wambolt, R.B.; Barr, R.L. An Imbalance between Glycolysis and Glucose Oxidation Is a Possible Explanation for the Detrimental Effects of High Levels of Fatty Acids during Aerobic Reperfusion of Ischemic Hearts. *J. Pharmacol. Exp. Ther.* **1993**, *264*, 135–144. [PubMed]
46. Diakos, N.A.; Navankasattusas, S.; Abel, E.D.; Rutter, J.; McCreath, L.; Ferrin, P.; McKellar, S.H.; Miller, D.V.; Park, S.Y.; Richardson, R.S.; et al. Evidence of Glycolysis Up-Regulation and Pyruvate Mitochondrial Oxidation Mismatch During Mechanical Unloading of the Failing Human Heart: Implications for Cardiac Reloading and Conditioning. *JACC Basic Transl. Sci.* **2016**, *1*, 432–444. [CrossRef] [PubMed]
47. Paolisso, G.; Gambardella, A.; Galzerano, D.; D'Amore, A.; Rubino, P.; Verza, M.; Teasuro, P.; Varricchio, M.; D'Onofrio, F. Total-Body and Myocardial Substrate Oxidation in Congestive Heart Failure. *Metabolism* **1994**, *43*, 174–179. [CrossRef]
48. Zhang, L.; Jaswal, J.S.; Ussher, J.R.; Sankaralingam, S.; Wagg, C.; Zaugg, M.; Lopaschuk, G.D. Cardiac Insulin-Resistance and Decreased Mitochondrial Energy Production Precede the Development of Systolic Heart Failure after Pressure-Overload Hypertrophy. *Circ. Heart Fail.* **2013**, *6*, 1039–1048. [CrossRef]
49. Zhabyeyev, P.; Gandhi, M.; Mori, J.; Basu, R.; Kassiri, Z.; Clanachan, A.; Lopaschuk, G.D.; Oudit, G.Y. Pressure-Overload-Induced Heart Failure Induces a Selective Reduction in Glucose Oxidation at Physiological Afterload. *Cardiovasc. Res.* **2013**, *97*, 676–685. [CrossRef] [PubMed]
50. Funada, J.; Betts, T.R.; Hodson, L.; Humphreys, S.M.; Timperley, J.; Frayn, K.N.; Karpe, F. Substrate Utilization by the Failing Human Heart by Direct Quantification Using Arterio-Venous Blood Sampling. *PLoS ONE* **2009**, *4*, e7533. [CrossRef]
51. Mori, J.; Alrob, O.A.; Wagg, C.S.; Harris, R.A.; Lopaschuk, G.D.; Oudit, G.Y. ANG II Causes Insulin Resistance and Induces Cardiac Metabolic Switch and Inefficiency: A Critical Role of PDK4. *Am. J. Physiol. Heart Circ. Physiol.* **2013**, *304*, H1103–H1113. [CrossRef] [PubMed]
52. Karwi, Q.G.; Uddin, G.M.; Ho, K.L.; Lopaschuk, G.D. Loss of Metabolic Flexibility in the Failing Heart. *Front. Cardiovasc. Med.* **2018**, *5*, 68. [CrossRef]
53. Cotter, D.G.; Schugar, R.C.; Crawford, P.A. Ketone Body Metabolism and Cardiovascular Disease. *Am. J. Physiol. Heart Circ. Physiol.* **2013**, *304*, H1060–H1076. [CrossRef] [PubMed]
54. Fukao, T.; Lopaschuk, G.D.; Mitchell, G.A. Pathways and Control of Ketone Body Metabolism: On the Fringe of Lipid Biochemistry. *Prostaglandins Leukot. Essent. Fatty Acids* **2004**, *70*, 243–251. [CrossRef]
55. Aubert, G.; Martin, O.J.; Horton, J.L.; Lai, L.; Vega, R.B.; Leone, T.C.; Koves, T.; Gardell, S.J.; Krüger, M.; Hoppel, C.L.; et al. The Failing Heart Relies on Ketone Bodies as a Fuel. *Circulation* **2016**, *133*, 698–705. [CrossRef] [PubMed]
56. Ho, K.L.; Zhang, L.; Wagg, C.; Al Batran, R.; Gopal, K.; Levasseur, J.; Leone, T.; Dyck, J.R.B.; Ussher, J.R.; Muoio, D.M.; et al. Increased Ketone Body Oxidation Provides Additional Energy for the Failing Heart without Improving Cardiac Efficiency. *Cardiovasc. Res.* **2019**, *115*, 1606–1616. [CrossRef]
57. Bedi, K.C.; Snyder, N.W.; Brandimarto, J.; Aziz, M.; Mesaros, C.; Worth, A.J.; Wang, L.L.; Javaheri, A.; Blair, I.A.; Margulies, K.B.; et al. Evidence for Intramyocardial Disruption of Lipid Metabolism and Increased Myocardial Ketone Utilization in Advanced Human Heart Failure. *Circulation* **2016**, *133*, 706–716. [CrossRef]
58. Obokata, M.; Negishi, K.; Sunaga, H.; Ishida, H.; Ito, K.; Ogawa, T.; Iso, T.; Ando, Y.; Kurabayashi, M. Association Between Circulating Ketone Bodies and Worse Outcomes in Hemodialysis Patients. *J. Am. Heart Assoc.* **2017**, *6*. [CrossRef]
59. Yokokawa, T.; Sugano, Y.; Shimouchi, A.; Shibata, A.; Jinno, N.; Nagai, T.; Kanzaki, H.; Aiba, T.; Kusano, K.; Shirai, M.; et al. Exhaled Acetone Concentration Is Related to Hemodynamic Severity in Patients with Non-Ischemic Chronic Heart Failure. *Circ. J.* **2016**, *80*, 1178–1186. [CrossRef]
60. Taegtmeyer, H. Failing Heart and Starving Brain: Ketone Bodies to the Rescue. *Circulation* **2016**, *134*, 265–266. [CrossRef]
61. Baquet, A.; Lavoinne, A.; Hue, L. Comparison of the Effects of Various Amino Acids on Glycogen Synthesis, Lipogenesis and Ketogenesis in Isolated Rat Hepatocytes. *Biochem. J.* **1991**, *273 Pt 1*, 57–62. [CrossRef]
62. Sansbury, B.E.; DeMartino, A.M.; Xie, Z.; Brooks, A.C.; Brainard, R.E.; Watson, L.J.; DeFilippis, A.P.; Cummins, T.D.; Harbeson, M.A.; Brittian, K.R.; et al. Metabolomic Analysis of Pressure-Overloaded and Infarcted Mouse Hearts. *Circ. Heart Fail.* **2014**, *7*, 634–642. [CrossRef]
63. Sun, H.; Olson, K.C.; Gao, C.; Prosdocimo, D.A.; Zhou, M.; Wang, Z.; Jeyaraj, D.; Youn, J.-Y.; Ren, S.; Liu, Y.; et al. Catabolic Defect of Branched-Chain Amino Acids Promotes Heart Failure. *Circulation* **2016**, *133*, 2038–2049. [CrossRef]
64. Uddin, G.M.; Zhang, L.; Shah, S.; Fukushima, A.; Wagg, C.S.; Gopal, K.; Al Batran, R.; Pherwani, S.; Ho, K.L.; Boisvenue, J.; et al. Impaired Branched Chain Amino Acid Oxidation Contributes to Cardiac Insulin Resistance in Heart Failure. *Cardiovasc. Diabetol.* **2019**, *18*, 86. [CrossRef]

65. Wang, W.; Zhang, F.; Xia, Y.; Zhao, S.; Yan, W.; Wang, H.; Lee, Y.; Li, C.; Zhang, L.; Lian, K.; et al. Defective Branched Chain Amino Acid Catabolism Contributes to Cardiac Dysfunction and Remodeling Following Myocardial Infarction. *Am. J. Physiol. Heart Circ. Physiol.* **2016**, *311*, H1160–H1169. [CrossRef] [PubMed]
66. Shiojima, I.; Sato, K.; Izumiya, Y.; Schiekofer, S.; Ito, M.; Liao, R.; Colucci, W.S.; Walsh, K. Disruption of Coordinated Cardiac Hypertrophy and Angiogenesis Contributes to the Transition to Heart Failure. *J. Clin. Investig.* **2005**, *115*, 2108–2118. [CrossRef] [PubMed]
67. Riehle, C.; Wende, A.R.; Sena, S.; Pires, K.M.; Pereira, R.O.; Zhu, Y.; Bugger, H.; Frank, D.; Bevins, J.; Chen, D.; et al. Insulin Receptor Substrate Signaling Suppresses Neonatal Autophagy in the Heart. *J. Clin. Investig.* **2013**, *123*, 5319–5333. [CrossRef]
68. Liesa, M.; Palacín, M.; Zorzano, A. Mitochondrial Dynamics in Mammalian Health and Disease. *Physiol. Rev.* **2009**, *89*, 799–845. [CrossRef] [PubMed]
69. Bayeva, M.; Gheorghiade, M.; Ardehali, H. Mitochondria as a Therapeutic Target in Heart Failure. *J. Am. Coll. Cardiol.* **2013**, *61*, 599–610. [CrossRef] [PubMed]
70. Hoppel, C.L.; Tandler, B.; Fujioka, H.; Riva, A. Dynamic Organization of Mitochondria in Human Heart and in Myocardial Disease. *Int. J. Biochem. Cell Biol.* **2009**, *41*, 1949–1956. [CrossRef] [PubMed]
71. Bers, D.M. Calcium Cycling and Signaling in Cardiac Myocytes. *Annu. Rev. Physiol.* **2008**, *70*, 23–49. [CrossRef] [PubMed]
72. Sharma, V.K.; Ramesh, V.; Franzini-Armstrong, C.; Sheu, S.S. Transport of Ca2+ from Sarcoplasmic Reticulum to Mitochondria in Rat Ventricular Myocytes. *J. Bioenerg. Biomembr.* **2000**, *32*, 97–104. [CrossRef] [PubMed]
73. Dorn, G.W.; Vega, R.B.; Kelly, D.P. Mitochondrial Biogenesis and Dynamics in the Developing and Diseased Heart. *Genes Dev.* **2015**, *29*, 1981–1991. [CrossRef] [PubMed]
74. Morciano, G.; Patergnani, S.; Bonora, M.; Pedriali, G.; Tarocco, A.; Bouhamida, E.; Marchi, S.; Ancora, G.; Anania, G.; Wieckowski, M.R.; et al. Mitophagy in Cardiovascular Diseases. *J. Clin. Med.* **2020**, *9*, 892. [CrossRef] [PubMed]
75. Gupta, A.; Gupta, S.; Young, D.; Das, B.; McMahon, J.; Sen, S. Impairment of Ultrastructure and Cytoskeleton during Progression of Cardiac Hypertrophy to Heart Failure. *Lab. Investig.* **2010**, *90*, 520–530. [CrossRef]
76. Sharov, V.G.; Goussev, A.; Lesch, M.; Goldstein, S.; Sabbah, H.N. Abnormal Mitochondrial Function in Myocardium of Dogs with Chronic Heart Failure. *J. Mol. Cell Cardiol.* **1998**, *30*, 1757–1762. [CrossRef] [PubMed]
77. Sharov, V.G.; Todor, A.V.; Silverman, N.; Goldstein, S.; Sabbah, H.N. Abnormal Mitochondrial Respiration in Failed Human Myocardium. *J. Mol. Cell Cardiol.* **2000**, *32*, 2361–2367. [CrossRef] [PubMed]
78. Sabbah, H.N.; Sharov, V.; Riddle, J.M.; Kono, T.; Lesch, M.; Goldstein, S. Mitochondrial Abnormalities in Myocardium of Dogs with Chronic Heart Failure. *J. Mol. Cell Cardiol.* **1992**, *24*, 1333–1347. [CrossRef]
79. Marín-García, J.; Akhmedov, A.T. Mitochondrial Dynamics and Cell Death in Heart Failure. *Heart Fail. Rev.* **2016**, *21*, 123–136. [CrossRef]
80. Dorn, G.W.; Kitsis, R.N. The Mitochondrial Dynamism-Mitophagy-Cell Death Interactome: Multiple Roles Performed by Members of a Mitochondrial Molecular Ensemble. *Circ. Res.* **2015**, *116*, 167–182. [CrossRef]
81. Youle, R.J.; Karbowski, M. Mitochondrial Fission in Apoptosis. *Nat. Rev. Mol. Cell Biol.* **2005**, *6*, 657–663. [CrossRef] [PubMed]
82. Chaanine, A.H.; Joyce, L.D.; Stulak, J.M.; Maltais, S.; Joyce, D.L.; Dearani, J.A.; Klaus, K.; Nair, K.S.; Hajjar, R.J.; Redfield, M.M. Mitochondrial Morphology, Dynamics, and Function in Human Pressure Overload or Ischemic Heart Disease with Preserved or Reduced Ejection Fraction. *Circ. Heart Fail.* **2019**, *12*, e005131. [CrossRef]
83. Ren, M.; Phoon, C.K.L.; Schlame, M. Metabolism and Function of Mitochondrial Cardiolipin. *Prog. Lipid Res.* **2014**, *55*, 1–16. [CrossRef]
84. Dudek, J. Role of Cardiolipin in Mitochondrial Signaling Pathways. *Front. Cell Dev. Biol.* **2017**, *5*, 90. [CrossRef]
85. John, G.B.; Shang, Y.; Li, L.; Renken, C.; Mannella, C.A.; Selker, J.M.L.; Rangell, L.; Bennett, M.J.; Zha, J. The Mitochondrial Inner Membrane Protein Mitofilin Controls Cristae Morphology. *Mol. Biol. Cell* **2005**, *16*, 1543–1554. [CrossRef] [PubMed]
86. Sabbah, H.N.; Gupta, R.C.; Singh-Gupta, V.; Zhang, K.; Lanfear, D.E. Abnormalities of Mitochondrial Dynamics in the Failing Heart: Normalization Following Long-Term Therapy with Elamipretide. *Cardiovasc. Drugs Ther.* **2018**, *32*, 319–328. [CrossRef]
87. Paradies, G.; Paradies, V.; De Benedictis, V.; Ruggiero, F.M.; Petrosillo, G. Functional Role of Cardiolipin in Mitochondrial Bioenergetics. *Biochim. Biophys. Acta* **2014**, *1837*, 408–417. [CrossRef] [PubMed]
88. Riehle, C.; Wende, A.R.; Zaha, V.G.; Pires, K.M.; Wayment, B.; Olsen, C.; Bugger, H.; Buchanan, J.; Wang, X.; Moreira, A.B.; et al. PGC-1β Deficiency Accelerates the Transition to Heart Failure in Pressure Overload Hypertrophy. *Circ. Res.* **2011**, *109*, 783–793. [CrossRef] [PubMed]
89. Bugger, H.; Schwarzer, M.; Chen, D.; Schrepper, A.; Amorim, P.A.; Schoepe, M.; Nguyen, T.D.; Mohr, F.W.; Khalimonchuk, O.; Weimer, B.C.; et al. Proteomic Remodelling of Mitochondrial Oxidative Pathways in Pressure Overload-Induced Heart Failure. *Cardiovasc. Res.* **2010**, *85*, 376–384. [CrossRef]
90. Arany, Z.; Novikov, M.; Chin, S.; Ma, Y.; Rosenzweig, A.; Spiegelman, B.M. Transverse Aortic Constriction Leads to Accelerated Heart Failure in Mice Lacking PPAR-Gamma Coactivator 1alpha. *Proc. Natl. Acad. Sci. USA* **2006**, *103*, 10086–10091. [CrossRef]
91. Pisano, A.; Cerbelli, B.; Perli, E.; Pelullo, M.; Bargelli, V.; Preziuso, C.; Mancini, M.; He, L.; Bates, M.G.D.; Lucena, J.R.; et al. Impaired Mitochondrial Biogenesis Is a Common Feature to Myocardial Hypertrophy and End-Stage Ischemic Heart Failure. *Cardiovasc. Pathol.* **2016**, *25*, 103–112. [CrossRef]
92. Scarpulla, R.C.; Vega, R.B.; Kelly, D.P. Transcriptional Integration of Mitochondrial Biogenesis. *Trends Endocrinol. Metab.* **2012**, *23*, 459–466. [CrossRef]

93. Karamanlidis, G.; Nascimben, L.; Couper, G.S.; Shekar, P.S.; del Monte, F.; Tian, R. Defective DNA Replication Impairs Mitochondrial Biogenesis in Human Failing Hearts. *Circ. Res.* **2010**, *106*, 1541–1548. [CrossRef] [PubMed]
94. Sabbah, H.N. Targeting the Mitochondria in Heart Failure: A Translational Perspective. *JACC Basic Transl. Sci.* **2020**, *5*, 88–106. [CrossRef]
95. Bray, A.W.; Ballinger, S.W. Mitochondrial DNA Mutations and Cardiovascular Disease. *Curr. Opin. Cardiol.* **2017**, *32*, 267–274. [CrossRef] [PubMed]
96. Finsterer, J.; Kothari, S. Cardiac Manifestations of Primary Mitochondrial Disorders. *Int. J. Cardiol.* **2014**, *177*, 754–763. [CrossRef]
97. Wahbi, K.; Bougouin, W.; Béhin, A.; Stojkovic, T.; Bécane, H.M.; Jardel, C.; Berber, N.; Mochel, F.; Lombès, A.; Eymard, B.; et al. Long-Term Cardiac Prognosis and Risk Stratification in 260 Adults Presenting with Mitochondrial Diseases. *Eur. Heart J.* **2015**, *36*, 2886–2893. [CrossRef]
98. Malfatti, E.; Laforêt, P.; Jardel, C.; Stojkovic, T.; Behin, A.; Eymard, B.; Lombès, A.; Benmalek, A.; Bécane, H.-M.; Berber, N.; et al. High Risk of Severe Cardiac Adverse Events in Patients with Mitochondrial m.3243A>G Mutation. *Neurology* **2013**, *80*, 100–105. [CrossRef] [PubMed]
99. Lehman, J.J.; Barger, P.M.; Kovacs, A.; Saffitz, J.E.; Medeiros, D.M.; Kelly, D.P. Peroxisome Proliferator-Activated Receptor Gamma Coactivator-1 Promotes Cardiac Mitochondrial Biogenesis. *J. Clin. Investig.* **2000**, *106*, 847–856. [CrossRef] [PubMed]
100. St-Pierre, J.; Drori, S.; Uldry, M.; Silvaggi, J.M.; Rhee, J.; Jäger, S.; Handschin, C.; Zheng, K.; Lin, J.; Yang, W.; et al. Suppression of Reactive Oxygen Species and Neurodegeneration by the PGC-1 Transcriptional Coactivators. *Cell* **2006**, *127*, 397–408. [CrossRef] [PubMed]
101. Lin, J.; Wu, P.-H.; Tarr, P.T.; Lindenberg, K.S.; St-Pierre, J.; Zhang, C.-Y.; Mootha, V.K.; Jäger, S.; Vianna, C.R.; Reznick, R.M.; et al. Defects in Adaptive Energy Metabolism with CNS-Linked Hyperactivity in PGC-1alpha Null Mice. *Cell* **2004**, *119*, 121–135. [CrossRef] [PubMed]
102. Russell, L.K.; Mansfield, C.M.; Lehman, J.J.; Kovacs, A.; Courtois, M.; Saffitz, J.E.; Medeiros, D.M.; Valencik, M.L.; McDonald, J.A.; Kelly, D.P. Cardiac-Specific Induction of the Transcriptional Coactivator Peroxisome Proliferator-Activated Receptor Gamma Coactivator-1alpha Promotes Mitochondrial Biogenesis and Reversible Cardiomyopathy in a Developmental Stage-Dependent Manner. *Circ. Res.* **2004**, *94*, 525–533. [CrossRef]
103. Arany, Z.; He, H.; Lin, J.; Hoyer, K.; Handschin, C.; Toka, O.; Ahmad, F.; Matsui, T.; Chin, S.; Wu, P.-H.; et al. Transcriptional Coactivator PGC-1 Alpha Controls the Energy State and Contractile Function of Cardiac Muscle. *Cell Metab.* **2005**, *1*, 259–271. [CrossRef]
104. Leone, T.C.; Lehman, J.J.; Finck, B.N.; Schaeffer, P.J.; Wende, A.R.; Boudina, S.; Courtois, M.; Wozniak, D.F.; Sambandam, N.; Bernal-Mizrachi, C.; et al. PGC-1alpha Deficiency Causes Multi-System Energy Metabolic Derangements: Muscle Dysfunction, Abnormal Weight Control and Hepatic Steatosis. *PLoS Biol.* **2005**, *3*, e101. [CrossRef]
105. Morciano, G.; Pedriali, G.; Sbano, L.; Iannitti, T.; Giorgi, C.; Pinton, P. Intersection of Mitochondrial Fission and Fusion Machinery with Apoptotic Pathways: Role of Mcl-1. *Biol. Cell* **2016**, *108*, 279–293. [CrossRef]
106. Ishihara, T.; Ban-Ishihara, R.; Maeda, M.; Matsunaga, Y.; Ichimura, A.; Kyogoku, S.; Aoki, H.; Katada, S.; Nakada, K.; Nomura, M.; et al. Dynamics of Mitochondrial DNA Nucleoids Regulated by Mitochondrial Fission Is Essential for Maintenance of Homogeneously Active Mitochondria during Neonatal Heart Development. *Mol. Cell Biol.* **2015**, *35*, 211–223. [CrossRef]
107. Song, M.; Mihara, K.; Chen, Y.; Scorrano, L.; Dorn, G.W. Mitochondrial Fission and Fusion Factors Reciprocally Orchestrate Mitophagic Culling in Mouse Hearts and Cultured Fibroblasts. *Cell Metab.* **2015**, *21*, 273–286. [CrossRef] [PubMed]
108. Chen, Y.; Liu, Y.; Dorn, G.W. Mitochondrial Fusion Is Essential for Organelle Function and Cardiac Homeostasis. *Circ. Res.* **2011**, *109*, 1327–1331. [CrossRef]
109. Kageyama, Y.; Hoshijima, M.; Seo, K.; Bedja, D.; Sysa-Shah, P.; Andrabi, S.A.; Chen, W.; Höke, A.; Dawson, V.L.; Dawson, T.M.; et al. Parkin-Independent Mitophagy Requires Drp1 and Maintains the Integrity of Mammalian Heart and Brain. *EMBO J.* **2014**, *33*, 2798–2813. [CrossRef] [PubMed]
110. Qiu, Z.; Wei, Y.; Song, Q.; Du, B.; Wang, H.; Chu, Y.; Hu, Y. The Role of Myocardial Mitochondrial Quality Control in Heart Failure. *Front. Pharmacol.* **2019**, *10*, 1404. [CrossRef] [PubMed]
111. Dorn, G.W. Mitochondrial Dynamism and Heart Disease: Changing Shape and Shaping Change. *EMBO Mol. Med.* **2015**, *7*, 865–877. [CrossRef]
112. Menezes, T.N.; Ramalho, L.S.; Bechara, L.R.G.; Ferreira, J.C.B. Targeting Mitochondrial Fission-Fusion Imbalance in Heart Failure. *Curr. Tissue Microenviron. Rep.* **2020**, *1*, 239–247. [CrossRef]
113. Vásquez-Trincado, C.; García-Carvajal, I.; Pennanen, C.; Parra, V.; Hill, J.A.; Rothermel, B.A.; Lavandero, S. Mitochondrial Dynamics, Mitophagy and Cardiovascular Disease. *J. Physiol.* **2016**, *594*, 509–525. [CrossRef]
114. Chappie, J.S.; Dyda, F. Building a Fission Machine—Structural Insights into Dynamin Assembly and Activation. *J. Cell Sci.* **2013**, *126*, 2773–2784. [CrossRef] [PubMed]
115. Filadi, R.; Pendin, D.; Pizzo, P. Mitofusin 2: From Functions to Disease. *Cell Death Dis.* **2018**, *9*, 330. [CrossRef]
116. Koshiba, T.; Detmer, S.A.; Kaiser, J.T.; Chen, H.; McCaffery, J.M.; Chan, D.C. Structural Basis of Mitochondrial Tethering by Mitofusin Complexes. *Science* **2004**, *305*, 858–862. [CrossRef] [PubMed]
117. Formosa, L.E.; Ryan, M.T. Mitochondrial Fusion: Reaching the End of Mitofusin's Tether. *J. Cell Biol.* **2016**, *215*, 597–598. [CrossRef] [PubMed]
118. Dorn, G.W. Mitofusins as Mitochondrial Anchors and Tethers. *J. Mol. Cell Cardiol.* **2020**, *142*, 146–153. [CrossRef]

119. Song, Z.; Ghochani, M.; McCaffery, J.M.; Frey, T.G.; Chan, D.C. Mitofusins and OPA1 Mediate Sequential Steps in Mitochondrial Membrane Fusion. *Mol. Biol. Cell* **2009**, *20*, 3525–3532. [CrossRef] [PubMed]
120. Sauvanet, C.; Arnauné-Pelloquin, L.; David, C.; Belenguer, P.; Rojo, M. Mitochondrial morphology and dynamics: Actors, mechanisms and functions. *Med. Sci.* **2010**, *26*, 823–829. [CrossRef]
121. Qin, Y.; Li, A.; Liu, B.; Jiang, W.; Gao, M.; Tian, X.; Gong, G. Mitochondrial Fusion Mediated by Fusion Promotion and Fission Inhibition Directs Adult Mouse Heart Function toward a Different Direction. *FASEB J.* **2020**, *34*, 663–675. [CrossRef]
122. Chen, H.; Chomyn, A.; Chan, D.C. Disruption of Fusion Results in Mitochondrial Heterogeneity and Dysfunction. *J. Biol. Chem.* **2005**, *280*, 26185–26192. [CrossRef]
123. Chen, L.; Gong, Q.; Stice, J.P.; Knowlton, A.A. Mitochondrial OPA1, Apoptosis, and Heart Failure. *Cardiovasc. Res.* **2009**, *84*, 91–99. [CrossRef]
124. Eisner, V.; Cupo, R.R.; Gao, E.; Csordás, G.; Slovinsky, W.S.; Paillard, M.; Cheng, L.; Ibetti, J.; Chen, S.R.W.; Chuprun, J.K.; et al. Mitochondrial Fusion Dynamics Is Robust in the Heart and Depends on Calcium Oscillations and Contractile Activity. *Proc. Natl. Acad. Sci. USA* **2017**, *114*, E859–E868. [CrossRef]
125. Smirnova, E.; Griparic, L.; Shurland, D.L.; van der Bliek, A.M. Dynamin-Related Protein Drp1 Is Required for Mitochondrial Division in Mammalian Cells. *Mol. Biol. Cell* **2001**, *12*, 2245–2256. [CrossRef]
126. Yoon, Y.; Krueger, E.W.; Oswald, B.J.; McNiven, M.A. The Mitochondrial Protein HFis1 Regulates Mitochondrial Fission in Mammalian Cells through an Interaction with the Dynamin-like Protein DLP1. *Mol. Cell Biol.* **2003**, *23*, 5409–5420. [CrossRef]
127. Suzuki, M.; Jeong, S.Y.; Karbowski, M.; Youle, R.J.; Tjandra, N. The Solution Structure of Human Mitochondria Fission Protein Fis1 Reveals a Novel TPR-like Helix Bundle. *J. Mol. Biol.* **2003**, *334*, 445–458. [CrossRef] [PubMed]
128. Losón, O.C.; Song, Z.; Chen, H.; Chan, D.C. Fis1, Mff, MiD49, and MiD51 Mediate Drp1 Recruitment in Mitochondrial Fission. *Mol. Biol. Cell* **2013**, *24*, 659–667. [CrossRef]
129. Jin, J.-Y.; Wei, X.-X.; Zhi, X.-L.; Wang, X.-H.; Meng, D. Drp1-Dependent Mitochondrial Fission in Cardiovascular Disease. *Acta Pharmacol. Sin.* **2021**, *42*, 655–664. [CrossRef] [PubMed]
130. Kim, E.Y.; Zhang, Y.; Beketaev, I.; Segura, A.M.; Yu, W.; Xi, Y.; Chang, J.; Wang, J. SENP5, a SUMO Isopeptidase, Induces Apoptosis and Cardiomyopathy. *J. Mol. Cell Cardiol.* **2015**, *78*, 154–164. [CrossRef] [PubMed]
131. Shirakabe, A.; Zhai, P.; Ikeda, Y.; Saito, T.; Maejima, Y.; Hsu, C.-P.; Nomura, M.; Egashira, K.; Levine, B.; Sadoshima, J. Drp1-Dependent Mitochondrial Autophagy Plays a Protective Role Against Pressure Overload-Induced Mitochondrial Dysfunction and Heart Failure. *Circulation* **2016**, *133*, 1249–1263. [CrossRef]
132. Shires, S.E.; Gustafsson, Å.B. Mitophagy and Heart Failure. *J. Mol. Med.* **2015**, *93*, 253–262. [CrossRef]
133. Jin, S.M.; Lazarou, M.; Wang, C.; Kane, L.A.; Narendra, D.P.; Youle, R.J. Mitochondrial Membrane Potential Regulates PINK1 Import and Proteolytic Destabilization by PARL. *J. Cell Biol.* **2010**, *191*, 933–942. [CrossRef]
134. Narendra, D.; Tanaka, A.; Suen, D.-F.; Youle, R.J. Parkin Is Recruited Selectively to Impaired Mitochondria and Promotes Their Autophagy. *J. Cell Biol.* **2008**, *183*, 795–803. [CrossRef]
135. Kazlauskaite, A.; Kondapalli, C.; Gourlay, R.; Campbell, D.G.; Ritorto, M.S.; Hofmann, K.; Alessi, D.R.; Knebel, A.; Trost, M.; Muqit, M.M.K. Parkin Is Activated by PINK1-Dependent Phosphorylation of Ubiquitin at Ser65. *Biochem. J.* **2014**, *460*, 127–139. [CrossRef]
136. Koyano, F.; Okatsu, K.; Kosako, H.; Tamura, Y.; Go, E.; Kimura, M.; Kimura, Y.; Tsuchiya, H.; Yoshihara, H.; Hirokawa, T.; et al. Ubiquitin Is Phosphorylated by PINK1 to Activate Parkin. *Nature* **2014**, *510*, 162–166. [CrossRef] [PubMed]
137. Kirkin, V.; Lamark, T.; Sou, Y.-S.; Bjørkøy, G.; Nunn, J.L.; Bruun, J.-A.; Shvets, E.; McEwan, D.G.; Clausen, T.H.; Wild, P.; et al. A Role for NBR1 in Autophagosomal Degradation of Ubiquitinated Substrates. *Mol. Cell* **2009**, *33*, 505–516. [CrossRef]
138. Lamark, T.; Kirkin, V.; Dikic, I.; Johansen, T. NBR1 and P62 as Cargo Receptors for Selective Autophagy of Ubiquitinated Targets. *Cell Cycle* **2009**, *8*, 1986–1990. [CrossRef] [PubMed]
139. Tanaka, A.; Cleland, M.M.; Xu, S.; Narendra, D.P.; Suen, D.-F.; Karbowski, M.; Youle, R.J. Proteasome and P97 Mediate Mitophagy and Degradation of Mitofusins Induced by Parkin. *J. Cell Biol.* **2010**, *191*, 1367–1380. [CrossRef] [PubMed]
140. Billia, F.; Hauck, L.; Konecny, F.; Rao, V.; Shen, J.; Mak, T.W. PTEN-Inducible Kinase 1 (PINK1)/Park6 Is Indispensable for Normal Heart Function. *Proc. Natl. Acad. Sci. USA* **2011**, *108*, 9572–9577. [CrossRef]
141. Kubli, D.A.; Zhang, X.; Lee, Y.; Hanna, R.A.; Quinsay, M.N.; Nguyen, C.K.; Jimenez, R.; Petrosyan, S.; Murphy, A.N.; Gustafsson, A.B. Parkin Protein Deficiency Exacerbates Cardiac Injury and Reduces Survival Following Myocardial Infarction. *J. Biol. Chem.* **2013**, *288*, 915–926. [CrossRef] [PubMed]
142. Hoshino, A.; Mita, Y.; Okawa, Y.; Ariyoshi, M.; Iwai-Kanai, E.; Ueyama, T.; Ikeda, K.; Ogata, T.; Matoba, S. Cytosolic P53 Inhibits Parkin-Mediated Mitophagy and Promotes Mitochondrial Dysfunction in the Mouse Heart. *Nat. Commun.* **2013**, *4*, 2308. [CrossRef] [PubMed]
143. Siddall, H.K.; Yellon, D.M.; Ong, S.-B.; Mukherjee, U.A.; Burke, N.; Hall, A.R.; Angelova, P.R.; Ludtmann, M.H.R.; Deas, E.; Davidson, S.M.; et al. Loss of PINK1 Increases the Heart's Vulnerability to Ischemia-Reperfusion Injury. *PLoS ONE* **2013**, *8*, e62400. [CrossRef]
144. Wang, B.; Nie, J.; Wu, L.; Hu, Y.; Wen, Z.; Dong, L.; Zou, M.-H.; Chen, C.; Wang, D.W. AMPKα2 Protects Against the Development of Heart Failure by Enhancing Mitophagy via PINK1 Phosphorylation. *Circ. Res.* **2018**, *122*, 712–729. [CrossRef] [PubMed]
145. Nah, J.; Miyamoto, S.; Sadoshima, J. Mitophagy as a Protective Mechanism against Myocardial Stress. *Compr. Physiol.* **2017**, *7*, 1407–1424. [CrossRef]

146. Zhou, H.; Zhu, P.; Wang, J.; Zhu, H.; Ren, J.; Chen, Y. Pathogenesis of Cardiac Ischemia Reperfusion Injury Is Associated with CK2α-Disturbed Mitochondrial Homeostasis via Suppression of FUNDC1-Related Mitophagy. *Cell Death Differ.* **2018**, *25*, 1080–1093. [CrossRef]
147. Saito, T.; Nah, J.; Oka, S.-I.; Mukai, R.; Monden, Y.; Maejima, Y.; Ikeda, Y.; Sciarretta, S.; Liu, T.; Li, H.; et al. An Alternative Mitophagy Pathway Mediated by Rab9 Protects the Heart against Ischemia. *J. Clin. Investig.* **2019**, *129*, 802–819. [CrossRef]
148. Bonora, M.; Patergnani, S.; Ramaccini, D.; Morciano, G.; Pedriali, G.; Kahsay, A.E.; Bouhamida, E.; Giorgi, C.; Wieckowski, M.R.; Pinton, P. Physiopathology of the Permeability Transition Pore: Molecular Mechanisms in Human Pathology. *Biomolecules* **2020**, *10*,.998. [CrossRef]
149. Morciano, G.; Bonora, M.; Giorgi, C.; Pinton, P. Other Bricks for the Correct Construction of the Mitochondrial Permeability Transition Pore Complex. *Cell Death Dis.* **2017**, *8*, e2698. [CrossRef]
150. Morciano, G.; Giorgi, C.; Bonora, M.; Punzetti, S.; Pavasini, R.; Wieckowski, M.R.; Campo, G.; Pinton, P. Molecular Identity of the Mitochondrial Permeability Transition Pore and Its Role in Ischemia-Reperfusion Injury. *J. Mol. Cell Cardiol.* **2015**, *78*, 142–153. [CrossRef]
151. Morciano, G.; Pedriali, G.; Bonora, M.; Pavasini, R.; Mikus, E.; Calvi, S.; Bovolenta, M.; Lebiedzinska-Arciszewska, M.; Pinotti, M.; Albertini, A.; et al. A naturally occurring mutation in ATP synthase subunit c is associated with increased damage following hypoxia/reoxygenation in STEMI patients. *Cell Rep.* **2021**, *35*, 108983. [CrossRef]
152. Algieri, C.; Trombetti, F.; Pagliarani, A.; Ventrella, V.; Bernardini, C.; Fabbri, M.; Forni, M.; Nesci, S. Mitochondrial Ca2+ -Activated F1 FO -ATPase Hydrolyzes ATP and Promotes the Permeability Transition Pore. *Ann. N. Y. Acad. Sci.* **2019**, *1457*, 142–157. [CrossRef]
153. Nesci, S. Mitochondrial Permeability Transition, F1FO-ATPase and Calcium: An Enigmatic Triangle. *EMBO Rep.* **2017**. [CrossRef] [PubMed]
154. Nesci, S.; Trombetti, F.; Ventrella, V.; Pagliarani, A. From the Ca2+-Activated F1FO-ATPase to the Mitochondrial Permeability Transition Pore: An Overview. *Biochimie* **2018**, *152*, 85–93. [CrossRef] [PubMed]
155. Penzo, D.; Tagliapietra, C.; Colonna, R.; Petronilli, V.; Bernardi, P. Effects of Fatty Acids on Mitochondria: Implications for Cell Death. *Biochim. Biophys. Acta* **2002**, *1555*, 160–165. [CrossRef]
156. Moon, S.H.; Liu, X.; Cedars, A.M.; Yang, K.; Kiebish, M.A.; Joseph, S.M.; Kelley, J.; Jenkins, C.M.; Gross, R.W. Heart Failure-Induced Activation of Phospholipase IPLA2γ Generates Hydroxyeicosatetraenoic Acids Opening the Mitochondrial Permeability Transition Pore. *J. Biol. Chem.* **2018**, *293*, 115–129. [CrossRef]
157. Marcil, M.; Ascah, A.; Matas, J.; Bélanger, S.; Deschepper, C.F.; Burelle, Y. Compensated Volume Overload Increases the Vulnerability of Heart Mitochondria without Affecting Their Functions in the Absence of Stress. *J. Mol. Cell Cardiol.* **2006**, *41*, 998–1009. [CrossRef]
158. Sharov, V.G.; Todor, A.; Khanal, S.; Imai, M.; Sabbah, H.N. Cyclosporine A Attenuates Mitochondrial Permeability Transition and Improves Mitochondrial Respiratory Function in Cardiomyocytes Isolated from Dogs with Heart Failure. *J. Mol. Cell Cardiol.* **2007**, *42*, 150–158. [CrossRef]
159. Kang, P.T.; Chen, C.-L.; Ohanyan, V.; Luther, D.J.; Meszaros, J.G.; Chilian, W.M.; Chen, Y.-R. Overexpressing Superoxide Dismutase 2 Induces a Supernormal Cardiac Function by Enhancing Redox-Dependent Mitochondrial Function and Metabolic Dilation. *J. Mol. Cell Cardiol.* **2015**, *88*, 14–28. [CrossRef] [PubMed]
160. Van Remmen, H.; Williams, M.D.; Guo, Z.; Estlack, L.; Yang, H.; Carlson, E.J.; Epstein, C.J.; Huang, T.T.; Richardson, A. Knockout Mice Heterozygous for Sod2 Show Alterations in Cardiac Mitochondrial Function and Apoptosis. *Am. J. Physiol. Heart Circ. Physiol.* **2001**, *281*, H1422–H1432. [CrossRef]
161. Kisucka, J.; Chauhan, A.K.; Patten, I.S.; Yesilaltay, A.; Neumann, C.; Van Etten, R.A.; Krieger, M.; Wagner, D.D. Peroxiredoxin1 Prevents Excessive Endothelial Activation and Early Atherosclerosis. *Circ. Res.* **2008**, *103*, 598–605. [CrossRef]
162. Ardanaz, N.; Yang, X.-P.; Cifuentes, M.E.; Haurani, M.J.; Jackson, K.W.; Liao, T.-D.; Carretero, O.A.; Pagano, P.J. Lack of Glutathione Peroxidase 1 Accelerates Cardiac-Specific Hypertrophy and Dysfunction in Angiotensin II Hypertension. *Hypertension* **2010**, *55*, 116–123. [CrossRef] [PubMed]
163. Kiermayer, C.; Northrup, E.; Schrewe, A.; Walch, A.; de Angelis, M.H.; Schoensiegel, F.; Zischka, H.; Prehn, C.; Adamski, J.; Bekeredjian, R.; et al. Heart-Specific Knockout of the Mitochondrial Thioredoxin Reductase (Txnrd2) Induces Metabolic and Contractile Dysfunction in the Aging Myocardium. *J. Am. Heart Assoc.* **2015**, *4*. [CrossRef] [PubMed]
164. Dey, S.; DeMazumder, D.; Sidor, A.; Foster, D.B.; O'Rourke, B. Mitochondrial ROS Drive Sudden Cardiac Death and Chronic Proteome Remodeling in Heart Failure. *Circ. Res.* **2018**, *123*, 356–371. [CrossRef] [PubMed]
165. Zorov, D.B.; Filburn, C.R.; Klotz, L.O.; Zweier, J.L.; Sollott, S.J. Reactive Oxygen Species (ROS)-Induced ROS Release: A New Phenomenon Accompanying Induction of the Mitochondrial Permeability Transition in Cardiac Myocytes. *J. Exp. Med.* **2000**, *192*, 1001–1014. [CrossRef] [PubMed]
166. Zorov, D.B.; Juhaszova, M.; Sollott, S.J. Mitochondrial Reactive Oxygen Species (ROS) and ROS-Induced ROS Release. *Physiol. Rev.* **2014**, *94*, 909–950. [CrossRef] [PubMed]
167. Giorgi, C.; Danese, A.; Missiroli, S.; Patergnani, S.; Pinton, P. Calcium Dynamics as a Machine for Decoding Signals. *Trends Cell Biol.* **2018**, *28*, 258–273. [CrossRef]
168. Giorgi, C.; Marchi, S.; Pinton, P. The Machineries, Regulation and Cellular Functions of Mitochondrial Calcium. *Nat. Rev. Mol. Cell Biol.* **2018**, *19*, 713–730. [CrossRef] [PubMed]

169. Gambardella, J.; Trimarco, B.; Iaccarino, G.; Santulli, G. New Insights in Cardiac Calcium Handling and Excitation-Contraction Coupling. *Adv. Exp. Med. Biol.* **2018**, *1067*, 373–385. [CrossRef] [PubMed]
170. Nánási, P.P.; Magyar, J.; Varró, A.; Ördög, B. Beat-to-Beat Variability of Cardiac Action Potential Duration: Underlying Mechanism and Clinical Implications. *Can. J. Physiol. Pharmacol.* **2017**, *95*, 1230–1235. [CrossRef]
171. Eisner, D.A.; Caldwell, J.L.; Kistamás, K.; Trafford, A.W. Calcium and Excitation-Contraction Coupling in the Heart. *Circ. Res.* **2017**, *121*, 181–195. [CrossRef] [PubMed]
172. Simoes, I.C.M.; Morciano, G.; Lebiedzinska-Arciszewska, M.; Aguiari, G.; Pinton, P.; Potes, Y.; Wieckowski, M.R. The Mystery of Mitochondria-ER Contact Sites in Physiology and Pathology: A Cancer Perspective. *Biochim. Biophys. Acta Mol. Basis Dis.* **2020**, *1866*, 165834. [CrossRef]
173. Santulli, G.; Monaco, G.; Parra, V.; Morciano, G. Editorial: Mitochondrial Remodeling and Dynamic Inter-Organellar Contacts in Cardiovascular Physiopathology. *Front. Cell Dev. Biol.* **2021**, *9*, 679725. [CrossRef]
174. Robert, V.; Gurlini, P.; Tosello, V.; Nagai, T.; Miyawaki, A.; Di Lisa, F.; Pozzan, T. Beat-to-Beat Oscillations of Mitochondrial [Ca2+] in Cardiac Cells. *EMBO J.* **2001**, *20*, 4998–5007. [CrossRef]
175. Bravo-Sagua, R.; Parra, V.; López-Crisosto, C.; Díaz, P.; Quest, A.F.G.; Lavandero, S. Calcium Transport and Signaling in Mitochondria. *Compr. Physiol.* **2017**, *7*, 623–634. [CrossRef] [PubMed]
176. Bround, M.J.; Wambolt, R.; Luciani, D.S.; Kulpa, J.E.; Rodrigues, B.; Brownsey, R.W.; Allard, M.F.; Johnson, J.D. Cardiomyocyte ATP Production, Metabolic Flexibility, and Survival Require Calcium Flux through Cardiac Ryanodine Receptors in Vivo. *J. Biol. Chem.* **2013**, *288*, 18975–18986. [CrossRef] [PubMed]
177. Santulli, G.; Xie, W.; Reiken, S.R.; Marks, A.R. Mitochondrial Calcium Overload Is a Key Determinant in Heart Failure. *Proc. Natl. Acad. Sci. USA* **2015**, *112*, 11389–11394. [CrossRef]
178. Gutiérrez, T.; Parra, V.; Troncoso, R.; Pennanen, C.; Contreras-Ferrat, A.; Vasquez-Trincado, C.; Morales, P.E.; Lopez-Crisosto, C.; Sotomayor-Flores, C.; Chiong, M.; et al. Alteration in Mitochondrial Ca(2+) Uptake Disrupts Insulin Signaling in Hypertrophic Cardiomyocytes. *Cell Commun. Signal.* **2014**, *12*, 68. [CrossRef]
179. Fernandez-Sanz, C.; Ruiz-Meana, M.; Miro-Casas, E.; Nuñez, E.; Castellano, J.; Loureiro, M.; Barba, I.; Poncelas, M.; Rodriguez-Sinovas, A.; Vázquez, J.; et al. Defective Sarcoplasmic Reticulum-Mitochondria Calcium Exchange in Aged Mouse Myocardium. *Cell Death Dis.* **2014**, *5*, e1573. [CrossRef]
180. Kubo, H.; Margulies, K.B.; Piacentino, V.; Gaughan, J.P.; Houser, S.R. Patients with End-Stage Congestive Heart Failure Treated with Beta-Adrenergic Receptor Antagonists Have Improved Ventricular Myocyte Calcium Regulatory Protein Abundance. *Circulation* **2001**, *104*, 1012–1018. [CrossRef] [PubMed]
181. Marx, S.O.; Reiken, S.; Hisamatsu, Y.; Jayaraman, T.; Burkhoff, D.; Rosemblit, N.; Marks, A.R. PKA Phosphorylation Dissociates FKBP12.6 from the Calcium Release Channel (Ryanodine Receptor): Defective Regulation in Failing Hearts. *Cell* **2000**, *101*, 365–376. [CrossRef]
182. Baker, D.L.; Hashimoto, K.; Grupp, I.L.; Ji, Y.; Reed, T.; Loukianov, E.; Grupp, G.; Bhagwhat, A.; Hoit, B.; Walsh, R.; et al. Targeted Overexpression of the Sarcoplasmic Reticulum Ca2+-ATPase Increases Cardiac Contractility in Transgenic Mouse Hearts. *Circ. Res.* **1998**, *83*, 1205–1214. [CrossRef] [PubMed]
183. Samuel, T.J.; Rosenberry, R.P.; Lee, S.; Pan, Z. Correcting Calcium Dysregulation in Chronic Heart Failure Using SERCA2a Gene Therapy. *Int. J. Mol. Sci.* **2018**, *19*, 1086. [CrossRef]
184. Ferrandi, M.; Barassi, P.; Tadini-Buoninsegni, F.; Bartolommei, G.; Molinari, I.; Tripodi, M.G.; Reina, C.; Moncelli, M.R.; Bianchi, G.; Ferrari, P. Istaroxime Stimulates SERCA2a and Accelerates Calcium Cycling in Heart Failure by Relieving Phospholamban Inhibition. *Br. J. Pharmacol.* **2013**, *169*, 1849–1861. [CrossRef] [PubMed]
185. Schröder, F.; Handrock, R.; Beuckelmann, D.J.; Hirt, S.; Hullin, R.; Priebe, L.; Schwinger, R.H.; Weil, J.; Herzig, S. Increased Availability and Open Probability of Single L-Type Calcium Channels from Failing Compared with Nonfailing Human Ventricle. *Circulation* **1998**, *98*, 969–976. [CrossRef] [PubMed]
186. Winslow, R.L.; Rice, J.; Jafri, S.; Marbán, E.; O'Rourke, B. Mechanisms of Altered Excitation-Contraction Coupling in Canine Tachycardia-Induced Heart Failure, II: Model Studies. *Circ. Res.* **1999**, *84*, 571–586. [CrossRef]
187. Pieske, B.; Maier, L.S.; Bers, D.M.; Hasenfuss, G. Ca2+ Handling and Sarcoplasmic Reticulum Ca2+ Content in Isolated Failing and Nonfailing Human Myocardium. *Circ. Res.* **1999**, *85*, 38–46. [CrossRef] [PubMed]
188. Sipido, K.R.; Volders, P.G.A.; Vos, M.A.; Verdonck, F. Altered Na/Ca Exchange Activity in Cardiac Hypertrophy and Heart Failure: A New Target for Therapy? *Cardiovasc. Res.* **2002**, *53*, 782–805. [CrossRef]
189. Piacentino, V.; Weber, C.R.; Chen, X.; Weisser-Thomas, J.; Margulies, K.B.; Bers, D.M.; Houser, S.R. Cellular Basis of Abnormal Calcium Transients of Failing Human Ventricular Myocytes. *Circ. Res.* **2003**, *92*, 651–658. [CrossRef]
190. Alegría-Torres, J.A.; Baccarelli, A.; Bollati, V. Epigenetics and Lifestyle. *Epigenomics* **2011**, *3*, 267–277. [CrossRef]
191. Peixoto, P.; Cartron, P.-F.; Serandour, A.A.; Hervouet, E. From 1957 to Nowadays: A Brief History of Epigenetics. *Int. J. Mol. Sci.* **2020**, *21*, 7571. [CrossRef]
192. Handy, D.E.; Castro, R.; Loscalzo, J. Epigenetic Modifications: Basic Mechanisms and Role in Cardiovascular Disease. *Circulation* **2011**, *123*, 2145–2156. [CrossRef]
193. Antos, C.L.; McKinsey, T.A.; Dreitz, M.; Hollingsworth, L.M.; Zhang, C.-L.; Schreiber, K.; Rindt, H.; Gorczynski, R.J.; Olson, E.N. Dose-Dependent Blockade to Cardiomyocyte Hypertrophy by Histone Deacetylase Inhibitors. *J. Biol. Chem.* **2003**, *278*, 28930–28937. [CrossRef]

194. Zhang, C.L.; McKinsey, T.A.; Chang, S.; Antos, C.L.; Hill, J.A.; Olson, E.N. Class II Histone Deacetylases Act as Signal-Responsive Repressors of Cardiac Hypertrophy. *Cell* **2002**, *110*, 479–488. [CrossRef]
195. Yanazume, T.; Hasegawa, K.; Morimoto, T.; Kawamura, T.; Wada, H.; Matsumori, A.; Kawase, Y.; Hirai, M.; Kita, T. Cardiac P300 Is Involved in Myocyte Growth with Decompensated Heart Failure. *Mol. Cell Biol.* **2003**, *23*, 3593–3606. [CrossRef]
196. Kaneda, R.; Takada, S.; Yamashita, Y.; Choi, Y.L.; Nonaka-Sarukawa, M.; Soda, M.; Misawa, Y.; Isomura, T.; Shimada, K.; Mano, H. Genome-Wide Histone Methylation Profile for Heart Failure. *Genes Cells* **2009**, *14*, 69–77. [CrossRef] [PubMed]
197. Gilsbach, R.; Preissl, S.; Grüning, B.A.; Schnick, T.; Burger, L.; Benes, V.; Würch, A.; Bönisch, U.; Günther, S.; Backofen, R.; et al. Dynamic DNA Methylation Orchestrates Cardiomyocyte Development, Maturation and Disease. *Nat. Commun.* **2014**, *5*, 5288. [CrossRef] [PubMed]
198. Xiao, D.; Dasgupta, C.; Chen, M.; Zhang, K.; Buchholz, J.; Xu, Z.; Zhang, L. Inhibition of DNA Methylation Reverses Norepinephrine-Induced Cardiac Hypertrophy in Rats. *Cardiovasc. Res.* **2014**, *101*, 373–382. [CrossRef] [PubMed]
199. Dorn, L.E.; Tual-Chalot, S.; Stellos, K.; Accornero, F. RNA Epigenetics and Cardiovascular Diseases. *J. Mol. Cell Cardiol.* **2019**, *129*, 272–280. [CrossRef] [PubMed]
200. Mohammed, S.A.; Ambrosini, S.; Lüscher, T.; Paneni, F.; Costantino, S. Epigenetic Control of Mitochondrial Function in the Vasculature. *Front. Cardiovasc. Med.* **2020**, *7*, 28. [CrossRef] [PubMed]
201. Takahashi, H.; McCaffery, J.M.; Irizarry, R.A.; Boeke, J.D. Nucleocytosolic Acetyl-Coenzyme a Synthetase Is Required for Histone Acetylation and Global Transcription. *Mol. Cell* **2006**, *23*, 207–217. [CrossRef]
202. Zhao, S.; Xu, W.; Jiang, W.; Yu, W.; Lin, Y.; Zhang, T.; Yao, J.; Zhou, L.; Zeng, Y.; Li, H.; et al. Regulation of Cellular Metabolism by Protein Lysine Acetylation. *Science* **2010**, *327*, 1000–1004. [CrossRef]
203. Matsuhashi, T.; Hishiki, T.; Zhou, H.; Ono, T.; Kaneda, R.; Iso, T.; Yamaguchi, A.; Endo, J.; Katsumata, Y.; Atsushi, A.; et al. Activation of Pyruvate Dehydrogenase by Dichloroacetate Has the Potential to Induce Epigenetic Remodeling in the Heart. *J. Mol. Cell Cardiol.* **2015**, *82*, 116–124. [CrossRef]
204. Chriett, S.; Dąbek, A.; Wojtala, M.; Vidal, H.; Balcerczyk, A.; Pirola, L. Prominent Action of Butyrate over β-Hydroxybutyrate as Histone Deacetylase Inhibitor, Transcriptional Modulator and Anti-Inflammatory Molecule. *Sci. Rep.* **2019**, *9*, 742. [CrossRef]
205. Cai, Y.; Yu, S.-S.; Chen, S.-R.; Pi, R.-B.; Gao, S.; Li, H.; Ye, J.-T.; Liu, P.-Q. Nmnat2 Protects Cardiomyocytes from Hypertrophy via Activation of SIRT6. *FEBS Lett.* **2012**, *586*, 866–874. [CrossRef] [PubMed]
206. Hsu, C.-P.; Oka, S.; Shao, D.; Hariharan, N.; Sadoshima, J. Nicotinamide Phosphoribosyltransferase Regulates Cell Survival through NAD+ Synthesis in Cardiac Myocytes. *Circ. Res.* **2009**, *105*, 481–491. [CrossRef] [PubMed]
207. Imai, S.; Armstrong, C.M.; Kaeberlein, M.; Guarente, L. Transcriptional Silencing and Longevity Protein Sir2 Is an NAD-Dependent Histone Deacetylase. *Nature* **2000**, *403*, 795–800. [CrossRef] [PubMed]
208. Tanno, M.; Kuno, A.; Horio, Y.; Miura, T. Emerging Beneficial Roles of Sirtuins in Heart Failure. *Basic Res. Cardiol.* **2012**, *107*, 273. [CrossRef]
209. Kietzmann, T.; Petry, A.; Shvetsova, A.; Gerhold, J.M.; Görlach, A. The Epigenetic Landscape Related to Reactive Oxygen Species Formation in the Cardiovascular System. *Br. J. Pharmacol.* **2017**, *174*, 1533–1554. [CrossRef]
210. Kim, S.Y.; Morales, C.R.; Gillette, T.G.; Hill, J.A. Epigenetic Regulation in Heart Failure. *Curr. Opin. Cardiol.* **2016**, *31*, 255–265. [CrossRef]

 life

Review

Interplay between Mitochondrial Protein Import and Respiratory Complexes Assembly in Neuronal Health and Degeneration

Hope I. Needs [1], Margherita Protasoni [2], Jeremy M. Henley [1,3], Julien Prudent [2], Ian Collinson [1,*] and Gonçalo C. Pereira [2,*]

1 School of Biochemistry, University of Bristol, Bristol BS8 1TD, UK; hope-isobel.needs@bristol.ac.uk (H.I.N.); j.m.henley@bristol.ac.uk (J.M.H.)
2 Medical Research Council-Mitochondrial Biology Unit, University of Cambridge, Cambridge CB2 0XY, UK; mp802@mrc-mbu.cam.ac.uk (M.P.); julien.prudent@mrc-mbu.cam.ac.uk (J.P.)
3 Centre for Neuroscience and Regenerative Medicine, Faculty of Science, University of Technology Sydney, Ultimo, NSW 2007, Australia
* Correspondence: ian.collinson@bristol.ac.uk (I.C.); g.pereira@mrc-mbu.cam.ac.uk (G.C.P.)

Citation: Needs, H.I.; Protasoni, M.; Henley, J.M.; Prudent, J.; Collinson, I.; Pereira, G.C. Interplay between Mitochondrial Protein Import and Respiratory Complexes Assembly in Neuronal Health and Degeneration. *Life* **2021**, *11*, 432. https://doi.org/10.3390/life11050432

Academic Editors: Giorgio Lenaz and Salvatore Nesci

Received: 25 March 2021
Accepted: 2 May 2021
Published: 11 May 2021

Abstract: The fact that >99% of mitochondrial proteins are encoded by the nuclear genome and synthesised in the cytosol renders the process of mitochondrial protein import fundamental for normal organelle physiology. In addition to this, the nuclear genome comprises most of the proteins required for respiratory complex assembly and function. This means that without fully functional protein import, mitochondrial respiration will be defective, and the major cellular ATP source depleted. When mitochondrial protein import is impaired, a number of stress response pathways are activated in order to overcome the dysfunction and restore mitochondrial and cellular proteostasis. However, prolonged impaired mitochondrial protein import and subsequent defective respiratory chain function contributes to a number of diseases including primary mitochondrial diseases and neurodegeneration. This review focuses on how the processes of mitochondrial protein translocation and respiratory complex assembly and function are interlinked, how they are regulated, and their importance in health and disease.

Keywords: protein import; mitochondrial dysfunction; respiratory complex assembly; supercomplexes; neurodegeneration; mitochondrial proteostasis

1. Introduction

Mitochondria provide the main source of cellular energy in the form of ATP. This is particularly important in high energy consuming cells such as cardiac and muscle cells as well as in neurons. On top of this vital role in ATP synthesis, mitochondria have a plethora of other roles, including regulation of cellular metabolism, calcium storage and signalling, reactive oxygen species (ROS) signalling, damage-associated molecular patterns (DAMPs) production in inflammation and immunity, and programmed cell death [1]. The presence of key enzymes and proteins in different submitochondrial compartments is indispensable for these roles. Consequently, protein translocation becomes a fundamental process for efficient mitochondrial physiology. Due to their diverse proteome, mitochondria have distinct import pathways, which must be fully operational to maintain a healthy organelle [2]. The first section of this review will cover how cytoplasmic translated proteins are imported into mitochondria, as well as how mitochondrial-encoded proteins are translocated from the matrix to the inner mitochondrial membrane (IMM). We then explore the special case of respiratory complexes, which are multimeric proteins assembled from subunits encoded by both the nuclear and mitochondrial genomes to highlight the importance of the import machineries.

Due to the fundamental importance of mitochondrial homeostasis for the regulation of multiple central processes and pathways, it is not surprising that mitochondrial defects, and more specifically mitochondrial import defects, have been implicated in several diseases [3]. These include most neurodegenerative diseases [4], as well as mitochondrial diseases associated with deficiencies of the respiratory complexes due to mutations affecting the import machineries [5]. These will be discussed later, followed by a discussion of the recent advances therein and with respect to the most common mitochondrial stress response and proteostatic pathways thought to counteract import dysfunction.

2. Protein Translocation

All but 13 of the estimated >1000 human mitochondrial proteins [6] required to perform key mitochondrial functions are encoded by the nucleus and synthesised on cytoplasmic ribosomes and thus must be imported into mitochondria through highly conserved protein translocation pathways (Figure 1). Owing to the double membrane bound structure of mitochondria, these multistep protein translocation pathways involve numerous protein complexes (Figure 1 and Table 1). Moreover, their proteome consists of soluble, membrane-bound, and transmembrane proteins with different mitochondrial sub-localisations. Therefore, specialised import machinery has evolved to successfully import all classes of proteins.

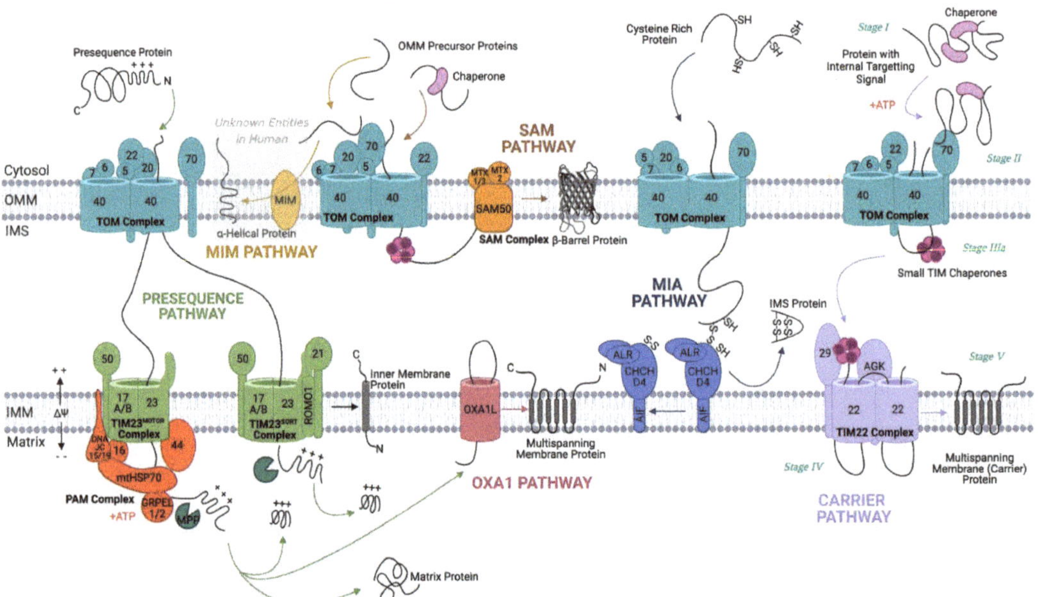

Figure 1. Overview of human mitochondrial protein import pathways. The TOM complex acts as the central entry gate for precursor proteins to enter the IMS, where they are diverted into one of five pathways, depending on their structure, function, and target destination. The MIM pathway (only currently understood in yeast) is an exception in that proteins usually do not cross the Tom40 channel. Instead, OMM α-helical proteins are recognised by Tom70 and transferred through MIM to be inserted into the OMM. The five major pathways proteins take after crossing the TOM channel are the following. The presequence pathway: Presequences containing precursor proteins are transported via the presequence pathway. Of these proteins, proteins with a hydrophobic sorting sequence are inserted into the IMM by the TIM23^{SORT} complex, whereas hydrophilic matrix proteins are pulled through the TIM23^{MOTOR} complex, with the help of the PAM complex and

ATP hydrolysis cycles. The presequences of both these groups of proteins are cleaved by MPP on the matrix side. The OXA1 pathway: N-terminally inserted multispanning membrane proteins, once passed through TIM23^{MOTOR} and cleaved by MPP, are passed to OXA1L, which inserts them into the IMM in the N-terminal formation. OXA1L is also responsible for the insertion of mtDNA encoded proteins into the IMM. The SAM pathway: β-barrel proteins are transported to the TOM complex by cytoplasmic chaperones. They are then passed through the TOM complex and received by small TIM chaperones on the other side for insertion into the OMM by the SAM complex. The MIA pathway: Cysteine-rich proteins in an unfolded, reduced state are passed via the TOM complex to the MIA complex, which inserts disulphide bonds in them, allowing them to reside in a folded, oxidised state in the IMS. Carrier pathway: Proteins with internal targeting signals are protected in the cytosol by cytosolic chaperones (Stage I), which pass them to the TOM complex (Stage II). They are received on the IMS side by small TIM chaperones (Stage III), which transfer them through the IMS to the TIM22 complex (Stage IV) for insertion into the IMM (Stage IV).

Table 1. Structure and function of subunits of the mitochondrial translocase complexes in humans and their yeast counterparts.

<table>
<tr><th>Pathway</th><th colspan="2">Complex</th><th>Subunit (Mammalian)</th><th>Yeast Homolog</th><th>Main Function</th><th>Topology</th></tr>
<tr><td rowspan="7">TOM</td><td rowspan="7">TOM-Holo Complex</td><td rowspan="5">Core complex</td><td>TOM40 and TOM40L</td><td>Tom40</td><td>Channel protein</td><td>β-barrel (19 β strands) and one N-terminal α-helical segment located inside pore</td></tr>
<tr><td>TOM22</td><td>Tom22</td><td>Receptor protein. Located at the dimer interface between TOM40 pores.</td><td>α-helical (single TMD); C_{in}-N_{out}</td></tr>
<tr><td>TOM5</td><td>Tom5</td><td>Complex assembly/stability</td><td>α-helical (single TMD); C_{in}-N_{out}</td></tr>
<tr><td>TOM6</td><td>Tom6</td><td>Complex assembly/stability</td><td>α-helical (single TMD); C_{in}-N_{out}</td></tr>
<tr><td>TOM7</td><td>Tom7</td><td>Complex assembly/stability</td><td>α-helical (single TMD); C_{in}-N_{out}</td></tr>
<tr><td rowspan="2">Receptors</td><td>TOM70</td><td>Tom70</td><td>Receptor for carrier precursors</td><td>α-helical (single TMD); N-terminally inserted</td></tr>
<tr><td>TOM20</td><td>Tom20</td><td>Receptor for presequence precursors</td><td>α-helical (single TMD); N-terminally inserted</td></tr>
<tr><td rowspan="3">SAM</td><td colspan="2" rowspan="3">SAM Complex</td><td>SAM50</td><td>Sam50</td><td>Core subunit responsible for β-barrel protein insertion</td><td>β-barrel (16 β-strands)</td></tr>
<tr><td>MTX1 and MTX3</td><td>Sam37</td><td>Accessory subunit</td><td>N/A</td></tr>
<tr><td>MTX2</td><td>Sam35</td><td>Accessory subunit</td><td>N/A</td></tr>
<tr><td rowspan="2">MIM</td><td colspan="2" rowspan="2">MIM Complex</td><td>Unknown</td><td>Mim1</td><td>Biogenesis of α-helical OMM proteins</td><td>-</td></tr>
<tr><td>Unknown</td><td>Mim2</td><td>Biogenesis of α-helical OMM proteins</td><td>-</td></tr>
<tr><td rowspan="4">MIA</td><td colspan="2" rowspan="4">MIA Complex</td><td>CHCHD4</td><td>Mia40</td><td>Oxidoreductase</td><td>Helix-loop-helix attached to a flexible helical arm</td></tr>
<tr><td>ALR</td><td>Erv1</td><td>Reoxidises Mia40</td><td>α-helical (a1-5) bundle</td></tr>
<tr><td>Cytochrome C/ETC</td><td>Cytochrome C/ETC</td><td>Final electron acceptor</td><td>Class I of the c type cytochrome</td></tr>
<tr><td>AIF</td><td>-</td><td>Anchors CHCHD4 to the IMM</td><td>One C-terminal TMD; N_{in}, C_{out}</td></tr>
</table>

Table 1. *Cont.*

Pathway	Complex		Subunit (Mammalian)	Yeast Homolog	Main Function	Topology
TIM23/ Presequence	TIM23^{SORT} Complex		TIM21	Tim21	Recognition/direction of precursor proteins to TIM23	α-helical (single TMD) with a large IMS domain; N_{in}-C_{out}
			ROMO1	Mgr2	Lateral release of proteins into the IMM	Two α-helical TMDs, joined by a basic loop
		TIM23^{MOTOR} Complex	TIM17A/B	Tim17	Channel forming	4 TMDs and a small IMS domain
			TIM23	Tim23	Channel forming	Multiple TMDs, and IMS exposed hydrophilic domain
			TIM50	Tim50	Receptor Protein	Single TMD, large IMS exposed C-terminal domain
	PAM Complex		TIM44	Tim44	Scaffold for complex & binding emerging precursor	Peripheral membrane protein on matrix side
			mtHSP70 (Mortalin)	SSC1 (mtHsp70)	ATPase	β-sheet and α-helical domains
			DNAJC15 and DNAJC19	Pam18 (Tim14)	Stimulates ATPase activity of mHsp70	Single α-helical TMD, with large C-terminal matrix domain and small N-terminal IMS domain
			TIM16	Pam16 (Tim16)	Inhibits Pam18 stimulatory effect on ATPase activity of mHsp70	Three α-helices forming an antiparallel hairpin
			GrpEL1/2	Mgel	Regeneration of mtHSP70	Long N-terminal α-helical region, small helical bundle region, and a C-teminal β-sheet domain
			Unknown	Pam17	Binds precursor:chaperone complex in matrix	-
TIM22/Carrier	TIM22 Complex		TIM22	Tim22	Channel	4 TMs that form a curved surface; IMS-facing N-helix
			TIM29	-	Scaffold	Matrix-facing N-helix, single TM and an IMS domain
			AGK	-	Assembly and function	N-terminally inserted with an IMS α/β motif
			TIM9	Tim9	Chaperone	Donut-shaped hexamer structure
			TIM10A	Tim10	Chaperone	
			TIM10B	Tim12	Chaperone	
			-	Tim54	Holds chaperone ring in tilted conformation	N-terminally inserted with an IMS α/β motif
			-	Tim18	Docking platform for chaperones	3 TMs and an amphipathic helix on the IMS side
			-	Sdh3	Docking platform for chaperones	3 TMs and an amphipathic helix on the IMS side
OXA Pathway	OXA Complex		OXA1L	OXA1	Insertion of mtDNA encoded proteins and N-terminal insertion of nuclear encoded proteins into the IMM	5 TM helices and a large internal C-terminal domain; N_{out}-C_{in}

2.1. Crossing the Outer Membrane

All proteins destined to the mitochondria must first cross the outer mitochondrial membrane (OMM), which they gain access to via the translocase of the outer membrane (TOM) complex (Figure 1). The TOM core complex (TOM-CC) consists of five components: TOM40, TOM22, TOM7, TOM6, and TOM5. The TOM holo-complex is formed following weak association of the TOM-CC with an additional two subunits: TOM20 and TOM70 [2,7]. These subunits are highly conserved between humans and yeast (Table 1); however, we refer to the yeast translocases in the following section, since this was the first organism it was discovered in. Precursor proteins are recognised by the receptor proteins Tom20, which recognises proteins with a mitochondrial targeting sequence (MTS), i.e., presequence proteins [8,9], and Tom70, which specifically recognises precursors with internal targeting signals, such as those belonging to the solute carrier family (SLC25) [10,11]. Proteins are then passed to the Tom40 pore via another receptor component, Tom22, which has also been shown to assist in the assembly of the TOM complex [12–14]. Tom22 physically interacts with Tom40 via its transmembrane segment, whilst its cytosolic domain has been suggested to act as a docking site for the other receptor proteins, Tom70 and Tom20. Recently, the OMM porin metabolite channel (also known as the voltage-dependent anion channel) has been reported to regulate Tom22 integration into the TOM complex in yeast, thus regulating the assembly and stability of the TOM complex [15,16]. Por1, the major yeast isoform of Porin, binds newly imported Tom22 and integrates it into the TOM complex, promoting formation of the mature trimeric form of the TOM complex required for import of precursor proteins [15]. Por1 also sequesters dissociated Tom22, stabilising the dimeric TOM complex under situations where this is preferable, i.e., for the import of proteins destined for the mitochondrial intermembrane space (IMS) assembly (MIA) pathway [15]. Porin is also thought to cooperate with Tom6 in regulating trimeric TOM assembly and stability and thereby modulating protein import during the cell cycle [15,17].

The different oligomeric states of the TOM complex and the nature of these different states remains unclear. Whilst it had generally been accepted that the mature form of TOM complex exists as a trimer [18–21], a cryogenic electron microscopy (cryo-EM) study in *Neurospora crassa* showed the TOM complex in a dimeric form [22]. More recently, high resolution cryo-EM studies in *Saccharomyces cerevisiae* showed that the TOM-CC exists as dimers and tetramers. The latter is essentially a dimer of the dimeric form of TOM-CC, achieved by lateral stacking of the dimeric TOM complex [7]. Due to the dynamic properties of the TOM complex, it may be proposed that the trimeric complex is formed by dissociation of a monomer from the tetrameric form.

Of note, the only protein of the TOM complex with a significant IMS domain is Tom22, which is important for its role in directing emerging precursor proteins to the Tim50 receptor of the translocase of the inner membrane 23 (TIM23) complex for further translocation [23]. Structural analysis of the interactions between these differing structural subunits showed that association is mainly mediated by hydrophobic interactions, along with high surface complementarity between the transmembrane domains [7].

2.2. Biogenesis of OMM Proteins

Evidence has also shown that, in yeast, Tom40 may simultaneously act as an insertase, assisting in the lateral release and insertion of proteins destined for the OMM. However, this is highly dependent on specific determining factors within the precursor sequence and is not yet fully understood [24]. Although this initial observation was monitored using an artificial import substrate, it has since been suggested that a similar process might be responsible for the accumulation of high-molecular weight PINK1 in the OMM in a TOM7-dependent manner in human cells [25,26].

Recently, it has been proposed that in addition to their role in quality control, PINK1/Parkin also regulate protein import under physiological conditions where mitochondrial function remains normal [27]. It is proposed that 'local dysfunction', as in mitochondrial membrane potential ($\Delta\psi$) depolarisation or import efficiency, is sensed by the PINK1/Parkin pair, which phosphorylates several subunits of the TOM complex, namely Tom20, Tom70 and Tom22, facilitating the import of presequence precursors. Importantly, the ubiquitylation pattern under this condition is significantly different from the PINK1/Parkin activation experienced from global mitochondrial dysfunction. Conversely, the mitochondrial ubiquitylase USP30 antagonises these effects [27–29]. Additionally, USP30 was shown to work in a reciprocal manner to MARCH5, a E3 ubiquitin-protein ligase of the OMM, under basal conditions, for deubiquitinating presequence substrates during translocation, facilitating their import. For other regulatory mechanisms of protein import, please see [30].

2.2.1. Insertion of β-Barrel Proteins in the OMM

Precursors of β-barrel proteins destined to be inserted into the OMM are passed via small TIM chaperone proteins to the sorting and assembly machinery (SAM) complex, for insertion into the OMM [31,32]. The SAM pathway has been described in detail in another review [33]. The human SAM complex consists of accessory subunits MTX2 (yeast Sam35), MTX1, and MTX3 (yeast Sam37) and OMM associated β-barrel core subunit SAM50 (yeast Sam50; Table 1) [34]. In yeast, β-barrel precursor proteins are translocated through the TOM complex, where they are bound by small TIM chaperones and transferred through the IMS to the SAM complex (Figure 1). Substrate proteins are recognised by Sam35, which interacts with the β-signal located in the last strand of the substrate protein. This initiates insertion into Sam50, which is responsible for folding and inserting substrates into the OMM [32]. Sam37 is required for substrate release and has also recently been proposed to assist in the formation of a SAM-TOM supercomplex, mediated by physical interaction of Sam37 and Tom22 on the cytosolic side of the OMM [35]. This SAM–TOM interaction has been shown to be essential for coupling of the two OMM complexes and promoting efficient precursor transfer [35]. Though not a part of the core SAM complex, Mdm10 is thought to associate with the SAM complex and have an important role in Tom40 assembly into the TOM complex [36]. This pathway is very similar to that observed for β-barrel proteins of the outer membrane in bacteria, which are folded and inserted into the outer membrane by the bacterial assembly machinery (BAM) complex, the *E.Coli* homolog of SAM [34].

2.2.2. Incorporation of α-Helical Anchors in the OMM

Over 90% of integral OMM proteins contain α-helical membrane anchors, yet the import pathway undertaken by these proteins is still relatively poorly understood, particularly in humans [37]. In yeast, the majority of these proteins are recognised by the Tom70 receptor of the TOM complex and passed on to the insertase of the outer mitochondrial membrane (MIM) complex, which aids in their insertion into the OMM (Figure 1 and Table 1) [38,39]. Multiple copies of Mim1 arrange themselves in such a way that, when reconstituted into the lipid bilayer, a channel is formed, and along with a couple of copies of Mim2, this establishes the MIM complex [40,41].

There are, however, known exceptions to this rule, whereby these α-helical proteins are passed through the Tom40 channel into the IMS prior to insertion into the OMM, aided by the MIM complex [42,43]. Interestingly, one of these proteins, yeast Om45, has been shown to require the TOM, TIM23, and MIM complexes for insertion into the OMM, where it is anchored by its N-terminal signal sequence with the bulk of the protein exposed to the IMS [42]. The final topology of Om45 is thus opposite to the N_{in}-C_{out} topology typical of MIM pathway proteins. The other known exception, yeast Mcp3, is also directed via TOM and TIM23, but is then processed by the inner mitochondrial membrane protease (IMP) before being transferred via MIM and inserted into the OMM with a final topology of N_{out}-C_{out} [43]. Notably, whilst both proteins interact with components of the TIM23 complex and are dependent on $\Delta\psi$, they do not cross or interact with the IMM [42,43].

2.3. Co- and Post-Translational Translocation

Importantly, preproteins must be unfolded in the cytosol and subsequently stabilised, in an ATP dependent process, by molecular chaperones of the heat shock protein (hsp) families Hsp70 and Hsp90, to then be efficiently imported [44,45]. Conversely, the subsequent translocation of these unfolded preproteins through the TOM channel occurs independently of ATP and $\Delta\psi$, and instead relies on an indirect driving force. That is the increased affinity of the presequences for the *trans* over *cis* side of the TOM channel, allowing transport of the preproteins across the channel where the presequence is bound by TIM50 [46]. This transport is also thought to rely on the sequential binding of the presequence to acidic domains of receptor proteins in what is known as the 'acid chain' hypothesis [47].

Interestingly, whilst the majority of preproteins are synthesised in the cytosol and must be unfolded prior to insertion into the TOM complex, others are unable to be imported into mitochondria post-translationally, and instead must undergo co-translational translocation whereby cytosolic ribosomes associate with mitochondria [48]. For this subgroup of proteins, it is thought that signals within the 3′-untranslated region (UTR) and coding regions of their mRNAs mediate their targeting to the cytosolic side of the OMM [49–51], where cytosolic ribosomes have also been observed [52,53].

2.4. Staying in the Intermembrane Space—The Disulfide Relay System

Proteins destined for the IMS take the route of the MIA pathway (Figure 1 and Table 1), which has been reviewed in great detail previously [54]. This class of proteins lack an MTS, are generally small, and share a conserved coiled coil-helix1-coiled coil-helix 2 domain (CHCHD). These cysteine-rich proteins contain two pairs of cysteines separated by three or nine amino acid residues (Cx_3C or Cx_9C) in the helices [54]. The small TIM chaperones of the IMS, important for translocase of the inner mitochondrial membrane 22 (TIM22)-dependent translocation described below, and assembly factors of IMM proteins, such as the respiratory complexes (see below and Table S1), are some examples of MIA substrates. The substrates are also relatively unstable and prone to degradation prior to their reduction by the relay system [55]. These cysteine-rich proteins undergo oxidation driven import whereby, upon passing through the TOM complex in an unfolded, reduced state, they form transient disulphide bonds with Mia40 [56,57]. CHCHD4 is the human ortholog of yeast Mia40 and shows high conservation despite the smaller size (16 vs. 40 kDa, respectively), lack of MTS, and no transmembrane anchor domain [58]. Instead, the human CHCHD4 interacts with the apoptosis inducing factor (AIF) and its cofactor NADH for association with the IMM [58].

The second player in the MIA pathway is ALR (Erv1 in yeast), a FAD-linked sulfhydryl oxidase that enables new rounds of precursor import and oxidation by re-oxidising reduced CHCHD4 after it has carried out its role as an oxidoreductase, thus allowing the cycle to continue [59]. Similarly, reduced ALR can relay its electrons to cytochrome *c* and, afterwards, to CIV of the respiratory chain [60]. Therefore, despite not requiring ATP or $\Delta\psi$ to operate, the MIA pathway still depends on a functional electron transport chain (ETC) to successfully oxidise its substrates.

2.5. Crossing or Insertion in the Inner Membrane

Proteins that are destined elsewhere within the mitochondria, namely the matrix or its membrane, must subsequently pass through or into the IMM (Figure 1). This membrane crossing (or insertion) import event is facilitated by one of two translocase complexes, the translocase of the inner mitochondrial membrane 23 (TIM23) complex, or the TIM22 complex.

2.5.1. TIM23 Complex (Presequence Pathway)

Precursor proteins destined for the mitochondrial matrix, along with some IMM sorted proteins, containing an N-terminal presequence (i.e., MTS), are passed directly from the TOM complex to the TIM23 complex [2,30]. The MTS is a cleavable region of 15 to 50 amino acids that precedes the mature protein and which is rich in hydrophobic, hydroxylated, and basic residues, with an overrepresentation of arginine residues and a near absence of acidic residues, forming a positively charged, amphipathic α-helix [61]. Interestingly, it has recently been suggested that preproteins may also contain additional internal MTS-like signal sequences (iMTS), located in the mature region of the preprotein, which act similarly to presequences and mediate the binding of the preprotein to Tom70, increasing the efficiency of protein import via the presequence pathway [62].

The TIM23 complex is anchored to the IMM and exists as a hetero-oligomeric complex, composed of various subunits (Table 1). It consists of an integral membrane embedded core complex as well as an import motor [63]. The core complex contains three essential subunits: TIM17A/B, TIM23, and TIM50 (Tim17, Tim23, and Tim50 in yeast) [46,64–66]. Additionally, the membrane-embedded part has two non-essential subunits: TIM21 and ROMO1 (Tim21 and Mgr2 in yeast) [12,67]. The import motor, also known as the presequence translocase-associated motor (PAM) complex, drives translocation across the IMM, aided by ATP hydrolysis, and consists of TIM44, mtHSP70, DNAJC15/19, TIM16, and GRPEL1/2. In yeast, the homologs are Tim44, SSC1 (also known as mtHsp70), Tim16 (also known as Pam16), Tim14 (also known as Pam18), and Mge1, as well as Pam17, which is not known to have a human homolog [68–73].

In yeast, precursor proteins released from the TOM complex and destined for the presequence pathway are recognised by Tim50 and the IMS region of Tim23, which act as receptor proteins for the incoming precursors [63]. This is achieved by binding of the hydrophilic, IMS-exposed part of the Tim23 subunit and the IMS-extending part of the Tim50 subunit in the IMS [46,65,66,74]. The Tim23 pore acts as a voltage gated channel and is ~13 Å wide, thus wide enough for only one α-helix to pass through at a time [75,76]. The pore is formed by the hydrophobic, C-terminal membrane domain of Tim23, and Tim17, which has been shown in the yeast model to be important for formation of the twin-pore structure, since it is unable to form in Tim17-depleted mitochondria [77]. In the handover of proteins from the TOM complex to the TIM23 complex, Tim50 also interacts with various partner proteins, including Tom22 and Tom21, which are necessary for the correct recognition and direction of precursor proteins across the IMS to the TIM23 channel [74,78–80]. Notably, it has recently been shown that phosphorylation/dephosphorylation of mammalian TIM50 is required for regulation of import activity, that is, phosphorylation of TIM50 reduces mitochondrial import, whilst its dephosphorylation by human phosphatase PPTC7 enhances it [81]. TIM50/Tim50 is phosphorylated on its matrix-facing segment in both mouse and yeast (T33 and S103, respectively) [81], but the identity of the kinase(s) responsible for this effect is still unknown. Furthermore, various matrix proteins were found to have phosphorylation sites around their MTS, the dephosphorylation of which is also thought to be important for enhancing their import and processing within the matrix [81]. This study highlights the importance of further work to dissect the currently unclear mechanisms regulating translocation.

The crossing of precursor proteins across the import channel of the IMM is driven by a number of forces: the proton motive force, i.e., $\Delta\psi$ and ΔpH, the affinity of the presequence for the *cis* side over the *trans* side of the membrane, and ATP hydrolysis [63,82]. As mentioned above, the higher affinity of presequences towards Tim50 initiates the handover from TOM to TIM23 complex. Additionally, the positively charged MTS means that the $\Delta\psi$ across the IMM exerts an electrophoretic effect on the proteins, facilitating the threading through TIM23.

As soon as the precursor emerges from the channel, it immediately interacts with Tim44. Importantly, it was shown that the affinity of presequences is higher for Tim44 compared to Tim50 [78], strengthening the directionality of presequence movement across the IMM. Additionally, Tim44 is known to act as a scaffold and to recruit the PAM complex (Table 1) [83]. In this model, one arm of Tim44 is anchored to Tim23 while another arm is dynamic and interacts with mtHsp70, Tim16 and, indirectly, Tim14, controlling the active:inactive state of the motor [76]. A typical cycle would involve the recruitment of ATP-bound mtHsp70 followed by a loose binding to the emerging precursor. Then, Tim14 would stimulate the ATPase activity of mtHsp70, trapping the bound polypeptide and consequently releasing the chaperone from Tim44, allowing the sliding of the precursor:chaperone complex into the matrix. The binding of Mge1 to this complex in the matrix allows the release of ADP and subsequent binding of a new ATP molecule coupled with the release of bound precursor [84]. The presequences are cleaved off by mitochondrial processing peptidase (MPP), leading to protein folding and maturation [12].

Nonetheless, not all precursors that are passed to the TIM23 complex are destined for the matrix. In fact, TIM23 is also responsible for the sorting and lateral insertion of membrane proteins into the IMM. These proteins contain a stop transfer signal, a region adjacent to the presequence of ~20 amino acids, which is rich in hydrophobic residues flanked by charged resides, also known as a sorting signal sequence, which targets them for this pathway of insertion [85]. The assembled TIM23 complex responsible for protein insertion into the IMM differs from the motor associated TIM23 in that it contains TIM21 (Tim21) and ROMO1 (Mgr2) and lacks the PAM complex [12], since it does not require the motor activity, but is instead driven supposedly solely by $\Delta\psi$ [85,86]. For these reasons, the motor-associated TIM23 complex is known as TIM23^{MOTOR} complex, whilst the lateral release TIM23 complex is known as the TIM23^{SORT} complex. Tim21 is important in regulating the lateral release of IMM proteins [87,88]. Furthermore, Mgr2 is important in aiding the binding of Tim21 to the TIM23^{SORT} complex, as well as in the lateral release of proteins into the IMM [89]. The ability of Mgr2 to be crosslinked to precursors in transit suggests that it may make up part of the channel [67].

2.5.2. TIM22 Complex (Carrier Pathway)

In the previous section, we described how proteins resident in the IMM, containing a single transmembrane domain and a mitochondrial targeting sequence, use the TIM23 complex for insertion. However, some hydrophobic proteins destined for the IMM are synthesised without a presequence and comprise multiple transmembrane domains and consequently, require a different import pathway named TIM22 or carrier pathway [90–92]. The majority of these proteins belong to the solute carrier family, typically containing six α-helical domains with multiple internal targeting sequences within the mature protein [90,93]. However, the exact mechanism by which these internal targeting sequences target carrier proteins to the IMM remains to be fully elucidated. The carrier pathway is particularly important for mitochondrial protein translocation as a whole since some of its substrates include translocase subunits Tim17, Tim22, and Tim23 [94].

Recent cryo-EM studies have determined the structure of the human TIM22 complex at 3.7 Å from overexpression in HEK293T cells [95] and yeast TIM22 at 3.8 Å resolution from endogenous protein levels [96]. The obtained models revealed notable structural differences between the two. Human TIM22 is a complex of ~440 kDa, and the cryo-EM structure (approx. 100 Å height and 160 Å width) revealed six subunits: TIM22, TIM29, acylglycerol kinase (AGK), TIM9, TIM10A, and TIM10B (Table 1) [95]. This structure shows the complex mainly extending into the IMS, along with a transmembrane region consisting of four transmembranes of TIM22 and one transmembrane of TIM29 and AGK. TIM29 acts as a scaffold, holding both TIM9-TIM10A-TIM10B and AGK in proximity to the TIM22 channel. The human TIM22 structure showed the chaperone ring to be tilted at a 45° angle [95]. It is also thought that TIM29 links the TIM22 and TOM complexes, mediating transfer of the carrier protein, a link that has not yet been shown in yeast [97,98]. Recent studies have revealed that AGK, which is involved in lipid biosynthesis, is important for TIM22 assembly and function [99,100].

In yeast, the TIM22 complex is ~300 kDa and consists of seven subunits: Tim22, Tim18, Tim54, Sdh3, Tim9, Tim10, and Tim12 (Table 1) [96]. The yeast structure showed that the small TIM subunits (Tim9–Tim10–Tim12) sit on the membrane in a hexameric ring formation and are anchored to the rest of the TIM22 complex via a docking platform consisting of Tim18-Sdh3 and Tim22. Tim54 is also required to hold Tim9–Tim10–Tim12 in a tilted conformation, like in humans, at around 45°, allowing them to receive substrates and pass them to the Tim22 channel [96]. Interestingly, Sdh3 is also a component of respiratory Complex II [101]. However, there is no evidence to suggest that the human Sdh3 homolog SDHC associates with the TIM22 complex.

Overall, the TIM22 carrier import pathway can be divided into five distinct and consecutive stages (Figure 1) with different energy requirements, producing perceivable transport intermediates to be monitored in vitro [102]. The stages are described in yeast below but are thought to be very similar in humans. In Stage I, the recently translated precursor is found in a soluble chaperone-bound form (chaperones of the Hsp70/Hsp90 families) not associated with mitochondria.

Then, during Stage II, the precursor–chaperone complex is passed on to the Tom70 receptor in an energy-independent manner, driven solely by the affinity of the receptor towards the precursor and the tetratricopeptide repeats in the chaperone. The Tom70 molecules contain two binding sites, one for the precursor and one for the chaperones [103], and aid in the transfer of the protein to Tom22 for insertion into the Tom40 channel [104,105]. More recently, the biological significance of Tom70 has been challenged, and it is suggested that the receptor acts as a general interface between cytosolic chaperones and the mitochondrial import machinery, and not as a specific receptor for carrier precursors [106]. In this regard, Tom70 would play a key role in reducing precursor-induced proteostasis stress. Next, ATP binding to the cytosolic chaperone triggers the release of the precursor and progression through the Tom40 channel. Importantly, the precursor can be arrested in Stage II by ATP depletion [102]. Interestingly, it is thought that carrier proteins are inserted into the Tom40 channel with both termini remaining in the cytosol, in a loop-like formation [107].

During Stage III, the precursor emerges from the IMS-facing side of the Tom40 channel, binding the small TIM chaperones (Tim9–Tim10), which tend to exist as hetero-hexameric complexes, for handover to the TIM22 complex. However, experimental data where $\Delta\psi$ was dissipated showed the accumulation of two distinct populations, suggesting that the following stages, namely insertion, are $\Delta\psi$-dependent, and that Stage III is further divided in two sub-stages. Stage IIIa represents the precursor deeply inserted in the TOM complex and protected from exogenous proteases [102]. Stage IIIb represents a fully translocated precursor across the OMM, tethered to the TIM22-bound TIM chaperone complex (Tim9–Tim10–Tim12) via hydrophobic interactions [102]. Tim12 is bound to the TIM22 complex, and thus aids in passing chaperoned carrier proteins to the Tim22 channel via the Tim54 docking site. Recently, it has been shown in yeast that Porins can assist the translocation by recruiting and interacting with the TIM22 complex, forming contact sites between OMM/IMM, to spatially coordinate inner and outer membrane transport steps [108]. However, others have identified that these juxtapositions are maintained by the interaction of TIM22 with the mitochondrial contact site and cristae organising system (MICOS) complex in humans [109]. Conversely, MICOS is found in association with the TIM23 complex in yeast [109].

Interestingly, the last two stages of the translocation of carrier precursors show differential dependence on $\Delta\psi$, confirmed experimentally through the use of ionophores [92]. Stage IV, also known as docking, can occur in a partially depolarised membrane (-120 to -60 mV) whereby the precursor is in full association with the TIM22 complex and one of its loops is inserted in the Tim22 channel [92]. Despite the low $\Delta\psi$, the electrophoretic effect experienced by the positive charges on the matrix loops of the carrier precursor is apparently sufficient to drive its partial translocation into the complex. Finally, Stage V requires a fully energised membrane (>-120 mV) to successfully insert the carrier precursor into the IMM after lateral opening of TIM22 [92,102].

Recently, the canonical even-numbered paired transmembrane helices with N_{out}- and C_{out}-terminal rule for TIM22 substrates has been challenged [110]. In this report, authors observed that the yeast mitochondrial pyruvate carrier, which has an odd number of transmembrane segments and a matrix-facing N-terminus, was imported specifically via the TIM22 complex. Similarly, it has been recently reported that human sideroflexins, a class of IMM proteins that contain five transmembrane domains and that do not belong to the SLC25 family, are imported via TIM22 [111]. Therefore, we can assume that the TIM22 substrate spectrum is less intransigent and contains proteins with paired and non-paired transmembrane domains.

2.5.3. Oxa1

Despite the endosymbiotic character of mitochondria, the organelle lacks a SecY-like translocon and possesses instead an import machinery that more closely relates to the bacterial membrane insertase YidC [112]. The so called IMM protein oxidase assembly protein 1 (OXA1L, OXA1 in yeast; Figure 1 and Table 1) is highly conserved from bacteria to mammals and plants [113].

OXA1 is nuclear-encoded, translated in the cytosol, and imported into the mitochondria by the TOM/TIM23 pathway via its N-terminal MTS in an mtHsp70- and ATP-dependent manner [114]. Interestingly, recently imported OXA1 is first observed in the matrix and then uses endogenous OXA1 to successfully insert itself into the IMM [114]. Mature OXA1 (36 kDa) is known to form oligomers, although its behaviour is still controversial. For example, in *Neurospora crassa*, it exists as a homo-tetramer [115], while human OXA1L has an apparent mass of 600–700 kDa, suggesting a hetero-oligomeric complex of unknown identity [116].

Since the majority of mtDNA-encoded proteins are highly hydrophobic, it is predictable that OXA1L interacts with mito-ribosomes for a co-translation process, whereby nascent chains associate with the insertase to suppress possible aggregation of the polypeptide in the matrix. This interaction occurs via the long C-terminus of OXA1L/OXA1, in both humans and yeast [117]. Recently, a cryo-EM structure showed an association between human OXA1L and mitochondrial ribosomes in a native state, coupling protein synthesis and membrane delivery [118].

In addition to its role in the insertion of mtDNA-encoded proteins, OXA1 is also responsible for N-terminal insertion of some nuclear-encoded proteins [119]. In these cases, proteins with N-terminal MTS are not arrested during import via $TIM23^{SORT}$ but are fully imported into the matrix via $TIM23^{MOTOR}$ and thereafter sorted for export from the matrix via OXA1 after cleavage of the MTS [119]. Similarly, multispanning proteins such as the ABC transporter Mdl1 can cooperatively make use of the stop-transfer (TIM23) and conservative (OXA1) sorting for integration into the IMM [120]. In regard to yeast Mdl1, the insertion topology occurs as follows: transmembranes 1 and 2 are imported via stop-transfer; the subsequent transmembranes 3 and 4 are imported into the matrix in an mtHSP60/ATP-dependent manner, and exported into the IMM via OXA1; transmembranes 5 and 6 are OXA1-independent and probably use the stop-transfer mechanism. Interestingly, the middle two TM helices 2 and 3 (of Mdl1), dependent on Oxa1 for their insertion, are not particularly hydrophobic. This ties in well with the noted evolutionary conservation and striking structural similarity of the Oxa1/YidC family with EMC3 of the ER membrane complex [121–123]. Given their common mechanism for membrane protein insertion, it is perhaps significant that the EMC is also recruited for the incorporation of TM helices with reduced hydrophobicity [124]. Therefore, the possibility that OXA1 assists more widely in the insertion of less-hydrophobic TM helices, such as those possessed by transporters (like Mdl1), proton translocators and carriers, is worthy of further investigation.

In regard to energy dependence, OXA1 does not require ATP for protein insertion, similarly to TIM22; however, its dependence on $\Delta\psi$ is not as obvious. For example, export of the N-terminus of nuclear-encoded proteins requires an energised membrane [125], as is the case for the mtDNA-encoded Cox2 yeast protein [126], but not for yeast Cox1, Cox3, or cytochrome *b* [126]. Interestingly, this same correlation is observed in regard to negative charges, i.e., substrates with negatively charged N-terminus and/or IMS loops are $\Delta\psi$-dependent, while those with less negative or neutral character are not [127], suggesting that the content of charged residues in an IMM protein determines its dependence on the OXA1 translocase.

3. The Respiratory Chain and Supercomplexes

The IMM is extremely rich in protein content and accommodates among other classes a vital group of proteins known as the ETC. Under physiological conditions, the respiratory complexes forming the ETC can exist as individual entities and/or in association with one another to form high-order structures, known as supercomplexes (SC) [128]. Interestingly, it has been suggested that an important role of SC is to participate in the assembly and/or stability of single respiratory complexes [129–131]. In fact, defects in one complex can lead to multi-complex deficiencies [132–134]. Additionally, CI and CIII intermediates were found to bind CIII and CIV subunits before maturation of the respiratory complex [135].

There are numerous reported interactions between components of the ETC and the import machinery. For example, Tim21 and the two regulatory PAM subunits Pam16 and Pam18, all part of TIM23, were found to interact with $SCIII_2IV$ in yeast [86,136]. Moreover, human TIM21 was co-purified with CI assembly intermediates and identified as a CI interactor by complexome profiling [137]. Other import-related proteins have also been found to associate with respiratory complex subunits in yeast, such as mHsp70, which was found to interact with Mss51, an MTCO1 mRNA-specific translation activator [138], and also with CIV subunit Cox4 [139]. These interactions suggest a functional interdependence between the import machinery and the respiration complexes, which still needs to be clarified. Hypothetically, a direct interaction with the translocase system might favour a faster and more efficient regulation of the ETC complexes assembly, possibly in response to cellular signalling. Alternatively, the import machinery in the direct vicinity of proton-pumping respiratory complexes could benefit from the higher $\Delta\psi$ [140] required for protein import.

3.1. *Respiratory Complexes Assembly*

As mentioned earlier, the OXPHOS machinery is composed of both nuclear and mitochondrial-encoded subunits, requiring the synchronisation of a series of pathways and cellular machineries (Figure 2). Firstly, nuclear and mitochondrial gene expression must be coordinated. This process has been observed in yeast [141], but the underlying mechanisms are still poorly understood. It is believed that the translation of mtDNA-encoded mRNAs is regulated by a series of translational activators acting on the 5′-UTR, while other translational activators could interact with ribosomes or play a role in transcript stabilisation [142–145]. Moreover, feedback regulation mechanisms linking respiratory complex subunits' expression with the state of complexes assembly have been described for CIII [146,147], CIV [148–151], and CV [152,153].

Interestingly, the route of import can vary for different OXPHOS subunits as well as for assembly factors (Table S1). It has long been known that the mRNA encoding nuclear proteins targeted to mitochondria can form polysomes with several ribosomes and localise to the surface of the OMM where it is translated and imported, a phenomenon known as co-translation [154–156]. This mechanism, observed for example for the CV subunit *ATP2* in yeast [50], is thought to promote import and assembly efficiency and requires specific nucleotide signals in the mRNA 3′-UTRs in addition to the MTS [50,157]. However, other ETC subunits, such as CIV COX4 [158], are encoded in a different type of polysomes, known as 'free polysomes', which are not attached to the organelle membrane [51,159–162]. Moreover, detailed observations revealed that evolutionary ancient proteins are mainly synthesised at the mitochondrial surface, the core subunits (bacterial orthologs), or proteins involved in the synthesis of metal and heme co-factors, while eukaryotic-specific supernumerary subunits are more likely to be produced in free polysomes [163].

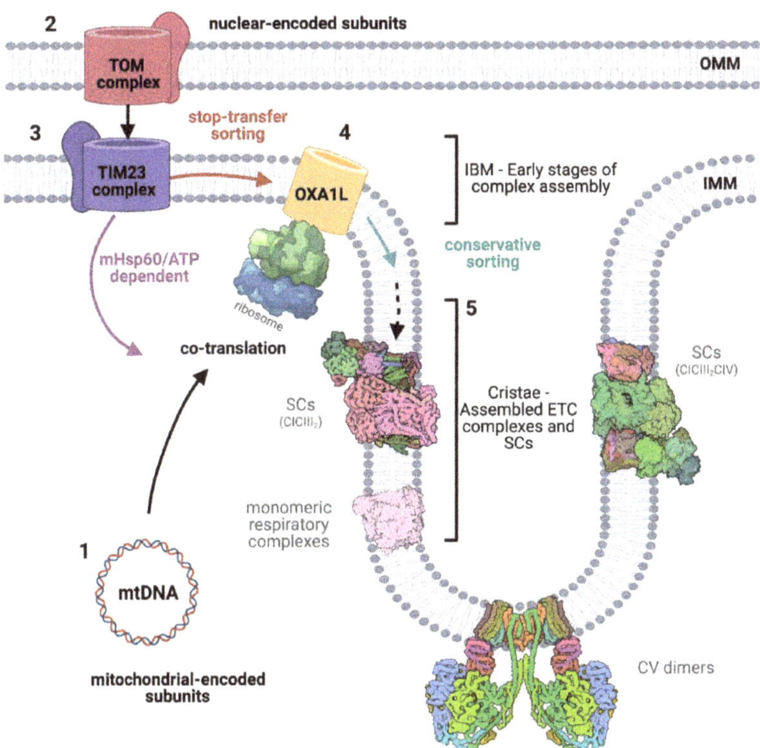

Figure 2. Spatial orchestration of mitochondrial respiratory complexes import and assembly and their organisation in the IMM. ETC complexes I, III, IV, and V are composed of both mitochondrial and nuclear-encoded subunits. Transcripts from the mitochondrial genome (**1**) are co-translated by mitochondrial ribosomes (here depicted as a simple arrow for clarity) and proteins inserted in the IMM via OXA1L. These newly synthesised proteins are then assembled together with the nuclear-encoded subunits, which are imported primarily through the TOM/TIM23 (**2**,**3**) complex. Additionally, proteins carrying a hydrophobic segment downstream of the MTS are arrested in the Tim23 channel and laterally inserted into the IMM through a stop-transfer sorting mechanism acquiring a N_{in}/C_{out} topology. Proteins with a N_{out}/C_{in} topology are instead fully imported and inserted into the IMM from the matrix side through a process known as conservative sorting, involving OXA1L (**4**). The import and insertion of these subunits in the IMM take place predominantly in IBM, a section of the IMM that runs parallel to the OMM. Then, the ETC subunits undergo a series of post-translational modifications and are incorporated in a nascent enzyme, often due to the interaction with assembly factors or chaperons. This process can occur in the monomeric enzymes and/or in the high-order SCs. Fully assembled enzymes and SCs are enriched in the cristae region of the IMM (**5**). Note: the size of monomeric respiratory complexes, supercomplexes, import machineries, and ribosomes are not to scale.

Quantitative immunogold electron microscopy studies in both isolated mammalian mitochondria and yeast cells showed that fully assembled Complexes I-IV and SC are found preferentially in cristae membranes [164,165]. However, it has been observed that complexes intermediates might localise in specific regions of the IMM during different maturation stages [166]. Therefore, respiratory complex assembly requires the temporal and spatial coordination of two independent protein synthesis machineries. In this regard, it was initially proposed that mitochondrial-encoded subunits translated in the matrix are inserted into the cristae membrane and that imported nuclear-encoded subunits are primarily inserted in the inner boundary membrane (IBM) for later incorporation in the nascent enzymes [165].

Using super-resolution microscopy and quantitative cryo-immunogold-EM a group of authors addressed this issue in yeast and concluded that although mature CIII, CIV, and CV localise mainly in the cristae, the early stages of assembly are enriched in the IBM [166]. Nonetheless, the complete assembly pathway of CV appears to develop specifically at IMM invaginations [166]. Mature CV is known to reside at the tip of these invaginations and to play a role in membrane curvature and cristae organisation [167,168].

Lastly, ETC subunits can undergo a series of post-translational modifications, in particular cleavages and insertion of prosthetic groups, such as heme groups, copper centres, and iron/sulphur (Fe/S) clusters, that are incorporated in the nascent enzymes, as extensively reviewed [169–171]. The insertion of these subunits occurs in a precise order and might require the involvement of assembly factors, as described more in detail in the following sections.

3.1.1. CI Assembly

Complex I (CI) is composed of 44 different subunits in mammals [172], organised in three structural domains: the *P-module*, inserted in the IMM, and the *N-* and *Q-modules*, protruding into the mitochondrial matrix (Figure 3a and Table S1). While the N- and Q-modules are formed exclusively by nuclear-encoded subunits, the P-module contains seven mitochondrial-encoded proteins (NDs; Figure 3a) [173]. CI assembly requires the formation of six independent modules, N, Q, ND1/P_P-a, ND2/P_P-b, ND4/$P_{D\text{-}a}$, and ND5/$P_{D\text{-}b}$, and the incorporation of each of them in a specific order [174]. The ND2 module is generated first [137] and is stabilised by its interaction with numerous assembly factors: ACAD9, ECSIT, TMEM126B, NDUFAF1, COA1, and the putative assembly factor TMEM186, which form the mitochondrial CI intermediate assembly (MCIA) [175,176]. Then, ND3, ND6, and ND4L are added to this intermediate. At least two of these assembly factors, namely TMEM126B and NDUFAF1, have been recently reported to be imported in a TIM22-dependent manner [111].

In parallel, a Q-module intermediate starts emerging through the binding of the assembly factor TIMMDC1 and the subunits ND1, NDUFA3, NDUFA8, and NDUFA13 (Figure 3a) [137]. In the intermediate phase of CI assembly, the ND4 module is formed, involving the assembly factors FOXRED1, ATP5SL, and TMEM70, followed by the ND5 module, the distal extremity of the membrane arm, which binds the assembly factor DMAC1/TMEM261 [177]. Finally, the N-module, composed of NDUFV1, NDUFV2, NDUFS1, and NDUFA2, is incorporated, generating the functional enzyme (Figure 3a).

It is worth noting that all mtDNA-enconded subunits, i.e., the NDs, are inserted in the IMM via OXA1. Still, most CI subunits are synthesised in the cytosol and then targeted to mitochondria by their N-terminal MTS, and thus preferentially use the TIM23 route [178]. Contrarily, NDUFS5, NDUFB7, NDUFB10, and NDUFA8 have been shown to be imported to the IMS using the MIA pathway [179,180]. The remaining 12 subunits (NDUFA5, NDUFS5, NDUFC2, NDUFB10, NDUFB6, NDUFB9, NDUFB3, NDUFA3, NDUFA8, NDUFA13, NDUFB1, NDUFB4) [177] and three CI assembly factors (TMEM126B, FOXRED1, and TIMMDC1) [181] do not contain a cleavable MTS and are imported into the organelle as a result of uncharacterised internal signals. Notwithstanding, it has been recently demonstrated that NDUFB10, TIMMDC1, and TMEM126B are imported via TIM22 [111], possibly suggesting a similar import route for this class of proteins that requires further investigation. This pathway is also used by NDUFA11, a supernumerary subunit of CI that is conspicuously located at the interface between CI and CIII in SCs. Interestingly, TIMMDC1 and NDUFA11 belong to the Tim17 family [182], providing another link between protein import and respiratory/SC assembly and function. A direct involvement can be observed in plants where in this case the NDUFA11 homolog B14.7 is directly associated with TIM23 complex [183].

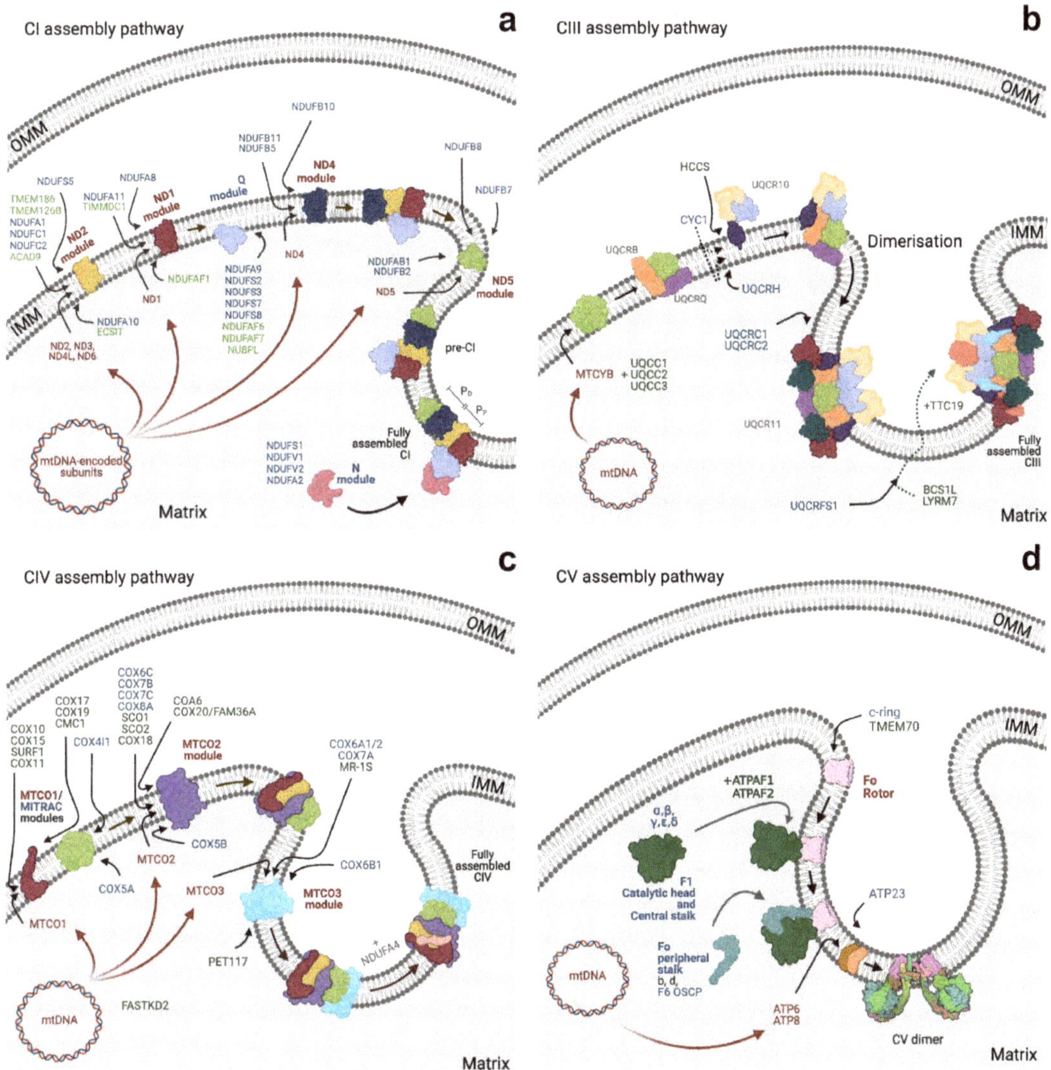

Figure 3. Assembly pathways for the different respiratory complexes. For the purpose of simplification, only a small portion of subunits and assembly factors are depicted in the figure, namely those for which the import route is known or expected (see Supplementary Table S1). For detailed view, please see [174]. Panel (**a**)-Complex I (CI) assembly pathway is modular and takes place in the IMM. After the synthesis, import, and maturation of both the mitochondrial- and the nuclear-encoded subunits, six subassemblies are independently formed: ND2, ND1, ND4, ND5, Q, and N-module. The preassembled modules associate with each other in a precise order. Initially, the central structure of the enzyme is formed, starting from the ND2-module and continuing with the subsequential addition of the ND1, Q, and then ND4 modules. Secondly, the extremities of the enzyme are incorporated, starting from the ND5-module, the last part of the membrane domain of CI, and finishing with the N-module, the FMN-containing intermediate that binds NADH and completes the assembly of the functional enzyme. mtDNA-encoded subunits (NDs) are indicated in red, while modules containing only nuclear-encoded subunits are indicated in blue. Panel (**b**)-Complex III (CIII) assembly starts with the maturation and insertion in the IMM of the single mitochondrial-encoded subunit, cytochrome *b* (MTCYB). The remaining nine subunits (in

blue or gray, for those with known and unknown import routes) are incorporated on top of this 'seed', following a precise order. For CYC1, the two debated import routes are shown as dashed line. Few assembly factors (in green) are known for CIII and are involved in the maturation of MTCYB and the Rieske protein (UQCRFS1). Different assembly factors are also shown in green. Mature CIII dimerisation is required for full activity and competence. Panel (**c**)-Complex IV (CIV) assembly is modular and is initiated by the parallel formation of the MTCO1 and of the COX4/COX5A modules. The MTCO1 module associates with a variety of assembly factors including Tim21, forming the MITRAC complex. One of the last subunits to be added is NDUFA4, which was initially misattributed to Complex I. Structural subunits are shown in red/blue and assembly factors in green colour. Panel (**d**)-Complex V (CV) is comprised of three modules: F1, Fo, and the peripheral stalk. The mtDNA subunits ATP6 and ATP8 (in red) together with other nuclear-encoded subunits (in blue), including the c-ring, form the Fo domain inserted in the IMM. The F1 domain is the matrix-facing part of the enzyme. The peripheral stalk is important for the stability of the complex and also contains key subunits required for the dimerisation of mature CV.

Once imported, several core subunits need further maturation and insertion of prosthetic groups, such as Fe/S clusters. To date, only one assembly factor is known to be involved in the incorporation of 4Fe/4S clusters in the peripheral arm: NUBPL, a member of the Mrp/NBP35 ATP-binding proteins family [184,185]. However, it is expected that several other unidentified proteins play a role in this process.

3.1.2. CII Assembly

CII subunits (SDHA-D) are all nuclear-encoded and imported into mitochondria post-translationally. The hydrophilic membrane domain (SDHC and SDHD) contains a heme *b* group and two ubiquinone binding sites [186] and forms an intermediate subcomplex [187]. In contrast, the mature forms of SDHA and SDHB are produced and inserted independently.

A flavine adenine dinucleotide (FAD) cofactor is inserted in SDHA via the interaction with the assembly factor SDHAF2/Sdh5 [188], while SDHAF1, assisted by SDHAF3, promotes the insertion of SDHB Fe/S clusters ([2Fe-2S], [4Fe-4S], and [3Fe-4S]) [189–192].

Interestingly, the yeast ortholog of SDHC, Sdh3, was found to form a subcomplex with Tim18 and participate in the biogenesis and assembly of the TIM22 complex [193]. However, Tim18 arose from duplication of the Sdh3 gene and does not have an ortholog in mammals. Similarly, the mammalian SDHC subunit has never been detected interacting with the TIM22 complex, suggestive of divergent mechanisms for the formation of the translocase between the two organisms [98], as discussed in the previous section.

3.1.3. CIII Assembly

Complex III (CIII) is composed of 10 subunits in both yeast and mammals. All CIII subunits are encoded by nuclear DNA except cytochrome *b* [194,195].

CIII assembly (Figure 3b) begins with the synthesis and insertion of the mtDNA-encoded subunit cytochrome *b* in the IMM, in both yeast and in mammals. The insertion of the yeast subunit has been shown to occur via Oxa1 [126], while in mammals, the depletion of OXA1L only marginally affects the biogenesis and function of the enzyme, suggesting a possible alternative route or compensatory mechanisms [196]. In yeast, this process is highly coordinated with the synthesis of nuclear-encoded proteins as a result of translational activators that regulate the expression of mitochondrial genes and their own expression in relation to mitochondrial respiration [197]; however, the same mechanism has not been observed in mammals yet. Four translational activators are known to be involved in the stability and translation of cytochrome *b* mRNA: Cbp1, Cbs1, Cbs2, and the complex Cbp3/Cbp6 [198]. However, only three factors are known in mammals, the ubiquinol–cytochrome *c* reductase complex assembly factors 1 and 2 (UQCC1 and UQCC2) [199], orthologs of Cbp3/Cbp6, and UQCC3, ortholog of Cbp4 (Table S1) [200].

In both yeast and mammals, the second step of CIII maturation involves the insertion of the subunits Qcr7 and Qcr8 (UQCRB and UQCRQ in mammals) [201], whilst the following proceedings differ between the two model organisms. In yeast, for example, an independent subassembly module containing the two large structural core subunits, Cor1 and Cor2, and the catalytic subunit cytochrome *c1* is formed [202]. This intermediate is then incorporated into the nascent enzyme together with the Qcr6 subunit [202–204]. In humans, however, CYC1 forms sub-assemblies with UQCR10 and potentially UQCRH (Qcr6 in yeast), without interacting directly with the core proteins UQCRC1 and UQCRC2 [131]. Moreover, this intermediate can be found in association with CIV subunits [131], suggesting that CIII might use CIV or modules of CIV as a structural scaffold during biogenesis. An alternative hypothesis is that CIII intermediates sequester CIV subunits when SC formation is impaired [131]. Importantly, it has been shown that dimerisation of two nascent CIII occurs during this stage [205].

During the intermediate assembly process, another player is added to the CIII complex. Mature Cyt1 contains a single heme centre and is anchored to the IMM via a single transmembrane segment near its C-terminus with its mature N-terminus soluble in the cytosol [206]. The full mechanism for the insertion and maturation of this atypical topology of an MTS-containing protein is still under debate. Nonetheless, both models share the initial steps with the Cyt1 precursor being translated in the cytosol and imported via TOM/TIM23 complexes into the mitochondria, where its N-terminal bipartite presequence is cleaved into two independent processes [207]. According to one proposed mechanism, the whole protein is fully imported into the mitochondrial matrix, and only after the first cleavage is performed, the hydrophobic sequence is re-located into the membrane, allowing the second proteolytic cleavage in the IMS [208]. However, earlier studies showed that depletion of matrix ATP has no impact on the import and maturation of Cyt1, suggesting that the precursor does not cross the IMM completely during translocation [209]. Therefore, the second model suggests that although the positively charged portion of the MTS reaches the mitochondrial matrix, the internal hydrophobic signal halts import, allowing for the lateral release (stop-transfer) of Cyt1 from the TIM23 channel and its insertion in the membrane [198]. The positive-charged MTS is then cleaved by MPP in the matrix, while the C-terminal α-helix is inserted into the IMM and the heme group is subsequently added to the protein by the holocytochrome c_1 synthetase (Cyt2 or HCCS1 in mammals) [210]. This modification leads to a conformational change that exposes the second hydrophobic targeting sequence for cleavage by Imp2, leaving the N-terminal of the mature Cyt1 soluble in the IMS [211].

The last stages of CIII assembly involve the incorporation of Qcr9 (UQCR10 in mammals), Qcr10 (UQCR11), and Rip1 (UQCRFS1) [212,213]. Once more, minor differences in the import of the nuclear-encoded Rip1/UQCRFS1 subunit exist between yeast and mammals. In yeast, the subunit is imported into the matrix via the TOM/TIM23 route [214] and subsequently cleaved by matrix proteases, MPP and mitochondrial intermediate peptidase (MIP), for complete removal of the MTS [215]. Next, Rip1 is translocated back across the IMM to the IMS, where it is incorporated in the complex. In this regard, Bcs1 is thought to be involved in the export of the Rieske Fe/S domain from the matrix into the IMS [216], since it is able to recognise the correctly folded Rieske protein and act as a protein translocase. In fact, cryo-EM structures of Bcs1 in yeast [217] and mouse [218] suggest the formation of an airlock-like mechanism for Rip1/UQCRFS1 translocation. Conversely, Mzm1 (LYRM7 or MZM1L in humans) stabilises Rip1 in the matrix before translocation to the IMS [219,220].

In contrast to yeast, the mammalian UQCRFS1 N-terminal import signal is not cleaved during import but rather after successful incorporation of the subunit, and in a single cleavage step. Importantly, the cleaved segment remains attached to the enzyme as an extra subunit [221]. Then, mammal-specific assembly factor TTC19 binds to fully assembled CIII for clearance of UQCRFS1 fragments, converting it to a fully functional and competent respiratory complex [222,223].

3.1.4. CIV Assembly

Mammalian CIV is composed of 14 subunits, 11 of which are nuclear-encoded and the remaining three are encoded by mtDNA (MTCO1, MTCO2, and MTCO3) [224]. Similar to CI, CIV assembly also occurs in a modular fashion (Figure 3c), and the first subassembly structure formed during CIV biogenesis contains two nuclear-encoded subunits, COX4I1 and COX5A [225], as well as an assembly factor, HIGD1A [226].

Next, the MTCO1 module, also known as 'MITRAC' (MItochondrial Translation Regulation Assembly intermediate of Cytochrome *c* oxidase) [227,228], is generated. This subassembly is composed of the mitochondrial-encoded subunit MTCO1 and a series of assembly factors necessary for its insertion into the IMM (including COX14/C12ORF62, COA3/CCDC56/MITRAC12, and OXA1 [126,227,229–231]) and for its maturation (including COX10, COX15, and SURF1 involved in the heme group biosynthesis and insertion [145,232,233] and COX11, COX17, and COX19 involved in the incorporation of the CuB group [234–236]). Interestingly, Tim21, a subunit of TIM23, was found to be associated with the MITRAC complex and seems to shuttle imported CIV subunits from the TIM23 translocase to the nascent enzyme [227]. Tim21 was also suggested to play a possible role in CI biogenesis [227].

Following the formation of MITRAC, the next assembly step requires the incorporation of the MTCO2 module (Figure 3c). The mtDNA-encoded subunit MTCO2 is inserted into the IMM via OXA1L together with the assembly factors COX18, COX20/FAM36A, and TMEM177, which are required for the export of MTCO2 C-terminal domain [237–239]. Afterwards, the copper-binding proteins COX17, SCO1, and SCO2 [240–242] together with COA6 [243,244] and COX16 help with the insertion of the Cu_A centre [245,246]. In the meantime, the nuclear-encoded subunits that form this module (COX5B, COX6C, COX7C, COX8A, and, most probably, COX7B) are incorporated. Finally, the MTCO3 module (MTCO3, COX6A1, COX6B1, COX7A2) is formed and added to the nascent enzyme [225], followed by NDUFA4, initially described as a CI subunit and later assigned to CIV [145].

3.1.5. CV Assembly

Complex V (CV), also known as ATP synthase, is organised in two domains: the entirely nuclear-encoded F_1 domain facing the mitochondrial matrix, and the F_o domain embedded in the IMM, containing both nuclear- and mtDNA-encoded subunits, namely ATP6 and ATP8 [247,248].

The formation of CV occurs through three independent sub-assembly steps (Figure 3d) [249]. First, the F1 subcomplex is formed through the interaction of chaperones ATPAF1/ATP11 and ATPAF2/ATP12 with the subunits ATP5B and ATP5A1, respectively [250]. The c-ring module is then assembled independently via mechanisms that remain unclear. Finally, the peripheral stalk is incorporated in two steps: first there is the inclusion of b/ATP5F1, d/ATPH, F6/ATP5J, and OSCP/ATP5O and then the addition of e/ATP5I, g/ATP5L, and f/ATPJ2 [251,252].

Interestingly, the mammalian c-subunit is encoded by three nuclear genes (*ATP5G1*, *ATP5G2*, and *ATP5G3*), which differ in their cleavable N-terminal targeting sequences but give rise to identical mature proteins [253,254]. The mechanisms behind the import and insertion of mammalian c-subunit(s) in the IMM are still unclear. However, studies in *Neurospora crassa* suggest that the MTS and the first transmembrane region could be initially translocated to the matrix via the TIM23 complex. Then, following the removal of the presequence, the transmembrane domain would be inserted into the membrane and the N-terminus exported to the IMS [125]. In bacteria, the insertase YidC, homolog of OXA1, facilitates the membrane insertion of the c-subunit [255]. The second transmembrane domain, instead, might be imported in a follow up step through a stop-transfer mechanism via the TIM23 complex as previously described for Cox2 in plants [256]. Interestingly, two assembly factors previously known for being involved in CI assembly, TMEM70 and TMEM242, were found acting as a scaffold for c-ring assembly [257,258].

4. Pathologies with Underlying Mitochondrial Import Defects

As mentioned earlier, TIM23 is responsible for the import of the vast majority of matrix and IMM proteins. Therefore, it is no surprise that TIM23 knockout in mice is embryonic lethal, even prior to implantation [259]. Similarly, TIM23 haploinsufficiency displays neurological defects and premature aging phenotype, further demonstrating the importance of protein import to maintain mitochondrial function and body health [259].

However, there are other occasions where the dysfunction might result from precursors clogging the channel or impaired ETC unable to provide driving force energy, rather than issues with the translocase *per se*. The cell has developed methods to detect and try to repair these problems, discussed in more detail in the next section. However, whenever this repair system fails or become overwhelmed, it creates cell and tissue stress, leading to general mitochondrial dysfunction, cytosolic toxicity, and disease. In regard to neurodegeneration, whilst mitochondrial dysfunction, amongst other effects, has long been recognised as a contributing factor in the pathogenic mechanisms of neurodegenerative diseases, the involvement of mitochondrial protein import, be it in a causative or consequential manner, is just beginning to emerge more recently. These defects have been summarised in Table 2.

Table 2. Summary of import defects associated with neurodegenerative diseases and their consequences on respiratory complexes.

Pathology	Import Defect(s)	Known Consequence(s)	Model Organism/System	Reference
Alzheimer's Disease	APP accumulation in Tom40 and Tim23 channels, with higher levels in AD susceptible brain regions.	Inhibition of import of CIV 4 and 5b, and subsequent reduction in CIV activity, leading to increased ROS.	Human AD brains.	[260]
	Chronic, sub-lethal Aβ exposure induces a significant reduction in mitochondrial protein import.	Reduction in Δψ, altered mitochondrial morphology, and increased ROS production.	PC12 cells.	[261]
	Tau accumulation in OMM and IMS, and interactions between N-terminal Tau fragment with OPA1 and Mfn1.	N/A	HEK293T cells, HeLa cells.	[262,263]
Parkinson's Disease	α-syn localises to and accumulates within mitochondria, mediated by a cryptic non-canonical MTS, in an ATP and Δψ dependent manner	N/A	Human dopaminergic neuronal cultures, PD brains.	[264]
	A53T version of α-syn is imported more efficiently than wildtype variant.	May account for faster development of cellular abnormalities seen in cells expressing the A53T version of α-syn compared to the wildtype.	Human dopaminergic neuronal cultures, PD brains, A53T mutant alpha-synuclein-inducible PC12 cell lines.	[265,266]
	Mitochondrial α-syn accumulates at IMM and interacts with CI.	Reduction in CI activity, increase in ROS production, inducing oxidative stress.	Human dopaminergic neuronal cultures, PD brains, rat *SN* neurons, human neuroblastoma cell line (SK-N-MC cells).	[266]
	S129 phosphorylated α-syn binds tightly to Tom20, inducing loss in Tom20-Tom22 interaction.	Impaired protein import, loss of Δψ, reduced respiratory capacity, and increased oxidative stress. Rescued by in vivo knockdown of endogenous α-syn, and by in vitro Tom20 overexpression.	SH-SY5Y cells and dopaminergic neurons from *SN* of post-mortem PD patient brains.	[267]

Table 2. *Cont.*

Pathology	Import Defect(s)	Known Consequence(s)	Model Organism/System	Reference
Parkinson's Disease	Tom40 downregulation, corresponding with α-syn accumulation in PD brains.	N/A	Midbrain of PD patients and α-syn transgenic mice.	[268]
	Excessively low levels of mitochondrial import in cells from *PINK1*- and *PARK2*-linked PD patients.	N/A Import defects reversed by phosphomimetic ubiquitin in cells with residual Parkin activity.	Cells from *PINK1*- and *PARK2*-linked PD patients.	[27]
Huntington's Disease	Disease variant Htt localises to mitochondria and directly interacts with the TIM23 complex.	Inhibited import and subsequent respiratory dysfunction, triggering cell death, rescued by TIM23 overexpression.	Isolated mitochondria from human HD brains, primary neurons expressing Htt variant, forebrain synaptosomal mitochondria in HD mice at early stages of HD.	[269]
	Dysfunctions in MIA pathway associated with mutant Htt: reduced levels and ratio of Erv1 and Mia40.	Reduced import of MIA pathway precursors, CIV assembly defects, deficient respiration, alterations in mtDNA, altered mitochondrial morphology.	Neuronal cell lines.	[270]
Amyotrophic Lateral Sclerosis	Variants of SOD1 accumulate in IMS, matrix, and OMM, and interact with OMM proteins.	Excessive ROS production, mitochondrial dysfunction, and toxic effects on the cells rescued by selective IMS targeting of wildtype SOD1.	Transgenic mouse models, spinal cord mitochondria.	[271–273]
	Increased levels of TOM subunits Tom20, Tom22, and Tom40. Overall reduction in import efficiency by 30%.	Changes in CI related protein expression levels.	Rat spinal cord of ALS-linked variant $SOD1^{G93A}$.	[274]
	Novel CHCHD10 mutant, *Q108P*, discovered in a patient with rapidly progressing ALS, almost completely abolishes its import.	Reduced mitochondrial respiratory capacity, an effect that is rescued by Mia40 overexpression.	HeLa cells and primary rat embryonic neurons transduced with genomic DNA from a young ALS patient.	[275]

4.1. *Mitochondrial Diseases*

Mitochondrial diseases are generally provoked by genetic mutations in complexes' subunits or assembly factors, as extensively discussed in numerous other reviews [276–280]. However, defects in the machinery involved in the import of these subunits have also been identified as the cause of mitochondrial pathologies, leading to different clinical features.

Mutations in *TIMM50*, which encodes a subunit of TIM23 complex, have been associated with severe lactic acidosis and seizures, linked to defects in import of ETC proteins, alterations in SC formation and a general respiratory deficiency [281–283]. Biochemical analysis of a patient with impaired TIM23 complex due to compound heterozygous mutations in *TIMM50* revealed reduced steady state levels of CI, II, and IV, but interestingly not CIII and V [282]. These results suggest the possibility of alternative import routes for certain ETC subunits or different interactions with the import machinery for subunits transported into the matrix or inserted into the IMM. Another study with patients with two homozygous missense mutations in the *TIMM50* gene, however, produced opposite results, showing normal activities of Complex I–IV and defective activity of CV [281]. A possible explanation for this phenotype is that, as shown in yeast, the import of subunits 9 and β of CV is highly dependent on TIM50, whilst import of CIII subunit CYC1 and CIV subunit COX5A import is only mildly affected by TIM50 defects [66]. Finally, another subject carrying compound heterozygous mutations within the IMS domain of *TIMM50* exhibited 3-ethylglutaconic aciduria, symptoms of Leigh syndrome, and dilated hyper-cardiomyopathy, associated with altered mitochondrial morphology [283]. Additionally, patients also experienced a general decrease in the levels of fully assembled complexes and their activity alongside a reduction in SC formation and a drop in the maximum respiratory capacity [283].

Interestingly, defects in other TIM23 complex subunits do not result in impaired mitochondrial respiration. For example, alterations in the formation of the DDP1/TIMM8a–TIMM13 complex found in a patient with deafness/dystonia and with a *de novo* mutation in DDP1(C66W) did not lead to defects in the activity of any of the respiratory complexes [284]. More recently, another reported case of neuromuscular presentation of mitochondrial disease was found to be associated with compound heterozygous mutations in *TIMM22* [285]. However, while cellular respiration was reduced in the patient cells, no evident defect in respiratory complexes or SC assembly was found. In a different report, where TIM22 assembly and activity was impaired by the removal of AGK, a mild CI assembly defect and respiration impairment was observed [111]. In contrast to the compound heterozygous variant, this observation suggests a direct involvement of the TIM22 complex in the import of CI subunits and/or factors required for CI assembly that requires further investigation.

Regarding defects at the TOM complex level, a patient with severe anaemia, lactic acidosis, and developmental delay were identified with compound heterozygous variants in the *TOMM70* gene [286]. Interestingly, this patient presented with respiratory complex deficiencies with a primarily marked defect in CIV, including decreased steady state levels of fully assembled enzyme, activity, and a reduction in CIV-containing SC, while SCI:$CIII_2$ species appeared unaffected.

Finally, defects in OXA1L, essential for the re-localisation of newly imported nuclear-encoded proteins in the matrix into the IMM, could have an impact on complexes' assembly and activity. One patient was identified with mutations in *OXA1L* and tissue-specific combined respiratory complex deficiencies, which led to severe encephalopathy, hypotonia, and developmental delay [196]. Interestingly, skeletal muscle biopsy from this patient showed defects primarily at the CIV and CV level with only milder defects in CI despite the fact that neuropathology experiments indicated an isolated CI deficiency in the central nervous system. Although the tissue-specificity observed in this patient remains unclear, it suggests a possible differential expression of OXA1L isoforms in different tissues or the presence of alternative insertases in human mitochondria.

4.2. Alzheimer's Disease

Alzheimer's disease (AD) is the most commonly occurring form of neurodegeneration, and growing evidence is linking it to mitochondrial dysfunction at all levels of AD neuropathology. AD is characterised by the death or loss of neurons in specific, susceptible areas of the brain, as well as by the presence of two pathological hallmarks: extracellular senile plaques and neurofibrillary tangles (NFTs) [287].

Senile plaques are deposits of accumulated amyloid-beta peptide (Aβ), a 40–42 amino acid peptide that is produced by specific, sequential proteolytic cleavages of amyloid precursor protein (APP). The biology of APP processing and its relevance in AD is reviewed in great detail in a previous review [287]. In a study carried out in mitochondria from human AD brains, APP has been found to accumulate in the TOM40 channel, forming a stable complex of ~480 kDa (Table 2) [260]. It also accumulates with both TOM40 and TIM23 to form a supercomplex of ~620 kDa [260]. Interestingly, mitochondrial APP levels varied both among patients, corresponding to the severity of AD, as well as across brain regions, with higher levels displayed in the regions of the brain that are more vulnerable to AD: the cortex, hippocampus, and amygdala [260]. Furthermore, the levels of APP accumulation in the mitochondria of AD brains directly correlates with mitochondrial dysfunction [260], suggesting that APP-mitochondrial translocase complex formation and aggregation may in fact be a causative factor in AD progression.

Furthermore, a study in PC12 cells showed that chronic, sub-lethal Aβ exposure induces a significant reduction in mitochondrial protein import, and that this, when sustained over long periods, leads to mitochondrial dysfunction highlighted by a reduction in $\Delta\psi$, altered mitochondrial morphology, and increased ROS production (Table 2) [261]. This consequential negative impact on mitochondrial function is likely due to the loss of important proteins that are usually imported via TOM40, such as proteins necessary for respiratory complex activity and assembly, as well as ROS scavenging proteins.

The second characteristic hallmark of AD, neurofibrillary tangles, insoluble aggregations made up primarily of hyperphosphorylated Tau protein, is a symptom of not only AD, but of all tauopathies. Cell line studies have shown that various forms of aggregation-prone Tau (wildtype, hyperphosphorylated, or caspase cleaved N-terminal fragment) are imported into mitochondria and localised to the IMS and OMM (Table 2) [262,263]. Whilst, to the best of our knowledge, no studies have specifically looked at the impact of Tau on mitochondrial protein import efficiency, the body of evidence highlighting Tau accumulation in mitochondria suggests this would be worth investigating.

4.3. Parkinson's Disease

Parkinson's disease (PD) is very closely associated with mitochondrial dysfunction, owing to consistent evidence suggesting reductions in CI activity in PD patient brains and other tissues [288,289], in addition to genetic links between familial PD and mitochondrial dysfunction [290]. These well characterised mitochondrial abnormalities in PD and potential therapeutic strategies to target them have been reviewed extensively previously [291].

Lewy bodies, which form in the *SN*, are the main pathological hallmark of PD and are made up mainly of aggregated alpha-synuclein (α-syn), an abundant presynaptic molecule [292,293]. Alpha-synuclein is a 140 amino acid molecule, which is thought to play a role in neuronal plasticity and synaptic function [292,294,295]. The aggregation of α-syn is highly neurotoxic, and studies of transgenic mice overexpressing α-syn have shown that its accumulation can lead to a PD-like phenotype, consisting of the formation of prominent intraneuronal inclusion bodies, loss of dopamine neuron terminals, and motor deficits [296]. Intriguingly, much evidence has suggested that neuronal injury caused by α-syn may be mediated by mitochondrial dysfunction and degeneration [264,266,297–301].

Multiple studies have shown that α-syn localises to, and accumulates within, mitochondria (Table 2) [264,266,300,301]. This is thought to be mediated by a cryptic, non-canonical MTS within the N-terminal 32 amino acids of α-syn [266]. The transport of α-syn into mitochondria does not occur in the presence of oligomycin, which inhibits ATP synthase and thus depletes mitochondrial ATP, or, carbonyl cyanide-*m*-lorophenylhydrazone (CCCP), which disrupts the mitochondrial $\Delta\psi$, highlighting that its import is dependent on both ATP and $\Delta\psi$, consistent with the import requirements for known mitochondrial proteins [266]. The A53T point mutation that occurs in rare familial PD cases is also imported into mitochondria, but with significantly higher efficiency than the wildtype protein [266], which may account for the faster development of cellular abnormalities seen in cells expressing the A53T version of α-syn compared to the wildtype [265].

It has been previously shown by electron microscopy that the majority of mitochondrial α-syn accumulates at the IMM and that it interacts with CI [266]. This causes a significant reduction in CI activity, as well as an increase in ROS production, inducing oxidative stress [266], which may account for some of the toxic effects on dopaminergic neurons. Importantly, α-syn lacking the N-terminal MTS failed to localise to mitochondria and did not exhibit any of the mitochondrial dysfunctions seen in the wildtype [266].

A study carried out in cell models of PD showed that in vitro treatment with rotenone leads to an increase in S129 phosphorylation of α-syn [267]. The resulting post-translationally modified α-syn species were observed to bind with high affinity to TOM20 molecules, leading to a loss of the critical interaction between TOM20 and TOM22 (Table 2) [267]. Consequently, mitochondria have impaired protein import and widespread mitochondrial dysfunction, displayed by a loss of $\Delta\psi$, reduced respiratory capacity, and increased oxidative stress in SH-SY5Y cells [267]. This α-syn/ TOM20 interaction and subsequent loss of import were also detected in the dopaminergic neurons from the *SN* of post-mortem brains of PD patients [267]. The authors highlighted mechanisms for rescuing this disorder, namely by in vivo knockdown of endogenous α-syn and by in vitro TOM20 overexpression, both of which preserve mitochondrial import and thus present potential therapeutic strategies for further investigation [267,302].

It has been shown that the core component of the TOM complex, TOM40, is downregulated in the midbrain of PD patients as well as in α-syn transgenic mice (Table 2) [268]. Importantly, levels of TOM20 remained the same, suggesting that this is a specific effect of TOM40, rather than a general reduction in mitochondrial proteins. Furthermore, this reduction in TOM40 levels corresponded with α-syn accumulation in PD brains, inferring a further functional link between α-syn aggregation and mitochondrial import dysfunction [268].

A recent study showed that, in addition to the key roles in mitochondrial quality control and biogenesis already established [303–308], Parkin, an E3 ubiquitin ligase, also plays a part in stimulating mitochondrial protein import, whilst stimulation of import is not achieved by disease-causing Parkin variants (Table 2) [27]. Furthermore, the results of this study showed that this effect relies on PINK1-mediated Parkin activation and results in ubiquitylation of TOM40 subunits, as well as an increase in K11 ubiquitin chains on mitochondria [27]. The importance of PINK1-Parkin regulation of mitochondrial import is highlighted by data showing excessively low levels of mitochondrial import in cells from PINK1- and PARK2-linked PD patients. This effect may be reversed by phosphomimetic ubiquitin in cells with residual Parkin activity, probably by bypassing the need for PINK1-dependent Parkin activation or by enhancing Parkin activity [27].

4.4. Huntington's Disease

Huntington's disease (HD) is an autosomal dominant neurological disorder characterised by neuronal loss in the striatal and cortical regions of the brain. The genetic cause of HD is an abnormal expansion of polyglutamine repeats (encoded by the CAG codon) in the huntingtin gene (HTT) [309].

N-terminal fragments of variant Huntingtin proteins, which form cytotoxic aggregates [310,311], have been shown to interact directly with mitochondria in cell and mouse models of HD (Table 2) [312,313]. Furthermore, a study showed that the variant Huntingtin localises to mitochondria from human HD brains isolated mitochondria, and that it directly interacts with the TIM23 complex, inhibiting import as a result (Table 2) [269]. These import defects were consistent in primary neurons expressing Huntingtin variant as well as in forebrain synaptosomal mitochondria in HD mice at early stages of the disease [269]. Notably, these import defects were not found in liver mitochondria from the same mice, suggesting that the import defects are specific to neurons [269]. Additionally, the inhibition of import preceded mitochondrial respiratory dysfunction and acted as a trigger for cell death, which was rescued upon augmentation of mitochondrial import by overexpression of TIM23 complex subunits, highlighting this pathway as a potential therapeutic strategy against HD [269].

Considering the early detection of impaired import in HD mice [269], it suggests that import defects precede the other mitochondrial insults described in HD models, namely decreased $\Delta\psi$ [314], reduced respiratory capacity and ATP levels [315,316], defective calcium buffering function [317], and altered mitochondrial morphology and number [318]. A plausible explanation is that inhibition of import would prevent key respiratory complex proteins from being imported and carrying out their functions, resulting in widespread mitochondrial damage.

Mutant Huntingtin has been linked to dysfunctions in the MIA pathway (Table 2) [270]. In neuronal cell lines, the expression of proteins of the MIA pathway were found to be significantly different to levels in control cells [270]. More specifically, ALR and CHCHD4 levels were reduced, and the ratio altered compared to control cells, whilst cytochrome *c* levels were increased, compared to the control group. Proteins that require the MIA pathway for import also displayed reduced expression levels, whilst CIV proteins not imported via this route, such as MTCO3, were unchanged, highlighting that this effect is specific to MIA substrates rather than a CIV effect [270]. In cells with a homozygous variant, however, levels of MTCO3 were also reduced [270], suggesting that there may be some CIV assembly defects. The observed effects on the MIA pathway were accompanied by deficient respiration, alterations in mtDNA, and changes in mitochondrial morphology [270]. These effects are consistent with what has been shown previously in both HD models and MIA deficient models [319–324].

4.5. Amyotrophic Lateral Sclerosis

Amyotrophic lateral sclerosis (ALS) is a rare motor neuron disease, strongly associated with mutations in *SOD1*, a ROS scavenging enzyme [325,326]. Characteristic features of mitochondrial dysfunction have been observed across ALS patients, and respiratory chain impairment has been highlighted as a common feature in the muscles of ALS patients, even prior to neuronal deficits being found [327–329]. This finding is consistent across both patient samples and experimental model systems and has highlighted mitochondrial dysfunction as a major pathological feature in ALS [330].

A small proportion of wildtype SOD1 is known to localise to the IMS under physiological conditions in both yeast and mammals [331,332]. Its antioxidant role in detoxifying ROS species produced by the ETC (mainly CI and CIII) is well established [332,333]. Disease associated variants of SOD1, however, have been shown to accumulate not only in the IMS but also within the matrix and the OMM, where it aggregates and interacts with OMM proteins (Table 2) [272,273]. This mislocalisation of SOD1 variants lead to excessive ROS production and subsequent mitochondrial dysfunction and toxic effects on the cells, which can be rescued by selective targeting of wildtype SOD1 to the IMS [271]. Evidence also shows alterations in activity of the respiratory complexes and in mitochondrial calcium buffering capacity associated with disease-causing SOD1 variants [317,334].

A proteomic screen of protein level changes in mitochondria from rat spinal cord of ALS-linked variant $SOD1^{G93A}$ showed vast changes in mitochondrial import and CI related proteins compared to $SOD1^{WT}$ mitochondria (Table 2) [274]. Levels of TOM subunits TOM20, TOM22, and TOM40 were increased in the affected mitochondria although, surprisingly, in vitro import assays highlighted a 30% reduction in protein import levels in these mitochondria compared to wildtype [274].

Furthermore, variants of mitochondrial IMS protein CHCHD10, which is crucial for cristae remodelling, have been linked to progression of ALS as well as frontotemporal dementia [275]. The native version of this protein is imported via the MIA pathway, where disulphide bonds are formed within the CHCHD of the protein [275]. A novel CHCHD10 variant, Q108P, discovered in a patient with rapidly progressing ALS, has been shown to almost completely abolish its import, resulting in reduced mitochondrial respiratory capacity, an effect that is rescued by overexpression of CHCHD4 (Table 2) [335]. Interestingly, the C9orf72 protein, which is often mutated in cases of ALS and frontotemporal dementia, has recently been shown to be an IMM protein vital for the assembly and stabilisation of CI, and its translocation occurs via the MIA pathway [336]. These studies demonstrate the importance of mitochondrial protein import and proper respiratory function in the prevention of motor neuron diseases such as ALS, highlighting import pathways as interesting potential targets for treatment.

5. Repair Pathways

In order to maintain the integrity and function of the mitochondria, a complex hierarchy of quality control mechanisms exists. This consists of repair mechanisms at the molecular, organelle, and cellular levels via a plethora of complex systems including mitochondrial chaperones and proteases, mitochondrial dynamics and distribution, mitochondrial-derived vesicles (MDVs), mitophagy, and apoptosis [337]. In addition to the emergence of links between mitochondrial import defects and neurodegenerative diseases, there is also evidence implicating stress response pathways in neurodegeneration. This evidence suggests that the pathways may have either a protective or exacerbating role in disease progression in different models. This section will discuss some of the stress response pathways that cells have developed in response to mitochondrial dysfunction for restoration of mitochondrial import function, respiratory capacity, and mitochondrial and cytosolic proteostasis.

5.1. *UPR^{mt}*

The mitochondrial unfolded protein response (UPR^{mt}) is known to be directly activated in response to impaired proteostasis in the mitochondrial matrix and has been extensively studied in *Caenorhabditis elegans*, where it was first identified [338]. The UPR^{mt} is a transcriptional response pathway that eliminates proteotoxic stress and fine-tunes mitochondrial respiration [339,340].

The sensor for this pathway is stress activated transcription factor (ATFS-1, ATF5 in mammals), which contains both a weak N-terminal MTS and a strong C-terminal nuclear localisation sequence (NLS) [341]. Proteotoxic mitochondrial stress, caused by a variety of mitochondrial stressors including: impairment of the import machinery (*timm23* or *tomm40*(RNAi) or paraquat application via CI inhibition), loss of ETC quality control (*spg-7*(RNAi)), or mtDNA depletion (ethidium bromide application) [341], results in retargeting of ATFS-1 primarily to the nucleus. There, ATFS-1 acts with transcriptional regulators DVE-1 and UBL-5 to induce the production of mitochondrial chaperone proteins HSP-6 and HSP-60, as well as proteases CLPP-1, LONP-1, SPG-7, and YMEL-1, metabolic genes GPD-2 and SKN-1, and core component of the TIM23 complex, TIM17 [339,342,343]. ATFS-1 is also responsible for repressing the expression of ETC genes, thus shifting expression capacity to increase mitochondrial protein folding and reducing the proteotoxic stress from mistargeted proteins in the cytosol [344].

Importantly, the localisation of ATFS-1 is mediated by HAF-1, the previously identified UPR^{mt} regulator and general attenuator of mitochondrial protein import during stress [345]. In the absence of HAF-1, ATFS-1 is unable to transition to the nucleus under stress conditions, thus failing to activate the UPR^{mt} [341,345]. It is important to note that ATFS-1 has a relatively weak MTS, meaning that minor effects on mitochondrial protein import efficiency, such as partially depolarised mitochondria, can trigger the stress response pathway, even though some mitochondrial proteins with stronger targeting sequences may still be imported successfully under these conditions [346].

In mammalian cells, the UPR^{mt} is thought to act in a similar way to that described above for *C. elegans*, where transcription factor ATF5 is regulated and triggers a stress response very similar to that described for the *C. elegans* homolog ATFS-1 [347]. However, studies have shown that integrated stress response (ISR) factor ATF4 is also involved in the transcriptional reprogramming of the mammalian UPR^{mt} [348,349]. It is also thought that the heat shock response (HSR) is activated alongside the UPR^{mt} in what is known as the mitochondrial to cytosolic stress response (MCSR) [338]. The HSR is activated by dysfunctional ETC activity or complex assembly and restores cytosolic proteostasis via transcription factor HSF-1 [350].

Given the vast mitochondrial dysfunction described in neurodegeneration, it is not surprising that there is an emerging body of evidence linking the UPR^{mt} to neurodegeneration. In PD, variants of *C. elegans* PINK1 and Parkin orthologs PINK-1 and PDR-1 lead to increased activation of the UPR^{mt}, which mitigates mitochondrial dysfunction caused by the corresponding mutations, subsequently increasing dopaminergic neuron survival [351]. However, a study in *C. elegans* showed that prolonged UPR^{mt} activation can in fact exacerbate mitochondrial dysfunction and dopaminergic cell death by favouring retention of dysfunctional mitochondria [352], which is important to note given the long-term and progressive nature of neurodegenerative diseases. Furthermore, the HSR has been shown to be activated in mouse and cell models of PD [353,354], and studies have also highlighted heat shock protein overexpression as an attenuator of α-syn aggregation and subsequent dopaminergic cell death [355].

In vivo studies also reveal that the accumulation of ALS SOD1 variant SOD1^{G93A} in the IMS leads to activation of the UPR^{mt} [356], consistent with other studies showing that activation of the UPR^{mt} precedes disease onset and increases throughout disease progression in ALS mutant mice [357]. Similarly, in AD, accumulation of Aβ has been shown to activate the UPR^{mt} [358], and there are high levels of UPR^{mt} marker genes in post-mortem brain samples from AD patients [359]. Interestingly, the inhibition of UPR^{mt} by knockdown of genes coding for key UPR^{mt} proteins HSP-6, HSP-60, and DVE-1 exacerbates AD phenotypes in *C. elegans* [360], suggesting that the UPR^{mt} may play a protective role in AD progression.

Recently, evidence has shown that an earlier form of the UPR^{mt} precedes the classical UPR^{mt}, and is activated by the accumulation of unprocessed precursor proteins inside mitochondria, due to impaired processing by MPP [361]. In this case, yeast nuclear transcription factor Rox1 is relocalised to mitochondria, binding to mtDNA and regulating mtDNA transcription and translation, and maintenance of mitochondrial respiratory and import functions [361].

5.2. *UPR^{am}*

The 'UPR activated by the mistargeting of proteins' (UPR^{am}) is another major stress response pathway that responds to mitochondrial import defects via the TIM23 or MIA pathways [362]. It has been well characterised in yeast, and there is some evidence that suggests that it also takes place in mammalian cells [362,363]. In yeast, the trigger for this is not the lack of import of a sensor protein, like ATFS-1 in the UPR^{mt}, but instead the accumulation of cytosolic precursor proteins [362]. This accumulation of cytosolic precursors leads to increased proteasome assembly, triggered by increased activity of proteasome assembly factors Irc25 and Poc4, and subsequent proteasomal degradation of the accumulated cytosolic precursor proteins [362]. This is accompanied by an inhibition of protein synthesis, which acts to prevent further accumulation of mistargeted proteins in the cytosol [362].

The UPR^{am} pathway is in part identical to the UPR^{mt} and is probably activated simultaneously alongside the UPR^{mt}; however, they differ in that the UPR^{mt} acts by regulating the abundance of mitochondrial chaperones and proteases, whilst the UPR^{am} regulates the expression of all mitochondrial proteins, as well as activating the proteasome to clear aggregated proteins [214,364].

To the best of our knowledge, there have been no studies thus far directly implicating the UPR^{am} pathway in neurodegeneration. However, proteasomal degradation via the ubiquitin–proteasome system is known to be downregulated in the affected neurons of many neurodegenerative diseases including AD, PD, HD, and ALS, and it is thought that this is mainly caused by the accumulation of cytotoxic protein aggregates [365–367]. For example, in AD, aggregated, ubiquitinated Tau can block entry of unfolded proteins to the 19S catalytic subunit of the proteasome by binding to the recognition site, resulting in impaired proteasomal degradation and enhancing the accumulation of precursor proteins [368].

5.3. *mPOS*

The mitochondrial precursor over-accumulation stress (mPOS) pathway is a mechanism of mitochondria mediated cell death, and has been characterised in yeast [369]. mPOS is usually triggered by any dysfunction that leads to over-accumulation of precursor proteins in the cytoplasm. Usually, this accumulation would occur as a consequence of import dysfunctions, but it can also be related to other mitochondrial damage, particularly damage that alters IMM integrity such as misfolding of IMM proteins [369]. mPOS is thought to lead to cell degeneration due to the toxic cytosolic accumulation of misfolded proteins exceeding the cells' capacity to remove these proteins [369]. However, there is a large network of genes responsible for suppressing mPOS and thus promoting cell survival by means of modulating ribosomal biogenesis, translation of specific transcripts, increasing protein chaperones and turnover, and decapping mRNA [369]. Of these proteins in yeast, Gis2 and Nog2 are particularly important in encouraging cell survival. Gis2 is involved in promoting cap-independent translation whilst Nog2 inhibits the nuclear export of the 60S RNA subunit of the ribosome, promoting cell survival and attenuating mPOS [369–371]. Furthermore, the mPOS pathway can trigger additional stress response pathways within the cell, including the ISR, which restores cellular homeostasis by reducing global protein synthesis, triggered by phosphorylation of eukaryotic translation initiation factor 2 alpha (eIF2α) [372].

Though there have been no specific examples of mPOS in neurodegeneration as of yet, it may have extremely important implications, especially given the mutations in genes of the anti-degenerative network seen in some neurodegenerative diseases such as ALS [373] and PD [374], which have been implicated in suppressing mPOS. The potential association of mPOS in neurodegeneration has been discussed in detail in a recent review [375].

5.4. mitoCPR

The mitochondrial compromised protein import response (mitoCPR) pathway was discovered in yeast and is activated when a mitochondrial protein is stalled in the Tom40 channel, inducing mitochondrial import stress and accumulation of proteins on the mitochondrial surface [376]. In yeast, the mitoCPR is activated and transcription factor Pdr3 induces the expression of *CIS1*. Cytosolic protein Cis1 binds to Tom70 and recruits the AAA+ ATPase Msp1, which removes stalled precursor proteins from mitochondrial channels and targets them for proteasomal degradation [376]. This allows mitochondria to maintain their functions under import stress conditions. This is interesting in the context of AD, especially given that APP, the precursor protein responsible for the production of toxic amyloid plaques in Alzheimer's brains, was shown to accumulate within TOM channels, driving mitochondrial dysfunction in AD [260]. This indicates that the mitoCPR pathway may be defective under these conditions, or may not be sufficient to rescue mitochondrial dysfunction associated with APP-TOM aggregation [377].

5.5. mitoTAD

The mitochondrial protein translocation-associated degeneration (mitoTAD) pathway differs from those described already in that it is a quality control pathway that occurs constitutively under non-stress conditions [378]. In yeast, it is triggered by precursor proteins trapped in the Tom40 channel, sensed by Ubx2, which consistently interacts with the TOM complex under normal conditions, monitoring protein import through Tom40 [378]. If Ubx2 senses that a precursor protein is arrested within the TOM complex, a pool of Ubx2 binds to TOM and recruits the AAA+ ATPase Cdc48 for removal of arrested precursor proteins from the Tom40 channel [378]. The mitoTAD pathway was discovered in yeast, and interestingly, shows similarities to a quality control pathway in the ER, which involves Ubx2 exporting unfolded proteins from the ER [379,380]. No examples of the mitoTAD pathway have been described in models of neurodegeneration as of yet; however, as discussed above for the mitoCPR pathway, it is intriguing in the context of studies showing accumulation of proteins in the mitochondria during neurodegeneration, and further research into this link would be most interesting.

6. Concluding Remarks

Over recent years, remarkable progress has been made towards understanding the processes of mitochondrial protein import and respiratory complexes assembly. Recent advances in structural biology have begun to further elucidate the different structural properties of the mitochondrial translocases in high resolution, and this sheds further light on the various processes of mitochondrial protein import for specific protein classes. Whilst progress has been made, there remain areas of uncertainty regarding the organisation and dynamic action of the translocase complexes. Advances in import assay methods, such as the one recently developed [381], are also of paramount importance to dissect the mechanism of the import process and its kinetics. Hopefully, a revamped in cell assay will allow one to perform drug and phenotypic screenings, allowing for the easy identification of new players and modulators as well as small molecules that target this biological pathway.

Here, we highlight the body of evidence surrounding how closely interlinked mitochondrial protein translocation pathways are with the assembly of respiratory complexes and their function. Recent advances have begun clarifying exactly which translocation pathways are taken by nuclear-encoded respiratory complex proteins, though much more is yet to be done. Elucidating this link between import and respiratory function will be of vital importance, especially since in cases of mitochondrial disease, the importance of import pathways for respiratory complexes assembly and function is now clear. This suggests that targeting import pathways in cases of mitochondrial disease may become a credible therapeutic strategy.

Whilst mitochondrial dysfunction has long been recognised as a key factor in neurodegenerative diseases, mitochondrial protein import is now being implicated as a key factor in this dysfunction, across all levels of neurodegenerative disease models from the simple cell line setup right up to animal models and patient samples. Interestingly, the findings from these studies suggest that dysfunctional mitochondrial import is a driving force for the prominent mitochondrial irregularities observed in these diseases. This therefore represents an important target for further research to address the major outstanding questions. Namely, are the links between mitochondrial import defects and disease causative or consequential? What exact role might such defects play in disease progression?

This review has also outlined the various stress response pathways that have been shown to be activated in response to mitochondrial protein import defects. We highlight their importance in maintaining cell proteostasis and fine-tuning respiratory processes that rescue mitochondrial function. It is thought that these pathways are interlinked with one another; for example, the UPR^{am} and UPR^{mt}, the two most well characterised pathways to date, are thought to be activated simultaneously, despite having different triggers [364]. Interestingly, the UPR^{am} and mPOS pathways both share a common trigger, that is, they become activated by accumulation of precursor proteins in the cytosol, yet thus far no evidence has shown their simultaneous activation. Interestingly, the mitoCPR and mitoTAD pathways also share the same trigger. It is thought that the mitoTAD pathway is active under non-stress conditions, whilst the mitoCPR pathway is activated only under stress conditions. Does this mean that mitoCPR is activated only when the mitoTAD pathway has failed? Since these stress response pathways are relatively new concepts, much more research is required. As more evidence emerges, enlightening the precise mechanisms of these pathways, they may generate therapeutically interesting targets for interventions against neurodegenerative diseases.

Supplementary Materials: The following are available online at https://www.mdpi.com/article/10.3390/life11050432/s1, Table S1: Supporting Table for Main Figure 3 on the Assembly of human Respiratory Complexes.

Author Contributions: Conceptualization, G.C.P. and H.I.N.; Writing, H.I.N., M.P. and G.C.P.; Figures, H.I.N., M.P. and G.C.P.; Revisions before submission, I.C. and J.P.; Revisions after submission, H.I.N., M.P., G.C.P. and J.M.H. All authors have read and agreed to the published version of the manuscript.

Funding: HN is supported by the Wellcome Trust Dynamic Molecular Cell Biology PhD programme (215317/Z/19/Z). MP is recipient of an MRC funded PhD scholarship. Research in JMH laboratory is supported the BBSRC (BB/R00787X/1), Wellcome Trust Investigator Award (220799/Z/20/Z), and Leverhulme Trust (RPG-2019-191). Research in the JP laboratory is supported by the Medical Research Council, UK (MC_UU_00015/7). Research in the IC laboratory is supported by the Wellcome Trust: Investigator Award (104632/Z/14/Z). GCP is supported by the Swiss National Science Foundation (Synergia project CRSII5_180326). The APC was funded by University of Bristol Open Access via IC.

Institutional Review Board Statement: Not applicable.

Informed Consent Statement: Not applicable.

Data Availability Statement: Not applicable.

Conflicts of Interest: The authors declare no conflict of interest.

Abbreviations

$\Delta\psi$: mitochondrial transmembrane potential; α-syn: alpha-synuclein; Aβ: amyloid-beta peptide; AD: Alzheimer's Disease; AGK: acylglycerol kinase; AIF: apoptosis inducing factor; ALS: Amyotrophic lateral sclerosis; APP: amyloid precursor protein; CCCP: carbonyl cyanide-*m*-lorophenylhydrazone; CHCHD: coiled-coil-helix-coiled-coil-helix domain; CI-V: mitochondrial respiratory complexes one to five; cryo-EM: cryogenic electron microscopy; DAMPs: damage-associated molecular patterns; Drp1: mitochondrial fission GTPase dynamin-related protein 1; eIF2α: eukaryotic translation initiation factor 2 alpha; ER: endoplasmic reticulum; ETC: electron transport chain; Fe/S: iron/sulphur; HD: Huntington's disease; HSR: heat shock response; Hsp: heat-shock protein; HTT: Huntingtin; IBM: inner boundary membrane; IMM: inner mitochondrial membrane; IMS: intermembrane space; iMTS: internal mitochondrial targeting sequence like signal sequence; ISR: integrated stress response; LHON: Leber hereditary optic neuropathy; MCIA: mitochondrial complex I intermediate assembly; MCSR: mitochondrial to cytosolic stress response; MDVs: mitochondrial-derived vesicles; MIA: mitochondrial IMS assembly; MICOS: mitochondrial contact site and cristae organising system; MIM: insertase of the outer mitochondrial membrane; MIP: mitochondrial intermediate peptidase; mitoCPR: mitochondrial compromised protein import response; mitoTAD: mitochondrial protein translocation-associated degeneration; MITRAC: Mitochondrial Translation Regulation Assembly intermediate of Cytochrome *c* oxidase; mPOS: mitochondrial precursor over-accumulation stress; MPP: mitochondrial processing peptidase; mtDNA: mitochondrial DNA; MTS: mitochondrial targeting sequence; NFTs: neurofibrillary tangles; NLS: nuclear localisation sequence; OMM: outer mitochondrial membrane; OXA1: mitochondrial oxidase assembly protein 1; PAM: presequence translocase-associated motor; PD: Parkinson's disease; ROS: reactive oxygen species; SAM: sorting and assembly machinery; SC: supercomplexes; *SN*: *Substantia Nigra;* TIM22: translocase of the inner membrane 23; TIM23: translocase of the inner membrane 23; TOM: translocase of the outer membrane; TOM-CC: translocase of the outer membrane core complex; UQCC1 and UQCC2: ubiquinol-cytochrome *c* reductase complex assembly factors 1 and 2; UPR^{am}: UPR activated by the mistargeting of proteins; UPR^{mt}: mitochondrial unfolded protein response; UTR: untranslated region.

References

1. Osellame, L.D.; Blacker, T.S.; Duchen, M.R. Cellular and molecular mechanisms of mitochondrial function. *Best Pract. Res. Clin. Endoc. Metab.* **2012**, *26*, 711–723. [CrossRef] [PubMed]
2. Chacinska, A.; Koehler, C.M.; Milenkovic, D.; Lithgow, T.; Pfanner, N. Importing Mitochondrial Proteins: Machineries and Mechanisms. *Cell* **2009**, *138*, 628–644. [CrossRef] [PubMed]
3. MacKenzie, J.A.; Payne, R.M. Mitochondrial protein import and human health and disease. *Biochim. Biophys. Acta Mol. Basis Dis.* **2007**, *1772*, 509–523. [CrossRef]

4. Briston, T.; Hicks, A.R. Mitochondrial dysfunction and neurodegenerative proteinopathies: Mechanisms and prospects for therapeutic intervention. *Biochem. Soc. Trans.* **2018**, *46*, 829–842. [CrossRef] [PubMed]
5. Jackson, T.D.; Palmer, C.S.; Stojanovski, D. Mitochondrial diseases caused by dysfunctional mitochondrial protein import. *Biochem. Soc. Trans.* **2018**, *46*, 1225–1238. [CrossRef]
6. Rath, S.; Sharma, R.; Gupta, R.; Ast, T.; Chan, C.; Durham, T.J.; Goodman, R.P.; Grabarek, Z.; Haas, M.E.; Hung, W.H.W.; et al. MitoCarta3.0: An updated mitochondrial proteome now with sub-organelle localization and pathway annotations. *Nucleic Acids Res.* **2021**, *49*, D1541–D1547. [CrossRef]
7. Tucker, K.; Park, E. Cryo-EM structure of the mitochondrial protein-import channel TOM complex at near-atomic resolution. *Nat. Struct. Mol. Biol.* **2019**, *26*, 1158–1166. [CrossRef]
8. Sollner, T.; Griffiths, G.; Pfaller, R.; Pfanner, N.; Neupert, W. Mom19, an import receptor for mitochondrial precursor proteins. *Cell* **1989**, *59*, 1061–1070. [CrossRef]
9. Moczko, M.; Dietmeier, K.; Sollner, T.; Segui, B.; Steger, H.F.; Neupert, W.; Pfanner, N. Identification of the mitochondrial receptor complex in saccharomyces-cerevisiae. *FEBS Lett.* **1992**, *310*, 265–268. [CrossRef]
10. Hines, V.; Brandt, A.; Griffiths, G.; Horstmann, H.; Brutsch, H.; Schatz, G. Protein import into yeast mitochondria is accelerated by the outer-membrane protein mas70. *Embo J.* **1990**, *9*, 3191–3200. [CrossRef]
11. Sollner, T.; Pfaller, R.; Griffiths, G.; Pfanner, N.; Neupert, W. A mitochondrial import receptor for the adp/atp carrier. *Cell* **1990**, *62*, 107–115. [CrossRef]
12. Chacinska, A.; Lind, M.; Frazier, A.E.; Dudek, J.; Meisinger, C.; Geissler, A.; Sickmann, A.; Meyer, H.E.; Truscott, K.N.; Guiard, B.; et al. Mitochondrial presequence translocase: Switching between TOM tethering and motor recruitment involves Tim21 and Tim17. *Cell* **2005**, *120*, 817–829. [CrossRef] [PubMed]
13. Van Wilpe, S.; Ryan, M.T.; Hill, K.; Maarse, A.C.; Meisinger, C.; Brix, J.; Dekker, P.J.; Moczko, M.; Wagner, R.; Meijer, M.; et al. Tom22 is a multifunctional organizer of the mitochondrial preprotein translocase. *Nature* **1999**, *401*, 485–489. [CrossRef]
14. Lithgow, T.; Junne, T.; Suda, K.; Gratzer, S.; Schatz, G. The mitochondrial outer membrane protein Mas22p is essential for protein import and viability of yeast. *Proc. Natl. Acad. Sci. USA* **1994**, *91*, 11973–11977. [CrossRef] [PubMed]
15. Sakaue, H.; Shiota, T.; Ishizaka, N.; Kawano, S.; Tamura, Y.; Tan, K.S.; Imai, K.; Motono, C.; Hirokawa, T.; Taki, K.; et al. Porin Associates with Tom22 to Regulate the Mitochondrial Protein Gate Assembly. *Mol. Cell* **2019**, *73*, 1044–1055. [CrossRef] [PubMed]
16. Grevel, A.; Becker, T. Porins as helpers in mitochondrial protein translocation. *Biol. Chem.* **2020**, *401*, 699–708. [CrossRef] [PubMed]
17. Harbauer, A.B.; Opalinska, M.; Gerbeth, C.; Herman, J.S.; Rao, S.; Schonfisch, B.; Guiard, B.; Schmidt, O.; Pfanner, N.; Meisinger, C. Cell cycle-dependent regulation of mitochondrial preprotein translocase. *Science* **2014**, *346*, 1109–1113. [CrossRef]
18. Kunkele, K.P.; Heins, S.; Dembowski, M.; Nargang, F.E.; Benz, R.; Thieffry, M.; Walz, J.; Lill, R.; Nussberger, S.; Neupert, W. The preprotein translocation channel of the outer membrane of mitochondria. *Cell* **1998**, *93*, 1009–1019. [CrossRef]
19. Shiota, T.; Imai, K.; Qiu, J.; Hewitt, V.L.; Tan, K.; Shen, H.H.; Sakiyama, N.; Fukasawa, Y.; Hayat, S.; Kamiya, M.; et al. Molecular architecture of the active mitochondrial protein gate. *Science* **2015**, *349*, 1544–1548. [CrossRef]
20. Model, K.; Prinz, T.; Ruiz, T.; Radermacher, M.; Krimmer, T.; Kuhlbrandt, W.; Pfanner, N.; Meisinger, C. Protein translocase of the outer mitochondrial membrane: Role of import receptors in the structural organization of the TOM complex. *J. Mol. Biol.* **2002**, *316*, 657–666. [CrossRef] [PubMed]
21. Model, K.; Meisinger, C.; Kuhlbrandt, W. Cryo-Electron Microscopy Structure of a Yeast Mitochondrial Preprotein Translocase. *J. Mol. Biol.* **2008**, *383*, 1049–1057. [CrossRef] [PubMed]
22. Bausewein, T.; Mills, D.J.; Langer, J.D.; Nitschke, B.; Nussberger, S.; Kuhlbrandt, W. Cryo-EM Structure of the TOM Core Complex from Neurospora crassa. *Cell* **2017**, *170*, 693–700. [CrossRef] [PubMed]
23. Moczko, M.; Bomer, U.; Kubrich, M.; Zufall, N.; Honlinger, A.; Pfanner, N. The intermembrane space domain of mitochondrial Tom22 functions as a trans binding site for properties with N-terminal targeting sequences. *Mol. Cell. Biol.* **1997**, *17*, 6574–6584. [CrossRef] [PubMed]
24. Harner, M.; Neupert, W.; Deponte, M. Lateral release of proteins from the TOM complex into the outer membrane of mitochondria. *Embo J.* **2011**, *30*, 3232–3241. [CrossRef]
25. Sekine, S.; Youle, R.J. PINK1 import regulation; a fine system to convey mitochondrial stress to the cytosol. *BMC Biol.* **2018**, *16*, 2. [CrossRef]
26. Hasson, S.A.; Kane, L.A.; Yamano, K.; Huang, C.H.; Sliter, D.A.; Buehler, E.; Wang, C.X.; Heman-Ackah, S.M.; Hessa, T.; Guha, R.; et al. High-content genome-wide RNAi screens identify regulators of parkin upstream of mitophagy. *Nature* **2013**, *504*, 291–295. [CrossRef]
27. Jacoupy, M.; Hamon-Keromen, E.; Ordureau, A.; Erpapazoglou, Z.; Coge, F.; Corvol, J.C.; Nosjean, O.; la Cour, C.M.; Millan, M.J.; Boutin, J.A.; et al. The PINK1 kinase-driven ubiquitin ligase Parkin promotes mitochondrial protein import through the presequence pathway in living cells. *Sci. Rep.* **2019**, *9*, 11829. [CrossRef]
28. Phu, L.; Rose, C.M.; Tea, J.S.; Wall, C.E.; Verschueren, E.; Cheung, T.K.; Kirkpatrick, D.S.; Bingol, B. Dynamic Regulation of Mitochondrial Import by the Ubiquitin System. *Mol. Cell* **2020**, *77*, 1107–1123. [CrossRef]
29. Ordureau, A.; Paulo, J.A.; Zhang, J.C.; An, H.; Swatek, K.N.; Cannon, J.R.; Wan, Q.Q.; Komander, D.; Harper, J.W. Global Landscape and Dynamics of Parkin and USP30-Dependent Ubiquitylomes in iNeurons during Mitophagic Signaling. *Mol. Cell* **2020**, *77*, 1124–1142. [CrossRef]

30. Harbauer, A.B.; Zahedi, R.P.; Sickmann, A.; Pfanner, N.; Meisinger, C. The Protein Import Machinery of Mitochondria-A Regulatory Hub in Metabolism, Stress, and Disease. *Cell Metab.* **2014**, *19*, 357–372. [CrossRef]
31. Paschen, S.A.; Waizenegger, T.; Stan, T.; Preuss, M.; Cyrklaff, M.; Hell, K.; Rapaport, D.; Neupert, W. Evolutionary conservation of biogenesis of beta-barrel membrane proteins. *Nature* **2003**, *426*, 862–866. [CrossRef]
32. Kutik, S.; Stojanovski, D.; Becker, L.; Becker, T.; Meinecke, M.; Kruger, V.; Prinz, C.; Meisinger, C.; Guiard, B.; Wagner, R.; et al. Dissecting membrane insertion of mitochondrial beta-barrel proteins. *Cell* **2008**, *132*, 1011–1024. [CrossRef] [PubMed]
33. Becker, T.; Voegtle, F.N.; Stojanovski, D.; Meisinger, C. Sorting and assembly of mitochondrial outer membrane proteins. *Biochim. Biophys. Acta-Bioenerg.* **2008**, *1777*, 557–563. [CrossRef]
34. Diederichs, K.A.; Ni, X.D.; Rollauer, S.E.; Botos, I.; Tan, X.F.; King, M.S.; Kunji, E.R.S.; Jiang, J.S.; Buchanan, S.K. Structural insight into mitochondrial beta-barrel outer membrane protein biogenesis. *Nat. Commun.* **2020**, *11*, 3290. [CrossRef] [PubMed]
35. Wenz, L.S.; Ellenrieder, L.; Qiu, J.; Bohnert, M.; Zufall, N.; van der Laan, M.; Pfanner, N.; Wiedemann, N.; Becker, T. Sam37 is crucial for formation of the mitochondrial TOM-SAM supercomplex, thereby promoting beta-barrel biogenesis. *J. Cell Biol.* **2015**, *210*, 1047–1054. [CrossRef] [PubMed]
36. Meisinger, C.; Rissler, M.; Chacinska, A.; Szklarz, L.K.S.; Milenkovic, D.; Kozjak, V.; Schonfisch, B.; Lohaus, C.; Meyer, H.E.; Yaffe, M.P.; et al. The mitochondrial morphology protein Mdm10 functions in assembly of the preprotein translocase of the outer membrane. *Dev. Cell* **2004**, *7*, 61–71. [CrossRef] [PubMed]
37. Doan, K.N.; Grevel, A.; Martensson, C.U.; Ellenrieder, L.; Thornton, N.; Wenz, L.S.; Opalinski, L.; Guiard, B.; Pfanner, N.; Becker, T. The Mitochondrial Import Complex MIM Functions as Main Translocase for alpha-Helical Outer Membrane Proteins. *Cell Rep.* **2020**, *31*, 107567. [CrossRef]
38. Becker, T.; Wenz, L.S.; Kruger, V.; Lehmann, W.; Muller, J.M.; Goroncy, L.; Zufall, N.; Lithgow, T.; Guiard, B.; Chacinska, A.; et al. The mitochondrial import protein Mim1 promotes biogenesis of multispanning outer membrane proteins. *J. Cell Biol.* **2011**, *194*, 387–395. [CrossRef] [PubMed]
39. Krumpe, K.; Frumkin, I.; Herzig, Y.; Rimon, N.; Ozbalci, C.; Brugger, B.; Rapaport, D.; Schuldiner, M. Ergosterol content specifies targeting of tail-anchored proteins to mitochondrial outer membranes. *Mol. Biol. Cell* **2012**, *23*, 3927–3935. [CrossRef]
40. Kruger, V.; Becker, T.; Becker, L.; Montilla-Martinez, M.; Ellenrieder, L.; Vogtle, F.N.; Meyer, H.E.; Ryan, M.T.; Wiedemann, N.; Warscheid, B.; et al. Identification of new channels by systematic analysis of the mitochondrial outer membrane. *J. Cell Biol.* **2017**, *216*, 3485–3495. [CrossRef]
41. Dimmer, K.S.; Papic, D.; Schumann, B.; Sperl, D.; Krumpe, K.; Walther, D.M.; Rapaport, D. A crucial role for Mim2 in the biogenesis of mitochondrial outer membrane proteins. *J. Cell Sci.* **2012**, *125*, 3464–3473. [CrossRef]
42. Wenz, L.S.; Opalinski, L.; Schuler, M.H.; Ellenrieder, L.; Ieva, R.; Bottinger, L.; Qiu, J.; van der Laan, M.; Wiedemann, N.; Guiard, B.; et al. The presequence pathway is involved in protein sorting to the mitochondrial outer membrane. *Embo Rep.* **2014**, *15*, 678–685. [CrossRef] [PubMed]
43. Sinzel, M.; Tan, T.; Wendling, P.; Kalbacher, H.; Ozbalci, C.; Chelius, X.; Westermann, B.; Brugger, B.; Rapaport, D.; Dimmer, K.S. Mcp3 is a novel mitochondrial outer membrane protein that follows a unique IMP-dependent biogenesis pathway. *Embo Rep.* **2016**, *17*, 965–981. [CrossRef] [PubMed]
44. Mokranjac, D.; Neupert, W. Energetics of protein translocation into mitochondria. *Biochim. Biophys. Acta Bioenerg.* **2008**, *1777*, 758–762. [CrossRef]
45. Komiya, T.; Rospert, S.; Schatz, G.; Mihara, K. Binding of mitochondrial precursor proteins to the cytoplasmic domains of the import receptors Tom70 and Tom20 is determined by cytoplasmic chaperones. *Embo J.* **1997**, *16*, 4267–4275. [CrossRef] [PubMed]
46. Yamamoto, H.; Esaki, M.; Kanamori, T.; Tamura, Y.; Nishikawa, S.; Endo, T. Tim50 is a subunit of the TIM23 complex that links protein translocation across the outer and inner mitochondrial membranes. *Cell* **2002**, *111*, 519–528. [CrossRef]
47. Komiya, T.; Rospert, S.; Koehler, C.; Looser, R.; Schatz, G.; Mihara, K. Interaction of mitochondrial targeting signals with acidic receptor domains along the protein import pathway: Evidence for the 'acid chain' hypothesis. *Embo J.* **1998**, *17*, 3886–3898. [CrossRef]
48. Dudek, J.; Rehling, P.; van der Laan, M. Mitochondrial protein import: Common principles and physiological networks. *Biochim. Biophys. Acta-Mol. Cell Res.* **2013**, *1833*, 274–285. [CrossRef]
49. Garcia, M.; Delaveau, T.; Goussard, S.; Jacq, C. Mitochondrial presequence and open reading frame mediate asymmetric localization of messenger RNA. *Embo Rep.* **2010**, *11*, 285–291. [CrossRef] [PubMed]
50. Margeot, A.; Blugeon, C.; Sylvestre, J.; Vialette, S.; Jacq, C.; Corral-Debrinski, M. In Saccharomyces cerevisiae, ATP2 mRNA sorting to the vicinity of mitochondria is essential for respiratory function. *Embo J.* **2002**, *21*, 6893–6904. [CrossRef]
51. Corral-Debrinski, M.; Blugeon, C.; Jacq, C. In yeast, the 3′ untranslated region or the presequence of ATM1 is required for the exclusive localization of its mRNA to the vicinity of mitochondria. *Mol Cell Biol* **2000**, *20*, 7881–7892. [CrossRef]
52. George, R.; Walsh, P.; Beddoe, T.; Lithgow, T. The nascent polypeptide-associated complex (NAC) promotes interaction of ribosomes with the mitochondrial surface in vivo. *FEBS Lett.* **2002**, *516*, 213–216. [CrossRef]
53. MacKenzie, J.A.; Payne, R.M. Ribosomes specifically bind to mammalian mitochondria via protease-sensitive proteins on the outer membrane. *J. Biol. Chem.* **2004**, *279*, 9803–9810. [CrossRef] [PubMed]
54. Edwards, R.; Eaglesfield, R.; Tokatlidis, K. The mitochondrial intermembrane space: The most constricted mitochondrial sub-compartment with the largest variety of protein import pathways. *Open Biol.* **2021**, *11*, 210002. [CrossRef]

55. Peleh, V.; Cordat, E.; Herrmann, J.M. Mia40 is a trans-site receptor that drives protein import into the mitochondrial intermembrane space by hydrophobic substrate binding. *eLife* **2016**, *5*, e16177. [CrossRef] [PubMed]
56. Chacinska, A.; Pfannschmidt, S.; Wiedemann, N.; Kozjak, V.; Szklarz, L.K.S.; Schulze-Specking, A.; Truscott, K.N.; Guiard, B.; Meisinger, C.; Pfanner, N. Essential role of Mia40 in import and assembly of mitochondrial intermembrane space proteins. *Embo J.* **2004**, *23*, 3735–3746. [CrossRef]
57. Banci, L.; Bertini, I.; Cefaro, C.; Ciofi-Baffoni, S.; Gallo, A.; Martinelli, M.; Sideris, D.P.; Katrakili, N.; Tokatlidis, K. MIA40 is an oxidoreductase that catalyzes oxidative protein folding in mitochondria. *Nat. Struct. Mol. Biol.* **2009**, *16*, 198–206. [CrossRef] [PubMed]
58. Reinhardt, C.; Arena, G.; Nedara, K.; Edwards, R.; Brenner, C.; Tokatlidis, K.; Modjtahedi, N. AIF meets the CHCHD4/Mia40-dependent mitochondrial import pathway. *Biochim. Biophys. Acta Mol. Basis Dis.* **2020**, *1866*, 165746. [CrossRef] [PubMed]
59. Bien, M.; Longen, S.; Wagener, N.; Chwalla, I.; Herrmann, J.M.; Riemer, J. Mitochondrial Disulfide Bond Formation Is Driven by Intersubunit Electron Transfer in Erv1 and Proofread by Glutathione. *Mol. Cell* **2010**, *37*, 516–528. [CrossRef]
60. Allen, S.; Balabanidou, V.; Sideris, D.P.; Lisowsky, T.; Tokatlidis, K. Erv1 mediates the Mia40-dependent protein import pathway and provides a functional link to the respiratory chain by shuttling electrons to cytochrome c. *J. Mol. Biol.* **2005**, *353*, 937–944. [CrossRef]
61. Neupert, W. Protein import into mitochondria. *Annu. Rev. Biochem.* **1997**, *66*, 863–917. [CrossRef] [PubMed]
62. Backes, S.; Hess, S.; Boos, F.; Woellhaf, M.W.; Godel, S.; Jung, M.; Muhlhaus, T.; Herrmann, J.M. Tom70 enhances mitochondrial preprotein import efficiency by binding to internal targeting sequences. *J. Cell Biol.* **2018**, *217*, 1369–1382. [CrossRef]
63. Demishtein-Zohary, K.; Azem, A. The TIM23 mitochondrial protein import complex: Function and dysfunction. *Cell Tissue Res.* **2017**, *367*, 33–41. [CrossRef] [PubMed]
64. Maarse, A.C.; Blom, J.; Keil, P.; Pfanner, N.; Meijer, M. Identification of the essential yeast protein mim17, an integral mitochondrial inner membrane-protein involved in protein import. *FEBS Lett.* **1994**, *349*, 215–221. [CrossRef]
65. Dekker, P.J.T.; Keil, P.; Rassow, J.; Maarse, A.C.; Pfanner, N.; Meijer, M. Identification of mim23, a putative component of the protein import machinery of the mitochondrial inner membrane. *FEBS Lett.* **1993**, *330*, 66–70. [CrossRef]
66. Geissler, A.; Chacinska, A.; Truscott, K.N.; Wiedemann, N.; Brandner, K.; Sickmann, A.; Meyer, H.E.; Meisinger, C.; Pfanner, N.; Rehling, P. The mitochondrial presequence translocase: An essential role of Tim50 in directing preproteins to the import channel. *Cell* **2002**, *111*, 507–518. [CrossRef]
67. Gebert, M.; Schrempp, S.G.; Mehnert, C.S.; Heisswolf, A.K.; Oeljeklaus, S.; Ieva, R.; Bohnert, M.; von der Malsburg, K.; Wiese, S.; Kleinschroth, T.; et al. Mgr2 promotes coupling of the mitochondrial presequence translocase to partner complexes. *J. Cell Biol.* **2012**, *197*, 595–604. [CrossRef] [PubMed]
68. Maarse, A.C.; Blom, J.; Grivell, L.A.; Meijer, M. Mpi1, an essential gene encoding a mitochondrial-membrane protein, is possibly involved in protein import into yeast mitochondria. *Embo J.* **1992**, *11*, 3619–3628. [CrossRef]
69. Kang, P.J.; Ostermann, J.; Shilling, J.; Neupert, W.; Craig, E.A.; Pfanner, N. Requirement for hsp70 in the mitochondrial matrix for translocation and folding of precursor proteins. *Nature* **1990**, *348*, 137–143. [CrossRef] [PubMed]
70. Frazier, A.E.; Dudek, J.; Guiard, B.; Voos, W.; Li, Y.F.; Lind, M.; Meisinger, C.; Geissler, A.; Sickmann, A.; Meyer, H.E.; et al. Pam16 has an essential role in the mitochondrial protein import motor. *Nat. Struct. Mol. Biol.* **2004**, *11*, 226–233. [CrossRef]
71. Truscott, K.N.; Voos, W.; Frazier, A.E.; Lind, M.; Li, Y.F.; Geissler, A.; Dudek, J.; Muller, H.; Sickmann, A.; Meyer, H.E.; et al. A J-protein is an essential subunit of the presequence translocase-associated protein import motor of mitochondria. *J. Cell Biol.* **2003**, *163*, 707–713. [CrossRef] [PubMed]
72. Van der Laan, M.; Chacinska, A.; Lind, M.; Perschil, I.; Sickmann, A.; Meyer, H.E.; Guiard, B.; Meisinger, C.; Pfanner, N.; Rehling, P. Pam17 is required for architecture and translocation activity of the mitochondrial protein import motor. *Mol. Cell. Biol.* **2005**, *25*, 7449–7458. [CrossRef] [PubMed]
73. Laloraya, S.; Gambill, B.D.; Craig, E.A. A role for a eukaryotic grpe-related protein, mge1p, in protein translocation. *Proc. Natl. Acad. Sci. USA* **1994**, *91*, 6481–6485. [CrossRef]
74. Tamura, Y.; Harada, Y.; Shiota, T.; Yamano, K.; Watanabe, K.; Yokota, M.; Yamamoto, H.; Sesaki, H.; Endo, T. Tim23-Tim50 pair coordinates functions of translocators and motor proteins in mitochondrial protein import. *J. Cell Biol.* **2009**, *184*, 129–141. [CrossRef]
75. Schwartz, M.P.; Matouschek, A. The dimensions of the protein import channels in the outer and inner mitochondrial membranes. *Proc. Natl. Acad. Sci. USA* **1999**, *96*, 13086–13090. [CrossRef] [PubMed]
76. Truscott, K.N.; Kovermann, P.; Geissler, A.; Merlin, A.; Meijer, M.; Driessen, A.J.M.; Rassow, J.; Pfanner, N.; Wagner, R. A presequence- and voltage-sensitive channel of the mitochondrial preprotein translocase formed by Tim23. *Nat. Struct. Biol.* **2001**, *8*, 1074–1082. [CrossRef]
77. Martinez-Caballero, S.; Grigoriev, S.M.; Herrmann, J.M.; Campo, M.L.; Kinnally, K.W. Tim17p regulates the twin pore structure and voltage gating of the mitochondrial protein import complex TIM23. *J. Biol. Chem.* **2007**, *282*, 3584–3593. [CrossRef] [PubMed]
78. Chacinska, A.; Rehling, P.; Guiard, B.; Frazier, A.E.; Schulze-Specking, A.; Pfanner, N.; Voos, W.; Meisinger, C. Mitochondrial translocation contact sites: Separation of dynamic and stabilizing elements in formation of a TOM-TIM-preprotein supercomplex. *Embo J.* **2003**, *22*, 5370–5381. [CrossRef]
79. Waegemann, K.; Popov-Celeketic, D.; Neupert, W.; Azem, A.; Mokranjac, D. Cooperation of TOM and TIM23 complexes during translocation of proteins into mitochondria. *J Mol. Biol.* **2015**, *427*, 1075–1084. [CrossRef]

80. Shiota, T.; Mabuchi, H.; Tanaka-Yamano, S.; Yamano, K.; Endo, T. In vivo protein-interaction mapping of a mitochondrial translocator protein Tom22 at work. *Proc. Natl. Acad. Sci. USA* **2011**, *108*, 15179–15183. [CrossRef]
81. Niemi, N.M.; Wilson, G.M.; Overmyer, K.A.; Vogtle, F.N.; Myketin, L.; Lohman, D.C.; Schueler, K.L.; Attie, A.D.; Meisinger, C.; Coon, J.J.; et al. Pptc7 is an essential phosphatase for promoting mammalian mitochondrial metabolism and biogenesis. *Nat. Commun.* **2019**, *10*, 3197. [CrossRef]
82. Marom, M.; Dayan, D.; Demishtein-Zohary, K.; Mokranjac, D.; Neupert, W.; Azem, A. Direct Interaction of Mitochondrial Targeting Presequences with Purified Components of the TIM23 Protein Complex. *J. Biol. Chem.* **2011**, *286*, 43809–43815. [CrossRef] [PubMed]
83. Ting, S.Y.; Yan, N.L.; Schilke, B.A.; Craig, E.A. Dual interaction of scaffold protein Tim44 of mitochondrial import motor with channel-forming translocase subunit Tim23. *eLife* **2017**, *6*, e23609. [CrossRef] [PubMed]
84. Neupert, W.; Brunner, M. The protein import motor of mitochondria. *Nat. Rev. Mol. Cell Biol.* **2002**, *3*, 555–565. [CrossRef] [PubMed]
85. Gruhler, A.; Arnold, I.; Seytter, T.; Guiard, B.; Schwarz, E.; Neupert, W.; Stuart, R.A. N-terminal hydrophobic sorting signals of preproteins confer mitochondrial hsp70 independence for import into mitochondria. *J. Biol. Chem.* **1997**, *272*, 17410–17415. [CrossRef] [PubMed]
86. Van der Laan, M.; Wiedemann, N.; Mick, D.U.; Guiard, B.; Rehling, P.; Pfanner, N. A role for Tim21 in membrane-potential-dependent preprotein sorting in mitochondria. *Curr. Biol.* **2006**, *16*, 2271–2276. [CrossRef]
87. Albrecht, R.; Rehling, P.; Chacinska, A.; Brix, J.; Cadamuro, S.A.; Volkmer, R.; Guiard, B.; Pfanner, N.; Zeth, K. The Tim21 binding domain connects the preprotein translocases of both mitochondrial membranes. *Embo Rep.* **2006**, *7*, 1233–1238. [CrossRef]
88. Mokranjac, D.; Popov-Celeketic, D.; Hell, K.; Neupert, W. Role of Tim21 in mitochondrial translocation contact sites. *J. Biol. Chem.* **2005**, *280*, 23437–23440. [CrossRef] [PubMed]
89. Ieva, R.; Schrempp, S.G.; Opalinski, L.; Wollweber, F.; Hoss, P.; Heisswolf, A.K.; Gebert, M.; Zhang, Y.; Guiard, B.; Rospert, S.; et al. Mgr2 Functions as Lateral Gatekeeper for Preprotein Sorting in the Mitochondrial Inner Membrane. *Mol. Cell* **2014**, *56*, 641–652. [CrossRef]
90. Endres, M.; Neupert, W.; Brunner, M. Transport of the ADP ATP carrier of mitochondria from the TOM complex to the TIM22.54 complex. *Embo J.* **1999**, *18*, 3214–3221. [CrossRef]
91. Curran, S.P.; Leuenberger, D.; Oppliger, W.; Koehler, C.M. The Tim9p-Tim10p complex binds to the transmembrane domains of the ADP/ATP carrier. *Embo J.* **2002**, *21*, 942–953. [CrossRef]
92. Rehling, P.; Model, K.; Brandner, K.; Kovermann, P.; Sickmann, A.; Meyer, H.E.; Kuhlbrandt, W.; Wagner, R.; Truscott, K.N.; Pfanner, N. Protein insertion into the mitochondrial inner membrane by a twin-pore translocase. *Science* **2003**, *299*, 1747–1751. [CrossRef] [PubMed]
93. Brix, J.; Rudiger, S.; Bukau, B.; Schneider-Mergener, J.; Pfanner, N. Distribution of binding sequences for the mitochondrial import receptors Tom20, Tom22, and Tom70 in a presequence-carrying preprotein and a non-cleavable preprotein. *J. Biol. Chem.* **1999**, *274*, 16522–16530. [CrossRef] [PubMed]
94. Curran, S.P.; Leuenberger, D.; Schmidt, E.; Koehler, C.M. The role of the Tim8p-Tim13p complex in a conserved import pathway for mitochondrial polytopic inner membrane proteins. *J. Cell Biol.* **2002**, *158*, 1017–1027. [CrossRef] [PubMed]
95. Qi, L.B.; Wang, Q.; Guan, Z.Y.; Wu, Y.; Shen, C.C.; Hong, S.X.; Cao, J.B.; Zhang, X.; Yan, C.Y.; Yin, P. Cryo-EM structure of the human mitochondrial translocase TIM22 complex. *Cell Res.* **2021**, *31*, 369–372. [CrossRef] [PubMed]
96. Zhang, Y.; Ou, X.; Wang, X.; Sun, D.; Zhou, X.; Wu, X.; Li, Q.; Li, L. Structure of the mitochondrial TIM22 complex from yeast. *Cell Res* **2021**, *31*, 366–368. [CrossRef]
97. Callegari, S.; Richter, F.; Chojnacka, K.; Jans, D.C.; Lorenzi, I.; Pacheu-Grau, D.; Jakobs, S.; Lenz, C.; Urlaub, H.; Dudek, J.; et al. TIM29 is a subunit of the human carrier translocase required for protein transport. *Febs Lett.* **2016**, *590*, 4147–4158. [CrossRef] [PubMed]
98. Kang, Y.L.; Baker, M.J.; Liem, M.; Louber, J.; McKenzie, M.; Atukorala, I.; Ang, C.S.; Keerthikumar, S.; Mathivanan, S.; Stojanovski, D. Tim29 is a novel subunit of the human TIM22 translocase and is involved in complex assembly and stability. *Elife* **2016**, *5*, e17463. [CrossRef]
99. Vukotic, M.; Nolte, H.; Konig, T.; Saita, S.; Ananjew, M.; Kruger, M.; Tatsuta, T.; Langer, T. Acylglycerol Kinase Mutated in Sengers Syndrome Is a Subunit of the TIM22 Protein Translocase in Mitochondria. *Mol. Cell* **2017**, *67*, 471–483. [CrossRef]
100. Kang, Y.L.; Stroud, D.A.; Baker, M.J.; De Souza, D.P.; Frazier, A.E.; Liem, M.; Tull, D.; Mathivanan, S.; McConville, M.J.; Thorburn, D.R.; et al. Sengers Syndrome-Associated Mitochondrial Acylglycerol Kinase Is a Subunit of the Human TIM22 Protein Import Complex. *Mol. Cell* **2017**, *67*, 457–470. [CrossRef]
101. Wiedemann, N.; Pfanner, N. Mitochondrial Machineries for Protein Import and Assembly. In *Annual Review of Biochemistry*; Kornberg, R.D., Ed.; Annual Reviews: Palo Alto, CA, USA, 2017; Volume 86, pp. 685–714.
102. Rehling, P.; Brandner, K.; Pfanner, N. Mitochondrial import and the twin-pore translocase. *Nat. Rev. Mol. Cell Biol.* **2004**, *5*, 519–530. [CrossRef]
103. Wu, Y.K.; Sha, B.D. Crystal structure of yeast mitochondrial outer membrane translocon member Tom70p. *Nat. Struct. Mol. Biol.* **2006**, *13*, 589–593. [CrossRef]
104. Young, J.C.; Hoogenraad, N.J.; Hartl, F.U. Molecular chaperones Hsp90 and Hsp70 deliver preproteins to the mitochondrial import receptor Tom70. *Cell* **2003**, *112*, 41–50. [CrossRef]

105. Bhangoo, M.K.; Tzankov, S.; Fan, A.C.Y.; Dejgaard, K.; Thomas, D.Y.; Young, J.C. Multiple 40-kDa heat-shock protein chaperones function in Tom70-dependent mitochondrial import. *Mol. Biol. Cell* **2007**, *18*, 3414–3428. [CrossRef]
106. Backes, S.; Bykov, Y.S.; Räschle, M.; Zhou, J.; Lenhard, S.; Krämer, L.; Mühlhaus, T.; Bibi, C.; Jann, C.; Smith, J.D.; et al. The mitochondrial surface receptor Tom70 protects the cytosol against mitoprotein-induced stress. *bioRxiv* **2020**. [CrossRef]
107. Wiedemann, N.; Pfanner, N.; Ryan, M.T. The three modules of ADP/ATP carrier cooperate in receptor recruitment and translocation into mitochondria. *Embo J.* **2001**, *20*, 951–960. [CrossRef] [PubMed]
108. Ellenrieder, L.; Dieterle, M.P.; Doan, K.N.; Martensson, C.U.; Floerchinger, A.; Campo, M.L.; Pfanner, N.; Becker, T. Dual Role of Mitochondrial Porin in Metabolite Transport across the Outer Membrane and Protein Transfer to the Inner Membrane. *Mol. Cell* **2019**, *73*, 1056–1065. [CrossRef] [PubMed]
109. Callegari, S.; Muller, T.; Schulz, C.; Lenz, C.; Jans, D.C.; Wissel, M.; Opazo, F.; Rizzoli, S.O.; Jakobs, S.; Urlaub, H.; et al. A MICOS-TIM22 Association Promotes Carrier Import into Human Mitochondria. *J. Mol. Biol.* **2019**, *431*, 2835–2851. [CrossRef]
110. Rampelt, H.; Sucec, I.; Bersch, B.; Horten, P.; Perschil, I.; Martinou, J.C.; van der Laan, M.; Wiedemann, N.; Schanda, P.; Pfanner, N. The mitochondrial carrier pathway transports non-canonical substrates with an odd number of transmembrane segments. *BMC Biol.* **2020**, *18*, 2. [CrossRef]
111. Jackson, T.D.; Hock, D.H.; Fujihara, K.M.; Palmer, C.S.; Frazier, A.E.; Low, Y.C.; Kang, Y.; Ang, C.-S.; Clemons, N.J.; Thorburn, D.R.; et al. The TIM22 complex mediates the import of Sideroflexins and is required for efficient mitochondrial one-carbon metabolism. *Mol. Biol. Cell* **2021**, *32*, 475–491. [CrossRef] [PubMed]
112. Luirink, J.; Samuelsson, T.; de Gier, J.W. YidC/Oxa1p/Alb3: Evolutionarily conserved mediators of membrane protein assembly. *FEBS Lett.* **2001**, *501*, 1–5. [CrossRef]
113. Bonnefoy, N.; Chalvet, F.; Hamel, P.; Slonimski, P.P.; Dujardin, G. OXA1, a saccharomyces-cerecconserved from prokaryotes to eukaryotes controls cytochrome-oxidase biogenesis. *J. Mol. Biol.* **1994**, *239*, 201–212. [CrossRef]
114. Herrmann, J.M.; Neupert, W.; Stuart, R.A. Insertion into the mitochondrial inner membrane of a polytopic protein, the nuclear-encoded Oxa1p. *Embo J.* **1997**, *16*, 2217–2226. [CrossRef]
115. Nargang, F.E.; Preuss, M.; Neupert, W.; Herrmann, J.M. The oxal protein forms a homooligomeric complex and is an essential part of the mitochondrial export translocase in Neurospora crassa. *J. Biol. Chem.* **2002**, *277*, 12846–12853. [CrossRef] [PubMed]
116. Stiburek, L.; Fornuskova, D.; Wenchich, L.; Pejznochova, M.; Hansikova, H.; Zeman, J. Knockdown of human Oxa1l impairs the biogenesis of F1Fo-ATP synthase and NADH: Ubiquinone oxidoreductase. *J. Mol. Biol.* **2007**, *374*, 506–516. [CrossRef]
117. Haque, M.E.; Elmore, K.B.; Tripathy, A.; Koc, H.; Koc, E.C.; Spremulli, L.L. Properties of the C-terminal Tail of Human Mitochondrial Inner Membrane Protein Oxa1L and Its Interactions with Mammalian Mitochondrial Ribosomes. *J. Biol. Chem.* **2010**, *285*, 28353–28362. [CrossRef] [PubMed]
118. Itoh, Y.; Andrell, J.; Choi, A.; Richter, U.; Maiti, P.; Best, R.B.; Barrientos, A.; Battersby, B.J.; Amunts, A. Mechanism of membrane-tethered mitochondrial protein synthesis. *Science* **2021**, *371*, 846–849. [CrossRef]
119. Stuart, R.A. Insertion of proteins into the inner membrane of mitochondria: The role of the Oxa1 complex. *Biochim. Biophys. Acta Mol. Cell Res.* **2002**, *1592*, 79–87. [CrossRef]
120. Bohnert, M.; Rehling, P.; Guiard, B.; Herrmann, J.M.; Pfanner, N.; van der Laan, M. Cooperation of Stop-Transfer and Conservative Sorting Mechanisms in Mitochondrial Protein Transport. *Curr. Biol.* **2010**, *20*, 1227–1232. [CrossRef]
121. Anghel, S.A.; McGilvray, P.T.; Hegde, R.S.; Keenan, R.J. Identification of Oxa1 Homologs Operating in the Eukaryotic Endoplasmic Reticulum. *Cell Rep.* **2017**, *21*, 3708–3716. [CrossRef]
122. Pleiner, T.; Tomaleri, G.P.; Januszyk, K.; Inglis, A.J.; Hazu, M.; Voorhees, R.M. Structural basis for membrane insertion by the human ER membrane protein complex. *Science* **2020**, *369*, 433–436. [CrossRef]
123. Bai, L.; You, Q.L.; Feng, X.; Kovach, A.; Li, H.L. Structure of the ER membrane complex, a transmembrane-domain insertase. *Nature* **2020**, *584*, 475–478. [CrossRef] [PubMed]
124. Chitwood, P.J.; Juszkiewicz, S.; Guna, A.; Shao, S.C.; Hegde, R.S. EMC Is Required to Initiate Accurate Membrane Protein Topogenesis. *Cell* **2018**, *175*, 1507–1519. [CrossRef] [PubMed]
125. Rojo, E.E.; Stuart, R.A.; Neupert, W. Conservative sorting of f-0-atpase subunit-9—export from matrix requires delta-ph across inner membrane and matrix atp. *Embo J.* **1995**, *14*, 3445–3451. [CrossRef] [PubMed]
126. Hell, K.; Neupert, W.; Stuart, R.A. Oxa1p acts as a general membrane insertion machinery for proteins encoded by mitochondrial DNA. *Embo J.* **2001**, *20*, 1281–1288. [CrossRef]
127. Herrmann, J.M.; Bonnefoy, N. Protein export across the inner membrane of mitochondria—The nature of translocated domains determines the dependence on the Oxa1 translocase. *J. Biol. Chem.* **2004**, *279*, 2507–2512. [CrossRef]
128. Schagger, H.; Pfeiffer, K. Supercomplexes in the respiratory chains of yeast and mammalian mitochondria. *Embo J.* **2000**, *19*, 1777–1783. [CrossRef] [PubMed]
129. Acín-Perez, R.; Bayona-Bafaluy, M.P.; Fernandez-Silva, P.; Moreno-Loshuertos, R.; Perez-Martos, A.; Bruno, C.; Moraes, C.T.; Enriquez, J.A. Respiratory complex III is required to maintain complex I in mammalian mitochondria. *Mol. Cell* **2004**, *13*, 805–815. [CrossRef]
130. Diaz, F.; Fukui, H.; Garcia, S.; Moraes, C.T. Cytochrome c oxidase is required for the assembly/stability of respiratory complex I in mouse fibroblasts. *Mol. Cell Biol.* **2006**, *26*, 4872–4881. [CrossRef]

131. Protasoni, M.; Pérez-Pérez, R.; Lobo-Jarne, T.; Harbour, M.E.; Ding, S.; Peñas, A.; Diaz, F.; Moraes, C.T.; Fearnley, I.M.; Zeviani, M.; et al. Respiratory supercomplexes act as a platform for complex III-mediated maturation of human mitochondrial complexes I and IV. *Embo J.* **2020**, *39*, e102817. [CrossRef]
132. Blakely, E.L.; Mitchell, A.L.; Fisher, N.; Meunier, B.; Nijtmans, L.G.; Schaefer, A.M.; Jackson, M.J.; Turnbull, D.M.; Taylor, R.W. A mitochondrial cytochrome b mutation causing severe respiratory chain enzyme deficiency in humans and yeast. *FEBS J.* **2005**, *272*, 3583–3592. [CrossRef] [PubMed]
133. Andreu, A.L.; Hanna, M.G.; Reichmann, H.; Bruno, C.; Penn, A.S.; Tanji, K.; Pallotti, F.; Iwata, S.; Bonilla, E.; Lach, B.; et al. Exercise intolerance due to mutations in the cytochrome b gene of mitochondrial DNA. *N. Engl. J. Med.* **1999**, *341*, 1037–1044. [CrossRef] [PubMed]
134. Lamantea, E.; Carrara, F.; Mariotti, C.; Morandi, L.; Tiranti, V.; Zeviani, M. A novel nonsense mutation (Q352X) in the mitochondrial cytochrome b gene associated with a combined deficiency of complexes I and III. *Neuromuscul. Disord. NMD* **2002**, *12*, 49–52. [CrossRef]
135. Moreno-Lastres, D.; Fontanesi, F.; Garcia-Consuegra, I.; Martin, M.A.; Arenas, J.; Barrientos, A.; Ugalde, C. Mitochondrial complex I plays an essential role in human respirasome assembly. *Cell Metab.* **2012**, *15*, 324–335. [CrossRef]
136. Wiedemann, N.; van der Laan, M.; Hutu, D.P.; Rehling, P.; Pfanner, N. Sorting switch of mitochondrial presequence translocase involves coupling of motor module to respiratory chain. *J. Cell Biol.* **2007**, *179*, 1115–1122. [CrossRef]
137. Guerrero-Castillo, S.; Baertling, F.; Kownatzki, D.; Wessels, H.J.; Arnold, S.; Brandt, U.; Nijtmans, L. The Assembly Pathway of Mitochondrial Respiratory Chain Complex I. *Cell Metab.* **2017**, *25*, 128–139. [CrossRef]
138. Fontanesi, F.; Soto, I.C.; Horn, D.; Barrientos, A. Mss51 and Ssc1 facilitate translational regulation of cytochrome c oxidase biogenesis. *Mol. Cell Biol.* **2010**, *30*, 245–259. [CrossRef]
139. Böttinger, L.; Guiard, B.; Oeljeklaus, S.; Kulawiak, B.; Zufall, N.; Wiedemann, N.; Warscheid, B.; van der Laan, M.; Becker, T. A complex of Cox4 and mitochondrial Hsp70 plays an important role in the assembly of the cytochrome c oxidase. *Mol. Biol. Cell* **2013**, *24*, 2609–2619. [CrossRef] [PubMed]
140. Rieger, B.; Junge, W.; Busch, K.B. Lateral pH gradient between OXPHOS complex IV and F0F1 ATP-synthase in folded mitochondrial membranes. *Nat. Commun.* **2014**, *5*, 3103. [CrossRef]
141. Couvillion, M.T.; Soto, I.C.; Shipkovenska, G.; Churchman, L.S. Synchronized mitochondrial and cytosolic translation programs. *Nature* **2016**, *533*, 499–503. [CrossRef]
142. Herrmann, J.M.; Woellhaf, M.W.; Bonnefoy, N. Control of protein synthesis in yeast mitochondria: The concept of translational activators. *Biochim. Biophys. Acta* **2013**, *1833*, 286–294. [CrossRef]
143. Ott, M.; Amunts, A.; Brown, A. Organization and Regulation of Mitochondrial Protein Synthesis. *Annu. Rev. Biochem.* **2016**, *85*, 77–101. [CrossRef] [PubMed]
144. Dennerlein, S.; Wang, C.; Rehling, P. Plasticity of Mitochondrial Translation. *Trends Cell Biol.* **2017**, *27*, 712–721. [CrossRef] [PubMed]
145. Timón-Gómez, A.N.E.; Abriata, L.A.; Vila, A.J.; Hosler, J.; Barrientos, A. Mitochondrial cytochrome c oxidase biogenesis: Recent developments. *Semin Cell Dev. Biol.* **2018**, *76*, 163–178. [CrossRef]
146. Gruschke, S.; Römpler, K.; Hildenbeutel, M.; Kehrein, K.; Kühl, I.; Bonnefoy, N.; Ott, M. The Cbp3-Cbp6 complex coordinates cytochrome b synthesis with bc(1) complex assembly in yeast mitochondria. *J. Cell Biol.* **2012**, *199*, 137–150. [CrossRef] [PubMed]
147. Hildenbeutel, M.; Hegg, E.L.; Stephan, K.; Gruschke, S.; Meunier, B.; Ott, M. Assembly factors monitor sequential hemylation of cytochrome b to regulate mitochondrial translation. *J. Cell Biol.* **2014**, *205*, 511–524. [CrossRef]
148. García-Villegas, R.; Camacho-Villasana, Y.; Shingú-Vázquez, M.; Cabrera-Orefice, A.; Uribe-Carvajal, S.; Fox, T.D.; Pérez-Martínez, X. The Cox1 C-terminal domain is a central regulator of cytochrome c oxidase biogenesis in yeast mitochondria. *J. Biol. Chem.* **2017**, *292*, 10912–10925. [CrossRef]
149. Perez-Martinez, X.; Butler, C.A.; Shingu-Vazquez, M.; Fox, T.D. Dual functions of Mss51 couple synthesis of Cox1 to assembly of cytochrome c oxidase in Saccharomyces cerevisiae mitochondria. *Mol. Biol Cell* **2009**, *20*, 4371–4380. [CrossRef]
150. Tavares-Carreón, F.; Camacho-Villasana, Y.; Zamudio-Ochoa, A.; Shingú-Vázquez, M.; Torres-Larios, A.; Pérez-Martínez, X. The pentatricopeptide repeats present in Pet309 are necessary for translation but not for stability of the mitochondrial COX1 mRNA in yeast. *J. Biol. Chem.* **2008**, *283*, 1472–1479. [CrossRef]
151. Zamudio-Ochoa, A.; Camacho-Villasana, Y.; García-Guerrero, A.E.; Pérez-Martínez, X. The Pet309 pentatricopeptide repeat motifs mediate efficient binding to the mitochondrial COX1 transcript in yeast. *RNA Biol.* **2014**, *11*, 953–967. [CrossRef]
152. Godard, F.; Tetaud, E.; Duvezin-Caubet, S.; di Rago, J.P. A genetic screen targeted on the FO component of mitochondrial ATP synthase in Saccharomyces cerevisiae. *J. Biol. Chem.* **2011**, *286*, 18181–18189. [CrossRef]
153. Helfenbein, K.G.; Ellis, T.P.; Dieckmann, C.L.; Tzagoloff, A. ATP22, a nuclear gene required for expression of the F0 sector of mitochondrial ATPase in Saccharomyces cerevisiae. *J. Biol. Chem.* **2003**, *278*, 19751–19756. [CrossRef]
154. Kellems, R.E.; Allison, V.F.; Butow, R.A. Cytoplasmic type 80S ribosomes associated with yeast mitochondria. IV. Attachment of ribosomes to the outer membrane of isolated mitochondria. *J. Cell Biol.* **1975**, *65*, 1–14. [CrossRef] [PubMed]
155. Ades, I.Z.; Butow, R.A. The transport of proteins into yeast mitochondria. Kinetics and pools. *J. Biol. Chem.* **1980**, *255*, 9925–9935. [CrossRef]
156. Suissa, M.; Schatz, G. Import of proteins into mitochondria. Translatable mRNAs for imported mitochondrial proteins are present in free as well as mitochondria-bound cytoplasmic polysomes. *J. Biol. Chem.* **1982**, *257*, 13048–13055. [CrossRef]

157. Hwang, S.T.; Wachter, C.; Schatz, G. Protein import into the yeast mitochondrial matrix. A new translocation intermediate between the two mitochondrial membranes. *J. Biol. Chem.* **1991**, *266*, 21083–21089. [CrossRef]
158. Garcia, M.D.X.; Devaux, F.; Singer, R.H.; Jacq, C. *Yeast Mitochondrial Transcriptomics*; Humana Press: Totowa, NJ, USA, 2007; Volume 372.
159. Marc, P.; Margeot, A.; Devaux, F.; Blugeon, C.; Corral-Debrinski, M.; Jacq, C. Genome-wide analysis of mRNAs targeted to yeast mitochondria. *EMBO Rep.* **2002**, *3*, 159–164. [CrossRef] [PubMed]
160. Garcia, M.; Darzacq, X.; Delaveau, T.; Jourdren, L.; Singer, R.H.; Jacq, C. Mitochondria-associated yeast mRNAs and the biogenesis of molecular complexes. *Mol. Biol. Cell* **2007**, *18*, 362–368. [CrossRef] [PubMed]
161. Saint-Georges, Y.; Garcia, M.; Delaveau, T.; Jourdren, L.; Le Crom, S.; Lemoine, S.; Tanty, V.; Devaux, F.; Jacq, C. Yeast mitochondrial biogenesis: A role for the PUF RNA-binding protein Puf3p in mRNA localization. *PLoS ONE* **2008**, *3*, e2293. [CrossRef] [PubMed]
162. Gadir, N.; Haim-Vilmovsky, L.; Kraut-Cohen, J.; Gerst, J.E. Localization of mRNAs coding for mitochondrial proteins in the yeast Saccharomyces cerevisiae. *RNA* **2011**, *17*, 1551–1565. [CrossRef]
163. Fox, T. Mitochondrial Protein Synthesis, Import, and Assembly. *Genetics* **2012**, *192*, 1203–1234. [CrossRef] [PubMed]
164. Gilkerson, R.W.; Selker, J.M.; Capaldi, R.A. The cristal membrane of mitochondria is the principal site of oxidative phosphorylation. *FEBS Lett.* **2003**, *546*, 355–358. [CrossRef]
165. Vogel, F.; Bornhövd, C.; Neupert, W.; Reichert, A.S. Dynamic subcompartmentalization of the mitochondrial inner membrane. *J. Cell Biol.* **2006**, *175*, 237–247. [CrossRef]
166. Stoldt, S.; Wenzel, D.; Kehrein, K.; Riedel, D.; Ott, M.; Jakobs, S. Spatial orchestration of mitochondrial translation and OXPHOS complex assembly. *Nat. Cell Biol.* **2018**, *20*, 528–534. [CrossRef]
167. Paumard, P.; Vaillier, J.; Coulary, B.; Schaeffer, J.; Soubannier, V.; Mueller, D.M.; Brethes, D.; di Rago, J.P.; Velours, J. The ATP synthase is involved in generating mitochondrial cristae morphology. *Embo J.* **2002**, *21*, 221–230. [CrossRef] [PubMed]
168. Spikes, T.E.; Montgomery, M.G.; Walker, J.E. Interface mobility between monomers in dimeric bovine ATP synthase participates in the ultrastructure of inner mitochondrial membranes. *Proc. Natl. Acad. Sci. USA* **2021**, *118*, e2021012118. [CrossRef]
169. Kim, H.J.; Khalimonchuk, O.; Smith, P.M.; Winge, D.R. Structure, function, and assembly of heme centers in mitochondrial respiratory complexes. *Biochim. Biophys. Acta Mol. Cell Res.* **2012**, *1823*, 1604–1616. [CrossRef]
170. Stram, A.R.; Payne, R.M. Post-translational modifications in mitochondria: Protein signaling in the powerhouse. *Cell. Mol. Life Sci.* **2016**, *73*, 4063–4073. [CrossRef] [PubMed]
171. Cardenas-Rodriguez, M.; Chatzi, A.; Tokatlidis, K. Iron-sulfur clusters: From metals through mitochondria biogenesis to disease. *J. Biol. Inorg. Chem.* **2018**, *23*, 509–520. [CrossRef]
172. Vinothkumar, K.R.; Zhu, J.; Hirst, J. Architecture of mammalian respiratory complex I. *Nature* **2014**, *515*, 80–84. [CrossRef]
173. Carroll, J.; Ding, S.; Fearnley, I.M.; Walker, J.E. Post-translational modifications near the quinone binding site of mammalian complex I. *J. Biol. Chem.* **2013**, *288*, 24799–24808. [CrossRef] [PubMed]
174. Signes, A.; Fernandez-Vizarra, E. Assembly of mammalian oxidative phosphorylation complexes I-V and supercomplexes. *Essays Biochem.* **2018**, *62*, 255–270. [CrossRef] [PubMed]
175. Formosa, L.E.; Muellner-Wong, L.; Reljic, B.; Sharpe, A.J.; Jackson, T.D.; Beilharz, T.H.; Stojanovski, D.; Lazarou, M.; Stroud, D.A.; Ryan, M.T. Dissecting the Roles of Mitochondrial Complex I Intermediate Assembly Complex Factors in the Biogenesis of Complex I. *Cell Rep.* **2020**, *31*, 107541. [CrossRef]
176. Andrews, B.; Carroll, J.; Ding, S.J.; Fearnley, I.M.; Walker, J.E. Assembly factors for the membrane arm of human complex I. *Proc. Natl. Acad. Sci. USA* **2013**, *110*, 18934–18939. [CrossRef] [PubMed]
177. Stroud, D.A.; Surgenor, E.E.; Formosa, L.E.; Reljic, B.; Frazier, A.E.; Dibley, M.G.; Osellame, L.D.; Stait, T.; Beilharz, T.H.; Thorburn, D.R.; et al. Accessory subunits are integral for assembly and function of human mitochondrial complex I. *Nature* **2016**, *538*, 123–126. [CrossRef]
178. Claros, M.G.; Vincens, P. Computational method to predict mitochondrially imported proteins and their targeting sequences. *Eur. J. Biochem.* **1996**, *241*, 779–786. [CrossRef] [PubMed]
179. Szklarczyk, R.; Wanschers, B.F.J.; Nabuurs, S.B.; Nouws, J.; Nijtmans, L.G.; Huynen, M.A. NDUFB7 and NDUFA8 are located at the intermembrane surface of complex I. *FEBS Lett.* **2011**, *585*, 737–743. [CrossRef]
180. Friederich, M.W.; Erdogan, A.J.; Coughlin, C.R.; Elos, M.T.; Jiang, H.; O'Rourke, C.P.; Lovell, M.A.; Wartchow, E.; Gowan, K.; Chatfield, K.C.; et al. Mutations in the accessory subunit NDUFB10 result in isolated complex I deficiency and illustrate the critical role of intermembrane space import for complex I holoenzyme assembly. *Hum. Mol. Genet.* **2017**, *26*, 702–716. [CrossRef]
181. Sánchez-Caballero, L.; Ruzzenente, B.; Bianchi, L.; Assouline, Z.; Barcia, G.; Metodiev, M.D.; Rio, M.; Funalot, B.; van den Brand, M.A.; Guerrero-Castillo, S.; et al. Mutations in Complex I Assembly Factor TMEM126B Result in Muscle Weakness and Isolated Complex I Deficiency. *Am. J. Hum. Genet.* **2016**, *99*, 208–216. [CrossRef]
182. Zarsky, V.; Dolezal, P. Evolution of the Tim17 protein family. *Biol. Direct* **2016**, *11*, 54. [CrossRef]
183. Wang, Y.; Carrie, C.; Giraud, E.; Elhafez, D.; Narsai, R.; Duncan, O.; Whelan, J.; Murcha, M.W. Dual Location of the Mitochondrial Preprotein Transporters B14.7 and Tim23-2 in Complex I and the TIM17:23 Complex in Arabidopsis Links Mitochondrial Activity and Biogenesis. *Plant Cell* **2012**, *24*, 2675–2695. [CrossRef] [PubMed]
184. Sheftel, A.D.; Stehling, O.; Pierik, A.J.; Netz, D.J.; Kerscher, S.; Elsasser, H.P.; Wittig, I.; Balk, J.; Brandt, U.; Lill, R. Human ind1, an iron-sulfur cluster assembly factor for respiratory complex I. *Mol. Cell Biol.* **2009**, *29*, 6059–6073. [CrossRef] [PubMed]

185. Bych, K.; Kerscher, S.; Netz, D.J.; Pierik, A.J.; Zwicker, K.; Huynen, M.A.; Lill, R.; Brandt, U.; Balk, J. The iron-sulphur protein Ind1 is required for effective complex I assembly. *Embo J.* **2008**, *27*, 1736–1746. [CrossRef] [PubMed]
186. Sun, F.; Huo, X.; Zhai, Y.; Wang, A.; Xu, J.; Su, D.; Bartlam, M.; Rao, Z. Crystal structure of mitochondrial respiratory membrane protein complex II. *Cell* **2005**, *121*, 1043–1057. [CrossRef] [PubMed]
187. Van Vranken, J.; Na, U.; Winge, D.R.; Rutter, J. Protein-mediated assembly of succinate dehydrogenase and its cofactors. *Crit. Rev. Biochem. Mol. Biol.* **2015**, *50*, 168–180. [CrossRef]
188. Hao, H.-X.; Khalimonchuk, O.; Schraders, M.; Dephoure, N.; Bayley, J.-P.; Kunst, H.; Devilee, P.; Cremers, C.W.R.J.; Schiffman, J.D.; Bentz, B.G.; et al. SDH5, a gene required for flavination of succinate dehydrogenase, is mutated in paraganglioma. *Science* **2009**, *325*, 1139–1142. [CrossRef]
189. Cecchini, G. Function and structure of complex II of the respiratory chain. *Annu Rev. Biochem.* **2003**, *72*, 77–109. [CrossRef] [PubMed]
190. Ghezzi, D.; Goffrini, P.; Uziel, G.; Horvath, R.; Klopstock, T.; Lochmüller, H.; D'Adamo, P.; Gasparini, P.; Strom, T.M.; Prokisch, H.; et al. SDHAF1, encoding a LYR complex-II specific assembly factor, is mutated in SDH-defective infantile leukoencephalopathy. *Nat. Genet.* **2009**, *41*, 654–656. [CrossRef]
191. Maio, N.; Ghezzi, D.; Verrigni, D.; Rizza, T.; Bertini, E.; Martinelli, D.; Zeviani, M.; Singh, A.; Carrozzo, R.; Rouault, T.A. Disease-Causing SDHAF1 Mutations Impair Transfer of Fe-S Clusters to SDHB. *Cell Metab.* **2016**, *23*, 292–302. [CrossRef]
192. Na, U.; Yu, W.; Cox, J.; Bricker, D.K.; Brockmann, K.; Rutter, J.; Thummel, C.S.; Winge, D.R. The LYR factors SDHAF1 and SDHAF3 mediate maturation of the iron-sulfur subunit of succinate dehydrogenase. *Cell Metab.* **2014**, *20*, 253–266. [CrossRef] [PubMed]
193. Gebert, N.; Gebert, M.; Oeljeklaus, S.; von der Malsburg, K.; Stroud, D.A.; Kulawiak, B.; Wirth, C.; Zahedi, R.P.; Dolezal, P.; Wiese, S.; et al. Dual Function of Sdh3 in the Respiratory Chain and TIM22 Protein Translocase of the Mitochondrial Inner Membrane. *Mol. Cell* **2011**, *44*, 811–818. [CrossRef]
194. Schagger, H.; Link, T.A.; Engel, W.D.; von Jagow, G. Isolation of the eleven protein subunits of the bc1 complex from beef heart. *Methods Enzymol.* **1986**, *126*, 224–237.
195. Iwata, S.; Lee, J.W.; Okada, K.; Lee, J.K.; Iwata, M.; Rasmussen, B.; Link, T.A.; Ramaswamy, S.; Jap, B.K. Complete structure of the 11-subunit bovine mitochondrial cytochrome bc1 complex. *Science* **1998**, *281*, 64–71. [CrossRef] [PubMed]
196. Thompson, K.; Mai, N.; Oláhová, M.; Scialó, F.; Formosa, L.E.; Stroud, D.A.; Garrett, M.; Lax, N.Z.; Robertson, F.M.; Jou, C.; et al. OXA1L mutations cause mitochondrial encephalopathy and a combined oxidative phosphorylation defect. *EMBO Mol. Med.* **2018**, *10*, e9060. [CrossRef]
197. Cogliati, S.; Lorenzi, I.; Rigoni, G.; Caicci, F.; Soriano, M.E. Regulation of Mitochondria! Electron Transport Chain Assembly. *J. Mol. Biol.* **2018**, *430*, 4849–4873. [CrossRef] [PubMed]
198. Ndi, M.; Marin-Buera, L.; Salvatori, R.; Singh, A.P.; Ott, M. Biogenesis of the bc1 Complex of the Mitochondrial Respiratory Chain. *J. Mol. Biol.* **2018**, *430*, 3892–3905. [CrossRef] [PubMed]
199. Tucker, E.J.; Wanschers, B.F.; Szklarczyk, R.; Mountford, H.S.; Wijeyeratne, X.W.; van den Brand, M.A.; Leenders, A.M.; Rodenburg, R.J.; Reljic, B.; Compton, A.G.; et al. Mutations in the UQCC1-interacting protein, UQCC2, cause human complex III deficiency associated with perturbed cytochrome b protein expression. *PLoS Genet.* **2013**, *9*, e1004034. [CrossRef] [PubMed]
200. Wanschers, B.F.J.; Szklarczyk, R.; van den Brand, M.A.M.; Jonckheere, A.; Suijskens, J.; Smeets, R.; Rodenburg, R.J.; Stephan, K.; Helland, I.B.; Elkamil, A.; et al. A mutation in the human CBP4 ortholog UQCC3 impairs complex III assembly, activity and cytochrome b stability. *Hum. Mol. Genet.* **2014**, *23*, 6356–6365. [CrossRef]
201. Zara, V.; Palmisano, I.; Conte, L.; Trumpower, B.L. Further insights into the assembly of the yeast cytochrome bc1 complex based on analysis of single and double deletion mutants lacking supernumerary subunits and cytochrome b. *Eur. J. Biochem.* **2004**, *271*, 1209–1218. [CrossRef]
202. Zara, V.; Conte, L.; Trumpower, B.L. Identification and characterization of cytochrome bc(1) subcomplexes in mitochondria from yeast with single and double deletions of genes encoding cytochrome bc(1) subunits. *FEBS J.* **2007**, *274*, 4526–4539. [CrossRef]
203. Zara, V.; Conte, L.; Trumpower, B.L. Biogenesis of the yeast cytochrome bc1 complex. *Biochim. Biophys. Acta* **2009**, *1793*, 89–96. [CrossRef] [PubMed]
204. Zara, V.; Conte, L.; Trumpower, B.L. Evidence that the assembly of the yeast cytochrome bc1 complex involves the formation of a large core structure in the inner mitochondrial membrane. *FEBS J.* **2009**, *276*, 1900–1914. [CrossRef]
205. Stephan, K.; Ott, M. Timing of dimerization of the bc(1) complex during mitochondrial respiratory chain assembly. *Biochim. Et Biophys. Acta. Bioenerg.* **2020**, *1861*, 148177. [CrossRef] [PubMed]
206. Xia, D.; Yu, C.A.; Kim, H.; Xia, J.Z.; Kachurin, A.M.; Zhang, L.; Yu, L.; Deisenhofer, J. Crystal structure of the cytochrome bc1 complex from bovine heart mitochondria. *Science* **1997**, *277*, 60–66. [CrossRef]
207. Sadler, I.; Suda, K.; Schatz, G.; Kaudewitz, F.; Haid, A. Sequencing of the nuclear gene for the yeast cytochrome c1 precursor reveals an unusually complex amino-terminal presequence. *Embo J.* **1984**, *3*, 2137–2143. [CrossRef] [PubMed]
208. Arnold, I.; Folsch, H.; Neupert, W.; Stuart, R.A. Two distinct and independent mitochondrial targeting signals function in the sorting of an inner membrane protein, cytochrome c1. *J. Biol. Chem.* **1998**, *273*, 1469–1476. [CrossRef] [PubMed]
209. Wachter, C.; Schatz, G.; Glick, B.S. Role of atp in the intramitochondrial sorting of cytochrome-C(1) and the adenine-nucleotide translocator. *Embo J.* **1992**, *11*, 4787–4794. [CrossRef]
210. Zollner, A.; Rödel, G.; Haid, A. Molecular cloning and characterization of the Saccharomyces cerevisiae CYT2 gene encoding cytochrome-c1–heme lyase. *FEBS J.* **1992**, *207*, 1093–1100. [CrossRef] [PubMed]

211. Van Loon, A.; Brandli, A.W.; Pesold-Hurt, B.; Blank, D.; Schatz, G. Transport of proteins to the mitochondrial intermembrane space: The 'matrix-targeting' and the 'sorting' domains in the cytochrome c1 presequence. *Embo J.* **1987**, *6*, 2433–2439. [CrossRef]
212. Phillips, J.; Schmitt, M.E.; Brown, T.A.; Beckmann, J.D.; Trumpower, B.L. Isolation and characterization of QCR9, a nuclear gene encoding the 7.3-kDa subunit 9 of the Saccharomyces cerevisiae ubiquinol-cytochrome c oxidoreductase complex. An intron-containing gene with a conserved sequence occurring in the intron of COX4. *J. Biol. Chem.* **1990**, *265*, 20813–20821. [CrossRef]
213. Brandt, U.; Uribe, S.; Schägger, H.; Trumpower, B.L. Isolation and characterization of QCR10, the nuclear gene encoding the 8.5-kDa subunit 10 of the Saccharomyces cerevisiae cytochrome bc1 complex. *J. Biol. Chem.* **1994**, *269*, 12947–12953. [CrossRef]
214. Wasilewski, M.; Chojnacka, K.; Chacinska, A. Protein trafficking at the crossroads to mitochondria. *Biochim. Biophys. Acta. Mol. Cell Res.* **2017**, *1864*, 125–137. [CrossRef]
215. Fu, W.; Japa, S.; Beattie, D.S. Import of the iron-sulfur protein of the cytochrome b.c1 complex into yeast mitochondria. *J. Biol. Chem.* **1990**, *265*, 16541–16547. [CrossRef]
216. Wagener, N.; Ackermann, M.; Funes, S.; Neupert, W. A pathway of protein translocation in mitochondria mediated by the AAA-ATPase Bcs1. *Mol. Cell.* **2011**, *44*, 191–202. [CrossRef] [PubMed]
217. Kater, L.; Wagener, N.; Berninghausen, O.; Becker, T.; Neupert, W.; Beckmann, R. Structure of the Bcs1 AAA-ATPase suggests an airlock-like translocation mechanism for folded proteins. *Nat. Struct. Mol. Biol.* **2020**, *27*, 142–149. [CrossRef]
218. Tang, W.K.; Borgnia, M.J.; Hsu, A.L.; Esser, L.; Fox, T.; de Val, N.; Xia, D. Structures of AAA protein translocase Bcs1 suggest translocation mechanism of a folded protein. *Nat. Struct. Mol. Biol.* **2020**, *27*, 202–209. [CrossRef] [PubMed]
219. Cui, T.Z.; Smith, P.M.; Fox, J.L.; Khalimonchuk, O.; Winge, D.R. Late-Stage Maturation of the Rieske Fe/S Protein: Mzm1 Stabilizes Rip1 but Does Not Facilitate Its Translocation by the AAA ATPase Bcs1. *Mol. Cell. Biol.* **2012**, *32*, 4400–4409. [CrossRef] [PubMed]
220. Sanchez, E.; Lobo, T.; Fox, J.L.; Zeviani, M.; Winge, D.R.; Fernandez-Vizarra, E. LYRM7/MZM1L is a UQCRFS1 chaperone involved in the last steps of mitochondrial Complex III assembly in human cells. *Biochim. Biophys. Acta Bioenerg.* **2013**, *1827*, 285–293. [CrossRef] [PubMed]
221. Brandt, U.; Yu, L.; Yu, C.A.; Trumpower, B.L. The mitochondrial targeting presequence of the Rieske iron-sulfur protein is processed in a single step after insertion into the cytochrome bc1 complex in mammals and retained as a subunit in the complex. *J. Biol. Chem.* **1993**, *268*, 8387–8390. [CrossRef]
222. Bottani, E.; Cerutti, R.; Harbour, M.E.; Ravaglia, S.; Dogan, S.A.; Giordano, C.; Fearnley, I.M.; D'Amati, G.; Viscomi, C.; Fernandez-Vizarra, E.; et al. TTC19 Plays a Husbandry Role on UQCRFS1 Turnover in the Biogenesis of Mitochondrial Respiratory Complex III. *Mol. Cell* **2017**, *67*, 96–105. [CrossRef]
223. Ghezzi, D.; Arzuffi, P.; Zordan, M.; Da Re, C.; Lamperti, C.; Benna, C.; D'Adamo, P.; Diodato, D.; Costa, R.; Mariotti, C.; et al. Mutations in TTC19 cause mitochondrial complex III deficiency and neurological impairment in humans and flies. *Nat. Genet.* **2011**, *43*, 259–263. [CrossRef]
224. Capaldi, R.A. Structure and function of cytochrome c oxidase. *Annu. Rev. Biochem.* **1990**, *59*, 569–596. [CrossRef] [PubMed]
225. Vidoni, S.H.M.E.; Guerrero-Castillo, S.; Signes, A.; Ding, S.; Fearnley, I.M.; Taylor, R.W.; Tiranti, V.; Arnold, S.; Fernandez-Vizarra, E.; Zeviani, M. MR-1S Interacts with PET100 and PET117 in Module-Based Assembly of Human Cytochrome c Oxidase. *Cell Rep.* **2017**, *18*, 1727–1738. [CrossRef]
226. Hayashi, T.; Asano, Y.; Shintani, Y.; Aoyama, H.; Kioka, H.; Tsukamoto, O.; Hikita, M.; Shinzawa-Itoh, K.; Takafuji, K.; Higo, S.; et al. Higd1a is a positive regulator of cytochrome c oxidase. *Proc. Natl. Acad. Sci. USA* **2015**, *112*, 1553–1558. [CrossRef]
227. Mick, D.; Dennerlein, S.; Wiese, H.; Reinhold, R.; Pacheu-Grau, D.; Lorenzi, I.; Sasarman, F.; Weraarpachai, W.; Shoubridge, E.A.; Warscheid, B.; et al. MITRAC links mitochondrial protein translocation to respiratory-chain assembly and translational regulation. *Cell* **2012**, *15*, 1528–1541. [CrossRef] [PubMed]
228. Dennerlein, S.; Oeljeklaus, S.; Jans, D.; Hellwig, C.; Bareth, B.; Jakobs, S.; Deckers, M.; Warscheid, B.; Rehling, P. MITRAC7 Acts as a COX1-Specific Chaperone and Reveals a Checkpoint during Cytochrome c Oxidase Assembly. *Cell Rep.* **2015**, *12*, 1644–1655. [CrossRef] [PubMed]
229. Szklarczyk, R.; Wanschers, B.F.; Cuypers, T.D.; Esseling, J.J.; Riemersma, M.; van den Brand, M.A.; Gloerich, J.; Lasonder, E.; van den Heuvel, L.P.; Nijtmans, L.G.; et al. Iterative orthology prediction uncovers new mitochondrial proteins and identifies C12orf62 as the human ortholog of COX14, a protein involved in the assembly of cytochrome c oxidase. *Genome Biol.* **2012**, *13*, R12. [CrossRef] [PubMed]
230. Clemente, P.; Peralta, S.; Cruz-Bermudez, A.; Echevarría, L.; Fontanesi, F.; Barrientos, A.; Fernandez-Moreno, M.A.; Garesse, R. hCOA3 stabilizes cytochrome c oxidase 1 (COX1) and promotes cytochrome c oxidase assembly in human mitochondria. *J. Biol. Chem.* **2013**, *288*, 8321–8331. [CrossRef] [PubMed]
231. Mick, D.; Vukotic, M.; Piechura, H.; Meyer, H.E.; Warscheid, B.; Deckers, M.; Rehling, P. Coa3 and Cox14 are essential for negative feedback regulation of COX1 translation in mitochondria. *J. Cell Biol.* **2010**, *191*, 141–154. [CrossRef] [PubMed]
232. Antonicka, H.; Leary, S.C.; Guercin, G.H.; Agar, J.N.; Horvath, R.; Kennaway, N.G.; Harding, C.O.; Jaksch, M.; Shoubridge, E.A. Mutations in COX10 result in a defect in mitochondrial heme A biosynthesis and account for multiple, early-onset clinical phenotypes associated with isolated COX deficiency. *Hum. Mol. Genet.* **2003**, *12*, 2693–2702. [CrossRef]
233. Diaz, F.; Thomas, C.K.; Garcia, S.; Hernandez, D.; Moraes, C.T. Mice lacking COX10 in skeletal muscle recapitulate the phenotype of progressive mitochondrial myopathies associated with cytochrome c oxidase deficiency. *Hum. Mol. Genet.* **2005**, *14*, 2737–2748. [CrossRef]

234. Hiser, L.; Di Valentin, M.; Hamer, A.G.; Hosler, J.P. Cox11p is required for stable formation of the Cu(B) and magnesium centers of cytochrome c oxidase. *J. Biol. Chem.* **2000**, *275*, 619–623. [CrossRef] [PubMed]
235. Glerum, D.; Shtanko, A.; Tzagoloff, A. Characterization of COX17, a yeast gene involved in copper metabolism and assembly of cytochrome oxidase. *J. Biol. Chem.* **1996**, *271*, 14504–14509. [CrossRef]
236. Bode, M.; Woellhaf, M.W.; Bohnert, M.; van der Laan, M.; Sommer, F.; Jung, M.; Zimmermann, R.; Schroda, M.; Herrmann, J.M. Redox-regulated dynamic interplay between Cox19 and the copper-binding protein Cox11 in the intermembrane space of mitochondria facilitates biogenesis of cytochrome c oxidase. *Mol. Biol. Cell* **2015**, *26*, 2385–2401. [CrossRef] [PubMed]
237. Mansilla, N.; Racca, S.; Gras, D.E.; Gonzalez, D.H.; Welchen, E. The Complexity of Mitochondrial Complex IV: An Update of Cytochrome c Oxidase Biogenesis in Plants. *Int. J. Mol. Sci.* **2018**, *19*, 662. [CrossRef] [PubMed]
238. Bourens, M.; Boulet, A.; Leary, S.C.; Barrientos, A. Human COX20 cooperates with SCO1 and SCO2 to mature COX2 and promote the assembly of cytochrome c oxidase. *Hum. Mol. Genet.* **2014**, *23*, 2901–2913. [CrossRef]
239. Lorenzi, I.; Oeljeklaus, S.; Aich, A.; Ronsör, C.; Callegari, S.; Dudek, J.; Warscheid, B.; Dennerlein, S.; Rehling, P. The mitochondrial TMEM177 associates with COX20 during COX2 biogenesis. *Biochim. Biophys. Acta Mol. Cell Res.* **2018**, *1865*, 323–333. [CrossRef] [PubMed]
240. Leary, S.C.; Sasarman, F.; Nishimura, T.; Shoubridge, E.A. Human SCO2 is required for the synthesis of CO II and as a thiol-disulphide oxidoreductase for SCO1. *Hum. Mol. Genet.* **2009**, *18*, 2230–2240. [CrossRef]
241. Leary, S.; Cobine, P.A.; Kaufman, B.A.; Guercin, G.H.; Mattman, A.; Palaty, J.; Lockitch, G.; Winge, D.R.; Rustin, P.; Horvath, R.; et al. The human cytochrome c oxidase assembly factors SCO1 and SCO2 have regulatory roles in the maintenance of cellular copper homeostasis. *Cell Metab.* **2007**, *5*, 9–20. [CrossRef] [PubMed]
242. Leary, S.; Kaufman, B.A.; Pellecchia, G.; Guercin, G.H.; Mattman, A.; Jaksch, M.; Shoubridge, E.A. Human SCO1 and SCO2 have independent, cooperative functions in copper delivery to cytochrome c oxidase. *Hum. Mol. Genet.* **2004**, *13*, 1839–1848. [CrossRef]
243. Stroud, D.; Maher, M.J.; Lindau, C.; Vögtle, F.N.; Frazier, A.E.; Surgenor, E.; Mountford, H.; Singh, A.P.; Bonas, M.; Oeljeklaus, S.; et al. COA6 is a mitochondrial complex IV assembly factor critical for biogenesis of mtDNA-encoded COX2. *Hum. Mol. Genet.* **2015**, *24*, 5404–5415. [CrossRef]
244. Pacheu-Grau, D.; Bareth, B.; Dudek, J.; Juris, L.; Vögtle, F.N.; Wissel, M.; Leary, S.C.; Dennerlein, S.; Rehling, P.; Deckers, M. Cooperation between COA6 and SCO2 in COX2 maturation during cytochrome c oxidase assembly links two mitochondrial cardiomyopathies. *Cell Metab.* **2015**, *21*, 823–833. [CrossRef] [PubMed]
245. Aich, A.; Wang, C.; Chowdhury, A.; Ronsör, C.; Pacheu-Grau, D.; Richter-Dennerlein, R.; Dennerlein, S.; Rehling, P. COX16 promotes COX2 metallation and assembly during respiratory complex IV biogenesis. *eLife* **2018**, *7*, e32572. [CrossRef] [PubMed]
246. Cerqua, C.; Morbidoni, V.; Desbats, M.A.; Doimo, M.; Frasson, C.; Sacconi, S.; Baldoin, M.C.; Sartori, G.; Basso, G.; Salviati, L.; et al. COX16 is required for assembly of cytochrome c oxidase in human cells and is involved in copper delivery to COX2. *Biochim. Biophys. Acta* **2018**, *1859*, 244–252. [CrossRef] [PubMed]
247. Jonckheere, A.I.; Smeitink, J.A.; Rodenburg, R.J. Mitochondrial ATP synthase: Architecture, function and pathology. *J. Inherit. Metab. Dis.* **2012**, *35*, 211–225. [CrossRef]
248. Abrahams, J.P.; Leslie, A.G.; Lutter, R.; Walker, J.E. Structure at 2.8 A resolution of F1-ATPase from bovine heart mitochondria. *Nature* **1994**, *370*, 621–628. [CrossRef]
249. Wittig, I.; Schagger, H. Structural organization of mitochondrial ATP synthase. *Biochim. Biophys. Acta* **2008**, *1777*, 592–598. [CrossRef] [PubMed]
250. Ackerman, S.H.; Tzagoloff, A. Identification of two nuclear genes (ATP11, ATP12) required for assembly of the yeast F1-ATPase. *Proc. Natl. Acad. Sci. USA* **1990**, *87*, 4986–4990. [CrossRef]
251. He, J.; Ford, H.C.; Carroll, J.; Ding, S.; Fearnley, I.M.; Walker, J.E. Persistence of the mitochondrial permeability transition in the absence of subunit c of human ATP synthase. *Proc. Natl. Acad. Sci. USA* **2017**, *114*, 3409–3414. [CrossRef]
252. Walker, J.E. The ATP synthase: The understood, the uncertain and the unknown. *Biochem. Soc. Trans.* **2013**, *41*, 1–16. [CrossRef]
253. Dyer, M.R.; Walker, J.E. Sequences of members of the human gene family for the c subunit of mitochondrial ATP synthase. *Biochem. J.* **1993**, *293 Pt 1*, 51–64. [CrossRef]
254. Yan, W.L.; Lerner, T.J.; Haines, J.L.; Gusella, J.F. Sequence analysis and mapping of a novel human mitochondrial ATP synthase subunit 9 cDNA (ATP5G3). *Genomics* **1994**, *24*, 375–377. [CrossRef]
255. Van Bloois, E.; Haan, G.J.; de Gier, J.W.; Oudega, B.; Luirink, J. F1F0 ATP synthase subunit c is targeted by the SRP to YidC in the E. coli inner membrane. *FEBS Lett.* **2004**, *576*, 97–100. [CrossRef] [PubMed]
256. Kolli, R.; Soll, J.; Carrie, C. Plant Mitochondrial Inner Membrane Protein Insertion. *Int. J. Mol. Sci.* **2018**, *19*, 641. [CrossRef]
257. Bahri, H.; Buratto, J.; Rojo, M.; Dompierre, J.P.; Salin, B.; Blancard, C.; Cuvellier, S.; Rose, M.; Ben Ammar Elgaaied, A.; Tetaud, E.; et al. TMEM70 forms oligomeric scaffolds within mitochondrial cristae promoting in situ assembly of mammalian ATP synthase proton channel. *Biochim. Biophys. Acta. Mol. Cell Res.* **2021**, *1868*, 118942. [CrossRef] [PubMed]
258. Carroll, J.; He, J.; Ding, S.; Fearnley, I.M.; Walker, J.E. TMEM70 and TMEM242 help to assemble the rotor ring of human ATP synthase and interact with assembly factors for complex I. *Proc. Natl. Acad. Sci. USA* **2021**, *118*, e2100558118. [CrossRef] [PubMed]
259. Ahting, U.; Floss, T.; Uez, N.; Schneider-Lohmar, I.; Becker, L.; Kling, E.; Iuso, A.; Bender, A.; de Angelis, M.H.; Gailus-Durner, V.; et al. Neurological phenotype and reduced lifespan in heterozygous Tim23 knockout mice, the first mouse model of defective mitochondrial import. *Biochim. Biophys. Acta Bioenerg.* **2009**, *1787*, 371–376. [CrossRef] [PubMed]

260. Devi, L.; Prabhu, B.M.; Galati, D.F.; Avadhani, N.G.; Anandatheerthavarada, H.K. Accumulation of amyloid precursor protein in the mitochondrial import channels of human Alzheimer's disease brain is associated with mitochondrial dysfunction. *J. Neurosci.* **2006**, *26*, 9057–9068. [CrossRef] [PubMed]
261. Sirk, D.; Zhu, Z.; Wadia, J.S.; Shulyakova, N.; Phan, N.; Fong, J.; Mills, L.R. Chronic exposure to sub-lethal beta-amyloid (A beta) inhibits the import of nuclear-encoded proteins to mitochondria in differentiated PC12 cells. *J. Neurochem.* **2007**, *103*, 1989–2003. [CrossRef] [PubMed]
262. Cieri, D.; Vicario, M.; Vallese, F.; D'Orsi, B.; Berto, P.; Grinzato, A.; Catoni, C.; De Stefani, D.; Rizzuto, R.; Brini, M.; et al. Tau localises within mitochondrial sub-compartments and its caspase cleavage affects ER-mitochondria interactions and cellular Ca2+ handling. *Biochim. Et Biophys. Acta-Mol. Basis Dis.* **2018**, *1864*, 3247–3256. [CrossRef]
263. Hu, Y.; Li, X.C.; Wang, Z.H.; Luo, Y.; Zhang, X.N.; Liu, X.P.; Feng, Q.; Wang, Q.; Yue, Z.Y.; Chen, Z.; et al. Tau accumulation impairs mitophagy via increasing mitochondrial membrane potential and reducing mitochondrial Parkin. *Oncotarget* **2016**, *7*, 17356–17368. [CrossRef] [PubMed]
264. Parihar, M.S.; Parihar, A.; Fujita, M.; Hashimoto, M.; Ghafourifar, P. Mitochondrial association of alpha-synuclein causes oxidative stress. *Cell. Mol. Life Sci.* **2008**, *65*, 1272–1284. [CrossRef] [PubMed]
265. Smith, W.W.; Jiang, H.B.; Pei, Z.; Tanaka, Y.; Morita, H.; Sawa, A.; Dawson, V.L.; Dawson, T.M.; Ross, C.A. Endoplasmic reticulum stress and mitochondrial cell death pathways mediate A53T mutant alpha-synuclein-induced toxicity. *Hum. Mol. Genet.* **2005**, *14*, 3801–3811. [CrossRef] [PubMed]
266. Devi, L.; Raghavendran, V.; Prabhu, B.M.; Avadhani, N.G.; Anandatheerthavarada, H.K. Mitochondrial import and accumulation of alpha-synuclein impair complex I in human dopaminergic neuronal cultures and Parkinson disease brain. *J. Biol. Chem.* **2008**, *283*, 9089–9100. [CrossRef]
267. Di Maio, R.; Barrett, P.J.; Hoffman, E.K.; Barrett, C.W.; Zharikov, A.; Borah, A.; Hu, X.P.; McCoy, J.; Chu, C.T.; Burton, E.A.; et al. Alpha-Synuclein binds to TOM20 and inhibits mitochondrial protein import in Parkinson's disease. *Sci. Transl. Med.* **2016**, *8*, 342ra78. [CrossRef]
268. Bender, A.; Desplats, P.; Spencer, B.; Rockenstein, E.; Adame, A.; Elstner, M.; Laub, C.; Mueller, S.; Koob, A.O.; Mante, M.; et al. TOM40 Mediates Mitochondrial Dysfunction Induced by alpha-Synuclein Accumulation in Parkinson's Disease. *PLoS ONE* **2013**, *8*, e62277. [CrossRef]
269. Yano, H.; Baranov, S.V.; Baranova, O.V.; Kim, J.; Pan, Y.C.; Yablonska, S.; Carlisle, D.L.; Ferrante, R.J.; Kim, A.H.; Friedlander, R.M. Inhibition of mitochondrial protein import by mutant huntingtin. *Nat. Neurosci.* **2014**, *17*, 822–831. [CrossRef]
270. Napoli, E.; Wong, S.; Hung, C.; Ross-Inta, C.; Bomdica, P.; Giulivi, C. Defective mitochondrial disulfide relay system, altered mitochondrial morphology and function in Huntingtons disease. *Hum. Mol. Genet.* **2013**, *22*, 989–1004. [CrossRef]
271. Fischer, L.R.; Igoudjil, A.; Magrane, J.; Li, Y.J.; Hansen, J.M.; Manfredi, G.; Glass, J.D. SOD1 targeted to the mitochondrial intermembrane space prevents motor neuropathy in the Sod1 knockout mouse. *Brain* **2011**, *134*, 196–209. [CrossRef]
272. Vijayvergiya, C.; Beal, M.F.; Buck, J.; Manfredi, G. Mutant superoxide dismutase 1 forms aggregates in the brain mitochondrial matrix of amyotrophic lateral sclerosis mice. *J. Neurosci.* **2005**, *25*, 2463–2470. [CrossRef]
273. Pasinelli, P.; Belford, M.E.; Lennon, N.; Bacskai, B.J.; Hyman, B.T.; Trotti, D.; Brown, R.H. Amyotrophic lateral sclerosis-associated SOD1 mutant proteins bind and aggregate with Bcl-2 in spinal cord mitochondria. *Neuron* **2004**, *43*, 19–30. [CrossRef]
274. Li, Q.A.; Vande Velde, C.; Israelson, A.; Xie, J.; Bailey, A.O.; Dong, M.Q.; Chun, S.J.; Roy, T.; Winer, L.; Yates, J.R.; et al. ALS-linked mutant superoxide dismutase 1 (SOD1) alters mitochondrial protein composition and decreases protein import. *Proc. Natl. Acad. Sci. USA* **2010**, *107*, 21146–21151. [CrossRef]
275. Bannwarth, S.; Ait-El-Mkadem, S.; Chaussenot, A.; Genin, E.C.; Lacas-Gervais, S.; Fragaki, K.; Berg-Alonso, L.; Kageyama, Y.; Serre, V.; Moore, D.G.; et al. A mitochondrial origin for frontotemporal dementia and amyotrophic lateral sclerosis through CHCHD10 involvement. *Brain* **2014**, *137*, 2329–2345. [CrossRef] [PubMed]
276. Fernandez-Vizarra, E.; Zeviani, M. Mitochondrial disorders of the OXPHOS system. *FEBS Lett.* **2021**, *595*, 1062–1106. [CrossRef] [PubMed]
277. Brischigliaro, M.; Zeviani, M. Cytochrome c oxidase deficiency. *Biochim. Biophys. Acta Bioenerg.* **2021**, *1862*, 148335. [CrossRef] [PubMed]
278. Ghezzi, D.; Zeviani, M. Human diseases associated with defects in assembly of OXPHOS complexes. In *Mitochondrial Diseases*; Garone, C., Minczuk, M., Eds.; Portland Press Ltd: London, UK, 2018; Volume 62, pp. 271–286.
279. Gorman, G.S.; Chinnery, P.F.; DiMauro, S.; Hirano, M.; Koga, Y.; McFarland, R.; Suomalainen, A.; Thorburn, D.R.; Zeviani, M.; Turnbull, D.M. Mitochondrial diseases. *Nat. Rev. Dis. Primers* **2016**, *2*, 16080. [CrossRef]
280. Fiedorczuk, K.; Sazanov, L.A. Mammalian Mitochondrial Complex I Structure and Disease-Causing Mutations. *Trends Cell Biol.* **2018**, *28*, 835–867. [CrossRef] [PubMed]
281. Shahrour, M.A.; Staretz-Chacham, O.; Dayan, D.; Stephen, J.; Weech, A.; Damseh, N.; Pri Chen, H.; Edvardson, S.; Mazaheri, S.; Saada, A.; et al. Mitochondrial epileptic encephalopathy, 3-methylglutaconic aciduria and variable complex V deficiency associated with TIMM50 mutations. *Clin. Genet.* **2017**, *91*, 690–696. [CrossRef] [PubMed]
282. Reyes, A.; Melchionda, L.; Burlina, A.; Robinson, A.J.; Ghezzi, D.; Zeviani, M. Mutations in TIMM50 compromise cell survival in OxPhos-dependent metabolic conditions. *EMBO Mol. Med.* **2018**, *10*, e8698. [CrossRef]

283. Tort, F.; Ugarteburu, O.; Texidó, L.; Gea-Sorlí, S.; García-Villoria, J.; Ferrer-Cortès, X.; Arias, Á.; Matalonga, L.; Gort, L.; Ferrer, I.; et al. Mutations in TIMM50 cause severe mitochondrial dysfunction by targeting key aspects of mitochondrial physiology. *Hum. Mutat.* **2019**, *40*, 1700–1712. [CrossRef]
284. Roesch, K.; Curran, S.P.; Tranebjaerg, L.; Koehler, C.M. Human deafness dystonia syndrome is caused by a defect in assembly of the DDP1/TIMM8a-TIMM13 complex. *Hum. Mol. Genet.* **2002**, *11*, 477–486. [CrossRef]
285. Pacheu-Grau, D.; Callegari, S.; Emperador, S.; Thompson, K.; Aich, A.; Topol, S.E.; Spencer, E.G.; McFarland, R.; Ruiz-Pesini, E.; Torkamani, A.; et al. Mutations of the mitochondrial carrier translocase channel subunit TIM22 cause early-onset mitochondrial myopathy. *Hum. Mol. Genet.* **2018**, *27*, 4135–4144. [CrossRef] [PubMed]
286. Wei, X.; Du, M.; Xie, J.; Luo, T.; Zhou, Y.; Zhang, K.; Li, J.; Chen, D.; Xu, P.; Jia, M.; et al. Mutations in TOMM70 lead to multi-OXPHOS deficiencies and cause severe anemia, lactic acidosis, and developmental delay. *J. Hum. Genet.* **2020**, *65*, 231–240. [CrossRef] [PubMed]
287. Zheng, H.; Koo, E.H. Biology and pathophysiology of the amyloid precursor protein. *Mol. Neurodegener.* **2011**, *6*, 27. [CrossRef] [PubMed]
288. Mann, V.M.; Cooper, J.M.; Daniel, S.E.; Srai, K.; Jenner, P.; Marsden, C.D.; Schapira, A.H.V. Complex-i, iron, and ferritin in parkinsons-disease substantia-nigra. *Ann. Neurol.* **1994**, *36*, 876–881. [CrossRef]
289. Bindoff, L.A.; Birchmachin, M.A.; Cartlidge, N.E.F.; Parker, W.D.; Turnbull, D.M. Respiratory-chain abnormalities in skeletal-muscle from patients with parkinsons-disease. *J. Neurol. Sci.* **1991**, *104*, 203–208. [CrossRef]
290. Schapira, A.H.V. Mitochondria in the aetiology and pathogenesis of Parkinson's disease. *Parkinsonism Relat. Disord.* **1999**, *5*, 139–143. [CrossRef]
291. Macdonald, R.; Barnes, K.; Hastings, C.; Mortiboys, H. Mitochondrial abnormalities in Parkinson's disease and Alzheimer's disease: Can mitochondria be targeted therapeutically? *Biochem. Soc. Trans.* **2018**, *46*, 891–909. [CrossRef]
292. Iwai, A.; Masliah, E.; Yoshimoto, M.; Ge, N.F.; Flanagan, L.; Desilva, H.A.R.; Kittel, A.; Saitoh, T. The precursor protein of non-a-beta component of alzheimers-disease amyloid is a presynaptic protein of the central-nervous-system. *Neuron* **1995**, *14*, 467–475. [CrossRef]
293. Spillantini, M.G.; Schmidt, M.L.; Lee, V.M.Y.; Trojanowski, J.Q.; Jakes, R.; Goedert, M. alpha-synuclein in Lewy bodies. *Nature* **1997**, *388*, 839–840. [CrossRef]
294. Clayton, D.F.; George, J.M. The synucleins: A family of proteins involved in synaptic function, plasticity, neurodegeneration and disease. *Trends Neurosci.* **1998**, *21*, 249–254. [CrossRef]
295. Jakes, R.; Spillantini, M.G.; Goedert, M. Identification of two distinct synucleins from human brain. *FEBS Lett* **1994**, *345*, 27–32. [CrossRef]
296. Masliah, E.; Rockenstein, E.; Veinbergs, I.; Mallory, M.; Hashimoto, M.; Takeda, A.; Sagara, Y.; Sisk, A.; Mucke, L. Dopaminergic loss and inclusion body formation in alpha-synuclein mice: Implications for neurodegenerative disorders. *Science* **2000**, *287*, 1265–1269. [CrossRef] [PubMed]
297. Banerjee, R.; Starkov, A.A.; Beal, M.F.; Thomas, B. Mitochondrial dysfunction in the limelight of Parkinson's disease pathogenesis. *Biochim. Biophys. Acta Mol. Basis Dis.* **2009**, *1792*, 651–663. [CrossRef] [PubMed]
298. Martin, L.J.; Pan, Y.; Price, A.C.; Sterling, W.; Copeland, N.G.; Jenkins, N.A.; Price, D.L.; Lee, M.K. Parkinson's disease alpha-synuclein transgenic mice develop neuronal mitochondrial degeneration and cell death. *J. Neurosci.* **2006**, *26*, 41–50. [CrossRef] [PubMed]
299. Song, D.D.; Shults, C.W.; Sisk, A.; Rockenstein, E.; Masliah, E. Enhanced substantia nigra mitochondrial pathology in human alpha-synuclein transgenic mice after treatment with MPTP. *Exp. Neurol.* **2004**, *186*, 158–172. [CrossRef]
300. Li, W.W.; Yang, R.; Guo, J.C.; Ren, H.M.; Zha, X.L.; Cheng, J.S.; Cai, D.F. Localization of alpha-synuclein to mitochondria within midbrain of mice. *Neuroreport* **2007**, *18*, 1543–1546. [CrossRef] [PubMed]
301. Parihar, M.S.; Parihar, A.; Fujita, M.; Hashimoto, M.; Ghafourifar, P. Alpha-synuclein overexpression and aggregation exacerbates impairment of mitochondrial functions by augmenting oxidative stress in human neuroblastoma cells. *Int. J. Biochem. Cell Biol.* **2009**, *41*, 2015–2024. [CrossRef]
302. Tanner, C.M.; Kamel, F.; Ross, G.W.; Hoppin, J.A.; Goldman, S.M.; Korell, M.; Marras, C.; Bhudhikanok, G.S.; Kasten, M.; Chade, A.R.; et al. Rotenone, Paraquat, and Parkinson's Disease. *Environ. Health Perspect.* **2011**, *119*, 866–872. [CrossRef]
303. Pickrell, A.M.; Youle, R.J. The Roles of PINK1, Parkin, and Mitochondrial Fidelity in Parkinson's Disease. *Neuron* **2015**, *85*, 257–273. [CrossRef]
304. Kazlauskaite, A.; Kondapalli, C.; Gourlay, R.; Campbell, D.G.; Ritorto, M.S.; Hofmann, K.; Alessi, D.R.; Knebel, A.; Trost, M.; Muqit, M.M.K. Parkin is activated by PINK1-dependent phosphorylation of ubiquitin at Ser(65). *Biochem. J.* **2014**, *460*, 127–139. [CrossRef] [PubMed]
305. Kondapalli, C.; Kazlauskaite, A.; Zhang, N.; Woodroof, H.I.; Campbell, D.G.; Gourlay, R.; Burchell, L.; Walden, H.; Macartney, T.J.; Deak, M.; et al. PINK1 is activated by mitochondrial membrane potential depolarization and stimulates Parkin E3 ligase activity by phosphorylating Serine 65. *Open Biol.* **2012**, *2*, 120080. [CrossRef] [PubMed]
306. Lee, Y.; Stevens, D.A.; Kang, S.U.; Jiang, H.S.; Lee, Y.I.; Ko, H.S.; Scarffe, L.A.; Umanah, G.E.; Kang, H.; Ham, S.; et al. PINK1 Primes Parkin-Mediated Ubiquitination of PARIS in Dopaminergic Neuronal Survival. *Cell Rep.* **2017**, *18*, 918–932. [CrossRef] [PubMed]

307. Bertolin, G.; Jacoupy, M.; Traver, S.; Ferrando-Miguel, R.; Saint Georges, T.; Grenier, K.; Ardila-Osorio, H.; Muriel, M.P.; Takahashi, H.; Lees, A.J.; et al. Parkin maintains mitochondrial levels of the protective Parkinson's disease-related enzyme 17-beta hydroxysteroid dehydrogenase type 10. *Cell Death Differ.* **2015**, *22*, 1563–1576. [CrossRef] [PubMed]
308. Gehrke, S.; Wu, Z.H.; Klinkenberg, M.; Sun, Y.P.; Auburger, G.; Guo, S.; Lu, B.W. PINK1 and Parkin Control Localized Translation of Respiratory Chain Component mRNAs on Mitochondria Outer Membrane. *Cell Metab.* **2015**, *21*, 95–108. [CrossRef]
309. Macdonald, M.E.; Ambrose, C.M.; Duyao, M.P.; Myers, R.H.; Lin, C.; Srinidhi, L.; Barnes, G.; Taylor, S.A.; James, M.; Groot, N.; et al. A novel gene containing a trinucleotide repeat that is expanded and unstable on huntingtons-disease chromosomes. *Cell* **1993**, *72*, 971–983. [CrossRef]
310. Li, H.; Li, S.H.; Johnston, H.; Shelbourne, P.F.; Li, X.J. Amino-terminal fragments of mutant huntingtin show selective accumulation in striatal neurons and synaptic toxicity. *Nat. Genet.* **2000**, *25*, 385–389. [CrossRef]
311. DiFiglia, M.; Sapp, E.; Chase, K.O.; Davies, S.W.; Bates, G.P.; Vonsattel, J.P.; Aronin, N. Aggregation of huntingtin in neuronal intranuclear inclusions and dystrophic neurites in brain. *Science* **1997**, *277*, 1990–1993. [CrossRef] [PubMed]
312. Orr, A.L.; Li, S.H.; Wang, C.E.; Li, H.; Wang, J.J.; Rong, J.; Xu, X.S.; Mastroberardino, P.G.; Greenamyre, J.T.; Li, X.J. N-terminal mutant huntingtin associates with mitochondria and impairs mitochondrial trafficking. *J. Neurosci.* **2008**, *28*, 2783–2792. [CrossRef]
313. Yu, Z.X.; Li, S.H.; Evans, J.; Pillarisetti, A.; Li, H.; Li, X.J. Mutant huntingtin causes context-dependent neurodegeneration in mice with Huntington's disease. *J. Neurosci.* **2003**, *23*, 2193–2202. [CrossRef] [PubMed]
314. Bae, B.I.; Xu, H.; Igarashi, S.; Fujimuro, M.; Agrawal, N.; Taya, Y.; Hayward, S.D.; Moran, T.H.; Montell, C.; Ross, C.A.; et al. p53 mediates cellular dysfunction and behavioral abnormalities in Huntington's disease. *Neuron* **2005**, *47*, 29–41. [CrossRef] [PubMed]
315. Benchoua, A.; Trioulier, Y.; Zala, D.; Gaillard, M.C.; Lefort, N.; Dufour, N.; Saudou, F.; Elalouf, J.M.; Hirsch, E.; Hantraye, P.; et al. Involvement of mitochondrial complex II defects in neuronal death produced by N-terminus fragment of mutated Huntingtin. *Mol. Biol. Cell* **2006**, *17*, 1652–1663. [CrossRef] [PubMed]
316. Brennan, W.A.; Bird, E.D.; Aprille, J.R. Regional mitochondrial respiratory activity in huntingtons-disease brain. *J. Neurochem.* **1985**, *44*, 1948–1950. [CrossRef]
317. Damiano, M.; Starkov, A.A.; Petri, S.; Kipiani, K.; Kiaei, M.; Mattiazzi, M.; Beal, M.F.; Manfredi, G. Neural mitochondrial Ca2+ capacity impairment precedes the onset of motor symptoms in G93A Cu/Zn-superoxide dismutase mutant mice. *J. Neurochem.* **2006**, *96*, 1349–1361. [CrossRef]
318. Kim, J.; Moody, J.P.; Edgerly, C.K.; Bordiuk, O.L.; Cormier, K.; Smith, K.; Beal, M.F.; Ferrante, R.J. Mitochondrial loss, dysfunction and altered dynamics in Huntington's disease. *Hum. Mol. Genet.* **2010**, *19*, 3919–3935. [CrossRef] [PubMed]
319. Lisowsky, T. Erv1 is involved in the cell-division cycle and the maintenance of mitochondrial genomes in saccharomyces-cerevisiae. *Curr. Genet.* **1994**, *26*, 15–20. [CrossRef]
320. Becher, D.; Kricke, J.; Stein, G.; Lisowsky, T. A mutant for the yeast scERV1 gene displays a new defect in mitochondrial morphology and distribution. *Yeast* **1999**, *15*, 1171–1181. [CrossRef]
321. Di Fonzo, A.; Ronchi, D.; Lodi, T.; Fassone, E.; Tigano, M.; Lamperti, C.; Corti, S.; Bordoni, A.; Fortunato, F.; Nizzardo, M.; et al. The Mitochondrial Disulfide Relay System Protein GFER Is Mutated in Autosomal-Recessive Myopathy with Cataract and Combined Respiratory-Chain Deficiency. *Am. J. Hum. Genet.* **2009**, *84*, 594–604. [CrossRef]
322. Browne, S.E.; Bowling, A.C.; MacGarvey, U.; Baik, M.J.; Berger, S.C.; Muqit, M.M.K.; Bird, E.D.; Beal, M.F. Oxidative damage and metabolic dysfunction in Huntington's disease: Selective vulnerability of the basal ganglia. *Ann. Neurol.* **1997**, *41*, 646–653. [CrossRef]
323. Gu, M.; Gash, M.T.; Mann, V.M.; JavoyAgid, F.; Cooper, J.M.; Schapira, A.H.V. Mitochondrial defect in Huntington's disease on caudate nucleus. *Ann. Neurol.* **1996**, *39*, 385–389. [CrossRef]
324. Browne, S.E.; Beal, M.F. The energetics of Huntington's disease. *Neurochem. Res.* **2004**, *29*, 531–546. [CrossRef]
325. Valentine, J.S.; Doucette, P.A.; Potter, S.Z. Copper-zinc superoxide dismutase and amyotrophic lateral sclerosis. *Annu. Rev. Biochem* **2005**, *74*, 563–593. [CrossRef] [PubMed]
326. Carri, M.T.; Cozzolino, M. SOD1 and mitochondria in ALS: A dangerous liaison. *J. Bioenerg. Biomembr.* **2011**, *43*, 593–599. [CrossRef]
327. Cozzolino, M.; Rossi, S.; Mirra, A.; Carri, M.T. Mitochondrial dynamism and the pathogenesis of Amyotrophic Lateral Sclerosis. *Front. Cell. Neurosci.* **2015**, *9*, 31. [CrossRef]
328. Crugnola, V.; Lamperti, C.; Lucchini, V.; Ronchi, D.; Peverelli, L.; Prelle, A.; Sciacco, M.; Bordoni, A.; Fassone, E.; Fortunato, F.; et al. Mitochondrial Respiratory Chain Dysfunction in Muscle From Patients With Amyotrophic Lateral Sclerosis. *Arch. Neurol.* **2010**, *67*, 849–854. [CrossRef] [PubMed]
329. Corti, S.; Donadoni, C.; Ronchi, D.; Bordoni, A.; Fortunato, F.; Santoro, D.; Del Bo, R.; Lucchini, V.; Crugnola, V.; Papadimitriou, D.; et al. Amyotrophic lateral sclerosis linked to a novel SOD1 mutation with muscle mitochondrial dysfunction. *J. Neurol. Sci.* **2009**, *276*, 170–174. [CrossRef]
330. Schon, E.A.; Przedborski, S. Mitochondria: The Next (Neurode) Generation. *Neuron* **2011**, *70*, 1033–1053. [CrossRef] [PubMed]
331. Sturtz, L.A.; Diekert, K.; Jensen, L.T.; Lill, R.; Culotta, V.C. A fraction of yeast Cu,Zn-superoxide dismutase and its metallochaperone, CCS, localize to the intermembrane space of mitochondria—A physiological role for SOD1 in guarding against mitochondrial oxidative damage. *J. Biol. Chem.* **2001**, *276*, 38084–38089. [CrossRef] [PubMed]

332. Okado-Matsumoto, A.; Fridovich, I. Subcellular distribution of superoxide dismutases (SOD) in rat liver—Cu,Zn-SOD in mitochondria. *J. Biol. Chem.* **2001**, *276*, 38388–38393. [CrossRef] [PubMed]
333. Murphy, M.P. How mitochondria produce reactive oxygen species. *Biochem. J.* **2009**, *417*, 1–13. [CrossRef]
334. Hervias, I.; Beal, M.F.; Manfredi, G. Mitochondrial dysfunction and amyotrophic lateral sclerosis. *Muscle Nerve* **2006**, *33*, 598–608. [CrossRef]
335. Lehmer, C.; Schludi, M.H.; Ransom, L.; Greiling, J.; Junghanel, M.; Exner, N.; Riemenschneider, H.; van der Zee, J.; Van Broeckhoven, C.; Weydt, P.; et al. A novel CHCHD10 mutation implicates a Mia40-dependent mitochondrial import deficit in ALS. *EMBO Mol. Med.* **2018**, *10*, e8558. [CrossRef]
336. Wang, T.; Liu, H.; Itoh, K.; Oh, S.; Zhao, L.; Murata, D.; Sesaki, H.; Hartung, T.; Na, C.H.; Wang, J. C9orf72 regulates energy homeostasis by stabilizing mitochondrial complex I assembly. *Cell Metab.* **2021**, *33*, 531–546.e539. [CrossRef]
337. Baker, M.J.; Palmer, C.S.; Stojanovski, D. Mitochondrial protein quality control in health and disease. *Br. J. Pharm.* **2014**, *171*, 1870–1889. [CrossRef]
338. Samluk, L.; Chroscicki, P.; Chacinska, A. Mitochondrial protein import stress and signaling. *Curr. Opin. Physiol.* **2018**, *3*, 41–48. [CrossRef]
339. Quiros, P.M.; Mottis, A.; Auwerx, J. Mitonuclear communication in homeostasis and stress. *Nat. Rev. Mol. Cell Biol.* **2016**, *17*, 213–226. [CrossRef] [PubMed]
340. Haynes, C.M.; Ron, D. The mitochondrial UPR—Protecting organelle protein homeostasis. *J. Cell Sci.* **2010**, *123*, 3849–3855. [CrossRef] [PubMed]
341. Nargund, A.M.; Pellegrino, M.W.; Fiorese, C.J.; Baker, B.M.; Haynes, C.M. Mitochondrial Import Efficiency of ATFS-1 Regulates Mitochondrial UPR Activation. *Science* **2012**, *337*, 587–590. [CrossRef] [PubMed]
342. D'Amico, D.; Sorrentino, V.; Auwerx, J. Cytosolic Proteostasis Networks of the Mitochondrial Stress Response. *Trends Biochem Sci* **2017**, *42*, 712–725. [CrossRef] [PubMed]
343. Aldridge, J.E.; Horibe, T.; Hoogenraad, N.J. Discovery of genes activated by the mitochondrial unfolded protein response (mtUPR) and cognate promoter elements. *PLoS ONE* **2007**, *2*, e874. [CrossRef]
344. Nargund, A.M.; Fiorese, C.J.; Pellegrino, M.W.; Deng, P.; Haynes, C.M. Mitochondrial and Nuclear Accumulation of the Transcription Factor ATFS-1 Promotes OXPHOS Recovery during the UPRmt. *Mol. Cell* **2015**, *58*, 123–133. [CrossRef]
345. Haynes, C.M.; Yang, Y.; Blais, S.P.; Neubert, T.A.; Ron, D. The Matrix Peptide Exporter HAF-1 Signals a Mitochondrial UPR by Activating the Transcription Factor ZC376.7 in C. elegans. *Mol. Cell* **2010**, *37*, 529–540. [CrossRef] [PubMed]
346. Rolland, S.G.; Schneid, S.; Schwarz, M.; Rackles, E.; Fischer, C.; Haeussler, S.; Regmi, S.G.; Yeroslaviz, A.; Habermann, B.; Mokranjac, D.; et al. Compromised Mitochondrial Protein Import Acts as a Signal for UPRmt. *Cell Rep.* **2019**, *28*, 1659–1669. [CrossRef]
347. Fiorese, C.J.; Schulz, A.M.; Lin, Y.F.; Rosin, N.; Pellegrino, M.W.; Haynes, C.M. The Transcription Factor ATF5 Mediates a Mammalian Mitochondrial UPR. *Curr. Biol.* **2016**, *26*, 2037–2043. [CrossRef] [PubMed]
348. Kuhl, I.; Miranda, M.; Atanassov, I.; Kuznetsova, I.; Hinze, Y.; Mourier, A.; Filipovska, A.; Larsson, N.G. Transcriptomic and proteomic landscape of mitochondrial dysfunction reveals secondary coenzyme Q deficiency in mammals. *eLife* **2017**, *6*, e30952. [CrossRef]
349. Quiros, P.M.; Prado, M.A.; Zamboni, N.; D'Amico, D.; Williams, R.W.; Finley, D.; Gygi, S.P.; Auwerx, J. Multi-omics analysis identifies ATF4 as a key regulator of the mitochondrial stress response in mammals. *J. Cell Biol.* **2017**, *216*, 2027–2045. [CrossRef]
350. Labbadia, J.; Brielmann, R.M.; Neto, M.F.; Lin, Y.F.; Haynes, C.M.; Morimoto, R.I. Mitochondrial Stress Restores the Heat Shock Response and Prevents Proteostasis Collapse during Aging. *Cell Rep.* **2017**, *21*, 1481–1494. [CrossRef]
351. Cooper, J.F.; Machiela, E.; Dues, D.J.; Spielbauer, K.K.; Senchuk, M.M.; Van Raamsdonk, J.M. Activation of the mitochondrial unfolded protein response promotes longevity and dopamine neuron survival in Parkinson's disease models. *Sci Rep.* **2017**, *7*, 16441. [CrossRef] [PubMed]
352. Martinez, B.A.; Petersen, D.A.; Gaeta, A.L.; Stanley, S.P.; Caldwell, G.A.; Caldwell, K.A. Dysregulation of the Mitochondrial Unfolded Protein Response Induces Non-Apoptotic Dopaminergic Neurodegeneration in C-elegans Models of Parkinson's Disease. *J. Neurosci.* **2017**, *37*, 11085–11100. [CrossRef]
353. St Martin, J.L.; Klucken, J.; Outeiro, T.F.; Nguyen, P.; Keller-McGandy, C.; Cantuti-Castelvetri, I.; Grammatopoulos, T.N.; Standaert, D.G.; Hyman, B.T.; McLean, P.J. Dopaminergic neuron loss and up-regulation of chaperone protein mRNA induced by targeted over-expression of alpha-synuclein in mouse substantia nigra. *J. Neurochem.* **2007**, *100*, 1449–1457. [CrossRef] [PubMed]
354. Gorman, A.M.; Szegezdi, E.; Quigney, D.J.; Samali, A. Hsp27 inhibits 6-hydroxydopamine-induced cytochrome c release and apoptosis in PC12 cells. *Biochem. Biophys. Res. Commun.* **2005**, *327*, 801–810. [CrossRef]
355. Klucken, J.; Shin, Y.; Masliah, E.; Hyman, B.T.; McLean, P.J. Hsp70 reduces alpha-synuclein aggregation and toxicity. *J. Biol. Chem.* **2004**, *279*, 25497–25502. [CrossRef]
356. Riar, A.K.; Burstein, S.R.; Palomo, G.M.; Arreguin, A.; Manfredi, G.; Germain, D. Sex specific activation of the ER alpha axis of the mitochondrial UPR (UPRmt) in the G93A-SOD1 mouse model of familial ALS. *Hum. Mol. Genet.* **2017**, *26*, 1318–1327. [CrossRef] [PubMed]
357. Pharaoh, G.; Sataranatarajan, K.; Street, K.; Hill, S.; Gregston, J.; Ahn, B.; Kinter, C.; Kinter, M.; Van Remmen, H. Metabolic and Stress Response Changes Precede Disease Onset in the Spinal Cord of Mutant SOD1 ALS Mice. *Front. Neurosci.* **2019**, *13*, 487. [CrossRef]

358. Shen, Y.; Ding, M.; Xie, Z.H.; Liu, X.T.; Yang, H.; Jin, S.Q.; Xu, S.L.; Zhu, Z.Y.; Wang, Y.; Wang, D.W.; et al. Activation of Mitochondrial Unfolded Protein Response in SHSY5Y Expressing APP Cells and APP/PS1 Mice. *Front. Cell. Neurosci.* **2020**, *13*, 568. [CrossRef] [PubMed]
359. Beck, J.S.; Mufson, E.J.; Counts, S.E. Evidence for Mitochondrial UPR Gene Activation in Familial and Sporadic Alzheimer's Disease. *Curr. Alzheimer Res.* **2016**, *13*, 610–614. [CrossRef]
360. Regitz, C.; Fitzenberger, E.; Mahn, F.L.; Dussling, L.M.; Wenzel, U. Resveratrol reduces amyloid-beta (A beta(1-42))-induced paralysis through targeting proteostasis in an Alzheimer model of Caenorhabditis elegans. *Eur. J. Nutr.* **2016**, *55*, 741–747. [CrossRef]
361. Poveda-Huertes, D.; Matic, S.; Marada, A.; Habernig, L.; Licheva, M.; Myketin, L.; Gilsbach, R.; Tosal-Castano, S.; Papinski, D.; Mulica, P.; et al. An Early mtUPR: Redistribution of the Nuclear Transcription Factor Rox1 to Mitochondria Protects against Intramitochondrial Proteotoxic Aggregates. *Mol. Cell* **2020**, *77*, 180–188. [CrossRef]
362. Wrobel, L.; Topf, U.; Bragoszewski, P.; Wiese, S.; Sztolsztener, M.E.; Oeljeklaus, S.; Varabyova, A.; Lirski, M.; Chroscicki, P.; Mroczek, S.; et al. Mistargeted mitochondrial proteins activate a proteostatic response in the cytosol. *Nature* **2015**, *524*, 485–488. [CrossRef]
363. Papa, L.; Germain, D. Estrogen receptor mediates a distinct mitochondrial unfolded protein response. *J. Cell Sci.* **2011**, *124*, 1396–1402. [CrossRef] [PubMed]
364. Callegari, S.; Dennerlein, S. Sensing the Stress: A Role for the UPRmt and UPRam in the Quality Control of Mitochondria. *Front. Cell. Dev. Biol.* **2018**, *6*, 31. [CrossRef] [PubMed]
365. Hegde, A.N.; Upadhya, S.C. Role of ubiquitin-proteasome-mediated proteolysis in nervous system disease. *Biochim. Biophys. Acta Gene Regul. Mech.* **2011**, *1809*, 128–140. [CrossRef]
366. Dennissen, F.J.A.; Kholod, N.; van Leeuwen, F.W. The ubiquitin proteasome system in neurodegenerative diseases: Culprit, accomplice or victim? *Prog. Neurobiol.* **2012**, *96*, 190–207. [CrossRef] [PubMed]
367. Ciechanover, A.; Kwon, Y.T. Degradation of misfolded proteins in neurodegenerative diseases: Therapeutic targets and strategies. *Exp. Mol. Med.* **2015**, *47*, e147. [CrossRef]
368. Tai, H.C.; Serrano-Pozo, A.; Hashimoto, T.; Frosch, M.P.; Spires-Jones, T.L.; Hyman, B.T. The Synaptic Accumulation of Hyperphosphorylated Tau Oligomers in Alzheimer Disease Is Associated With Dysfunction of the Ubiquitin-Proteasome System. *Am. J. Pathol.* **2012**, *181*, 1426–1435. [CrossRef] [PubMed]
369. Wang, X.W.; Chen, X.J. A cytosolic network suppressing mitochondria-mediated proteostatic stress and cell death. *Nature* **2015**, *524*, 481–484. [CrossRef] [PubMed]
370. Gerbasi, V.R.; Link, A.J. The myotonic dystrophy type 2 protein ZNF9 is part of an ITAF complex that promotes cap-independent translation. *Mol. Cell. Proteom.* **2007**, *6*, 1049–1058. [CrossRef] [PubMed]
371. Matsuo, Y.; Granneman, S.; Thoms, M.; Manikas, R.G.; Tollervey, D.; Hurt, E. Coupled GTPase and remodelling ATPase activities form a checkpoint for ribosome export. *Nature* **2014**, *505*, 112–116. [CrossRef] [PubMed]
372. Pakos-Zebrucka, K.; Koryga, I.; Mnich, K.; Ljujic, M.; Samali, A.; Gorman, A.M. The integrated stress response. *EMBO Rep.* **2016**, *17*, 1374–1395. [CrossRef]
373. Elden, A.C.; Kim, H.J.; Hart, M.P.; Chen-Plotkin, A.S.; Johnson, B.S.; Fang, X.D.; Armakola, M.; Geser, F.; Greene, R.; Lu, M.M.; et al. Ataxin-2 intermediate-length polyglutamine expansions are associated with increased risk for ALS. *Nature* **2010**, *466*, 1069–1075. [CrossRef]
374. Chartier-Harlin, M.C.; Dachsel, J.C.; Vilarino-Guell, C.; Lincoln, S.J.; Lepretre, F.; Hulihan, M.M.; Kachergus, J.; Milnerwood, A.J.; Tapia, L.; Song, M.S.; et al. Translation Initiator EIF4G1 Mutations in Familial Parkinson Disease. *Am. J. Hum. Genet.* **2011**, *89*, 398–406. [CrossRef]
375. Coyne, L.P.; Chen, X.J. mPOS is a novel mitochondrial trigger of cell death—Implications for neurodegeneration. *FEBS Lett.* **2018**, *592*, 759–775. [CrossRef]
376. Weidberg, H.; Amon, A. MitoCPR-A surveillance pathway that protects mitochondria in response to protein import stress. *Science* **2018**, *360*, eaan4146. [CrossRef] [PubMed]
377. Tang, M.Z.; Luo, X.L.; Huang, Z.; Chen, L.X. MitoCPR: A novel protective mechanism in response to mitochondrial protein import stress. *Acta Biochim. Biophys. Sin.* **2018**, *50*, 1072–1074. [CrossRef]
378. Martensson, C.U.; Priesnitz, C.; Song, J.Y.; Ellenrieder, L.; Doan, K.N.; Boos, F.; Floerchinger, A.; Zufall, N.; Oeljeklaus, S.; Warscheid, B.; et al. Mitochondrial protein translocation-associated degradation. *Nature* **2019**, *569*, 679–683. [CrossRef]
379. Neuber, O.; Jarosch, E.; Volkwein, C.; Walter, J.; Sommer, T. Ubx2 links the Cdc48 complex to ER-associated protein degradation. *Nat. Cell Biol.* **2005**, *7*, 993–998. [CrossRef] [PubMed]
380. Schuberth, C.; Buchberger, A. Membrane-bound Ubx2 recruits Cdc48 to ubiquitin ligases and their substrates to ensure efficient ER-associated protein degradation. *Nat. Cell Biol.* **2005**, *7*, 999–1006. [CrossRef] [PubMed]
381. Pereira, G.C.; Allen, W.J.; Watkins, D.W.; Buddrus, L.; Noone, D.; Liu, X.; Richardson, A.P.; Chacinska, A.; Collinson, I. A High-Resolution Luminescent Assay for Rapid and Continuous Monitoring of Protein Translocation across Biological Membranes. *J. Mol. Biol.* **2019**, *431*, 1689–1699. [CrossRef]

Review

The Causal Role of Lipoxidative Damage in Mitochondrial Bioenergetic Dysfunction Linked to Alzheimer's Disease Pathology

Mariona Jové [1], Natàlia Mota-Martorell [1], Pascual Torres [1], Victoria Ayala [1], Manuel Portero-Otin [1], Isidro Ferrer [2,3,*] and Reinald Pamplona [1,*]

[1] Department of Experimental Medicine, Lleida Biomedical Research Institute (IRBLleida), Lleida University (UdL), 25198 Lleida, Spain; Mariona.jove@udl.cat (M.J.); natalia.mota@mex.udl.cat (N.M.-M.); pascual.torres@udl.cat (P.T.); victoria.ayala@udl.cat (V.A.); manuel.portero@udl.cat (M.P.-O.)

[2] Department of Pathology and Experimental Therapeutics, University of Barcelona, Bellvitge University Hospital/Bellvitge Biomedical Research Institute (IDIBELL), L'Hospitalet de Llobregat, 08907 Barcelona, Spain

[3] Center for Biomedical Research on Neurodegenerative Diseases (CIBERNED), ISCIII, 28220 Madrid, Spain

* Correspondence: 8082ifa@gmail.com (I.F.); reinald.pamplona@udl.cat (R.P.)

Abstract: Current shreds of evidence point to the entorhinal cortex (EC) as the origin of the Alzheimer's disease (AD) pathology in the cerebrum. Compared with other cortical areas, the neurons from this brain region possess an inherent selective vulnerability derived from particular oxidative stress conditions that favor increased mitochondrial molecular damage with early bioenergetic involvement. This alteration of energy metabolism is the starting point for subsequent changes in a multitude of cell mechanisms, leading to neuronal dysfunction and, ultimately, cell death. These events are induced by changes that come with age, creating the substrate for the alteration of several neuronal pathways that will evolve toward neurodegeneration and, consequently, the development of AD pathology. In this context, the present review will focus on description of the biological mechanisms that confer vulnerability specifically to neurons of the entorhinal cortex, the changes induced by the aging process in this brain region, and the alterations at the mitochondrial level as the earliest mechanism for the development of AD pathology. Current findings allow us to propose the existence of an altered allostatic mechanism at the entorhinal cortex whose core is made up of mitochondrial oxidative stress, lipid metabolism, and energy production, and which, in a positive loop, evolves to neurodegeneration, laying the basis for the onset and progression of AD pathology.

Keywords: aging; ATP synthase; energy metabolism; entorhinal cortex; lipoxidation-derived damage; mitochondrial dysfunction; neurodegeneration; oxidative damage

Citation: Jové, M.; Mota-Martorell, N.; Torres, P.; Ayala, V.; Portero-Otin, M.; Ferrer, I.; Pamplona, R. The Causal Role of Lipoxidative Damage in Mitochondrial Bioenergetic Dysfunction Linked to Alzheimer's Disease Pathology. *Life* **2021**, *11*, 388. https://doi.org/10.3390/life11050388

Academic Editors: Giorgio Lenaz and Salvatore Nesci

Received: 19 March 2021
Accepted: 21 April 2021
Published: 25 April 2021

1. Introduction

Population aging is a global phenomenon. Demographic projections from the United Nations Aging Program [1] and the Centers for Disease Control and Prevention [2] estimate that the number of people in the world aged 65+ will increase from 420 million in 2000 (7% of total population) to about 1 billion by 2030 (12% of the world population). Considering that aging represents the single biggest risk factor for Alzheimer's Disease (AD), it is clear that this dementing disorder presents a challenge to medicine, preventive medicine, public health, and elderly care systems in all countries across the world.

The global prevalence of dementia in the world is estimated at around 3.9% in people aged 60+ years but there are regional differences ranging from 1.6% in Africa to 6.4% in North America [3]. Furthermore, more than 25 million people in the world are currently affected by dementia with around 5 million new cases occurring every year [3–5]. The two most common forms of dementia around the world are AD (50–70%) and vascular dementia (15–25%), with AD the predominant form that affects the elderly. The data from population-based studies in Europe suggest that the age-standardized prevalence in people 65+ years old is 6.4% for dementia and 4.4% for AD [6]. In the US, a study evaluating people aged > 70 years showed a prevalence for AD of 9.7% [7]. The age-specific prevalence of AD almost doubles every 5 years after age 65. In industrialized countries, approximately 10% of people over 65 and more than 30% over 85 are affected by some degree of dementia [8,9]. Thus, AD is the most prevalent form of dementia predominantly affecting the aged.

In Europe, there is a rate of incidence of 19.4 per 1000 people every year in those over 65. In the US, data from two large-scale studies of people aged 65+ showed an incidence rate for AD of 15.0 (male, 13.0; female, 16.9) per 1000 person-years [10,11]. Furthermore, the rate of incidence of AD increases with increasing age until 85 years of age [6,11,12] and seems to decline in the early 90s and later [13] suggesting a resistance to this age-related disease in long-lived humans, as well as the possibility that the aging trajectory of long-lived humans differs from the 'normal' or pathological aging probably ascribed to biological differences. Currently, Alzheimer's disease has no cure, although intense research is being carried out aimed at improving understanding of the molecular and cellular mechanisms of the disease and early detection in preclinical stages of the disease, as well as identifying targets for specific treatments.

Mitochondria are evolutionarily conserved intracellular organelles present in all eukaryotic cells, and neurons are no exception. These organelles participate in a multitude of cellular physiological processes, particularly in cell bioenergetics and the biology of free radicals [14]. Both functions are relevant for neuronal cells, both to support their high energy demands in order to maintain neuronal homeostasis and survival, and to limit the potentially detrimental effects induced by the oxidative stress conditions which are inherently favored by the own neuronal structure and function [15–21]. However, neurons are long-lived post-mitotic cells with a lifespan identical to the organism itself, indicating the presence of efficient molecular mechanisms designed to ensure neuronal survival [22,23]. Other critical cellular processes in which mitochondria are also involved include biosynthesis of lipids, iron-sulfur clusters, heme groups, presynaptic neurotransmitters, calcium homeostasis, and cell death, among others [24]. To accomplish this diversity of functions, mitochondria show a great dynamism and establish a coordinated network and crosstalk among themselves and with other cell organelles including, for instance, endoplasmic reticulum and nucleus, in order to adapt their response to a continuously changing microenvironment under physiological and pathological conditions [25–27].

Dysfunction of mitochondria has been associated with several neurodegenerative diseases, including amyotrophic lateral sclerosis, Huntington's disease and Parkinson's disease [28–32]. Here we will review available evidence that demonstrates mitochondrial alterations in the brain of AD patients. Furthermore, we will show that mitochondrial defects in particular metabolic pathways of specific vulnerable neurons of the entorhinal cortex (EC) trigger the onset of AD pathology.

2. The Alzheimer's Disease Pathology

AD pathology is a human age-related biological process that causes progressive degeneration of the brain and is characterized clinically by cognitive impairment and dementia. At the cellular level, AD is characterized by a selective and progressive loss of nerve cells, spines and synapses, impaired neurotransmission, and progressive isolation of remaining nerve cells. Neuropathologically, AD is currently characterized by extracellular deposits of β-amyloid (Aβ40 and Aβ42 as the main species) in the brain parenchyma forming senile plaques (SPs) and giving rise to β-amyloid angiopathy around cerebral blood vessels, and intraneuronal deposits of hyper-phosphorylated, abnormally-conformed, and truncated tau configuring neurofibrillary tangles (NFTs), dystrophic neurites of senile plaques, and neuropil threads [33]. In this context, AD could be considered a pathology with abnormal accumulation of specific protein aggregates leading to neurodegeneration. Unfortunately, these two markers have attracted so much attention in biomedical research that a third potential hallmark, altered lipid metabolism, first described by Alzheimer, has gone practically unnoticed. Furthermore, we should note that studies in the late 1960s and 1970s also reported important changes at the mitochondrial level whose meaning was absolutely ignored until recently (see later).

In terms of disease progression and spread, and again based on SPs and NFTs as a neuropathological reference, early abnormal tau deposition appears in selected nuclei of the brainstem followed by the EC and olfactory bulb and tracts; later on, it extends to the hippocampal complex, basal forebrain, and limbic system, and eventually to the whole cerebral cortex and other regions such as the striatum and thalamus [34–38]. Systematic anatomical studies have allowed the categorization of stages of disease progression based on the accumulation of lesions in the brain. Thus, Braak and Braak stages I-II are manifested by NFTs in the olfactory bulb and tracts, and EC, followed by the transentorhinal cortex and initial CA1; stages III-IV show increased numbers of NFTs in the preceding regions and extension of NFTs to the whole CA1 region of the hippocampus, subiculum, temporal cortex, magnocellular nuclei of the basal forebrain including Meynert nucleus, amygdala, anterodorsal thalamic nuclei, and tubero-mammillary nucleus; stages V-VI entail, in addition to increased severity in the above-mentioned areas, the cortical association areas including the frontal and parietal cortices, the claustrum, reticular nuclei of the thalamus, and, finally, the primary sensory areas compromising the primary visual cortex [34,35,38] (Figure 1A). Regarding β-amyloid deposits, these first appear in the orbitofrontal cortex and temporal cortex, and they then progress to practically the whole cerebral cortex, diencephalic nuclei, and, lastly, the cerebellum [34,35,39] (Figure 1B). Regarding SPs, Braak and Braak stage A is characterized by plaques in the basal neocortex, particularly the orbitofrontal and temporal cortices; stage B involves, in addition, the association cortices; and stage C, the primary cortical areas [34,35].

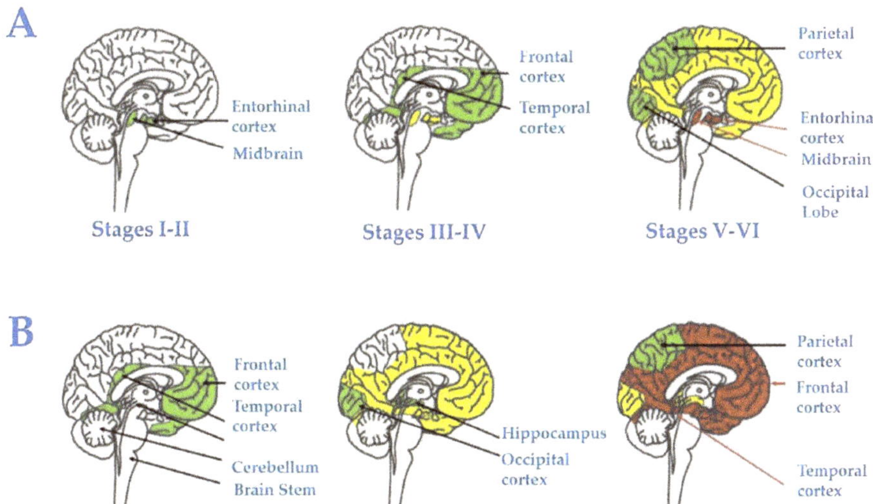

Figure 1. Progression and spread of Alzheimer's disease in humans based on tau (NFTs) and β-amyloid (SPs) as a neuropathological reference. (**A**) Progression of tau pathology (Braak staging), and (**B**) progression of β-amyloid pathology (β-amyloid deposits). For details, see main text. Reproduced and modified with permission from [40].

Importantly, cognitive impairment clinically categorized as mild or moderate may appear at stages III-IV, whereas dementia can occur in individuals with AD pathology at stages V-VI [41–43]. The delay between the first appearance of AD-related pathology and the development of cognitive decline and dementia has been estimated to be several decades in those individuals in whom dementia eventually occurs [44]. It is worth stressing that AD-related pathology, including stages I-II, is present in about 85% of individuals aged 65+ years [38,45,46]. Therefore, AD can be considered a very common and relatively well-tolerated degenerative process for a long time, depending on cognitive reserve, but it may have devastating effects once thresholds are exceeded [46]. This concept highlights the need for discovering early biomarkers and prompt interventional measures to maintain neuronal function. Discovering the triggering pathological events is relevant for the adoption of these interventions.

One of the reasons for the lack of knowledge concerning AD is the limited amount of information regarding early stages of the neurodegenerative process. This caveat in knowledge is secondary to the fact that the disease's clinical features—those that permit its recognition—occur at relatively advanced stages, once the degenerative process has reached thresholds that overwhelm the system's allostatic capacity as a whole to maintain essential neuronal functions. Thus, little is known about physiological aging changes, intertwined with the process's silent earliest stages (Braak stages I-II). Thus, attention should be given to those stages in which neurofibrillary pathology (one of the hallmarks of AD) is restricted to the EC, in order to improve understanding of early pathogenic mechanisms and identify possible targets for therapeutic intervention geared to curbing or retarding progression to clinical stages [46–48].

3. The Entorhinal Cortex as Starting Point

In mammals, the EC is located in the medial temporal lobe, adjacent to the hippocampus. This brain region presents two major divisions, the medial EC (MEC), which is located next to the pre- and parasubiculum, and the lateral EC (LEC), which is adjacent to the neocortex. The EC has five cell layers, in contrast to the neocortex which has six layers. There are three superficial layers (layers I, II, III), a relatively cell-free central layer (lamina dissecans), and two deep layers (layers V and VI). The principal cells of the EC are glutamatergic and include pyramidal neurons (located in layers II, III, V and VI), and stellate cells (located in layer II). GABAergic local circuit neurons (interneurons) are dispersed throughout the layers, similar to neocortex. From a neurophysiological perspective, the EC plays a critical role in memory consolidation [49].

3.1. *Selective Neuronal Vulnerability of the Human Entorhinal Cortex*

Human evolution is closely linked with rapid expansion of brain size and complexity, a prerequisite for the appearance of cognitive functions. These evolutionary changes have been associated with and supported by brain metabolism adaptations, especially concerning increased energy supply, and fatty acid uses [50–54]. The human brain's complexity is expressed at many levels, such as the organization in different regions, the diversity of functions (motor, sensory, regulatory, behavioral, and cognitive), and the morphological and functional diversity of neurons. This neuronal diversity requires specific gene expression profiles, in addition to the housekeeping genes required for the basic functions of every cell that in essence are linked to cell metabolism [55]. The gene expression profile supports a given proteomic profile which, in turn, configures a neuron-specific metabolomic fingerprint. Additionally, the fact that specific regions of the central nervous system exhibit differential vulnerabilities with respect to the aging process and neurodegenerative disease development indicates that neuronal responses to cell-damaging processes are heterogeneous [15,18,56]. To better understand the mechanisms involved in neuronal resistance/sensitivity to stress, disease, and death, it is crucial to define the different brain regions' cell vulnerability in physiological conditions.

In this context, developmental, structural, and functional traits suggest that neurons in EC, particularly neurons in layer II, have a selective vulnerability compared to other neurons. Thus, it has been observed that primates display a developmental precocity of the EC relative to other cortical regions [57]. This could be a factor in increased vulnerability of the EC to the effects of age [58] due to their neurons' high longevity. An additional factor of neuronal vulnerability is the morphological complexity and presence of long myelinated axons which carry a high energy cost to maintain their integrity [58–60] (Figure 2). Furthermore, the intraregional differences in the firing properties of the different neurons also introduce differential vulnerability based on the necessary adaptation of cellular pieces of machinery [58]. Other factors associated with a specific profile of layer II of the EC can be found at the gene expression profile level. For instance, high reelin expression has been described, involved in neuronal development and synaptic plasticity [61].

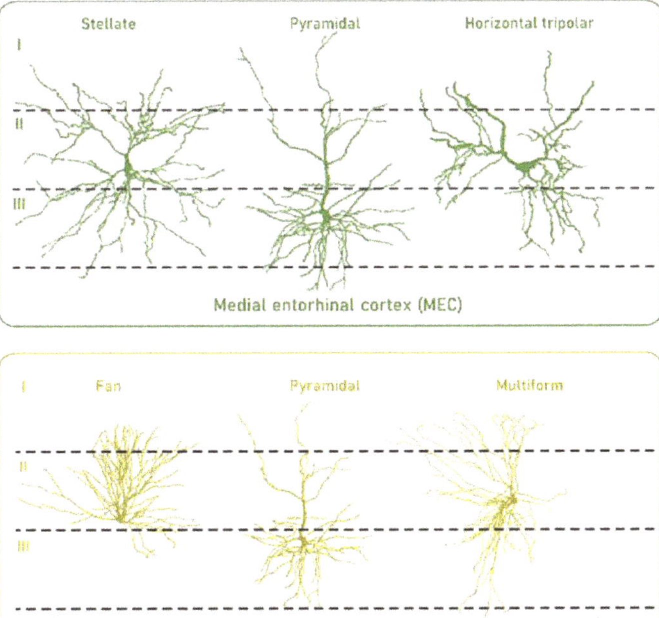

Figure 2. High longevity and morphological complexity are traits that confer selective vulnerability to neurons in layer II of the medial and lateral entorhinal cortex in humans. For details, see main text. Reproduced and modified with permission from [58].

Neurons in EC can also present a selective vulnerability based on specific metabolic traits. Thus, we find that a cross-regional comparative study designed to identify traits that characterize the selective neuronal vulnerability in three functionally and evolutionarily distinct brain regions—EC, hippocampus, and frontal cortex [19]—showed the existence of higher energy demand, mitochondrial stress, and higher one-carbon metabolism (particularly restricted to the methionine cycle) specifically in EC and hippocampus. These findings, along with the worse antioxidant capacity and higher mTOR signaling also seen in EC and hippocampus, suggest that these brain regions are especially vulnerable to stress compared to the frontal cortex, which is a more resistant region [19]. Therefore, specific, as–yet unknown, tradeoffs between energy metabolism, mTOR signaling, antioxidant capacities, and stress resilience could operate in these brain regions.

The one-carbon metabolism can be considered as an integrative network of nutrient status. Thus, inputs in the form of amino acids (which donate carbon units) enter the metabolic network, and are metabolized into intermediate metabolites that become a substrate for diverse biological functions, which include regulation of methylation reactions and redox status, biosynthesis of cell components, and regulation of nucleotide pools. The partitioning of carbon units into these different intermediate cellular metabolites involves three interrelated pathways: the folate cycle, the methionine cycle, and the transsulfuration pathway [62]. Several studies have shown that defects in one-carbon metabolism in the brain induce profound disturbances in cell physiology due to the relevant pathways where one-carbon metabolism is involved and, more importantly, through the toxic effects derived from the metabolites that shape the core of the methionine cycle. Thus, a connection has been established between high levels of homocysteine and cognitive function, from mild cognitive decline to vascular dementia and AD) [63]. In contrast, low methionine and derived metabolite content, either constitutively or induced by nutritional intervention, is associated with resistance to oxidative stress and greater longevity [64–66]. Hence, we may infer that the higher one-carbon metabolism observed in EC is a physiological adaptation that imparts vulnerability to stress in this region.

mTOR is a conserved serine/threonine kinase which regulates metabolism in response to nutrients, growth factors, and cellular energy conditions. Available evidence indicates that the mTOR signaling pathway is involved in brain aging and age-related neurodegenerative diseases [67–69]. In line with this, several studies show that increased mTOR signaling in the brain negatively affects multiple pathways including glucose and lipid metabolism, energy production, mitochondrial function, and autophagy. Conversely, attenuation of the mTOR signal, through nutritional or pharmacological intervention, is associated with a healthy lifespan, including improvement in brain function, and also increases longevity [67,69]. Consequently, we may also infer that the higher mTOR content observed in EC is a physiological adaptation which imparts vulnerability to stress in this region.

Lipids have played a determinant role in the human brain's evolution [54]. It is postulated that the morphological and functional diversity among the human central nervous system's neural cells is projected and achieved through the expression of particular lipid profiles [18,20]. A recent study evaluated the differential vulnerability to oxidative stress mediated by lipids through a cross-regional comparative approach [20]. To this end, the fatty acid profile and vulnerability to lipid peroxidation were determined in 12 brain regions of healthy adult subjects. Additionally, different components involved in polyunsaturated fatty acid (PUFA) biosynthesis, and adaptive defense mechanisms against lipid peroxidation, were measured. The results evidenced that the EC possesses a lipid profile that is highly vulnerable to oxidation due to the presence of a higher content of highly unsaturated fatty acids (UFAs) and a higher steady-state level of lipoxidation-derived protein damage compared to other cortical brain regions. Consequently, the lipid profile of the EC is prone to oxidative damage.

Globally, available information from a cross-regional comparative approach suggests that EC presents specific bioenergetic and lipid profile traits and signaling pathways that make this brain region more susceptible to damage compared to other cortical brain regions. Whether this fact is stochastic or is related to functional constraints is not known.

3.2. The Entorhinal Cortex during Physiological Aging

Very little is known about what occurs in the EC during normal aging (i.e., in the absence of dementia). Indeed, functional-structural studies with magnetic resonance imaging (MRI) in healthy subjects covering adult lifespan have demonstrated that the EC is a region relatively well preserved with aging [70–72]. Nevertheless, this does not exclude a number of changes that occur with age. Several studies have reported declines in EC volume and/or thickness with aging [71,73–80], even though the rate of change within this region is thought to be lower than other related regions such as the hippocampus [75,81]. Reinforcing these observations, recent studies using quantitative structural high resolution MRI applied to healthy subjects covering adult lifespan demonstrate that the trajectory of EC volume, thickness, and surface area initially increased with age, reaching a peak at about 32 years, 40 years, and 50 years of age, respectively, after which they decreased with age [79,82]. Importantly, it seems that there are right-left hemisphere differences, as well as differences derived from gender. Furthermore, there is a correlation between changes in

EC volume and cognitive decline in functions ascribed to the EC in aging [83,84]. On the whole, these studies suggest that across the healthy adult lifespan the EC suffers minor but significant morphological and functional changes with age affecting volume and thickness, as well as cognitive functions specifically ascribed to the EC.

At the cellular level, a recent study using stereological measurement of the neuronal soma demonstrated slight but significant increases in the neuron body size in layer II of EC in old subjects (65+, without signs of dementia) compared to younger healthy adult individuals [85]. Other findings include increased numbers of astrocytes with age, increased lipofuscin granule content (aggregates with oxidized lipids and a hallmark of aging), and nuclei that are rounder and more prominent than in younger subjects [86]. Diffuse SPs and NFT were not common traits [86]. Importantly, there is no significant neuronal loss with age in the EC [86,87], in contrast to a previous observation [88] reporting a highly significant correlation between loss of neurons and age in the EC. The biological meaning of these changes in neuronal soma is at present unknown.

At the biochemical level, age-related alterations of the phospholipid profile of mitochondrial and microsomal membranes from the EC of healthy humans ranging from 18 to 98 years have been also described [89]. Specifically, the proportion of total phosphatidylcholine (PC) of the mitochondrial fraction and the most abundant phospholipid present within the human brain, PC 16:0_18:1, increased in the EC with age. In contrast, the total mitochondrial phosphatidylethanolamine (PE) content decreased with age. Importantly, many specific mitochondrial PE molecular species containing docosahexaenoic acid (DHA) increased with age, although this did not translate into a generalized age-related increase in total mitochondrial DHA. When compared to other regions of the brain such as the hippocampus and frontal cortex, the phospholipid profile of the EC remains relatively stable in adults over the lifespan [89]. Nevertheless, the increase, although slight, in phospholipid species with highly unsaturated fatty acids like DHA, also highly susceptible to oxidative damage, can cause the mitochondrial membrane to become more vulnerable to oxidative damage which, in turn, may extend the molecular damage to other mitochondrial and cellular components. This observation may be relevant because it has been demonstrated in experimental animal models that very small variations in the degree of unsaturation of cell membranes in brain are translated into a magnified increase in the lipid peroxidation level and subsequent lipoxidation-derived molecular damage [90].

In consonance with this relative preservation of mitochondrial lipid profile during EC aging, the level of lipoxidation-derived protein adducts (neuroketal-protein adducts, and MDA-lysine adducts) is maintained, at least in total tissue, in old-aged individuals [91]. This maintenance suggests successful regulation of oxidative stress during aging, so that the potential increased damage is probably restricted to selective targets. In agreement with this idea, no differences in the content of COX-2 (enzyme involved in the generation of lipids with neuroinflammatory properties) or CYP2J2 (involved in the generation of neuroprotective lipid products) were observed in old-aged individuals [91]. In this line, comparing the expression of inflammatory mediators (complement system, colony stimulating factor receptors, toll-like receptors, and pro- and anti-inflammatory cytokines) in the brains of subjects without NFTs (mean age: 47.1 $\pm$ 5.7 years) with those with no neurological disease and neurofibrillary pathology stages I-II (mean age: 70.6 $\pm$ 6.3 years) revealed no differences either in EC or in any examined region [92]. Moreover, gene expression of significant anti-oxidative stress responses did not match neuroinflammation in aging or increased regional susceptibility to major neurodegenerative diseases [92].

Globally, these findings suggest minor but potentially relevant changes in neuronal vulnerability of EC with age. Importantly, mitochondrial membrane lipids prone to oxidative damage can offer the substrate for the alteration of several neuronal pathways through lipoxidation-derived molecular damage that could evolve, depending on its intensity, to a normal aging process or to become the basis of a neurogenerative process. Whether these findings can be attributed to specific neuron types present in the EC or whether all neurons are similarly affected are questions that remain to be answered.

4. The Entorhinal Cortex in AD

4.1. Early Bioenergetic Defects

Whereas the EC as a region is relatively resistant to detrimental aging phenomena, EC tissue loss with age is considered an important marker of early AD. Among all brain regions, EC volume and thickness loss (using neuroimaging) provides the best resolution between cognitively intact individuals and patients with mild cognitive impairment (MCI) or AD [93–98]. Additional studies have also demonstrated the presence of atrophy in the EC, hippocampus, and amygdala associated with clinical disease severity, and have even detected atrophy in the EC earlier than hippocampal and amygdala atrophy [99–103] and years before AD conversion [104,105]. More recent MRI studies have focused on evidence of atrophy that precede clinical symptoms [106–113], often detecting these smaller changes using time-series data analysis and survival analysis. In this line, a recent study detected a break point in the rate of atrophy in the EC 8–11 years prior to a diagnosis of MCI, and even earlier (9–14 years prior) in the transentorhinal cortex [103]. Interestingly, these observations are consistent with autopsy findings that verify neuronal changes in the EC. It may be inferred that these findings underestimate the time for the first events that trigger the AD pathology, based on the idea that the accumulation of the neuropathological hallmarks of AD (the NFTs and SPs) is an 'advanced step' in the origin of AD pathology.

The first electron microscopic (EM) studies revealed the structure of β-amyloid deposition and the presence of surrounding dystrophic neurites filled with altered mitochondria, residual bodies, and abnormal filaments in SPs, together with disorganization of the normal cytoskeleton and accumulation of paired helical filaments such as the subcellular substrates of NFTs [114–116]. In the 1970s, EM pictures of AD brains revealed altered mitochondrial structures [117,118]. Initial reports, though, offered little speculation as to the cause or significance of this basic finding. Later studies confirmed and extended the observation [119,120].

In the 1980s, fluorodeoxyglucose positron emission tomography (FDG PET) studies showed brains from AD patients used less glucose than those from control subjects [121–123]. PET studies designed to quantify brain oxygen consumption in vivo showed decreased oxygen consumption by AD brains [124,125]. These studies boosted interest in a potential metabolic component for this disease [126–129]. Later, different biochemical studies demonstrated activity deficiencies in bioenergetic-related enzymes such as cytochrome oxidase (COX), pointing to a mitochondrial compromise [59,119,130–133]. Furthermore, cytoplasmic hybrid cells in which mitochondria from sporadic cases of AD were fused with other cells also indicate a defect in mitochondria function in AD [134–136]. Reinforcing this idea, reduced mitochondrial activity was described as an early and persistent phenomenon in the EC [137,138]. Furthermore, the major abnormalities in mitochondrial structure and dynamics and derived oxidative molecular damage (to mitochondrial DNA, proteins, and lipids) are selective and restricted to the EC's vulnerable neurons [119], suggesting an intimate and early association between these features in AD. Therefore, accumulated data suggest that mitochondrial-bioenergetic dysfunction represents a fundamental AD event [128,139,140].

Among the myriad of reactions taking place in mitochondria, the principal and earliest AD-linked alteration is found in ATP synthase or complex V of the mitochondrial respiratory chain, which catalyzes the synthesis of ATP from adenosine diphosphate (ADP) and inorganic phosphate. ATP synthase is selectively damaged (lipoxidized) as a unique and prime target and its function is reduced in the EC as early as Braak stage I-II, exclusively affecting neuronal cells [137]. Interestingly, the modification of ATP synthase has also been verified in different regions of the human brain cortex at advanced Braak stages of AD [141]. This event is crucial because in addition to the defect in the energy metabolism, this loss-of-function can have a detrimental effect on electron transport chain activity leading to increased free radical (reactive oxygen species) production, with subsequent lipid oxidation and lipoxidation-derived molecular damage. Consequently, these alterations suggest an impairment in cellular oxidative stress conditions and later neuronal damage [46,142–146].

In this scenario, it is proposed that there is an altered allostatic mechanism in the EC whose core is made up of mitochondrial oxidative stress, lipid metabolism, and energy production, and that a positive loop or feedback leads to mitochondrial failure that becomes the onset of the AD pathology (see Figure 3).

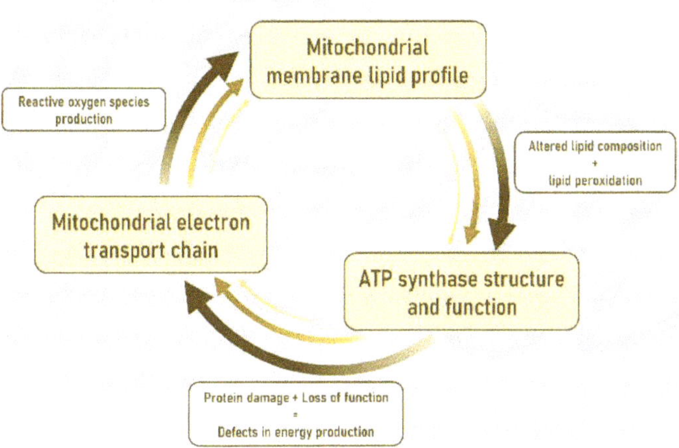

Figure 3. The triad formed by the activity of the mitochondrial electronic transport chain, the lipid profile, and the activity of ATP synthase constitutes the basic component whose alteration at the level of the neurons of the entorhinal cortex gives rise to the origin and subsequent progression of Alzheimer's disease pathology.

4.2. ATP Synthase as the Key Target of AD

ATP synthase is a macromolecular structure inserted in the inner mitochondrial membrane and is the last complex (complex V) of the electron transport chain with a key role in energy metabolism. Complex V has a central role in cellular energy (as ATP) supply. Figure 4 shows the structure of the mitochondrial ATP synthase in light of current knowledge [147,148].

Available evidence demonstrates that ATP synthase is a selectively damaged vital protein. This modification consists of the nonenzymatic modification of different complex V subunits by reactive compounds generated from lipid peroxidation (lipoxidation-derived protein damage) [18,141,147,149]. The lipoxidation damage mostly and preferentially affects the α and β subunits [149], although the specific residues targeted are still unknown. Whether this preferential and early modification of ATP synthase expresses a particular vulnerability of this complex in the individuals that will go on to develop AD is currently unknown. For this reason, proteomics, transcriptomics and genomics studies should be directed specifically to the analysis of the ATP synthase to detect or rule out particularities that may be underlying the susceptibility of an individual to developing AD. Why does the molecular onset of AD take place at the adult stage however? It is hypothesized that at the adult stage (40–50 years old), the changes that have been progressively implemented by age are the substrate that will determine a trajectory of normal (physiological) aging if these changes have been slight, or else there will be a bifurcation and change of trajectory to a pathological condition (AD) if the changes have crossed a threshold. We propose that the conditions of oxidative stress at the mitochondrial level reached at the adult age by the population susceptible to developing AD mark the critical point or threshold for modification of vulnerable ATP synthase and, thus, the onset of the pathological condition. The optimal conditions to trigger this modification are offered in specific neurons of the EC.

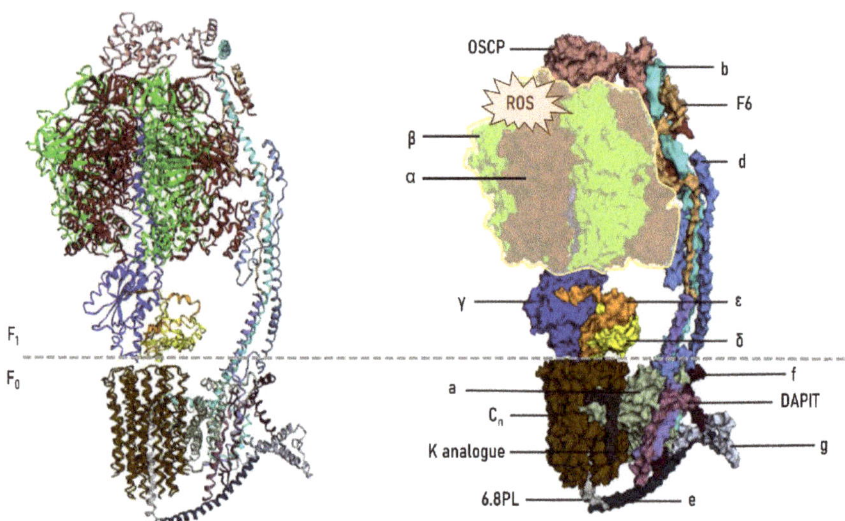

Figure 4. Structure of the mitochondrial F0F1 ATP synthase (complex V) in mammals. Human mitochondrial ATP synthase is a protein complex composed of 28 subunits and organized into two domains: a membrane-embedded F0 domain, and a membrane-extrinsic and matrix-oriented F1 catalytic domain. The two domains are connected by a peripheral and central stalk. For additional structural and functional details see [147,148]. Functionally, the F0 domain is a trans-membrane channel that translocates protons and F1, a synthase domain that binds to ADP and inorganic phosphate and synthesizes ATP. ATP synthase is also involved in mitochondrial cristae formation, as well as in the formation of the permeability transition pore (PTP), which triggers cell death. Importantly, some of the functional properties of ATP synthase and regulatory mechanisms are associated with the integrity of specific residues such as cysteine, arginine, and lysine [150]. Thus, for instance, specific post-translational modifications play important roles in ATP synthase regulation by modifying ε-amino groups of lysine, with the resulting conformational changes of the active sites and decreased enzymatic activity. Similarly, the integrity of structural lysine residues is a key point to promote interaction with membrane lipids, particularly cardiolipin, in order to ensure correct ATP synthase activity [150]. The oxidative and lipoxidative non-enzymatic modificationm—ediated by reactive oxygen species (ROS) and carbonyl species (RCS), respectively—of ATP synthase observed during early stages of AD pathology mostly and preferentially affects the α and β subunits. Reproduced and modified with permission from [147].

This selective and preferential damage of ATP synthase may be due to several factors such as functional characteristics, structural traits, and location, all of which influence this specificity [149,151]. Concerning structural traits, it has been demonstrated that the presence of alpha helices and loops, globular shapes that additionally form soluble coiled-shaped molecules with hydrophobic groups to the center and exposed hydrophilic groups, and the presence and exposure of amino acids like lysine which are particularly vulnerable, render ATP synthase susceptible to lipoxidative damage [149–151]. As to location, ATP synthase is a complex located inside the inner membrane and spatially oriented toward the mitochondrial matrix. Thus, ATP synthase is exposed to a potentially dangerous environment because of its proximity to the main generators of free radicals, mitochondrial complexes I and III [149], and its insertion in a lipid bilayer highly enriched in PUFAs.

4.3. *Advanced Effects in the Early Stage Resulting from Mitochondrial Dysfunction and Oxidative-Derived Damage*

The selective oxidative damage of ATP synthase leads to a progressive mitochondrial bioenergetic failure with the cell's deleterious effects postulated as the earliest molecular event from which AD pathology will progress in EC and spread to the rest of the brain. These deleterious effects, representing the first steps of disease progression, may be grouped as follows: (a) expansion of the modified proteins and functional consequences at the mitochondrial and cellular levels, and (b) mitochondrial functional alterations, cellular dysfunction, and neuronal death.

Thus, once the oxidative lesion of ATP synthase and its functional defect are established as an early marker, and given the persistence of free radical production and lipid peroxidation, there will be an increase in the diversity of modified proteins, particularly at the mitochondrial level, amplifying the functional defects associated with them. Thus, there is a significant number of selectively modified proteins described in different cortical regions of the brain at different stages of AD [18,141,144,146,149,152]. Interestingly, this apparent diversity of modified proteins can be grouped into highly restricted functional categories such as bioenergetics, proteostasis, neurotransmission, antioxidant, and ion channel [18,144,149]. Bioenergetics is the most affected functional category, deepening the bioenergetic worsening already initiated by the ATP synthase injury. Table 1 shows a list of modified proteins at the mitochondrial level by lipid peroxidation-derived compounds (lipoxidation reactions) identified with redox proteomics in different regions of the human cerebral cortex during different stages of AD. In this context, it may be hypothesized that this mitochondrial bioenergetic defect will end up affecting an additional number of mitochondrial functions that, in turn, will affect cellular functions beyond the mitochondria.

Thus, protein damage can be extended beyond mitochondria, increasing the pool of modified proteins. Effectively, several observations confirm a greater number of modified proteins, and these can again be restricted to very specific functional categories such as neurotransmission, cytoskeleton, and oxygen metabolism [144]. Importantly, synaptic proteins constitute another group of major deregulated AD targets [144,152–155]. Although mostly described in other brain regions such as hippocampus and frontal cortex, these targets probably are also modified in EC. However, this point needs to be confirmed.

Table 1. Mitochondrial lipoxidized proteins identified with redox proteomics in different cerebral cortex regions affected by aging and AD pathology in different stages.

ID (Entry Human)	Protein	Gene	Biological Process	Reference
Q99798	Aconitate hydratase	ACO2	Energy metabolism (TCA cycle)	[151,155–159]
P00367	Glutamate dehydrogenase 1	GLUD1	Energy metabolism (TCA cycle)	[141]
P40926	Malate dehydrogenase	MDH2	Energy metabolism (TCA cycle)	[156–159]
P20674	Cytochrome c oxidase subunit 5a	COX5A	Energy metabolism (ETC)	[159]
P09622	Dihydrolipoyl dehydrogenase	DLD	Energy metabolism (ETC)	[151]
O75489	NADH dehydrogenase (ubiquinone) iron-sulfur protein 3	NDUFS3	Energy metabolism (ETC)	[159]
P31930	Ubiquinol-cytochrome c reductase complex core protein 1	UQCRC1	Energy metabolism (ETC)	[141]
P25705	ATP synthase subunit alpha	ATP5F1A	Energy metabolism (OxPhos)	[137,151,155–160]
P06576	ATP synthase subunit beta	ATP5F1B	Energy metabolism (OxPhos)	[137,141,160]
O75947	ATP synthase subunit d	ATP5H	Energy metabolism (OxPhos)	[159]
P48047	ATP synthase subunit o	ATP5PO	Energy metabolism (OxPhos)	[159]
P12532	Creatine Kinase U-type	CKMT1A	Energy metabolism (energy transduction)	[159]
P15104	Glutamine synthetase	GLUL	Neurotransmission	[141,156–159]
P49411	Elongation factor Tu	TUFM	Proteostasis	[156–159]
P10809	Heat shock protein 60KDa	HSPD1	Proteostasis	[141,151,159]
Q99497	Protein/nucleic acid deglycase DJ-1	PARK7	Proteostasis	[151]
P04179	Manganese superoxide dismutase	SOD2	Antioxidants	[156–159]
P21796	Voltage-dependent anion-selective channel protein 1	VDAC1	Ion channel	[159]

Brain regions from which modified proteins have been identified are entorhinal cortex, cingulate gyrus, hippocampus, parietal cortex, temporal cortex, and frontal cortex, see also: [18,141,144,146,149,152].

As a consequence of the bioenergetic defects and loss-of-function of many mitochondrial proteins, it is feasible to postulate that additional mitochondrial activities, e.g., the machinery for the import of proteins/subunits of nuclear origin, and the communication between mitochondrion and nucleus, among others, can be affected, inducing a dysfunction in the gene expression of mitochondrial structural and regulatory components. In light of this bioenergetics compromise, we might also explain early EC changes in AD as the altered expression of several subunits of mitochondrial complexes and enzymes involved in energy metabolism [138,161–165], altered mitochondrial DNA methylation pattern [161,166], and alteration of the phosphorylation state of the mitochondrial channel VDAC (Voltage-dependent anion channel) [167]. Remarkably, bioinformatics processing has identified a large cluster of altered protein expression in the EC at relatively early stages of the disease, as shown in Figure 5 [144]. A dominant cluster is composed of mitochondrial proteins. This cumulative evidence points to an aggravation of mitochondrial dysfunction.

Other changes at the neuronal level are also described in the EC of AD-related pathology at initial stages includes deregulation of purine metabolism [168], alteration of pro-NGF [169], minor changes in microRNA expression [170], abnormal expression and distribution of metalloproteinase MMP2 [171], and alterations in the phosphorylation of the translation initiation factor 2 alpha (TIF2) [172].

An additional point that needs to be explored is related to lipid metabolism. The accrual of lipid granules was noted in the central studies of Alois Alzheimer. Recent studies have confirmed and provided further details of alteration in brain lipid metabolism in AD in general [18], and in the EC in particular [173–175]. These alterations include changes in lipidomic profiles, which must be added to the role as a prime target of oxidative damage for unsaturated acyl chains in the context of lipid peroxidation and lipoxidation-derived molecular damage previously treated in this review. This increased damage to PUFAs, along with potential alterations in biosynthesis pathways (currently unknown), could explain the reduced content of these fatty acids described in the EC and in the lipid rafts from EC at early stages I-II of AD [173,174]. This change in fatty acid profile is relevant for neuronal membrane properties (fluidity, thickness, curvature, packing, and activities of membrane-bound proteins) because the biophysical traits of polyunsaturated phospholipids do not favor the formation of highly ordered lamellar microdomains, whereas a relative increase in phospholipids containing short-chain saturated or monounsaturated fatty acids interacts favorably with cholesterol and sphingolipids in lipid rafts [176]. Thus, these features point to an increased propensity of neuronal membranes of EC in the earliest stages of AD (AD I/II) to form lipid rafts [173,174].

Furthermore, these lipid rafts show alterations in their profiles of lipid classes (increased content of phosphatidylcholine, sphingomyelins, and gangliosides) [173,174] which determine an increased membrane order and viscosity in these microdomains [174]. The physiopathological consequences of these changes in lipid profile in the onset and progression of AD in the EC are given credence by the specific accumulation of beta-secretase within AD subjects' lipid rafts even at the earliest stages. So, these findings provide a mechanistic connection between lipid alterations in these microdomains and amyloidogenic processing of amyloid precursor protein (APP) and subsequent cytotoxic effects [174]. Consequently, these changes in lipid metabolism seem to precede and play a causal role in forming SPs and probably NFTs, and thus represent an early event in the onset of AD at the EC. Similarly, the relevant role of amyloid fragments as cholesterol-binding proteins in cellular homeostasis of this vital lipid [177] and the upregulation of its subcellular transport towards mitochondrially associated membranes of endoplasmic reticulum in neurons [178] reinforces the relevance of lipid changes in the pathophysiology of AD.

We hypothesize the existence of a detrimental self-sustained loop between mitochondrial oxidative stress, lipid peroxidation-alteration of lipid metabolism, and bioenergetic defects in AD. We propose that this loop could be the basis of additional functional alteration at mitochondrial levels described in EC (and other brain regions) during AD. These secondary alterations would include altered mitochondrial genomic homeostasis, dysfunctional mitochondrial fusion and fission, deficits in mitochondrial axonal trafficking and distribution, impaired mitochondrial biogenesis, abnormal endoplasmic reticulum-mitochondrial interaction, and impaired mitophagy [140,179,180]. Overall, all these phenomena would lead to cell failure and eventual neuronal death. Table 2 shows a summary of alterations described in human EC during aging and AD.

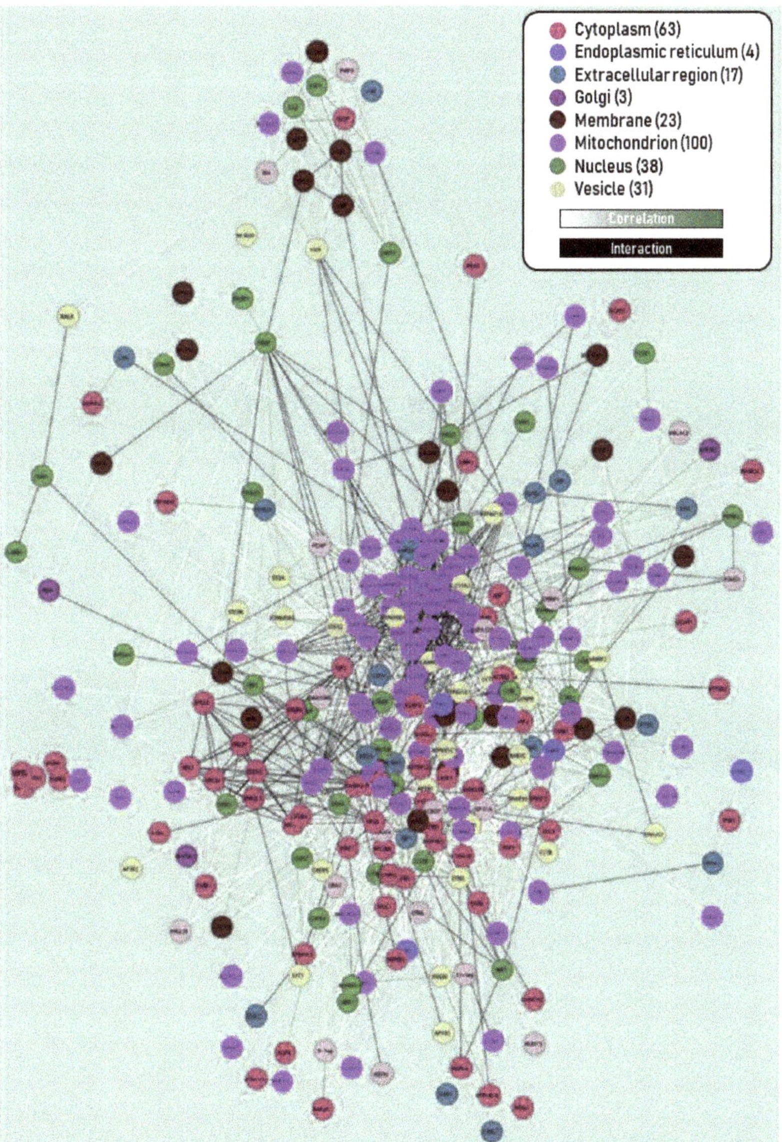

Figure 5. Interactome map of deregulated proteins in the EC of AD at III–IV stage. Edge colors: (i) protein level correlations of proteomic data obtained by Orbitrap Velos, and (ii) interactions retrieved from the public databases BIND, CCSB, DIP, GRID, PubMed, Reactome, KEGG, HPRD, IntAct, MDC, and MINT. Node colors indicate cellular components provided by GO. A large cluster in the center corresponds to deregulated mitochondria-related proteins. Reproduced and modified with permission from [144].

Table 2. Summary of changes reported in human entorhinal cortex during aging and AD.

Aging	Alzheimer's Disease
Minor loss of volume, thickness, and surface area	↓↓ volume and thickness
↑ neuron body size	Abnormalities in mitochondrial structure and dynamics
↑ number of astrocytes	Mitochondrial-bioenergetic failure
Minor changes in fatty acid profile	Loss-of-function of mitochondrial ATP synthase
Minor changes of the phospholipid profile of mitochondrial and microsomal membranes: ↑ phosphatidycholine content, and ↓ phosphatidylethanolamine (PE) content (but increase PE molecular species containing DHA)	Alterations in lipid metabolism and lipidomic profile of neuronal membrane
↑ lipoxidation-derived protein adducts	↑↑ lipid peroxidation and lipoxidation-derived molecular damage
↑ lipofuscin granule content	Expansion of molecular damage to components belonging to bioenergetics, neurotransmission, cytoskeleton, proteostasis, antioxidants, ion channel, and oxygen metabolism
No changes in the expression of inflammatory mediators	Alterations of several mitochondrial activities (import of proteins, fusion and fission, mitophagy, cross-talk with other cell compartments) and gene expression
No loss of neurons	Neuronal death

5. Conclusions

The origin of AD pathology seems to be associated with alterations located explicitly in the EC. Particular neurons from this brain region possess an inherent selective vulnerability prone to oxidative damage with early involvement of energy metabolism. This bioenergetic alteration is the onset for subsequent changes in a multitude of cell mechanisms leading to neuronal dysfunction and, eventually, cell death. Changes induced during physiological aging are the substrate for the emergence of dysfunctional mechanisms which will evolve toward neurodegeneration and, consequently, the development of the AD pathology. Current observations allow us to propose the existence of an altered allostatic mechanism in the EC whose central nucleus is made up of increased mitochondrial oxidative stress, lipid oxidation, and bioenergetic failure, and which in detrimental self-sustained feedback evolves to neurodegeneration, laying the basis for the onset and progression of the AD pathology. Since these alterations are already identified in middle-aged individuals, it seems reasonable to act upon the appropriate free radical-producing targets and lipid metabolism at the appropriate middle-age window. The present observations form a framework for further experiments and provide potential new targets for neuroprotective therapeutic interventions in AD pathology at the earliest stages. Curiously, the evolutionary traits that support human longevity are based on our vulnerability to age-related degenerative processes. Two key traits that define long-lived animal species (humans included) are the presence of cellular components resistant to oxidative stress and a low generation rate of molecular damage. These two characteristics are expressed at the biological level in the lipid profile and the mitochondrial production of free radicals. Curiously, both mechanisms, when altered, are the basis for a pathological condition, in this case AD, which limits human longevity.

Author Contributions: The manuscript was written by M.J., N.M.-M., P.T., V.A., M.P.-O., I.F. and R.P. and edited by R.P. All authors have read and agreed to the published version of the manuscript.

Funding: Research by the authors was supported by the Institute of Health Carlos III (FIS grants PI14/00757, PI14/00328, PI20/0155), the Spanish Ministry of Science, Innovation, and Universities (Ministerio de Ciencia, Innovación y Universidades, grant RTI2018-099200-B-I00), and the Generalitat of Catalonia: Agency for Management of University and Research Grants (2017SGR696) to M.P-O., I.F., and R.P. This study was co-financed by FEDER funds from the European Union ('A way to build Europe').

Institutional Review Board Statement: The study was conducted according to the guidelines of the Declaration of Helsinki, and approved by the Institutional Review Board (or Ethics Committee) of IDIBELL and IRBLleida.

Informed Consent Statement: Informed consent was obtained from all subjects involved in the study.

Acknowledgments: M.J. is a 'Serra-Hunter' Fellow. N.M.-M. and P.T. received predoctoral fellowships from the Generalitat of Catalonia (AGAUR, ref 2018FI_B2_00104) and the Spanish Ministry of Science, Innovation, and Universities (Ministerio de Ciencia, Innovación y Universidades, ref. FPU16/01446), respectively. We thank T. Yohannan for editorial help.

Conflicts of Interest: The authors declare no competing financial interest.

References

1. Organisation, U.N. World population ageing 1950–2050. *Popul. Dev. Rev.* **2002**, *XLIX*, 483p.
2. The Centers for Disease Control and Prevention. Public health and aging: Trends in aging-United States and worldwide. *JAMA* **2003**, *289*, 1371–1373. [CrossRef]
3. Ferri, C.P.; Prince, M.; Brayne, C.; Brodaty, H.; Fratiglioni, L.; Ganguli, M.; Hall, K.; Hasegawa, K.; Hendrie, H.; Huang, Y.; et al. Global prevalence of dementia: A Delphi consensus study. *Lancet* **2005**, *366*, 2112–2117. [CrossRef]
4. Wimo, A.; Winblad, B.; Aguero-Torres, H.; von Strauss, E. The magnitude of dementia occurrence in the world. *Alzheimer Dis. Assoc. Disord.* **2003**, *17*, 63–67. [CrossRef]
5. Brookmeyer, R.; Johnson, E.; Ziegler-Graham, K.; Arrighi, H.M. Forecasting the global burden of Alzheimer's disease. *Alzheimer's Dement.* **2007**, *3*, 186–191. [CrossRef]
6. Lobo, A.; Launer, L.J.; Fratiglioni, L.; Andersen, K.; Di Carlo, A.; Breteler, M.M.; Copeland, J.R.; Dartigues, J.F.; Jagger, C.; Martinez-Lage, J.; et al. Prevalence of dementia and major subtypes in Europe: A collaborative study of population-based cohorts. Neurologic diseases in the elderly research group. *Neurology* **2000**, *54*, S4–S9. [PubMed]
7. Plassman, B.L.; Langa, K.M.; Fisher, G.G.; Heeringa, S.G.; Weir, D.R.; Ofstedal, M.B.; Burke, J.R.; Hurd, M.D.; Potter, G.G.; Rodgers, W.L.; et al. Prevalence of dementia in the United States: The aging, demographics, and memory study. *Neuroepidemiology* **2007**, *29*, 125–132. [CrossRef]
8. von Strauss, E.; Viitanen, M.; De Ronchi, D.; Winblad, B.; Fratiglioni, L. Aging and the occurrence of dementia: Findings from a population-based cohort with a large sample of nonagenarians. *Arch. Neurol.* **1999**, *56*, 587. [CrossRef]
9. Corrada, M.M.; Brookmeyer, R.; Berlau, D.; Paganini-Hill, A.; Kawas, C.H. Prevalence of dementia after age 90: Results from the 90+ study. *Neurology* **2008**, *71*, 337–343. [CrossRef] [PubMed]
10. Kawas, C.; Gray, S.; Brookmeyer, R.; Fozard, J.; Zonderman, A. Age-specific incidence rates of Alzheimer's disease: The Baltimore longitudinal study of aging. *Neurology* **2000**, *54*, 2072–2077. [CrossRef] [PubMed]
11. Kukull, W.A.; Higdon, R.; Bowen, J.D.; McCormick, W.C.; Teri, L.; Schellenberg, G.D.; van Belle, G.; Jolley, L.; Larson, E.B. Dementia and Alzheimer disease incidence: A prospective cohort study. *Arch. Neurol.* **2002**, *59*, 1737. [CrossRef]
12. Jorm, A.F.; Jolley, D. The incidence of dementia: A meta-analysis. *Neurology* **1998**, *51*, 728–733. [CrossRef]
13. Miech, R.A.; Breitner, J.C.S.; Zandi, P.P.; Khachaturian, A.S.; Anthony, J.C.; Mayer, L. Incidence of AD may decline in the early 90s for men, later for women: The Cache Country study. *Neurology* **2002**, *58*, 209–218. [CrossRef]
14. Pamplona, R.; Jové, M.; Mota-Martorell, N.; Barja, G. Is the NDUFV2 subunit of the hydrophilic complex I domain a key determinant of animal longevity? *FEBS J.* **2021**, febs.15714. [CrossRef]
15. Mattson, M.P.; Magnus, T. Ageing and neuronal vulnerability. *Nat. Rev. Neurosci.* **2006**, *7*, 278–294. [CrossRef]
16. Wang, X.; Michaelis, E.K. Selective neuronal vulnerability to oxidative stress in the brain. *Front. Aging Neurosci.* **2010**, *2*, 12. [CrossRef]
17. Jové, M.; Portero-Otín, M.; Naudí, A.; Ferrer, I.; Pamplona, R. Metabolomics of humanbrain aging and age-related neurodegenerative diseases. *J. Neuropathol. Exp. Neurol.* **2014**, *73*, 640–657. [CrossRef] [PubMed]
18. Naudí, A.; Cabré, R.; Jové, M.; Ayala, V.; Gonzalo, H.; Portero-Otín, M.; Ferrer, I.; Pamplona, R. Lipidomics of human brain aging and Alzheimer's disease pathology. *Int. Rev. Neurobiol.* **2015**, *122*, 133–189. [PubMed]
19. Cabré, R.; Jové, M.; Naudí, A.; Ayala, V.; Piñol-Ripoll, G.; Gil-Villar, M.P.; Dominguez-Gonzalez, M.; Obis, È.; Berdun, R.; Mota-Martorell, N.; et al. Specific metabolomics adaptations define a differential regional vulnerability in the adult human cerebral cortex. *Front. Mol. Neurosci.* **2016**, *9*, 138. [CrossRef] [PubMed]
20. Naudí, A.; Cabré, R.; Ayala, V.; Jové, M.; Mota-Martorell, N.; Portero-Otín, M.; Pamplona, R. Region-specific vulnerability to lipid peroxidation and evidence of neuronal mechanisms for polyunsaturated fatty acid biosynthesis in the healthy adult human central nervous system. *Biochim. Biophys. Acta Mol. Cell Biol. Lipids* **2017**, *1862*. [CrossRef] [PubMed]
21. Cobley, J.N.; Fiorello, M.L.; Bailey, D.M. 13 reasons why the brain is susceptible to oxidative stress. *Redox Biol.* **2018**, *15*, 490–503. [CrossRef]
22. Lu, T.; Aron, L.; Zullo, J.; Pan, Y.; Kim, H.; Chen, Y.; Yang, T.-H.; Kim, H.-M.; Drake, D.; Liu, X.S.; et al. REST and stress resistance in ageing and Alzheimer's disease. *Nature* **2014**, *507*, 448–454. [CrossRef] [PubMed]
23. Cabré, R.; Naudí, A.; Dominguez-Gonzalez, M.; Ayala, V.; Jové, M.; Mota-Martorell, N.; Piñol-Ripoll, G.; Gil-Villar, M.P.; Rué, M.; Portero-Otín, M.; et al. Sixty years old is the breakpoint of human frontal cortex aging. *Free Radic. Biol. Med.* **2017**, *103*. [CrossRef] [PubMed]
24. Lin, M.T.; Beal, M.F. Mitochondrial dysfunction and oxidative stress in neurodegenerative diseases. *Nature* **2006**, *443*, 787–795. [CrossRef] [PubMed]

25. Desai, R.; East, D.A.; Hardy, L.; Faccenda, D.; Rigon, M.; Crosby, J.; Alvarez, M.S.; Singh, A.; Mainenti, M.; Hussey, L.K.; et al. Mitochondria form contact sites with the nucleus to couple prosurvival retrograde response. *Sci. Adv.* **2020**, *6*, eabc9955. [CrossRef] [PubMed]
26. Karakaidos, P.; Rampias, T. Mitonuclear interactions in the maintenance of mitochondrial integrity. *Life* **2020**, *10*, 173. [CrossRef] [PubMed]
27. Picard, M.; Sandi, C. The social nature of mitochondria: Implications for human health. *Neurosci. Biobehav. Rev.* **2021**, *120*, 595–610. [CrossRef]
28. Johri, A.; Beal, M.F. Mitochondrial dysfunction in neurodegenerative diseases. *J. Pharmacol. Exp. Ther.* **2012**, *342*, 619–630. [CrossRef]
29. Bose, A.; Beal, M.F. Mitochondrial dysfunction in Parkinson's disease. *J. Neurochem.* **2016**, *139*, 216–231. [CrossRef]
30. Trinh, D.; Israwi, A.R.; Arathoon, L.R.; Gleave, J.A.; Nash, J.E. The multi-faceted role of mitochondria in the pathology of Parkinson's disease. *J. Neurochem.* **2021**, *156*, 715–752. [CrossRef]
31. Dionísio, P.A.; Amaral, J.D.; Rodrigues, C.M.P. Oxidative stress and regulated cell death in Parkinson's disease. *Ageing Res. Rev.* **2021**, *67*, 101263. [CrossRef] [PubMed]
32. Malpartida, A.B.; Williamson, M.; Narendra, D.P.; Wade-Martins, R.; Ryan, B.J. Mitochondrial dysfunction and mitophagy in Parkinson's disease: From mechanism to therapy. *Trends Biochem. Sci.* **2021**, *46*, 329–343. [CrossRef] [PubMed]
33. Duyckaerts, C. Disentangling Alzheimer's disease. *Lancet Neurol.* **2011**, *10*, 774–775. [CrossRef]
34. Braak, H.; Braak, E. Neuropathological stageing of Alzheimer-related changes. *Acta Neuropathol.* **1991**, *82*, 239–259. [CrossRef]
35. Braak, E.; Griffing, K.; Arai, K.; Bohl, J.; Bratzke, H.; Braak, H. Neuropathology of Alzheimer's disease: What is new since A. Alzheimer? *Eur. Arch. Psychiatry Clin. Neurosci.* **1999**, *249*, S14–S22. [CrossRef]
36. Grinberg, L.T.; Rüb, U.; Ferretti, R.E.L.; Nitrini, R.; Farfel, J.M.; Polichiso, L.; Gierga, K.; Jacob-Filho, W.; Heinsen, H. Brazilian Brain Bank Study Group The dorsal raphe nucleus shows phospho-tau neurofibrillary changes before the transentorhinal region in Alzheimer's disease. A precocious onset? *Neuropathol. Appl. Neurobiol.* **2009**, *35*, 406–416. [CrossRef]
37. Simic, G.; Stanic, G.; Mladinov, M.; Jovanov-Milosevic, N.; Kostovic, I.; Hof, P.R. Does Alzheimer's disease begin in the brainstem? *Neuropathol. Appl. Neurobiol.* **2009**, *35*, 532–554. [CrossRef]
38. Braak, H.; Thal, D.R.; Ghebremedhin, E.; Del Tredici, K. Stages of the pathologic process in Alzheimer disease: Age categoriesfrom 1 to 100 years. *J. Neuropathol. Exp. Neurol.* **2011**, *70*, 960–969. [CrossRef]
39. Thal, D.R.; Rüb, U.; Orantes, M.; Braak, H. Phases of Aβ-deposition in the human brain and its relevance for the development of AD. *Neurology* **2002**, *58*, 1791–1800. [CrossRef]
40. Lewandowski, C.T.; Maldonado Weng, J.; LaDu, M.J. Alzheimer's disease pathology in APOE transgenic mouse models: The who, what, when, where, why, and how. *Neurobiol. Dis.* **2020**, *139*, 104811. [CrossRef]
41. Price, J.L.; McKeel, D.W.; Buckles, V.D.; Roe, C.M.; Xiong, C.; Grundman, M.; Hansen, L.A.; Petersen, R.C.; Parisi, J.E.; Dickson, D.W.; et al. Neuropathology of nondemented aging: Presumptive evidence for preclinical Alzheimer disease. *Neurobiol. Aging* **2009**, *30*, 1026–1036. [CrossRef] [PubMed]
42. Markesbery, W.R. Neuropathologic alterations in mild cognitive impairment: A review. *J. Alzheimer's Dis.* **2010**, *19*, 221–228. [CrossRef] [PubMed]
43. Nelson, P.T.; Alafuzoff, I.; Bigio, E.H.; Bouras, C.; Braak, H.; Cairns, N.J.; Castellani, R.J.; Crain, B.J.; Davies, P.; Tredici, K.; et al. Correlation of Alzheimer disease neuropathologic changes with cognitive status: A review of the literature. *J. Neuropathol. Exp. Neurol.* **2012**, *71*, 362–381. [CrossRef]
44. Ohm, T.G.; Müller, H.; Braak, H.; Bohl, J. Close-meshed prevalence rates of different stages as a tool to uncover the rate of Alzheimer's disease-related neurofibrillary changes. *Neuroscience* **1995**, *64*, 209–217. [CrossRef]
45. Braak, H.; Braak, E. Frequency of stages of Alzheimer-related lesions in different age categories. *Neurobiol. Aging* **1997**, *18*, 351–357. [CrossRef]
46. Ferrer, I. Defining Alzheimer as a common age-related neurodegenerative process not inevitably leading to dementia. *Prog. Neurobiol.* **2012**, *97*, 38–51. [CrossRef]
47. Selkoe, D.J. Preventing Alzheimer's disease. *Science* **2012**, *337*, 1488–1492. [CrossRef]
48. McGeer, P.L.; McGeer, E.G. The amyloid cascade-inflammatory hypothesis of Alzheimer disease: Implications for therapy. *Acta Neuropathol.* **2013**, *126*, 479–497. [CrossRef]
49. Canto, C.B.; Wouterlood, F.G.; Witter, M.P. What does the anatomical organization of the entorhinal cortex tell us? *Neural Plast.* **2008**, *2008*, 1–18. [CrossRef]
50. Mink, J.W.; Blumenschine, R.J.; Adams, D.B. Ratio of central nervous system to body metabolism in vertebrates: Its constancy and functional basis. *Am. J. Physiol. Integr. Comp. Physiol.* **1981**, *241*, R203–R212. [CrossRef]
51. Caceres, M.; Lachuer, J.; Zapala, M.A.; Redmond, J.C.; Kudo, L.; Geschwind, D.H.; Lockhart, D.J.; Preuss, T.M.; Barlow, C. Elevated gene expression levels distinguish human from non-human primate brains. *Proc. Natl. Acad. Sci. USA* **2003**, *100*, 13030–13035. [CrossRef]
52. Uddin, M.; Wildman, D.E.; Liu, G.; Xu, W.; Johnson, R.M.; Hof, P.R.; Kapatos, G.; Grossman, L.I.; Goodman, M. Sister grouping of chimpanzees and humans as revealed by genome-wide phylogenetic analysis of brain gene expression profiles. *Proc. Natl. Acad. Sci. USA* **2004**, *101*, 2957–2962. [CrossRef]

53. Fu, X.; Giavalisco, P.; Liu, X.; Catchpole, G.; Fu, N.; Ning, Z.-B.; Guo, S.; Yan, Z.; Somel, M.; Paabo, S.; et al. Rapid metabolic evolution in human prefrontal cortex. *Proc. Natl. Acad. Sci. USA* **2011**, *108*, 6181–6186. [CrossRef]
54. Somel, M.; Liu, X.; Khaitovich, P. Human brain evolution: Transcripts, metabolites and their regulators. *Nat. Rev. Neurosci.* **2013**, *14*, 112–127. [CrossRef]
55. Hawrylycz, M.J.; Lein, E.S.; Guillozet-Bongaarts, A.L.; Shen, E.H.; Ng, L.; Miller, J.A.; van de Lagemaat, L.N.; Smith, K.A.; Ebbert, A.; Riley, Z.L.; et al. An anatomically comprehensive atlas of the adult human brain transcriptome. *Nature* **2012**, *489*, 391–399. [CrossRef]
56. Jové, M.; Naudí, A.; Ramírez-Núñez, O.; Portero-Otín, M.; Selman, C.; Withers, D.J.; Pamplona, R. Caloric restriction reveals a metabolomic and lipidomic signature in liver of male mice. *Aging Cell* **2014**, *13*, 828–837. [CrossRef] [PubMed]
57. Rakic, P.; Nowakowski, R.S. The time of origin of neurons in the hippocampal region of the rhesus monkey. *J. Comp. Neurol.* **1981**, *196*, 99–128. [CrossRef] [PubMed]
58. Stranahan, A.M.; Mattson, M.P. Selective vulnerability of neurons in layer II of the entorhinal cortex during aging and Alzheimer's disease. *Neural Plast.* **2010**, *2010*, 108190. [CrossRef] [PubMed]
59. Hevner, R.F.; Wong-Riley, M.T. Entorhinal cortex of the human, monkey, and rat: Metabolic map as revealed by cytochrome oxidase. *J. Comp. Neurol.* **1992**, *326*, 451–469. [CrossRef]
60. Buckmaster, P.S.; Alonso, A.; Canfield, D.R.; Amaral, D.G. Dendritic morphology, local circuitry, and intrinsic electrophysiology of principal neurons in the entorhinal cortex of macaque monkeys. *J. Comp. Neurol.* **2004**, *470*, 317–329. [CrossRef]
61. Liang, W.S.; Dunckley, T.; Beach, T.G.; Grover, A.; Mastroeni, D.; Walker, D.G.; Caselli, R.J.; Kukull, W.A.; McKeel, D.; Morris, J.C.; et al. Gene expression profiles in anatomically and functionally distinct regions of the normal aged human brain. *Physiol. Genomics* **2007**, *28*, 311–322. [CrossRef] [PubMed]
62. Locasale, J.W. Serine, glycine and one-carbon units: Cancer metabolism in full circle. *Nat. Rev. Cancer* **2013**, *13*, 572–583. [CrossRef]
63. Miller, A.L. The methionine-homocysteine cycle and its effects on cognitive diseases. *Altern. Med. Rev.* **2003**, *8*, 7–19.
64. Pamplona, R.; Barja, G. Mitochondrial oxidative stress, aging and caloric restriction: The protein and methionine connection. *Biochim. Biophys. Acta Bioenerg.* **2006**, *1757*, 496–508. [CrossRef]
65. Pamplona, R.; Barja, G. An evolutionary comparative scan for longevity-related oxidative stress resistance mechanisms in homeotherms. *Biogerontology* **2011**, *12*, 409–435. [CrossRef]
66. Naudí, A.; Caro, P.; Jové, M.; Gómez, J.; Boada, J.; Ayala, V.; Portero-Otín, M.; Barja, G.; Pamplona, R. Methionine restriction decreases endogenous oxidative molecular damage and increases mitochondrial biogenesis and uncoupling protein 4 in rat brain. *Rejuvenation Res.* **2007**, *10*, 473–484. [CrossRef]
67. Garelick, M.G.; Kennedy, B.K. TOR on the brain. *Exp. Gerontol.* **2011**, *46*, 155–163. [CrossRef] [PubMed]
68. Bockaert, J.; Marin, P. mTOR in Brain Physiology and Pathologies. *Physiol. Rev.* **2015**, *95*, 1157–1187. [CrossRef]
69. Perluigi, M.; Di Domenico, F.; Butterfield, D.A. mTOR signaling in aging and neurodegeneration: At the crossroad between metabolism dysfunction and impairment of autophagy. *Neurobiol. Dis.* **2015**, *84*, 39–49. [CrossRef]
70. Salat, D.H. Thinning of the cerebral cortex in aging. *Cereb. Cortex* **2004**, *14*, 721–730. [CrossRef] [PubMed]
71. Lemaitre, H.; Goldman, A.L.; Sambataro, F.; Verchinski, B.A.; Meyer-Lindenberg, A.; Weinberger, D.R.; Mattay, V.S. Normal age-related brain morphometric changes: Nonuniformity across cortical thickness, surface area and gray matter volume? *Neurobiol. Aging* **2012**, *33*, 617.e1–617.e9. [CrossRef] [PubMed]
72. Feng, X.; Guo, J.; Sigmon, H.C.; Sloan, R.P.; Brickman, A.M.; Provenzano, F.A.; Small, S.A. Brain regions vulnerable and resistant to aging without Alzheimer's disease. *PLoS ONE* **2020**, *15*, e0234255. [CrossRef] [PubMed]
73. Morrison, J.H.; Hof, P. Life and death of neurons in the aging brain. *Science* **1997**, *278*, 412–419. [CrossRef]
74. Insausti, R.; Insausti, A.M.; Sobreviela, M.T.; Salinas, A.; Martínez-Peñuela, J.M. Human medial temporal lobe in aging: Anatomical basis of memory preservation. *Microsc. Res. Tech.* **1998**, *43*, 8–15. [CrossRef]
75. Raz, N.; Lindenberger, U.; Rodrigue, K.M.; Kennedy, K.M.; Head, D.; Williamson, A.; Dahle, C.; Gerstorf, D.; Acker, J.D. Regional brain changes in aging healthy adults: General trends, individual differences and modifiers. *Cereb. Cortex* **2005**, *15*, 1676–1689. [CrossRef]
76. Donix, M.; Burggren, A.C.; Scharf, M.; Marschner, K.; Suthana, N.A.; Siddarth, P.; Krupa, A.K.; Jones, M.; Martin-Harris, L.; Ercoli, L.M.; et al. APOE associated hemispheric asymmetry of entorhinal cortical thickness in aging and Alzheimer's disease. *Psychiatry Res. Neuroimaging* **2013**, *214*, 212–220. [CrossRef] [PubMed]
77. Jiang, J.; Sachdev, P.; Lipnicki, D.M.; Zhang, H.; Liu, T.; Zhu, W.; Suo, C.; Zhuang, L.; Crawford, J.; Reppermund, S.; et al. A longitudinal study of brain atrophy over two years in community-dwelling older individuals. *Neuroimage* **2014**, *86*, 203–211. [CrossRef]
78. Fjell, A.M.; Westlye, L.T.; Grydeland, H.; Amlien, I.; Espeseth, T.; Reinvang, I.; Raz, N.; Dale, A.M.; Walhovd, K.B. Accelerating cortical thinning: Unique to dementia or universal in aging? *Cereb. Cortex* **2014**, *24*, 919–934. [CrossRef]
79. Hasan, K.M.; Mwangi, B.; Cao, B.; Keser, Z.; Tustison, N.J.; Kochunov, P.; Frye, R.E.; Savatic, M.; Soares, J. Entorhinal cortex thickness across the human lifespan. *J. Neuroimaging* **2016**, *26*, 278–282. [CrossRef]
80. Kurth, F.; Cherbuin, N.; Luders, E. The impact of aging on subregions of the hippocampal complex in healthy adults. *Neuroimage* **2017**, *163*, 296–300. [CrossRef]
81. Raz, N.; Rodrigue, K.M.; Head, D.; Kennedy, K.M.; Acker, J.D. Differential aging of the medial temporal lobe: A study of a five-year change. *Neurology* **2004**, *62*, 433–438. [CrossRef]

82. Wang, S.-Y.; Wang, W.-J.; Liu, J.-Q.; Song, Y.-H.; Li, P.; Sun, X.-F.; Cai, G.-Y.; Chen, X.-M. Methionine restriction delays senescence and suppresses the senescence-associated secretory phenotype in the kidney through endogenous hydrogen sulfide. *Cell Cycle* **2019**, *18*, 1573–1587. [CrossRef]
83. Olsen, R.K.; Yeung, L.-K.; Noly-Gandon, A.; D'Angelo, M.C.; Kacollja, A.; Smith, V.M.; Ryan, J.D.; Barense, M.D. Human anterolateral entorhinal cortex volumes are associated with cognitive decline in aging prior to clinical diagnosis. *Neurobiol. Aging* **2017**, *57*, 195–205. [CrossRef] [PubMed]
84. Reagh, Z.M.; Noche, J.A.; Tustison, N.J.; Delisle, D.; Murray, E.A.; Yassa, M.A. Functional imbalance of anterolateral entorhinal cortex and hippocampal dentate/CA3 underlies age-related object pattern separation deficits. *Neuron* **2018**, *97*, 1187–1198.e4. [CrossRef] [PubMed]
85. Delgado-González, J.C.; Rosa-Prieto, C.; Tarruella-Hernández, D.L.; Vallejo-Calcerrada, N.; Cebada-Sánchez, S.; Insausti, R.; Artacho-Pérula, E. Neuronal volume of the hippocampal regions in ageing. *J. Anat.* **2020**, *237*, 301–310. [CrossRef] [PubMed]
86. Martínez-Pinilla, E.; Ordóñez, C.; del Valle, E.; Navarro, A.; Tolivia, J. Regional and gender study of neuronal density in brain during aging and in Alzheimer's disease. *Front. Aging Neurosci.* **2016**, *8*. [CrossRef]
87. Gómez-Isla, T.; Price, J.L.; McKeel, D.W., Jr.; Morris, J.C.; Growdon, J.H.; Hyman, B.T. Profound loss of layer II entorhinal cortex neurons occurs in very mild Alzheimer's disease. *J. Neurosci.* **1996**, *16*, 4491–4500. [CrossRef]
88. Simic, G.; Bexheti, S.; Kelovic, Z.; Kos, M.; Grbic, K.; Hof, P.R.; Kostovic, I. Hemispheric asymmetry, modular variability and age-related changes in the human entorhinal cortex. *Neuroscience* **2005**, *130*, 911–925. [CrossRef]
89. Hancock, S.E.; Friedrich, M.G.; Mitchell, T.W.; Truscott, R.J.W.; Else, P.L. The phospholipid composition of the human entorhinal cortex remains relatively stable over 80 years of adult aging. *GeroScience* **2017**, *39*, 73–82. [CrossRef]
90. Pamplona, R.; Portero-Otın, M.; Sanz, A.; Requena, J.; Barja, G. Modification of the longevity-related degree of fatty acid unsaturation modulates oxidative damage to proteins and mitochondrial DNA in liver and brain. *Exp. Gerontol.* **2004**, *39*, 725–733. [CrossRef]
91. Domínguez-González, M.; Puigpinós, M.; Jové, M.; Naudi, A.; Portero-Otín, M.; Pamplona, R.; Ferrer, I. Regional vulnerability to lipoxidative damage and inflammation in normal human brain aging. *Exp. Gerontol.* **2018**, *111*, 218–228. [CrossRef]
92. López-González, I.; Tebé Cordomí, C.; Ferrer, I. Regional gene expression of inflammation and oxidative stress responses does not predict neurodegeneration in aging. *J. Neuropathol. Exp. Neurol.* **2017**. [CrossRef]
93. Dickerson, B.C.; Bakkour, A.; Salat, D.H.; Feczko, E.; Pacheco, J.; Greve, D.N.; Grodstein, F.; Wright, C.I.; Blacker, D.; Rosas, H.D.; et al. The cortical signature of Alzheimer's disease: Regionally specific cortical thinning relates to symptom severity in very mild to mild AD dementia and is detectable in asymptomatic amyloid-positive individuals. *Cereb. Cortex* **2009**, *19*, 497–510. [CrossRef]
94. Whitwell, J.L.; Dickson, D.W.; Murray, M.E.; Weigand, S.D.; Tosakulwong, N.; Senjem, M.L.; Knopman, D.S.; Boeve, B.F.; Parisi, J.E.; Petersen, R.C.; et al. Neuroimaging correlates of pathologically defined subtypes of Alzheimer's disease: A case-control study. *Lancet Neurol.* **2012**, *11*, 868–877. [CrossRef]
95. Cho, Y.; Seong, J.-K.; Jeong, Y.; Shin, S.Y. Individual subject classification for Alzheimer's disease based on incremental learning using a spatial frequency representation of cortical thickness data. *Neuroimage* **2012**, *59*, 2217–2230. [CrossRef] [PubMed]
96. Guzman, V.A.; Carmichael, O.T.; Schwarz, C.; Tosto, G.; Zimmerman, M.E.; Brickman, A.M. White matter hyperintensities and amyloid are independently associated with entorhinal cortex volume among individuals with mild cognitive impairment. *Alzheimer's Dement.* **2013**, *9*, S124–S131. [CrossRef]
97. Mah, L.; Binns, M.A.; Steffens, D.C. Anxiety symptoms in amnestic mild cognitive impairment are associated with medial remporal atrophy and predict conversion to Alzheimer disease. *Am. J. Geriatr. Psychiatry* **2015**, *23*, 466–476. [CrossRef] [PubMed]
98. Varon, D.; Barker, W.; Loewenstein, D.; Greig, M.; Bohorquez, A.; Santos, I.; Shen, Q.; Harper, M.; Vallejo-Luces, T.; Duara, R. Visual rating and volumetric measurement of medial temporal atrophy in the Alzheimer's Disease Neuroimaging Initiative (ADNI) cohort: Baseline diagnosis and the prediction of MCI outcome. *Int. J. Geriatr. Psychiatry* **2015**, *30*, 192–200. [CrossRef]
99. Devanand, D.P.; Pradhaban, G.; Liu, X.; Khandji, A.; De Santi, S.; Segal, S.; Rusinek, H.; Pelton, G.H.; Honig, L.S.; Mayeux, R.; et al. Hippocampal and entorhinal atrophy in mild cognitive impairment: Prediction of Alzheimer disease. *Neurology* **2007**, *68*, 828–836. [CrossRef] [PubMed]
100. La Joie, R.; Perrotin, A.; Barre, L.; Hommet, C.; Mezenge, F.; Ibazizene, M.; Camus, V.; Abbas, A.; Landeau, B.; Guilloteau, D.; et al. Region-specific hierarchy between atrophy, hypometabolism, and β-Amyloid (Aβ) load in Alzheimer's disease dementia. *J. Neurosci.* **2012**, *32*, 16265–16273. [CrossRef]
101. Younes, L.; Albert, M.; Miller, M.I. Inferring changepoint times of medial temporal lobe morphometric change in preclinical Alzheimer's disease. *NeuroImage Clin.* **2014**, *5*, 178–187. [CrossRef]
102. Miller, M.I.; Ratnanather, J.T.; Tward, D.J.; Brown, T.; Lee, D.S.; Ketcha, M.; Mori, K.; Wang, M.-C.; Mori, S.; Albert, M.S.; et al. Network neurodegeneration in Alzheimer's disease via MRI based shape diffeomorphometry and high-field atlasing. *Front. Bioeng. Biotechnol.* **2015**, *3*. [CrossRef] [PubMed]
103. Kulason, S.; Xu, E.; Tward, D.J.; Bakker, A.; Albert, M.; Younes, L.; Miller, M.I. Entorhinal and transentorhinal atrophy in preclinical Alzheimer's disease. *Front. Neurosci.* **2020**, *14*. [CrossRef] [PubMed]
104. Atiya, M.; Hyman, B.T.; Albert, M.S.; Killiany, R. Structural magnetic resonance imaging in established and prodromal Alzheimer disease: A review. *Alzheimer Dis. Assoc. Disord.* **2003**, *17*, 177–195. [CrossRef] [PubMed]
105. Kantarci, K.; Jack, C.R. Neuroimaging in Alzheimer disease: An evidence-based review. *Neuroimaging Clin. N. Am.* **2003**, *13*, 197–209. [CrossRef]

106. Jack, C.R.; Shiung, M.M.; Gunter, J.L.; O'Brien, P.C.; Weigand, S.D.; Knopman, D.S.; Boeve, B.F.; Ivnik, R.J.; Smith, G.E.; Cha, R.H.; et al. Comparison of different MRI brain atrophy rate measures with clinical disease progression in AD. *Neurology* **2004**, *62*, 591–600. [CrossRef]
107. Csernansky, J.G.; Wang, L.; Swank, J.; Miller, J.P.; Gado, M.; McKeel, D.; Miller, M.I.; Morris, J.C. Preclinical detection of Alzheimer's disease: Hippocampal shape and volume predict dementia onset in the elderly. *Neuroimage* **2005**, *25*, 783–792. [CrossRef] [PubMed]
108. den Heijer, T.; Geerlings, M.I.; Hoebeek, F.E.; Hofman, A.; Koudstaal, P.J.; Breteler, M.M.B. Use of hippocampal and amygdalar volumes on magnetic resonance imaging to predict dementia in cognitively intact elderly people. *Arch. Gen. Psychiatry* **2006**, *63*, 57. [CrossRef] [PubMed]
109. Apostolova, L.G.; Mosconi, L.; Thompson, P.M.; Green, A.E.; Hwang, K.S.; Ramirez, A.; Mistur, R.; Tsui, W.H.; de Leon, M.J. Subregional hippocampal atrophy predicts Alzheimer's dementia in the cognitively normal. *Neurobiol. Aging* **2010**, *31*, 1077–1088. [CrossRef] [PubMed]
110. Dickerson, B.C.; Stoub, T.R.; Shah, R.C.; Sperling, R.A.; Killiany, R.J.; Albert, M.S.; Hyman, B.T.; Blacker, D.; DeToledo-Morrell, L. Alzheimer-signature MRI biomarker predicts AD dementia in cognitively normal adults. *Neurology* **2011**, *76*, 1395–1402. [CrossRef]
111. Miller, M.I.; Younes, L.; Ratnanather, J.T.; Brown, T.; Trinh, H.; Postell, E.; Lee, D.S.; Wang, M.-C.; Mori, S.; O'Brien, R.; et al. The diffeomorphometry of temporal lobe structures in preclinical Alzheimer's disease. *NeuroImage Clin.* **2013**, *3*, 352–360. [CrossRef]
112. Soldan, A.; Pettigrew, C.; Lu, Y.; Wang, M.; Selnes, O.; Albert, M.; Brown, T.; Ratnanather, J.T.; Younes, L.; Miller, M.I. Relationship of medial temporal lobe atrophy, APOE genotype, and cognitive reserve in preclinical Alzheimer's disease. *Hum. Brain Mapp.* **2015**, *36*, 2826–2841. [CrossRef]
113. Pettigrew, C.; Soldan, A.; Zhu, Y.; Wang, M.-C.; Moghekar, A.; Brown, T.; Miller, M.; Albert, M. Cortical thickness in relation to clinical symptom onset in preclinical AD. *NeuroImage Clin.* **2016**, *12*, 116–122. [CrossRef] [PubMed]
114. Kidd, M. Alzheimer's Disease—An electron microscopical study. *Brain* **1964**, *87*, 307–320. [CrossRef] [PubMed]
115. Luse, S.A.; Smith, K.R. The ultrastructure of senile plaques. *Am. J. Pathol.* **1964**, *44*, 553–563. [PubMed]
116. Terry, R.D.; Gonatas, N.K.; Weiss, M. Ultrastructural studies in Alzheimer's presenile demientia. *Am. J. Pathol.* **1964**, *44*, 269–297. [PubMed]
117. Johnson, A.B.; Blum, N.R. Nucleoside phosphatase activities associated with the tangles and plaques of Alzheimer's disease: A histochemical study of natural and experimental neurofibrillary tangles. *J. Neuropathol. Exp. Neurol.* **1970**, *29*, 463–478. [CrossRef] [PubMed]
118. Wiśniewski, H.; Terry, R.D.; Hirano, A. Neurofibrillary pathology. *J. Neuropathol. Exp. Neurol.* **1970**, *29*, 163–176. [CrossRef] [PubMed]
119. Hirai, K.; Aliev, G.; Nunomura, A.; Fujioka, H.; Russell, R.L.; Atwood, C.S.; Johnson, A.B.; Kress, Y.; Vinters, H.V.; Tabaton, M.; et al. Mitochondrial abnormalities in Alzheimer's disease. *J. Neurosci.* **2001**, *21*, 3017–3023. [CrossRef] [PubMed]
120. Baloyannis, S.J. Mitochondrial alterations in Alzheimer's disease. *J. Alzheimer's Dis.* **2006**, *9*, 119–126. [CrossRef] [PubMed]
121. Ferris, S.H.; de Leon, M.J.; Wolf, A.P.; Farkas, T.; Christman, D.R.; Reisberg, B.; Fowler, J.S.; MacGregor, R.; Goldman, A.; George, A.E.; et al. Positron emission tomography in the study of aging and senile dementia. *Neurobiol. Aging* **1980**, *1*, 127–131. [CrossRef]
122. Foster, N.L.; Chase, T.N.; Fedio, P.; Patronas, N.J.; Brooks, R.A.; Chiro, G.D. Alzheimer's disease: Focal cortical changes shown by positron emission tomography. *Neurology* **1983**, *33*, 961. [CrossRef] [PubMed]
123. de Leon, M.J.; George, A.E.; Ferris, S.H.; Rosenbloom, S.; Christman, D.R.; Gentes, C.I.; Reisberg, B.; Kricheff, I.I.; Wolf, A.P. Regional correlation of PET and CT in senile dementia of the Alzheimer type. *AJNR. Am. J. Neuroradiol.* **1983**, *4*, 553–556. [PubMed]
124. Frackowiak, R.S.J.; Pozilli, C.J.; Legg, N.D.; Boulay, G.H.; Marshall, J.; Lenzi, G.L.; Jones, T. Regional cerebral oxygen supply and utilization in dementia. A clinical and physiological study with oxygen-15 and positron tomography. *Brain* **1981**, *104*, 753–778. [CrossRef]
125. Fukuyama, H.; Ogawa, M.; Yamauchi, H.; Yamaguchi, S.; Kimura, J.; Yonekura, Y.; Konishi, J. Altered cerebral energy metabolism in Alzheimer's disease: A PET study. *J. Nucl. Med.* **1994**, *35*, 1–6.
126. Blass, J.P.; Zemcov, A. Alzheimer's disease. A metabolic systems degeneration? *Neurochem. Pathol.* **1984**, *2*, 103–114. [CrossRef] [PubMed]
127. Hoyer, S. Abnormalities in brain glucose utilization and its impact on cellular and molecular mechanisms in sporadic dementia of Alzheimer type. *Ann. N. Y. Acad. Sci.* **1993**, *695*, 77–80. [CrossRef] [PubMed]
128. Swerdlow, R.H. Mitochondria and mitochondrial cascades in Alzheimer's disease. *J. Alzheimer's Dis.* **2018**, *62*, 1403–1416. [CrossRef] [PubMed]
129. Beal, M.F. Aging, energy, and oxidative stress in neurodegenerative diseases. *Ann. Neurol.* **1995**, *38*, 357–366. [CrossRef]
130. Sorbi, S.; Bird, E.D.; Blass, J.P. Decreased pyruvate dehydrogenase complex activity in Huntington and Alzheimer brain. *Ann. Neurol.* **1983**, *13*, 72–78. [CrossRef]
131. Gibson, G.E.; Sheu, K.-F.R.; Blass, J.P.; Baker, A.; Carlson, K.C.; Harding, B.; Perrino, P. Reduced activities of thiamine-dependent enzymes in the brains and peripheral tissues of patients with Alzheimer's disease. *Arch. Neurol.* **1988**, *45*, 836–840. [CrossRef]
132. Parker, W.D.; Filley, C.M.; Parks, J.K. Cytochrome oxidase deficiency in Alzheimer's disease. *Neurology* **1990**, *40*, 1302. [CrossRef] [PubMed]

133. Gibson, G.E.; Starkov, A.; Blass, J.P.; Ratan, R.R.; Beal, M.F. Cause and consequence: Mitochondrial dysfunction initiates and propagates neuronal dysfunction, neuronal death and behavioral abnormalities in age-associated neurodegenerative diseases. *Biochim. Biophys. Acta Mol. Basis Dis.* **2010**, *1802*, 122–134. [CrossRef] [PubMed]
134. Ghosh, S.S.; Swerdlow, R.H.; Miller, S.W.; Sheeman, B.; Parker, W.D.; Davis, R.E. Use of cytoplasmic hybrid cell lines for elucidating the role of mitochondrial dysfunction in Alzheimer's disease and Parkinson's disease. *Ann. N. Y. Acad. Sci.* **1999**, *893*, 176–191. [CrossRef]
135. Khan, S.M.; Cassarino, D.S.; Abramova, N.N.; Keeney, P.M.; Borland, M.K.; Trimmer, P.A.; Krebs, C.T.; Bennett, J.C.; Parks, J.K.; Swerdlow, R.H.; et al. Alzheimer's disease cybrids replicate beta-amyloid abnormalities through cell death pathways. *Ann. Neurol.* **2000**, *48*, 148–155. [CrossRef]
136. Trimmer, P.A.; Swerdlow, R.H.; Parks, J.K.; Keeney, P.; Bennett, J.P.; Miller, S.W.; Davis, R.E.; Parker, W.D. Abnormal mitochondrial morphology in sporadic Parkinson's and Alzheimer's disease cybrid cell lines. *Exp. Neurol.* **2000**, *162*, 37–50. [CrossRef]
137. Terni, B.; Boada, J.; Portero-Otin, M.; Pamplona, R.; Ferrer, I. Mitochondrial ATP-synthase in the entorhinal cortex is a target of oxidative stress at stages I/II of Alzheimer's disease pathology. *Brain Pathol.* **2010**, *20*, 222–233. [CrossRef] [PubMed]
138. Armand-Ugon, M.; Ansoleaga, B.; Berjaoui, S.; Ferrer, I. Reduced mitochondrial activity is early and steady in the entorhinal cortex but it is mainly unmodified in the frontal cortex in Alzheimer's disease. *Curr. Alzheimer Res.* **2017**, *14*. [CrossRef] [PubMed]
139. Cadonic, C.; Sabbir, M.G.; Albensi, B.C. Mechanisms of mitochondrial dysfunction in Alzheimer's disease. *Mol. Neurobiol.* **2016**, *53*, 6078–6090. [CrossRef]
140. Wang, W.; Zhao, F.; Ma, X.; Perry, G.; Zhu, X. Mitochondria dysfunction in the pathogenesis of Alzheimer's disease: Recent advances. *Mol. Neurodegener.* **2020**, *15*, 30. [CrossRef]
141. Pamplona, R.; Dalfó, E.; Ayala, V.; Bellmunt, M.J.; Prat, J.; Ferrer, I.; Portero-Otín, M. Proteins in human brain cortex are modified by oxidation, glycoxidation, and lipoxidation. Effects of Alzheimer disease and identification of lipoxidation targets. *J. Biol. Chem.* **2005**, *280*, 21522–21530. [CrossRef]
142. Wang, X.; Wang, W.; Li, L.; Perry, G.; Lee, H.; Zhu, X. Oxidative stress and mitochondrial dysfunction in Alzheimer's disease. *Biochim. Biophys. Acta Mol. Basis Dis.* **2014**, *1842*, 1240–1247. [CrossRef]
143. Tramutola, A.; Lanzillotta, C.; Perluigi, M.; Butterfield, D.A. Oxidative stress, protein modification and Alzheimer disease. *Brain Res. Bull.* **2017**, *133*, 88–96. [CrossRef]
144. Ferrer, I. Proteomics and lipidomics in the human brain. In *Handbook of Clinical Neurology*; Elsevier: Amsterdam, The Netherlands, 2018; pp. 285–302.
145. Tobore, T.O. On the central role of mitochondria dysfunction and oxidative stress in Alzheimer's disease. *Neurol. Sci.* **2019**, *40*, 1527–1540. [CrossRef] [PubMed]
146. Butterfield, D.A.; Boyd-Kimball, D. Mitochondrial oxidative and nitrosative stress and Alzheimer disease. *Antioxidants* **2020**, *9*, 818. [CrossRef] [PubMed]
147. Ebanks, B.; Ingram, T.L.; Chakrabarti, L. ATP synthase and Alzheimer's disease: Putting a spin on the mitochondrial hypothesis. *Aging (Albany. NY).* **2020**, *12*, 16647–16662. [CrossRef] [PubMed]
148. Pinke, G.; Zhou, L.; Sazanov, L.A. Cryo-EM structure of the entire mammalian F-type ATP synthase. *Nat. Struct. Mol. Biol.* **2020**, *27*, 1077–1085. [CrossRef] [PubMed]
149. Jové, M.; Pradas, I.; Dominguez-Gonzalez, M.; Ferrer, I.; Pamplona, R. Lipids and lipoxidation in human brain aging. Mitochondrial ATP-synthase as a key lipoxidation target. *Redox Biol.* **2019**, *23*, 101082. [CrossRef] [PubMed]
150. Nesci, S.; Trombetti, F.; Ventrella, V.; Pagliarani, A. Post-translational modifications of the mitochondrial F1FO-ATPase. *Biochim. Biophys. Acta Gen. Subj.* **2017**, *1861*, 2902–2912. [CrossRef]
151. Domínguez, M.; de Oliveira, E.; Odena, M.A.; Portero, M.; Pamplona, R.; Ferrer, I. Redox proteomic profiling of neuroketal-adducted proteins in human brain: Regional vulnerability at middle age increases in the elderly. *Free Radic. Biol. Med.* **2016**, *95*, 1–15. [CrossRef] [PubMed]
152. Martínez, A.; Portero-Otin, M.; Pamplona, R.; Ferrer, I. Protein targets of oxidative damage in human neurodegenerative diseases with abnormal protein aggregates. *Brain Pathol.* **2010**, *20*, 281–297. [CrossRef] [PubMed]
153. Manavalan, A.; Mishra, M.; Feng, L.; Sze, S.K.; Akatsu, H.; Heese, K. Brain site-specific proteome changes in aging-related dementia. *Exp. Mol. Med.* **2013**, *45*, e39. [CrossRef] [PubMed]
154. Brinkmalm, A.; Brinkmalm, G.; Honer, W.G.; Moreno, J.A.; Jakobsson, J.; Mallucci, G.R.; Zetterberg, H.; Blennow, K.; Öhrfelt, A. Targeting synaptic pathology with a novel affinity mass spectrometry approach. *Mol. Cell. Proteom.* **2014**, *13*, 2584–2592. [CrossRef] [PubMed]
155. Chang, R.Y.K.; Etheridge, N.; Dodd, P.R.; Nouwens, A.S. Targeted quantitative analysis of synaptic proteins in Alzheimer's disease brain. *Neurochem. Int.* **2014**, *75*, 66–75. [CrossRef] [PubMed]
156. Reed, T.; Perluigi, M.; Sultana, R.; Pierce, W.M.; Klein, J.B.; Turner, D.M.; Coccia, R.; Markesbery, W.R.; Butterfield, D.A. Redox proteomic identification of 4-hydroxy-2-nonenal-modified brain proteins in amnestic mild cognitive impairment: Insight into the role of lipid peroxidation in the progression and pathogenesis of Alzheimer's disease. *Neurobiol. Dis.* **2008**, *30*, 107–120. [CrossRef] [PubMed]
157. Perluigi, M.; Sultana, R.; Cenini, G.; Di Domenico, F.; Memo, M.; Pierce, W.M.; Coccia, R.; Butterfield, D.A. Redox proteomics identification of 4-hydroxynonenal-modified brain proteins in Alzheimer's disease: Role of lipid peroxidation in Alzheimer's disease pathogenesis. *Proteomics. Clin. Appl.* **2009**, *3*, 682–693. [CrossRef]

158. Reed, T.T.; Pierce, W.M.; Markesbery, W.R.; Butterfield, D.A. Proteomic identification of HNE-bound proteins in early Alzheimer disease: Insights into the role of lipid peroxidation in the progression of AD. *Brain Res.* **2009**, *1274*, 66–76. [CrossRef] [PubMed]
159. Di Domenico, F.; Tramutola, A.; Butterfield, D.A. Role of 4-hydroxy-2-nonenal (HNE) in the pathogenesis of alzheimer disease and other selected age-related neurodegenerative disorders. *Free Radic. Biol. Med.* **2017**, *111*, 253–261. [CrossRef]
160. Ottis, P.; Koppe, K.; Onisko, B.; Dynin, I.; Arzberger, T.; Kretzschmar, H.; Requena, J.R.; Silva, C.J.; Huston, J.P.; Korth, C. Human and rat brain lipofuscin proteome. *Proteomics* **2012**, *12*, 2445–2454. [CrossRef]
161. Semick, S.A.; Bharadwaj, R.A.; Collado-Torres, L.; Tao, R.; Shin, J.H.; Deep-Soboslay, A.; Weiss, J.R.; Weinberger, D.R.; Hyde, T.M.; Kleinman, J.E.; et al. Integrated DNA methylation and gene expression profiling across multiple brain regions implicate novel genes in Alzheimer's disease. *Acta Neuropathol.* **2019**, *137*, 557–569. [CrossRef]
162. Ding, B.; Xi, Y.; Gao, M.; Li, Z.; Xu, C.; Fan, S.; He, W. Gene expression profiles of entorhinal cortex in Alzheimer's disease. *Am. J. Alzheimers. Dis. Other Demen.* **2014**, *29*, 526–532. [CrossRef]
163. Chandrasekaran, K.; Hatanpää, K.; Brady, D.R.; Stoll, J.; Rapoport, S.I. Downregulation of oxidative phosphorylation in Alzheimer disease: Loss of cytochrome oxidase subunit mRNA in the hippocampus and entorhinal cortex. *Brain Res.* **1998**, *796*, 13–19. [CrossRef]
164. Cenini, G.; Fiorini, A.; Sultana, R.; Perluigi, M.; Cai, J.; Klein, J.B.; Head, E.; Butterfield, D.A. An investigation of the molecular mechanisms engaged before and after the development of Alzheimer disease neuropathology in Down syndrome: A proteomics approach. *Free Radic. Biol. Med.* **2014**, *76*, 89–95. [CrossRef] [PubMed]
165. Butterfield, D.A.; Palmieri, E.M.; Castegna, A. Clinical implications from proteomic studies in neurodegenerative diseases: Lessons from mitochondrial proteins. *Expert Rev. Proteom.* **2016**, *13*, 259–274. [CrossRef] [PubMed]
166. Blanch, M.; Mosquera, J.L.; Ansoleaga, B.; Ferrer, I.; Barrachina, M. Altered mitochondrial DNA methylation pattern in Alzheimer disease–related pathology and in Parkinson disease. *Am. J. Pathol.* **2016**, *186*, 385–397. [CrossRef] [PubMed]
167. Fernandez-Echevarria, C.; Díaz, M.; Ferrer, I.; Canerina-Amaro, A.; Marin, R. Aβ promotes VDAC1 channel dephosphorylation in neuronal lipid rafts. Relevance to the mechanisms of neurotoxicity in Alzheimer's disease. *Neuroscience* **2014**, *278*, 354–366. [CrossRef]
168. Ansoleaga, B.; Jové, M.; Schlüter, A.; Garcia-Esparcia, P.; Moreno, J.; Pujol, A.; Pamplona, R.; Portero-Otín, M.; Ferrer, I. Deregulation of purine metabolism in Alzheimer's disease. *Neurobiol. Aging* **2015**, *36*, 68–80. [CrossRef]
169. Kichev, A.; Ilieva, E.V.; Piñol-Ripoll, G.; Podlesniy, P.; Ferrer, I.; Portero-Otín, M.; Pamplona, R.; Espinet, C. Cell death and learning impairment in mice caused by in vitro modified pro-NGF can be related to its increased oxidative modifications in Alzheimer disease. *Am. J. Pathol.* **2009**, *175*, 2574–2585. [CrossRef]
170. Llorens, F.; Thüne, K.; Andrés-Benito, P.; Tahir, W.; Ansoleaga, B.; Hernández-Ortega, K.; Martí, E.; Zerr, I.; Ferrer, I. MicroRNA expression in the locus coeruleus, entorhinal cortex, and hippocampus at early and middle stages of Braak neurofibrillary tangle pathology. *J. Mol. Neurosci.* **2017**, *63*, 206–215. [CrossRef]
171. Terni, B.; Ferrer, I. Abnormal expression and distribution of MMP2 at initial stages of Alzheimer's disease-related pathology. *J. Alzheimer's Dis.* **2015**, *46*, 461–469. [CrossRef]
172. Ferrer, I. Differential expression of phosphorylated translation initiation factor 2 alpha in Alzheimer's disease and Creutzfeldt-Jakob's disease. *Neuropathol. Appl. Neurobiol.* **2002**, *28*, 441–451. [CrossRef]
173. Chan, R.B.; Oliveira, T.G.; Cortes, E.P.; Honig, L.S.; Duff, K.E.; Small, S.A.; Wenk, M.R.; Shui, G.; Di Paolo, G. Comparative lipidomic analysis of mouse and human brain with Alzheimer disease. *J. Biol. Chem.* **2012**, *287*, 2678–2688. [CrossRef] [PubMed]
174. Fabelo, N.; Martín, V.; Marín, R.; Moreno, D.; Ferrer, I.; Díaz, M. Altered lipid composition in cortical lipid rafts occurs at early stages of sporadic Alzheimer's disease and facilitates APP/BACE1 interactions. *Neurobiol. Aging* **2014**, *35*, 1801–1812. [CrossRef] [PubMed]
175. Xia, J.; Rocke, D.M.; Perry, G.; Ray, M. Differential network analyses of Alzheimer's disease identify early events in Alzheimer's disease pathology. *Int. J. Alzheimers Dis.* **2014**, *2014*, 1–18. [CrossRef]
176. Wassall, S.R.; Stillwell, W. Polyunsaturated fatty acid–cholesterol interactions: Domain formation in membranes. *Biochim. Biophys. Acta Biomembr.* **2009**, *1788*, 24–32. [CrossRef] [PubMed]
177. Montesinos, J.; Pera, M.; Larrea, D.; Guardia-Laguarta, C.; Agrawal, R.R.; Velasco, K.R.; Yun, T.D.; Stavrovskaya, I.G.; Xu, Y.; Koo, S.Y.; et al. The Alzheimer's disease-associated C99 fragment of APP regulates cellular cholesterol trafficking. *EMBO J.* **2020**, *39*. [CrossRef] [PubMed]
178. Pera, M.; Larrea, D.; Guardia-Laguarta, C.; Montesinos, J.; Velasco, K.R.; Agrawal, R.R.; Xu, Y.; Chan, R.B.; Di Paolo, G.; Mehler, M.F.; et al. Increased localization of APP-C99 in mitochondria-associated ER membranes causes mitochondrial dysfunction in Alzheimer disease. *EMBO J.* **2017**, *36*, 3356–3371. [CrossRef]
179. Leng, K.; Li, E.; Eser, R.; Piergies, A.; Sit, R.; Tan, M.; Neff, N.; Li, S.H.; Rodriguez, R.D.; Suemoto, C.K.; et al. Molecular characterization of selectively vulnerable neurons in Alzheimer's disease. *Nat. Neurosci.* **2021**, *24*, 276–287. [CrossRef]
180. Olajide, O.J.; Suvanto, M.E.; Chapman, C.A. Molecular mechanisms of neurodegeneration in the entorhinal cortex that underlie its selective vulnerability during the pathogenesis of Alzheimer's disease. *Biol. Open* **2021**, *10*, bio056796. [CrossRef]

Review

Organization of the Respiratory Supercomplexes in Cells with Defective Complex III: Structural Features and Metabolic Consequences

Michela Rugolo, **Claudia Zanna** and **Anna Maria Ghelli** *

Department of Pharmacy and Biotechnology, University of Bologna, 40126 Bologna, Italy; michela.rugolo@unibo.it (M.R.); claudia.zanna@unibo.it (C.Z.)
* Correspondence: annamaria.ghelli@unibo.it

Abstract: The mitochondrial respiratory chain encompasses four oligomeric enzymatic complexes (complex I, II, III and IV) which, together with the redox carrier ubiquinone and cytochrome *c*, catalyze electron transport coupled to proton extrusion from the inner membrane. The protonmotive force is utilized by complex V for ATP synthesis in the process of oxidative phosphorylation. Respiratory complexes are known to coexist in the membrane as single functional entities and as supramolecular aggregates or supercomplexes (SCs). Understanding the assembly features of SCs has relevant biomedical implications because defects in a single protein can derange the overall SC organization and compromise the energetic function, causing severe mitochondrial disorders. Here we describe in detail the main types of SCs, all characterized by the presence of complex III. We show that the genetic alterations that hinder the assembly of Complex III, not just the activity, cause a rearrangement of the architecture of the SC that can help to preserve a minimal energetic function. Finally, the major metabolic disturbances associated with severe SCs perturbation due to defective complex III are discussed along with interventions that may circumvent these deficiencies.

Keywords: respiratory complexes; respiratory supercomplexes; oxidative stress; mitochondrial DNA; *MTCYB* mutations; cytochrome *b*; complex III; mitochondrial diseases

Citation: Rugolo, M.; Zanna, C.; Ghelli, A.M. Organization of the Respiratory Supercomplexes in Cells with Defective Complex III: Structural Features and Metabolic Consequences. *Life* **2021**, *11*, 351. https://doi.org/10.3390/life11040351

Academic Editors: Giorgio Lenaz and Salvatore Nesci

Received: 18 March 2021
Accepted: 14 April 2021
Published: 17 April 2021

1. Introduction

The mitochondria are cytosolic organelles of eukaryotic cells in charge of ATP production through the process of oxidative phosphorylation (OXPHOS). However, several other important pathways are associated with mitochondria, such as the citric acid cycle [1,2], the fatty acids oxidation [3] and lipid droplets formation [4], the iron–sulfur (Fe–S) protein biogenesis [5] and amino acids catabolism [6]. Furthermore, mitochondria are implicated in the buffering of cytosolic calcium concentration [7], in generation of reactive oxygen species (ROS) [8], and in regulation and execution of different types of cell death [9]. They are also involved in an array of adaptive responses triggered by perturbations of intracellular homeostasis [10], orchestrating anabolic and catabolic reactions, which are finely adjusted according to different cytosolic conditions. All these interconnected functions are sustained by the activity of the "mitochondrial proteome", estimated to contain at least 1000 (in yeast) [11] to 1500 (in humans) [12] different proteins, 15% of which are directly involved in energy metabolism and the OXPHOS system. Note that recent bioinformatics analysis in yeast provided evidence for more proteins than expected, cryptically localized inside mitochondria [13].

As typically described in the textbooks, mitochondria have two membranes, the outer membrane, which acts a barrier separating mitochondria from the cytoplasm, and the inner membrane surrounding the matrix, where soluble enzymes of intermediary metabolism, ribosomes and the mitochondrial genome (mtDNA) are hosted. The inner mitochondrial membrane is characterized by an extraordinarily high protein content and, in particular, encloses many copies of the respiratory chain components that together with the ATP synthase (named also complex V, CV) form the molecular machinery of OXPHOS, i.e., the ATP production from ADP and inorganic phosphate. The mitochondrial respiratory chain consists of four enzymatic multi-subunit complexes, namely the NADH-coenzyme Q reductase (Complex I, CI), the succinate-Coenzyme Q reductase (Complex II, CII), the ubiquinol-cytochrome *c* reductase (Complex III, CIII), and the cytochrome *c* oxidase (Complex IV, CIV). Two mobile redox-active compounds, the lipophilic coenzyme Q (CoQ) and the hydrophilic cytochrome *c*, connect the enzymatic complexes, thus allowing the electron transfer from soluble reducing equivalents (NADH and $FADH_2$) to molecular oxygen.

Unlike the oxidation of NADH which only occurs via CI, $FADH_2$ can be oxidized at the inner membrane mainly by CII, but also by other less abundant proteins such as the glycerol-3-phosphate dehydrogenase [14], the electron transfer flavoprotein dehydrogenases [15–17], the dihydroorotate dehydrogenase [18], the choline dehydrogenase [19], the sulfide CoQ reductase [20], and the proline dehydrogenase [21]. All these proteins are able to feed electrons to CoQ and in turn to CIII, which therefore can be considered the central collector delivering electrons through cytochrome *c* to CIV. The electron transport is coupled to proton extrusion from the matrix into the intermembrane space generating a transmembrane proton gradient at the level of CI, CIII and CIV, but not of CII. This later, together with other FAD-linked enzymes, does not contribute to energy conservation.

2. Mitochondrial Proteins Are Encoded by Two Genomes

Most mitochondrial proteins are encoded in the nucleus, synthesized in the cytosol and then imported into mitochondria by specific targeting mechanisms. However, these organelles are characterized by the presence of an independent genome, the mtDNA. This peculiarity is believed to be due to the evolutionary origin of mitochondria from alpha-proteobacteria integrated into proto-eukaryotic host, of which the details are still debated [22]. Most mitochondrial genes were then transferred to the nucleus, although a few of respiration-competent genes were conserved as an independent genome. The mtDNA is a circular double-stranded DNA, in humans of approximately 16.5 kb, encoding for thirteen polypeptides, all essential subunits of the OXPHOS system, and also for two ribosomal RNAs and twenty-two transfer RNAs, required for the intra-mitochondrial translation of the thirteen proteins. The evolutionary pathways involved in maintaining this transcriptionally active genome in addition to nuclear DNA are still poorly understood. New system biology and bioinformatics approaches have confirmed that the very high hydrophobicity of the proteins encoded by mtDNA is crucial to limit their translocation from the cytoplasm to the mitochondrial membrane and to favour mistargeting to the endoplasmic reticulum. In addition, the high CG content has been shown to increase the thermodynamic stability of the mtDNA, protecting from environmental insults [23]. Of note is that the high GC content might be related also to the surprisingly very high local temperature (about 50 °C) recently determined inside mitochondria [24]. Finally, the preferential encoding of components essential for organelle function in the mtDNA would allow localized control of gene expression and therefore the assembly of protein complexes [25].

3. Both Genomes Contribute to the Onset of Mitochondrial Diseases

Mitochondrial diseases are genetically heterogeneous disorders caused by mutations in nuclear genes encoding OXPHOS structural proteins or assembly factors, which are proteins required for the correct maturation of the complexes, but not contributing to the final structures. Mutations can also affect the molecular machineries of mtDNA replication and maintenance, of mitochondrial transcription and translation, as well as proteins involved in cristae shaping, network dynamics and quality control, composition of membrane lipids or mechanisms of antioxidant defences. Disorders can also be caused by mutations in the mtDNA, encoding structural OXPHOS subunits. In this case the genetic features are very peculiar, since the mtDNA inheritance mode follows the maternal lineage.

Furthermore, each cell presents multiple copies (100–1000) of this genome, so that mutated and non-mutated copies can co-exist in the same individual, generating a phenomenon called heteroplasmy. Accordingly, the clinical phenotype and the severity of biochemical dysfunctions are highly variable, and pathology only manifests when the percentage of mutated mtDNA exceeds a threshold, which is variable for each kind of mutation. For a recent exhaustive review on the genetic basis of primary mitochondrial diseases we refer to Fernandez-Vizarra and Zeviani, 2020 [26].

4. The OXPHOS System

The structures of individual mitochondrial respiratory chain complexes have been determined by X-ray crystallography [27–30] or electron cryo-microscopy [31–35]. The bovine or human mitochondrial CI contains 44 different subunits, forming an L-shaped structure. The minimal functional unit of CI, which is conserved from bacteria to mammals, comprises 14 subunits known as core subunits. Subunits ND1-ND6 and ND4L form the hydrophobic membrane arm, the other seven core subunits form the hydrophilic arm protruding into the matrix and comprising a flavin mononucleotide (FMN) and eight iron–sulfur clusters as redox active prosthetic groups. This latter contains the NADH binding and electron transfer sites, whereas the membrane arm performs the proton translocation. The other supernumerary subunits play significant roles in the assembly, stabilization and regulation of CI [35–37].

CII is composed of four subunits, forming a hydrophilic head, containing a FAD binding protein and an iron–sulfur protein, and the hydrophobic arm with two membrane-anchor proteins (CybL and CybS). Three kinds of prosthetic groups, FAD, heme and iron–sulfur clusters, were recognized in CII, coupled with two Q-binding sites [38,39].

The mammalian CIII monomer is composed of three respiratory subunits (cytochrome *b*/MTCYB, cytochrome *c1*/CYC1 and the Rieske iron–sulfur protein/UQCRFS1, two core proteins (UQCRC1, UQCRC2) and six low-molecular-weight proteins (UQCRH/QCR6, UQCRB/QCR7, UQCRQ/QCR8, UQCR10/QCR9, UQCR11/QCR10 and a cleavage product of UQCRFS1). CIII is present as dimer (approx 450 kDa), although it is still controversial whether the two monomers of $CIII_2$ function cooperatively or independently [40,41].

The CIV monomer has a mass of approximately 200 kDa and is believed to occur in the membrane both as a monomer and a dimer [42]. Each CIV monomer consists of 14 subunits [43], since NDUFA4, which was considered to be a subunit of CI, is actually a subunit of CIV [44]. The four redox-active metal centres constituting the electron transport pathway are heme *a3* and CuB, forming the binuclear centre that binds oxygen, and heme *a*, located in subunit COXI. The CuA center is incorporated in COXII [45]. The remaining subunits (COX4, 5A, 5B, 6A, 6B, 6C, 7A, 7B, 7C, 8A) are thought to have a structural role in the stabilization of the complex [46,47].

5. OXPHOS Optimization by the Inner Membrane Architecture

The general notion that the inner membrane is heavily folded to form the cristae has been reconsidered after 3D electron microscopy (EM) tomographic analysis in mitochondria from a wide variety of organisms, revealing that the cristae membrane opens toward the intermembrane space through narrow tubular structures, the crista junctions [48]. The molecular components of these cristae junctions have been subsequently identified, the most relevant being the mitochondrial contact site and cristae organizing system (MICOS) [49–51] and the high-molecular-weight GTPase Optic atrophy 1 (OPA1) [52–54] (Figure 1).

Furthermore, the observation of mitochondrial specimens in their native environment by means of the immuno-EM and cryo-EM tomography allowed to demonstrate that dimers of CV are arranged in long rows along the tightly curved ridges of the cristae membrane [55] (Figure 1). These ribbons of CV dimers are involved in the folding of the crista membrane [56]. It seems therefore that the intra-mitochondrial architecture is more complex than initially believed, with MICOS, OPA1 and CV dimers playing a structural role, together with the abundance of non-bilayer cone-shaped phospholipids, such as cardiolipin and phosphatidylethanolamine [53,57,58].

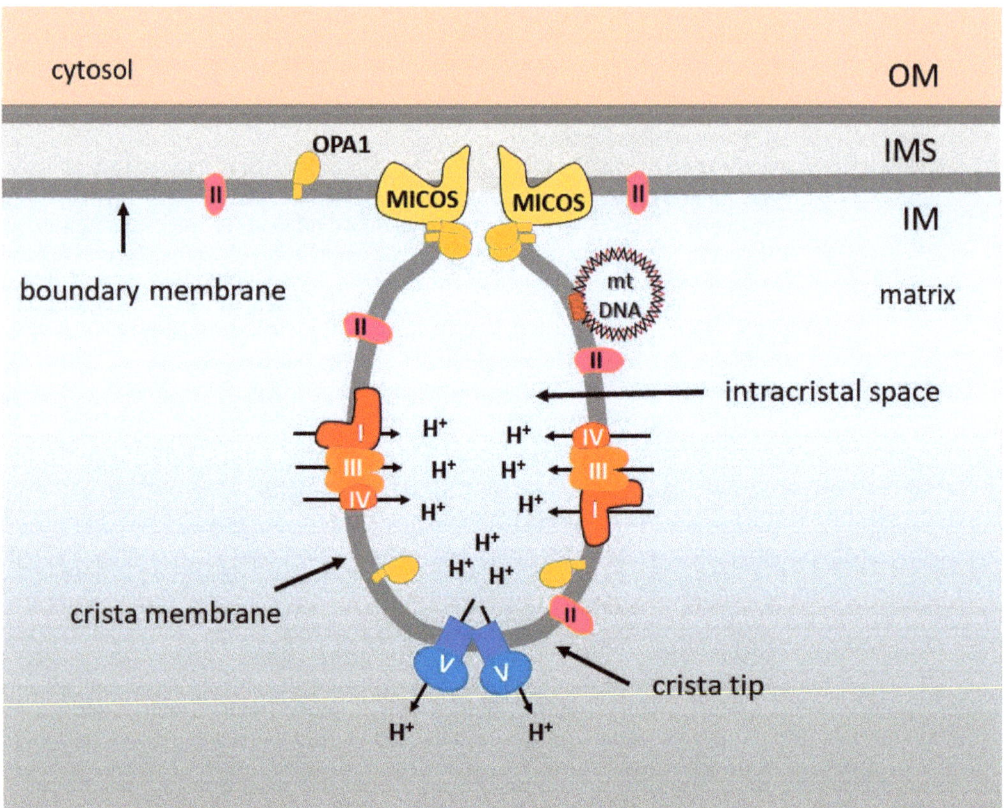

Figure 1. Architecture of the inner membrane and unequal distribution of OXPHOS complexes. CI, CIII and CIV are in the cristae membrane, CV dimers are at the cristae tip, whereas CII is also found in the inner boundary membrane. MICOS and OPA1 stabilize the cristae junctions, providing constrains to membrane mobility of complexes.

Of note is that the presence of MICOS and other proteins at the cristae junctions provides severe constraints to the mobility of protein complexes along the inner membrane, and therefore two sub-compartments with unequal protein distribution and functions can be envisaged: the cristae membrane, protruding in the matrix, and the inner boundary membrane, facing the outer membrane. The respiratory complexes are mainly found in the cristae membrane [56]. Further, sophisticated analysis of diffusion and directionality of movement of single complexes by super resolution microscopy revealed that all the complexes are trapped in the laminar part of cristae, except CII which is also found in the inner boundary membrane [59] (Figure 1). It has to be noticed that CII is a rather small complex and does not associate in supramolecular assemblies with other complexes. It follows that the specific geometry of the inner membrane, dictated by the cristae junction organization at the bottom and CV dimers on the top of crista, has a critical impact on the energetic function of mitochondria. As summarized in Figure 1, protons ejected by CI, CIII and CIV segregated on the laminar part of the cristae, are enriched in the intracristal space and flow back into the matrix through CV. This arrangement was predicted to be highly efficient in promoting energy conversion [56]. However, the intracristal membrane geometry has not been shown to influence the local pH gradients [60,61], consequently it has been proposed that the tight packing of the OXPHOS machinery in the cristae membranes favours the kinetic coupling between proton pumping and ATP synthesis [61]. The extraordinary power of super resolution live-cell imaging in combination with EM tomography and genome editing will further illuminate the functional details and dynamic aspects of this important microcompartment.

6. Supramolecular Organization of the Respiratory Complexes

The unequal protein complex distribution within the inner membrane subcompartments is also favoured by their size and their ability to interact with other complexes to form high-molecular-weight macromolecular aggregates. Differently form CII, which is always found alone, the proton-pumping CI, CIII and CIV can assemble in non-covalent associations defined as respiratory supercomplexes (SCs). Respiratory SCs were first identified by non-denaturing blue native gel electrophoresis (BN-PAGE) of mitochondrial membrane extracts, using the mild detergents Triton X-100 and digitonin, as high-molecular-weight gel bands, showing activity for CI, CIII and CIV [62,63]. The mammalian SCs containing CI, CIII dimer ($CIII_2$), and CIV with different stoichiometry are sometimes referred to as respirasomes, because they contain all the components required to transfer electrons from NADH to molecular oxygen [63–65]. In human cells, the respirasome comprises most of CI (>90%), approximately 40–50% of $CIII_2$ and 20–30% of CIV. Differently from CI, significant amounts of $CIII_2$ (50–60%) and CIV (70–80%) can be also found as isolated complexes within the membrane. The CIII+CIV SC represents about 5% of total amount of the complexes [66].

6.1. The CI+$CIII_2$+CIV SC or Respirasome

The structures of the CI+$CIII_2$+CIV SC isolated from different mammalian mitochondria have been determined by single-particle electron cryo-EM at resolutions ranging from ~30 to ~4 Å [67–70] or by electron cryotomography (cryo-ET) at ~30 Å resolution [71,72].

In the structure of respirasome, $CIII_2$ borders the concave arc of CI membrane arm, and CIV is located near $CIII_2$ at the distal end of the CI membrane arm, with cardiolipin molecules filling the gaps between the individual complexes [67,71]. In the respirasome, two distinct arrangements have been identified, a major "tight" and a minor "loose" form, which mainly diverge for the position of CIV. As illustrated in the cartoon of Figure 2, the most extensive and stable interactions take place between CI and $CIII_2$.

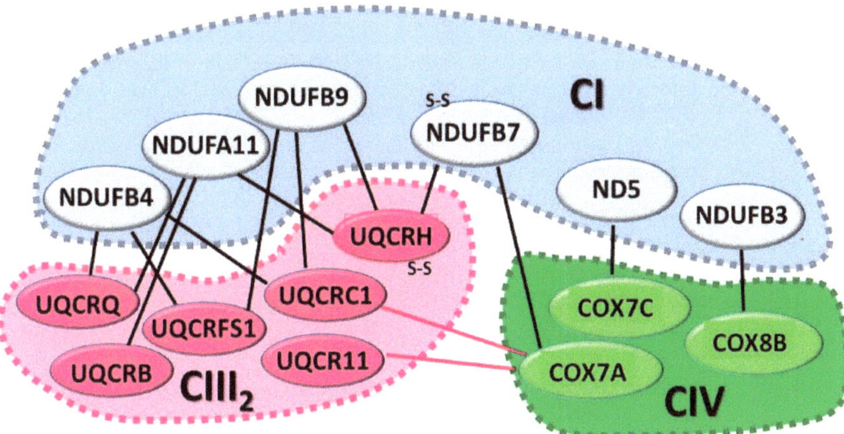

Figure 2. Proposed interactions between the respiratory complexes of the respirasome (according to [69]). Nomenclature of human subunits is indicated. The number of tight contacts between CI and CIII is greater than those between CI–CIV and CIII–CIV.

Two are the major interaction points: in the inner membrane between CI subunit NDUFA11 and the CIII subunits UQCRB, UQCRQ and UQCRH, and at the matrix between the CI subunits NDUFB4 and NDUFB9 and CIII subunits UQCRC1 and UQCRFS1 [69]. Another important interaction occurs between CI subunit NDUFB7 and subunit UQCRH on $CIII_2$. Of note is that both subunits contain disulphide bonds, suggesting that redox regulation might modulate the interactions between the respiratory complexes [69].

Few interactions occur between CI and CIV, the most important linking the CI ND5 subunit to COX7C at the interface between matrix and inner membrane, the other between NDUFB3 subunit and COX8B. The contacts between CIII and CIV mainly involve the CIV COX7A subunit with the UQCRC1 and UQCR11 subunits and COX5B subunits with the UQCRC1 subunit [69].

Interestingly, high-molecular-weight bands above the CI+$CIII_2$+CIV SC were previously described by BN-PAGE analysis [63]. More recently, mass spectrometry analysis has suggested that the main components of these bands are subunits of CI, CIII and CIV, and EM analysis detected a minor population of particles with circular arrangements. This led to proposing a higher oligomeric state named megacomplex CI_2+$CIII_2$+CIV_2. This assembled structure is shaped by a central $CIII_2$ surrounded by two copies each of CI and CIV. This arrangement may be an oligomerization form of respiratory complexes operating under the most efficient emergency conditions, because both monomers of the CIII dimer could receive $CoQH_2$ from each CI and pass reduced cytochrome *c* to one adjacent CIV. Further analysis by cryo-EM allowed to better define the architecture of the megacomplex [73], although these results were intensely debated, mainly due to limitations of cryo-ET technology in the reconstruction of supramolecular assemblies.

6.2. The CI+$CIII_2$ SC

CI can also assemble with the CIII dimer alone to form the CI+$CIII_2$ SC. Recently, a functional CI+$CIII_2$ SC has been isolated from ovine heart mitochondria and characterized by cryoEM, demonstrating that the contacts between CI and $CIII_2$ are evolutionarily conserved [72] and are similar to those of the respirasome, confirming the stabilizing role of $CIII_2$ on CI [69,74].

6.3. The $CIII_2$+CIV SC

Isolated $CIII_2$ and CIV coexists with the $CIII_2$+CIV SC, which has not been structurally characterized in mammalian tissues, likely due to its low relative abundance. Cryo-EM studies on yeast, which lacks CI, detected a CIII dimer at the core of the SC flanked by a CIV monomer on either side. The CIII–CIV interface revealed protein-protein interactions on either side of the membrane and with lipids within the membrane. The majority of interactions occur on the matrix side between Cor1/UQCRC1 and the N-terminus of COX5A, whereas the C-terminal domain of COX5A interacts with both Qcr6/UQCRH and cytochrome *c*1 on the intermembrane space. Within the membrane, COX5A contacts the N-terminal helix of Rip1/UQCRH and Qcr8/UQCRQ via a cardiolipin molecule and another lipid modelled as phosphocholine. Two other cardiolipins indirectly support the CIII–CIV interface highlighting again their crucial role in SC formation [75].

7. SCs Assembly Factors

Several molecules able to directly or indirectly promote mitochondrial SCs assembly and stability have been identified, such as cardiolipins, stomatin-like protein 2, prohibitin 1 and 2, and others (for a recent review, see [76]). Here we consider the protein factors involved in SCs assembly, the most studied being HIGD2A (hypoxia inducible domain family member 2A) [66,77–80] and COX7A2L [81]. HIGD2A mainly functions in CIV assembly [77,82] and mediates CIV integration within the respirasome, as well [66,80]. The mammalian homologous HIGD1A has been reported to assist the assembly of CIII and CIII-containing SCs biogenesis, having overlapping functions with HIGD2A [82].

These findings suggest the involvement of multiple pathways to assemble the respiratory complexes and to gather the SCs [82–84].

COX7A2L (also called SCAF1) has been the subject of various studies with contradictory results, due to the presence of two variants (long and short forms) differently expressed in mouse strains and tissues [81]. There is an agreement that long COX7A2L can bind to both $CIII_2$ and CIV and is required for formation of the $CIII_2$+IV SC [79,83,84], as well as for the assembly of the megacomplexes [73], but is dispensable for the respirasome [66,80].

Assembly factors specifically interacting with CIII were identified in yeast and their orthologs validated in human cells, namely, Cbp3 (UQCC1) and Cbp6 (UQCC2) [85] and Cbp4 (UQCC3) [86]. The best-characterized is UQCC3 (ubiquinol-cytochrome *c* reductase complex assembly factor 3; also known as C11orf83), which is involved in the early phase of CIII assembly and in the stabilization of CIII-containing SCs [86,87]. Interestingly, UQCC3 was reported to be indispensable for simultaneously maintaining both OXPHOS and glycolysis during hepatocarcinoma cells hypoxia adaption, suggesting a role in energetic reprogramming [88].

Table 1 summarizes the genetic alterations affecting CIII structural subunits and assembly factors, which will be considered in the following paragraphs, also indicating the redox activities and the assembly state of complexes/SCs.

8. Models of SC Organization

The question of the organization of the respiratory chain has been debated since the pioneering studies aimed at identifying the molecular components and catalytic mechanisms [89,90]. Hackenbrock et al. (1986) described the "fluid" or "random collision model" of electron transfer, where each complex acts as an individual entity, CoQ and cytochrome *c* freely diffuse within the lipid bilayer, and electron transfer occurs during random and transient collision events [91]. The alternative solid-state model derives from the early observations reporting that CI and CIII preferentially associated in the native membranes [92]. Since 2000, the solid-state model has received strong support from isolation of SCs by BN-PAGE and then by development of protein crystallography and cryo-EM approaches. According to the solid model, a unique multicomplex unit is able to execute all the steps of respiration. More recently, the two models have been merged in the so-called "plasticity" model [64,93], based on the coexistence of individual CIII and CIV complexes, with a variable combination of SCs. Dynamic SC association/dissociation can be triggered under physiological conditions by availability of different substrates (NADH and $FADH_2$), determining a variety of different structural options that allow to adapt the efficiency of the respiratory chain to metabolic demands. One consequence of the possibility to preferentially use NADH- or $FADH_2$-dependent substrates through isolated complexes and SCs is the presence of partially dedicated pools of the mobile electron carrier CoQ [81,94]. However, demonstrating the presence of two CoQ pools is experimentally difficult and therefore this issue has been disputed [95,96]. According to the working model proposed by the Enriquez group, under normal circumstances, the superassembly in the respirasome generates a CoQ fraction within the SCs functionally dedicated to NADH oxidation. Given that also individual $CIII_2$ co-exist with the SC, $CoQH_2$ generated by CII or by other $FADH_2$-dependent enzymes can be oxidized by the free CIII out of respirasome. Under conditions of block/lack of CIII or CIV, all CIII is associated with CI and $CoQH_2$ can diffuse out and be oxidized outside the SCs, and, on the other hand, the ubiquinol generated by CII can diffuse in and be oxidized by CI+$CIII_2$ SC [81,94]. Additionally, for cytochrome *c* the existence of a unique pool is also unlikely, given that this protein, besides transferring electrons between CIII and IV, can interact with many mitochondrial and non-mitochondrial components, exerting a variety of roles, among them the trigger of apoptotic cell death. SCAF1, which has been shown to be required for CIII and CIV interaction, plays an important role in the cytochrome *c* pool functional segmentation and likely in the efficient use of respiratory substrates. For a detailed discussion on CoQ and cytochrome *c* segmentation in SCs, we refer to [97].

9. Functional Roles of the SCs and CIII Involvement

Even if the evidence in support of SCs in detergent extracts is beyond question and the SCs in situ arrangement in the mitochondrial inner membrane has been defined, their physiological functional significance is still debated. The hypothesis of a catalytic advantage provided by SCs [67,81,98] has been questioned [65,72,95,96,99]. Catalytic advantage would imply substrate channelling, i.e., a defined conduit for the hydrophobic quinol from its reduction site in CI to CIII, where it is oxidized. However, no structural evidence for such a protein-defined conduit between CI and $CIII_2$ was obtained [100].

As alternative hypotheses, it has been proposed that SC organization could support CI assembly and stability [101,102], preventing dangerous casual protein aggregations within the membrane [100,103]. Finally, it has to be recalled that CI and $CIII_2$ are the main sources of reactive oxygen species (ROS) in the mitochondrial matrix and in the inner membrane [104–107], and that experimental dissociation of the SCs results in increased ROS production from CI [108]. It was, therefore, proposed that SCs organization could reduce ROS generation and subsequent oxidative damage to membrane components [108,109], although the molecular mechanism was not clearly defined. Analysis of recent structural data allowed to propose that in the respirasome, the interactions of $CIII_2$ with CI and CIV break the symmetry of $CIII_2$, ensuring efficient oxidation of QH_2 and allowing CI to operate at full rate, thus helping to reduce ROS formation at CI [100].

The relationship between SCs and ROS is intriguing, because it is difficult to dissect how important SCs are to prevent ROS production and how ROS production depends on SCs disassembly. There is evidence that the production of ROS due to CIII dysfunction plays a role in the stability of SCs. It has been reported that some pathogenic mutations in *BCS1L*, the chaperone protein needed to incorporate the UQCFRS1 protein in CIII, affected CIII activity, inducing ROS production and secondary CI and CIV activity/stability alteration [110]. Furthermore, in a mouse cellular model lacking UQCFRS1 protein, the increase of ROS production due to CIII deficit induced a general reduction in the assembly and stability of CI, CIV, and in turn of SCs; however, this latter defect was partially rescued in presence of antioxidants treatment or hypoxia [111].

In addition to a direct ROS production by CIII due to its defective activity, it has been shown that the lack of CIII, by increasing the $CoQH_2/CoQ$ ratio, could promote the backflow of electrons from $CoQH_2$ to CI by the reverse electron transfer (RET) reaction through CI. RET produced the oxidation of cysteine residues of CI, triggering its degradation and in turn hampering SC formation [94]. Although in this study the occurrence of RET and ROS production was not directly demonstrated, the ROS involvement was suggested by partial rescue of CI assembly after treatment with the CI inhibitor rotenone and under hypoxic conditions, all interventions that prevent RET. The overexpression of SOD2, however, was not effective [94]. An interesting finding has been recently reported, showing that Na^+ modulates ROS production during acute hypoxia through the regulation of inner membrane fluidity [112]. Noticeably, this increased ROS production was associated with reduction in combined CII+CIII enzymatic activity and respiratory capacity, whereas combined CI+CIII activity and respiration remained unchanged. This finding reinforces the general idea of that formation of the CI+$CIII_2$ SC limits the production of ROS.

Finally, in a human cellular model carrying an 18-bp frame deletion in *MTCYB* associated with a severe impairment of CIII, CI and CIV assembly and ROS production, treatment with the antioxidant N-acetyl cysteine partially rescued respirasome formation [113]. Taken together, these data indicate that ROS production may affect respiratory complexes and SCs assembly, thus, it is reasonable to speculate that the correct assembly of SCs could be useful to reduce ROS production in a healthy respiratory chain.

The fact that the mutual arrangement of CI and $CIII_2$ is essentially conserved from obligate aerobic yeast to mammals, and plants seems to favour the hypothesis that specifically these two complexes have a functional role in maintaining the respiratory chain stability reducing ROS production [51]. Further structural and biochemical work is needed, also considering that in some cases a slight increase of ROS was detected, but the assembly of respiratory complexes and SCs was normal, suggesting that ROS levels could modulate the response to this supermolecular organization [114–116]. The influence of tissue-specific subunit isoforms is also to be taken into account [83].

10. SCs Biogenesis and Role of CIII

An important issue to be elucidated concerns the SCs biogenesis, i.e., how CI, CIII and CIV interact in the respirasome formation. Currently, two models have been presented. The first model proposes that the close interplay among the three complexes favours better structural stability, implying that the isolated forms of respiratory complexes, in particular CI, are more prone to degradation. Accordingly, the complexes are supposed to follow separate assembly pathways to build mature individual complexes and to form SCs in a second step [93,117]. The second model suggests a central role of CI acting as a scaffold for the sequential incorporation of $CIII_2$ and CIV_n subassemblies to form mature SCs [102]. However, this latter model does not rule out the occurrence of a dynamic exchange of $CIII_2$ and CIV once the respirasome assembly has been completed, allowing the formation of the other SCs and the co-existence with free complexes [66]. Both models have been developed from data obtained in experiments in which SCs biogenesis was studied following the time-course of respiratory complexes assembly after mitochondrial translation inhibition, by using diverse antibiotics and different experimental conditions, and this may explain some conflicting results. Recently, interesting information obtained by combining the Stable Isotope Labelling by Amino acids in Cell culture (SILAC) and complexome profiling techniques, suggested a new main role of CIII in SCs formation as a structural and functional platform for the overall respiratory chain biogenesis [118,119]. In particular, Protasoni et al., 2020 [118] analyzed human cells bearing a 4-bp *MTCYB* deletion that induced a frameshift with the loss of the encoded protein and of $CIII_2$, associated with hampered CI biogenesis due to the stall in N-module incorporation and decreased CI stability. Furthermore, the CIV assembly was also defective because some CIV subunits were recruited within the accumulated CIII subassemblies. A similar analysis was carried out by Páleníková et al. (2021) [119], in cells bearing a 18-bp frame deletion in *MTCYB* gene that produced a protein shortened of six amino acids and induced a strong defect of $CIII_2$ activity/assembly as well as of CI and CIV [113,120,121]. These cells, differently from those with the 4-bp deletion, exhibited some amount of isolated $CIII_2$ and CI, that were associated in the enzymatically active CI+III_2 SC [113,121], in agreement with structural data demonstrating the existence of multiple interactions between the two complexes [74]. However, Páleníková et al. (2021) showed that several sub-complexes with mixed CIII, CI and CIV subunits were also present, indicating that the decreased $CIII_2$ assembly leads the formation of intermediates that trap other respiratory complex subunits impairing their assembly [119]. Taken together, these data support a cooperative assembly model in the respiratory chain structural and functional maturation and in SCs biogenesis, highlighting the central role of $CIII_2$ as scaffold for the ordered association with mixed CI and CIV subunits. Exhaustive studies focused on the mechanisms of SCs biogenesis in presence of CIII deficiency are still lacking; however, after reviewing literature it appears that mutations that disrupt CIII assembly also induce an impairment of CI and CIV and of SCs formation as well. Indeed, severe mutations in both CIII structural subunits or in early assembly factors strongly affect CIII structure and reduce CI and CIV stability [85,86,111,112,118,122–128] (Table 1). Unfortunately, several reported mutations that could be relevant for CIII structure, such as truncating mutations in cytochrome *b*, were poorly or not at all investigated for CIII assembly and related influence on CI, CIV and SCs structure (Table 1). However, it appears that most *MTCYB* missense mutations affect neither the assembly of CIII nor CI and CIV (Table 1), except for the p.Y278C mutation that was associated with a slight reduction of CI activity only and a reduced amount of the $CIII_2$ +CIV SC [129], and the p.E373K that disassembled CIII and CI, but not CIV [130]. Taken together, these findings indicate that the assembly of CI and CIV depends on the physical presence of assembled CIII species and not on their catalytic activity, although ROS production could play a role in specific cases [111].

However, the role of CIII structure in SCs biogenesis is far from being clarified; for instance, some authors suggested that the complete formation of the $CIII_2$+CIV SC is important to safely activate CI only when the respiratory chain is fully assembled [102], but some data show that CI could interact with a pre-CIII to form enough SC to ensure sufficient respiratory chain activity [115,131]. These data suggest that regardless of the mutation, there is a tendency to maintain respiratory complex stability and SC assembly to mitigate CIII dysfunction. This latter piece of evidence is supported by recent papers showing that missense mutations in *MTCYB* that induce defective CIII enzymatic activity when detected in the isolated complex, are mitigated when CIII activity is measured under conditions in which the respiratory complex is organized into SCs [129,132]. Although further work is clearly needed, recent structural details obtained from cryo-EM analysis of active SC particles from sheep mitochondria highlighted the specific involvement of cytochrome *b* in the crosstalk between CI and $CIII_2$, confirming its role in structural/functional interactions between the two complexes [74].

11. Metabolic Disturbances and Treatment Options

Defective CIII and associated perturbation of the supramolecular organization of CIII-containing SCs result in several significant metabolic alterations, which are briefly summarized in Figure 3 and discussed in detail below, providing also some hints on experimental treatments specifically aimed at improving these metabolic disturbances. For an extensive recent review of the therapeutic strategies to treat mitochondrial disease, we refer to [133].

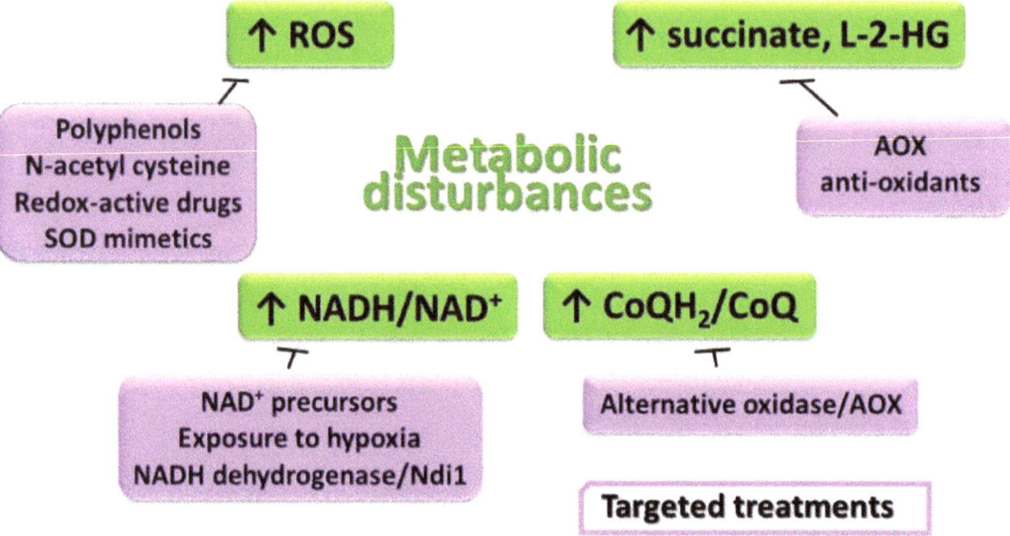

Figure 3. Major metabolic alterations associated with perturbed supramolecular organization of CIII-containing SCs and proposed treatments.

11.1. Unbalanced Intracellular Redox Homeostasis

It is widely accepted that even under physiological conditions the electron flow through the mitochondrial respiratory chain results in mild ROS production [134,135], with CI and CIII being the main redox components responsible for molecular oxygen reduction to superoxide anions [104–107,136]. As mentioned above (point 9), one of the proposed roles of the supramolecular organization into SCs is to limit ROS formation, avoiding the diffusion of free radicals with damaging effects at protein and lipid levels [100,137,138].

Accordingly, different experimental conditions causing disruption or prevention of the association between CI and CIII were shown to increase ROS production, supporting the view that dissociation of SCs may strictly link oxidative stress and energy failure [108]. Furthermore, pathological conditions leading to dismantling the SCs organization, such those described in cells with 4- and 18-bp *MTCYB* deletions, were also shown to enhance ROS generation [94], and, due to up-regulation of intracellular efficient antioxidant defences in both cytosol and mitochondrial compartments, to cause a significant unbalance in the redox homeostasis [113,121,139]. Of note, mild to moderate increase in oxidative stress associated with parallel SCs depletion was detected in different brain areas of neuron-specific mice KO for the UQCRFS1 gene, demonstrating that ROS can modulate the SCs architecture to cope with a high level of ROS [140]. This conclusion is further supported by previous finding that a superoxide dismutase mimetic compound and SOD2 overexpression induced a partial increase in SCs in the UQCRFS1 KO cells [111,140]. Moreover, the prolonged treatment of cells bearing the 18-bp *MTCYB* deletion with N-acetyl cysteine (NAC) significantly increased the rate of ATP synthesis driven by CI substrates as well as the amount of free CI, CIII and CIV and of the respirasome. It is likely that NAC may provide optimal redox conditions for respiratory complexes interactions and SCs re-organization [113], also considering that both CI subunit NDUFB7 and CIII subunit UQCRH contain disulphide bonds [69].

11.2. Accumulation of the Reduced Form of Pyridine Nucleotides and CoQ

Direct consequences of defective CIII are energy failure and metabolic derangements, as indicated by a huge number of case reports describing lactic acidosis and hypoglycaemia as recurrent clinical phenotypes of patients bearing mutations in different CIII-related genes, i.e., *MTCYB* [120,141], *UQCRC2* [127,142], *UQCC3* [86] and others. Metabolomics analyses in liver of the mouse model of CIII dysfunction ($Bcs1l^{c.232A>G}$ mutant) revealed a decrease in carbohydrate intermediates, demonstrating an increase in glycolysis to compensate for the reduced mitochondrial ATP production [143]. Subsequently, targeted metabolomics detected increases in glucogenic and ketogenic amino acids in circulation, supporting a starvation-like condition [144]. Of note is that this mouse model, despite the severe CIII dysfunction, does not present significant perturbations in the SCs organization [115].

On the other hand, when the assembly of CIII-containing SCs is compromised, in addition to CIII, the amount of CI collapses as well, leading to elevation of the cellular ratio of reduced and oxidized pyridine nucleotides ($NADH/NAD^+$). The inability to oxidize NADH in the mitochondrial matrix affects not only the efficiency of OXPHOS but also the flux of metabolites through the Krebs cycle. As a consequence, the cells become heavily dependent on aerobic glycolysis for survival. The glycolytic flux relies on the activity of glyceraldehyde-3-phosphate dehydrogenase which requires NAD^+, generated from NADH oxidation by the cytosolic lactate dehydrogenase enzyme. In agreement with this notion, we found that the amount of lactate released into the growth medium by the homoplasmic cells bearing the 18-bp *MTCYB* deletion was significantly greater than WT cells. Noticeably, cells bearing the p.278Y>C *MTCYB* mutation impairing CIII activity without affecting SCs organization failed to increase lactate release [121], in accord with the clinical phenotype of patients, presenting lactic acidosis in the patient bearing the 18-bp *MTCYB* deletion [120], but not in that with the p.278Y>C mutation [145]. The molecular mechanism underlining this metabolic switch is unknown, although the possible role for UQCC3 may be worth investigating [88].

Interestingly, previous studies described benefits in lifespan and energetic function of defective CI by interventions targeting NADH elevation, such as supplementation with NAD-precursor [146] inhibition of mTOR [147] and of mitochondrial serine catabolism [148], as well as hypoxia treatment [149]. To circumvent the CI deficiency, some studies took advantage of the xenotopic expression of the single-subunit yeast enzyme NADH dehydrogenase (Ndi1) [150–152]. In yeast, Ndi1 catalyses the oxidation of NADH in the matrix like CI, but is unable to restore the proton pumping. Ndi1 protein expression in human cultured cells lacking CI restored the NADH-dependent respiration as well as the growth in glucose-free medium containing galactose [153,154]. Recently, McEllroy et al. (2020) generated a mouse that conditionally expresses Ndi1, confirming that its expression dramatically prolong lifespan, but was unable to significantly improve motor function in a mouse model of Leigh syndrome due to loss of the NDUFS4 CI subunit [155]. In the absence of structural data showing Ndi1 association with SCs, it is reasonable to speculate that the ability of Ndi1 to ameliorate the cell viability does not depend on association with other respiratory complexes, rather it depends on the restoration of NADH oxidation allowing for a compensatory increase in glycolysis and sufficient metabolite flux in the Krebs cycle.

At cellular level, the primary consequence of the specific drop/lack of CIII is the blockade of $CoQH_2$ oxidation, preventing the NADH and $FADH_2$ oxidation by CI and CII. As mentioned above, elevation of $CoQH_2$/CoQ ratio causes reverse electron transport through CI, with local generation of superoxide, triggering CI subunits degradation and tuning the amount of this complex [94]. This is in agreement with previous data showing that CIII can be released from CI-containing SCs under metabolic conditions (e.g., starvation) when electron flux from FAD overwhelms the oxidation of CoQ, supporting the plasticity model of SCs organization [81]. The alternative oxidase (AOX) is a single-protein electron transport system present in bacteria, lower eukaryotes and plants that can perform $CoQH_2$ oxidation instead of CIII and CIV, by transferring electrons directly from quinols to oxygen without proton translocation [156]. The xenotopic expression of tunicated AOX in mouse was recently investigated, failing to show any association of AOX with SCs. This finding supports the notion that xenotopically expressed AOX acts as a freely diffusible redox partner [157]. Of note is that the expression of AOX from *Emericella nidulans* in *MTCYB* KO cells induced $CoQH_2$ oxidation, thus reducing the oxidative stress and inhibiting CI degradation. Despite the increased CI amount, the SCs were not restored due to the lack of CIII [94], further corroborating the central role of CIII as a scaffold for incorporation of CI and CIV [118]. Furthermore, AOX was reported to provide a full functional rescue of the cardiomyopathy of the *Bcs1l*$^{c.232A>G}$ mutant mice, by restoring respiration to wild-type level. Noticeably, the CIII and CI+$CIII_2$+CIV assembly was partially rescued in cardiac mitochondria, likely secondary to the general improvement in mitochondrial structure and function [144].

The observation that expression of a single enzyme, such as AOX, can bypass defective oxidative reactions carried out by dozens of proteins is intriguing. Besides representing a useful tool for detecting the contribution of ATP requirement from the NAD^+ and CoQ regeneration, the oxidase may be of potential use for respiratory chain deficiencies, although a gene therapy approach seems quite problematic at present. In fact, it has to be considered that correction of the mutated gene in affected tissue of monogenic diseases by CRISPR/CAS9 genome editing, already shown to be promising in animal models, requires sophisticated gene-specific tools and is still under development for mtDNA interventions. Conversely, the expression of one protein such AOX might be, in theory, beneficial for restoring most of the metabolic stress induced by OXPHOS impairment caused by a wide variety of mutations.

11.3. Elevation of Succinate and Effects on Gene Expression Regulation

The lack of CIII and the extremely limited availability of oxidized CoQ results in the inability of the Krebs cycle to progress from succinate to fumarate, as demonstrated by the markedly increased levels of succinate and reduction of fumarate and malate detected by us in cells with the 18-bp *MTCYB* deletion [121] and also in cells bearing the 4-bp *MTCYB* deletion [158]. In cells with the p.278Y>C *MTCYB* mutation, with normal SCs organization, we also detected a weak increase of succinate, but increased malate and normal fumarate, suggesting that some succinate can be oxidized to fumarate which in turn produces malate, likely as a consequence of CIII assembly in the respirasome, that can preserve significant electron transport [121]. Elevation of succinate and fumarate was reported in the liver of *Bcs1l*$^{c.232A>G}$ mutant mouse, in line with the blockade of the Krebs cycle flux [159].

Succinate accumulation in the cytosol has been shown to have a strong impact in gene expression regulation, by inhibiting the 2-oxoglutarate-dependent dioxygenases, which catalyse hydroxylation reactions on various types of substrates. In particular, succinate, competing with 2-oxoglutarate, inhibits the activity of prolyl hydroxylases, leading to stabilization of Hypoxia Inducible Factor-1α (HIF-1α) under normoxia, defined as a pseudo-hypoxic condition [160]. HIF-1α can then translocate from cytoplasm to the nucleus where it associates with HIF-1β, to activate transcription of HIF-1α-target genes, among which are those encoding glycolytic enzymes [161]. It is likely that the huge increase of succinate levels determined in cells with the 18-bp *MTCYB* deletion may be at least in part responsible for the glycolytic switch revealed by increased extracellular lactate production [121].

In addition to succinate and fumarate, the levels of the oncometabolite 2-hydroxyglutarate (L-2-HG) were increased in RISP/ UQCRFS1 KO cells [162] and also in cells bearing the 4-bp *MTCYB* deletion [158]. All these Krebs cycle metabolites competitively inhibit the activity of 2-oxoglutarate-dependent dioxygenases, including also JmJC domain-containing histone lysine demethylases and ten-eleven translocation TETs family of 5-methlycytosine hydroxylases, involved in oxidizing 5-methylcytosine into 5-hydroxymethylcytosine [163]. Accordingly, the DNA and histone methylation were increased upon loss of RISP in fetal hematopoietic stem cells, impairing their differentiation and maintenance of stemness [162]. The increase in these metabolites, through inhibition of the histone and DNA demethylases, can therefore represent a very important factor affecting the epigenetic landscape of the cells and causing wide-ranging effects on cell physiology (for a review on Krebs metabolites and epigenetics, see [164]).

12. Conclusions and Perspectives

The development of cryo-EM technology has provided a powerful tool to analyse at atomic level the specific associations of CI, CIII and CIV into the respirasome, the CI+$CIII_2$ and the $CIII_2$+$CIV_{1\text{-}2}$ SCs. However, these techniques suffer from some limitations, mainly associated with the mitochondrial purification procedures and the type/amount of detergent. In this regard, the development of the "in situ" reconstruction of SCs in eukaryotic cells in vivo by using the proximity-dependent labeling followed by mass spectrometry will open the possibility to identify potentially interacting proteins and their subcellular spatial localization [165,166]. Dynamic rearrangements between individual complexes and SCs have been demonstrated to occur (Table 1). It is necessary to identify the functional consequences of these arrangements as well as their implications in the regulation of the respiratory function under different physiological conditions. More biochemical and biophysical experiments, in combination with advances in super-resolution light microscopy are needed for clarify the functional mechanism of the SCs. This information will be crucial to elucidate the pathogenic mechanisms underlying the mitochondrial disorders associated with both nuclear and mtDNA mutations, and, hopefully, to identify effective treatments.

Table 1. Disease genes encoding structural subunits and assembly factors associated with CIII deficiency.

Mutated Gene	Mutation	Enzymatic Activity			Isolated Complexes Assembly			Supercomplexes Assembly			Refs.
	CIII	CIII	CI	CIV	CIII	CI	CIV	CI+CIII+CIV	CI+CIII	CIII+CIV	
					Structural subunits						
MTCYB	p.14I> *	↓	↓	↓	↓	↓	↓	↓	↓	↓	[118,167,168]
	p.34G>S	↓	mild ↓	=	n.d.	n.d.	n.d.	n.d.	n.d.	n.d.	[169]
	p.35S>P	↓	n.d.	n.d	n.d.	n.d.	n.d.	n.d.	n.d.	n.d.	[170]
	p.40C>R	mild ↓	mild ↓	mild ↓	n.d.	n.d.	n.d.	n.d.	n.d.	n.d.	[171]
	p.113W> *	↓	n.d.	n.d	n.d.	n.d.	n.d.	n.d.	n.d.	n.d.	[169]
	p.135W> *	↓	n.d.	n-d	↓	n.d.	n.d.	n.d.	n.d.	n.d.	[172]
	p.141W> *	↓	=	=	n.d.	n.d.	n.d.	n.d.	n.d.	n.d.	[169]
	p.142G> *	↓	↓	n.d.	n.d.	n.d.	n.d.	n.d.	n.d.	n.d.	[173]
	p.151S>P	↓	n.d.	n.d.	↓	n.d.	n.d.	n.d.	n.d.	n.d.	[172]
	p.166G> *	↓	n.d.	n.d.	n.d.	n.d.	n.d.	n.d.	n.d.	n.d.	[174]
	p.166G>E	↓	n.d.	n.d.	n.d.	n.d.	n.d.	n.d.	n.d.	n.d.	[175]
	p.Δ251-258	↓	mild ↓	=	n.d.	n.d.	n.d.	n.d.	n.d.	n.d.	[169]
	p.271E>K	mild ↓	=	=	=	=	=	=	=	=	[132]
	p.278Y>C	↓	mild ↓	n.d.	=	=	=	mild↑	mild↑	↓	[129,145]
	p.290G>D	↓	n.d.	n.d	n.d.	n.d.	n.d.	n.d.	n.d.	n.d.	[176,177]
	p.297S>P	↓	=	=	↓	=	=	n.d.	n.d.	n.d.	[178]
	p.Δ300-305	↓	↓	↓	↓	↓	↓	↓	↓	↓	[113,120]
	p.318K>P	↓	↓	=	↓	↓	=	n.d.	n.d.	n.d.	[122]
	p.326W> *	↓	n.d.	n.d	n.d.	n.d.	n.d.	n.d.	n.d.	n.d.	[169]
	p.339G> *	↓	n.d.	n.d	n.d.	n.d.	n.d.	n.d.	n.d.	n.d.	[179]
	p.339G>E	↓	n.d.	n.d	n.d.	n.d.	n.d.	n.d.	n.d.	n.d.	[180]
	p.352Q> *	↓	↓	↓	n.d.	n.d.	n.d.	n.d.	n.d.	n.d.	[123]
	p.373E>K	↓	↓	=	↓	↓	=	n.d.	n.d.	n.d.	[130]
UQCRB	Change at C-term	↓	↓	↓	n.d.	n.d.	n.d.	n.d.	n.d.	n.d.	[124]
UQCRQ	p.45S>F; p.45S>F	↓	↓	=	n.d.	n.d.	n.d.	n.d.	n.d.	n.d.	[125]

Table 1. *Cont.*

Mutated Gene	Mutation	Enzymatic Activity			Isolated Complexes Assembly			Supercomplexes Assembly			Refs.
	CIII	CIII	CI	CIV	CIII	CI	CIV	CI+CIII+CIV	CI+CIII	CIII+CIV	
CYC1	p.96W>C; p.215L>F	↓	↓	↓	n.d.	n.d.	n.d.	n.d.	n.d.	n.d.	[126]
UQCRC2	p.183R>W; p.183R>W	↓	↑	↓	↓	↓	=	↓	↓	n.d.	[127]
	p.183R>W; p183R>W	↓	↓	=	↓	↓	n.d.	n.d.	n.d.	n.d.	[142]
UQCRFS1	p.14V>D; p.204R> *	reduced overall respiration			↓	n.d.	n.d.	n.d.	n.d.	n.d.	[181]
	p.72V>T81del10; p.72V>T81del10	reduced overall respiration			↓	n.d.	n.d.	n.d.	n.d.	n.d.	[181]
	mouse KO	↓	↓	↓	↓	↓	↓	↓	↓	↓	[111]
Assembly Factors											
BCS1L	p.35G>R; p.184R>C	↓	n.d.	n.d.	↓	n.d.	n.d.	=	n.d.	↓	[182]
	p.45R>C; p.56R> *	↓	↓	=	↓	↓	↓	n.d.	n.d.	n.d.	[110,183]
	p.50T>A; p.50T>A	mild ↓	n.d.	n.d.	mild↓	=	n.d.	n.d.	n.d.	n.d.	[110,184]
	p.R56 *; g1181A>G/g1164C>C	↓	=	↓	↓	=	↓	n.d.	n.d.	n.d.	[110,183]
	p.R56 *; p.327V>A	=	=	=	n.d.	n.d.	n.d.	n.d.	n.d.	n.d.	[185]
	p.R56 *; p.69R>C	=	=	=	↓	=	=	n.d.	n.d.	n.d.	[186]
	p.73R>C; p.368F>I	↓	=	=	=	=	=	=	=	=	[187]
	p.78S>G; p.144R>Q	=	=	=	n.d.	n.d.	n.d.	n.d.	n.d.	n.d.	[185]
	p.99P>L; p.99P>L	↓	↓	↓	↓	↓	↓	n.d.	n.d.	n.d.	[110,188,189]
	p.109R>W; p.109R>W	=	=	=	↓	=	↓	n.d.	n.d.	n.d.	[186]
	p.129G>R; p.129G>R	↓	n.d.	n.d.	↓	n.d.	n.d.	n.d.	n.d.	n.d.	[190,191]
	p.155R>P; p.353V>M	↓	n.d.	=	n.d.	n.d.	n.d.	n.d.	n.d.	n.d.	[188]
	p.183R>C; p.184R>C	↓	=	=	=	=	=	=	=	=	[187]
	p.184R>C; g1892A>G	↓	=	=	=	mild ↓	=	=	n.d.	↓	[110]
	p.184R>C; p.280L>F	n.d.	n.d.	n.d.	n.d.	n.d.	n.d.	n.d.	n.d.	n.d.	[192]
	p.277S>N; p.277S>N	↓	n.d.	=	n.d.	n.d.	n.d.	n.d.	n.d.	n.d.	[188]
	decreased levels BCS1L	↓	n.d.	↓	↓	=	=	n.d.	n.d.	n.d.	[193]
	mouse p.78S>G: p.78S>G	↓	=	=	↓	=	=	mild ↓	mild ↓	mild ↓	[115]
	mouse KO										[116]

Table 1. *Cont.*

Mutated Gene	Mutation	Enzymatic Activity			Isolated Complexes Assembly			Supercomplexes Assembly			Refs.
	CIII	CIII	CI	CIV	CIII	CI	CIV	CI+CIII+CIV	CI+CIII	CIII+CIV	
TTC19	p.54P>A *	↓	n.d.	n.d.	n.d.	n.d.	n.d.	n.d.	n.d.	n.d.	[194]
	p.77Q>R *; p.77Q>R *	↓	n.d.	n.d.	=	n.d.	n.d.	n.d.	n.d.	n.d.	[195]
	p.173Q> *	↓	=	=	↓	n.d.	n.d.	n.d.	n.d.	n.d.	[196]
	p.185L>P	n.d.	n.d.	n.d.	=	n.d.	n.d.	n.d.	n.d.	n.d.	[197]
	p.186W> *; p.322G>M *	↓	n.d.	↓	n.d.	n.d.	n.d.	n.d.	n.d.	n.d.	[198]
	p.194R>N *	n.d.	n.d.	n.d.	n.d.	n.d.	n.d.	n.d.	n.d.	n.d.	[199]
	p.219L> *	↓	=	=	↓	n.d.	n.d.	n.d.	n.d.	n.d.	[196]
	p.261E>G *; p.261A>G *	n.d.	n.d.	n.d.	n.d.	n.d.	n.d.	n.d.	n.d.	n.d.	[200]
	p.277Q> *; p.277Q> *	n.d.	n.d.	n.d.	n.d.	n.d.	n.d.	n.d.	n.d.	n.d.	[201]
	p.313Q> *	↓	n.d.	n.d.	n.d.	n.d.	n.d.	n.d.	n.d.	n.d.	[202]
	p.321A> *; p.321A> *	↓	n.d.	n.d.	↓	n.d.	n.d.	n.d.	n.d.	↓	[114]
	p.324L>P	n.d.	n.d.	n.d.	=	n.d.	n.d.	n.d.	n.d.	n.d.	[197]
	mouse and human KO	↓	=	=	mild ↓	=	=	=	=	=	[203]
UQCC2	Protein absent	↓	↓	↓	↓	mild ↓	mild↑	↓	↓	↓	[85]
	p.[8R>P;10L>F];[8R>P;10L>F]	↓	↓	=	↓	↓	=	↓	↓	↓	[128]
UQCC3	p.20V>E; p.20V>E	↓	mild ↓	=	↓	↓	=	n.d.	n.d.	n.d.	[86]
LYRM7	p.13T>H *; p.13T>H *	↓	n.d.	n.d.	n.d.	n.d.	n.d.	n.d.	n.d.	n.d.	[204]
LYRM7	p.18R>D *; p.18R>A *	↓	mild ↓	mild ↓	↓	mild ↓	mild ↓	n.d.	n.d.	n.d.	[205]
	p.25D>N; p25D>N	↓	n.d.	n.d.	↓	n.d.	=	n.d.	n.d.	n.d.	[206]
	p.25D>N; p25D>N	↓	n.d.	n.d.	n.d.	n.d.	n.d.	n.d.	n.d.	n.d.	[204]
	p.72Q> *; p72Q> *	↓	n.d.	n.d.	n.d.	n.d.	n.d.	n.d.	n.d.	n.d.	[204]
	p.82K>N *; p.82K>N *	↓	n.d.	n.d.	↓	n.d.	n.d.	n.d.	n.d.	n.d.	[204]
	Protein absent	↓	n.d.	n.d.	n.d.	n.d.	n.d.	n.d.	n.d.	n.d.	[207]

*, a stop in the protein synthesis; ↓, decrease; ↑, increase; =, no change; n.d., not determined.

Author Contributions: Conceptualization, A.M.G. and M.R.; writing, original draft and figures preparation, M.R.; writing and editing, A.M.G.; supervision, C.Z. All authors have read and agreed to the published version of the manuscript.

Funding: This research received no external funding.

Institutional Review Board Statement: Not applicable.

Informed Consent Statement: Not applicable.

Acknowledgments: We thank Erika Fernandez-Vizarra, University of Glasgow, Scotland, UK, for kindly sharing some SILAC and complexome profiling results. This article is dedicated to the memory of Francisca Diaz, University of Miami, USA, who significant contributed to the study of SCs alteration in mitochondrial dysfunctions.

Conflicts of Interest: The authors declare no conflict of interest.

References

1. Raimundo, N.; Baysal, B.E.; Shadel, G.S. Revisiting the TCA cycle: Signaling to tumor formation. *Trends Mol. Med.* **2011**, *17*, 641–649. [CrossRef]
2. Martínez-Reyes, I.; Diebold, L.P.; Kong, H.; Schieber, M.; Huang, H.; Hensley, C.T.; Mehta, M.M.; Wang, T.; Santos, J.H.; Woychik, R.; et al. TCA Cycle and Mitochondrial Membrane Potential Are Necessary for Diverse Biological Functions. *Mol. Cell* **2016**, *61*, 199–209. [CrossRef]
3. Kastaniotis, A.J.; Autio, K.J.; Kerätär, J.M.; Monteuuis, G.; Mäkelä, A.M.; Nair, R.R.; Pietikäinen, L.P.; Shvetsova, A.; Chen, Z.; Hiltunen, J.K. Mitochondrial fatty acid synthesis, fatty acids and mitochondrial physiology. *Biochim. Biophys. Acta Mol. Cell Biol. Lipids* **2017**, *1862*, 39–48. [CrossRef]
4. Benador, I.Y.; Veliova, M.; Liesa, M.; Shirihai, O.S. Mitochondria Bound to Lipid Droplets: Where Mitochondrial Dynamics Regulate Lipid Storage and Utilization. *Cell Metab.* **2019**, *29*, 827–835. [CrossRef] [PubMed]
5. Lill, R.; Broderick, J.B.; Dean, D.R. Special issue on iron-sulfur proteins: Structure, function, biogenesis and diseases. *Biochim. Biophys. Acta Bioenerg.* **2015**, *1853*, 1251–1252. [CrossRef]
6. Minton, D.R.; Nam, M.; McLaughlin, D.J.; Shin, J.; Bayraktar, E.C.; Alvarez, S.W.; Sviderskiy, V.O.; Papagiannakopoulos, T.; Sabatini, D.M.; Birsoy, K.; et al. Serine Catabolism by SHMT2 Is Required for Proper Mitochondrial Translation Initiation and Maintenance of Formylmethionyl-tRNAs. *Mol. Cell* **2018**, *69*, 610–621.e5. [CrossRef] [PubMed]
7. Gherardi, G.; Monticelli, H.; Rizzuto, R.; Mammucari, C. The Mitochondrial Ca^{2+} Uptake and the Fine-Tuning of Aerobic Metabolism. *Front. Physiol.* **2020**, *11*, 554904. [CrossRef] [PubMed]
8. Sena, L.A.; Chandel, N.S. Physiological Roles of Mitochondrial Reactive Oxygen Species. *Mol. Cell* **2012**, *48*, 158–167. [CrossRef] [PubMed]
9. Bock, F.J.; Tait, S.W.G. Mitochondria as multifaceted regulators of cell death. *Nat. Rev. Mol. Cell Biol.* **2020**, *21*, 85–100. [CrossRef]
10. Liesa, M.; Shirihai, O.S. Mitochondrial Dynamics in the Regulation of Nutrient Utilization and Energy Expenditure. *Cell Metab.* **2013**, *17*, 491–506. [CrossRef]
11. Morgenstern, M.; Stiller, S.B.; Lübbert, P.; Peikert, C.D.; Dannenmaier, S.; Drepper, F.; Weill, U.; Höß, P.; Feuerstein, R.; Gebert, M.; et al. Definition of a High-Confidence Mitochondrial Proteome at Quantitative Scale. *Cell Rep.* **2017**, *19*, 2836–2852. [CrossRef] [PubMed]
12. Kulak, N.A.; Pichler, G.; Paron, I.; Nagaraj, N.; Mann, M. Minimal, encapsulated proteomic-sample processing applied to copy-number estimation in eukaryotic cells. *Nat. Methods* **2014**, *11*, 319–324. [CrossRef]
13. Monteuuis, G.; Miścicka, A.; Świrski, M.; Zenad, L.; Niemitalo, O.; Wrobel, L.; Alam, J.; Chacinska, A.; Kastaniotis, A.J.; Kufel, J. Non-canonical translation initiation in yeast generates a cryptic pool of mitochondrial proteins. *Nucleic Acids Res.* **2019**, *47*, 5777–5791. [CrossRef]
14. Harding, J.W.; Pyeritz, E.A.; Copeland, E.S.; White, H.B. Role of glycerol 3-phosphate dehydrogenase in glyceride metabolism. Effect of diet on enzyme activities in chicken liver. *Biochem. J.* **1975**, *146*, 223–229. [CrossRef]
15. Watmough, N.J.; Frerman, F.E. The electron transfer flavoprotein: Ubiquinone oxidoreductases. *Biochim. Biophys. Acta Bioenerg.* **2010**, *1797*, 1910–1916. [CrossRef]
16. Alcázar-Fabra, M.; Navas, P.; Brea-Calvo, G. Coenzyme Q biosynthesis and its role in the respiratory chain structure. *Biochim. Biophys. Acta Bioenerg.* **2016**, *1857*, 1073–1078. [CrossRef] [PubMed]
17. Henriques, B.J.; Olsen, R.K.J.; Gomes, C.M.; Bross, P. Electron transfer flavoprotein and its role in mitochondrial energy metabolism in health and disease. *Gene* **2021**, *776*, 145407. [CrossRef]
18. Evans, D.R.; Guy, H.I. Mammalian Pyrimidine Biosynthesis: Fresh Insights into an Ancient Pathway. *J. Biol. Chem.* **2004**, *279*, 33035–33038. [CrossRef] [PubMed]
19. Salvi, F.; Gadda, G. Human choline dehydrogenase: Medical promises and biochemical challenges. *Arch. Biochem. Biophys.* **2013**, *537*, 243–252. [CrossRef]

20. Hildebrandt, T.M.; Grieshaber, M.K. Three enzymatic activities catalyze the oxidation of sulfide to thiosulfate in mammalian and invertebrate mitochondria. *FEBS J.* **2008**, *275*, 3352–3361. [CrossRef]
21. Hancock, C.N.; Liu, W.; Alvord, W.G.; Phang, J.M. Co-regulation of mitochondrial respiration by proline dehydrogenase/oxidase and succinate. *Amino Acids* **2016**, *48*, 859–872. [CrossRef] [PubMed]
22. Zachar, I.; Boza, G. Endosymbiosis before eukaryotes: Mitochondrial establishment in protoeukaryotes. *Cell. Mol. Life Sci.* **2020**, *77*, 3503–3523. [CrossRef]
23. Johnston, I.G.; Williams, B.P. Evolutionary Inference across Eukaryotes Identifies Specific Pressures Favoring Mitochondrial Gene Retention. *Cell Syst.* **2016**, *2*, 101–111. [CrossRef] [PubMed]
24. Chrétien, D.; Bénit, P.; Ha, H.-H.; Keipert, S.; El-Khoury, R.; Chang, Y.-T.; Jastroch, M.; Jacobs, H.T.; Rustin, P.; Rak, M. Mitochondria are physiologically maintained at close to 50 °C. *PLoS Biol.* **2018**, *16*, e2003992. [CrossRef] [PubMed]
25. Allen, J.F. Why chloroplasts and mitochondria retain their own genomes and genetic systems: Colocation for redox regulation of gene expression. *Proc. Natl. Acad. Sci. USA* **2015**, *112*, 10231–10238. [CrossRef]
26. Fernandez-Vizarra, E.; Zeviani, M. Mitochondrial disorders of the OXPHOS system. *FEBS Lett.* **2020**. [CrossRef]
27. Iwata, S.; Lee, J.W.; Okada, K.; Lee, J.K.; Iwata, M.; Rasmussen, B.; Link, T.A.; Ramaswamy, S.; Jap, B.K. Complete Structure of the 11-Subunit Bovine Mitochondrial Cytochrome *bc*1 Complex. *Science* **1998**, *281*, 64–71. [CrossRef]
28. Tsukihara, T.; Aoyama, H.; Yamashita, E.; Tomizaki, T.; Yamaguchi, H.; Shinzawa-Itoh, K.; Nakashima, R.; Yaono, R.; Yoshikawa, S. The Whole Structure of the 13-Subunit Oxidized Cytochrome *c* Oxidase at 2.8 A. *Science* **1996**, *272*, 1136–1144. [CrossRef]
29. Sun, F.; Huo, X.; Zhai, Y.; Wang, A.; Xu, J.; Su, D.; Bartlam, M.; Rao, Z. Crystal Structure of Mitochondrial Respiratory Membrane Protein Complex II. *Cell* **2005**, *121*, 1043–1057. [CrossRef]
30. Zickermann, V.; Wirth, C.; Nasiri, H.; Siegmund, K.; Schwalbe, H.; Hunte, C.; Brandt, U. Mechanistic insight from the crystal structure of mitochondrial complex I. *Science* **2015**, *347*, 44–49. [CrossRef]
31. Zhu, J.; Vinothkumar, K.R.; Hirst, J.Z.J. Structure of mammalian respiratory complex I. *Nat. Cell Biol.* **2016**, *536*, 354–358. [CrossRef] [PubMed]
32. Blaza, J.N.; Vinothkumar, K.R.; Hirst, J. Structure of the Deactive State of Mammalian Respiratory Complex I. *Structure* **2018**, *26*, 312–319.e3. [CrossRef] [PubMed]
33. Agip, A.-N.A.; Blaza, J.N.; Bridges, H.R.; Viscomi, C.; Rawson, S.; Muench, S.P.; Hirst, J. Cryo-EM structures of complex I from mouse heart mitochondria in two biochemically defined states. *Nat. Struct. Mol. Biol.* **2018**, *25*, 548–556. [CrossRef]
34. Fiedorczuk, K.; Letts, J.A.; Degliesposti, G.; Kaszuba, K.; Skehel, G.D.M.; Sazanov, L.A. Atomic structure of the entire mammalian mitochondrial complex I. *Nat. Cell Biol.* **2016**, *538*, 406–410. [CrossRef]
35. Kampjut, D.; Sazanov, L.A. The coupling mechanism of mammalian respiratory complex I. *Science* **2020**, *370*, eabc4209. [CrossRef]
36. Vinothkumar, K.R.; Zhu, J.; Hirst, J. Architecture of mammalian respiratory complex I. *Nat. Cell Biol.* **2014**, *515*, 80–84. [CrossRef]
37. Yoga, E.G.; Angerer, H.; Parey, K.; Zickermann, V. Respiratory complex I—Mechanistic insights and advances in structure determination. *Biochim. Biophys. Acta Bioenerg.* **2020**, *1861*, 148153. [CrossRef]
38. Cecchini, G. Function and Structure of Complex II of the Respiratory Chain. *Annu. Rev. Biochem.* **2003**, *72*, 77–109. [CrossRef]
39. Bezawork-Geleta, A.; Rohlena, J.; Dong, L.; Pacak, K.; Neuzil, J. Mitochondrial Complex II: At the Crossroads. *Trends Biochem. Sci.* **2017**, *42*, 312–325. [CrossRef]
40. Sarewicz, M.; Pintscher, S.; Pietras, R.; Borek, A.; Bujnowicz, Ł.; Hanke, G.; Cramer, W.A.; Finazzi, G.; Osyczka, A. Catalytic Reactions and Energy Conservation in the Cytochrome *bc*1 and b6f Complexes of Energy-Transducing Membranes. *Chem. Rev.* **2021**, *121*, 2020–2108. [CrossRef] [PubMed]
41. Guo, R.; Gu, J.; Wu, M.; Yang, M. Amazing structure of respirasome: Unveiling the secrets of cell respiration. *Protein Cell* **2016**, *7*, 854–865. [CrossRef]
42. Hakvoort, T.B.; Moolenaar, K.; Lankvelt, A.H.; Sinjorgo, K.M.; Dekker, H.L.; Muijsers, A.O. Separation, stability and kinetics of monomeric and dimeric bovine heart cytochrome *c* oxidase. *Biochim. Biophys. Acta Bioenerg.* **1987**, *894*, 347–354. [CrossRef]
43. Zong, S.; Wu, M.; Gu, J.; Liu, T.; Guo, R.; Yang, M. Structure of the intact 14-subunit human cytochrome *c* oxidase. *Cell Res.* **2018**, *28*, 1026–1034. [CrossRef] [PubMed]
44. Balsa, E.; Marco, R.; Perales-Clemente, E.; Szklarczyk, R.; Calvo, E.; Landázuri, M.O.; Enríquez, J.A. NDUFA4 Is a Subunit of Complex IV of the Mammalian Electron Transport Chain. *Cell Metab.* **2012**, *16*, 378–386. [CrossRef] [PubMed]
45. Hill, B.C. The sequence of electron carriers in the reaction of cytochrome *c* oxidase with oxygen. *J. Bioenerg. Biomembr.* **1993**, *25*, 115–120. [CrossRef] [PubMed]
46. Kadenbach, B.; Hüttemann, M. The subunit composition and function of mammalian cytochrome *c* oxidase. *Mitochondrion* **2015**, *24*, 64–76. [CrossRef] [PubMed]
47. Ishigami, I.; Zatsepin, N.A.; Hikita, M.; Conrad, C.E.; Nelson, G.; Coe, J.D.; Basu, S.; Grant, T.D.; Seaberg, M.H.; Sierra, R.G.; et al. Crystal structure of CO-bound cytochrome *c* oxidase determined by serial femtosecond X-ray crystallography at room temperature. *Proc. Natl. Acad. Sci. USA* **2017**, *114*, 8011–8016. [CrossRef] [PubMed]
48. Frey, T.G.; Mannella, C.A. The internal structure of mitochondria. *Trends Biochem. Sci.* **2000**, *25*, 319–324. [CrossRef]
49. Harner, M.; Körner, C.; Walther, D.; Mokranjac, D.; Kaesmacher, J.; Welsch, U.; Griffith, J.; Mann, M.; Reggiori, F.; Neupert, W. The mitochondrial contact site complex, a determinant of mitochondrial architecture. *EMBO J.* **2011**, *30*, 4356–4370. [CrossRef]
50. Rampelt, H.; Zerbes, R.M.; van der Laan, M.; Pfanner, N. Role of the mitochondrial contact site and cristae organizing system in membrane architecture and dynamics. *Biochim. Biophys. Acta Bioenerg.* **2017**, *1864*, 737–746. [CrossRef] [PubMed]

51. Stephan, T.; Brüser, C.; Deckers, M.; Steyer, A.M.; Balzarotti, F.; Barbot, M.; Behr, T.S.; Heim, G.; Hübner, W.; Ilgen, P.; et al. MICOS assembly controls mitochondrial inner membrane remodeling and crista junction redistribution to mediate cristae formation. *EMBO J.* **2020**, *39*, 104105. [CrossRef] [PubMed]
52. Frezza, C.; Cipolat, S.; De Brito, O.M.; Micaroni, M.; Beznoussenko, G.V.; Rudka, T.; Bartoli, D.; Polishuck, R.S.; Danial, N.N.; De Strooper, B.; et al. OPA1 Controls Apoptotic Cristae Remodeling Independently from Mitochondrial Fusion. *Cell* **2006**, *126*, 177–189. [CrossRef] [PubMed]
53. Quintana-Cabrera, R.; Quirin, C.; Glytsou, C.; Corrado, M.; Urbani, A.; Pellattiero, A.; Calvo, E.; Vázquez, J.; Enríquez, J.A.; Gerle, C.; et al. The cristae modulator Optic atrophy 1 requires mitochondrial ATP synthase oligomers to safeguard mitochondrial function. *Nat. Commun.* **2018**, *9*, 3399. [CrossRef]
54. Del Dotto, V.; Fogazza, M.; Carelli, V.; Rugolo, M.; Zanna, C. Eight human OPA1 isoforms, long and short: What are they for? *Biochim. Biophys. Acta Bioenerg.* **2018**, *1859*, 263–269. [CrossRef] [PubMed]
55. Paumard, P.; Vaillier, J.; Coulary, B.; Schaeffer, J.; Soubannier, V.; Mueller, D.M.; Brèthes, D.; Di Rago, J.-P.; Velours, J. The ATP synthase is involved in generating mitochondrial cristae morphology. *EMBO J.* **2002**, *21*, 221–230. [CrossRef]
56. Davies, K.M.; Strauss, M.; Daum, B.; Kief, J.H.; Osiewacz, H.D.; Rycovska, A.; Zickermann, V.; Kühlbrandt, W. Macromolecular organization of ATP synthase and complex I in whole mitochondria. *Proc. Natl. Acad. Sci. USA* **2011**, *108*, 14121–14126. [CrossRef]
57. Rampelt, H.; Van Der Laan, M. The Yin & Yang of Mitochondrial Architecture—Interplay of MICOS and F1Fo-ATP synthase in cristae formation. *Microb. Cell* **2017**, *4*, 236–239. [CrossRef]
58. Kondadi, A.K.; Anand, R.; Reichert, A.S. Cristae Membrane Dynamics—A Paradigm Change. *Trends Cell Biol.* **2020**, *30*, 923–936. [CrossRef]
59. Busch, K.B. Inner mitochondrial membrane compartmentalization: Dynamics across scales. *Int. J. Biochem. Cell Biol.* **2020**, *120*, 105694. [CrossRef]
60. Rieger, B.; Junge, W.; Busch, K.B. Lateral pH gradient between OXPHOS complex IV and F0F1 ATP-synthase in folded mitochondrial membranes. *Nat. Commun.* **2014**, *5*, 3103. [CrossRef]
61. Toth, A.; Meyrat, A.; Stoldt, S.; Santiago, R.; Wenzel, D.; Jakobs, S.; Von Ballmoos, C.; Ott, M. Kinetic coupling of the respiratory chain with ATP synthase, but not proton gradients, drives ATP production in cristae membranes. *Proc. Natl. Acad. Sci. USA* **2020**, *117*, 2412–2421. [CrossRef] [PubMed]
62. Stuart, R.A.; Cruciat, C.-M.; Brunner, S.; Baumann, F.; Neupert, W. The Cytochrome *bc* 1 and Cytochrome *c* Oxidase Complexes Associate to Form a Single Supracomplex in Yeast Mitochondria. *J. Biol. Chem.* **2000**, *275*, 18093–18098. [CrossRef] [PubMed]
63. Schägger, H. Supercomplexes in the respiratory chains of yeast and mammalian mitochondria. *EMBO J.* **2000**, *19*, 1777–1783. [CrossRef]
64. Acín-Pérez, R.; Fernández-Silva, P.; Peleato, M.L.; Pérez-Martos, A.; Enriquez, J.A. Respiratory Active Mitochondrial Supercomplexes. *Mol. Cell* **2008**, *32*, 529–539. [CrossRef]
65. Milenkovic, D.; Blaza, J.N.; Larsson, N.-G.; Hirst, J. The Enigma of the Respiratory Chain Supercomplex. *Cell Metab.* **2017**, *25*, 765–776. [CrossRef] [PubMed]
66. Lobo-Jarne, T.; Ugalde, C. Respiratory chain supercomplexes: Structures, function and biogenesis. *Semin. Cell Dev. Biol.* **2018**, *76*, 179–190. [CrossRef]
67. Althoff, T.; Mills, D.J.; Popot, J.-L.; Kühlbrandt, W. Arrangement of electron transport chain components in bovine mitochondrial supercomplex I1III2IV1. *EMBO J.* **2011**, *30*, 4652–4664. [CrossRef]
68. Sousa, J.S.; Mills, D.J.; Vonck, J.; Kühlbrandt, W. Functional asymmetry and electron flow in the bovine respirasome. *eLife* **2016**, *5*, e21290. [CrossRef]
69. Letts, J.A.; Fiedorczuk, K.; Sazanov, J.A.L.K.F.L.A. The architecture of respiratory supercomplexes. *Nat. Cell Biol.* **2016**, *537*, 644–648. [CrossRef]
70. Gu, J.; Wu, M.; Guo, R.; Yan, K.; Lei, J.; Gao, N.; Yang, M. The architecture of the mammalian respirasome. *Nat. Cell Biol.* **2016**, *537*, 639–643. [CrossRef]
71. Dudkina, N.V.; Kudryashev, M.; Stahlberg, H.; Boekema, E.J. Interaction of complexes I, III, and IV within the bovine respirasome by single particle cryoelectron tomography. *Proc. Natl. Acad. Sci. USA* **2011**, *108*, 15196–15200. [CrossRef] [PubMed]
72. Davies, K.M.; Blum, T.B.; Kühlbrandt, W. Conserved in situ arrangement of complex I and III_2 in mitochondrial respiratory chain supercomplexes of mammals, yeast, and plants. *Proc. Natl. Acad. Sci. USA* **2018**, *115*, 3024–3029. [CrossRef] [PubMed]
73. Guo, R.; Zong, S.; Wu, M.; Gu, J.; Yang, M. Architecture of Human Mitochondrial Respiratory Megacomplex I2III2IV2. *Cell* **2017**, *170*, 1247–1257.e12. [CrossRef] [PubMed]
74. Letts, J.A.; Fiedorczuk, K.; Degliesposti, G.; Skehel, M.; Sazanov, L.A. Structures of Respiratory Supercomplex I+III_2 Reveal Functional and Conformational Crosstalk. *Mol. Cell* **2019**, *75*, 1131–1146.e6. [CrossRef] [PubMed]
75. Hartley, A.M.; Lukoyanova, N.; Zhang, Y.; Cabrera-Orefice, A.; Arnold, S.; Meunier, B.; Pinotsis, N.; Maréchal, A. Structure of yeast cytochrome *c* oxidase in a supercomplex with cytochrome *bc*1. *Nat. Struct. Mol. Biol.* **2019**, *26*, 78–83. [CrossRef]
76. Azuma, K.; Ikeda, K.; Inoue, S. Functional Mechanisms of Mitochondrial Respiratory Chain Supercomplex Assembly Factors and Their Involvement in Muscle Quality. *Int. J. Mol. Sci.* **2020**, *21*, 3182. [CrossRef]
77. Chen, Y.-C.; Taylor, E.B.; Dephoure, N.; Heo, J.-M.; Tonhato, A.; Papandreou, I.; Nath, N.; Denko, N.C.; Gygi, S.P.; Rutter, J. Identification of a Protein Mediating Respiratory Supercomplex Stability. *Cell Metab.* **2012**, *15*, 348–360. [CrossRef]

78. Ikeda, K.; Shiba, S.; Horie-Inoue, K.; Shimokata, K.; Inoue, S. A stabilizing factor for mitochondrial respiratory supercomplex assembly regulates energy metabolism in muscle. *Nat. Commun.* **2013**, *4*, 2147. [CrossRef]
79. Pérez-Pérez, R.; Lobo-Jarne, T.; Milenkovic, D.; Mourier, A.; Bratic, A.; García-Bartolomé, A.; Fernández-Vizarra, E.; Cadenas, S.; Delmiro, A.; García-Consuegra, I.; et al. COX7A2L Is a Mitochondrial Complex III Binding Protein that Stabilizes the III_2+IV Supercomplex without Affecting Respirasome Formation. *Cell Rep.* **2016**, *16*, 2387–2398. [CrossRef] [PubMed]
80. Lobo-Jarne, T.; Pérez-Pérez, R.; Fontanesi, F.; Timon-Gomez, A.; Wittig, I.; Peñas, A.; Serrano-Lorenzo, P.; García-Consuegra, I.; Arenas, J.; Martín, M.A.; et al. Multiple pathways coordinate assembly of human mitochondrial complex IV and stabilization of respiratory supercomplexes. *EMBO J.* **2020**, *39*, e103912. [CrossRef]
81. Lapuente-Brun, E.; Moreno-Loshuertos, R.; Acín-Pérez, R.; Latorre-Pellicer, A.; Colás, C.; Balsa, E.; Perales-Clemente, E.; Quirós, P.M.; Calvo, E.; Rodríguez-Hernández, Á.; et al. Supercomplex Assembly Determines Electron Flux in the Mitochondrial Electron Transport Chain. *Science* **2013**, *340*, 1567–1570. [CrossRef]
82. Timón-Gómez, A.; Garlich, J.; Stuart, R.A.; Ugalde, C.; Barrientos, A. Distinct Roles of Mitochondrial HIGD1A and HIGD2A in Respiratory Complex and Supercomplex Biogenesis. *Cell Rep.* **2020**, *31*, 107607. [CrossRef]
83. Cogliati, S.; Calvo, E.; Loureiro, M.; Guaras, A.M.; Nieto-Arellano, R.; Garcia-Poyatos, C.; Ezkurdia, I.; Mercader, C.G.-P.N.; Vázquez, J.; Enriquez, J.A. Mechanism of super-assembly of respiratory complexes III and IV. *Nat. Cell Biol.* **2016**, *539*, 579–582. [CrossRef]
84. Zhang, K.; Wang, G.; Zhang, X.; Hüttemann, P.P.; Qiu, Y.; Liu, J.; Mitchell, A.; Lee, I.; Zhang, C.; Lee, J.-S.; et al. COX7AR is a Stress-inducible Mitochondrial COX Subunit that Promotes Breast Cancer Malignancy. *Sci. Rep.* **2016**, *6*, 31742. [CrossRef]
85. Tucker, E.J.; Wanschers, B.F.J.; Szklarczyk, R.; Mountford, H.S.; Wijeyeratne, X.W.; Brand, M.A.M.V.D.; Leenders, A.M.; Rodenburg, R.J.; Reljić, B.; Compton, A.G.; et al. Mutations in the UQCC1-Interacting Protein, UQCC2, Cause Human Complex III Deficiency Associated with Perturbed Cytochrome *b* Protein Expression. *PLoS Genet.* **2013**, *9*, e1004034. [CrossRef]
86. Wanschers, B.F.; Szklarczyk, R.; Brand, M.A.V.D.; Jonckheere, A.; Suijskens, J.; Smeets, R.; Rodenburg, R.J.; Stephan, K.; Helland, I.B.; Elkamil, A.; et al. A mutation in the human CBP4 ortholog UQCC3 impairs complex III assembly, activity and cytochrome *b* stability. *Hum. Mol. Genet.* **2014**, *23*, 6356–6365. [CrossRef] [PubMed]
87. Desmurs, M.; Foti, M.; Raemy, E.; Vaz, F.M.; Martinou, J.-C.; Bairoch, A.; Lane, L. C11orf83, a Mitochondrial Cardiolipin-Binding Protein Involved inbc1Complex Assembly and Supercomplex Stabilization. *Mol. Cell. Biol.* **2015**, *35*, 1139–1156. [CrossRef]
88. Yang, Y.; Zhang, G.; Guo, F.; Li, Q.; Luo, H.; Shu, Y.; Shen, Y.; Gan, J.; Xu, L.; Yang, H. Mitochondrial UQCC3 Modulates Hypoxia Adaptation by Orchestrating OXPHOS and Glycolysis in Hepatocellular Carcinoma. *Cell Rep.* **2020**, *33*, 108340. [CrossRef] [PubMed]
89. Chance, B.; Williams, G.R. A Method for the Localization of Sites for Oxidative Phosphorylation. *Nat. Cell Biol.* **1955**, *176*, 250–254. [CrossRef]
90. Green, D.E.; Tzagoloff, A. The mitochondrial electron transfer chain. *Arch. Biochem. Biophys.* **1966**, *116*, 293–304. [CrossRef]
91. Hackenbrock, C.R.; Chazotte, B.; Gupte, S.S. The random collision model and a critical assessment of diffusion and collision in mitochondrial electron transport. *J. Bioenerg. Biomembr.* **1986**, *18*, 331–368. [CrossRef] [PubMed]
92. Hatefi, Y.; Haavik, A.; Griffiths, D. Studies on the Electron Transfer System. *J. Biol. Chem.* **1962**, *237*, 1676–1680. [CrossRef]
93. Acin-Perez, R.; Enriquez, J.A. The function of the respiratory supercomplexes: The plasticity model. *Biochim. Biophys. Acta Bioenerg.* **2014**, *1837*, 444–450. [CrossRef] [PubMed]
94. Guarás, A.; Perales-Clemente, E.; Calvo, E.; Acín-Pérez, R.; Loureiro-Lopez, M.; Pujol, C.; Martínez-Carrascoso, I.; Nuñez, E.; García-Marqués, F.; Rodríguez-Hernández, M.A.; et al. The CoQH2/CoQ Ratio Serves as a Sensor of Respiratory Chain Efficiency. *Cell Rep.* **2016**, *15*, 197–209. [CrossRef] [PubMed]
95. Blaza, J.N.; Serreli, R.; Jones, A.J.Y.; Mohammed, K.; Hirst, J. Kinetic evidence against partitioning of the ubiquinone pool and the catalytic relevance of respiratory-chain supercomplexes. *Proc. Natl. Acad. Sci. USA* **2014**, *111*, 15735–15740. [CrossRef] [PubMed]
96. Fedor, J.G.; Hirst, J. Mitochondrial Supercomplexes Do Not Enhance Catalysis by Quinone Channeling. *Cell Metab.* **2018**, *28*, 525–531.e4. [CrossRef]
97. Hernansanz-Agustín, P.; Enríquez, J.A. Functional segmentation of CoQ and cyt c pools by respiratory complex superassembly. *Free Radic. Biol. Med.* **2021**, *167*, 232–242. [CrossRef]
98. Bianchi, C.; Genova, M.L.; Castelli, G.P.; Lenaz, G. The Mitochondrial Respiratory Chain Is Partially Organized in a Supercomplex Assembly. *J. Biol. Chem.* **2004**, *279*, 36562–36569. [CrossRef]
99. Trouillard, M.; Meunier, B.; Rappaport, F. Questioning the functional relevance of mitochondrial supercomplexes by time-resolved analysis of the respiratory chain. *Proc. Natl. Acad. Sci. USA* **2011**, *108*, E1027–E1034. [CrossRef]
100. Letts, J.A.; Sazanov, L.A. Clarifying the supercomplex: The higher-order organization of the mitochondrial electron transport chain. *Nat. Struct. Mol. Biol.* **2017**, *24*, 800–808. [CrossRef]
101. Schägger, H.; de Coo, R.; Bauer, M.F.; Hofmann, S.; Godinot, C.; Brandt, U. Significance of Respirasomes for the Assembly/Stability of Human Respiratory Chain Complex I. *J. Biol. Chem.* **2004**, *279*, 36349–36353. [CrossRef] [PubMed]
102. Moreno-Lastres, D.; Fontanesi, F.; García-Consuegra, I.; Martín, M.A.; Arenas, J.; Barrientos, A.; Ugalde, C. Mitochondrial Complex I Plays an Essential Role in Human Respirasome Assembly. *Cell Metab.* **2012**, *15*, 324–335. [CrossRef] [PubMed]
103. Hirst, J. Open questions: Respiratory chain supercomplexes—why are they there and what do they do? *BMC Biol.* **2018**, *16*, 111. [CrossRef]

104. Muller, F.L.; Liu, Y.; Van Remmen, H. Complex III Releases Superoxide to Both Sides of the Inner Mitochondrial Membrane. *J. Biol. Chem.* **2004**, *279*, 49064–49073. [CrossRef]
105. Kussmaul, L.; Hirst, J. The mechanism of superoxide production by NADH: Ubiquinone oxidoreductase (complex I) from bovine heart mitochondria. *Proc. Natl. Acad. Sci. USA* **2006**, *103*, 7607–7612. [CrossRef]
106. Murphy, M.P. How mitochondria produce reactive oxygen species. *Biochem. J.* **2009**, *417*, 1–13. [CrossRef] [PubMed]
107. Pryde, K.R.; Hirst, J. Superoxide Is Produced by the Reduced Flavin in Mitochondrial Complex I. *J. Biol. Chem.* **2011**, *286*, 18056–18065. [CrossRef]
108. Maranzana, E.; Barbero, G.; Falasca, A.I.; Lenaz, G.; Genova, M.L. Mitochondrial Respiratory Supercomplex Association Limits Production of Reactive Oxygen Species from Complex I. *Antioxid. Redox Signal.* **2013**, *19*, 1469–1480. [CrossRef]
109. Lopez-Fabuel, I.; Le Douce, J.; Logan, A.; James, A.M.; Bonvento, G.; Murphy, M.P.; Almeida, A.; Bolaños, J.P. Complex I assembly into supercomplexes determines differential mitochondrial ROS production in neurons and astrocytes. *Proc. Natl. Acad. Sci. USA* **2016**, *113*, 13063–13068. [CrossRef]
110. Morán, M.; Marín-Buera, L.; Gil-Borlado, M.C.; Rivera, H.; Blázquez, A.; Seneca, S.; Vázquez-López, M.; Arenas, J.; Martín, M.A.; Ugalde, C. Cellular pathophysiological consequences of BCS1L mutations in mitochondrial complex III enzyme deficiency. *Hum. Mutat.* **2010**, *31*, 930–941. [CrossRef] [PubMed]
111. Diaz, F.; Enriquez, J.A.; Moraes, C.T. Cells Lacking Rieske Iron-Sulfur Protein Have a Reactive Oxygen Species-Associated Decrease in Respiratory Complexes I and IV. *Mol. Cell. Biol.* **2012**, *32*, 415–429. [CrossRef] [PubMed]
112. Hernansanz-Agustín, P.; Choya-Foces, C.; Carregal-Romero, S.; Ramos, E.; Oliva, T.; Villa-Piña, T.; Moreno, L.; Izquierdo-Álvarez, A.; Cabrera-García, J.D.; Cortés, A.; et al. Na+ controls hypoxic signalling by the mitochondrial respiratory chain. *Nature* **2020**, *586*, 287–291. [CrossRef] [PubMed]
113. Tropeano, C.V.; Aleo, S.J.; Zanna, C.; Roberti, M.; Scandiffio, L.; Polosa, P.L.; Fiori, J.; Porru, E.; Roda, A.; Carelli, V.; et al. Fine-tuning of the respiratory complexes stability and supercomplexes assembly in cells defective of complex III. *Biochim. Biophys. Acta Bioenerg.* **2020**, *1861*, 148133. [CrossRef]
114. Nogueira, C.; Barros, J.; Sá, M.J.; Azevedo, L.; Taipa, R.; Torraco, A.; Meschini, M.C.; Verrigni, D.; Nesti, C.; Rizza, T.; et al. Novel TTC19 mutation in a family with severe psychiatric manifestations and complex III deficiency. *Neurogenetics* **2013**, *14*, 153–160. [CrossRef] [PubMed]
115. Davoudi, M.; Kotarsky, H.; Hansson, E.; Fellman, V. Complex I Function and Supercomplex Formation Are Preserved in Liver Mitochondria Despite Progressive Complex III Deficiency. *PLoS ONE* **2014**, *9*, e86767. [CrossRef] [PubMed]
116. Davoudi, M.; Kotarsky, H.; Hansson, E.; Kallijärvi, J.; Fellman, V. COX7A2L/SCAFI and Pre-Complex III Modify Respiratory Chain Supercomplex Formation in Different Mouse Strains with a Bcs1l Mutation. *PLoS ONE* **2016**, *11*, e0168774. [CrossRef] [PubMed]
117. Guerrero-Castillo, S.; Baertling, F.; Kownatzki, D.; Wessels, H.J.; Arnold, S.; Brandt, U.; Nijtmans, L. The Assembly Pathway of Mitochondrial Respiratory Chain Complex I. *Cell Metab.* **2017**, *25*, 128–139. [CrossRef]
118. Protasoni, M.; Pérez-Pérez, R.; Lobo-Jarne, T.; Harbour, M.E.; Ding, S.; Peñas, A.; Diaz, F.; Moraes, C.T.; Fearnley, I.M.; Zeviani, M.; et al. Respiratory supercomplexes act as a platform for complex III -mediated maturation of human mitochondrial complexes I and IV. *EMBO J.* **2020**, *39*, e102817. [CrossRef]
119. Páleníková, P.; Harbour, M.E.; Prodi, F.; Minczuk, M.; Zeviani, M.; Ghelli, A.; Fernández-Vizarra, E. Duplexing complexome profiling with SILAC to study human respiratory chain assembly defects. *Biochim. Biophys. Acta Bioenerg.* **2021**, *1862*, 148395. [CrossRef]
120. Carossa, V.; Ghelli, A.; Tropeano, C.V.; Valentino, M.L.; Iommarini, L.; Maresca, A.; Caporali, L.; La Morgia, C.; Liguori, R.; Barboni, P.; et al. A Novel in-Frame 18-bp Microdeletion in *MTCYB* Causes a Multisystem Disorder with Prominent Exercise Intolerance. *Hum. Mutat.* **2014**, *35*, 954–958. [CrossRef]
121. Tropeano, C.V.; Fiori, J.; Carelli, V.; Caporali, L.; Daldal, F.; Ghelli, A.M.; Rugolo, M. Complex II phosphorylation is triggered by unbalanced redox homeostasis in cells lacking complex III. *Biochim. Biophys. Acta Bioenerg.* **2018**, *1859*, 182–190. [CrossRef]
122. Blakely, E.L.; Mitchell, A.L.; Fisher, N.; Meunier, B.; Nijtmans, L.G.; Schaefer, A.M.; Jackson, M.J.; Turnbull, D.M.; Taylor, R.W. A mitochondrial cytochrome *b* mutation causing severe respiratory chain enzyme deficiency in humans and yeast. *FEBS J.* **2005**, *272*, 3583–3592. [CrossRef]
123. Lamantea, E.; Carrara, F.; Mariotti, C.; Morandi, L.; Tiranti, V.; Zeviani, M. A novel nonsense mutation (Q352X) in the mitochondrial cytochrome *b* gene associated with a combined deficiency of complexes I and III. *Neuromuscul. Disord.* **2002**, *12*, 49–52. [CrossRef]
124. Haut, S.; Brivet, M.; Touati, G.; Rustin, P.; Lebon, S.; Garcia-Cazorla, A.; Saudubray, J.M.; Boutron, A.; Legrand, A.; Slama, A. A deletion in the human QP-C gene causes a complex III deficiency resulting in hypoglycaemia and lactic acidosis. *Qual. Life Res.* **2003**, *113*, 118–122. [CrossRef]
125. Barel, O.; Shorer, Z.; Flusser, H.; Ofir, R.; Narkis, G.; Finer, G.; Shalev, H.; Nasasra, A.; Saada, A.; Birk, O.S. Mitochondrial Complex III Deficiency Associated with a Homozygous Mutation in UQCRQ. *Am. J. Hum. Genet.* **2008**, *82*, 1211–1216. [CrossRef]
126. Gaignard, P.; Menezes, M.; Schiff, M.; Bayot, A.; Rak, M.; de Baulny, H.O.; Su, C.-H.; Gilleron, M.; Lombes, A.; Abida, H.; et al. Mutations in CYC1, Encoding Cytochrome *c*1 Subunit of Respiratory Chain Complex III, Cause Insulin-Responsive Hyperglycemia. *Am. J. Hum. Genet.* **2013**, *93*, 384–389. [CrossRef] [PubMed]

127. Miyake, N.; Yano, S.; Sakai, C.; Hatakeyama, H.; Matsushima, Y.; Shiina, M.; Watanabe, Y.; Bartley, J.; Abdenur, J.E.; Wang, R.Y.; et al. Mitochondrial Complex III Deficiency Caused by a HomozygousUQCRC2Mutation Presenting with Neonatal-Onset Recurrent Metabolic Decompensation. *Hum. Mutat.* **2013**, *34*, 446–452. [CrossRef] [PubMed]
128. Feichtinger, R.G.; Brunner-Krainz, M.; Alhaddad, B.; Wortmann, S.B.; Kovacs-Nagy, R.; Stojakovic, T.; Erwa, W.; Resch, B.; Windischhofer, W.; Verheyen, S.; et al. Combined Respiratory Chain Deficiency and UQCC2 Mutations in Neonatal Encephalomyopathy: Defective Supercomplex Assembly in Complex III Deficiencies. *Oxidative Med. Cell. Longev.* **2017**, *2017*, 7202589. [CrossRef] [PubMed]
129. Ghelli, A.; Tropeano, C.V.; Calvaruso, M.A.; Marchesini, A.; Iommarini, L.; Porcelli, A.M.; Zanna, C.; De Nardo, V.; Martinuzzi, A.; Wibrand, F.; et al. The cytochrome *b* p.278Y>C mutation causative of a multisystem disorder enhances superoxide production and alters supramolecular interactions of respiratory chain complexes. *Hum. Mol. Genet.* **2013**, *22*, 2141–2151. [CrossRef]
130. Acín-Pérez, R.; Bayona-Bafaluy, M.P.; Fernández-Silva, P.; Moreno-Loshuertos, R.; Pérez-Martos, A.; Bruno, C.; Moraes, C.T.; Enríquez, J.A. Respiratory complex III is required to maintain complex I in mammalian mitochondria. *Mol. Cell.* **2004**, *13*, 805–815. [CrossRef]
131. Spinazzi, M.; Radaelli, E.; Horré, K.; Arranz, A.M.; Gounko, N.V.; Agostinis, P.; Maia, T.M.; Impens, F.; Morais, V.A.; Lopez-Lluch, G.; et al. PARL deficiency in mouse causes Complex III defects, coenzyme Q depletion, and Leigh-like syndrome. *Proc. Natl. Acad. Sci. USA* **2019**, *116*, 277–286. [CrossRef] [PubMed]
132. Iommarini, L.; Ghelli, A.; Leone, G.; Tropeano, C.V.; Kurelac, I.; Amato, L.B.; Gasparre, G.; Porcelli, A.M. Mild phenotypes and proper supercomplex assembly in human cells carrying the homoplasmic m.15557G > A mutation in cytochrome *b* gene. *Hum. Mutat.* **2018**, *39*, 92–102. [CrossRef] [PubMed]
133. Bottani, E.; Lamperti, C.; Prigione, A.; Tiranti, V.; Persico, N.; Brunetti, D. Therapeutic Approaches to Treat Mitochondrial Diseases: "One-Size-Fits-All" and "Precision Medicine" Strategies. *Pharmaceutics* **2020**, *12*, 1083. [CrossRef] [PubMed]
134. Ježek, P.; Hlavatá, L. Mitochondria in homeostasis of reactive oxygen species in cell, tissues, and organism. *Int. J. Biochem. Cell Biol.* **2005**, *37*, 2478–2503. [CrossRef]
135. Raha, S.; Robinson, B.H. Mitochondria, oxygen free radicals, disease and ageing. *Trends Biochem. Sci.* **2000**, *25*, 502–508. [CrossRef]
136. Wong, H.-S.; Dighe, P.A.; Mezera, V.; Monternier, P.-A.; Brand, M.D. Production of superoxide and hydrogen peroxide from specific mitochondrial sites under different bioenergetic conditions. *J. Biol. Chem.* **2017**, *292*, 16804–16809. [CrossRef]
137. Seelert, H.; Dani, D.; Dante, S.; Hauß, T.; Krause, F.; Schafer, E.; Frenzel, M.; Poetsch, A.; Rexroth, S.; Schwaßmann, H.; et al. From protons to OXPHOS supercomplexes and Alzheimer's disease: Structure–dynamics–function relationships of energy-transducing membranes. *Biochim. Biophys. Acta Bioenerg.* **2009**, *1787*, 657–671. [CrossRef]
138. Genova, M.L.; Lenaz, G. Functional role of mitochondrial respiratory supercomplexes. *Biochim. Biophys. Acta Bioenerg.* **2014**, *1837*, 427–443. [CrossRef]
139. Malferrari, M.; Ghelli, A.; Roggiani, F.; Valenti, G.; Paolucci, F.; Rugolo, M.; Rapino, S. Reactive Oxygen Species Produced by Mutated Mitochondrial Respiratory Chains of Entire Cells Monitored Using Modified Microelectrodes. *ChemElectroChem* **2019**, *6*, 627–633. [CrossRef]
140. Anwar, M.R.; Saldana-Caboverde, A.; Garcia, S.; Diaz, F. The Organization of Mitochondrial Supercomplexes is Modulated by Oxidative Stress In Vivo in Mouse Models of Mitochondrial Encephalopathy. *Int. J. Mol. Sci.* **2018**, *19*, 1582. [CrossRef]
141. Mori, M.; Goldstein, J.; Young, S.P.; Bossen, E.H.; Shoffner, J.; Koeberl, D.D. Complex III deficiency due to an in-frame *MTCYB* deletion presenting as ketotic hypoglycemia and lactic acidosis. *Mol. Genet. Metab. Rep.* **2015**, *4*, 39–41. [CrossRef]
142. Gaignard, P.; Eyer, D.; Lebigot, E.; Oliveira, C.; Therond, P.; Boutron, A.; Slama, A. UQCRC2 mutation in a patient with mitochondrial complex III deficiency causing recurrent liver failure, lactic acidosis and hypoglycemia. *J. Hum. Genet.* **2017**, *62*, 729–731. [CrossRef] [PubMed]
143. Levéen, P.; Kotarsky, H.; Mörgelin, M.; Karikoski, R.; Elmér, E.; Fellman, V. The GRACILE mutation introduced into Bcs1l causes postnatal complex III deficiency: A viable mouse model for mitochondrial hepatopathy. *Hepatology* **2011**, *53*, 437–447. [CrossRef]
144. Rajendran, J.; Tomašić, N.; Kotarsky, H.; Hansson, E.; Velagapudi, V.; Kallijärvi, J.; Fellman, V. Effect of High-Carbohydrate Diet on Plasma Metabolome in Mice with Mitochondrial Respiratory Chain Complex III Deficiency. *Int. J. Mol. Sci.* **2016**, *17*, 1824. [CrossRef] [PubMed]
145. Wibrand, F.; Ravn, K.; Schwartz, M.; Rosenberg, T.; Horn, N.; Vissing, J. Multisystem disorder associated with a missense mutation in the mitochondrial cytochrome *b* gene. *Ann. Neurol.* **2001**, *50*, 540–543. [CrossRef] [PubMed]
146. Lee, C.F.; Caudal, A.; Abell, L.; Gowda, G.A.N.; Tian, R. Targeting NAD^+ Metabolism as Interventions for Mitochondrial Disease. *Sci. Rep.* **2019**, *9*, 3073. [CrossRef]
147. Johnson, S.C.; Yanos, M.E.; Kayser, E.-B.; Quintana, A.; Sangesland, M.; Castanza, A.; Uhde, L.; Hui, J.; Wall, V.Z.; Gagnidze, A.; et al. mTOR Inhibition Alleviates Mitochondrial Disease in a Mouse Model of Leigh Syndrome. *Science* **2013**, *342*, 1524–1528. [CrossRef]
148. Yang, L.; Canaveras, J.C.G.; Chen, Z.; Wang, L.; Liang, L.; Jang, C.; Mayr, J.A.; Zhang, Z.; Ghergurovich, J.M.; Zhan, L.; et al. Serine Catabolism Feeds NADH when Respiration Is Impaired. *Cell Metab.* **2020**, *31*, 809–821.e6. [CrossRef]
149. Jain, I.H.; Zazzeron, L.; Goli, R.; Alexa, K.; Schatzman-Bone, S.; Dhillon, H.; Goldberger, O.; Peng, J.; Shalem, O.; Sanjana, N.E.; et al. Hypoxia as a therapy for mitochondrial disease. *Science* **2016**, *352*, 54–61. [CrossRef]

150. Seo, B.B.; Kitajima-Ihara, T.; Chan, E.K.L.; Scheffler, I.E.; Matsuno-Yagi, A.; Yagi, T. Molecular remedy of complex I defects: Rotenone-insensitive internal NADH-quinone oxidoreductase of Saccharomyces cerevisiae mitochondria restores the NADH oxidase activity of complex I-deficient mammalian cells. *Proc. Natl. Acad. Sci. USA* **1998**, *95*, 9167–9171. [CrossRef]
151. Perales-Clemente, E.; Bayona-Bafaluy, M.P.; Pérez-Martos, A.; Barrientos, A.; Fernández-Silva, P.; Enriquez, J.A. Restoration of electron transport without proton pumping in mammalian mitochondria. *Proc. Natl. Acad. Sci. USA* **2008**, *105*, 18735–18739. [CrossRef] [PubMed]
152. Wheaton, W.W.; Weinberg, S.E.; Hamanaka, R.B.; Soberanes, S.; Sullivan, L.B.; Anso, E.; Glasauer, A.; Dufour, E.; Mutlu, G.M.; Budigner, G.S.; et al. Metformin inhibits mitochondrial complex I of cancer cells to reduce tumorigenesis. *eLife* **2014**, *3*, e02242. [CrossRef]
153. Bai, Y.; Hájek, P.; Chomyn, A.; Chan, E.; Seo, B.B.; Matsuno-Yagi, A.; Yagi, T.; Attardi, G. Lack of Complex I Activity in Human Cells Carrying a Mutation in MtDNA-encoded ND4 Subunit Is Corrected by the Saccharomyces cerevisiae NADH-Quinone Oxidoreductase (NDI1) Gene. *J. Biol. Chem.* **2001**, *276*, 38808–38813. [CrossRef] [PubMed]
154. Yagi, T.; Seo, B.B.; Nakamaru-Ogiso, E.; Marella, M.; Barber-Singh, J.; Yamashita, T.; Matsuno-Yagi, A. Possibility of trans-kingdom gene therapy for Complex I diseases. *Biochim. Biophys. Acta Bioenerg.* **2006**, *1757*, 708–714. [CrossRef] [PubMed]
155. McElroy, G.S.; Reczek, C.R.; Reyfman, P.A.; Mithal, D.S.; Horbinski, C.M.; Chandel, N.S. NAD^+ Regeneration Rescues Lifespan, but Not Ataxia, in a Mouse Model of Brain Mitochondrial Complex I Dysfunction. *Cell Metab.* **2020**, *32*, 301–308.e6. [CrossRef] [PubMed]
156. El-Khoury, R.; Kemppainen, K.K.; Dufour, E.; Szibor, M.; Jacobs, H.T.; Rustin, P. Engineering the alternative oxidase gene to better understand and counteract mitochondrial defects: State of the art and perspectives. *Br. J. Pharmacol.* **2014**, *171*, 2243–2249. [CrossRef]
157. Szibor, M.; Gainutdinov, T.; Fernandez-Vizarra, E.; Dufour, E.; Gizatullina, Z.; Debska-Vielhaber, G.; Heidler, J.; Wittig, I.; Viscomi, C.; Gellerich, F.; et al. Bioenergetic consequences from xenotopic expression of a tunicate AOX in mouse mitochondria: Switch from RET and ROS to FET. *Biochim. Biophys. Acta Bioenerg.* **2020**, *1861*, 148137. [CrossRef]
158. Mullen, A.R.; Wheaton, W.W.; Jin, E.S.; Chen, P.-H.; Sullivan, L.B.; Cheng, T.; Yang, Y.; Linehan, W.M.; Chandel, N.S.; DeBerardinis, R.J. Reductive carboxylation supports growth in tumour cells with defective mitochondria. *Nat. Cell Biol.* **2011**, *481*, 385–388. [CrossRef]
159. Kotarsky, H.; Keller, M.; Davoudi, M.; Leveen, P.; Karikoski, R.; Enot, D.P.; Fellman, V. Metabolite Profiles Reveal Energy Failure and Impaired Beta-Oxidation in Liver of Mice with Complex III Deficiency Due to a BCS1L Mutation. *PLoS ONE* **2012**, *7*, e41156. [CrossRef]
160. Selak, M.A.; Armour, S.M.; MacKenzie, E.D.; Boulahbel, H.; Watson, D.G.; Mansfield, K.D.; Pan, Y.; Simon, M.; Thompson, C.B.; Gottlieb, E. Succinate links TCA cycle dysfunction to oncogenesis by inhibiting HIF-α prolyl hydroxylase. *Cancer Cell* **2005**, *7*, 77–85. [CrossRef] [PubMed]
161. Semenza, G.; Roth, P.; Fang, H.; Wang, G. Transcriptional regulation of genes encoding glycolytic enzymes by hypoxia-inducible factor 1. *J. Biol. Chem.* **1994**, *269*, 23757–23763. [CrossRef]
162. Ansó, E.; Weinberg, S.E.; Diebold, L.P.; Thompson, B.J.; Malinge, S.; Schumacker, P.T.; Liu, X.; Zhang, Y.; Shao, Y.Z.Z.; Steadman, M.; et al. The mitochondrial respiratory chain is essential for haematopoietic stem cell function. *Nat. Cell Biol.* **2017**, *19*, 614–625. [CrossRef] [PubMed]
163. Chowdhury, R.; Yeoh, K.K.; Tian, Y.; Hillringhaus, L.; Bagg, E.A.; Rose, N.R.; Leung, I.K.H.; Li, X.S.; Woon, E.C.Y.; Yang, M.; et al. The oncometabolite 2-hydroxyglutarate inhibits histone lysine demethylases. *EMBO Rep.* **2011**, *12*, 463–469. [CrossRef]
164. Martínez-Reyes, I.; Chandel, N.S. Mitochondrial TCA cycle metabolites control physiology and disease. *Nat. Commun.* **2020**, *11*, 102. [CrossRef] [PubMed]
165. Roux, K.J. Marked by association: Techniques for proximity-dependent labeling of proteins in eukaryotic cells. *Cell. Mol. Life Sci.* **2013**, *70*, 3657–3664. [CrossRef]
166. Varnaitė, R.; MacNeill, S.A. Meet the neighbors: Mapping local protein interactomes by proximity-dependent labeling with BioID. *Proteomics* **2016**, *16*, 2503–2518. [CrossRef] [PubMed]
167. De Coo, I.F.M.; Renier, W.O.; Ruitenbeek, W.; Ter Laak, H.J.; Bakker, M.; Schägger, H.; Van Oost, B.A.; Smeets, H.J.M. A 4–base pair deletion in the mitochondrial cytochrome *b* gene associated with parkinsonism/MELAS overlap syndrome. *Ann. Neurol.* **1999**, *45*, 130–133. [CrossRef]
168. Rana, M.; De Coo, I.; Diaz, F.; Smeets, H.; Moraes, C.T. An out-of-frame cytochrome *b* gene deletion from a patient with parkinsonism is associated with impaired complex III assembly and an increase in free radical production. *Ann. Neurol.* **2000**, *48*, 774–781. [CrossRef]
169. Andreu, A.L.; Hanna, M.G.; Reichmann, H.; Bruno, C.; Penn, A.S.; Tanji, K.; Pallotti, F.; Iwata, S.; Bonilla, E.; Lach, B.; et al. Exercise Intolerance Due to Mutations in the Cytochrome *b* Gene of Mitochondrial DNA. *N. Engl. J. Med.* **1999**, *341*, 1037–1044. [CrossRef]
170. Schuelke, M.; Krude, H.; Finckh, B.; Mayatepek, E.; Janssen, A.; Schmelz, M.; Trefz, F.; Trijbels, F.; Smeitink, J. Septo-optic dysplasia associated with a new mitochondrial cytochrome *b* mutation. *Ann. Neurol.* **2002**, *51*, 388–392. [CrossRef]
171. Emmanuele, V.; Sotiriou, E.; Rios, P.G.; Ganesh, J.; Ichord, R.; Foley, A.R.; Akman, H.O.; DiMauro, S. A Novel Mutation in the Mitochondrial DNA Cytochrome *b* Gene (*MTCYB*) in a Patient with Mitochondrial Encephalomyopathy, Lactic Acidosis, and Strokelike Episodes Syndrome. *J. Child Neurol.* **2013**, *28*, 236–242. [CrossRef]

172. Legros, F.; Chatzoglou, E.; Frachon, P.; De Baulny, H.O.; Laforêt, P.; Jardel, C.; Godinot, C.; Lombès, A. Functional characterization of novel mutations in the human cytochrome *b* gene. *Eur. J. Hum. Genet.* **2001**, *9*, 510–518. [CrossRef]
173. Bruno, C.; Santorelli, F.M.; Assereto, S.; Tonoli, E.; Tessa, A.; Traverso, M.; Scapolan, S.; Bado, M.; Tedeschi, S.; Minetti, C. Progressive exercise intolerance associated with a new muscle-restricted nonsense mutation (G142X) in the mitochondrial cytochrome *b* gene. *Muscle Nerve* **2003**, *28*, 508–511. [CrossRef]
174. Keightley, J.A.; Anitori, R.; Burton, M.D.; Quan, F.; Buist, N.R.; Kennaway, N.G. Mitochondrial Encephalomyopathy and Complex III Deficiency Associated with a Stop-Codon Mutation in the Cytochrome *b* Gene. *Am. J. Hum. Genet.* **2000**, *67*, 1400–1410. [CrossRef]
175. Valnot, I.; Kassis, J.; Chretien, D.; De Lonlay, P.; Parfait, B.; Munnich, A.; Kachaner, J.; Rustin, P.; Rötig, A. A mitochondrial cytochrome *b* mutation but no mutations of nuclearly encoded subunits in ubiquinol cytochrome *c* reductase (complex III) deficiency. *Qual. Life Res.* **1999**, *104*, 460–466. [CrossRef]
176. Dumoulin, R.; Mandon, G.; Collombet, J.M.; Blond, J.L.; Carrier, H.; Godinot, C.; Flocard, F.; Villard, J.; Guibaud, P.; Mathieu, M.; et al. Human cultured myoblasts: A model for the diagnosis of mitochondrial diseases. *J. Inherit. Metab. Dis.* **1993**, *16*, 545–547. [CrossRef]
177. Bouzidi, M.F.; Carrier, H.; Godinot, C. Antimycin resistance and ubiquinol cytochrome *c* reductase instability associated with a human cytochrome *b* mutation. *Biochim. Biophys. Acta Mol. Basis Dis.* **1996**, *1317*, 199–209. [CrossRef]
178. Fragaki, K.; Procaccio, V.; Bannwarth, S.; Serre, V.; O'Hearn, S.; Potluri, P.; Augé, G.; Casagrande, F.; Caruba, C.; Lambert, J.C.; et al. A neonatal polyvisceral failure linked to a de novo homoplasmic mutation in the mitochondrially encoded cytochrome *b* gene. *Mitochondrion* **2009**, *9*, 346–352. [CrossRef]
179. Mancuso, M.; Filosto, M.; Stevens, J.C.; Patterson, M.; Shanske, S.; Krishna, S.; DiMauro, S. Mitochondrial myopathy and complex III deficiency in a patient with a new stop-codon mutation (G339X) in the cytochrome *b* gene. *J. Neurol. Sci.* **2003**, *209*, 61–63. [CrossRef]
180. Andreu, A.L.; Bruno, C.; Shanske, S.; Shtilbans, A.; Hirano, M.; Krishna, S.; Hayward, L.; Systrom, D.S.; Brown, R.H.; DiMauro, S. Missense mutation in the mtDNA cytochrome *b* gene in a patient with myopathy. *Neurology* **1998**, *51*, 1444–1447. [CrossRef]
181. Gusic, M.; Schottmann, G.; Feichtinger, R.G.; Du, C.; Scholz, C.; Wagner, M.; Mayr, J.A.; Lee, C.-Y.; Yépez, V.A.; Lorenz, N.; et al. Bi-Allelic UQCRFS1 Variants Are Associated with Mitochondrial Complex III Deficiency, Cardiomyopathy, and Alopecia Totalis. *Am. J. Hum. Genet.* **2020**, *106*, 102–111. [CrossRef]
182. Hinson, J.T.; Fantin, V.R.; Schönberger, J.; Breivik, N.; Siem, G.; McDonough, B.; Sharma, P.; Keogh, I.; Godinho, R.; Santos, F.; et al. Missense Mutations in the BCS1LGene as a Cause of the Björnstad Syndrome. *N. Engl. J. Med.* **2007**, *356*, 809–819. [CrossRef]
183. De Meirleir, L.; Seneca, S.; Damis, E.; Sepulchre, B.; Hoorens, A.; Gerlo, E.; Silva, M.T.G.; Hernandez, E.M.; Lissens, W.; Van Coster, R. Clinical and diagnostic characteristics of complex III deficiency due to mutations in the BCS1Lgene. *Am. J. Med. Genet. Part A* **2003**, *121A*, 126–131. [CrossRef] [PubMed]
184. Blázquez, A.; Gil-Borlado, M.C.; Morán, M.; Verdú, A.; Cazorla-Calleja, M.R.; Martín, M.A.; Arenas, J.; Ugalde, C. Infantile mitochondrial encephalomyopathy with unusual phenotype caused by a novel BCS1L mutation in an isolated complex III-deficient patient. *Neuromuscul. Disord.* **2009**, *19*, 143–146. [CrossRef]
185. Visapää, I.; Fellman, V.; Vesa, J.; Dasvarma, A.; Hutton, J.L.; Kumar, V.; Payne, G.S.; Makarow, M.; Van Coster, R.; Taylor, R.W.; et al. GRACILE Syndrome, a Lethal Metabolic Disorder with Iron Overload, Is Caused by a Point Mutation in BCS1L. *Am. J. Hum. Genet.* **2002**, *71*, 863–876. [CrossRef] [PubMed]
186. Oláhová, M.; Berti, C.C.; Collier, J.J.; Alston, C.L.; Jameson, E.; Jones, S.A.; Edwards, N.; He, L.; Chinnery, P.F.; Horvath, R.; et al. Molecular genetic investigations identify new clinical phenotypes associated with BCS1L-related mitochondrial disease. *Hum. Mol. Genet.* **2019**, *28*, 3766–3776. [CrossRef] [PubMed]
187. Fernandez-Vizarra, E.; Bugiani, M.; Goffrini, P.; Carrara, F.; Farina, L.; Procopio, E.; Donati, A.; Uziel, G.; Ferrero, I.; Zeviani, M. Impaired complex III assembly associated with BCS1L gene mutations in isolated mitochondrial encephalopathy. *Hum. Mol. Genet.* **2007**, *16*, 1241–1252. [CrossRef]
188. De Lonlay, P.; Valnot, I.; Barrientos, A.; Gorbatyuk, M.; Tzagoloff, A.; Taanman, J.-W.; Benayoun, E.; Chrétien, D.; Kadhom, N.; Lombès, A.; et al. A mutant mitochondrial respiratory chain assembly protein causes complex III deficiency in patients with tubulopathy, encephalopathy and liver failure. *Nat. Genet.* **2001**, *29*, 57–60. [CrossRef]
189. Ezgu, F.; Senaca, S.; Gunduz, M.; Tumer, L.; Hasanoglu, A.; Tiras, U.; Unsal, R.; Bakkaloglu, S.A. Severe renal tubulopathy in a newborn due to BCS1L gene mutation: Effects of different treatment modalities on the clinical course. *Gene* **2013**, *528*, 364–366. [CrossRef]
190. Tuppen, H.A.; Fehmi, J.; Czermin, B.; Goffrini, P.; Meloni, F.; Ferrero, I.; He, L.; Blakely, E.L.; McFarland, R.; Horvath, R.; et al. Long-term survival of neonatal mitochondrial complex III deficiency associated with a novel BCS1L gene mutation. *Mol. Genet. Metab.* **2010**, *100*, 345–348. [CrossRef]
191. Al-Owain, M.; Colak, D.; AlBakheet, A.; Al-Younes, B.; Al-Humaidi, Z.; Al-Sayed, M.; Al-Hindi, H.; Al-Sugair, A.; Al-Muhaideb, A.; Rahbeeni, Z.; et al. Clinical and biochemical features associated with BCS1L mutation. *J. Inherit. Metab. Dis.* **2013**, *36*, 813–820. [CrossRef]
192. Baker, R.A.; Priestley, J.R.C.; Wilstermann, A.M.; Reese, K.J.; Mark, P.R. Clinical spectrum of BCS1L Mitopathies and their underlying structural relationships. *Am. J. Med. Genet. Part A* **2019**, *179*, 373–380. [CrossRef]

193. Gil-Borlado, M.C.; González-Hoyuela, M.; Blázquez, A.; García-Silva, M.T.; Gabaldón, T.; Manzanares, J.; Vara, J.; Martín, M.A.; Seneca, S.; Arenas, J.; et al. Pathogenic mutations in the 5′ untranslated region of BCS1L mRNA in mitochondrial complex III deficiency. *Mitochondrion* **2009**, *9*, 299–305. [CrossRef]
194. Kunii, M.; Doi, H.; Higashiyama, Y.; Kugimoto, C.; Ueda, N.; Hirata, J.; Tomita-Katsumoto, A.; Kashikura-Kojima, M.; Kubota, S.; Taniguchi, M.; et al. A Japanese case of cerebellar ataxia, spastic paraparesis and deep sensory impairment associated with a novel homozygous TTC19 mutation. *J. Hum. Genet.* **2015**, *60*, 187–191. [CrossRef] [PubMed]
195. Melchionda, L.; Damseh, N.S.; Abu Libdeh, B.Y.; Nasca, A.; Elpeleg, O.; Zanolini, A.; Ghezzi, D. A novel mutation in TTC19 associated with isolated complex III deficiency, cerebellar hypoplasia, and bilateral basal ganglia lesions. *Front. Genet.* **2014**, *5*, 397. [CrossRef]
196. Ghezzi, D.; Arzuffi, P.; Zordan, M.; Da Re, C.; Lamperti, C.; Benna, C.; D'Adamo, P.; Diodato, D.; Costa, R.; Mariotti, C.; et al. Mutations in TTC19 cause mitochondrial complex III deficiency and neurological impairment in humans and flies. *Nat. Genet.* **2011**, *43*, 259–263. [CrossRef]
197. Koch, J.; Freisinger, P.; Feichtinger, R.G.; Zimmermann, F.A.; Rauscher, C.; Wagentristl, H.P.; Konstantopoulou, V.; Seidl, R.; Haack, T.B.; Prokisch, H.; et al. Mutations in TTC19: Expanding the molecular, clinical and biochemical phenotype. *Orphanet J. Rare Dis.* **2015**, *10*, 40. [CrossRef]
198. Atwal, P.S.; Zschocke, J.; Gibson, K.M. Mutations in the Complex III Assembly Factor Tetratricopeptide 19 Gene TTC19 Are a Rare Cause of Leigh Syndrome. *JIMD Rep. Vol. 14* **2014**, *14*, 43–45. [CrossRef]
199. Habibzadeh, P.; Inaloo, S.; Silawi, M.; Dastsooz, H.; Fard, M.A.F.; Sadeghipour, F.; Faghihi, Z.; Rezaeian, M.; Yavarian, M.; Böhm, J.; et al. A Novel TTC19 Mutation in a Patient with Neurological, Psychological, and Gastrointestinal Impairment. *Front. Neurol.* **2019**, *10*, 944. [CrossRef]
200. Ardissone, A.; Granata, T.; Legati, A.; Diodato, D.; Melchionda, L.; Lamantea, E.; Garavaglia, B.; Ghezzi, D.; Moroni, I.; Zschocke, J. Mitochondrial Complex III Deficiency Caused by TTC19 Defects: Report of a Novel Mutation and Review of Literature. *JIMD Rep.* **2015**, *22*, 115–120. [CrossRef]
201. Morino, H.; Miyamoto, R.; Ohnishi, S.; Maruyama, H.; Kawakami, H. Exome sequencing reveals a novel TTC19 mutation in an autosomal recessive spinocerebellar ataxia patient. *BMC Neurol.* **2014**, *14*, 5. [CrossRef] [PubMed]
202. Mordaunt, D.A.; Jolley, A.; Balasubramaniam, S.; Thorburn, D.R.; Mountford, H.S.; Compton, A.G.; Nicholl, J.; Manton, N.; Clark, D.; Bratkovic, D.; et al. Phenotypic variation of TTC19-deficient mitochondrial complex III deficiency: A case report and literature review. *Am. J. Med. Genet. Part A* **2015**, *167*, 1330–1336. [CrossRef] [PubMed]
203. Bottani, E.; Cerutti, R.; Harbour, M.E.; Ravaglia, S.; Dogan, S.A.; Giordano, C.; Fearnley, I.M.; D'Amati, G.; Viscomi, C.; Fernandez-Vizarra, E.; et al. TTC19 Plays a Husbandry Role on UQCRFS1 Turnover in the Biogenesis of Mitochondrial Respiratory Complex III. *Mol. Cell* **2017**, *67*, 96–105.e4. [CrossRef] [PubMed]
204. Dallabona, C.; Abbink, T.E.M.; Carrozzo, R.; Torraco, A.; Legati, A.; Van Berkel, C.G.M.; Niceta, M.; Langella, T.; Verrigni, D.; Rizza, T.; et al. LYRM7 mutations cause a multifocal cavitating leukoencephalopathy with distinct MRI appearance. *Brain* **2016**, *139*, 782–794. [CrossRef]
205. Kremer, L.S.; L'Hermitte-Stead, C.; Lesimple, P.; Gilleron, M.; Filaut, S.; Jardel, C.; Haack, T.B.; Strom, T.M.; Meitinger, T.; Azzouz, H.; et al. Severe respiratory complex III defect prevents liver adaptation to prolonged fasting. *J. Hepatol.* **2016**, *65*, 377–385. [CrossRef]
206. Invernizzi, F.; Varanese, S.; Thomas, A.M.; Carrara, F.; Onofrj, M.; Zeviani, M. Two novel POLG1 mutations in a patient with progressive external ophthalmoplegia, levodopa-responsive pseudo-orthostatic tremor and parkinsonism. *Neuromuscul. Disord.* **2008**, *18*, 460–464. [CrossRef]
207. Hempel, M.; Kremer, L.S.; Tsiakas, K.; Alhaddad, B.; Haack, T.B.; Löbel, U.; Feichtinger, R.G.; Sperl, W.; Prokisch, H.; Mayr, J.A.; et al. LYRM7—Associated complex III deficiency: A clinical, molecular genetic, MR tomographic, and biochemical study. *Mitochondrion* **2017**, *37*, 55–61. [CrossRef]

 life

Review

Mitochondrial Dynamics, ROS, and Cell Signaling: A Blended Overview

Valentina Brillo †, Leonardo Chieregato †, Luigi Leanza *,‡, Silvia Muccioli ‡ and Roberto Costa ‡

Department of Biology, University of Padova, 35121 Padova, Italy; valentina.brillo@studenti.unipd.it (V.B.); leonardo.chieregato@studenti.unipd.it (L.C.); silvia.muccioli@studenti.unipd.it (S.M.); roberto.costa@unipd.it (R.C.)

* Correspondence: luigi.leanza@unipd.it; Tel.: +39-049-827-6343

† These two authors contributed equally.

‡ These authors share senior authorship.

Abstract: Mitochondria are key intracellular organelles involved not only in the metabolic state of the cell, but also in several cellular functions, such as proliferation, Calcium signaling, and lipid trafficking. Indeed, these organelles are characterized by continuous events of fission and fusion which contribute to the dynamic plasticity of their network, also strongly influenced by mitochondrial contacts with other subcellular organelles. Nevertheless, mitochondria release a major amount of reactive oxygen species (ROS) inside eukaryotic cells, which are reported to mediate a plethora of both physiological and pathological cellular functions, such as growth and proliferation, regulation of autophagy, apoptosis, and metastasis. Therefore, targeting mitochondrial ROS could be a promising strategy to overcome and hinder the development of diseases such as cancer, where malignant cells, possessing a higher amount of ROS with respect to healthy ones, could be specifically targeted by therapeutic treatments. In this review, we collected the ultimate findings on the blended interplay among mitochondrial shaping, mitochondrial ROS, and several signaling pathways, in order to contribute to the dissection of intracellular molecular mechanisms involved in the pathophysiology of eukaryotic cells, possibly improving future therapeutic approaches.

Keywords: mitochondrial dynamic; cell signaling; ROS; cancer

Citation: Brillo, V.; Chieregato, L.; Leanza, L.; Muccioli, S.; Costa, R. Mitochondrial Dynamics, ROS, and Cell Signaling: A Blended Overview. *Life* **2021**, *11*, 332. https://doi.org/10.3390/life11040332

Academic Editors: Giorgio Lenaz and Salvatore Nesci

Received: 2 March 2021
Accepted: 7 April 2021
Published: 10 April 2021

1. Introduction

Mitochondria are intracellular organelles present in eukaryotic cells that evolutionarily originated from symbiotic resident proteobacteria [1]. These organelles are involved in many cellular functions, such as oxidative phosphorylation, the regulation of cell proliferation, differentiation, and death. Their different roles in several cellular processes are largely dependent on ATP and reactive oxygen species (ROS) production, both generated during oxidative phosphorylation [2]. Indeed, targeting mitochondrial metabolism with molecules able to specifically disrupt mitochondrial fitness and trigger cell death has become a promising strategy against several diseases [3].

Importantly, mitochondria are physically interconnected with other subcellular organelles, such as endoplasmic reticulum (ER), lipid droplets, Golgi apparatus, lysosomes, melanosomes, and peroxisomes [4]. Indeed, mitochondria–organelle contact sites represent real signaling hubs that are involved in multiple cellular functions, such as lipid trafficking, mitochondrial dynamics, calcium (Ca^{2+}) flow, and ER stress, such that the contacts not only result in physical but also functional links that finely tune multiple signaling pathways.

Moreover, the capability to establish these interactions with other intracellular organelles is strongly dependent on mitochondria's high attitude to fuse and divide, leading to modification of the intracellular mitochondrial network [5].

In addition, ROS figure as byproducts of oxygen consumption and cellular metabolism, and 45% of their total amount is related to mitochondria, specifically to Complex I and

Complex III leakage of electrons, which is involved in superoxide (O_2^-) and hydrogen peroxide (H_2O_2) production [6]. While ROS production was at first believed to be only detrimental for the cell in physiological conditions, in the last two decades, it has been considered that its presence in a sublethal concentration could act as a secondary messenger that specifically modulates distinct cellular pathways [7] and mitochondrial dynamics and morphology [8]. As a consequence, ROS homeostasis is strictly regulated by enzymatic and nonenzymatic mechanisms, with the aim of maintaining balance among ROS production and scavenging [9]. Once this critical equilibrium is impaired, ROS overload is one of the main players in the onset of a plethora of different diseases, including cancer [6], where it exerts a dual regulation, influencing cell survival and oxidative stress, leading to cell death, as well as mediating redox signaling events beneficial for the progression of the disease [9].

In this review, we describe the known signaling pathways mediated by mitochondrial structure rearrangements or by mitochondrial ROS release, focusing also on possible therapeutic targets against disease formation.

2. Mitochondrial Dynamics: A Multiplayer Regulation

Proper mitochondrial integrity and physiology is essential for cell homeostasis. Mitochondrial fusion and fission dynamics, organelle transport, mitophagy, interaction with other organelles, such as the endoplasmic reticulum (ER) and the cytoskeleton, and genomic mitochondrial control are only a few of the several mechanisms involved in the fitness of these fundamental organelles. Thus, the improper surveillance on mitochondrial dynamics and partitioning on daughter cells can give rise to a wide spectrum of syndromes and diseases [5].

Focusing on the fusion and fission processes, these are mainly regulated by proteins belonging to the dynamin-related family of large GTPases that utilize GTP hydrolysis to drive mechanical work on biological membranes (Figure 1) [10].

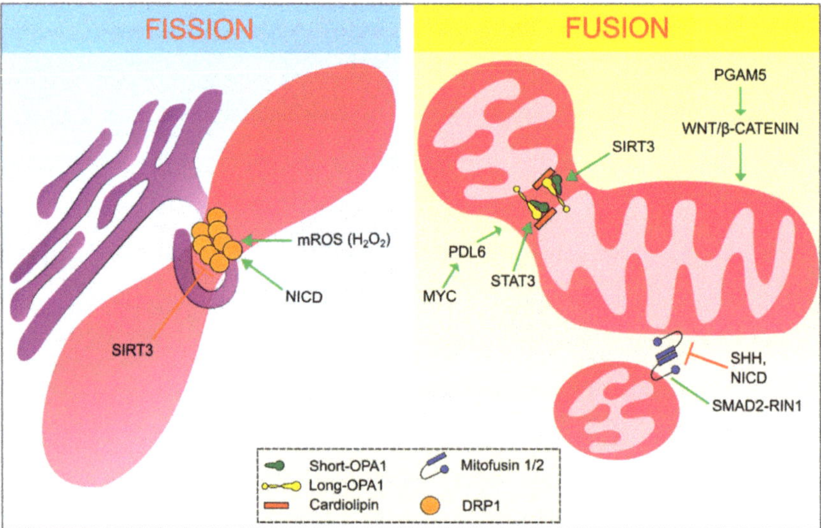

Figure 1. Modulation of fusion and fission processes. Left panel: the endoplasmic reticulum is wrapping the mitochondria in the site of fission, where polymers of dynamin-related protein 1 (DRP1) (main interactor in the fission process) are present. Right panel: the two different events of outer mitochondrial membrane (OMM) fusion and inner mitochondrial membrane (IMM) fusion are separately shown. Essentials components for OMM fusion are Mitofusins. In the IMM fusion process, instead, the role of long and short optic atrophy protein 1 (OPA1) is fundamental, as well as their interaction with cardiolipins. Green arrows point out the positive regulators of these processes, whereas the red ones represent the negative modulators.

The membrane potential is also a crucial player in the mitochondrial fission process, by triggering dynamin-related protein 1 (DRP1) activity [11,12]. DRP1 translocates from the cytoplasm to the outer mitochondrial membrane (OMM) via a physical interaction with several adapter proteins, such as mitochondrial fission factor (MFF), mitochondrial dynamics protein 49 and 51 (MID49, MID51), and mitochondrial fission 1 protein (Fis1) [13]. Specifically, DRP1 firstly binds GTP through its GTP-binding site. This mediates a conformational change that allows DRP1 to interact with the OMM receptors and polymerize, encircling the mitochondria in a spiral fashion [14]. Once a complete turn is performed, contacts between the GTPase domains of the DRP1 polymers trigger the GTP hydrolysis causing the detachment of filaments from the OMM receptors and the constriction of the spiral [15]. Nevertheless, it seems that DRP1 itself is not sufficient to completely perform the fission process. In fact, it has been shown that Dynamin-2 is a fundamental component that works in concert with DRP1 in order to orchestrate sequential constriction events that ultimately lead to division [16]. Some authors suggested that the fission process occurs in sites that have been previously wrapped by a smooth ER protrusion. These ER–mitochondria contact sites act as the fission starting point, so that DRP1 can cooperate with the actin-nucleating protein inverted formin 2 (INF2) causing the accumulation of actin in the site of fission. Actin filament accumulation can ultimately facilitate the formation of the initial constriction, supporting the subsequent DRP1 activity [13,17].

In particular, ER–mitochondria contact sites, also known as mitochondria–ER contacts (MERCs), are relevant for mitochondrial fitness and plasticity regulation. Therefore, the characterization of the proteins involved in MERCs revealed the presence of several players that allow the connection between the two organelles. In fact, the existence of dynamic bridges that consist of proteins inserted in the OMM, such as voltage-dependent anion channel (VDAC), physically connected to the ER membrane proteins (such as inositol 1,4,5-triphosphate receptor (IP3R)), by linker proteins (e.g., glucose-regulated protein (GRP75) and transglutaminase type 2 (TG2)), contribute to the modulation of many mitochondrial events such as lipid trafficking, Ca^{2+} homeostasis, and ER stress [16]. Moreover, the functional role of such contacts is highlighted by their involvement in several pathologies, such as diabetes, neurodegenerative diseases, and cancer [18]. Interestingly, very recently, it was demonstrated that downregulation of transglutaminase type 2, which is involved in ER–mitochondria contacts [19], is linked to a decrease in canonical Wnt signaling targets, such as β-catenin and Lymphoid enhancer binding factor 1 (LEF1), suggesting new possible ways of modulating Wnt-dependent proliferation, strongly associated with diseases development [20].

Since mitochondria are characterized by a double membrane, the fusion process requires the coordination of two separate events which occur almost simultaneously. Indeed, mitochondrial fusion is a mechanism mediated by MFN1 and MFN2 at the level of the OMM and by optic atrophy protein 1 (OPA1) in the inner mitochondrial membrane (IMM) [10,17]. While MFN1 and MFN2 must both be present in the OMM in order to mediate outer membrane fusion, it is enough for OPA1 to be present in only one of the IMMs to mediate inner membrane fusion [10]. Fusion onset is established by the docking of two Mitofusin molecules. This interaction mediates conformational changes that trigger Mitofusin-mediated GTP hydrolysis and the subsequent OMM fusion [17].

Concerning IMM fusion, the *opa1* gene encodes eight different long isoforms [21], and all of them are firstly inserted into the IMM thanks to the presence of the mitochondrial targeting sequence (MTS), which is subsequently cleaved by the matrix processing protease (MPP). Once inserted into the IMM, OPA1 isoforms can be processed by two inner membrane peptidases: zinc metallopeptidase (OMA1) and ATP-dependent zinc metalloprotease (YME1L1), thus forming the essential short form of OPA1. Up to now, it seems that both the isoforms are required in order to guarantee the correct and physiological levels of mitochondrial fusion [22]. Then, the heterodimer of one IMM, formed by long-OPA1 and short-OPA1, interacts with cardiolipin of the other IMM and mediates IMM fusion [17]. It is important also to remember that OPA1 is essential in cristae structure maintenance [10].

3. Mitochondrial Plasticity and Cell Signaling: A Two-Way Interaction

In recent years, numerous studies pointed out links between key oncogenic signaling pathways and mitochondria [13]. These networks among the two players have been deeply explored, revealing not only that mitochondrial plasticity could be influenced by distinct cellular pathways, but also that mitochondrial shaping could be crucial for the modulation of several signaling cascades.

Indeed, Malhotra et al. explored the role of Sonic Hedgehog (Shh) signaling in mitochondrial biogenesis regulation in cerebellar granule neuron precursors (CGNPs), the progenitor of Shh-associated medulloblastoma. Surprisingly, the authors observed a decreased mitochondrial membrane potential (MMP) and overall ATP production in CGNP cells upon Shh induction, along with an increase in glycolysis levels, which resulted in higher intracellular acidity leading to mitochondrial fragmentation and reduced cristae network formation. These results are quite controversial since decreased MMP is usually linked to a reduction in proliferation and apoptotic induction. Inversely, these cells are characterized by a rise in the proliferation rate and absence of cell differentiation. This effect seems to be derived from the Shh-mediated lowering of MFN1 and MFN2 activity. Really interesting is that the phenotype can be rescued by the inhibition of the Shh pathway, as well as through the downregulation of DRP1, remarking the importance of the delicate balance between fission and fusion mechanisms involved in mitochondrial biogenesis [23].

Canonical Wnt-signaling is involved in a plethora of different cellular functions, such as neuroprotection, stemness maintenance, self-renewal, and regulation of mitochondrial dynamic inside eukaryotic cells [24–26]. Strikingly, recent studies demonstrated a novel mechanism via which damaged mitochondria promote restoration of the mitochondrial pool through the activation of canonical Wnt-signaling via the Pgam5/β-catenin axis. Presenilins-associated rhomboid-like (PARL)-mediated cleavage of the mitochondrial Serine/threonine-protein phosphatase Pgam5 occurs in stressed mitochondria characterized by a decreased membrane potential. This cleaved and cytosolic form seems to be able to interact with Axin at the level of the β-catenin destruction complex. Such interaction counters the casein kinase 1 α (CK1α)- and glycogen synthase 3 (GSK3)-mediated phosphorylation of β-catenin, thus avoiding its degradation and leading to an increased transcriptional activity performed through the activation of the canonical Wnt/β-catenin axis. This cell-intrinsic stimulation of the Wnt/β-catenin cascade can trigger the mitophagy process, in order to degrade or recycle the old and dysfunctional mitochondria, thus supporting the process of mitochondrial biogenesis [27].

Several other pathways are involved in the regulation of mitochondrial dynamics. For instance, it has been shown that the activity of mothers against decapentaplegic homolog (Smad) proteins—mediators of the transforming growth factor β (TGF-β) signaling—are able to modulate mitochondrial fusion when present in their inactive and cytoplasmatic form. In fact, SMAD2 can promote mitochondrial fusion through its interaction with MFN2 at the level of the OMM. In particular, a model has been established in which SMAD2 colocalizes with Ras and Rab interactor 1 (RIN1) and MFN2 at the level of the OMM and stimulates the process of fusion. Problems or mutations at the level of any of these characters can interfere with the modulation of mitochondrial dynamics, thereby favoring the development of different kinds of diseases, such as carcinogenesis and metabolic issues [28].

Indeed, increased mitochondrial fission is often associated with tumor formation, e.g., lung cancers and breast cancers. In particular, the presence of a mitochondrial fission and Notch signaling positive feedback loop has been elucidated in triple-negative breast cancers (TNBCs). In fact, it seems that mitochondrial fragmentation is linked to an increased cytoplasmic Ca^{2+} level, causing the subsequent activation of calcineurin. Therefore, calcineurin activates Notch signaling, increasing the level of the cleaved and active Notch intracellular domain (NICD) inside the nucleus. In turn, Notch signaling promotes the upregulation of DRP1 and the downregulation of MFN1, thus establishing a vicious cycle. Moreover, this positive feedback loop enhances the apoptotic resistance and survival of tumor cells through the Notch-mediated upregulation of the inhibitor of apoptosis (IAP) protein Survivin [29].

Moreover, some types of human breast cancers are characterized by dysregulated Myc signaling. Overexpression of Myc leads, among all the other targets, to the overexpression of Phospholipase D Family Member 6 (PLD6)—a phospholipase of the OMM involved in the regulation of lipid metabolism, which is able to mediate mitochondrial fusion in order to improve lipid metabolism, but which also cooperates with the increased nucleotide demand during DNA synthesis. This Myc-mediated metabolic reprograming, in part caused by the overstimulated mitochondrial fusion derived by PLD6 activity, strains cellular energy resources and leads to 5′ adenosine monophosphate-activated protein kinase (AMPK) activation. AMPK is also able to phosphorylate and inhibit yes-associated protein (YAP), and YAP inactivation is characteristic of some types of Myc-dependent triple-negative mammary carcinomas. Another effect mediated by PLD6-dependent mitochondrial fusion is also the increase in the levels of glutaminolysis, an essential process for tumor survival since MYC-driven cell growth depends on glutamine [30].

Additionally, mitochondrial fission is also correlated to other diseases, such as unilateral unilateral obstruction (UUO)-induced renal tubulointerstitial fibrosis. Indeed, it has been shown that Honokiol (2-(4-hydroxy-3-prop-2-enyl-phenyl)-4-prop-2-enyl-phenol, HKL) is able to stimulate the activity of Sirtuin 3 (SIRT3), which sequentially mediates the activation of OPA1 and decreases DRP1 expression, restoring the correct mitochondrial fusion and fission dynamics and normal mitochondrial shape and function. Thus, targeting mitochondrial dynamics can be a novel therapeutic approach for the treatment of acute or chronic kidney diseases [31].

Recently, the importance of signal transducer and activator of transcription 3 (STAT3) was also elucidated in the regulation of mitochondrial dynamics. Indeed, Zhang et al. demonstrated in diabetic mice and in albumin-treated proximal tubular HK-2 cells how anomalies resulting from diabetes, e.g., hyperglycemia and ROS, can mediate the overexpression or overactivation of the dipeptidyl peptidase-4 (DPP4) enzyme, leading to DPP4-mediated cleavage of stromal cell-derived factor-1α (SDF-1α) and suppression of the SDF-1α/C-X-C Motif Chemokine Receptor 4 (CXCR4) phosphorylation of STAT3 at the level of serine-727. Thus, this impedes STAT3 translocation into the mitochondria and its interaction with OPA1, ultimately leading to increased mitochondrial fragmentation. This result highlights novel targets for managing diabetic kidney disease [32].

The connection between STAT3 and mitochondrial fusion protein OPA1 has also been described in a myocardial ischemia and reperfusion mouse model. In this study, it was demonstrated how κ-opioid receptor (κ-OR) activation mediates mitochondrial fusion through enhanced OPA1 expression. In particular, this suggests that κ-OR activation can stimulate STAT3 phosphorylation at the level of tyrosine-705, allowing its nuclear translocation where it can mediate OPA1 overexpression. This result allows novel insight into therapeutic strategies for myocardial ischemia and reperfusion injury [33].

Thus, STAT3 induction of mitochondrial fusion through the modulation of OPA1 seems to be quite clear. Nevertheless, a more in-depth investigation is still needed into the effective mechanism.

4. Mitochondrial ROS in the Modulation of Cell Signaling

ROS are small molecules that figure as byproducts of oxygen consumption and cellular metabolism, which derive from the partial reduction of molecular oxygen. The most known molecules among the ROS family are the highly unstable oxygen free radicals, superoxide (O_2^-) and hydroxyl (OH^-), which can be converted into more stable non-radical and diffusible forms, e.g., hydrogen peroxide (H_2O_2) or hypochlorous acid [6,34].

As it is well known, mitochondria represent one of the major contributors to ROS generation. In fact, it was recently demonstrated that, in resting C2C12 myoblasts, mitochondria account for the 45% of the total amount of reactive oxygen species produced inside the cell [35], and that up to 1% of the mitochondrial oxygen is utilized for superoxide production [36]. In addition, 11 distinct sites associated with substrate oxidation in the electron transport chain (ETC) in mammalian mitochondria resulted in the release of electrons involved in the production of superoxide (O_2^-) and hydrogen peroxide (H_2O_2). In particular, Complex I and Complex III are the main sources of ROS both in healthy and in pathological conditions, which are required for a plethora of biological processes such as cell differentiation and proliferation, oxygen cell sensing, and Hypoxia-inducible factor (HIF) activation [7,37–39].

Inside the mitochondria, mitochondrial ROS are mainly produced by Nicotinamide adenine dinucleotide phosphate (NADPH) oxidases (NOX) and, to a minor extent, by other enzymes such as cyclooxygenases (COX), lipoxygenases, xanthine oxidases, and cytochrome P450 enzymes [34,40–43]. Moreover, the electron transport chain is intrinsically leaky; indeed, even in physiological conditions, 0.2–2% of the electrons generated by the respiratory chain are not coupled with the production of ATP but contribute to the generation of superoxide anion (O_2^-) or hydrogen peroxide (H_2O_2) due to their premature interaction with oxygen. Thus, a minor percentage of ROS is physiologically released during all the respiratory processes, playing a crucial role in mitochondria and cell fate [44].

Actually, ROS generation is involved in the regulation and induction of both physiological and pathological cellular pathways. For a long time they were considered with a negative connotation in physiological conditions, being responsible for the induction of oxidative stress and consequent apoptosis and necrosis, ultimately resulting in alterations in cell survival rate [45]. Indeed, mitochondrial dysfunction in the ETC is strongly linked to an unregulated release of mitochondrial ROS, which causes both DNA and macromolecular oxidative damage, leading to the development of degenerative pathologies and biological aging [46]. For this reason, ROS homeostasis must be strictly regulated by enzymatic and non-enzymatic mechanisms, such as superoxide dismutases (SODs), catalases, glutathione peroxidases (GPX), peroxiredoxins (PRX), thioredoxins (TRX), and vitamins A, C, and E (for extended reviews, see [9,47–49]).

Accordingly, ROS are well recognized mediators of DNA damage, affecting the DNA damage response (DDR), and their accumulation can also induce mitochondrial DNA lesions, strand breaks, and degradation of mitochondrial DNA [50]. Specifically, ROS-induced DNA damage and the consequent inability to evoke the DNA repair system are responsible for Cellular tumor antigen p53 (p53) activation and mitochondrial-mediated apoptosis, a pathway that is elicited by different anticancer drugs, leading to the suggestion that ROS modulators could be promising for cancer targeting [51,52].

Interestingly, in the latest two decades, a new role for mitochondrial ROS emerged. In fact, sublethal concentrations of ROS could act as potential secondary messengers which could be used to specifically modulate distinct cellular pathways, introducing new possible therapeutic approaches [7,53]. Specifically, via the reversible oxidation of specific cysteine (and also methionine) residues within redox-sensitive proteins, ROS can modify a putative target protein activity or conformation, altering the signal transduction. In this regard, ROS can act on phosphatases, kinases, proteases, and transcription factors [54], regulating growth factor cascades, cell proliferation and differentiation, cellular oxygen sensing, and hypoxia (and the consequent angiogenic stimulation), while also finely tuning aging-related mechanisms, immunity responses, inflammation, and autophagy [9,44,45,53,55–57].

Moreover, a role for ROS in the modulation of mitochondrial dynamics was recently elucidated, suggesting a link between the redox homeostasis of the cell and the regulation of mitochondrial morphology [8]. In particular, high levels of ROS, if not counterbalanced by an efficient antioxidant system, promote mitochondrial fragmentation, swelling, or shortening, whereas a reduction in ROS leads to mitochondrial filamentation. In fact, exogenous concentrations of hydrogen peroxide induced dose-dependent mitochondrial fragmentation in human umbilical vein endothelial cells (HUVECs) and the expression of several fusion and fission-related genes [58]; in C2C12 myocytes, hydrogen peroxide induced mitochondrial membrane depolarization, stimulating fragmentation involving an increased DRP1 activity [59,60]. On the contrary, lowering ROS levels in fibroblasts triggered MFN2-dependent mitochondrial filamentation [61]. The redox regulation of fission and fusion proteins by ROS is mediated by post-translational modifications, such as phosphorylation, ubiquitination, and sumoylation, in addition to the *S*-glutathionylation and *S*-nitrosylation of their Cys residues [62]. Moreover, ROS also act at the transcriptional level, stimulating the expression of factors that are involved in both redox regulation and mitochondrial dynamics; an example is the peroxisome proliferator-activated receptor gamma coactivator (PGC1α/β), which is redox-sensitive and associated with MFN2 regulation [63]. Another important role in the link among mitochondrial dynamics and ROS is played by AMPK; once activated, it phosphorylates MFF and DRP1, necessary for mitochondrial fission [62].

5. Mitochondrial ROS Involvement in Cancer

Interestingly, when an imbalance among the production and the scavenging of ROS species occurs, impaired ROS homeostasis results in the onset and the progression of various pathologies, including neurodegenerative diseases [64], diabetes [65,66], cardiovascular diseases [67], and cancer [6,68]. More specifically, it is clear that mitochondrial ROS can act in a dual mode during the progression of these pathologies; as oxidants, at elevated concentrations, they influence cell survival and oxidative stress, ultimately leading to cell death, whereas, at lower concentrations, they act as signaling molecules which mediate redox signaling events beneficial for the progression of the disease [9,37].

In addition, cancer cells are characterized by increased ROS levels with respect to normal cells; this is due to their abnormal metabolism, which exploits normal cell machinery in a constitutive way in order to maximize cellular growth and proliferation, to enhance aerobic glycolysis (the so-called "Warburg effect") [9], and to promote altered expression of pro-tumorigenic networks (as for example, Kras and Myc overexpression [69,70]), as well as the inhibition of tumor suppressors [71]. Moreover, the accumulation of mutations in mitochondrial DNA (mtDNA), increased tumor-derived hypoxia, and mitochondrial shape changes, along with alterations in the antioxidant system and in cellular signaling pathways, all contribute to the increased ROS level in neoplastic cells [45].

High ROS levels have been demonstrated to be causative of a cascade of multiple events in cancer—perpetuating the tumorigenic transformation—including DNA damage [50], genetic instability, enhanced cell proliferation, cellular injury, cell death, and resistance to drugs [34]. Moreover, ROS species work as signaling intermediates in several pathways that are physiologically used by healthy cells in order to sustain both proliferation and cellular growth [71]. Crucial pathways that are activated by ROS accumulation are the mitogen activated-protein kinase (MAPK)/extracellular-regulated kinase 1/2 (ERK1/2) and phosphatidyl inositol 3 kinase (PI3K) signaling cascades that are mainly responsible for cell proliferation, growth, and survival. Indeed, ROS have been found to be involved in the inhibition of the phosphatase and tensin homolog (PTEN) via cysteine oxidation, thereby promoting Akt activity and positively regulating the PI3K pathway, which, in turn, results in higher proliferation rates [72].

Moreover, it was recently discovered that high concentrations of mitochondrial ROS in cancer stem cells (CSCs) promote cancer metastasis, via fatty acid β-oxidation, involving the activation of PI3K/AKT and ERK signaling, leading to epithelial-to-mesenchymal transition (EMT) [73]. In addition, Wang et al. demonstrated that, in colorectal cancer, elevated cholesterol levels increased ROS production, which, in turn, activated the MAPK signaling pathway, stimulating tumor progression (Figure 2) [74].

Even STAT3, which is activated in a plethora of cancers and controls the expression of multiple genes involved in tumor initiation, progression, and chemoresistance, has been proven to be regulated by mitochondrial ROS production. Normally, STAT3, in its inactive form, is present as a monomer in the cytoplasm, whereas, once activated by Janus activated kinases (JAKs), proto-oncogene tyrosine-protein kinase Src (Src), and MAP kinases, on its tyrosine-705 (Y705) and serine-727 (S727), it dimerizes, migrates into the nucleus, and regulates the transcription of several proliferative and antiapoptotic genes such as cyclins,

Bcl-2, Bcl-xl, and Survivin [75,76]. Moreover, a distinct pool of STAT3 resides in the mitochondria and is responsible for the control of ETC, modulation of ROS production, Ras transformation, cellular growth, and protection from ischemia/reperfusion injuries through the regulation of the mitochondrial permeability transition pore (MPTP) [77]. Very recently, Lee et al. elucidated a role in hepatoma cell invasiveness of ROS-induced STAT3 activation, which in turn promoted Nrf2 transcription and syntaxin 12 expression [78].

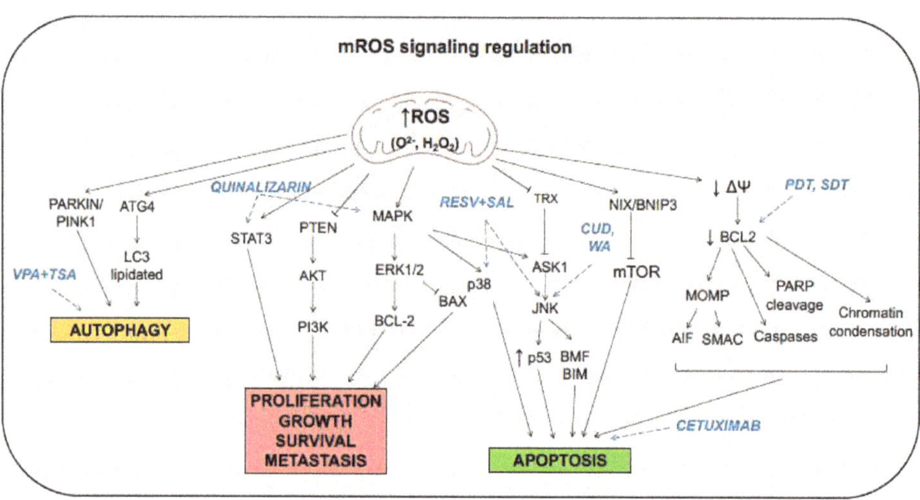

Figure 2. Mitochondrial reactive oxygen species (ROS) regulation of cellular signaling pathways. Many convergent signaling pathways that contribute to autophagy, proliferation, metastasis, and apoptosis are deeply modulated by an increase in mitochondrial ROS. In blue are depicted several drugs discussed in the text and reported in Table 1, which have been demonstrated to target key mediators of the pathways involved in ROS signaling.

Dysregulated mitochondrial dynamics have been reported in various diseases including cancer, and they can contribute to development, progression, and chemoresistance of tumors. Recent studies demonstrated that higher levels of ROS induce a DRP1-mediated mitochondrial fission in metastatic cancer and tumor-initiating cells, increasing migration and chemoresistance [79]. As an example, hypoxia-induced ROS in ovarian cancer cells are responsible for an increase in mitochondrial fission rate through the activation of DRP1 and downregulation of MFN1, leading to cisplatin resistance [80].

Generally, high levels of ROS production are counterbalanced by enhanced levels of antioxidant and scavenging activity, carefully maintaining a redox balance in order to avoid reaching a toxic amount of ROS which would lead to programmed cell death by apoptosis or necrosis. The most important way in which tumor cells potentiate their antioxidant system is through the activation of the nuclear factor erythroid 2-related factor 2 (Nrf2) [81]. Normally, this protein interacts with Kelch-like Enoyl CoA hydratase (ECH)-associated protein 1 (KEAP1) and ubiquitine ligase cullin3 (CUL3) and is targeted for proteasomal degradation. In elevated ROS condition, the oxidation of several cysteine residues in KEAP1 releases Nrf2, which translocates into the nucleus, associates with the MAF proteins, and binds to antioxidant-responsive elements (AREs) within the regulatory regions of several antioxidant genes [71], including those encoding Glutathione (GSH) S-transferase (GST), heme oxygenase 1 [53], and HIF1α (Figure 3) [82].

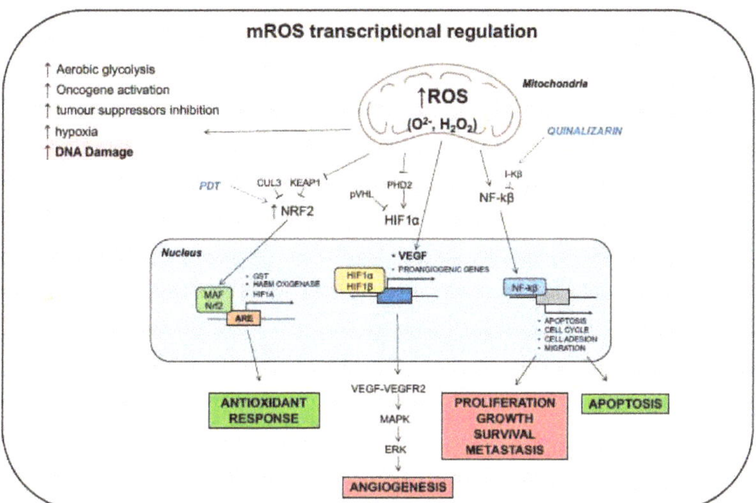

Figure 3. Mitochondrial ROS regulation of cellular processes at a transcriptional level. Antioxidant response, angiogenesis, proliferation, metastasis, and apoptosis are strictly regulated events by an increase in ROS production in the mitochondria (mROS). Indeed, mitochondrial ROS increase promotes the translocation into the nucleus of important factors that possess transcriptional activity, leading to the synthesis of genes related to these main events. In blue, several drugs that target intermediates of different signaling cascades are shown, as reported in Table 1.

Along with cell proliferation, other ROS-dependent signaling pathways are important for the adaptation of tumor cells to hypoxia-induced metabolic stress. Generally, in non-hypoxic conditions, hypoxia-inducible factors (HIFs) form heterodimers made up of two subunits: HIF1α and HIF1β. The oxygen-sensitive HIF1α is then hydroxylated by prolyl hydroxylases (PHDs) and targeted to proteasomal degradation due to its ubiquitylation by von Hippel–Lindau protein [83]. Instead, hypoxia stabilizes the HIFs, and the larger production of ROS inhibits PHD2 [84], thereby stabilizing HIF1α that, in turn, translocates into the nucleus, dimerizes with HIF1β, and regulates the expression of proangiogenic genes, including vascular endothelial growth factor (VEGF) [54]. Eventually, ROS are also able to directly enhance VEGF production at a transcriptional level. Finally, once bound to its receptor VEGFR2, VEGF promotes the proangiogenic signaling cascade, leading to activation of the ERK/MAPK pathway (Figure 3) [54,85].

An excessive level of ROS could give rise to apoptotic and autophagic responses, through the interaction with fundamental signaling molecules. Indeed, either extrinsic or intrinsic apoptosis has been demonstrated to be activated by mitochondrial ROS. For instance, ROS oxidation of thioredoxin (Trx) mediates the separation of Thx from Apoptosis signal-regulating kinase 1 (ASK1), a mitogen activated protein (MAP) kinase kinase kinase (MAPKKK) that upstream regulates c-Jun n-terminal kinase (JNK) pathways. ASK1 homo-oligomerization and activation by autophosphorylation phosphorylates JNK that, in turn, phosphorylates Bim and Bmf proteins, further activating Bcl-2-associated death promoter Bax and Bak, to initiate apoptosis. Moreover, JNK can increase p53 expression inducing apoptosis [68]. Additionally, other signaling cascades have been demonstrated to drive apoptosis through higher ROS levels, such as the mitogen-activated protein kinases (MAPKs), the signal transducer and activator of transcription-3 (STAT3), and the nuclear factor κB (NF-κB) signaling pathways. MAPK signaling includes extracellular-signal-regulated kinase (ERK), JNK, and p38, which regulate not only proliferation but also a variety of other cellular behaviors [86]. In fact, JNK (as previously reported) and p38 are considered mediators of apoptosis and are activated through phosphorylation by MAPK in response to several stress signals, including ROS. On the contrary, ERK, which is activated by growth factors, is considered pro-survival and oncogenic, and it antagonizes apoptosis by phosphorylating proapoptotic Bax and antiapoptotic Bcl-2 proteins, inhibiting and promoting their functions, respectively (Figure 2) [87]. Activated ROS production also plays a role in JAK2/STAT3 signaling suppression and subsequent apoptosis induction; for example, Cao et al. demonstrated that CYT997, a novel synthetic microtubule-disrupting agent, through the upregulation of mitochondrial ROS, triggers protective autophagy and inhibits the JAK2/STAT3 pathway, inducing gap 2 (G2)/mitosis (M) arrest and apoptosis in gastric cancer cells [88].

Furthermore, the activity of NF-κB, which is part of a family of signal-responsive transcription factors, has been shown to be modulated by ROS levels. In fact, in the classical pathway, NF-κB can be maintained inactive within the cytoplasm through interactions and binding to inhibitor of κB (i-κB) in normal cells, whereas it is constitutively activated in cancer cells; the phosphorylation of i-κB protein results in it being targeted by protease, releasing NF-κB that is translocated to the nucleus where it acts as transcription factor, leading to the expression of genes related to apoptosis, cell cycle control, adhesion, and migration [89]. All these processes are strictly related to tumor progression [90]. Chen et al. recently discovered that deferoxamine (DFO), an iron chelator and anticancer drug, was able to increase mitochondrial ROS and, in turn, elicit NF-κB and TGF-β pathways, promoting migration of a TNBC cell line MDA-MB-231 (Figure 2) [91].

Lastly, a complex interconnection among ROS and autophagy is present in cancer cells. Autophagy stands for the regulated self-degradative process in mammalian cells where unnecessary or dysfunctional cytoplasmic organelles are degraded in the lysosomes. This process has been demonstrated to be elicited by several anticancer drugs [92]. Autophagy driven by mitochondrial ROS possesses a double role; the first is to decrease the intracellular ROS level, mediating the mitophagy (degradation of damaged mitochondria) that contributes to oxidative stress. Mitophagy is achieved through the NIX/ B-Cell lymphoma 2 (BCL2) and adenovirus E1B 19kDa-interacting protein 3 like (BNIP3L) and ubiquitin-protein ligase PARKIN/PTEN induced putative kinase 1 (PINK1) molecular pathways [93,94]. On the other hand, elevated ROS levels contribute to defective autophagy in cancer cells, leading to autophagic cell death [95]. As an example, hydrogen peroxide, through the activation of BNIP3, inhibits mammalian target of rapamycin (mTOR) activity and induces autophagy in C6 glioma cells after sanguinarine treatment [96]. Moreover, under starvation conditions, autophagy related 4 (ATG4) protease becomes a target of mitochondrial produced hydrogen peroxide that oxidates its cysteine residue, mediating its inactivation, and promoting the lipidation of LC-3, starting the autophagosome formation process (Figure 2) [97] (for extended reviews on ROS control of autophagy, see [98,99]).

Table 1. Novel pharmaceutical treatments based on mitochondrial ROS exploitation which proved to be effective in cancer management.

Pharmacological Treatments	Cancer Types	Cell Lines	Mechanism of Action	Reference
Resveratrol + salinomycin	Breast cancer	MCF-7	↑ ROS impairs mitochondrial membrane potential; decreased Bcl2 expression, activation of caspases 7,8,9, chromatin condensation, PARP cleavage, apoptosis	[100]
Resveratrol + salinomycin	Breast cancer	MCF-7	↑ ROS activates MAPK pathway, phosphorylates JNK and p38, leading to apoptosis	[100]
Withaferin A *(WA)*	Colorectal cancer	HCT-116, RKO	↑ ROS reduces mitochondrial membrane potential, decreasing Bcl-2/Bax ratio, activating caspase 3–9, leading to apoptosis, and activating JNK pathway	[101]
Carnosic Acid *(CA)*	Colon cancer	HCT-116	↑ ROS diminishes STAT3 phosphorylation, decreasing STAT3 gene products	[76]
Quinalizarin	Breast cancer	MCF-7	↑ ROS affects MAPK, STAT3, and NF-κB signaling pathways, inducing cell-cycle arrest and apoptosis	[102]

Table 1. *Cont.*

Pharmacological Treatments	Cancer Types	Cell Lines	Mechanism of Action	Reference
Quinalizarin	Lung cancer	A549	↑ ROS affects MAPK, STAT3, and NF-κB signaling pathways, inducing cell-cycle arrest and apoptosis	[103]
Cucurbitacin *(CuD)*	Pancreatic cancer	Capan-1	↑ ROS induces G2/M cell-cycle arrest and mediates p38/MAPK pathway, promoting cell death	[104]
Imiquimod *(IMQ)*	Skin cancer	BCC/KMC-1, B16F10 and A375	↑ ROS causes mitochondrial membrane potential loss, mitochondrial fission, and mitophagy	[105]
Isorhamnetin *(IH)* + chloroquine *(CQ)*	Breast cancer	MDA-MB-231, MCF-7, BT549, MCF-10A	ROS-mediated phosphorylation of CaMKII/Drp1 promotes Bax translocation and release of cytochrome c, mitochondrial fission, caspase activation, and apoptosis	[106]
Cetuximab + oridonin	Laryngeal cancer	Hep-2, Tu212	↑ ROS, through NF-κB, PI3K/Akt, and JAK2/STAT3, induces apoptosis	[107]
Valproic acid *(VPA)* + Trichostatin A *(TSA)*	Pancreatic cancer	PANC1, PaCa44	↑ ROS triggers autophagy	[108]

Legend: ↑: increase; ROS: reactive oxygen species; Bcl-2: B-cell lymphoma 2; PARP: poly adenosine phosphate-ribose polymerase; MAPK: mitogen activated protein kinase; JNK: c-Jun N-terminal kinase; Bcl-2/Bax; B-cell lymphoma 2/Bcl-2-associated X protein; STAT3: signal transducer and activator of transcription; NF-κB: nuclear factor kappa-light-chain-enhancer of activated B cells; CaMKII/Drp1: Ca^{2+}/calmodulin-dependent protein kinase II/Dynamin-1-like protein; PI3K/Akt: Phosphatidylinositol 3 Kinase/Protein Kinase B; JAK2: Janus kinase 2.

6. Hints for Anticancer Therapy: Exploitation of Mitochondrial ROS

As clearly stated in the previous paragraphs, tumor cells generate and maintain high levels of ROS to preserve pro-tumorigenic signaling cascades, granting proliferation, growth, and metabolic adaptation. However, their level must be tightly regulated by the antioxidant system of the cell, in order to not exceed the toxic threshold ROS level, preventing cell death due to oxidative stress. This duality represents the specific challenge in the effort to find an effective ROS therapy in cancer.

Indeed, manipulating ROS in the context of cancer treatment is a promising approach recently developed, either by decreasing or by increasing their levels in cancer cells. The first approach relies on trying to decrease ROS levels while increasing antioxidant systems, in order to diminish the pro-tumorigenic activity of ROS. The reduction in ROS levels not only decreases cell survival and proliferation but also reduces DNA damage and genetic instability, lowering the pro-tumorigenic signaling and the exacerbation of the tumorigenicity. A great variety of studies aimed at investigating the effects of a range of antioxidants on tumor growth and yielded different outcomes, from no effect to, in some cases, increased cancer-related mortality [109]. On the other hand, metformin, a pleiotropic drug that targets mitochondrial complex I with antineoplastic functions, seems to suppress ROS production, decreasing ROS levels and inhibiting inflammatory signaling and metastatic progression in breast cancer [110]; moreover, metformin decreased the viability of Mia PaCa and PANC1 pancreatic ductal adenocarcinoma cell lines through the reduction in intracellular ROS, increasing MnSOD and decreasing NOX2 and NOX4 [111].

The second approach consists of pushing the ROS concentration over the threshold of toxicity, selectively killing tumor cells by disabling antioxidants and activating different cell death processes such as apoptosis, necrosis, and autophagy-mediated cell death. Necrosis, for example, is a programmed cell death characterized by organelle swelling and membrane rupture. As apoptosis, it involves a controlled signaling cascade which requires the receptor-interacting protein kinase 1 (RIP1)/ receptor-interacting protein kinase 3 (RIP3) complex, whose formation was proven to be regulated by mitochondrial ROS [112]. A novel type of cell death is ferroptosis, an iron-dependent programmed cell death occurring when the intracellular levels of lipid reactive oxygen species exceed the activity of glutathione peroxidase 4 (GPX4), leading to the collapse of redox homeostasis. Mitochondria are focal hubs for iron metabolism and homeostasis; moreover, the free and redox active iron pool has been demonstrated to participate in the accumulation of mitochondrial ROS, which can interact with polyunsaturated fatty acids, leading to lipid peroxidation, initiating ferroptosis in cancer and healthy cells [113]. Lastly, pyroptosis could also be an option. This mechanism is mediated by the gasdermin family, accompanied by inflammatory and immune responses; in the last few years, it has been considered a potential cancer treatment strategy [114]. One of the latest updates in ROS-exploiting cancer therapy, in fact, identifies iron as an amplifier of ROS signaling to induce pyroptosis (a lytic programmed cell death initiated by inflammasomes), via the Tom20/Bax/caspase-3-cleaved gasdermin E (GSDME) pathway in melanoma cells [115].

Chemotherapy, the most common treatment in cancer, in the majority of cases, elevates intracellular levels of ROS, in general pushing the cancer cell over a threshold to induce cell death; this is one of the proposed mechanisms via which chemotherapeutics provoke tumor regression. There are two causes for the increase in ROS level in the tumor cell: mitochondrial ROS generation and inhibition of the antioxidant system [116]. Intracellular ROS increase promotes a series of signaling cascades, including the activation of MAPK and NF-κB pathways; moreover, DNA damage induced by ROS accumulation can promote p53 accumulation, activating the p53/Bax pathway and resulting in apoptosis [117].

The combinatorial therapy against breast cancer using resveratrol (RESV)—a natural polyphenol having antiproliferative activity against breast cancer cells—and salinomycin (SAL)—a monocarboxylic polyether ionophore—in MCF7 cell lines has been observed to elicit an apoptotic response through the enhancement of mitochondrial ROS, because of mitochondrial impairment. In fact, after the combinatorial treatment, ROS increase induced mitochondrial membrane potential disruption, decreasing the expression of Bcl-2. This led to the activation of caspases 7,8,9, chromatin condensation, and Poly adenosine diphosphate (ADP)-ribose polymerase (PARP) cleavage, inducing apoptosis. In addition, ROS activated the MAPK pathway, which responds to cellular stress and metabolism by phosphorylating JNK and p38 and leading to apoptosis [100]. In the same direction, Xia et al. studied for the first time on colorectal cancer cells the effect on tumor cells of Withaferin A (WA), an active steroidal lactone derived from *Withania somnifera* that exhibits antitumor activity in several cancers, including breast cancer, lung cancer, and pancreatic cancer, via ROS production. They validated the hypothesis that ROS production, driven by mitochondrial dysfunction, inhibited cell growth and increased apoptosis; the reduction in mitochondrial membrane potential started the traditional apoptotic cascade (decrease on Bcl-2/Bax ratio, subsequent activation of caspase 3–9) and activated of the JNK pathway [101]. Carnosic acid (CA), an antioxidant compound derived from *Rosmarinus officinalis*, was able to induce apoptosis in HCT116 colon cancer cell line via ROS generation and inactivation of STAT3 signaling. Specifically, treatment with CA, generating ROS, diminished the phosphorylation of STAT3, JAK2, and Src kinases (it is likely that ROS may cause oxidative modification of Cys residues on these proteins), decreasing also the expression of STAT3 gene products, such as D-cyclins and survivin [76]. Quinalizarin, an anthraquinone component isolated from Rubiaceae, has been demonstrated to link ROS generation to MAPK, STAT3, and mitochondrial dynamics and inheritance during cell division, as well as the development and disease NF-κB signaling pathways, leading the MCF7 breast cancer cell line and A549 lung cancer cell line to cell-cycle arrest and caspase-dependent apoptosis [102,103]. Cucurbitacin (CuD), a common phytochemical derived from *Trichosanthes kirilowii*, was used in Capan-1 pancreatic cancer cell line, demonstrating that the drug-induced ROS production induced G2/M cell-cycle arrest and mediated the p38/MAPK pathway, promoting cell death (Figures 2 and 3) [104].

More studies, instead, are needed to understand the exact mechanism and correlation among mitochondrial ROS production and mitochondrial dynamics in cancer, to utilize these findings for therapeutic purposes, in order to overcome chemoresistance and/or to improve patient prognosis [79]. Meanwhile, Chuang et al. very recently demonstrated that imiquimod (IMQ), a Toll-like receptor (TLR) 7 ligand, induced severe ROS production that in turn caused mitochondrial membrane potential loss, mitochondrial fission, and mitophagy in skin cancer cells [105]. Moreover, isorhamnetin (IH), a flavonoid that is present in plants of the Polygonaceae family, in combination with chloroquine (CQ), was able to induce apoptosis in triple-negative breast cancer cells, via an ROS-mediated phosphorylation of CaMKII/Drp1, leading to Bax translocation and release of cytochrome c, mitochondrial fission, caspase activation, and apoptosis [106].

In the last few years, the role of ROS in cancer therapy, especially the increase in ROS levels elicited by targeted therapy, has received more and more attention; monoclonal antibodies and small-molecule inhibitors, which specifically target tyrosine kinases, have been demonstrated to show ROS-mediated anticancer effects, eliciting signaling cascades that provoke apoptosis [68]. Moreover, other targeted therapies such as proteasome inhibitors, histone deacetylase (HDAC) inhibitors (HDACi), and STAT3 inhibitors have been shown to sensitize tumor cells by increasing the level of ROS. Cetuximab, in combination with oridonin, inducing ROS production, enhanced mitochondrial apoptosis through the NF-κB, PI3K/Akt, and JAK2/STAT3 pathways in laryngeal squamous carcinoma cells [107]. Moreover, histone deacetylase inhibitors valproic acid (VPA) and trichostatin A (TSA) in PANC1 and PaCa44 pancreatic cancer-derived cell lines triggered autophagy through ROS production [108].

Photodynamic therapy (PDT) is a method for the treatment of tumors, based on a photochemical reaction between a photosensitizer (PS) and molecular oxygen. These three apparently harmless components, taken together, result in the formation of ROS [118]. When the PS, after intravenous, intraperitoneal, or topical administration, is exposed to light with a precise wavelength, it changes from a ground (singlet) state to an excited (triplet) state. The excited state can undergo two kinds of reactions; it can react directly with substrates in the cells, such as the membrane or a molecule, transferring an electron or a proton to form radical anion or cation species (type I reactions), whereby these radicals react with oxygen to form oxidizing free radicals and singlet oxygen [119]. Alternatively, excited PS can be restored to the ground state, which then releases energy inducing the conversion of oxygen to the excited state singlet oxygen. Both species produced exert a cytotoxic effect on the cell, as they both interact with lipids, proteins, and nucleic acids. The irradiation of the tumor can selectively activate the photosensitive drug in situ, triggering a photochemical reaction and tumor destruction, via three different mechanisms: (1) PDT can kill the malignant cells directly, through ROS generation; (2) PDT can damage the tumor-associated vasculature, leading to tumor infarction; (3) PDT can activate an inflammatory and immune response against tumor cells [120]. Focusing on the first mechanism, PDT can evoke apoptosis, necrosis, and autophagy-associated cell death pathways. As an example, mitochondria-associated PSs leading to the photodamage of Bcl-2 is a permissive signal for mitochondrial outer membrane permeabilization (MOMP), mediating the release of caspase activators cytochrome c and Smac/DIABLO or proapoptotic molecules such as apoptosis-inducing factor (AIF) [121]. Moreover, other nonapoptotic pathways could be elicited, including the necrosis signaling cascade [122] and autophagy that can have both a cryoprotective and a pro-death role, depending on the PDT doses [123,124]. However, it has been demonstrated that cancer cells exploit their antioxidant activity to neutralize ROS derived for PDT, as an increase in SODs and other antioxidant enzymes has been observed following PDT [125]. Moreover, PDT induces the expression of proteins that are related to signaling pathways such as apoptosis [126] or are responsible for cell survival mechanisms, in order to cope with the oxidative stress and damage. Transcription factors such as Nrf2, activator protein 1 (AP-1), HIF1, and NF-κB are among the factors that are expressed, in addition to those that mediate the proteotoxic stress response [127,128]. New combinatorial approaches to increase the efficacy of the therapy are now being studied, while also integrating chemotherapeutic drugs and PSs into nanocarriers [129,130]. Developed on the basis of PDT, sonodynamic therapy (SDT) is a novel noninvasive approach for use against solid tumors, with low-intensity ultrasound and sonosensitizers [131,132], inducing an excess of ROS, thereby promoting cell death pathways via downregulation of Bcl-2 family proteins [133]. Lastly, new ways to improve traditional PDT are being developed; since PDT has limited killing capacity due to hypoxia in the tumor niche, strategies are taken into consideration not only to increase the ROS killing effect, but also to inhibit ROS defense systems (Figures 2 and 3) [134].

7. Conclusions and Future Perspectives

In this review, we discussed the currently known intracellular pathways mediated either by mitochondrial structure rearrangements or by mitochondrial ROS production and release. In particular, we demonstrated how finely tuned the regulation of mitochondrial shaping is, reporting the presence of a two-way modulation of mitochondrial dynamics by several pathways and the existence of a vice versa axis [23,27–30]. Interestingly, these two players can also establish more intricated positive feedback loops or vicious cycles, directly responsible for the maintenance of physiological states or contributing to pathological conditions. For instance, we reported that mitochondria can restore their own biogenesis in normal tissues through an upregulation of canonical Wnt via the Pgam5/β-catenin interaction, which stimulates mitophagy and organelle remodeling [27], while a mitochondria/Notch cascade alters mitochondrial fusion and fission rates, ultimately supporting tumor proliferation [29].

Moreover, we addressed the possibility of exploring the functional role of "contactology" in cell signaling modulation, especially to unravel possible links with disease formation and development. Nevertheless, we believe that mitochondrial biology is now evolving into "organellar biology", via which several different organelles work together to regulate important intracellular pathways. In this regard, further studies may be helpful to more deeply investigate the role of ER/mitochondria in cell signaling modulation, but further experiments will be necessary to address this issue.

The existence of a direct link between mitochondrial ROS and cell signaling was also reported in this review, resulting in the modulation of important cellular functions such as proliferation, autophagy, and apoptosis, also acting on a transcriptional level, as summarized in Figures 2 and 3. This leads to the possibility of taking advantage of mitochondrial ROS production for anticancer treatment in multiple ways, by both lowering and enhancing mitochondrial ROS levels inside the cells, resulting in the promotion of cell death via, for instance, DNA damage or mitochondrial impairment, which ultimately provokes the block of tumor progression [68]. To support the idea of the efficacy of this strategy, we collected the novel findings on mitochondrial ROS-targeting drugs (Table 1) which proved to be useful in in vitro studies and could be possibly employed for future clinical trials. The presence of innovative approaches, such as the introduction of photodynamics [118–121] and sonodynamics [131–133], to specifically activate mitochondrial ROS targeting pharmaceuticals to treat cancer supports the relevance of the exploitation of this molecular species, underlining the importance of dissecting cell signaling cascades in which they are involved.

In conclusion, it is clear that mitochondrial physiology has a fundamental role in tuning intracellular functions, leading to the possibility to target these organelles to treat several human diseases. Further work will be necessary to improve drug selectivity to preferentially hit pathological cells while sparing healthy ones.

Author Contributions: Writing—original draft preparation, V.B., L.C., S.M., R.C., and L.L.; writing—review and editing, S.M., R.C., and L.L.; supervision, L.L.; funding acquisition, L.L. All authors read and agreed to the published version of the manuscript.

Funding: The research leading to these results received funding from the Italian Association for Cancer Research (AIRC) under MFAG 2019—ID. 23271 project—P.I. Leanza Luigi. L.L. is also grateful for the PRID 2017 grant (n. BIRD162511) from the University of Padova.

Conflicts of Interest: The authors declare no conflict of interest. The funders had no role in the design of the study; in the collection, analyses, or interpretation of data; in the writing of the manuscript, or in the decision to publish the results.

References

1. Dyall, S.D.; Brown, M.T.; Johnson, P.J. Ancient Invasions: From Endosymbionts to Organelles. *Science (80-.)* **2004**, *304*, 253–257. [CrossRef]
2. Youle, R.J. Mitochondria-Striking a balance between host and endosymbiont. *Science* **2019**, *365*, eaaw9855. [CrossRef]
3. Leanza, L.; Romio, M.; Becker, K.A.; Azzolini, M.; Trentin, L.; Managò, A.; Venturini, E.; Zaccagnino, A.; Mattarei, A.; Carraretto, L.; et al. Direct Pharmacological Targeting of a Mitochondrial Ion Channel Selectively Kills Tumor Cells In Vivo. *Cancer Cell* **2017**, *31*, 516–531.e10. [CrossRef]
4. Xia, M.; Zhang, Y.; Jin, K.; Lu, Z.; Zeng, Z.; Xiong, W. Communication between mitochondria and other organelles: A brand-new perspective on mitochondria in cancer. *Cell Biosci.* **2019**, *9*, 27. [CrossRef]
5. Mishara, P.; Chan, D.C. Mitochondrial dynamics and inheritance during cell division, development and disease. *Nat. Rev. Mol. Cell Biol.* **2014**, *15*, 634–646. [CrossRef]
6. Kalyanaraman, B.; Cheng, G.; Hardy, M.; Ouari, O.; Bennett, B.; Zielonka, J. Teaching the basics of reactive oxygen species and their relevance to cancer biology: Mitochondrial reactive oxygen species detection, redox signaling, and targeted therapies. *Redox Biol.* **2018**, *15*, 347–362. [CrossRef] [PubMed]
7. Brand, M.D. Mitochondrial generation of superoxide and hydrogen peroxide as the source of mitochondrial redox signaling. *Free Radic. Biol. Med.* **2016**, *100*, 14–31. [CrossRef]
8. Willems, P.H.G.M.; Rossignol, R.; Dieteren, C.E.J.; Murphy, M.P.; Koopman, W.J.H. Redox Homeostasis and Mitochondrial Dynamics. *Cell Metab.* **2015**, *22*, 207–218. [CrossRef]
9. Sena, L.A.; Chandel, N.S. Physiological Roles of Mitochondrial Reactive Oxygen Species. *Mol. Cell* **2012**, *48*, 158–167. [CrossRef] [PubMed]
10. Chan, D.C. Mitochondrial Dynamics and Its Involvement in Disease. *Annu. Rev. Pathol. Mech. Dis.* **2020**, *15*, 235–259. [CrossRef] [PubMed]
11. Ishihara, N.; Jofuku, A.; Eura, Y.; Mihara, K. Regulation of mitochondrial morphology by membrane potential, and DRP1-dependent division and FZO1-dependent fusion reaction in mammalian cells. *Biochem. Biophys. Res. Commun.* **2003**, *301*, 891–898. [CrossRef]
12. Meyer, J.N.; Leuthner, T.C.; Luz, A.L. Mitochondrial fusion, fission, and mitochondrial toxicity. *Toxicology* **2017**, *391*, 42–53. [CrossRef]
13. Kashatus, D.F. The regulation of tumor cell physiology by mitochondrial dynamics. *Biochem. Biophys. Res. Commun.* **2018**, *500*, 9–16. [CrossRef] [PubMed]
14. Kalia, R.; Wang, R.Y.R.; Yusuf, A.; Thomas, P.V.; Agard, D.A.; Shaw, J.M.; Frost, A. Structural basis of mitochondrial receptor binding and constriction by DRP1. *Nature* **2018**, *558*, 401–405. [CrossRef] [PubMed]
15. Van der Bliek, A.M.; Shen, Q.; Kawajiri, S. Mechanisms of mitochondrial fission and fusion. *Cold Spring Harb. Perspect. Biol.* **2013**, *5*, 1–16. [CrossRef]
16. Lee, J.E.; Westrate, L.M.; Wu, H.; Page, C.; Voeltz, G.K. Multiple dynamin family members collaborate to drive mitochondrial division. *Nature* **2016**, *540*, 139–143. [CrossRef]
17. Giacomello, M.; Pyakurel, A.; Glytsou, C.; Scorrano, L. The cell biology of mitochondrial membrane dynamics. *Nat. Rev. Mol. Cell Biol.* **2020**, *21*, 204–224. [CrossRef] [PubMed]
18. Peruzzo, R.; Costa, R.; Bachmann, M.; Leanza, L.; Szabò, I. Mitochondrial metabolism, contact sites and cellular calcium signaling: Implications for tumorigenesis. *Cancers (Basel)* **2020**, *12*, 2574. [CrossRef]
19. D'Eletto, M.; Rossin, F.; Occhigrossi, L.; Farrace, M.G.; Faccenda, D.; Desai, R.; Marchi, S.; Refolo, G.; Falasca, L.; Antonioli, M.; et al. Transglutaminase Type 2 Regulates ER-Mitochondria Contact Sites by Interacting with GRP75. *Cell Rep.* **2018**, *25*, 3573–3581.e4. [CrossRef]
20. Rossin, F.; Costa, R.; Bordi, M.; D'Eletto, M.; Occhigrossi, L.; Farrace, M.G.; Barlev, N.; Ciccosanti, F.; Muccioli, S.; Chieregato, L.; et al. Transglutaminase Type 2 regulates the Wnt/β-catenin pathway in vertebrates. *Cell Death Dis.* **2021**, *12*, 249. [CrossRef] [PubMed]
21. Delettre, C.; Griffoin, J.M.; Kaplan, J.; Dollfus, H.; Lorenz, B.; Faivre, L.; Lenaers, G.; Belenguer, P.; Hamel, C.P. Mutation spectrum and splicing variants in the OPA1 gene. *Hum. Genet.* **2001**, *109*, 584–591. [CrossRef] [PubMed]
22. Song, Z.; Chen, H.; Fiket, M.; Alexander, C.; Chan, D.C. OPA1 processing controls mitochondrial fusion and is regulated by mRNA splicing, membrane potential, and Yme1L. *J. Cell Biol.* **2007**, *178*, 749–755. [CrossRef]
23. Malhotra, A.; Dey, A.; Prasad, N.; Kenney, A.M. Sonic Hedgehog signaling drives mitochondrial fragmentation by suppressing mitofusins in cerebellar granule neuron precursors and medulloblastoma. *Mol. Cancer Res.* **2016**, *14*, 114–124. [CrossRef] [PubMed]
24. Arrázola, M.S.; Silva-Alvarez, C.; Inestrosa, N.C. How the Wnt signaling pathway protects from neurodegeneration: The mitochondrial scenario. *Front. Cell. Neurosci.* **2015**, *9*, 1–13. [CrossRef] [PubMed]
25. Bengoa-Vergniory, N.; Kypta, R.M. Canonical and noncanonical Wnt signaling in neural stem/progenitor cells. *Cell. Mol. Life Sci.* **2015**, *72*, 4157–4172. [CrossRef]
26. Zhao, H.; Luo, Y.; Chen, L.; Zhang, Z.; Shen, C.; Li, Y.; Xu, R. Sirt3 inhibits cerebral ischemia-reperfusion injury through normalizing Wnt/β-catenin pathway and blocking mitochondrial fission. *Cell Stress Chaperones* **2018**, *23*, 1079–1092. [CrossRef] [PubMed]

27. Bernkopf, D.B.; Jalal, K.; Brückner, M.; Knaup, K.X.; Gentzel, M.; Schambony, A.; Behrens, J. Pgam5 released from damaged mitochondria induces mitochondrial biogenesis via Wnt signaling. *J. Cell Biol.* **2018**, *217*, 1383–1394. [CrossRef] [PubMed]
28. Shah, N.; Lee, N.Y. Regulation of gene expression and mitochondrial dynamics by SMAD. *Mol. Cell. Oncol.* **2016**, *3*, 1–3. [CrossRef] [PubMed]
29. Chen, L.; Zhang, J.; Lyu, Z.; Chen, Y.; Ji, X.; Cao, H.; Jin, M.; Zhu, J.; Yang, J.; Ling, R.; et al. Positive feedback loop between mitochondrial fission and Notch signaling promotes survivin-mediated survival of TNBC cells. *Cell Death Dis.* **2018**, *9*, 1050. [CrossRef]
30. Von Eyss, B.; Jaenicke, L.A.; Kortlever, R.M.; Royla, N.; Wiese, K.E.; Letschert, S.; McDuffus, L.A.; Sauer, M.; Rosenwald, A.; Evan, G.I.; et al. A MYC-Driven Change in Mitochondrial Dynamics Limits YAP/TAZ Function in Mammary Epithelial Cells and Breast Cancer. *Cancer Cell* **2015**, *28*, 743–757. [CrossRef]
31. Quan, Y.; Park, W.; Jin, J.; Kim, W.; Park, S.K.; Kang, K.P. Sirtuin 3 Activation by Honokiol Decreases Unilateral Ureteral Obstruction-Induced Renal Inflammation and Fibrosis via Regulation of Mitochondrial Dynamics and the Renal NF-κB-TGF-β1/Smad Signaling Pathway. *Int. J. Mol. Sci.* **2020**, *21*, 402. [CrossRef] [PubMed]
32. Zhang, Q.; He, L.; Dong, Y.; Fei, Y.; Wen, J.; Li, X.; Guan, J.; Liu, F.; Zhou, T.; Li, Z.; et al. Sitagliptin ameliorates renal tubular injury in diabetic kidney disease via STAT3-dependent mitochondrial homeostasis through SDF-1α/CXCR4 pathway. *FASEB J.* **2020**, *34*, 7500–7519. [CrossRef] [PubMed]
33. Wang, K.; Liu, Z.; Zhao, M.; Zhang, F.; Wang, K.; Feng, N.; Fu, F.; Li, J.; Li, J.; Liu, Y.; et al. κ-opioid receptor activation promotes mitochondrial fusion and enhances myocardial resistance to ischemia and reperfusion injury via STAT3-OPA1 pathway. *Eur. J. Pharmacol.* **2020**, *874*, 172987. [CrossRef]
34. Moloney, J.N.; Cotter, T.G. ROS signalling in the biology of cancer. *Semin. Cell Dev. Biol.* **2018**, *80*, 50–64. [CrossRef]
35. Wong, H.-S.; Benoit, B.; Brand, M.D. Mitochondrial and cytosolic sources of hydrogen peroxide in resting C2C12 myoblasts. *Free Radic. Biol. Med.* **2019**, *130*, 140–150. [CrossRef]
36. Quinlan, C.L.; Treberg, J.R.; Perevoshchikova, I.V.; Orr, A.L.; Brand, M.D. Native rates of superoxide production from multiple sites in isolated mitochondria measured using endogenous reporters. *Free Radic. Biol. Med.* **2012**, *53*, 1807–1817. [CrossRef] [PubMed]
37. Yang, S.; Lian, G. ROS and diseases: Role in metabolism and energy supply. *Mol. Cell. Biochem.* **2020**, *467*, 1–12. [CrossRef] [PubMed]
38. Mazat, J.-P.; Devin, A.; Ransac, S. Modelling mitochondrial ROS production by the respiratory chain. *Cell. Mol. Life Sci.* **2020**, *77*, 455–465. [CrossRef] [PubMed]
39. Dröse, S.; Brandt, U. Molecular Mechanisms of Superoxide Production by the Mitochondrial Respiratory Chain. In *Advances in Experimental Medicine and Biology;* Springer: New York, NY, USA, 2012; pp. 145–169. ISBN 9781461435723.
40. Burtenshaw, D.; Hakimjavadi, R.; Redmond, E.; Cahill, P. Nox, Reactive Oxygen Species and Regulation of Vascular Cell Fate. *Antioxidants* **2017**, *6*, 90. [CrossRef] [PubMed]
41. Aviello, G.; Knaus, U.G. NADPH oxidases and ROS signaling in the gastrointestinal tract. *Mucosal Immunol.* **2018**, *11*, 1011–1023. [CrossRef]
42. Brandes, R.P.; Weissmann, N.; Schröder, K. Nox family NADPH oxidases: Molecular mechanisms of activation. *Free Radic. Biol. Med.* **2014**, *76*, 208–226. [CrossRef]
43. Conteh, A.M.; Reissaus, C.A.; Hernandez-Perez, M.; Nakshatri, S.; Anderson, R.M.; Mirmira, R.G.; Tersey, S.A.; Linnemann, A.K. Platelet-type 12-lipoxygenase deletion provokes a compensatory 12/15-lipoxygenase increase that exacerbates oxidative stress in mouse islet β cells. *J. Biol. Chem.* **2019**, *294*, 6612–6620. [CrossRef]
44. Zhao, R.; Jiang, S.; Zhang, L.; Yu, Z. Mitochondrial electron transport chain, ROS generation and uncoupling (Review). *Int. J. Mol. Med.* **2019**, *44*, 3–15. [CrossRef]
45. Delierneux, C.; Kouba, S.; Shanmughapriya, S.; Potier-Cartereau, M.; Trebak, M.; Hempel, N. Mitochondrial Calcium Regulation of Redox Signaling in Cancer. *Cells* **2020**, *9*, 432. [CrossRef]
46. Federico, A.; Cardaioli, E.; Da Pozzo, P.; Formichi, P.; Gallus, G.N.; Radi, E. Mitochondria, oxidative stress and neurodegeneration. *J. Neurol. Sci.* **2012**, *322*, 254–262. [CrossRef]
47. Zou, X.; Ratti, B.A.; O'Brien, J.G.; Lautenschlager, S.O.; Gius, D.R.; Bonini, M.G.; Zhu, Y. Manganese superoxide dismutase (SOD2): Is there a center in the universe of mitochondrial redox signaling? *J. Bioenerg. Biomembr.* **2017**, *49*, 325–333. [CrossRef]
48. Arteel, G.E.; Sies, H. The biochemistry of selenium and the glutathione system. *Environ. Toxicol. Pharmacol.* **2001**, *10*, 153–158. [CrossRef]
49. Cao, Z.; Lindsay, J.G. The Peroxiredoxin Family: An Unfolding Story. *Subcell. Biochem.* **2017**, *83*, 127–147. [CrossRef] [PubMed]
50. Srinivas, U.S.; Tan, B.W.Q.; Vellayappan, B.A.; Jeyasekharan, A.D. ROS and the DNA damage response in cancer. *Redox Biol.* **2019**, *25*, 101084. [CrossRef] [PubMed]
51. Khan, S.; Zafar, A.; Naseem, I. Copper-redox cycling by coumarin-di(2-picolyl)amine hybrid molecule leads to ROS-mediated DNA damage and apoptosis: A mechanism for cancer chemoprevention. *Chem. Biol. Interact.* **2018**, *290*, 64–76. [CrossRef] [PubMed]
52. Kumar, K.; Mishra, J.P.N.; Singh, R.P. Usnic acid induces apoptosis in human gastric cancer cells through ROS generation and DNA damage and causes up-regulation of DNA-PKcs and γ-H2A.X phosphorylation. *Chem. Biol. Interact.* **2020**, *315*, 108898. [CrossRef]

53. Holmström, K.M.; Finkel, T. Cellular mechanisms and physiological consequences of redox-dependent signalling. *Nat. Rev. Mol. Cell Biol.* **2014**, *15*, 411–421. [CrossRef]
54. Diebold, L.; Chandel, N.S. Mitochondrial ROS regulation of proliferating cells. *Free Radic. Biol. Med.* **2016**, *100*, 86–93. [CrossRef] [PubMed]
55. D'Autréaux, B.; Toledano, M.B. ROS as signalling molecules: Mechanisms that generate specificity in ROS homeostasis. *Nat. Rev. Mol. Cell Biol.* **2007**, *8*, 813–824. [CrossRef] [PubMed]
56. Del Pilar Sosaldelchik, M.; Begley, U.; Begley, T.J.; Melendez, J.A. Mitochondrial ROS control of cancer. *Semin. Cancer Biol.* **2017**, *47*, 57–66. [CrossRef]
57. Zorov, D.B.; Juhaszova, M.; Sollott, S.J. Mitochondrial Reactive Oxygen Species (ROS) and ROS-Induced ROS Release. *Physiol. Rev.* **2014**, *94*, 909–950. [CrossRef] [PubMed]
58. Jendrach, M.; Mai, S.; Pohl, S.; Vöth, M.; Bereiter-Hahn, J. Short- and long-term alterations of mitochondrial morphology, dynamics and mtDNA after transient oxidative stress. *Mitochondrion* **2008**, *8*, 293–304. [CrossRef] [PubMed]
59. Fan, X.; Hussien, R.; Brooks, G.A. H2O2-induced mitochondrial fragmentation in C2C12 myocytes. *Free Radic. Biol. Med.* **2010**, *49*, 1646–1654. [CrossRef] [PubMed]
60. Iqbal, S.; Hood, D.A. Oxidative stress-induced mitochondrial fragmentation and movement in skeletal muscle myoblasts. *Am. J. Physiol. Physiol.* **2014**, *306*, C1176–C1183. [CrossRef] [PubMed]
61. Distelmaier, F.; Valsecchi, F.; Liemburg-Apers, D.C.; Lebiedzinska, M.; Rodenburg, R.J.; Heil, S.; Keijer, J.; Fransen, J.; Imamura, H.; Danhauser, K.; et al. Mitochondrial dysfunction in primary human fibroblasts triggers an adaptive cell survival program that requires AMPK-α. *Biochim. Biophys. Acta - Mol. Basis Dis.* **2015**, *1852*, 529–540. [CrossRef]
62. Trewin, A.; Berry, B.; Wojtovich, A. Exercise and Mitochondrial Dynamics: Keeping in Shape with ROS and AMPK. *Antioxidants* **2018**, *7*, 7. [CrossRef]
63. Liesa, M.; Borda-d'Água, B.; Medina-Gómez, G.; Lelliott, C.J.; Paz, J.C.; Rojo, M.; Palacín, M.; Vidal-Puig, A.; Zorzano, A. Mitochondrial Fusion Is Increased by the Nuclear Coactivator PGC-1β. *PLoS ONE* **2008**, *3*, e3613. [CrossRef]
64. Angelova, P.R.; Abramov, A.Y. Role of mitochondrial ROS in the brain: From physiology to neurodegeneration. *FEBS Lett.* **2018**, *592*, 692–702. [CrossRef]
65. Rendra, E.; Riabov, V.; Mossel, D.M.; Sevastyanova, T.; Harmsen, M.C.; Kzhyshkowska, J. Reactive oxygen species (ROS) in macrophage activation and function in diabetes. *Immunobiology* **2019**, *224*, 242–253. [CrossRef] [PubMed]
66. Ahmad, W.; Ijaz, B.; Shabbiri, K.; Ahmed, F.; Rehman, S. Oxidative toxicity in diabetes and Alzheimer's disease: Mechanisms behind ROS/ RNS generation. *J. Biomed. Sci.* **2017**, *24*, 76. [CrossRef] [PubMed]
67. Incalza, M.A.; D'Oria, R.; Natalicchio, A.; Perrini, S.; Laviola, L.; Giorgino, F. Oxidative stress and reactive oxygen species in endothelial dysfunction associated with cardiovascular and metabolic diseases. *Vascul. Pharmacol.* **2018**, *100*, 1–19. [CrossRef]
68. Zou, Z.; Chang, H.; Li, H.; Wang, S. Induction of reactive oxygen species: An emerging approach for cancer therapy. *Apoptosis* **2017**, *22*, 1321–1335. [CrossRef]
69. Yang, Y.; Karakhanova, S.; Hartwig, W.; D'Haese, J.G.; Philippov, P.P.; Werner, J.; Bazhin, A.V. Mitochondria and Mitochondrial ROS in Cancer: Novel Targets for Anticancer Therapy. *J. Cell. Physiol.* **2016**, *231*, 2570–2581. [CrossRef] [PubMed]
70. Hamanaka, R.B.; Chandel, N.S. Mitochondrial reactive oxygen species regulate cellular signaling and dictate biological outcomes. *Trends Biochem. Sci.* **2010**, *35*, 505–513. [CrossRef] [PubMed]
71. Schieber, M.; Chandel, N.S. ROS Function in Redox Signaling and Oxidative Stress. *Curr. Biol.* **2014**, *24*, R453–R462. [CrossRef] [PubMed]
72. Jiang, N.; Dai, Q.; Su, X.; Fu, J.; Feng, X.; Peng, J. Role of PI3K/AKT pathway in cancer: The framework of malignant behavior. *Mol. Biol. Rep.* **2020**, *47*, 4587–4629. [CrossRef] [PubMed]
73. Wang, C.; Shao, L.; Pan, C.; Ye, J.; Ding, Z.; Wu, J.; Du, Q.; Ren, Y.; Zhu, C. Elevated level of mitochondrial reactive oxygen species via fatty acid β-oxidation in cancer stem cells promotes cancer metastasis by inducing epithelial–mesenchymal transition. *Stem Cell Res. Ther.* **2019**, *10*, 175. [CrossRef] [PubMed]
74. Wang, C.; Li, P.; Xuan, J.; Zhu, C.; Liu, J.; Shan, L.; Du, Q.; Ren, Y.; Ye, J. Cholesterol Enhances Colorectal Cancer Progression via ROS Elevation and MAPK Signaling Pathway Activation. *Cell. Physiol. Biochem.* **2017**, *42*, 729–742. [CrossRef]
75. Kim, B.-H.; Yi, E.H.; Ye, S.-K. Signal transducer and activator of transcription 3 as a therapeutic target for cancer and the tumor microenvironment. *Arch. Pharm. Res.* **2016**, *39*, 1085–1099. [CrossRef]
76. Kim, D.H.; Park, K.W.; Chae, I.G.; Kundu, J.; Kim, E.H.; Kundu, J.K.; Chun, K.S. Carnosic acid inhibits STAT3 signaling and induces apoptosis through generation of ROS in human colon cancer HCT116 cells. *Mol. Carcinog.* **2016**. [CrossRef]
77. Meier, J.A.; Hyun, M.; Cantwell, M.; Raza, A.; Mertens, C.; Raje, V.; Sisler, J.; Tracy, E.; Torres-Odio, S.; Gispert, S.; et al. Stress-induced dynamic regulation of mitochondrial STAT3 and its association with cyclophilin D reduce mitochondrial ROS production. *Sci. Signal.* **2017**, *10*, eaag2588. [CrossRef]
78. Lee, Y.-K.; Kwon, S.M.; Lee, E.; Kim, G.-H.; Min, S.; Hong, S.-M.; Wang, H.-J.; Lee, D.M.; Choi, K.S.; Park, T.J.; et al. Mitochondrial Respiratory Defect Enhances Hepatoma Cell Invasiveness via STAT3/NFE2L1/STX12 Axis. *Cancers (Basel)* **2020**, *12*, 2632. [CrossRef]
79. Kim, B.; Song, Y.S. Mitochondrial dynamics altered by oxidative stress in cancer. *Free Radic. Res.* **2016**, *50*, 1065–1070. [CrossRef]
80. Han, Y.; Kim, B.; Cho, U.; Park, I.S.; Kim, S.I.; Dhanasekaran, D.N.; Tsang, B.K.; Song, Y.S. Mitochondrial fission causes cisplatin resistance under hypoxic conditions via ROS in ovarian cancer cells. *Oncogene* **2019**, *38*, 7089–7105. [CrossRef]

81. Taguchi, K.; Yamamoto, M. The KEAP1–NRF2 System in Cancer. *Front. Oncol.* **2017**, *7*. [CrossRef] [PubMed]
82. Lacher, S.E.; Levings, D.C.; Freeman, S.; Slattery, M. Identification of a functional antioxidant response element at the HIF1A locus. *Redox Biol.* **2018**, *19*, 401–411. [CrossRef]
83. Schroedl, C.; McClintock, D.S.; Budinger, G.R.S.; Chandel, N.S. Hypoxic but not anoxic stabilization of HIF-1α requires mitochondrial reactive oxygen species. *Am. J. Physiol. Cell. Mol. Physiol.* **2002**, *283*, L922–L931. [CrossRef] [PubMed]
84. Bell, E.L.; Klimova, T.A.; Eisenbart, J.; Moraes, C.T.; Murphy, M.P.; Budinger, G.R.S.; Chandel, N.S. The Qo site of the mitochondrial complex III is required for the transduction of hypoxic signaling via reactive oxygen species production. *J. Cell Biol.* **2007**, *177*, 1029–1036. [CrossRef]
85. Pastukh, V.; Roberts, J.T.; Clark, D.W.; Bardwell, G.C.; Patel, M.; Al-Mehdi, A.-B.; Borchert, G.M.; Gillespie, M.N. An oxidative DNA "damage" and repair mechanism localized in the VEGF promoter is important for hypoxia-induced VEGF mRNA expression. *Am. J. Physiol. Cell. Mol. Physiol.* **2015**, *309*, L1367–L1375. [CrossRef] [PubMed]
86. Sun, Y.; Liu, W.-Z.; Liu, T.; Feng, X.; Yang, N.; Zhou, H.-F. Signaling pathway of MAPK/ERK in cell proliferation, differentiation, migration, senescence and apoptosis. *J. Recept. Signal Transduct.* **2015**, *35*, 600–604. [CrossRef]
87. Yang, R.; Piperdi, S.; Gorlick, R. Activation of the RAF/Mitogen-Activated Protein/Extracellular Signal-Regulated Kinase Kinase/Extracellular Signal-Regulated Kinase Pathway Mediates Apoptosis Induced by Chelerythrine in Osteosarcoma. *Clin. Cancer Res.* **2008**, *14*, 6396–6404. [CrossRef]
88. Cao, Y.; Wang, J.; Tian, H.; Fu, G.-H. Mitochondrial ROS accumulation inhibiting JAK2/STAT3 pathway is a critical modulator of CYT997-induced autophagy and apoptosis in gastric cancer. *J. Exp. Clin. Cancer Res.* **2020**, *39*, 119. [CrossRef] [PubMed]
89. Capece, D.; Verzella, D.; Di Francesco, B.; Alesse, E.; Franzoso, G.; Zazzeroni, F. NF-κB and mitochondria cross paths in cancer: Mitochondrial metabolism and beyond. *Semin. Cell Dev. Biol.* **2020**, *98*, 118–128. [CrossRef] [PubMed]
90. Dolcet, X.; Llobet, D.; Pallares, J.; Matias-Guiu, X. NF-kB in development and progression of human cancer. *Virchows Arch.* **2005**, *446*, 475–482. [CrossRef] [PubMed]
91. Chen, C.; Wang, S.; Liu, P. Deferoxamine Enhanced Mitochondrial Iron Accumulation and Promoted Cell Migration in Triple-Negative MDA-MB-231 Breast Cancer Cells via a ROS-Dependent Mechanism. *Int. J. Mol. Sci.* **2019**, *20*, 4952. [CrossRef]
92. Zou, Z.; Yuan, Z.; Zhang, Q.; Long, Z.; Chen, J.; Tang, Z.; Zhu, Y.; Chen, S.; Xu, J.; Yan, M.; et al. Aurora kinase A inhibition-induced autophagy triggers drug resistance in breast cancer cells. *Autophagy* **2012**, *8*, 1798–1810. [CrossRef]
93. Feng, D.; Liu, L.; Zhu, Y.; Chen, Q. Molecular signaling toward mitophagy and its physiological significance. *Exp. Cell Res.* **2013**, *319*, 1697–1705. [CrossRef]
94. Matsuda, N.; Sato, S.; Shiba, K.; Okatsu, K.; Saisho, K.; Gautier, C.A.; Sou, Y.; Saiki, S.; Kawajiri, S.; Sato, F.; et al. PINK1 stabilized by mitochondrial depolarization recruits Parkin to damaged mitochondria and activates latent Parkin for mitophagy. *J. Cell Biol.* **2010**, *189*, 211–221. [CrossRef]
95. Chen, Y.; McMillan-Ward, E.; Kong, J.; Israels, S.J.; Gibson, S.B. Oxidative stress induces autophagic cell death independent of apoptosis in transformed and cancer cells. *Cell Death Differ.* **2008**, *15*, 171–182. [CrossRef] [PubMed]
96. Pallichankandy, S.; Rahman, A.; Thayyullathil, F.; Galadari, S. ROS-dependent activation of autophagy is a critical mechanism for the induction of anti-glioma effect of sanguinarine. *Free Radic. Biol. Med.* **2015**, *89*, 708–720. [CrossRef]
97. Scherz-Shouval, R.; Shvets, E.; Fass, E.; Shorer, H.; Gil, L.; Elazar, Z. Reactive oxygen species are essential for autophagy and specifically regulate the activity of Atg4. *EMBO J.* **2007**, *26*, 1749–1760. [CrossRef] [PubMed]
98. Li, L.; Tan, J.; Miao, Y.; Lei, P.; Zhang, Q. ROS and Autophagy: Interactions and Molecular Regulatory Mechanisms. *Cell. Mol. Neurobiol.* **2015**, *35*, 615–621. [CrossRef]
99. Scherz-Shouval, R.; Elazar, Z. Regulation of autophagy by ROS: Physiology and pathology. *Trends Biochem. Sci.* **2011**, *36*, 30–38. [CrossRef]
100. Dewangan, J.; Tandon, D.; Srivastava, S.; Verma, A.K.; Yapuri, A.; Rath, S.K. Novel combination of salinomycin and resveratrol synergistically enhances the anti-proliferative and pro-apoptotic effects on human breast cancer cells. *Apoptosis* **2017**, *22*, 1246–1259. [CrossRef] [PubMed]
101. Xia, S.; Miao, Y.; Liu, S. Withaferin A induces apoptosis by ROS-dependent mitochondrial dysfunction in human colorectal cancer cells. *Biochem. Biophys. Res. Commun.* **2018**, *503*, 2363–2369. [CrossRef] [PubMed]
102. Zang, Y.Q.; Feng, Y.Y.; Luo, Y.H.; Zhai, Y.Q.; Ju, X.Y.; Feng, Y.C.; Sheng, Y.N.; Wang, J.R.; Yu, C.Q.; Jin, C.H. Quinalizarin induces ROS-mediated apoptosis via the MAPK, STAT3 and NF-κB signaling pathways in human breast cancer cells. *Mol. Med. Rep.* **2019**, *20*, 4576–4586. [CrossRef]
103. Meng, L.; Liu, C.; Luo, Y.; Piao, X.; Wang, Y.; Zhang, Y.; Wang, J.; Wang, H.; Xu, W.; Liu, Y.; et al. Quinalizarin exerts an anti-tumour effect on lung cancer A549 cells by modulating the Akt, MAPK, STAT3 and p53 signalling pathways. *Mol. Med. Rep.* **2017**. [CrossRef] [PubMed]
104. Kim, M.-S.; Lee, K.; Ku, J.M.; Choi, Y.-J.; Mok, K.; Kim, D.; Cheon, C.; Ko, S.-G. Cucurbitacin D Induces G2/M Phase Arrest and Apoptosis via the ROS/p38 Pathway in Capan-1 Pancreatic Cancer Cell Line. *Evid.-Based Complement. Altern. Med.* **2020**, *2020*, 1–10. [CrossRef]
105. Chuang, K.-C.; Chang, C.-R.; Chang, S.-H.; Huang, S.-W.; Chuang, S.-M.; Li, Z.-Y.; Wang, S.-T.; Kao, J.-K.; Chen, Y.-J.; Shieh, J.-J. Imiquimod-induced ROS production disrupts the balance of mitochondrial dynamics and increases mitophagy in skin cancer cells. *J. Dermatol. Sci.* **2020**, *98*, 152–162. [CrossRef] [PubMed]

106. Hu, J.; Zhang, Y.; Jiang, X.; Zhang, H.; Gao, Z.; Li, Y.; Fu, R.; Li, L.; Li, J.; Cui, H.; et al. ROS-mediated activation and mitochondrial translocation of CaMKII contributes to Drp1-dependent mitochondrial fission and apoptosis in triple-negative breast cancer cells by isorhamnetin and chloroquine. *J. Exp. Clin. Cancer Res.* **2019**, *38*, 225. [CrossRef] [PubMed]
107. Kang, N.; Cao, S.; Jiang, B.; Zhang, Q.; Donkor, P.O.; Zhu, Y.; Qiu, F.; Gao, X. Cetuximab enhances oridonin-induced apoptosis through mitochondrial pathway and endoplasmic reticulum stress in laryngeal squamous cell carcinoma cells. *Toxicol. In Vitro* **2020**, *67*, 104885. [CrossRef]
108. Gilardini Montani, M.S.; Granato, M.; Santoni, C.; Del Porto, P.; Merendino, N.; D'Orazi, G.; Faggioni, A.; Cirone, M. Histone deacetylase inhibitors VPA and TSA induce apoptosis and autophagy in pancreatic cancer cells. *Cell. Oncol.* **2017**, *40*, 167–180. [CrossRef] [PubMed]
109. Bjelakovic, G.; Nikolova, D.; Gluud, L.L.; Simonetti, R.G.; Gluud, C. Antioxidant supplements for prevention of mortality in healthy participants and patients with various diseases. *Cochrane Database Syst. Rev.* **2012**. [CrossRef] [PubMed]
110. Schexnayder, C.; Broussard, K.; Onuaguluchi, D.; Poché, A.; Ismail, M.; McAtee, L.; Llopis, S.; Keizerweerd, A.; McFerrin, H.; Williams, C. Metformin Inhibits Migration and Invasion by Suppressing ROS Production and COX2 Expression in MDA-MB-231 Breast Cancer Cells. *Int. J. Mol. Sci.* **2018**, *19*, 3692. [CrossRef]
111. Cheng, G.; Lanza-Jacoby, S. Metformin decreases growth of pancreatic cancer cells by decreasing reactive oxygen species: Role of NOX4. *Biochem. Biophys. Res. Commun.* **2015**, *465*, 41–46. [CrossRef] [PubMed]
112. Zhang, Y.; Su, S.S.; Zhao, S.; Yang, Z.; Zhong, C.-Q.; Chen, X.; Cai, Q.; Yang, Z.-H.; Huang, D.; Wu, R.; et al. RIP1 autophosphorylation is promoted by mitochondrial ROS and is essential for RIP3 recruitment into necrosome. *Nat. Commun.* **2017**, *8*, 14329. [CrossRef] [PubMed]
113. Battaglia, A.M.; Chirillo, R.; Aversa, I.; Sacco, A.; Costanzo, F.; Biamonte, F. Ferroptosis and Cancer: Mitochondria Meet the "Iron Maiden" Cell Death. *Cells* **2020**, *9*, 1505. [CrossRef] [PubMed]
114. Xia, X.; Wang, X.; Cheng, Z.; Qin, W.; Lei, L.; Jiang, J.; Hu, J. The role of pyroptosis in cancer: Pro-cancer or pro-"host"? *Cell Death Dis.* **2019**, *10*, 650. [CrossRef] [PubMed]
115. Zhou, B.; Zhang, J.; Liu, X.; Chen, H.; Ai, Y.; Cheng, K.; Sun, R.; Zhou, D.; Han, J.; Wu, Q. Tom20 senses iron-activated ROS signaling to promote melanoma cell pyroptosis. *Cell Res.* **2018**, *28*, 1171–1185. [CrossRef]
116. Yang, H.; Villani, R.M.; Wang, H.; Simpson, M.J.; Roberts, M.S.; Tang, M.; Liang, X. The role of cellular reactive oxygen species in cancer chemotherapy. *J. Exp. Clin. Cancer Res.* **2018**, *37*, 266. [CrossRef]
117. Ray, T.; Chakrabarti, M.K.; Pal, A. Hemagglutinin protease secreted by V. cholerae induced apoptosis in breast cancer cells by ROS mediated intrinsic pathway and regresses tumor growth in mice model. *Apoptosis* **2016**, *21*, 143–154. [CrossRef]
118. Dolmans, D.E.J.G.J.; Fukumura, D.; Jain, R.K. Photodynamic therapy for cancer. *Nat. Rev. Cancer* **2003**, *3*, 380–387. [CrossRef]
119. Agostinis, P.; Berg, K.; Cengel, K.A.; Foster, T.H.; Girotti, A.W.; Gollnick, S.O.; Hahn, S.M.; Hamblin, M.R.; Juzeniene, A.; Kessel, D.; et al. Photodynamic therapy of cancer: An update. *CA. Cancer J. Clin.* **2011**, *61*, 250–281. [CrossRef]
120. Van Straten, D.; Mashayekhi, V.; de Bruijn, H.; Oliveira, S.; Robinson, D. Oncologic Photodynamic Therapy: Basic Principles, Current Clinical Status and Future Directions. *Cancers (Basel)* **2017**, *9*, 19. [CrossRef]
121. Buytaert, E.; Dewaele, M.; Agostinis, P. Molecular effectors of multiple cell death pathways initiated by photodynamic therapy. *Biochim. Biophys. Acta - Rev. Cancer* **2007**, *1776*, 86–107. [CrossRef]
122. Vanlangenakker, N.; Berghe, T.; Krysko, D.; Festjens, N.; Vandenabeele, P. Molecular Mechanisms and Pathophysiology of Necrotic Cell Death. *Curr. Mol. Med.* **2008**, *8*, 207–220. [CrossRef] [PubMed]
123. Reiners, J.J.; Agostinis, P.; Berg, K.; Oleinick, N.L.; Kessel, D.H. Assessing autophagy in the context of photodynamic therapy. *Autophagy* **2010**, *6*, 7–18. [CrossRef]
124. Inguscio, V.; Panzarini, E.; Dini, L. Autophagy Contributes to the Death/Survival Balance in Cancer PhotoDynamic Therapy. *Cells* **2012**, *1*, 464–491. [CrossRef]
125. Nowis, D.; Szokalska, A.; Makowski, M.; Winiarska, M.; Golab, J. Improvement of anti-tumor activity of photodynamic therapy through inhibition of cytoprotective mechanism in tumor cells. In Proceedings of the Photodynamic Therapy: Back to the Future, Seattle, WA, USA, 13 July 2009; p. 73804F.
126. Oleinick, N.L.; Morris, R.L.; Belichenko, I. The role of apoptosis in response to photodynamic therapy: What, where, why, and how. *Photochem. Photobiol. Sci.* **2002**, *1*, 1–21. [CrossRef]
127. Broekgaarden, M.; Weijer, R.; van Gulik, T.M.; Hamblin, M.R.; Heger, M. Tumor cell survival pathways activated by photodynamic therapy: A molecular basis for pharmacological inhibition strategies. *Cancer Metastasis Rev.* **2015**, *34*, 643–690. [CrossRef]
128. Matroule, J.-Y.; Bonizzi, G.; Morlière, P.; Paillous, N.; Santus, R.; Bours, V.; Piette, J. Pyropheophorbide-a Methyl Ester-mediated Photosensitization Activates Transcription Factor NF-κB through the Interleukin-1 Receptor-dependent Signaling Pathway. *J. Biol. Chem.* **1999**, *274*, 2988–3000. [CrossRef]
129. Zhang, Q.; Li, L. Photodynamic combinational therapy in cancer treatment. *J. BUON.* **2018**, *23*, 561–567. [PubMed]
130. Sun, C.-Y.; Cao, Z.; Zhang, X.-J.; Sun, R.; Yu, C.-S.; Yang, X. Cascade-amplifying synergistic effects of chemo-photodynamic therapy using ROS-responsive polymeric nanocarriers. *Theranostics* **2018**, *8*, 2939–2953. [CrossRef]
131. Rengeng, L.; Qianyu, Z.; Yuehong, L.; Zhongzhong, P.; Libo, L. Sonodynamic therapy, a treatment developing from photodynamic therapy. *Photodiagnosis Photodyn. Ther.* **2017**, *19*, 159–166. [CrossRef]

132. Yang, Y.; Tu, J.; Yang, D.; Raymond, J.L.; Roy, R.A.; Zhang, D. Photo- and Sono-Dynamic Therapy: A Review of Mechanisms and Considerations for Pharmacological Agents Used in Therapy Incorporating Light and Sound. *Curr. Pharm. Des.* **2019**. [CrossRef] [PubMed]
133. Chen, H.; Gao, W.; Yang, Y.; Guo, S.; Wang, H.; Wang, W.; Zhang, S.; Zhou, Q.; Xu, H.; Yao, J.; et al. Inhibition of VDAC1 prevents Ca2+-mediated oxidative stress and apoptosis induced by 5-aminolevulinic acid mediated sonodynamic therapy in THP-1 macrophages. *Apoptosis* **2014**, *19*, 1712–1726. [CrossRef] [PubMed]
134. Ming, L.; Cheng, K.; Chen, Y.; Yang, R.; Chen, D. Enhancement of tumor lethality of ROS in photodynamic therapy. *Cancer Med.* **2021**, *10*, 257–268. [CrossRef] [PubMed]

 *life*

Review

The ATP Synthase Deficiency in Human Diseases

Chiara Galber [1,2], Stefania Carissimi [1], Alessandra Baracca [2] and Valentina Giorgio [1,2,*]

1 Consiglio Nazionale delle Ricerche, Institute of Neuroscience, I-35121 Padova, Italy; chiara.galber@phd.unipd.it (C.G.); stefania.carissimi@unipd.it (S.C.)
2 Department of Biomedical and Neuromotor Sciences, University of Bologna, I-40126 Bologna, Italy; alessandra.baracca@unibo.it
* Correspondence: valentina.giorgio4@unibo.it

Abstract: Human diseases range from gene-associated to gene-non-associated disorders, including age-related diseases, neurodegenerative, neuromuscular, cardiovascular, diabetic diseases, neurocognitive disorders and cancer. Mitochondria participate to the cascades of pathogenic events leading to the onset and progression of these diseases independently of their association to mutations of genes encoding mitochondrial protein. Under physiological conditions, the mitochondrial ATP synthase provides the most energy of the cell via the oxidative phosphorylation. Alterations of oxidative phosphorylation mainly affect the tissues characterized by a high-energy metabolism, such as nervous, cardiac and skeletal muscle tissues. In this review, we focus on human diseases caused by altered expressions of ATP synthase genes of both mitochondrial and nuclear origin. Moreover, we describe the contribution of ATP synthase to the pathophysiological mechanisms of other human diseases such as cardiovascular, neurodegenerative diseases or neurocognitive disorders.

Keywords: ATP synthase; human disease; mitochondria

Citation: Galber, C.; Carissimi, S.; Baracca, A.; Giorgio, V. The ATP Synthase Deficiency in Human Diseases. *Life* **2021**, *11*, 325. https://doi.org/10.3390/life11040325

Academic Editor: Gopal J. Babu; Giorgio Lenaz and Salvatore Nesci

Received: 8 March 2021
Accepted: 3 April 2021
Published: 8 April 2021

1. Introduction

Mitochondria support aerobic respiration and produce the majority of cellular ATP by oxidative phosphorylation (OXPHOS) [1]. Electrons derived from the oxidation of fatty acids, carbohydrates and amino acids are shuttled to oxygen along the respiratory chain complexes (I–IV) embedded in the inner mitochondrial membrane (IMM), producing water and releasing the energy necessary to pump protons from the mitochondrial matrix to the intermembrane space (IMS). This results in the formation of a transmembrane electrochemical gradient across the IMM, which enables the ATP synthase to produce ATP from ADP and inorganic phosphate [2]. A reverse catalytic process can occur under anoxia, a condition in which ATP synthase couples ATP hydrolysis to the generation of a transmembrane potential [3,4]. The mitochondrial OXPHOS is the only metabolic pathway that is under dual genetic control. It is therefore possible to distinguish genetic defects caused by (i) alterations in mitochondrial DNA (mtDNA), ~15%, e.g., Neuropathy, Ataxia, Retinitis Pigmentosa (NARP), Maternally Inherited Leigh's Syndrome (MILS) and Leber's Hereditary Optic Neuropathy (LHON) [5] and (ii) nuclear DNA (nDNA) mutations, which are inherited as Mendelian disorders. A recent review provided an update on the contribution of nuclear genes that impair mitochondrial respiration in patients and have been characterized in yeast [6]. More than 150 distinct genetic mitochondrial dysfunction syndromes characterized by a diminished OXPHOS capacity have been described [5,7–11]. Typical clinical traits include visual/hearing defects, encephalopathies, cardiomyopathies, myopathies, diabetes, liver and renal dysfunctions [12–14]. In other cases, mitochondria participate to the cascades of pathogenic events leading to the onset of several diseases, but they are not linked to their genetic origins. Mitochondria are damaged during the reperfusion of ischemic heart, age-related diseases and all the major neurodegenerative diseases—Parkinson's (PD), Alzheimer's (AD) and motor neuron diseases such as Amyotrophic Lateral Sclerosis (ALS).

In this scenario, ATP synthase has been shown to participate to the pathogenesis of different human diseases. The mitochondrial enzyme occupies the IMM of the organelle and forms dimers. Each monomeric unit (Figure 1), as shown in the latest dimeric mammalian enzyme by electron cryo-microscopy [15], is an assembly of 28 polypeptide chains of 17 different subunits organized into a catalytic globular domain, which is attached to an intrinsic membrane domain by a central stalk and a peripheral stalk [16]. This membrane-bound enzyme is a rotary machine. The membrane-bound rotor consists of eight identical c subunits (c- ring) in close association with a single a (or ATP6) subunit and is attached to the asymmetrical central stalk (subunits γ, δ and e) [17,18], which extends from the membrane domain and penetrates into the extrinsic globular catalytic domain along its central axis. As the central stalk rotates, it causes structural changes in the three catalytic sites, found mainly in each of the three β subunits, which alternate with three α subunits in the spherical extrinsic domain [16,19]. These structural changes lead to the enzyme's catalytic activity. The peripheral stalk, composed of the subunits oligomycin sensitivity conferral protein (OSCP), b, d, F6 and the membrane extrinsic region of A6L (or ATP8), links the external surface of the catalytic domain (F_1) to the a subunit in the membrane domain (F_o) [20,21]. The subunits e, f, g, A6L and 6.8 proteolipid also contribute to the membrane domain of the peripheral stalk [22–24], and in the dimeric complex, some of them are involved in forming the interface between monomers [24]. Another subunit, previously known as diabetes-associated protein in insulin sensitive-tissues (DAPIT) [23], may be involved in the formation of links between dimer units in the rows of dimers [25]. In this review, we mainly focus on mutations in mitochondrial and nuclear genes encoding ATP synthase subunits and factors important for their association to human diseases. Moreover, we describe the contribution of this enzyme to the pathogenic mechanisms of cardiovascular, neurodegenerative diseases and neurodevelopmental disorders. Due to space constraints, the modulation and regulation of the ATP synthase in cancer has not been addressed in this review; however, for details, see [26,27].

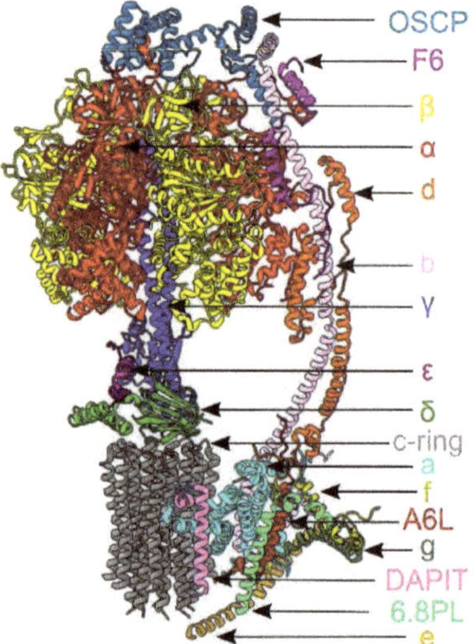

Figure 1. Subunit composition of the bovine ATP synthase monomer is mapped according to [15], (Protein Data Bank (PDB): 6ZQM). In the upper part, the subunits $\alpha(3)$ and $\beta(3)$ of the catalytic domain are red and yellow, respectively; the three central stalk subunits γ, δ and ε, are blue, indigo and green. In the lower part, the membrane domain is composed of the c8-ring and the a subunit (dark grey and light blue); the supernumerary subunits e, f, g, A6L, 6.8PL and DAPIT are khaki, straw yellow, forest green, brick red, lime green and dark pink. On the right of the model, the peripheral stalk subunits OSCP, b, d and F6 are teal, light pink, orange and magenta.

2. Gene Mutations of ATP Synthase and Its Assembly Factors in Human Disease

Disorders caused by ATP synthase deficiencies can be classified depending on the mitochondrial or nuclear genetic origin (Table 1). These diseases are often severe encephalo- or cardiomyopathies and manifest shortly after birth. Interestingly, they are less frequent than other OXPHOS-related diseases [28].

Table 1. Human pathogenic mutations occurring in ATP synthase subunits and assembly factors. The mutant subunits, or the mutant assembly factors of the ATP synthase found in human diseases are listed. Their specific nucleotide and amino acid substitutions and the related consequences on ATP synthase activity, assembly or mitochondrial morphology are summarized (nd, not defined).

ATP Synthase Subunit or Assembly Factor	mtDNA or nDNA Mutation	Protein Mutation	ATP Synthase			References
			Activity	*Assembly*	*Mitochondrial morphology*	
ATP6 (a subunit)	m.8993T>G	p.Leu156Arg	decreased	normal	nd	[29–33]
	m.8993T>C	p.Leu156Pro	decreased	nd	nd	[29–31]
	m.9176T>G	p.Leu217Arg	decreased	impaired	altered cristae	[28,34]
	m.9176T>C	p.Leu217Pro	decreased	impaired	altered cristae	[28,35]
	m.9035T>C	p.Leu170Pro	decreased	nd	nd	[36]
	m.9185T>C	p.Leu220Pro	decreased	nd	nd	[37–39]
	m.9191T>C	p.Leu222Pro	decreased	impaired (in the yeast model)	nd	[37,39]
	m.8969G>A	p.Ser148Asn	decreased	nd	nd	[40–42]
	m.8611_8612 insC	p.Leu29Profs*36	decreased	impaired	distorted mitochondria, aberrant cristae formation	[43]
ATP6 (a subunit) and ATP8 (A6L subunit)	m.8528T>C	a p.Met1Thr + A6L p.Trp55Arg	decreased	impaired	nd	[44,45]
	m.8529G>A	a p.Met1Ile + A6L p.Trp55 *	decreased	impaired	nd	[46]
	m.8561C>G	a p.Pro12Arg + A6L p.Pro66Ala	decreased	impaired	nd	[47]
	m.8561C>T	a p.Pro12Leu + A6L p.Pro66Ser	decreased	impaired	nd	[48]
ATP5F1E (ε subunit)	c.35A>G	p.Tyr12Cys	decreased	impaired	nd	[49]
ATP5F1A (α subunit)	c.985C>T	p.Arg329Cys	decreased	impaired	nd	[50]
	c.962A>G	p.Tyr321Cys	decreased	nd	nd	[51]
ATP5F1D (δ subunit)	c.245C>T	p.Pro82Leu	decreased	impaired	decreased number of cristae	[52]
	c.317T>G	p.Val106Gly	decreased	impaired	nd	[52]
ATP5MK (DAPIT subunit)	c.87+1G>C	/	decreased	impaired	altered cristae shape	[53]
ATPAF2	c.280T>A	p.Trp94Arg	decreased	Impaired	normal	[54,55]
TMEM70	c.317–2A>G	/	decreased	impaired	different alterations including swollen, giant or small mitochondria; or irregularly shaped mitochondria (with concentric, fragmented or aggregated cristae)	[56–59]

Note. (*) indicates a STOP codon.

2.1. Mitochondrial Gene Mutations of ATP Synthase

The better characterized ATP synthase diseases are caused by mutations in the mtDNA *ATP6* and *ATP8* genes [28], encoding for the human a and A6L subunits, respectively. The open reading frame of these two subunits overlap for 46 nucleotides; thus, changes occurring in this region can affect the expression of both subunits [44–48]. The majority mutations that cause defects of ATP synthase involve the *ATP6* gene. We focused our attention here on ATP synthase mutations, at the level of the *ATP6* gene, leading to altered enzyme functions that have been characterized, while we refer to other detailed reviews [11,28,60,61] and www.mitomap.org (accessed on 7 April 2021) for updated lists for the other *ATP6* mutations.

The most common of these mutations are the m.8993T>G/C (p.Leu156Arg/Pro) and the m.9176T>G/C (p.Leu217Arg/Pro) substitutions, which cause different clinical phenotypes varying from NARP to MILS, depending on mtDNA heteroplasmy [28]. These four mutations compromise mitochondrial ATP production with different degrees of severity and have also been modeled in yeast, in order to better clarify their role in the ATP synthase activity and assembly [34,35,62,63]. Not surprisingly, mutations in the yeast mitochondrial *ATP6* gene impair the ATP synthase function, since the a subunit is involved in the formation of the proton channel at the interface with the c-ring, which is formed in the F_o sector of the enzyme and is fundamental for the catalytic activity [15,64]. Another common mutation is the m.9035T>C, reported in studies of large patient cohorts [36,65,66]. Although a high level of heteroplasmy in patients has been shown to be required to develop a phenotype [36,66], the m.9035T>C mutation causes lower ATP levels, decreased ATP hydrolysis and increased reactive oxygen species (ROS) in patient tissues [42].

As clearly reported, the mtDNA 8993T>G mutation is associated with a more severe NARP/MILS clinical phenotype than the 8993T>C mutation [28]. Biochemical studies aimed at elucidating the pathogenic mechanisms of the two mutations showed that, in NARP/MILS patient cells harboring a high mutant load (>80%), the ATP synthase activity was drastically reduced (about 70%) and only slightly affected (about 20%) compared to the controls, when the mutations were the 8993T>G and 8993T>C, respectively [29–31]. Although, both mutations lead to cellular energy deficiency and increased ROS levels, the latter was reported as a major contributor to the pathogenesis of the NARP/MILS associated to the 8993T>C mutation [30]. In addition, a high percentage of 8993T>G mutation did not significantly affect either the ATP hydrolytic activity or the ATP-driven proton transport in mitochondria of patient cells, excluding that the mutation affects the assembly of the ATP synthase complex [32,33]. However, biochemical analyses in NARP/MILS lymphocytes revealed that the Leu156Arg mutant a subunit slightly affected the proton translocation through the enzyme, suggesting that the coupling between proton translocation through F_o and ATP synthesis on F_1 was altered in the mutant ATP synthase complex [29]. These studies also suggested a close relationship of biochemical defect and tissue heteroplasmy. In addition, the clinical phenotype associated to mutations at both 8993 and 9176 nucleotides was found to be worsened by defects in respiratory complex function and assembly [67]. These findings elucidate plausible factors that might contribute to the difference in severity of the clinical phenotype associated with MILS and NARP, which the alteration in ATP synthase alone was unable to explain. The rescue of the energy deficiency that characterizes the cells of NARP and MILS patients has been positively targeted by both genetic and biochemical approaches providing different tools for the development of therapeutic strategies for patient treatment [68–70].

The m.9185T>C or m.9191T>C mutations of *ATP6* are variants of the early described NARP-MILS clinical spectrum. In both cases, leucine is changed into proline, at position 220 or 222 in humans, respectively, near the C-terminus of the protein. The first mutation was reported in many patients with a mild clinical phenotype [37,38] and was associated with decreased Mg^{2+}-ATPase activity in isolated muscle mitochondria but normal respiratory chain enzyme activity [37,38]. The second mutation was instead discovered in a two-year-old patient who died presenting a severe clinical phenotype. This second mutation caused a severe reduction in Mg^{2+}-ATPase activity accompanied by a decrease in the mitochondrial respiration rate, indicating a possible reduction also in ATP synthesis [53].

The *Saccharomyces cerevisiae* yeast equivalent of the m.9185T>C mutation (p.Ser250Pro, corresponding to human p.Leu220Pro) partially impaired the yeast ATP synthase activity with a 30% decrease in mitochondrial ATP production without any evidence of a proton leak [39]. The equivalent of the human m.9191T>C mutation (p.Leu252Pro in *S. cerevisiae*) instead caused a more severe dysfunction in terms of a >95% decrease in the ATP synthesis rate accompanied by a defective ATP synthase assembly. Subcomplexes of the ATP synthase and free F_1 were detected by BN-PAGE analysis, suggesting that the mutant a subunit was not stably incorporated in the enzyme complex and, therefore, degraded. Since the proline amino acid is indeed a well-known α-helix breaker residue, a possible explanation of the described impaired ATP synthase assembly might reside in the fact that this mutation prevents the correctly folded structure of the a subunit and alters its proper interaction with the c-ring [39].

Another less frequent mutation is the *ATP6* m.8969G>A transition, which leads to the replacement of a highly conserved serine residue at position 148 of the human sequence with asparagine. This mutation was found in a six-year old male with Mitochondrial Myopathy, Lactic Acidosis and Sideroblastic Anemia (MLASA) [40], and in a 14-year old female with a severe nephropathy, carrying a high mutation level (>89%) in the kidney [57]. Biochemical investigations of mutant yeast and human cells revealed a decreased basal and oligomycin-sensitive respiration [40,41], indicating that the substitution of this serine into an asparagine severely compromised the ATP synthase activity, with a block of the proton transfer through the F_o [41]. Later on, it was shown that these detrimental consequences are caused by the amino acid substitution for asparagine. According to the authors, the asparagine (Asn175 in yeast), which replaces the serine of the normal sequence, binds (with the hydrogen bond) and neutralizes the nearby glutamate (Glu172 in yeast), which is critical for the proton flux in yeast ATP synthase [42].

A novel frameshift mutation (m.8611_8612insC) in the *ATP6* gene was discovered in 2017 in a patient with ataxia and encephalopathy symptoms [43]. A biochemical analysis revealed impaired assembly and accumulation of subcomplexes of ATP synthase, decrease in the enzyme activity and altered mitochondrial ultrastructure with aberrant cristae formation. All these features were attributed to an aberrant a subunit translation, with the consequent formation of a truncated form [43].

Mutations in the *ATP8* gene are less common. The mutations that cause ATP synthase deficiency are those occurring in the overlapping region of the *ATP6* and *ATP8* genes, thus interfering with the synthesis of both subunits. Two different nucleotide substitutions, m.8528T>C and m.8529G>A, both affecting the first amino acid residue in the human a subunit, and the amino acid residue in position 55 of the human A6L subunit (m.8528T>C (a subunit p.Met1Thr and A6L subunit p.Trp55Arg), m.8529G>A (a subunit p.Met1Ile and A6L subunit p.Trp55*)), were identified in patients suffering from severe cardiomyopathies [44–46]. The other two mutations, occurring in the overlapping region in the nucleotide m.8561 caused amino acid residue changes in position 12 and 66 of the human a and A6L subunits, respectively (m.8561C>G (a subunit p.Pro12Arg and A6L subunit p.Pro66Ala) and m.8561C>T (a subunit p.Pro12Arg and A6L subunit p.Pro66Ser)). These mutations were detected in individuals who had also early-onset ataxia and severe neurological signs [47,48]. In all these reported cases, the ATP synthase deficiency was due to an altered enzyme assembly that causes a consequent increase in the F_1 subcomplex [46–48] and a decrease in ATP production [44–48].

2.2. Nuclear Gene Mutations of ATP Synthase and Its Assembly Factors

Mutations in the nDNA genes encoding for ATP synthase subunits are very rare. Only a few different cases have been discovered over the years. The first mutation was reported in the *ATP5F1E* gene, which encodes ε subunit in the central stalk of the enzyme [49]. More recently, different mutations were found in the *ATP5F1A* (α subunit) [50,51], in the *ATP5F1D* (δ subunit) [63] or in the *ATP5MK* (DAPIT subunit) [64] genes. These nDNA mutations cause a similar and marked decrease in the content of fully assembled ATP synthase complexes, with a consequent decrease in their activity [49–52] and are detailed below.

The first case, the mutation causing Tyr12Cys amino acid residue substitution in the ε subunit, was described in a 22-year old woman presenting a neonatal-onset lactic acidosis, 3-methylglutaconic aciduria, mild mental retardation and developed peripheral neuropathy [49]. An analysis on patient fibroblasts with the homozygous missense mutation c.35A>G showed a decrease in the mitochondrial ATP synthase activity (both in ATP synthesis and hydrolysis) and assembly, caused by a reduction in the enzyme subunit level. Unlike the expression of the other subunits, the c subunit was found to accumulate and aggregate in a detergent-insoluble form, an accumulation that was also described in other disorders, such as Batten disease and fragile X syndrome, which will be discussed below. Overall, these findings suggest that the ε subunit is important for proper biosynthesis and assembly of the ATP synthase and for the proper incorporation of the c subunit into the rotor structure [49]. In line with these results, the downregulation of the ε subunit in HEK293 cells [71] or in yeast [72], caused a decrease in mitochondrial content and the activity of ATP synthase, with an effect on c subunit accumulation in HEK293 cells depleted of the ε subunit [71].

One of the other mutations (c.985C>T) of the nuclear *ATP5F1A* gene encoding the α subunit mentioned above was found in two different siblings who died in the first weeks of life [50]. The severity of this phenotype depends on the fact that the wild-type allele of the mother was not expressed in the siblings. Patient fibroblasts showed a reduction in the oxygen consumption rate, possibly caused by impairment of ATP synthase assembly and function. These cells displayed, in line with a decreased enzyme assembly, a decreased content of the subunits α, β, OSCP or d, which are important for the enzyme catalysis [50]. The authors showed that the expression of the wild-type gene encoding the α subunit in patient fibroblasts rescued the ATP synthase complex content and activity [50]. The possible explanation of the functional effects of the described mutation proposed by the authors implies that the substitution of arginine 329 for cysteine abolishes the three α-β interactions in the catalytic core of the enzyme, with the consequent loss of stability of the entire ATP synthase complex [50]. Another case is the homozygous c.962A>G mutation in the *ATP5F1A* gene that was described in two sisters born from consanguineous first-cousin parents. They both died early after birth with microcephaly, pulmonary hypertension and heart failure [51]. Patient muscle tissue showed OXPHOS deficiency and mtDNA depletion. Additionally, in this case, the p.Tyr321Cys mutation involved a highly conserved residue. The expression of the analogous yeast variant (*ATP1*: p.Tyr315Cys) in an *ATP1* knockout strain reflected the same severe phenotype with mtDNA loss, decrease of mitochondrial membrane potential and petite phenotype [51]

Other examples of nuclear mutations known to cause a mitochondrial dysfunction involved the δ subunit and its *ATP5F1D* gene. Two different homozygous mutations, c.245C>T (p.Pro82Leu) and c.317T>G (p.Val106Gly), were found in two unrelated individuals with a metabolic disorder [52]. Cultured skin fibroblasts from these individuals showed an impaired ATP synthase assembly, as revealed through BN-PAGE, and decreased enzyme activity. Moreover, in both subjects, the amount of the δ subunit was unchanged but not that of other subunits like α, β or OSCP, which were decreased in abundance. Through in silico modeling, the authors found that each of the amino acid substitutions induces changes in the predicted structure of the protein. According to these data, they hypothesized that these changes can alter the ability of the δ subunit to bind and interact with the F_1 subunits and thus affect the proper assembly of the enzyme. The pathogenicity of the two *ATP5F1D* variants was corroborated by studies performed in Drosophila. Indeed, both the mutated proteins were unable to complement the phenotypic defects caused by the δ subunit depletion in Drosophila, whereas the human wild-type subunit did [52]. Interestingly, and in line with the effect on the enzyme assembly, fibroblasts from the patient with a c.245C>T mutation showed significant decrease in mitochondrial cristae content [63], a fact consistent with the role of ATP synthase dimers in maintaining normal mitochondrial cristae morphology [73,74].

Recently, a novel homozygous splice-site mutation (c.87+1G>C) in the ATP synthase *ATP5MK* gene (encoding the DAPIT subunit) was described in three unrelated Ashkenazi Jewish families. The mutation negatively affects enzyme dimerization and ATP synthesis rate. Rescue with wild-type *ATP5MK* cDNA in patient fibroblasts restored the DAPIT protein levels, and enhanced ATP synthase dimers and their activity [53].

The biosynthesis of the eukaryotic ATP synthase is a highly organized process that requires the action of specific assembly factors [56,75–79]. It was shown that mutations in some of these "chaperone" proteins, named ATPA12 and transmembrane protein 70 (TMEM70), can be responsible for secondary ATP synthase deficiencies [80], leading to altered assembly and compromised activity of the enzyme.

The ATP12 protein is known to interact with the unassembled α subunit and is essential for its incorporation into the ATP synthase complex [78]. In a genetic study, De Meirleir et al. discovered a homozygous T>A missense mutation in exon 3 of the *ATPAF2* gene in a girl [54]. This mutation caused the amino acid substitution of a conserved tryptophan to an arginine at position 94 (p.Trp94Arg), which decreased the solubility of the protein with a tendency to aggregate [55]. The consequence is a severe decrease in the ATP synthase complex assembly and activity [54], even if no alteration in the mitochondrial morphology was observed in the fibroblasts derived from the patient carrying this mutation [55].

The other important regulatory protein in ATP synthase assembly is TMEM70, localized in the inner mitochondrial membrane [77]. Different mutations have been found over the years for *TMEM70*, with a broad spectrum of phenotypes and severity. The most common features of the syndrome caused by *TMEM70* mutations are a severe neonatal lactic acidosis, 3-methlyglutaconinc aciduria, cardiomyopathy, facial dysmorphism and mental retardation [7,81]. Early evidence for the role of TMEM70 in the enzyme assembly came in 2008 [56] and was later confirmed when it was shown that TMEM70 promotes the ATP synthase assembly by interacting with subunit c. This interaction facilitates the incorporation of the c subunit into the rotor structure of the enzyme within the inner mitochondrial membrane [76,77]. The c.317–2A>G mutation, which was firstly reported at the end of the second intron of the *TMEM70* gene, resulted in aberrant splicing and the loss of the transcript. As a consequence, low ATP synthase activity and assembly were observed. Fibroblast carrying this mutation were complemented with the wild-type *TMEM70*, which rescued structural and functional changes of ATP synthase, suggesting, for the first time, the importance of TMEM70 in the enzyme assembly [56]. All patients affected by this common mutation that were later diagnosed and exhibited ATP synthase deficiencies similar to the aforementioned case [57–59]. Other, less-common mutations on the *TMEM70* gene were found in many patients from different ethnic groups, with various phenotypes ranging from the absence of TMEM70 protein due to the premature stop codon, to the synthesis of an incomplete truncated form of the factor, lacking functional or structural domains [7,58,81–84]. As expected, a mitochondrial defect characterized by a decrease in ATP synthase assembly and activity was described in these patients [58,83,84]. Mitochondrial ultrastructural analysis in some *TMEM70* mutant samples showed a fragmented mitochondrial network and impaired mitochondrial morphology, swollen mitochondria or altered and concentric cristae [57–59,84] in line with the role of properly assembled ATP synthase in cristae shaping [73,74]. However, it has been shown that the disrupted mitochondrial cristae architecture in some patients also impairs the activity and localization of other OXPHOS complexes, increasing the severity of the disease [57,58,84]. Importantly, the mitochondrial effects caused by *TMEM70* mutations could be completely restored by complementation with the wild-type gene [84].

3. ATP Synthase Dysfunctions in other Human Diseases

The ATP synthase dysfunctions involved in the pathogenic events, leading to cardiovascular, neurodegenerative and neurocognitive diseases are described below and shown in Figure 2.

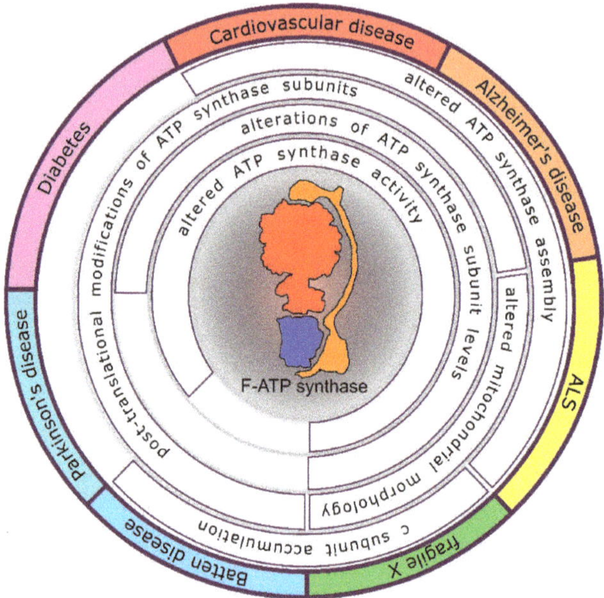

Figure 2. Schematic representation of the ATP synthase modifications involved in the progression of human diseases. Cardiovascular, Alzheimer's, Amyotrophic Lateral Sclerosis (ALS), fragile X, Batten and Parkinson's diseases are shown with different colors in the external perimeter. Changes in ATP synthase catalytic activity, assembly, subunit levels or subunit post-translational modifications and their consequence on mitochondrial morphology are listed inside the circle in correspondence of each related disease.

3.1. Cardiovascular Disease and Cardio-Protection

Myocardial cell death due to ischemia–reperfusion is a major cause of morbidity and mortality in western nations. Rouslin has first demonstrated downregulation of the mitochondrial ATP synthase activity in ischemic heart tissue from different animals [85,86]. Under ischemic condition, the ATP synthase works in reverse hydrolyzing ATP. Thus, inhibition of the ATP synthase hydrolytic activity under these conditions conserves cellular ATP levels. The membrane potential prevents uncontrolled influx of ATP into the mitochondrial matrix via the electrogenic ATP/ADP translocator, thus limiting ATP hydrolysis. Furthermore, it is stated that during ischemia, the mitochondrial ATPase inhibitor protein (IF1) binds to and inhibits the mitochondrial ATPase, thereby conserving ATP [85,87–89]. IF1 can also contribute to the myocardial ischemic preconditioning, reducing the mitochondria damage during early reperfusion [90]. Cyclophilin D (CyPD), the permeability transition pore modulator, which also inhibits the ATPase catalytic activity [91], may contribute in preventing ATP dissipation as an additional mechanism of ATP synthase modulation.

The reintroduction of oxygen during reperfusion allows the generation of ATP, but the damage to the electron transport chain results in increased mitochondrial generation of ROS. The catalytic activity of ATP synthase has been shown to be regulated in the presence of ROS in several cardiovascular studies, a fact which might be caused by oxidation of specific enzyme residues [92]. Mitochondrial Ca^{2+} overload and increased ROS can result in opening of the mitochondrial permeability transition pore, which further compromises cellular energetics and induces cell death. Apart from permeability transition-directed therapies [93], the cardioprotective strategy of ischemic preconditioning (PC), first described in 1986, provided an indication of the magnitude of the possible cardioprotective effect [94]. The rate of ATP consumption during ischemia is slower in PC hearts [95,96]. The full mechanism by which PC reduces ATP break down during ischemia remains still unknown. Interestingly, cardiac-specific overexpression of the antiapoptotic protein Bcl-2, overexpression of the cardioprotective PKC-ε and adenosine pretreatment have all been shown to slow the rate of ATP breakdown during ischemia [97–99]. Di Lisa et al. used the fluorescent membrane potential-sensitive dye JC1 to measure mitochondrial $\Delta\psi$ in anoxic rat cardiomyocytes and showed a biphasic decline in $\Delta\psi$ [100]. These authors showed that glycolytically generated ATP was used to maintain $\Delta\psi$, since $\Delta\psi$ was shown to decline more rapidly during ischemia in the presence of oligomycin, an ATP synthase inhibitor. Leyssens et al. obtained similar results using JC1 to measure $\Delta\psi$ in rat cardiomyocytes metabolically inhibited with cyanide and 2-deoxyglucose [101]. These data support the conclusion that ATP synthase is a major consumer of ATP during ischemia and/or metabolic inhibition, and they further demonstrated that the consumption of glycolytic ATP is used to maintain $\Delta\psi$ [87].

It was proposed that PC promoted earlier binding of the IF1 to ATP synthase. However, studies by two different groups using submitochondrial particles found no evidence supporting inhibition of the ATPase in PC hearts [102,103]. Other groups, on the other hand, have reported that PC and diazoxide enhance the IF1 binding to ATP synthase [104–106]. A previous study has also reported that pharmacological PC with adenosine results in increased phosphorylation of the β subunit of the ATP synthase [107], although the functional effects of phosphorylation on the ATPase activity were not addressed. Subsequent studies aimed at generating different yeast mutants to better define the role of the β subunit phosphorylation, demonstrated its impact on enzyme assembly/stability and catalytic activity [108].

Additionally, changes in the amount of ATP synthase subunits have been shown in other cardiac patients. During an inflammatory cardiomyopathy occurring in patients affected by chronic Chagas disease, the most characteristic histopathological lesions are consistent with inflammation and a myocardial remodeling process such as T-cell/macrophage-rich myocarditis, hypertrophy and fibrosis [109]. Protein analysis showed a 20% decrease in the ATP synthase subunits α and β in the myocardium of chronic Chagas patients compared to myocardial samples from individuals without cardiomyopathies [110]. Since the analysis of the mRNA levels did not show significant differences [110], it seems plausible that a modulation at the level of subunit turnover or enzyme assembly might occur in chronic Chagas patients.

3.2. *Neurodegenerative Diseases*

The neurodegenerative diseases include AD, PD, ALS and Multiple Sclerosis (MS), injury to the central nervous system (CNS) through chronic low-grade hypoxia, the rarer Huntington's disease (HD), Wilson's disease and Freidreich's Ataxia. In all these diseases, impaired ATP generation causes a failure of cellular homeostasis, with a number of consequences, including the ionic imbalance, altered Ca^{2+}-dependent transmission of information in the CNS and ultimately, necrotic or apoptotic cell death, depending on ATP depletion.

The first study to implicate ATP synthase in AD etiology found decreased level of the entire complex in the hippocampal tissue of AD patients through BN-PAGE analysis [111]. Whether the impaired complex stability in AD patients during detergent extraction also reflects changes in the catalytic activity of ATP synthase is still debated in the literature. An early study found no significant decrease in the ATP synthase catalytic activity in isolated mitochondria from AD patient hippocampal tissue, motor cortex and platelets [112]. However, there is evidence of post-translational modifications occurring in AD with consequences on the ATP synthase activity. In the hippocampus of AD patients, the α subunit was shown to be excessively nitrated in comparison to age-matched control brains [113], a post-translational modification that has been shown to inhibit the ATP synthase catalysis [92]. The α subunit can be subjected to glycosylation with O-linked β-N-acetylglucosamine (O-GlcNAcylated) on the Thr432 residue. This modification was reduced in brains of AD patients, Tg AD mice and in Aβ-treated mammalian cell cultures–resulting in reduced ATP levels [114]. Molecular modelling and co-IP experiments with deletion mutants of the α and β subunits showed that Aβ directly blocks the O-GlcNAcylation of this Thr432 residue.

Interestingly, the O-GlcNAcylation of Thr432 that had been inhibited by Aβ was rescued by treatment with the O-GlcNAcase inhibitor. These findings are particularly noteworthy as the authors demonstrated a chemical mechanism for the interaction of the Aβ peptide with mitochondrial ATP synthase, which could provide a potential therapeutic target for AD [114]. Another post-translational modification was shown in the presence of a lipid peroxidation product, 4-hydroxy-2-nonenal (4-HNE) [115]. This product was shown to modify the α subunit of ATP synthase. It caused a 35%-decrease of ATP hydrolysis in the hippocampal tissue of early stage AD individuals with mild cognitive impairment [116] and a 30%-decreased ATP synthase activity in the entorhinal cortex [117], suggesting that oxidative stress precedes the presence of Aβ in the affected tissue.

Multiple studies have pointed to a decrease in the levels of ATP synthase subunits in AD models. A decreased expression was observed in several of the nuclear ATP synthase genes in the posterior cingulate cortex, hippocampal field CA1, middle temporal gyrus, entorhinal cortex and posterior cingulate neurons [118]. A study using induced pluripotent stem cell (iPSC) -derived hippocampal neuronal cells, with familial associated presenilin 1 (PS1) mutation M146, observed a decreased level of the ATP synthase complex while PS1 expression was kept at physiological levels [119]. Neuroblastoma cells expressing the *ApoE4* allele of the *ApoE* gene, the major genetic risk factor for sporadic AD, showed a reduction in the levels of all detected ATP synthase subunits [120]. A reduced expression of the catalytic β subunit mRNA levels by over 50% was found in the mid-temporal cortex of AD patient brains [121]. In another study linking Aβ peptides with ATP synthase in AD, rats receiving a bilateral intrahippocampal injection of Aβ showed a significant decrease in the levels of β subunit of ATP synthase [122]. Gene expression analysis of the entorhinal cortex of AD patient brains showed reduced expression of γ, δ, c, and β subunit genes [123]. The subunits β, d, e, and F6 were down-regulated in the early-onset AD, as revealed by the iTRAQ quantitative mass spectrometric technique [124], whereas another proteomics analysis of hippocampal subcellular fractions from a murine AD model showed a decreased level of the peripheral stalk subunit d [125]. Interestingly, the d subunit gene was firstly thought to be genetic risk factor for AD in a genome wide association study [126].

On the contrary, increased expression was found for the ATP synthase α subunit gene in a Transgenic Swedish APP mouse (Tg2576) model for AD, with increased levels of amyloid plaque formation in the brain [127]. Moreover, a transgenic mice line (J20 Tg) producing a mutant form of APP, corresponding to the Swedish and Indiana familial forms of AD, showed a 12.2-fold increase in the α subunit level in a whole brain homogenate [128]. One might speculate that the increased subunit levels are due to an adaptive response only occurring in the AD animal models. Yet, in 2004 a study in AD patients by Manczak et al. showed increased mRNA levels for the mitochondrial *ATP6* and *ATP8* genes in brains, while increased levels of the δ subunit in the frontal cortex by immunofluorescence analysis [129].

ATP synthase is further controlled by possible interactions with other proteins modulating its activity and influencing neurodegeneration. Selective loss of the peripheral stalk subunit of ATP synthase, OSCP, was found in the brains of AD individuals and in an AD mouse model [130]. OSCP loss and complex interactions with Aβ leads to reduced ATP production, elevated oxidative stress and activated permeability transition [130]. The authors suggested that the restoration of OSCP ameliorates Aβ-mediated mouse and human neuronal mitochondrial impairments, including the effects on ATP synthesis and the resultant synaptic injury [130]. This finding demonstrating the OSCP involvement in AD is of particular interest, given that this subunit is the molecular-binding site for the CypD, the matrix prolyl-cis-trans-isomerase, which has been shown to modulate ATP synthase catalytic activity [91] and the ATP synthase transition to the permeability transition pore [131]. A previous study in an AD mouse model showed that neuronal and synaptic stress due to the interaction of CypD with mitochondrial Aβ are attenuated in CypD-deficient cortical mitochondria. CypD deficiency protected neurons from Aβ- and oxidative stress-induced cell death, in a mechanism involving the permeability transition pore [132], which was also confirmed in other AD transgenic mouse models [133]. Moreover, CypD levels, which increased in aging mice, have been shown to decrease ATP synthase activity and to promote mitochondrial dysfunction [134]. Compared with non-synaptic mitochondria, the synaptic mitochondria showed a greater degree of age-dependent accumulation of Aβ and deficits in mitochondrial function, as shown by increased mitochondrial permeability transition and decline in respiration [135]. In an AD animal model (5$\times$FAD mice) the genetic depletion of CypD mitigates OSCP loss via ubiquitin-dependent OSCP degradation and attenuates OSCP/Aβ interaction preserving the ATP synthase function, mitochondrial bioenergetics and improved mouse cognition [136]. The authors' interpretation is that CypD is a critical mediator that promotes OSCP deficits in AD-related conditions, providing a promising therapeutic strategy to correct mitochondrial dysfunction for AD therapy.

The most frequent form of neurodegenerative disorder affecting movement, PD, is caused by death of dopaminergic neurons in the mesencephalic region called substantia nigra pars compacta. In astrocytes derived from *PINK1*-knockout mice, proliferation defects were associated with a decrease in mitochondrial mass, membrane potential and ATP production as well as an increase in cellular ROS. Treatment of wild-type astrocytes with the ATP synthase inhibitor oligomycin was sufficient to mimic the proliferation phenotype observed in *PINK1*-deficient murine cells [137]. Protein aggregation and mitochondrial dysfunction are two central pathogenic processes in both familial and sporadic PD. However, the way in which these two processes converge to cause neurodegeneration was only recently proposed. Protein aggregation causes α-synuclein to switch from its physiological role to a pathological toxic gain of function form. Under physiological conditions, monomeric α-synuclein improves ATP synthase efficiency [138]. On the other hand, aggregation of α-synuclein monomers generates beta sheet-rich oligomers localized in the mitochondria in close proximity to several mitochondrial proteins including ATP synthase. Oligomers induce selective oxidation of the ATP synthase β subunit and mitochondrial lipid peroxidation. These oxidation events are proposed to increase the probability of permeability transition pore opening, triggering mitochondrial swelling and, ultimately, cell death [139]. Moreover, the protein DJ-1, linked to early onset PD, if defective, binds the ATP synthase β subunit. The interaction with the wild-type form of DJ-1 decreased the mitochondrial uncoupling and enhanced ATP production, while mutations in *PARK7* gene encoding DJ-1 (or *PARK7*-knockout) increased mitochondrial uncoupling and depolarized neuronal mitochondria [140]. The Authors suggested that this observation may depend on the presence of a leak at the level of the c-ring of ATP synthase in the membrane, which may be closed by pharmacological treatment [140].

ALS is an appalling neurodegenerative disease characterized by the loss of spinal motor neurons, which is rapidly progressive and lethal [141]. The most common genetic form of the disease is caused by GGGGCC repeat expansion in the *C9ORF72* gene. It was found that poly(GR) preferentially binds to the ATP synthase α subunit and promotes its degradation via the ubiquitin-proteasome pathway. Moreover, inducing the expression of *ATP5F1A* gene in poly(GR)-expressing neurons or reducing the poly(GR) level in adult mice after disease onset, rescued poly(GR)-induced neurotoxicity [142].

In a small subset of patients, the disease is caused by mutations in superoxide dismutase 1 [143]. Transgenic mice producing the mutant protein display mitochondrial alterations, including swelling, respiratory inhibition and an elevated generation of ROS [144,145]. Another form of ALS is caused by mutant forms of fused in sarcoma or translocated in liposarcoma (FUS), which is a multifunctional DNA/RNA-binding protein associated with neurodegeneration. In both cellular and animal models, the expression of wild-type or an ALS-associated mutant (p.Pro525Leu) FUS disrupts the formation of the mitochondrial ATP synthase supercomplexes and suppresses the activity of ATP synthase, resulting in mitochondrial cristae loss followed by mitochondrial fragmentation. Expression of FUS increases levels of the β subunit which is not properly assembled, and importantly, the downregulation of this subunit by RNA interference partially rescues neurodegenerative phenotypes [146]. In other studies, fibroblasts from patients affected by frontotemporal dementia and ALS presented mitochondrial ultrastructural alterations and fragmentation of the mitochondrial network together with respiratory chain deficiency. A missense mutation was identified (c.176C>T; p.Ser59Leu) in the *CHCHD10* gene which encodes a mitochondrial coiled-coil helix protein, whose function is unknown. Blue native-PAGE analysis of patient muscles revealed altered ATP synthase assembly which might contribute to the described abnormal organization of cristae morphology in these patients [147].

3.3. *The c Subunit of ATP Synthase and Neurodevelopmental Disorders*

Fragile X-related disorders are due to a dynamic mutation of the CGG repeat in the *FMR1* gene on chromosome X, encoding for the RNA-binding protein FMRP. In patients these disorders are associated to mental retardation and neurocognitive deficits. In primary human-derived fibroblasts, mitochondrial morphology is altered and displays "donut-shaped" organelles [148]. The mouse model of fragile X syndrome, the knockout of *FMR1*, resembled the human phenotype. Mouse brain mitochondria displayed a decreased ATP synthesis, while showing higher activities of the isolated respiratory chain complexes than in controls, suggesting a possible defect at the level of ATP synthase [149]. More recently, Elizabeth Jonas and coworkers showed that fragile X-affected neurons from mouse synthesized lower levels of cellular ATP [150]. These authors observed for the first time an increased level of the ATP synthase β subunit, an accumulation of the c subunit in insoluble aggregates in brain mitochondria and the presence of a proton leak in the F_o sector of the enzyme [150]. This finding can explain the decreased ATP synthesis in spite of a fully active respiratory chain shown in *FMR1*- knockout mouse [149]. The presence of c subunit aggregates was also revealed in the group of neuronal ceroid-lipofuscinoses diseases that are linked by common clinical and pathological features falling under the description of Batten disease. Although the ceroid-lipofuscinoses present pathologically as lysosomal storage diseases, there is severe but selective neurodegeneration that leads to the clinical signs of dementia, blindness, seizures, and premature death. It was the accumulation of subunit c of mitochondrial ATP synthase in lysosomes of ovine tissues that are models of ceroidlipofuscinosis that first drew attention to the possible association of mitochondrial dysfunction with the pathogenesis of these diseases. It was reasoned that accumulation of storage bodies containing the c subunit within lysosomes was the consequence of a defect in its catabolic pathway [151]. This hypothesis was extended to include its initial disassembly from the "F_o complex domain" of ATP synthase in the inner mitochondrial membrane [152]. This would be the initial step in the catabolic pathway and could depend on enzymes such as phospholipases rather than a protease [152]. Ca^{2+} causes a decrease of ATPase activity in isolated liver mitochondria from the ovine model of the disease, in comparison to controls displaying higher ATPase activity in the presence of Ca^{2+} [152], a fact that might be explained by proton leak but awaits further studies to be clarified.

4. Conclusions

The mitochondrial ATP synthase is a multi-subunit complex fundamental for the mitochondrial function and ATP synthesis under physiological conditions. In this review, we gave an update on the involvement of this mitochondrial enzyme in human diseases, such as encephalo- and cardiomyopathies of mitochondrial or nuclear origin, cardiovascular, neurodegenerative diseases or neurocognitive disorders, ranging from those that are caused by specific ATP synthase gene mutations to those that are instead initiated by other factors but are promoted by dysfunctions in the enzyme assembly and catalytic activity.

We here analyzed altered expressions of ATP synthase genes, enzyme subunit composition, post-translational modifications and interactions as causes of altered ATP synthase complex assembly and activity in human diseases, leading to mitochondrial morphology alterations and cell death.

Author Contributions: Conceptualization, V.G., A.B. and C.G.; writing—original draft preparation, V.G. and A.B.; writing—review and editing, V.G., A.B., C.G. and S.C. and funding acquisition, V.G. All authors have read and agreed to the published version of the manuscript.

Funding: The research in the authors' laboaroy received funding from AIRC under the MFAG 2017-ID. 20316 and PRIN 2017–ID. 201789LFKB projects, P.I. Giorgio Valentina.

Institutional Review Board Statement: Not applicable.

Informed Consent Statement: Not applicable.

Conflicts of Interest: The authors declare no conflict of interest. The funders had no role in the design of the study; in the collection, analyses or interpretation of data or in the writing of the manuscript.

References

1. Saraste, M. Oxidative Phosphorylation at the Fin de Siecle. *Science* **1999**, *283*, 1488–1493. [CrossRef] [PubMed]
2. Boyer, P.D. The Atp Synthase—A Splendid Molecular Machine. *Annu. Rev. Biochem.* **1997**, *66*, 717–749. [CrossRef] [PubMed]
3. Rees, D.M.; Montgomery, M.G.; Leslie, A.G.W.; Walker, J.E. Structural Evidence of a New Catalytic Intermediate in the Pathway of ATP Hydrolysis by F1-ATPase from Bovine Heart Mitochondria. *Proc. Natl. Acad. Sci. USA* **2012**, *109*, 11139–11143. [CrossRef] [PubMed]
4. Sgarbi, G.; Barbato, S.; Costanzini, A.; Solaini, G.; Baracca, A. The Role of the ATPase Inhibitor Factor 1 (IF1) in Cancer Cells Adaptation to Hypoxia and Anoxia. *Biochim. Biophys. Acta Bioenerg.* **2018**, *1859*, 99–109. [CrossRef] [PubMed]
5. Wallace, D.C. Mitochondrial DNAMutations in Disease and Aging. *Environ. Mol. Mutagen.* **2010**, *51*, 440–450. [CrossRef] [PubMed]
6. Franco, L.V.R.; Bremner, L.; Barros, M.H. Human Mitochondrial Pathologies of the Respiratory Chain and ATP Synthase: Contributions from Studies of *Saccharomyces Cerevisiae*. *Life* **2020**, *10*, 304. [CrossRef]
7. Hejzlarová, K.; Mráček, T.; Vrbacký, M.; Kaplanová, V.; Karbanová, V.; Nůsková, H.; Pecina, P.; Houštěk, J. Nuclear Genetic Defects of Mitochondrial ATP Synthase. *Physiol. Res.* **2014**, *63*, 57–71. [CrossRef] [PubMed]
8. Chinnery, P.F. Mitochondrial Disease in Adults: What's Old and What's New? *EMBO Mol. Med.* **2015**, *7*, 1503–1512. [CrossRef]
9. Ng, Y.S.; Turnbull, D.M. Mitochondrial Disease: Genetics and Management. *J. Neurol.* **2016**, *263*, 179–191. [CrossRef]
10. Stewart, J.B.; Chinnery, P.F. The Dynamics of Mitochondrial DNA Heteroplasmy: Implications for Human Health and Disease. *Nat. Rev. Genet.* **2015**, *16*, 530–542. [CrossRef]
11. Xu, T.; Pagadala, V.; Mueller, D.M. Understanding Structure, Function, and Mutations in the Mitochondrial ATP Synthase. *Microb. Cell* **2015**, *2*, 105–125. [CrossRef]
12. Dimauro, S.; Schon, E.A. Mitochondrial Respiratory-Chain Diseases. *N. Engl. J. Med.* **2003**, *348*, 2656–2668. [CrossRef] [PubMed]
13. Zeviani, M.; Carelli, V. Mitochondrial Disorders. *Curr. Opin. Neurol.* **2007**, *20*, 564–571. [CrossRef] [PubMed]
14. Vafai, S.B.; Mootha, V.K. Mitochondrial Disorders as Windows into an Ancient Organelle. *Nature* **2012**, *491*, 374–383. [CrossRef]
15. Spikes, T.E.; Montgomery, M.G.; Walker, J.E. Structure of the Dimeric ATP Synthase from Bovine Mitochondria. *Proc. Natl. Acad. Sci. USA* **2020**, *117*, 23519–23526. [CrossRef]
16. Walker, J.E. The ATP Synthase: The Understood, the Uncertain and the Unknown. *Biochem. Soc. Trans.* **2013**, *41*, 1–16. [CrossRef] [PubMed]
17. Watt, I.N.; Montgomery, M.G.; Runswick, M.J.; Leslie, A.G.W.; Walker, J.E. Bioenergetic Cost of Making an Adenosine Triphosphate Molecule in Animal Mitochondria. *Proc. Natl. Acad. Sci. USA* **2010**, *107*, 16823–16827. [CrossRef] [PubMed]
18. Zhou, A.; Rohou, A.; Schep, D.G.; Bason, J.V.; Montgomery, M.G.; Walker, J.E.; Grigorieffniko, N.; Rubinstein, J.L. Structure and Conformational States of the Bovine Mitochondrial ATP Synthase by Cryo-EM. *Elife* **2015**, *4*, 1–15. [CrossRef]
19. Walker, J.E. ATP Synthesis by Rotary Catalysis (Nobel Lecture). *Angew. Chem. Int. Ed.* **1998**, *37*, 2308–2319. [CrossRef]
20. Collinson, I.R.; van RaaiJ, M.J.; Runswick, J.M.; Fearnley, I.M.; Skehel, J.M.; Orriss, G.L.; Miroux, B.; Walker, J.E. ATP Synthase from Bovine Heart Mitochondria: In Vitro Assembly of a Stalk Complex in the Presence of F1-ATPase and in Its Absence. *J. Mol. Biol.* **1994**, *242*, 408–421. [CrossRef]
21. Dickson, V.K.; Silvester, J.A.; Fearnley, I.M.; Leslie, A.G.W.; Walker, J.E. On the Structure of the Stator of the Mitochondrial ATP Synthase. *EMBO J.* **2006**, *25*, 2911–2918. [CrossRef]
22. Collinson, I.R.; Skehel, J.M.; Fearnley, L.M.; Runswick, M.J.; Walker, J.E. The F1F0-ATPase Complex from Bovine Heart Mitochondria: The Molar Ratio of the Subunits in the Stalk Region Linking the F1 and F0 Domains. *Biochemistry* **1996**, *35*, 12640–12646. [CrossRef]
23. Chen, R.; Runswick, M.J.; Carroll, J.; Fearnley, I.M.; Walker, J.E. Association of Two Proteolipids of Unknown Function with ATP Synthase from Bovine Heart Mitochondria. *FEBS Lett.* **2007**, *581*, 3145–3148. [CrossRef]
24. Guo, H.; Bueler, S.; Rubinstein, J. Atomic Model for the Dimeric FO Region of Mitochondrial ATP Synthase. *Science* **2017**, *358*, 936–940. [CrossRef] [PubMed]
25. He, J.; Ford, H.C.; Carroll, J.; Douglas, C.; Gonzales, E.; Ding, S.; Fearnley, I.M.; Walker, J.E. Assembly of the Membrane Domain of ATP Synthase in Human Mitochondria. *Proc. Natl. Acad. Sci. USA* **2018**, *115*, 2988–2993. [CrossRef] [PubMed]
26. Galber, C.; Acosta, M.J.; Minervini, G.; Giorgio, V. The Role of Mitochondrial ATP Synthase in Cancer. *Biol. Chem.* **2020**, *401*, 1199–1214. [CrossRef] [PubMed]
27. Solaini, G.; Sgarbi, G.; Baracca, A. Oxidative Phosphorylation in Cancer Cells. *Biochim. Biophys. Acta Bioenerg.* **2011**, *1807*, 534–542. [CrossRef] [PubMed]

28. Dautant, A.; Meier, T.; Hahn, A.; Tribouillard-Tanvier, D.; di Rago, J.P.; Kucharczyk, R. ATP Synthase Diseases Of Mitochondrial Genetic Origin. *Front. Physiol.* **2018**, *9*, 1–16. [CrossRef] [PubMed]
29. Sgarbi, G.; Baracca, A.; Lenaz, G.; Valentino, L.M.; Carelli, V.; Solaini, G. Inefficient Coupling between Proton Transport and ATP Synthesis May Be the Pathogenic Mechanism for NARP and Leigh Syndrome Resulting from the T8993G Mutation in MtDNA. *Biochem. J.* **2006**, *395*, 493–500. [CrossRef] [PubMed]
30. Baracca, A.; Sgarbi, G.; Mattiazzi, M.; Casalena, G.; Pagnotta, E.; Valentino, M.L.; Moggio, M.; Lenaz, G.; Carelli, V.; Solaini, G. Biochemical Phenotypes Associated with the Mitochondrial ATP6 Gene Mutations at Nt8993. *Biochim. Biophys. Acta Bioenerg.* **2007**, *1767*, 913–919. [CrossRef]
31. Solaini, G.; Harris, D.A.; Lenaz, G.; Sgarbi, G.; Baracca, A. The Study of the Pathogenic Mechanism of Mitochondrial Diseases Provides Information on Basic Bioenergetics. *Biochim. Biophys. Acta Bioenerg.* **2008**, *1777*, 941–945. [CrossRef]
32. Baracca, A.; Barogi, S.; Carelli, V.; Lenaz, G.; Solaini, G. Catalytic Activities of Mitochondrial ATP Synthase in Patients with Mitochondrial DNA T8993G Mutation in the ATPase 6 Gene Encoding Subunit A. *J. Biol. Chem.* **2000**, *275*, 4177–4182. [CrossRef]
33. Carelli, V.; Baracca, A.; Barogi, S.; Pallotti, F.; Valentino, M.L.; Montagna, P.; Zeviani, M.; Pini, A.; Lenaz, G.; Baruzzi, A.; et al. Biochemical-Clinical Correlation in Patients with Different Loads of the Mitochondrial DNA T8993G Mutation. *Arch. Neurol.* **2002**, *59*, 264–270. [CrossRef] [PubMed]
34. Kucharczyk, R.; Salin, B.; Di Rago, J.P. Introducing the Human Leigh Syndrome Mutation T9176G into Saccharomyces Cerevisiae Mitochondrial DNA Leads to Severe Defects in the Incorporation of Atp6p into the ATP Synthase and in the Mitochondrial Morphology. *Hum. Mol. Genet.* **2009**, *18*, 2889–2898. [CrossRef] [PubMed]
35. Kucharczyk, R.; Ezkurdia, N.; Couplan, E.; Procaccio, V.; Ackerman, S.H.; Blondel, M.; di Rago, J.P. Consequences of the Pathogenic T9176C Mutation of Human Mitochondrial DNA on Yeast Mitochondrial ATP Synthase. *Biochim. Biophys. Acta Bioenerg.* **2010**, *1797*, 1105–1112. [CrossRef] [PubMed]
36. Ganetzky, R.D.; Stendel, C.; McCormick, E.M.; Zolkipli-Cunningham, Z.; Goldstein, A.C.; Klopstock, T.; Falk, M.J. MT-ATP6 Mitochondrial Disease Variants: Phenotypic and Biochemical Features Analysis in 218 Published Cases and Cohort of 14 New Cases. *Hum. Mutat.* **2019**, *40*, 499–515. [CrossRef] [PubMed]
37. Moslemi, A.R.; Darin, N.; Tulinius, M.; Oldfors, A.; Holme, E. Two New Mutations in the MTATP6 Gene Associated with Leigh Syndrome. *Neuropediatrics* **2005**, *36*, 314–318. [CrossRef]
38. Castagna, A.E.; Addis, J.; McInnes, R.R.; Clarke, J.T.R.; Ashby, P.; Blaser, S.; Robinson, B.H. Late Onset Leigh Syndrome and Ataxia Due to a T to C Mutation at Bp 9185 of Mitochondrial DNA. *Am. J. Med. Genet.* **2007**, *143*, 808–816. [CrossRef] [PubMed]
39. Kabala, A.M.; Lasserre, J.P.; Ackerman, S.H.; Di Rago, J.P.; Kucharczyk, R. Defining the Impact on Yeast ATP Synthase of Two Pathogenic Human Mitochondrial DNA Mutations, T9185C and T9191C. *Biochimie* **2014**, *100*, 200–206. [CrossRef]
40. Burrage, L.C.; Tang, S.; Wang, J.; Donti, T.R.; Walkiewicz, M.; Luchak, J.M.; Chen, L.C.; Schmitt, E.S.; Niu, Z.; Erana, R.; et al. Mitochondrial Myopathy, Lactic Acidosis, and Sideroblastic Anemia (MLASA) plus Associated with a Novel de Novo Mutation (m.8969G>A) in the Mitochondrial Encoded ATP6 Gene. *Mol. Genet. Metab.* **2014**, *113*, 207–212. [CrossRef]
41. Wen, S.; Niedzwiecka, K.; Zhao, W.; Xu, S.; Liang, S.; Zhu, X.; Xie, H.; Tribouillard-Tanvier, D.; Giraud, M.F.; Zeng, C.; et al. Identification of G8969>A in Mitochondrial ATP6 Gene That Severely Compromises ATP Synthase Function in a Patient with IgA Nephropathy. *Sci. Rep.* **2016**, *6*, 1–12. [CrossRef]
42. Skoczeń, N.; Dautant, A.; Binko, K.; Godard, F.; Bouhier, M.; Su, X.; Lasserre, J.P.; Giraud, M.F.; Tribouillard-Tanvier, D.; Chen, H.; et al. Molecular Basis of Diseases Caused by the MtDNA Mutation m.8969G>A in the Subunit a of ATP Synthase. *Biochim. Biophys. Acta Bioenerg.* **2018**, *1859*, 602–611. [CrossRef] [PubMed]
43. Jackson, C.B.; Hahn, D.; Schröter, B.; Richter, U.; Battersby, B.J.; Schmitt-Mechelke, T.; Marttinen, P.; Nuoffer, J.M.; Schaller, A. A Novel Mitochondrial ATP6 Frameshift Mutation Causing Isolated Complex V Deficiency, Ataxia and Encephalomyopathy. *Eur. J. Med. Genet.* **2017**, *60*, 345–351. [CrossRef] [PubMed]
44. Ware, S.M.; El-Hassan, N.; Kahler, S.G.; Zhang, Q.; Ma, Y.W.; Miller, E.; Wong, B.; Spicer, R.L.; Craigen, W.J.; Kozel, B.A.; et al. Infantile Cardiomyopathy Caused by a Mutation in the Overlapping Region of Mitochondrial ATPase 6 and 8 Genes. *J. Med. Genet.* **2009**, *46*, 308–314. [CrossRef] [PubMed]
45. Imai, A.; Fujita, S.; Kishita, Y.; Kohda, M.; Tokuzawa, Y.; Hirata, T.; Mizuno, Y.; Harashima, H.; Nakaya, A.; Sakata, Y.; et al. Rapidly Progressive Infantile Cardiomyopathy with Mitochondrial Respiratory Chain Complex v Deficiency Due to Loss of ATPase 6 and 8 Protein. *Int. J. Cardiol.* **2016**, *207*, 203–205. [CrossRef]
46. Jonckheere, A.I.; Hogeveen, M.; Nijtmans, L.G.J.; van den Brand, M.A.M.; Janssen, A.J.M.; Diepstra, J.H.S.; van den Brandt, F.C.A.; van den Heuvel, L.P.; Hol, F.A.; Hofste, T.G.J.; et al. A Novel Mitochondrial ATP8 Gene Mutation in a Patient with Apical Hypertrophic Cardiomyopathy and Neuropathy. *J. Med. Genet.* **2008**, *45*, 129–133. [CrossRef]
47. Kytövuori, L.; Lipponen, J.; Rusanen, H.; Komulainen, T.; Martikainen, M.H.; Majamaa, K. A Novel Mutation m.8561C>G in MT-ATP6/8 Causing a Mitochondrial Syndrome with Ataxia, Peripheral Neuropathy, Diabetes Mellitus, and Hypergonadotropic Hypogonadism. *J. Neurol.* **2016**, *263*, 2188–2195. [CrossRef] [PubMed]
48. Fragaki, K.; Chaussenot, A.; Serre, V.; Acquaviva, C.; Bannwarth, S.; Rouzier, C.; Chabrol, B.; Paquis-Flucklinger, V. A Novel Variant m.8561C>T in the Overlapping Region of MT-ATP6 and MT-ATP8 in a Child with Early-Onset Severe Neurological Signs. *Mol. Genet. Metab. Rep.* **2019**, *21*, 1–3. [CrossRef]

49. Mayr, J.A.; Havlíčková, V.; Zimmermann, F.; Magler, I.; Kaplanová, V.; Ješina, P.; Pecinová, A.; Nůsková, H.; Koch, J.; Sperl, W.; et al. Mitochondrial ATP Synthase Deficiency Due to a Mutation in the ATP5E Gene for the F1 ε Subunit. *Hum. Mol. Genet.* **2010**, *19*, 3430–3439. [CrossRef] [PubMed]
50. Jonckheere, A.I.; Herma Renkema, G.; Bras, M.; van den Heuvel, L.P.; Hoischen, A.; Gilissen, C.; Nabuurs, S.B.; Huynen, M.A.; de Vries, M.C.; Smeitink, J.A.M.; et al. A Complex v ATP5A1 Defect Causes Fatal Neonatal Mitochondrial Encephalopathy. *Brain* **2013**, *136*, 1544–1554. [CrossRef]
51. Lieber, D.S.; Calvo, S.E.; Shanahan, K.; Slate, N.G.; Liu, S.; Hershman, S.G.; Gold, N.B.; Chapman, B.A.; Thorburn, D.R.; Berry, G.T.; et al. Targeted Exome Sequencing of Suspected Mitochondrial Disorders. *Neurology* **2013**, *80*, 1762–1770. [CrossRef]
52. Oláhová, M.; Yoon, W.H.; Thompson, K.; Jangam, S.; Fernandez, L.; Davidson, J.M.; Kyle, J.E.; Grove, M.E.; Fisk, D.G.; Kohler, J.N.; et al. Biallelic Mutations in ATP5F1D, Which Encodes a Subunit of ATP Synthase, Cause a Metabolic Disorder. *Am. J. Hum. Genet.* **2018**, *102*, 494–504. [CrossRef] [PubMed]
53. Barca, E.; Ganetzky, R.D.; Potluri, P.; Juanola-Falgarona, M.; Gai, X.; Li, D.; Jalas, C.; Hirsch, Y.; Emmanuele, V.; Tadesse, S.; et al. USMG5 Ashkenazi Jewish Founder Mutation Impairs Mitochondrial Complex V Dimerization and ATP Synthesis. *Hum. Mol. Genet.* **2018**, *27*, 3305–3312. [CrossRef] [PubMed]
54. De Meirleir, L.; Seneca, S.; Lissens, W.; De Clercq, I.; Eyskens, F.; Gerlo, E.; Smet, J.; Van Coster, R. Respiratory Chain Complex V Deficiency Due to a Mutation in the Assembly Gene ATP12. *J. Med. Genet.* **2004**, *41*, 120–124. [CrossRef] [PubMed]
55. Meulemans, A.; Seneca, S.; Pribyl, T.; Smet, J.; Alderweirldt, V.; Waeytens, A.; Lissens, W.; van Coster, R.; De Meirleir, L.; Di Rago, J.P.; et al. Defining the Pathogenesis of the Human Atp12p W94R Mutation Using a Saccharomyces Cerevisiae Yeast Model. *J. Biol. Chem.* **2010**, *285*, 4099–4109. [CrossRef]
56. Čížková, A.; Stránecký, V.; Mayr, J.A.; Tesařová, M.; Havlíčková, V.; Paul, J.; Ivánek, R.; Kuss, A.W.; Hansíková, H.; Kaplanová, V.; et al. TMEM70 Mutations Cause Isolated ATP Synthase Deficiency and Neonatal Mitochondrial Encephalocardiomyopathy. *Nat. Genet.* **2008**, *40*, 1288–1290. [CrossRef] [PubMed]
57. Cameron, J.M.; Levandovskiy, V.; MacKay, N.; Ackerley, C.; Chitayat, D.; Raiman, J.; Halliday, W.H.; Schulze, A.; Robinson, B.H. Complex V TMEM70 Deficiency Results in Mitochondrial Nucleoid Disorganization. *Mitochondrion* **2011**, *11*, 191–199. [CrossRef] [PubMed]
58. Diodato, D.; Invernizzi, F.; Lamantea, E.; Fagiolari, G.; Parini, R.; Menni, F.; Parenti, G.; Bollani, L.; Pasquini, E.; Donati, M.A.; et al. Common and Novel TMEM70 Mutations in a Cohort of Italian Patients with Mitochondrial Encephalocardiomyopathy. *JIMD Rep.* **2014**, *4*, 71–78. [CrossRef]
59. Braczynski, A.K.; Vlaho, S.; Müller, K.; Wittig, I.; Blank, A.E.; Tews, D.S.; Drott, U.; Kleinle, S.; Abicht, A.; Horvath, R.; et al. ATP Synthase Deficiency Due to TMEM70 Mutation Leads to Ultrastructural Mitochondrial Degeneration and Is Amenable to Treatment. *Biomed Res. Int.* **2015**, 1–10. [CrossRef]
60. Kucharczyk, R.; Zick, M.; Bietenhader, M.; Rak, M.; Couplan, E.; Blondel, M.; Caubet, S.D.; di Rago, J.P. Mitochondrial ATP Synthase Disorders: Molecular Mechanisms and the Quest for Curative Therapeutic Approaches. *Biochim. Biophys. Acta Mol. Cell Res.* **2009**, *1793*, 186–199. [CrossRef]
61. Jonckheere, A.I.; Smeitink, J.A.M.; Rodenburg, R.J.T. Mitochondrial ATP Synthase: Architecture, Function and Pathology. *J. Inherit. Metab. Dis.* **2012**, *35*, 211–225. [CrossRef] [PubMed]
62. Rak, M.; Tetaud, E.; Duvezin-Caubet, S.; Ezkurdia, N.; Bietenhader, M.; Rytka, J.; Di Rago, J.P. A Yeast Model of the Neurogenic Ataxia Retinitis Pigmentosa (NARP) T8993G Mutation in the Mitochondrial ATP Synthase-6 Gene. *J. Biol. Chem.* **2007**, *282*, 34039–34047. [CrossRef] [PubMed]
63. Kucharczyk, R.; Rak, M.; di Rago, J.P. Biochemical Consequences in Yeast of the Human Mitochondrial DNA 8993T > C Mutation in the ATPase6 Gene Found in NARP/MILS Patients. *Biochim. Biophys. Acta Mol. Cell Res.* **2009**, *1793*, 817–824. [CrossRef] [PubMed]
64. Kühlbrandt, W.; Davies, K.M. Rotary ATPases: A New Twist to an Ancient Machine. *Trends Biochem. Sci.* **2016**, *41*, 106–116. [CrossRef] [PubMed]
65. Stendel, C.; Neuhofer, C.; Floride, E.; Yuqing, S.; Ganetzky, R.D.; Park, J.; Freisinger, P.; Kornblum, C.; Kleinle, S.; Schöls, L.; et al. Delineating MT-ATP6 -Associated Disease: From Isolated Neuropathy to Early Onset Neurodegeneration. *Neurol. Genet.* **2020**, *6*, e395–e406. [CrossRef]
66. Ng, Y.S.; Martikainen, M.H.; Gorman, G.S.; Blain, A.; Bugiardini, E.; Bunting, A.; Schaefer, A.M.; Alston, C.L.; Blakely, E.L.; Sharma, S.; et al. Pathogenic Variants in MT-ATP6: A United Kingdom–Based Mitochondrial Disease Cohort Study. *Ann. Neurol.* **2019**, *86*, 310–315. [CrossRef]
67. D'Aurelio, M.; Vives-Bauza, C.; Davidson, M.M.; Manfredi, G. Mitochondrial DNA Background Modifies the Bioenergetics of NARP/MILS ATP6 Mutant Cells. *Hum. Mol. Genet.* **2010**, *19*, 374–386. [CrossRef] [PubMed]
68. Manfredi, G.; Fu, J.; Ojaimi, J.; Sadlock, J.E.; Kwong, J.Q.; Guy, J.; Schon, E.A. Rescue of a Deficiency in ATP Synthesis by Transfer of MTATP6, a Mitochondrial DNA-Encoded Gene, to the Nucleus. *Nat. Genet.* **2002**, *30*, 394–399. [CrossRef] [PubMed]
69. Alexeyev, M.F.; Venediktova, N.; Pastukh, V.; Shokolenko, I.; Bonilla, G.; Wilson, G.L. Selective Elimination of Mutant Mitochondrial Genomes as Therapeutic Strategy for the Treatment of NARP and MILS Syndromes. *Gene Ther.* **2008**, *15*, 516–523. [CrossRef]
70. Sgarbi, G.; Casalena, G.A.; Baracca, A.; Lenaz, G.; DiMauro, S.; Solaini, G. Human NARP Mitochondrial Mutation Metabolism Corrected with α-Ketoglutarate/Aspartate. *Arch. Neurol.* **2009**, *66*, 951–957. [CrossRef] [PubMed]

71. HavlíČková, V.; Kaplanová, V.; NŮsková, H.; Drahota, Z.; Houštěk, J. Knockdown of F1 Epsilon Subunit Decreases Mitochondrial Content of ATP Synthase and Leads to Accumulation of Subunit C. *Biochim. Biophys. Acta Bioenerg.* **2010**, *1797*, 1124–1129. [CrossRef] [PubMed]
72. Tetaud, E.; Godard, F.; Giraud, M.F.; Ackerman, S.H.; Di Rago, J.P. The Depletion of F1 Subunit ε in Yeast Leads to an Uncoupled Respiratory Phenotype That Is Rescued by Mutations in the Proton-Translocating Subunits of F0. *Mol. Biol. Cell* **2014**, *25*, 791–799. [CrossRef] [PubMed]
73. Strauss, M.; Hofhaus, G.; Schröder, R.R.; Kühlbrandt, W. Dimer Ribbons of ATP Synthase Shape the Inner Mitochondrial Membrane. *EMBO J.* **2008**, *27*, 1154–1160. [CrossRef] [PubMed]
74. Davies, K.M.; Strauss, M.; Daum, B.; Kief, J.H.; Osiewacz, H.D.; Rycovska, A.; Zickermann, V.; Kühlbrandt, W. Macromolecular Organization of ATP Synthase and Complex I in Whole Mitochondria. *Proc. Natl. Acad. Sci. USA* **2011**, *108*, 14121–14126. [CrossRef] [PubMed]
75. Ackerman, S.H.; Tzagoloff, A. Identification of Two Nuclear Genes (ATPJ1, ATP12) Required for Assembly of the Yeast F1-ATPase. *Proc. Natl. Acad. Sci. USA* **1990**, *87*, 4986–4990. [CrossRef] [PubMed]
76. Kovalčíková, J.; Vrbacký, M.; Pecina, P.; Tauchmannová, K.; Nůsková, H.; Kaplanová, V.; Brázdová, A.; Alán, L.; Eliáš, J.; Čunátová, K.; et al. TMEM70 Facilitates Biogenesis of Mammalian ATP Synthase by Promoting Subunit c Incorporation into the Rotor Structure of the Enzyme. *FASEB J.* **2019**, *33*, 14103–14117. [CrossRef]
77. Bahri, H.; Buratto, J.; Rojo, M.; Dompierre, J.P.; Salin, B.; Blancard, C.; Cuvellier, S.; Rose, M.; Elgaaied, A.B.A.; Tetaud, E.; et al. TMEM70 Promotes ATP Synthase Assembly within Cristae via Transient Interactions with Subunit C. *bioRxiv Cell Biol.* **2020**. [CrossRef]
78. Wang, Z.G.; White, P.S.; Ackerman, S.H. Atp11p and Atp12p Are Assembly Factors for the F1-ATPase in Human Mitochondria. *J. Biol. Chem.* **2001**, *276*, 30773–30778. [CrossRef]
79. Vrbacký, M.; Kovalčiková, J.; Chawengsaksophak, K.; Beck, I.M.; Mráček, T.; Nůsková, H.; Sedmera, D.; Papoušek, F.; Kolář, F.; Sobol, M.; et al. Knockout of Tmem70 Alters Biogenesis of ATP Synthase and Leads to Embryonal Lethality in Mice. *Hum. Mol. Genet.* **2016**, *25*, 4674–4685. [CrossRef]
80. Ghezzi, D.; Zeviani, M. Human Diseases Associated with Defects in Assembly of OXPHOS Complexes. *Essays Biochem.* **2018**, *62*, 271–286. [CrossRef] [PubMed]
81. Magner, M.; Dvorakova, V.; Tesarova, M.; Mazurova, S.; Hansikova, H.; Zahorec, M.; Brennerova, K.; Bzduch, V.; Spiegel, R.; Horovitz, Y.; et al. TMEM70 Deficiency: Long-Term Outcome of 48 Patients. *J. Inherit. Metab. Dis.* **2015**, *38*, 417–426. [CrossRef] [PubMed]
82. Honzík, T.; Tesařová, M.; Mayr, J.A.; Hansíková, H.; Ješina, P.; Bodamer, O.; Koch, J.; Magner, M.; Freisinger, P.; Huemer, M.; et al. Mitochondrial Encephalocardio-Myopathy with Early Neonatal Onset Due to TMEM70 Mutation. *Arch. Dis. Child.* **2010**, *95*, 296–301. [CrossRef] [PubMed]
83. Spiegel, R.; Khayat, M.; Shalev, S.A.; Horovitz, Y.; Mandel, H.; Hershkovitz, E.; Barghuti, F.; Shaag, A.; Saada, A.; Korman, S.H.; et al. TMEM70 Mutations Are a Common Cause of Nuclear Encoded ATP Synthase Assembly Defect: Further Delineation of a New Syndrome. *J. Med. Genet.* **2011**, *48*, 177–182. [CrossRef] [PubMed]
84. Jonckheere, A.I.; Huigsloot, M.; Lammens, M.; Jansen, J.; van den Heuvel, L.P.; Spiekerkoetter, U.; von Kleist-Retzow, J.C.; Forkink, M.; Koopman, W.J.H.; Szklarczyk, R.; et al. Restoration of Complex V Deficiency Caused by a Novel Deletion in the Human TMEM70 Gene Normalizes Mitochondrial Morphology. *Mitochondrion* **2011**, *11*, 954–963. [CrossRef] [PubMed]
85. Rouslin, W. Protonic Inhibition of the Mitochondrial Oligomycin-Sensitive Adenosine 5′-Triphosphatase in Ischemic and Autolyzing Cardiac Muscle. *J. Biol. Chem.* **1983**, *258*, 9657–9661. [CrossRef]
86. Rouslin, W. The Mitochondrial Adenosine 5′-Triphosphatase in Slow and Fast Heart Rate Hearts. *Am. J. Physiol.* **1987**, *252*, H622–H627. [CrossRef]
87. Solaini, G.; Harris, D.A. Biochemical Dysfunction in Heart Mitochondria Exposed to Ischaemia and Reperfusion. *Biochem. J.* **2005**, *390*, 377–394. [CrossRef]
88. Rouslin, W.; Erickson, J.L.; Solaro, R.J. Effects of Oligomycin and Acidosis on Rates of ATP Depletion in Ischemic Heart Muscle. *Am. J. Physiol. Hear. Circ. Physiol.* **1986**, *250*, H503–H508. [CrossRef]
89. Rouslin, W.; Broge, C.W. IF1 Function in Situ in Uncoupler-Challenged Ischemic Rabbit, Rat, and Pigeon Hearts. *J. Biol. Chem.* **1996**, *271*, 23638–23641. [CrossRef] [PubMed]
90. Bosetti, F.; Baracca, A.; Lenaz, G.; Solaini, G. Increased State 4 Mitochondrial Respiration and Swelling in Early Post-Ischemic Reperfusion of Rat Heart. *FEBS Lett.* **2004**, *563*, 161–164. [CrossRef]
91. Giorgio, V.; Bisetto, E.; Soriano, M.E.; Dabbeni-Sala, F.; Basso, E.; Petronilli, V.; Forte, M.A.; Bernardi, P.; Lippe, G. Cyclophilin D Modulates Mitochondrial F0F1-ATP Synthase by Interacting with the Lateral Stalk of the Complex. *J. Biol. Chem.* **2009**, *284*, 33982–33988. [CrossRef]
92. Kaludercic, N.; Giorgio, V. The Dual Function of Reactive Oxygen/Nitrogen Species in Bioenergetics and Cell Death: The Role of ATP Synthase. *Oxidative Med. Cell. Longev.* **2016**, *2016*, 1–17. [CrossRef]
93. Di Lisa, F.; Carpi, A.; Giorgio, V.; Bernardi, P. The Mitochondrial Permeability Transition Pore and Cyclophilin D in Cardioprotection. *Biochim. Biophys. Acta* **2011**, *1813*, 1316–1322. [CrossRef] [PubMed]
94. Murry, C.E.; Jennings, R.B.; Reimer, K.A. Preconditioning with Ischemia: A Delay of Lethal Cell Injury in Ischemic Myocardium. *Circulation* **1986**, *74*, 1124–1136. [CrossRef] [PubMed]

95. Murry, C.E.; Richard, V.J.; Reimer, K.A.; Jennings, R.B. Ischemic Preconditioning Slows Energy Metabolism and Delays Ultrastructural Damage during a Sustained Ischemic Episode. *Circ. Res.* **1990**, *66*, 913–931. [CrossRef] [PubMed]
96. Steenbergen, C.; Perlman, M.E.; London, R.E.; Murphy, E. Mechanism of Preconditioning: Ionic Alterations. *Circ. Res.* **1993**, *72*, 112–125. [CrossRef]
97. Cross, H.R.; Murphy, E.; Bolli, R.; Ping, P.; Steenbergen, C. Expression of Activated PKC Epsilon (PKCε) Protects the Ischemic Heart, without Attenuating Ischemic H+ Production. *J. Mol. Cell. Cardiol.* **2002**, *34*, 361–367. [CrossRef] [PubMed]
98. Fralix, T.A.; Murphy, E.; London, R.E.; Steenbergen, C. Protective Effects of Adenosine in the Perfused Rat Heart: Changes in Metabolism and Intracellular Ion Homeostasis. *Am. J. Physiol. Cell Physiol.* **1993**, *264*, 986–994. [CrossRef] [PubMed]
99. Imahashi, K.; Schneider, M.D.; Steenbergen, C.; Murphy, E. Transgenic Expression of Bcl-2 Modulates Energy Metabolism, Prevents Cytosolic Acidification during Ischemia, and Reduces Ischemia/Reperfusion Injury. *Circ. Res.* **2004**, *95*, 734–741. [CrossRef] [PubMed]
100. Di Lisa, F.; Blank, P.S.; Colonna, R.; Gambassi, G.; Silverman, H.S.; Stern, M.D.; Hansford, R.G. Mitochondrial Membrane Potential in Single Living Adult Rat Cardiac Myocytes Exposed to Anoxia or Metabolic Inhibition. *J. Physiol.* **1995**, *486*, 1–13. [CrossRef]
101. Leyssens, A.; Nowicky, A.V.; Patterson, L.; Crompton, M.; Duchen, M.R. The Relationship between Mitochondrial State, ATP Hydrolysis, [Mg2+]i and [Ca2+]i Studied in Isolated Rat Cardiomyocytes. *J. Physiol.* **1996**, *496*, 111–128. [CrossRef] [PubMed]
102. Green, D.W.; Murray, H.N.; Sleph, P.G.; Wang, F.L.; Baird, A.J.; Rogers, W.L.; Grover, G.J. Preconditioning in Rat Hearts Is Independent of Mitochondrial F1F0 ATPase Inhibition. *Am. J. Physiol. Hear. Circ. Physiol.* **1998**, *274*, 90–97. [CrossRef] [PubMed]
103. Heide, R.S.V.; Hill, M.L.; Reimer, K.A.; Jennings, R.B. Effect of Reversible Ischemia on the Activity of the Mitochondrial ATPase: Relationship to Ischemic Preconditioning. *J. Mol. Cell. Cardiol.* **1996**, *28*, 103–112. [CrossRef] [PubMed]
104. Ala-Rämi, A.; Ylitalo, K.V.; Hassinen, I.E. Ischaemic Preconditioning and a Mitochondrial KATP Channel Opener Both Produce Cardioprotection Accompanied by F1F0-ATPase Inhibition in Early Ischaemia. *Basic Res. Cardiol.* **2003**, *98*, 250–258. [CrossRef] [PubMed]
105. Contessi, S.; Metelli, G.; Mavelli, I.; Lippe, G. Diazoxide Affects the IF1 Inhibitor Protein Binding to F 1 Sector of Beef Heart F0F1ATPsynthase. *Biochem. Pharmacol.* **2004**, *67*, 1843–1851. [CrossRef] [PubMed]
106. Comelli, M.; Metelli, G.; Mavelli, I. Downmodulation of Mitochondrial F0F1 ATP Synthase by Diazoxide in Cardiac Myoblasts: A Dual Effect of the Drug. *Am. J. Physiol. Hear. Circ. Physiol.* **2007**, *292*, 820–829. [CrossRef]
107. Arrell, D.K.; Elliott, S.T.; Kane, L.A.; Guo, Y.; Ko, Y.H.; Pedersen, P.L.; Robinson, J.; Murata, M.; Murphy, A.M.; Marbán, E.; et al. Proteomic Analysis of Pharmacological Preconditioning: Novel Protein Targets Converge to Mitochondrial Metabolism Pathways. *Circ. Res.* **2006**, *99*, 706–714. [CrossRef] [PubMed]
108. Kane, L.A.; Youngman, M.J.; Jensen, R.E.; Jennifer, E.; Van, E. Phosphorylation of the F1Fo ATP Synthase β Subunit: Functional and Structural Consequences Assessed in a Model System. *Circ. Res.* **2011**, *106*, 504–513. [CrossRef] [PubMed]
109. Higuchi, M.D.L.; De Morais, F.C.; Pereira Barreto, A.C.; Lopes, E.A.; Stolf, N.; Bellotti, G.; Pileggi, F. The Role of Active Myocarditis in the Development of Heart Failure in Chronic Chagas' Disease: A Study Based on Endomyocardial Biopsies. *Clin. Cardiol.* **1987**, *10*, 665–670. [CrossRef]
110. Teixeira, P.C.; Santos, R.H.B.; Fiorelli, A.I.; Bilate, A.M.B.; Benvenuti, L.A.; Stolf, N.A.; Kalil, J.; Cunha-Neto, E. Selective Decrease of Components of the Creatine Kinase System and ATP Synthase Complex in Chronic Chagas Disease Cardiomyopathy. *PLoS Negl. Trop. Dis.* **2011**, *5*, 1–9. [CrossRef]
111. Schägger, H.; Ohm, T.G. Human Diseases with Defects in Oxidative Phosphorylation. *Eur. J. Biochem.* **2008**, *227*, 916–921. [CrossRef]
112. Bosetti, F.; Brizzi, F.; Barogi, S.; Mancuso, M.; Siciliano, G.; Tendi, E.A.; Murri, L.; Rapoport, S.I.; Solaini, G. Cytochrome c Oxidase and Mitochondrial F1F0-ATPase (ATP Synthase) Activities in Platelets and Brain from Patients with Alzheimer's Disease. *Neurobiol. Aging* **2002**, *23*, 371–376. [CrossRef]
113. Sultana, R.; Poon, H.F.; Cai, J.; Pierce, W.M.; Merchant, M.; Klein, J.B.; Markesbery, W.R.; Butterfield, D.A. Identification of Nitrated Proteins in Alzheimer's Disease Brain Using a Redox Proteomics Approach. *Neurobiol. Dis.* **2006**, *22*, 76–87. [CrossRef] [PubMed]
114. Cha, M.Y.; Cho, H.J.; Kim, C.; Jung, Y.O.; Kang, M.J.; Murray, M.E.; Hong, H.S.; Choi, Y.J.; Choi, H.; Kim, D.K.; et al. Mitochondrial ATP Synthase Activity Is Impaired by Suppressed O-GlcNAcylation in Alzheimer's Disease. *Hum. Mol. Genet.* **2015**, *24*, 6492–6504. [CrossRef] [PubMed]
115. Reed, T.T. Lipid Peroxidation and Neurodegenerative Disease. *Free Radic. Biol. Med.* **2011**, *51*, 1302–1319. [CrossRef] [PubMed]
116. Reed, T.; Perluigi, M.; Sultana, R.; Pierce, W.M.; Klein, J.B.; Turner, D.M.; Coccia, R.; Markesbery, W.R.; Butterfield, D.A. Redox Proteomic Identification of 4-Hydroxy-2-Nonenal-Modified Brain Proteins in Amnestic Mild Cognitive Impairment: Insight into the Role of Lipid Peroxidation in the Progression and Pathogenesis of Alzheimer's Disease. *Neurobiol. Dis.* **2008**, *30*, 107–120. [CrossRef] [PubMed]
117. Terni, B.; Boada, J.; Portero-Otin, M.; Pamplona, R.; Ferrer, I. Mitochondrial ATP-Synthase in the Entorhinal Cortex Is a Target of Oxidative Stress at Stages I/II of Alzheimer's Disease Pathology. *Brain Pathol.* **2010**, *20*, 222–233. [CrossRef]
118. Liang, W.S.; Reiman, E.M.; Valla, J.; Dunckley, T.; Beach, T.G.; Grover, A.; Niedzielko, T.L.; Schneider, L.E.; Mastroeni, D.; Caselli, R.; et al. Alzheimer's Disease Is Associated with Reduced Expression of Energy Metabolism Genes in Posterior Cingulate Neurons. *Proc. Natl. Acad. Sci. USA* **2008**, *105*, 4441–4446. [CrossRef] [PubMed]

119. Martín-Maestro, P.; Sproul, A.; Martinez, H.; Paquet, D.; Gerges, M.; Noggle, S.; Starkov, A.A. Autophagy Induction by Bexarotene Promotes Mitophagy in Presenilin 1 Familial Alzheimer's Disease IPSC-Derived Neural Stem Cells. *Mol. Neurobiol.* **2019**, *56*, 8220–8236. [CrossRef] [PubMed]
120. Orr, A.L.; Kim, C.; Jimenez-Morales, D.; Newton, B.W.; Johnson, J.R.; Krogan, N.J.; Swaney, D.L.; Mahley, R.W. Neuronal Apolipoprotein E4 Expression Results in Proteome-Wide Alterations and Compromises Bioenergetic Capacity by Disrupting Mitochondrial Function. *J. Alzheimer's Dis.* **2019**, *68*, 991–1011. [CrossRef]
121. Chandrasekaran, K.; Hatanpää, K.; Rapoport, S.I.; Brady, D.R. Decreased Expression of Nuclear and Mitochondrial DNA-Encoded Genes of Oxidative Phosphorylation in Association Neocortex in Alzheimer Disease. *Mol. Brain Res.* **1997**, *44*, 99–104. [CrossRef]
122. Shi, X.; Lu, X.; Zhan, L.; Liu, L.; Sun, M.Z.; Gong, X.; Sui, H.; Niu, X.; Liu, S.; Zheng, L.; et al. Rat Hippocampal Proteomic Alterations Following Intrahippocampal Injection of Amyloid Beta Peptide (1–40). *Neurosci. Lett.* **2011**, *500*, 87–91. [CrossRef] [PubMed]
123. Ding, B.; Xi, Y.; Gao, M.; Li, Z.; Xu, C.; Fan, S.; He, W. Gene Expression Profiles of Entorhinal Cortex in Alzheimer's Disease. *Am. J. Alzheimers. Dis. Other Demen.* **2014**, *29*, 526–532. [CrossRef] [PubMed]
124. Adav, S.S.; Park, J.E.; Sze, S.K. Quantitative Profiling Brain Proteomes Revealed Mitochondrial Dysfunction in Alzheimer's Disease. *Mol. Brain* **2019**, *12*, 1–12. [CrossRef] [PubMed]
125. Yu, H.; Lin, X.; Wang, D.; Zhang, Z.; Guo, Y.; Ren, X.; Xu, B.; Yuan, J.; Liu, J.; Spencer, P.S.; et al. Mitochondrial Molecular Abnormalities Revealed by Proteomic Analysis of Hippocampal Organelles of Mice Triple Transgenic for Alzheimer Disease. *Front. Mol. Neurosci.* **2018**, *11*, 1–13. [CrossRef] [PubMed]
126. Boada, M.; Antúnez, C.; Ramírez-Lorca, R.; Destefano, A.L.; González-Pérez, A.; Gayán, J.; López-Arrieta, J.; Ikram, M.A.; Hernández, I.; Marín, J.; et al. ATP5H/KCTD2 Locus Is Associated with Alzheimer's Disease Risk. *Mol. Psychiatry* **2014**, *19*, 682–687. [CrossRef]
127. Carrette, O.; Burgess, J.A.; Burkhard, P.R.; Lang, C.; Côte, M.; Rodrigo, N.; Hochstrasser, D.F.; Sanchez, J.C. Changes of the Cortex Proteome and Apolipoprotein E in Transgenic Mouse Models of Alzheimer's Disease. *J. Chromatogr. B* **2006**, *840*, 1–9. [CrossRef] [PubMed]
128. Robinson, R.A.S.; Lange, M.B.; Sultana, R.; Galvan, V.; Fombonne, J.; Gorostiza, O.; Zhang, J.; Warrier, G.; Cai, J.; Pierce, W.M.; et al. Differential Expression and Redox Proteomics Analyses of an Alzheimer Disease Transgenic Mouse Model: Effects of the Amyloid-β Peptide of Amyloid Precursor Protein. *Neuroscience* **2011**, *177*, 207–222. [CrossRef] [PubMed]
129. Manczak, M.; Park, B.S.; Jung, Y.; Reddy, P.H. Differential Expression of Oxidative Phosphorylation Genes in Patients with Alzheimer's Disease. *NeuroMolecular Med.* **2004**, *5*, 147–162. [CrossRef]
130. Beck, S.J.; Guo, L.; Phensy, A.; Tian, J.; Wang, L.; Tandon, N.; Gauba, E.; Lu, L.; Pascual, J.M.; Kroener, S.; et al. Deregulation of Mitochondrial F1FO-ATP Synthase via OSCP in Alzheimer's Disease. *Nat. Commun.* **2016**, *7*, 1–16. [CrossRef] [PubMed]
131. Giorgio, V.; von Stockum, S.; Antoniel, M.; Fabbro, A.; Fogolari, F.; Forte, M.; Glick, G.D.; Petronilli, V.; Zoratti, M.; Szabó, I.; et al. Dimers of Mitochondrial ATP Synthase Form the Permeability Transition Pore. *Proc. Natl. Acad. Sci. USA* **2013**, *110*, 5887–5892. [CrossRef] [PubMed]
132. Du, H.; Guo, L.; Fang, F.; Chen, D.; Sosunov, A.A.; Mckhann, G.M.; Yan, Y.; Wang, C.; Zhang, H.; Molkentin, J.D.; et al. Cyclophilin D Deficiency Attenuates Mitochondrial and Neuronal Perturbation and Ameliorates Learning and Memory in Alzheimer's Disease. *Nat. Med.* **2009**, *14*, 1097–1105. [CrossRef] [PubMed]
133. Du, H.; Guo, L.; Zhang, W.; Rydzewska, M.; Yana, S. Cyclophilin D Deficiency Improves Mitochondrial Function and Learning/Memory in Aging Alzheimer Disease Mouse Mode. *Neurobiol. Aging* **2011**, *32*, 398–406. [CrossRef] [PubMed]
134. Gauba, E.; Guo, L.; Du, H. Cyclophilin D Promotes Brain Mitochondrial F1FO ATP Synthase Dysfunction in Aging Mice. *J. Alzheimer's Dis.* **2017**, *55*, 1351–1362. [CrossRef]
135. Du, H.; Guo, L.; Yan, S.; Sosunov, A.A.; Mckhann, G.M.; Shidu Yan, S. Early Deficits in Synaptic Mitochondria in an Alzheimer ' s Disease Mouse Model. *Proc. Natl. Acad. Sci. USA* **2010**, *107*, 18670–18675. [CrossRef] [PubMed]
136. Gauba, E.; Chen, H.; Guo, L.; Du, H. Cyclophilin D Deficiency Attenuates Mitochondrial F1Fo ATP Synthase Dysfunction via OSCP in Alzheimer's Disease. *Neurobiol. Dis.* **2019**, *121*, 138–147. [CrossRef]
137. Choi, I.; Kim, J.; Jeong, H.K.; Kim, B.; Jou, I.; Park, M.; Chen, L.; Kang, U.J.; Zhuang, X.; Joe, E.-h. PINK1 Deficiency Attenuates Astrocyte Proliferation through Mitochondrial Dysfunction, Reduced AKT and Increased P38 MAPK Activation, and Downregulation of EGFR. *Glia* **2013**, *61*, 800–812. [CrossRef] [PubMed]
138. Ludtmann, M.H.R.; Angelova, P.R.; Ninkina, N.N.; Gandhi, S.; Buchman, V.L.; Abramov, A.Y. Monomeric Alpha-Synuclein Exerts a Physiological Role on Brain ATP Synthase. *J. Neurosci.* **2016**, *36*, 10510–10521. [CrossRef] [PubMed]
139. Ludtmann, M.H.R.; Angelova, P.R.; Horrocks, M.H.; Choi, M.L.; Rodrigues, M.; Baev, A.Y.; Berezhnov, A.V.; Yao, Z.; Little, D.; Banushi, B.; et al. α-Synuclein Oligomers Interact with ATP Synthase and Open the Permeability Transition Pore in Parkinson's Disease. *Nat. Commun.* **2018**, *9*, 2293. [CrossRef] [PubMed]
140. Chen, R.; Park, H.; Mnatsakanyan, N.; Niu, Y.; Licznerski, P.; Wu, J.; Miranda, P.; Graham, M.; Tang, J.; Boon, A.J.W.; et al. Parkinson's Disease Protein DJ-1 Regulates ATP Synthase Protein Components to Increase Neuronal Process Outgrowth. *Cell Death Dis.* **2019**, *10*, 1–12. [CrossRef] [PubMed]
141. Pasinelli, P.; Brown, R.H. Molecular Biology of Amyotrophic Lateral Sclerosis: Insights from Genetics. *Nat. Rev. Neurosci.* **2006**, *7*, 710–723. [CrossRef] [PubMed]

142. Choi, S.Y.; Lopez-Gonzalez, R.; Krishnan, G.; Phillips, H.L.; Li, A.N.; Seeley, W.W.; Yao, W.D.; Almeida, S.; Gao, F.B. C9ORF72-ALS/FTD-Associated Poly(GR) Binds Atp5a1 and Compromises Mitochondrial Function in Vivo. *Nat. Neurosci.* **2019**, *22*, 851–862. [CrossRef] [PubMed]
143. Boillée, S.; Vande Velde, C.; Cleveland, D.W.W. ALS: A Disease of Motor Neurons and Their Nonneuronal Neighbors. *Neuron* **2006**, *52*, 39–59. [CrossRef] [PubMed]
144. Grosskreutz, J.; Van Den Bosch, L.; Keller, B.U. Calcium Dysregulation in Amyotrophic Lateral Sclerosis. *Cell Calcium* **2010**, *47*, 165–174. [CrossRef]
145. Kawamata, H.; Manfredi, G. Mitochondrial Dysfunction and Intracellular Calcium Dysregulation in ALS. *Mech. Ageing Dev.* **2010**, *131*, 517–526. [CrossRef] [PubMed]
146. Deng, J.; Wang, P.; Chen, X.; Cheng, H.; Liu, J.; Fushimi, K.; Zhu, L.; Wu, J.Y. FUS Interacts with ATP Synthase Beta Subunit and Induces Mitochondrial Unfolded Protein Response in Cellular and Animal Models. *Proc. Natl. Acad. Sci. USA* **2018**, *115*, E9678–E9686. [CrossRef] [PubMed]
147. Bannwarth, S.; Ait-El-Mkadem, S.; Chaussenot, A.; Genin, E.C.; Lacas-Gervais, S.; Fragaki, K.; Berg-Alonso, L.; Kageyama, Y.; Serre, V.; Moore, D.G.; et al. A Mitochondrial Origin for Frontotemporal Dementia and Amyotrophic Lateral Sclerosis through CHCHD10 Involvement. *Brain* **2014**, *137*, 2329–2345. [CrossRef] [PubMed]
148. Nobile, V.; Palumbo, F.; Lanni, S.; Ghisio, V.; Vitali, A.; Castagnola, M.; Marzano, V.; Maulucci, G.; De Angelis, C.; De Spirito, M.; et al. Altered Mitochondrial Function in Cells Carrying a Premutation or Unmethylated Full Mutation of the FMR1 Gene. *Hum. Genet.* **2020**, *139*, 227–245. [CrossRef]
149. D'Antoni, S.; De Bari, L.; Valenti, D.; Borro, M.; Bonaccorso, C.M.; Simmaco, M.; Vacca, R.A.; Catania, M.V. Aberrant Mitochondrial Bioenergetics in the Cerebral Cortex of the Fmr1 Knockout Mouse Model of Fragile X Syndrome. *Biol. Chem.* **2019**, *401*, 497–503. [CrossRef]
150. Licznerski, P.; Park, H.A.; Rolyan, H.; Chen, R.; Mnatsakanyan, N.; Miranda, P.; Graham, M.; Wu, J.; Cruz-Reyes, N.; Mehta, N.; et al. ATP Synthase C-Subunit Leak Causes Aberrant Cellular Metabolism in Fragile X Syndrome. *Cell* **2020**, *182*, 1170–1185. [CrossRef]
151. Fearnley, I.M.; Walker, J.E.; Martinus, R.D.; Jolly, R.D.; Kirkland, K.B.; Shaw, G.J.; Palmer, D.N. The Sequence of the Major Protein Stored in Ovine Ceroid Lipofuscinosis Is Identical with That of the Dicyclohexylcarbodi-Imide-Reactive Proteolipid of Mitochondrial ATP Synthase. *Biochem. J.* **1990**, *268*, 751–758. [CrossRef] [PubMed]
152. Jolly, R.D.; Brown, S.; Das, A.M.; Walkley, S.U. Mitochondrial Dysfunction in the Neuronal Ceroid-Lipofuscinoses (Batten Disease). *Neurochem. Int.* **2002**, *40*, 565–571. [CrossRef]

 life

MDPI

Review

Molecular and Supramolecular Structure of the Mitochondrial Oxidative Phosphorylation System: Implications for Pathology

Salvatore Nesci [1,*], Fabiana Trombetti [1], Alessandra Pagliarani [1,*], Vittoria Ventrella [1], Cristina Algieri [1], Gaia Tioli [2] and Giorgio Lenaz [2,*]

1 Department of Veterinary Medical Sciences, Alma Mater Studiorum University of Bologna, 40064 Ozzano Emilia, Italy; fabiana.trombetti@unibo.it (F.T.); vittoria.ventrella@unibo.it (V.V.); cristina.algieri2@unibo.it (C.A.)
2 Department of Biomedical and Neuromotor Sciences, Alma Mater Studiorum University of Bologna, 40138 Bologna, Italy; gaia.tioli2@unibo.it
* Correspondence: salvatore.nesci@unibo.it (S.N.); alessandra.pagliarani@unibo.it (A.P.); giorgio.lenaz@unibo.it (G.L.)

Abstract: Under aerobic conditions, mitochondrial oxidative phosphorylation (OXPHOS) converts the energy released by nutrient oxidation into ATP, the currency of living organisms. The whole biochemical machinery is hosted by the inner mitochondrial membrane (mtIM) where the protonmotive force built by respiratory complexes, dynamically assembled as super-complexes, allows the F_1F_O-ATP synthase to make ATP from ADP + Pi. Recently mitochondria emerged not only as cell powerhouses, but also as signaling hubs by way of reactive oxygen species (ROS) production. However, when ROS removal systems and/or OXPHOS constituents are defective, the physiological ROS generation can cause ROS imbalance and oxidative stress, which in turn damages cell components. Moreover, the morphology of mitochondria rules cell fate and the formation of the mitochondrial permeability transition pore in the mtIM, which, most likely with the F_1F_O-ATP synthase contribution, permeabilizes mitochondria and leads to cell death. As the multiple mitochondrial functions are mutually interconnected, changes in protein composition by mutations or in supercomplex assembly and/or in membrane structures often generate a dysfunctional cascade and lead to life-incompatible diseases or severe syndromes. The known structural/functional changes in mitochondrial proteins and structures, which impact mitochondrial bioenergetics because of an impaired or defective energy transduction system, here reviewed, constitute the main biochemical damage in a variety of genetic and age-related diseases.

Keywords: oxidative phosphorylation; respiratory supercomplexes; ROS; ATP synthase/hydrolase; mitochondrial dysfunction; mitochondrial permeability transition pore; *cristae*; cellular signaling

Citation: Nesci, S.; Trombetti, F.; Pagliarani, A.; Ventrella, V.; Algieri, C.; Tioli, G.; Lenaz, G. Molecular and Supramolecular Structure of the Mitochondrial Oxidative Phosphorylation System: Implications for Pathology. *Life* **2021**, *11*, 242. https://doi.org/10.3390/life11030242

Academic Editor: Angela Anna Messina

Received: 1 February 2021
Accepted: 11 March 2021
Published: 15 March 2021

Publisher's Note: MDPI stays neutral with regard to jurisdictional claims in published maps and institutional affiliations.

1. Introduction: Functions of Mitochondria. The Oxidative Phosphorylation System

The advancement of molecular medicine has pinpointed the role of mitochondria in the etiology and pathogenesis of most common chronic diseases [1–4], so much that the term "Mitochondrial Medicine" has been proposed [5] and then widely used [6–9].

The early biochemical studies on mitochondria were centered on their role in energy conservation. The energy-transducing membrane-bound enzyme complexes of inner mitochondrial membrane (mtIM) drive biochemical reactions involved in energy transformation or bioenergetics. Oxidation of substrate by respiratory complex and ATP production by ATP synthase are tightly coupled molecular mechanisms in the oxidative phosphorylation (OXPHOS) system as explained by Mitchell's chemiosmotic hypothesis [10].

Once the major aspects of the OXPHOS system were clarified, the interest in mitochondria somewhat decreased. However, in recent years it has been raised again due to a series of novel findings assigning new roles to mitochondria in molecular and cell biology, such as mitochondrial DNA and mitochondrial genetics, the role of mitochondria in generation of reactive oxygen species (ROS) and in cell signaling, in cellular quality control and apoptosis (programmed cell death). Nevertheless, these newly discovered functions are strictly intertwined with the central role of electron transfer and ATP synthesis.

This review intends to outline the major structural and functional aspects of mitochondrial bioenergetics that are at the basis of changes leading to pathology; in particular, we will deal with recent advances of the supramolecular structure of the respiratory chain complexes and of F_1F_O-ATP synthase/hydrolase; for this reason, this review will not analyze other important aspects in much detail such as the intimate mechanisms of electron transfer and proton translocation on one hand, and the analysis of the individual pathologies on the other.

2. The Respiratory Chain of Mitochondria

The major mechanism of energy conservation in eukaryotes is OXPHOS, performed by a multi-enzyme system embedded in the mtIM and constituted by two portions: the respiratory chain and the ATP synthase complex.

The membrane electron transfer system (membrane-ETS) is a series of enzymes that collects electrons which stem from the oxidations of intermediary metabolism and drives them downhill to oxygen molecules which are reduced. The free energy fall that accompanies electron flux creates an electrochemical proton gradient ($\Delta\mu_{H+}$) since H^+ are pumped from the mitochondrial the matrix, namely the compartment inside the mitochondrion, to the intermembrane space (IMS) localized between the inner and outer mitochondrial membranes [11]. The energy associated to the proton gradient is then largely used to synthesize ATP from ADP and Pi by the ATP synthase complex. The ATP synthesized is transferred to the cytoplasm in exchange with ADP by the ATP/ADP translocase, also exploiting the H^+ gradient.

As previously revised in [12], it was Hatefi et al. [13] who first isolated from mitochondria four enzyme multi-subunit complexes that concur on the oxidation of NADH and succinate, namely NADH-Coenzyme Q reductase (Complex I, CI), succinate-Coenzyme Q reductase (Complex II, CII), ubiquinol-cytochrome *c* reductase (Complex III, CIII or cytochrome bc_1 Complex) and cytochrome *c* oxidase (Complex IV, CIV) [14].

The connection among these enzyme complexes is ensured by two mobile transporters of electrons, i.e., Coenzyme Q (CoQ, ubiquinone) and cytochrome *c* (cyt. *c*) [14]. The former is a lipophilic quinone incorporated in the lipid bilayer of the mtIM, while cyt. *c* is a hydrophilic hemoprotein facing the mitochondrial IMS, in contact with the external surface of the mtIM. The membrane-ETS operates through the following sequence of the respiratory enzyme complexes: (Equations (1) and (2))

$$\text{NADH} \rightarrow \text{CI} \rightarrow \text{CoQ} \rightarrow \text{CIII} \rightarrow \text{cyt. } c \rightarrow \text{CIV} \rightarrow O_2 \tag{1}$$

or

$$\text{succinate} \rightarrow \text{CII} \rightarrow \text{CoQ} \rightarrow \text{CIII} \rightarrow \text{cyt. } c \rightarrow \text{CIV} \rightarrow O_2 \tag{2}$$

In addition, the membrane-ETS consists of other proteins having electron transfer activity [12] that converge at the CoQ junction. Glycerol-3-phosphate dehydrogenase is involved in a shuttle of reducing equivalents from cytosol to mitochondria [15], electron transfer flavoprotein (ETF) dehydrogenase is involved in fatty acid oxidation [16], dihydroorotate dehydrogenase catalyzes a step of pyrimidine nucleotide biosynthesis [17], choline dehydrogenase is important for the regulation of phospholipid metabolism [18], and sulfide dehydrogenase is involved in the disposal of sulfide [19]. A schematic representation of mammalian respiratory chain is found in Figure 1.

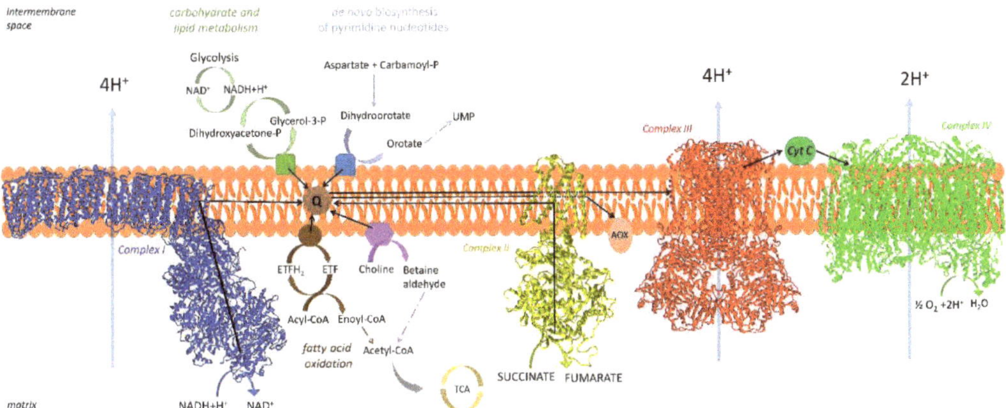

Figure 1. A schematic drawing of the respiratory chain depicting the protein complexes and their substrates. Complex I, Complex III and Complex IV are shown in their free form (modified PDB ID: 6YJ4, 2YBB, 1V54). Blue, CI, NADH-ubiquinone oxidoreductase; yellow, CII, succinate-ubiquinone oxidoreductase (modified PDB ID: 1ZOY); red, CIII, ubiquinol-cytochrome *c* oxidoreductase; green, CIV, cytochrome *c* oxidase; AOX, alternative oxidase; CoQ, Coenzyme Q (ubiquinone); Cytc, cytochrome *c*. Four enzymes that reduce CoQ are also shown together with an indication of their metabolic pathways: from the intermembrane space (IMS), glycerol-3.P dehydrogenase (green) and dihydroorotate dehydrogenase (blue); from the matrix, electron transfer flavoprotein (ETF) dehydrogenase (brown) and choline dehydrogenase (purple). (see text for details).

For reviews see, e.g., Sousa et al. [20] and, for individual complexes: Parey et al. [21] for CI, Bezawork-Geleta et al. [22] for CII, Xia et al. [23] for CIII, and Zong et al. [24] for CIV.

2.1. Organization of the Respiratory Chain: Historical Outline

The advancement of science is long and tortuous: it often happens that novel discoveries, rather than providing further clarifying information on the existing basic knowledge, bring doubts and contradictory issues. This happened for the discovery of the respiratory chain supercomplexes (SCs) and for the elucidation of their function. Therefore, we perform a concise historical survey of the respiratory chain, since some early findings came out before the present knowledge, but their significance was not understood.

As we mentioned in the previous section, Hatefi et al. [13] accomplished the systematic resolution and reconstitution of four functional respiratory complexes from mitochondria, and proposed that the overall electron transfer from substrates to oxygen results from both intra-complex and inter-complex redox reactions: intra-complex electron transfer takes place in the "solid" state of redox components (e.g., flavins, FeS clusters, cytochromes) having fixed steric relation, whereas inter-complex electron transfer operates by rapid diffusion of the mobile components acting as co-substrates, i.e., CoQ and cyt. *c* [14]. In the following years, this proposal was confirmed by the kinetic analysis of Kröger and Klingenberg [25,26], leading Hackenbrock et al. [27] to postulate the *Random Collision Model of Electron Transfer*:

"Electron transport is a diffusion-coupled kinetic process; Electron transport is a multicollisional, obstructed, long-range diffusional process; The rates of diffusion of the redox components have a direct influence on the overall kinetic process of electron transport and can be rate limiting, as in diffusion control ... It is concluded that mitochondrial electron transport is a diffusion-based random collision process, and that diffusion has an integral and controlling effect on electron transport."

The random collision model was accepted by most researchers in this field. Nevertheless, the accumulation of further experimental evidence obtained with newly developed techniques has led to the proposal of a different model of supramolecular organization based upon specific interactions between individual respiratory complexes [28].

Indeed, a supramolecular assembly of respiratory components had been already present in the pioneering studies of Chance and Williams [29], who described the respiratory chain as an aggregate of coenzymes (flavins and cytochromes) in solid-state, embedded in a protein matrix.

Surprisingly, evidence for a non-random arrangement of respiratory complexes also derived from the same early investigations of Hatefi's group, reporting isolation of CI+CIII units [30], suggesting that such units preferentially assemble in the native membrane. These aggregates were considered physiological by their discoverers, indicating the existence of supramolecular units of electron transfer. It is therefore clear that SCs, as well as the entire NADH oxidase (known as the respirasome, see later) had been isolated and reconstituted. The publication of the fluid mosaic model of biomembranes [31] was applied also to mitochondrial membranes, thus favoring the random collision model of Hackenbrock. Thus, Hatefi's idea of specific associations among complexes was overlooked.

The idea of fixed associations, however, was never completely abandoned: in fact, some authors reported the possible existence of specific associations between respiratory complexes, either fixed [32] or dynamic [33].

At the turning of the century, Schägger and Pfeiffer [34] applied the technique of Blue-Native Gel Electrophoresis (BN-PAGE) to digitonin-solubilized yeast and mammalian mitochondria and found electrophoretic bands of high molecular weight indicating specific associations among respiratory complexes. According to their discoverers, these associations represent the state of the respiratory chain under physiological conditions. The same study also provided evidence for a dimeric ATP synthase complex.

The supramolecular arrangement of the respiratory complexes is well established [35,36], although the functional role of respiratory SCs is still controversial and their relationship with a random distribution of the individual complexes partially remains to be clarified [37–40].

2.2. Distribution and Composition of Respiratory Supercomplexes

All SCs exhibit highly ordered structures, thus they are not the result of artificial protein–protein interaction due to the extraction and solubilization procedures [41].

In most studies, SCs appear to contain the three "core" respiratory complexes, i.e., CI, CIII, and CIV, whereas the other respiratory enzymes may be randomly distributed in the lipid bilayer [12]. These "core" complexes have the common feature of carrying out proton translocation and of having subunits encoded by mitochondrial DNA.

Respiratory SCs have been described and characterized in the mitochondria of several mammalian tissues and in many other organisms, also including plants, fungi and bacteria [42].

The supramolecular structural organization of the "core" respiratory complexes CI, CIII, and CIV is conserved in all higher eukaryotes. The SC $I_1III_2IV_{1-4}$ contains all three complexes required for the complete oxidation of NADH by molecular oxygen, and for this reason it was called "respirasome". SCs containing only two components are also normally found in large amounts, as the SC I_1III_2 in which CIV is not present, as well as SCs III_2IV. In bovine heart mitochondria, only a small aliquot of CI is present in free form in the presence of digitonin [43]; thus, it is believed that all CI is bound to CIII in the native membrane.

A higher level of complexity in the organization state of SCs may be due to the presence of additional protein components unrelated to electron transfer complexes (e.g., Shy1, SCAF1, Rcf2, HIGD2A) occasionally found by BN-PAGE [44].

Wang et al. [45] described a multifunctional mitochondrial fatty acid β-oxidation (FAO) complex that is physically associated with SCs. Analysis of genetic defects in both fatty acid oxidation and OXPHOS [46] revealed that some patients with primary FAO deficiencies exhibit secondary OXPHOS defects. These metabolic interrelations support the view that OXPHOS proteins and FAO are physically associated, and that these interactions are critical for both functions.

The other respiratory enzymes not comprising the "core" of the proton translocation machinery appear not to be associated in SCs [47]. In most studies of BN-PAGE, CII was not found associated with other complexes of the respiratory chain. In addition, mitochondrial glycerol phosphate dehydrogenase is absent in the respirasome.

Fang et al. [48] observed that mitochondrial function is hampered by dihydroorotate dehydrogenase deficiency; the enzyme was shown to physically interact with respiratory complexes CII and CIII by immunoprecipitation and BN/SDS/PAGE analysis.

Sulfide-quinone oxidoreductase and sulfite oxidase, involved in sulfide oxidation, which transfer electrons to the respiratory chain, were also described to be associated with SCs containing CIV [49].

3. Supercomplexes May Provide a Kinetic Advantage to Electron Transfer

The discovery of SCs led to the proposal that these aggregates form to improve substrate channeling or enhance catalysis in inter-complex electron transfer. Substrate channeling is the direct transfer of an intermediate between the active sites of two enzymes which catalyzes reactions which occur one after the other [50]; in the respiratory chain, this means that electrons are directly transferred between two consecutive enzymes by alternate reduction and re-oxidation of an intermediate which is not diffused in the medium. In such a case, inter-complex electron transfer cannot be distinguished from intra-complex electron transfer. Therefore, the mobile intermediates predicted to exhibit substrate-like behavior in the random collision model, i.e., CoQ and cyt. *c*, would be buried in the interface between two consecutive complexes within the SC.

In the CoQ region, the occurrence of electron transfer by channeling of CoQ between adjacent CI and CIII contrasts with the so-called CoQ "pool behavior", previously sustained by the kinetic analysis of Kröger and Klingenberg [25]. To better understand the following sections, we believe it necessary to briefly mention their kinetic analysis. These authors showed that steady-state respiration can be described as a simple two-enzyme system, where the first enzyme reduces ubiquinone (V_{red}) while the second one oxidizes ubiquinol (V_{ox}), kinetically behaving as a homogeneous pool. Thus, the total CoQ molecules must randomly link any number of the dehydrogenase(s) with any number of the oxidase(s); the overall steady-state activity (V_{obs}) is related to V_{red} and V_{ox} as shown by the so-called pool equation: (Equation (3))

$$Vobs = (Vred \times Vox)/(Vred + Vox) \quad (3)$$

It has been shown that cyt. *c* also exhibits "pool behavior " [51].

In the following sections, the relations existing between the two models and the ensuing controversial findings will be analyzed. The main challenge to study the physiological role of SCs is to selectively disrupt respiratory SCs in the cell [40].

3.1. Structural Evidence Suggesting Channeling

3.1.1. Molecular Structure of Supercomplexes

If channeling occurs between CI and CIII by the common intermediate CoQ, the redox groups involved in CoQ reduction by CI and $CoQH_2$ re-oxidation in CIII must be in close contact in order to form a driving pathway containing CoQ itself; similar reasoning applies to cyt. *c* between CIII and CIV. An essential requirement for this condition is the availability of a high-resolution map of the molecular structure of the SCs.

In the initial studies, purified SCs were analyzed by negative-stain electron microscopy and single-particle cryogenic electron microscopy (cryo-EM) [52,53]. Pseudo-atomic models of the respirasome (SC $I_1III_2IV_1$) were produced by fitting the known X-ray structures of the component complexes to the 3D maps of the respirasome. Evidence for channeling in the respirasome stems from the observed distinct arrangement of the three component complexes, in which the CoQ-binding sites in CI and in CIII face each other and are divided by a gap within the SC membrane core, which most likely contains lipids. Althoff et al. [52] proposed that CoQ may cover a trajectory through such a 13 nm-long gap.

Further high-resolution studies by cryo-EM of mammalian mitochondrial SCs [54–57] may be useful to predict the CoQ pathway between CI and CIII, as shown in Figure 2.

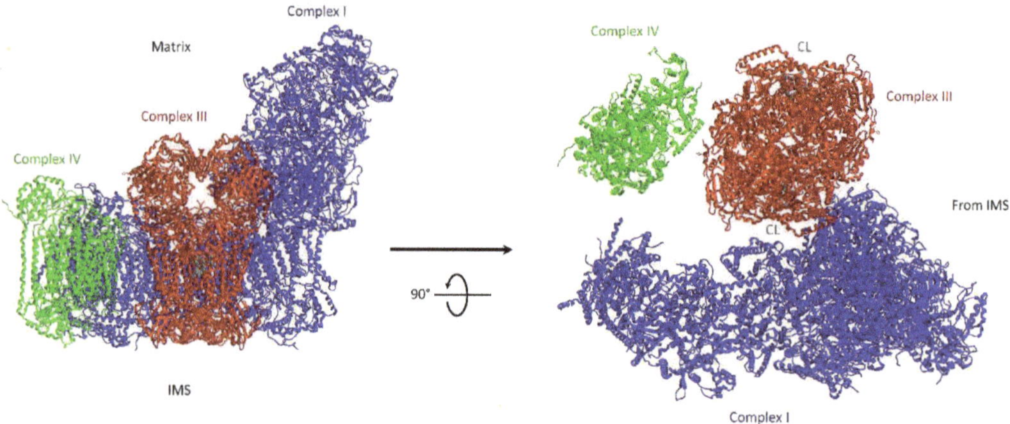

Figure 2. Bovine mitochondrial supercomplex (SC) $I_1III_2IV_1$. Fitted model by single particle cryo-EM. Side view (on the left) and view from the IMS (on the right) showing two CL (cardiolipin) molecules (in space filling mode) in the cavity of each monomer of CIII formed by cytochromes c_1 and *b*. Image modified from Mileykovskaya and W. Dowhan [58]. The enzyme subunits are drawn as ribbon representations obtained from modified PDB ID code: 2YBB. Blue, CI; red, CIII dimer; green, CIV.

In theory, channeling within the SC may occur through either close docking of the active sites with real tunneling of substrate among proteins, or covering relatively long distances presumably by substrate-restricted diffusion (microdiffusion) within the space between the two active sites; both alternatives share a channeling which necessarily occurs between two fixed sites of the same SC. Microdiffusion is kinetically distinguished from bulk diffusion (pool behavior), where the interaction of the substrate molecules stochastically occurs by collisional encounters with a large number of possible sites in distinct respiratory complexes attained by random diffusion.

Information on channeling within the SC molecular structure must be compatible with the known CoQ reduction mechanisms by CI and its re-oxidation by CIII dimers. Ubiquinol oxidation occurs via a cyclic mechanism known as the Q-cycle [59,60]. On considering the functional asymmetry of the CIII dimer [61], it is suggestive to consider that in the supercomplexed CIII dimer only one monomer is active for NADH-dependent respiration and CoQ-channeling, since the CoQ reducing site I CI and the Qo site where ubiquinol is reoxidized are close to each other. On the contrary, the other monomer, which lacks such constrictions, might be easily available to interact with the CoQ pool at the (distal) Q_i site.

The group of Kuhlbrandt [57] showed that the CoQ reduction site of CI in the SC is 11 nm far from the proximal CIII monomer, whereas the distal one is 18 nm away, thus suggesting that the latter is functionally inactive. The linear CI arrangement with the active CIII monomer and with CIV is strongly in favor of channeling for electron transfer between complexes in the respirasome by substrate microdiffusion.

On the other hand, the respirasome structure reported by Letts et al. [54] indicates that both CoQ interaction sites in CI and CIII are separated and easily accessible to the membrane, and are likely to provide no limit to free CoQ diffusion in the inner mitochondrial membrnae (IMM).

The SC structure, resolved at up to 3.8 Å in four distinct states, suggests that CoQ oxidation may be rate limiting because of unequal access of CoQ to the active sites of $CIII_2$ [62]. CI changes between "closed" and "open" conformations, while a key transmembrane helix rotates in a striking way. Moreover, the CI state influences the conformational flexibility within $CIII_2$, demonstrating crosstalk between the enzymes.

According to Letts et al. [62], under high turnover conditions, individual CIs may overcome the capacity for CoQ pool balance, so that the local $CoQH_2$ pool may attain higher concentration than in the bulk membrane. By ensuring close association of CI and $CIII_2$ in SCs, but still allowing free CoQ exchange with the bulk, the SCs would prevent any local accumulation of $CoQH_2$, without limiting CI's activity by making it entirely dependent on the adjacent $CIII_2$ for the $CoQH_2$ re-oxidation.

The structural analysis of SCs allows analogous considerations for cyt. *c* between CIII and CIV.

The structural evidence of purified SCs by cryo-EM adds structural reasons for the looseness of CIII–CIV interactions in mammalian mitochondria. In ovine mitochondria, Letts et al. [54] identified two distinct arrangements of SC $I_1III_2IV_1$: a major "tight" form and a minor "loose" form. In both structures, the density for CIV is weaker relative to that for CI and CIII, indicating the greater conformational flexibility of CIV. In the tight respirasome, CIV contacts both CI and CIII, whereas CIV in the loose respirasome has defined contacts only with CI. Tight and loose architectures may represent independent structures or may interconvert each other, according to distinct stages of assembly or disassembly. The presence of two respirasome forms with distinct CIV linkages was also detected by Sousa et al. [57] and ascribed to instability of the purified SC.

3.1.2. Supercomplexes Are Dynamic Structures: The Plasticity Model

Together with kinetic evidence by flux control analysis in favor of CoQ channeling (See Section 3.2.2), we also postulated [63] that the SC I_1III_2 may be in equilibrium with free CI and CIII. In fact, we found that NADH-cyt. *c* reductase activity in reconstituted proteoliposomes can occur either by channeling or by random diffusion, depending on the protein density in the lipid bilayer of the proteoliposomes.

Acín-Pérez et al. [64] in a study of purified SCs showed by BN-PAGE that other types of associations exist besides the respirasome, such as CI+CIII or CIII+CIV. They therefore suggested that a variety of associations between respiratory complexes is likely co-exists in vivo with free complexes and proposed an integrated model, the plasticity model, for the organization of the respiratory chain. According to their view, the previous contrasting models, solid vs. fluid, are only the two extremes of a dynamic range of molecular associations between respiratory complexes [64]. The plasticity model proposes that both CoQ and cyt. *c* are sequestered into distinct sub-domains and are both trapped into SCs and freely diffusible. This promiscuity can stem either from the dissociation of SCs or from the escape of CoQ or cyt. *c* from the supramolecular assemblies [36,65].

The plasticity model and the dynamics and stability of mitochondrial SCs [37] are still debated issues. The putative factors involved in association/dissociation of SCs, such as membrane potential and post-translational changes, are reviewed by Genova and Lenaz [42]. However, even if indirect evidence shoulders the SC dynamic assembly and disassembly, up to now no direct demonstration exists that SCs physiologically form and dissociate in vivo.

3.1.3. The Role of Lipids: Cardiolipin in Supercomplexes

SCs have 2–5 nm gaps at the transmembrane interfaces of the individual complexes (Figure 2) that are presumably filled with lipids. Wu et al. [56], in their high-resolution structure of the purified SC $I_1III_2IV_1$ from porcine heart, localized many phospholipid molecules including cardiolipin (CL) involved in the protein–protein interactions [66]. Substantial progress was made by studying Barth syndrome, a genetic syndrome characterized by severe cardiopathy, in which cardiolipin remodeling is altered due to the mutation of the gene *Tafazzin*; in Barth syndrome patients SCs are unstable, leading to a functional impairment characterizing the disease [67]. In immortalized lymphoblasts from Barth's syndrome patients the amount of SCs is decreased, as well as the amount of individual CI and CIV [68].

Direct involvement of CL in the formation of SC III–IV was demonstrated in genetically manipulated strains of *S. cerevisiae* in which the CL content can be regulated in vivo. Yeast mutants lacking CL have normal amounts of individual complexes, but lack the stable SC III_2IV_2 as observed in the wild type parental strain [69]. Since *S. cerevisiae* lacks CI, these studies could not provide information on SCs containing CI.

In contrast with CL depletion that destabilizes SCs, the loss of phosphatidylethanolamine (PE) favors the formation of larger SCs between CIII and CIV in *S. cerevisiae* mitochondria. The reason why CL and PE, even if both associate in non-bilayers, act in an opposite way on SC stability may depend on the different charge, since, at physiological pHs CL is anionic and PE is zwitterionic. Using yeast mutants of PE and phosphatidylcholine (PC) biosynthesis, Baker et al. [70] showed a specific requirement for mitochondrial PE, but not PC, in CIII and CIV activities, but not for their formation. Neither PE nor PC were involved in respiratory SC formation, emphasizing the specific requirement of CL in SC assembly.

Our laboratory has stressed the importance of CL in stabilizing CI-containing SCs. Reconstitution of binary CI/CIII proteoliposomes from bovine heart mitochondria at high lipid to protein ratio (30:1 w:w) prevents formation of SC I_1III_2 [71]. However, SC I_1III_2 and related NADH-cyt. *c* reductase activity are maintained if these high-lipid proteoliposomes are enriched with 20% CL (w:w), resembling the percent content of CL in the mitochondrial membrane (M. Kopuz, Y. Birinci, S. Nesci, G. Lenaz and M.L. Genova, unpublished data). Likewise, the dilution of native protein-bound CL in the excess exogenous phospholipids prevents SC assembly, but the effect is reversed by increasing the CL content of the liposomes, therefore shifting again the equilibrium to CL binding to protein.

Addition of reactive oxygen species (ROS) to mitochondria affects the respiratory activity due to cardiolipin peroxidation, since CL is required for the optimal activity of respiratory complexes [72,73]. We showed by flux control analysis (cf. Section 3.2.2) that the SC I_1III_2 in proteoliposomes disappears if lipid vesicles are peroxidized before protein reconstitution [71]. Evidently, the distortion of the lipid bilayer induced by peroxidation and the alteration of the phospholipid annulus originally present in the purified SC I_1III_2 determines its dissociation.

CL deficiency was reported to cause a defect in respiratory function and a decrease in ATP synthesis [74], an indirect demonstration of CL role in the structure of the SCs. Rieger et al. [75] investigated the effect of ALCAT1 overexpression. This enzyme catalyzes the incorporation of polyunsaturated fatty acids into CL. The resulting CL species are more susceptible to oxidative damage, leading to increased ROS production and mitochondrial dysfunction. Using galactose as sugar supply, cells must employ OXPHOS for ATP synthesis, shown as a significant increase in basal- and ADP-linked respiration; ALCAT1 overexpression prevents the galactose-induced increase in respiration.

Likewise, a genetic defect of CL synthase induces loss of SC assembly and of respiratory activity as well as other mitochondrial defects [76].

3.2. Kinetic Evidence for Channeling in the Coenzyme Q Region

3.2.1. Rate Advantage Imposed by SCs

The overall rate in a diffusion-coupled pathway is always less than that of the rate-limiting step and only approaches the latter when it is widely different (i.e., much slower) compared to the rate of the other step(s) in the pathway [25]. On the contrary, the rate is expected to be equal to the rate of the limiting step if channeling of the intermediate substrate(s) occurs.

Ragan and Heron [77] investigated in detail the reconstitution of CI in the respiratory chain; they showed that purified CI and CIII, when mixed as concentrated solutions, reversibly associate in a 1:1 molar ratio to form an SC I+III endowed with NADH-cyt.

c reductase activity. The study demonstrated a stoichiometric behavior for this activity, revealing the existence of an active SC formed by CI and CIII: electron transfer is fast within the SC and conversely very slow from the SC to free CIII via the CoQ pool.

On the other hand, prevailing CoQ-pool behavior could be induced and CI and CIII could be made to operate independently of each other if the SCs were functionally dissociated into "free" complexes, by raising the amount of phospholipid and ubiquinone in the concentrated mixture. Under this condition, electron transfer from CI to CIII is still ensured by CI+$CIII_2$ units that however dissociate and reform at rates exceeding the rates of electron transfer in the individual complexes.

According to Heron et al. [78], the mobility of CI and CIII in the membrane is lost and the complexes are frozen in their SC assembly when phospholipids are not in excess with respect to the content needed to form a lipid annulus around the protein. Such frozen state favors a stable orientation of the site of CoQ reduction by CI with respect to the site of oxidation by CIII. They also showed that protein-bound CoQ_{10} leaks out of the SC I+III in the lipid bilayer when an extra phospholipid is present in the proteoliposomes; the consequent decrease in activity could be reversed by adding additional quinone. As a corollary, we may deduce that a large amount of ubiquinone found in the natural membrane is needed to maintain the CoQ_{10} content in the SC unit when it is formed.

Our laboratory provided a direct evidence that the phospholipid amount affects the choice of channeling with respect to CoQ-pool behavior [79]. A proteoliposome system obtained by fusing a crude mitochondrial fraction enriched in CI and CIII with different amounts of phospholipids and CoQ_{10} allowed us to discriminate between these two mechanisms. The experimental NADH-cyt. *c* reductase activity was compared with the theoretical values obtained from the pool equation of Kroger and Klingenberg [25], showing overlapping results in the range from 1:10 to 1:40 (w/w) protein to lipid ratios. However, pool behavior was not shown at low (1:1 w/w) lipid to protein ratio. Moreover, the observed rates of NADH-cyt. *c* reductase were higher than the theoretical values; this 1:1 ratio corresponds to the mean nearest distance between respiratory complexes in mitochondria.

Several studies point out that SC disorganization by several causes is accompanied by decreased electron transfer and energetic efficiency (cf. also Section 3.1.3).

Recently Garcia-Poyatos et al. [80] in Enriquez's group investigated the physiological role of SCs, generating two null allele zebrafish lines for supercomplex assembly factor 1 (SCAF1). The *scaf1*$^{-/-}$ fish showed altered OXPHOS activity due to the disrupted interaction of C III and CIV.

Solsona-Vilarrasa et al. [81] studied the effects of increased cholesterol levels, as occurring in vivo in alcoholic and non-alcoholic fatty liver, or by in vitro cholesterol enrichment of mouse liver mitochondria. Cholesterol feeding caused oxidative stress and mitochondrial GSH (mGSH) depletion, which lead to liver steatosis and damage. The overload of cholesterol in mitochondria disrupts mitochondrial functional performance and the organization of respiratory SC assembly, thus potentially contributing to oxidative stress and liver injury.

Tomkova et al. [82] showed that Tamoxifen-resistant cells show a significant decrease in mitochondrial respiration accompanied by a decrease in mitochondrial respiratory SC and significantly increased levels of mitochondrial superoxide (See Section 5).

Balsa et al. [83] showed that endoplasmic reticulum (ER) stress and glucose deprivation stimulate mitochondrial OXPHOS and formation of respiratory SCs by action of protein kinase R-like ER kinase (PERK). Accordingly, PERK genetic ablation or pharmacological inhibition abolishes nutrient and ER stress-mediated increase in SC levels and reduce OXPHOS-dependent ATP synthesis.

These studies suggest that electron transfer between CI and CIII in NAD-linked respiration can take place either by CoQ channeling within the SC I_1III_2 or by a less efficient random diffusion behavior, depending on the state of the membrane lipids: nevertheless, it appears that channeling is preferred under physiological conditions.

3.2.2. Metabolic Flux Control Analysis

Our laboratory first exploited flux control analysis [84] to establish the functional association between individual respiratory complexes in beef heart mitochondria [63]. Using specific inhibitors to establish the extent of metabolic control exerted by each individual complex over the entire NADH-dependent respiration, we demonstrated that CI and CIII are both rate-limiting, since they exhibit flux control coefficients (FCC) close to 1, thus behaving as a single enzymatic unit (e.g., SC I_1III_2); we therefore concluded that electron transfer through CoQ is accomplished by channeling between CI and CIII [63]. Using the same method for succinate oxidation we found that CII, but not CIII, is rate-limiting, confirming the notion that CII does not form SCs and that the oxidation of succinate follows pool behavior.

A decade later, Blaza et al. [85], on the basis of flux control analysis, criticized the evidence for channeling in the respiratory chain: accordingly, rotenone competes with CoQ and, consequently the CI inhibition extent is affected by the additional quinone analog used as substrate in the CI assay, but absent in the NADH aerobic oxidation assay. Indeed, using rotenone and CoQ_1 as substrate they find an FCC for CI exceeding 1 that is certainly artefactual. However, in our study we used decyl-ubiquinone (DB) rather than CoQ_1. DB has much lower affinity than CoQ_1 for CI [86], thus exerting lower competition with rotenone and lower influence on the inhibition efficiency measured by us [63]. Moreover, the FCC values found by Blaza et al. [85] using alternative inhibitors of CI (e.g., piericidin and diphenyleneiodonium, not competitive with CoQ) are unrealistically low, indicating the presence of a rate-limiting step downhill. Most likely, the limiting amount of cyt. *c* in their sample would shift the main control of the chain to the cyt. *c* region. Significantly, in Blaza's study, addition of exogenous cyt. *c* to the mitochondrial samples enhanced the FCC from 0.19 to 0.67.

Overall, we must emphasize that our major clue [36] to demonstrate association of CI to CIII was not so much the high FCC of CI but the high FCC of CIII in NADH oxidation, which is incompatible with a CoQ-pool model where CI and CIII independently float in the membrane. Flux control of CIII was not performed in the study by Blaza et al. [85]. Such high FCC value of CIII over NADH-dependent respiration was measured by inhibitor titration with mucidin, and is clearly not artefactual because the corresponding FCC of CIII measured by using the same inhibitor in succinate oxidation is low (as expected from the rate-limiting role of CII and the lack of SC II+III). It is evident that the FCC of CIII is close to 1, indicating a rate-limiting step in CIII, only if the preceding COQ-reductase is assembled with CIII in an SC, which is the case only for CI, but not for CII. In addition, the high FCC that CIII shows over NADH-cyt. *c* reductase activity in reconstituted proteoliposomes is strongly lowered when the SC I_1III_2 is dissociated [71] by reconstitution in excess phospholipids or in peroxidized phospholipids.

The few other studies addressed to SCs using metabolic control analysis have confirmed the possibility that the respiratory chain is arranged as functionally relevant supramolecular structures [87–89].

3.2.3. Coenzyme Q Compartmentalization

The possibility that the CoQ pool may not be homogeneous has been advanced long time ago. Gutman and Silman [90] observed partial additivity and competition between succinate oxidation and NADH oxidation. On the basis of their observations (and those of others) [91], they proposed a compartmentalization of the CoQ pool, in which two sub-domains partially interact through a spill-over diffusion-mediated mechanism.

Gutman [92] also investigated NADH and succinate oxidation in submitochondrial particles (SMP), as well as the rates of energy-dependent reverse electron transfer from succinate to NAD^+ and of forward electron transfer from NADH to fumarate, concluding that:

"The electron flux from succinate dehydrogenase to oxygen (forward electron transfer towards Complex III) or to NADH dehydrogenase (reverse electron transfer) employs the same carrier and is controlled by the same reaction" whereas "the electron transfer from

NADH to oxygen does not share the same pathway through which electrons flow in the NADH-fumarate reductase".

In other words, CI and CII (reverse electron transfer) are linked by a different pathway with respect to CI and CIII (forward electron transfer), even if CoQ is a common intermediate for both pathways. We can clearly interpret this non-homogeneity of the CoQ pool with respect to succinate and NADH oxidation as deriving from sequestration of CoQ in the SC I+III in contrast with the free CoQ molecules that connect CII and CIII.

In past years, several other reports have suggested that the CoQ pool is not homogeneous and questioned its universal validity [93]. Similar observations were made by other authors [94] who observed different functional CoQ pools in mitochondria.

Lapuente-Brun et al. [95] demonstrated that the physical association of CI and CIII determines a preferential pathway for electrons mediated by a dedicated subset of CoQ molecules. This compartmentalization would prevent significant cross talk and mixing up between NADH oxidation (CI-dependent) and succinate oxidation (dependent on CII) or other flavoenzyme-dependent oxidations. Moreover, the CIII units within the SC I+III are exclusively dedicated to oxidation of NADH whereas free CIII is dedicated to oxidation of other substrates using the free CoQ pool (e.g., fatty acids through ETF, succinate, glycerol-3-phosphate and choline).

On the contrary, Blaza et al. [85] showed that the steady-state rates of aerobic NADH and succinate oxidation were only partly additive in bovine heart submitochondrial particles SMP; moreover, cytochromes b_H, b_L, c and c_1 in the same cyanide-inhibited particles were reduced to a similar extent by either NADH or succinate or a mixture of the two substrates. Blaza et al. [85] interpreted the results as a demonstration that a single homogeneous pool of CoQ molecules exists that receives electrons indifferently from CI and CII. Note that these data were obtained on the whole pathway of electrons from ubiquinol to oxygen and therefore also comprise the steps through cyt. *c* and CIV. It is therefore not possible to discriminate whether the pool behavior is due to the homogeneity of the CoQ-pool or of the cyt. *c* pool.

To overcome this uncertainty, we carried out experiments on succinate and NADH oxidation by exogenous cyt. *c* in KCN-inhibited mitochondria, thus shortening the electron route for both oxidation reactions by only including CoQ and CIII as redox partners [36]. Under such conditions, NADH and succinate oxidation by cyt. *c* was completely or almost completely additive and close to the theoretical sum of the two individual reactions, thus suggesting that two separate CoQ-compartments may exist.

Notably, we also obtained similar results (Tioli G, Falasca AI, Lenaz G and Genova ML, data communicated at EBEC2016-Session P4) using proteoliposomes enriched in CI, CII, and CIII. In these proteoliposomes, the additivity of NADH and succinate oxidation decreased at increasing CIII inhibition by mucidin, thus substantiating the hypothesis that the CoQ molecules in the SC I+III are in a dissociation equilibrium with the molecules in the pool [96], cf. next section.

Fedor and Hirst [97] in another study incorporated an alternative quinol oxidase (AOX) into mammalian heart mitochondrial membranes to have another competing pathway for ubiquinol oxidation. Since AOX strongly increased the rate of aerobic NADH oxidation, they concluded that the quinol generated in SCs by CI is more rapidly reoxidized in the CoQ pool by AOX than by CIII inside the SC. Based on these results, Hirst [98] concludes that quinone and quinol diffuse freely in and out of SC: no substrate channeling occurs, since it is not required to support respiration.

In our opinion, this criticism is inconsistent for the same reasons discussed above for concomitant NADH and succinate oxidation in presence of a CIII inhibitor, since Fedor and Hirst used KCN to block electron transfer through CIV: under this condition electrons from ubiquinol are forced to follow the alternative pathway by dissociation into the CoQ pool (cf. Section 3.3.1). It is intuitive that channeling is largely predominant during high turnover of the respiratory chain, whereas at low turnover, CoQ diffusion in the pool would preferentially take place.

Recently, our analysis was strongly supported by Szibor et al. [99] who expressed AOX in mouse mitochondria and observed that electron input by succinate (CII), but not by NADH (CI), almost completely reduces the Q pool (>90%) irrespective of the respiratory state. AOX enhances the forward electron flux from CII and decreases reverse electron transport. Interestingly, AOX does not act on CI substrates, except in the presence of a respiratory inhibitor (as proven by Fedor and Hirst [97]).

Recently, Enriquez's group [100] also presented results confirming that the SC of CI and CIII allows the partial segregation the CoQ pool and allows substrate channeling. Their results with different cellular models suggest that when CI is not superassembled, as in CIII-KO cells, CoQ occurs in a unique pool, whereas CI superassembly triggers the formation of two partially different CoQ pools.

3.3. *If Electron Transfer Occurs via Channeling, What Is the Role of Coenzyme Q Pool?*

Since a mobile pool of CoQ in the mtIM coexists with CoQ sequestered in SCs, we may ask whether this pool is a mere reservoir of an excess of CoQ molecules without a specific function or whether the CoQ pool is in any way required for physiological electron transfer and/or for additional functions.

3.3.1. Dissociation Equilibrium of Bound Coenzyme Q

Compelling evidence that electron transfer from CI to CIII occurs physiologically by channeling/restricted diffusion stems from the observation that CI is almost totally associated in an SC with CIII. However, we reasoned that an excess of entrapped CoQ in the pool is also required for channeling to occur [96]. In fact, the CoQ molecules bound in the SC, that ensure electron transfer directly from CI to CIII are in dissociation equilibrium with the CoQ pool. Therefore, the extent of CoQ binding to the SC depends on the equilibrium with free CoQ in the pool and on the size of the pool itself. The existence of this equilibrium is widely confirmed by a series of studies, such as the previously reported findings by Heron et al. [77] on CI to CIII association, by the saturation kinetics for CoQ exhibited by the integrated activity of CI and CIII (NADH-cyt. *c* oxidoreductase) [101] and by the decrease in respiration in mitochondria fused with phospholipids causing subsequent dilution of the CoQ pool [102].

In conclusion, besides acting as a diffusible substrate of several flavin reductases not involved in SC organization (cf. next section), free CoQ molecules are a reservoir for binding to the SC. Studies on respiration under pathological conditions [103,104] showed that SC assembly is required for proper respiration, even if activity of the individual complexes is normal. On the other hand, reconstitution in vitro as described in the previous Section 3.2, suggests that respiration can occur in both modes (channeling and diffusion), if CoQ is available. In a proteoliposome system where CI was reconstituted together with an alternative oxidase (AOX) and CoQ_{10}, and in absence of SC, Jones et al. [105] found high rates of NADH oxidation through AOX, demonstrating that CI can deliver electrons through the CoQ pool.

The dynamic character of CoQ bound within the SC, which is in dissociation equilibrium with the free pool of CoQ in the membrane, is the major point favoring controversy concerning channeling. We propose that the CoQ molecules trapped in a lipid microdomain within the SC and are channeled from CI to CIII during electron transfer at steady state, without significant exchange with the free pool; however, CoQ dissociation from the SC to the pool becomes significant when electron transfer in the respiratory chain is slow or blocked by an inhibitor. In this respect, the plasticity model would be a functional rather than structural feature of the respiratory chain, a possibility taken into account by Enriquez himself [37]. According to this idea, the SCs are stable structures that do not readily dissociate under physiological conditions, while CoQ is said to behave in a highly dynamic fashion: CoQ channeling and diffusion shall occur in proportions depending on the turnover rates of the respiratory chain.

3.3.2. Electron Transfer between Free Complexes

Several flavoenzymes reduce CoQ as outlined in Section 2. Among these, the most widely investigated is CII. All evidence converges in the statement that electron transfer from CII to CIII takes place only through the CoQ pool. Accordingly, succinate oxidation kinetically follows pool behavior after extraction and reconstitution [101] and in intact mitochondria [106] in accordance with the notion that CII does not participate in SC formation (see previous sections). For the same reason, also energy-dependent reverse electron transfer from succinate to NAD^+, taking place through sequential activity of CII and CI connected by CoQ, must take place by collisional interactions in the CoQ pool. The hyperbolic relation experimentally found by Gutman [92] between the rate of reverse electron transfer and succinate oxidase is in complete accordance with the pool equation.

Other CoQ reductases such as glycerol-3-phosphate dehydrogenase, ETF dehydrogenase, dihydroorotate dehydrogenase, choline dehydrogenase, sulfide dehydrogenase, that are present in minor amounts, are probably inserted in the respiratory chain by interaction through the CoQ pool [12], but kinetic evidence is poor.

Some knowledge exists for mitochondrial glycerol phosphate dehydrogenase (mtGPDH). A study [107] demonstrated that in brown adipose tissue (BAT) mitochondria the inhibition curve of glycerol phosphate-cyt. *c* reductase is sigmoidal in the presence of myxothiazol or antimycin A. Such type of inhibition suggests the occurrence of a homogeneous CoQ pool between mtGPDH and CIII [26]. CIII reduction by mtGPDH in human neutrophil mitochondria was also shown to occur in absence of SC organization and of NAD-linked respiration [103], providing further evidence that mtGPDH functions through the CoQ pool. Preliminary studies by BN-PAGE (M.L. Genova, M. and H. Rauchova, unpublished) showed that mtGPDH is not apparently linked to any of the respiratory complexes. Mráček et al. [108] demonstrated that mtGPDH associates into homo-oligomers but also in high molecular weight SCs of more than 1000 kDa whose composition is unknown, but not associated with CI, III, or CIV as shown by BN-PAGE analysis. No association was found between CI or other OXPHOS complexes and the electron transferring flavoprotein (ETF) that participates in fatty acid oxidation.

Indirect evidence, however, suggests that fatty acyl CoA oxidation proceeds through a separate CoQ pool with respect to NADH oxidation (cf. Section 3.5).

3.4. *Electron Transfer through Cytochrome c: Is There Evidence for Channeling?*

BN-PAGE clearly shows that a fraction of CIV, as well as CIII, participates in SC assembly (cf. Section 2). However, metabolic flux control analysis [63] showed that CIII has a high FCC, as well as CI, in NAD-linked respiration, whereas CIV has a low FCC; this means that CI and CIII behave as a functional enzyme unit, whereas CIV seems to be functionally independent.

These results cannot be easily explained [109]. It is true that most of CIV appears to be free in the BN-gels (cf. Section 2), and one might consider that the cytochrome *c* oxidase activity in the respirasome is masked by the large excess of active enzyme randomly distributed in the membrane. On the other hand, this would mean that channeling is not a major feature of electron transfer in the cyt. *c* region. The latter means that the free molecules of CIV are involved in electron transfer from NADH and implies that the molecules of CIV assembled in the respirasomes are not involved in the channeling of cyt. *c* [38,42], but do behave similarly to the free CIV units in electron transfer by random diffusion.

The conclusion that the absence of channeling in the cyt. *c* region of mammalian mitochondria is a physiological feature, is supported by the observation that the respirasome of potato tuber mitochondria is completely functional in cyt. *c* channeling, as detected by the same flux control analysis in mammalian mitochondria [71]. This difference may be ascribed to a tighter binding of cyt. *c* in the potato respirasome, as shown by BN-PAGE.

The structural evidence of purified SCs by cryo-EM adds structural reasons on the looseness of CIII–CIV interactions in mammalian mitochondria. In ovine mitochondria, Letts et al. [54] identified two distinct arrangements of SC $I_1III_2IV_1$: a major "tight" form and a minor "loose" form (resolved at the resolution of 5.8 Å and 6.7 Å, respectively). In both respirasome structures, CIV has a weaker density than the other complexes in the respirasome, indicating the greater conformational flexibility of CIV. In the tight respirasome, CIV contacts both CI and CIII, whereas CIV in the loose respirasome only interacts with CI. Tight and loose architectures may represent independent structural entities or may interconvert, indicating different stages of assembly or disassembly. The presence of two forms of the respirasome having distinct CIV linkages was also detected by Sousa et al. [57] and ascribed to instability of the purified SC.

In partial contrast with the above considerations, Lapuente-Brun et al. [95] demonstrated that at least part of CIV forms a functional SC with channeling of cyt. *c*, but the SC formation depends on the availability of the SC assembly factor SCAF1. They showed that when CIV participates in SCs due to the presence of SCAF1, a significant proportion of CIV activity is not utilized in cell respiratory activity. They also identified three CIV populations, one of which is dedicated exclusively to receive electrons from NADH oxidation (forming SC I+III+IV), presumably by cyt. *c*. channeling. The second population receives electrons from FAD-dependent enzymes (forming SC III+IV and presumably operating by cyt. *c* channeling), whereas the third major one is in free form and receives electrons from both NADH and $FADH_2$ oxidization, presumably using the free cyt. *c* pool. On the contrary, if CIV is maintained permanently detached from SCs by elimination of SCAF1, all CIV is in free form and electron transfer takes place via diffusion of a single pool of cyt. *c*.

Functional evidence for cyt. *c* channeling was also found in *S. cerevisiae* [110] mitochondria which are characterized by having all CIV bound to CIII in a supercomplex [111], thus preventing electron transfer through free CIV units. The structural evidence by single particle cryo-EM sustains channeling, since in the III_2IV_2 SC (there is no CI in this yeast species) the distance between the binding sites of cyt. *c*, i.e., cytochrome c_1 of CIII and the Cu_A-subunit II of CIV, is considerably shorter than that in bovine mitochondria [112]. At odds with this structural evidence, however, Trouillard et al. [113] showed that the time-resolved oxidation of cyt. *c* by CIV in yeast mitochondria is a random-collision process.

Rydström Lundin et al. [114] studied the kinetics of aerobic quinol oxidation in *S. cerevisiae* mitochondria as a function of the respiratory SC factors Rcf1 and Rcf2 that mediate supramolecular interactions between CIII and CIV forming SCs $III_2IV_{1\text{-}2}$. They demonstrated that Rcf1 promotes formation of a direct electron-transfer pathway from CIII to CIV via a tightly bound cyt. *c*. Accordingly, in these mitochondria under steady-state conditions in the presence of added homologous cyt. *c*, the direct electron transfer through the bound cyt. *c* (i.e., cyt. *c* channeling), is faster than the equilibration of electrons with the cyt. *c* pool. Interestingly, when using heterologous cyt. *c* (from horse heart), electron transfer between CIII and CIV occurs only via the cyt. *c* pool. Realistic theoretical assumptions indicate that electron transfer between CIII and CIV can become rate limiting. Hence, there is a kinetic advantage of bringing CIII and CIV together in the membrane to form SCs [115]. In a yeast mutant defective in SC formation, but containing fully functional individual complexes, the diffusion of cyt. *c* between the separated complexes is delayed, thus reducing electron transfer efficiency [116].

Moreover, other results indicate that in the absence of Rcf1, the interactions required for stability of the SC III+IV are disrupted [117]. Consistently, Rydström Lundin et al. [114] demonstrated that direct electron transfer does not take place in the absence of Rcf1 and electrons are only transferred via the cyt. *c* pool in the Rcf1Δ strain.

In the next Section 3.5, we will discuss the possible physiological reasons why there is no major channeling of cyt. *c* in mammalian mitochondria, contrary to yeast or plant mitochondria.

3.5. Supercomplexes and Regulation of Metabolic Fluxes

Does CoQ channeling occur under physiological conditions? As discussed in the previous Section 3.3, the presence of a bottleneck downhill CoQ might induce interaction of the NADH pathway (via channeled CoQ in SC $I_1III_2IV_n$) with the succinate pathway (CoQ pool behavior) by shifting the CoQ dissociation equilibrium. However, if a rate-limiting step is situated upstream, i.e., in or before the dehydrogenases, the reducing pressure of CoQ on its partner oxidases (particularly in the case of CIII assembled in the SC $I_1III_2IV_n$) may not be present and the two routes would take place independently. On the other hand, in mammalian mitochondria, it appears that the two fluxes mix up in the cyt. *c* region where most evidence excludes cyt. *c* channeling (Section 3.4).

The existence of two different functional CoQ compartments, one for NADH oxidation an another one for oxidation of succinate and other FAD-linked substrates, has deep implications for the metabolic adaptation to the feeding state. The flux of electrons in the respiratory chain is conditioned by the orientation of metabolism to use different fuels. The use of different fuel molecules generates different proportions of NADH and $FADH_2$, that require an optimal equilibrium between the corresponding routes of electron transfer in the respiratory chain. In a well-fed individual, the oxidative metabolism largely follows the glycolytic pathway and the Krebs cycle, so that the oxidation of NAD-linked substrates exceeds succinate oxidation (ca. 5:1) and the electron flux proceeds mainly through SC I_1III_2. On the other hand, metabolic adaptation of liver mitochondria to fasting forces fat mobilization and FAO [118]. In fact, an increase in the $FADH_2$-dependent respiration, as in FAO, induces saturation of the CoQ pool reoxidation capacity and promotes reverse electron transport from ubiquinol to CI [119]. The resulting local generation of superoxide triggers protein degradation of CI subunits by oxidative damage, and consequent disintegration of the complex, as well as lipid peroxidation, disassembly of SCs and further instability of their enzyme components. Thus, we can infer that CoQ redox status acts as a metabolic sensor that fine-tunes the configuration and supramolecular organization of the respiratory chain in order to match the prevailing substrate profile [120]. In order to avoid such saturation of the CoQ pool oxidation capacity, fasting conditions require proper disassembly of SCs to achieve adjustment of respiratory SC proportions favoring the FAD-linked route. Similar adaptations may occur when the increase in FAO results in response to high-fat diet. Impairment of this adaptation may be relevant to pathological processes associated to obesity.

As postulated by Lapuente-Brun et al. [95], at the metabolic level, the fast kinetic adjustments of the mitochondrial respiration are followed by gene expression changes in the level of individual respiratory complexes.

Why appreciable channeling may not exist at the level of cyt. *c* in mammalian mitochondria? (cf. Section 3.4). Mammalian mitochondria have several dehydrogenases directing electrons to CoQ, however they usually have only one oxidase (CIV) receiving electrons from CIII via cyt. *c*. For this reason, it may be worth speculating that, whilst it may be useful to separate the major NADH-dependent flux from CI from those departing from succinate, fatty acid oxidation and other metabolic pathways by separating the CoQ compartments, there is no such need for cyt. *c* that is by and large receiving electrons univocally from CIII. This assumption is reinforced by the fact that, contrary to what is found in mammalian mitochondria, CIII and CIV of plant mitochondria are functionally operating as an SC and, at the same time, cyt. *c* is tightly bound in the SC, as revealed by flux control analysis [63] and BN-PAGE [71]. Not surprisingly, plant mitochondria are characterized by a high branching of the electron transfer pathways feeding electrons directly to cyt. *c* [121], which requires adjustment of the different routes as a response to physiological needs, as it happens in mammalian mitochondria at the CoQ level. In plant mitochondria, segmentation might be achieved by regulating different compartments of free and bound cyt. *c*.

4. Overview of the F_1F_O-ATP Synthase/Hydrolase and Its Supramolecular Structure

The mitochondrial respiratory chain and its arrangement in SCs, detailed in the previous sections, has the main bioenergetic role to build the electrochemical gradient which allows the F_1F_O-ATP synthase to build ATP, namely, to accomplish the so-called energy transduction, which converts a transmembrane proton movement into chemical energy, which can be used for all the energy-consuming processes of the cell.

The rotary ATPase family, which embraces structurally similar membrane-bound enzyme complexes working as energy transduction mechanisms, consists of three subfamilies, A-, V- and F-type ATPases, which originate from a common evolutionary ancestor. A-type ATPases occur in Archaea and some bacteria, V-type ATPases are typical of eukaryotic vacuoles, while F-type ATPases occur in eukaryotic mitochondria, tylakoid membranes of chloroplasts and in bacterial cell membranes [122]. In mitochondria, the hydrophilic (F_1) and hydrophobic (F_O) ATPase domains are joined laterally by a single stalk stator and in the middle by a central stalk. The F-type ATPases can function as ATP synthesis or ion pumps coupled to ATP hydrolysis, a bifunctional mechanism unique in nature [123,124]. The F_1F_O-ATPase has a mushroom shape, in which F_O is membrane-embedded and F_1 protrudes outside the mtIM. The enzyme complex is universally known as the nano-machine that produces ATP, the "molecular energy currency" under aerobic conditions [125,126]. The hydrophilic enzyme domain basically consists of a spherical extrinsic hexamer formed by three catalytic β subunits alternated with three non-catalytic α subunits. The $(\alpha\beta)_3$ hexamer contains in the core the asymmetrical central stalk (in turn composed by γ, δ, and ε subunits). The central stalk is joined to the loop of each *c* subunit hairpin. The number of *c* subunits is species dependent: these subunits are arranged as a cylindric palisade to form the *c*-ring. This subunit assembly, namely the *c*-ring and the central stalk, makes up the rotor. The membrane-embedded *a* subunit with unusual "horizontal" hairpin helices named H5–H6 matches the concave barrel-like *c*-ring where the H^+ sites lie and perfectly fits the ring size in order to create the H^+ translocation pathway. A static structure peripheral to the rotor, which spans for the entire enzyme complex length, acts as a stator to prevent the $(\alpha\beta)_3$ rotation torque of the central stalk. The peripheral stalk is composed by various subunits, namely the hydrophilic oligomycin sensitivity-conferring protein (OSCP), F6, *b*, *d*, linked to F_1-catalytic domain and embedded in the mtIM by the hydrophobic portion of *b* and A6L subunits. The peripheral stalk is also associated to the supernumerary membrane subunits (*sms*) *e*, *f*, *g*, and 6.8-kDa proteolipid (6.8PL). However, 6.8PL was erroneously believed to be inserted in the central hole of the *c*-ring [127] and another supernumerary subunit, diabetes-associated protein in insulin-sensitive tissue (DAPIT), in the tetrameric F_1F_O-ATPase porcine model was misassigned, since recently it was reported as localized at the furthest edge of F_O domain [128] (Figure 3). So, the emerging advances in the enzyme knowledge make researchers continuously re-consider and re-evaluate the F_1F_O-ATPase structure and function.

In mammalian mitochondria, this astonishing enzyme complex has recently revealed versatile roles, in addressing cells to life or death, in the maintenance of the mitochondrial morphology and raised great expectations as a drug target [129,130]. Accordingly, the modulation of the enzyme functions, often compromised by mitochondrial dysfunctions associated with severe diseases, may contribute to innovative therapeutic strategies at the molecular level.

4.1. The F_1F_O-ATP Synthase from the Energy Production to Mitochondrial Morphology

In the energy-transducing mtIM, the synthesis of ATP is based on the coupling of the rotary mechanisms of the two main sectors, namely the hydrophobic F_O and the hydrophilic F_1 [131,132]. The protonmotive force (Δp), built by substrate oxidation in the respiratory chain, drives H^+ translocation through F_O domain by generating torque and allowing conformational changes in F_1 which leads to ADP phosphorylation to yield ATP. In mitochondria, the bi-functional enzyme can also work in reverse when the Δp drops. In this case, ATP hydrolysis by the F_1 domain fuels H^+ pumping by F_O which re-energizes the mtIM [133]. This energy-dissipating mechanism has also been associated with a variety of pathological conditions [134]. Both in the forward and in the reverse reaction, H^+ translocation across the mtIM is due to the reversible protonation/deprotonation of carboxylic sites, which also converts H^+ flux into F_O rotation. The disconnection between these two matched activities, namely H^+-translocation and catalysis, "uncouples" the F-ATP synthase and often leads to mitochondrial dysfunctions. The enzyme sensitivity to the antibiotic oligomycin, which by binding to the enzyme covers the H^+ binding sites and blocks the *c*-ring rotation and enzyme catalysis in either directions [135], is taken as a parameter of the efficient coupling between the two domains.

This amazing rotary enzyme complex offers a good example of how molecular structure and function are tightly linked features [130], which cannot work without each other.

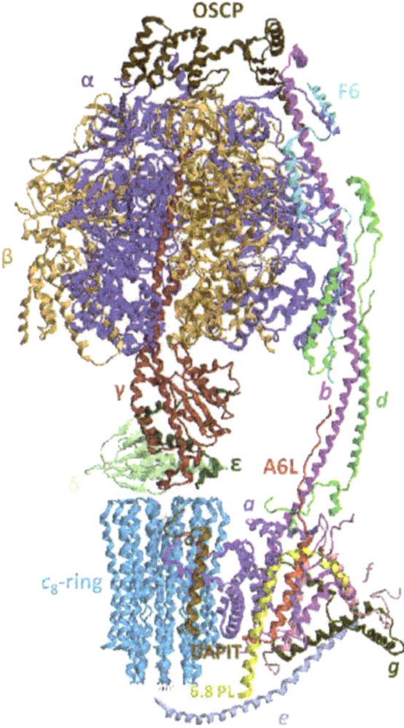

Figure 3. Structure of the mitochondrial F_1F_O-ATPase in mammals. The enzyme subunits are drawn as ribbon representations obtained from modified PDB ID code: 6TT7. The differently colored letters identify the subunits, drawn in the same color as the letter.

4.2. Structural Basis of the Bioenergetic Mechanism

Among all F-ATPases, the mitochondrial enzyme complex has the most complex subunit composition [136]. F_1 protrudes in the mitochondrial matrix as an asymmetric hexagonal globular assembly of α and β subunits, arranged as $(\alpha\beta)_3$ around the γ subunit of central stalk which connects F_O to F_1. The hexamer, in which α and β subunits alternate, hosts three catalytic and three non-catalytic sites. The three catalytic β-sites can adopt three conformations, namely "open", "closed", and "semi-closed", defined as β_E, which is empty, β_{TP}, which contains Mg-ATP or Mg-ADP and β_{DP}, which hosts Mg-ADP. Each of the three non-catalytic α subunits binds Mg-ATP [137] and also Mg-ADP [128]. According to the binding change mechanism, the β conformations interconvert each other as the *c*-ring rotates and transmits the rotation to central stalk [138], which when it meets the catalytic sites, changes their conformation and affinity for nucleotides. So, during a complete turn (360°) each catalytic site undergoes (stepwise) all the three different conformations leading to synthesize/hydrolyze three ATP molecules. Moreover, Mg-ATP bound to the non-catalytic sites allows ADP release from the βDP site and removes the Mg-ADP-driven enzyme inhibition during multiple ATP hydrolysis turnover [139]. Interestingly, the mammalian γ subunit shows two isozymes, a "heart isoform" and a "liver isoform", which differ by a single amino acid, an additional aspartate at the C-terminus of the liver enzyme.

The former is expressed in tissues with a high and variable energy demand such as heart, muscle, diaphragm, while the latter is expressed in brain, thyroid, spleen, pancreas, kidney, testis and liver, which, apart from brain, show a low ATP consumption and a steady energy demand. Even if the C-terminus Asp in the liver γ-subunit may form salt linkages and decrease the enzyme efficiency, no differences in enzyme kinetics were found, so the two γ isoforms probably only have a regulatory function [140].

Other than by the γ subunit, the functional coupling of the two main sectors F_O and F_1 in mammalian mitochondria is ensured by the peripheral stalk and especially by OSCP [134].

The membrane-embedded F_O domain, which allows H^+ flow across the mtIM, hosts the *a* subunit, which channels H^+, and the c_n-ring, namely the rotor, which consists of a cylindrical palisade of n subunits. The number n varies in the range 8–15 among the species, is constant in the same species and determines the ATP bioenergetic cost, which is obtained from the ratio between the number of translocated H^+ and the three ATP molecules constantly produced in a complete (360°) rotation of the rotor. Accordingly, the number of transported H^+ per complete rotation of the *c*-ring is related to the number of H^+ binding site(s) on each *c* subunit. In general, an increase in the *c*-ring size and consequently in the number of H^+-binding *c*-subunits, implies an increase in the bioenergetic cost of ATP, since more H^+ should flow downhill across the mtIM to produce one ATP molecule. Mammals which show a small *c*-ring (n = 8) are highly efficient by dissipating less H^+ to yield one ATP, namely they obtain ATP at a low bioenergetic cost [141]. So, the number of *c* subunits, and of H^+ binding sites, can be taken as a rotor efficiency parameter. In the OXPHOS system the transmembrane driving force Δp which links substrate oxidation to ADP phosphorylation to yield ATP, consists of an electrical $\Delta\Psi$ (membrane potential difference) and of a chemical component, ΔpH (pH gradient) between the positive and negative side of the mtIM. During evolution, species have adapted the *c*-ring size to the prevailing electrochemical parameter, $\Delta\Psi$ or ΔpH. Accordingly, when $\Delta\Psi$ prevails and the pH gradient is low, the *c*-ring is small [142]. This is the case of mammals, generally vertebrates and some invertebrates [143]. Conversely, a prevailing ΔpH is associated with the large *c*-rings (n = 11–15), typical of chloroplasts and bacteria [144,145].

The H^+ pathway within F_O only recently has been elucidated. It is built step by step within the *a* subunit by selected amino acid side chains, which form a quite unexpected route along the mtIM allowing membrane crossing in a perpendicular way to the two aqueous half-channels. The two half-channels are discontinuous and open one in the matrix and the other in the lumen and both contain two horizontal highly conserved α-helices, named H5 and H6, which lie along the mtIM and are juxtaposed to the *c* subunits, thus providing access to the *c*-ring H^+ binding sites [146]. This α-helix arrangement in the mtIM features all rotary A-, F- and V-type ATPases [147]. Each *c* subunit is hairpin-shaped and spans the mtIM. These gathered hairpins form a sort of hourglass, seen laterally from the membrane side, whose concavity hosts the H^+-binding sites on the outer C-terminal α-helices of *c* subunits. Therefore, the horizontal α-helix arrangement perfectly fits the rotor concavity [148].

The H^+ transfer across the mtIM in either direction requires subsequent protonation/deprotonation steps of H^+ binding sites that consist of amino acid side chains that interconvert between the protonated and deprotonated states. Interestingly, in mitochondria at the edge of the *cristae*, the pH value is higher than that in the IMS opposing the inner boundary membrane (IBM), as well as the *cristae* $\Delta\Psi$ generated by the respiratory chain [149] is higher than in the IBM [150]. The H^+ flow through F_O for ATP synthesis occurs from the lumen or *intracrista* space of the half-channel formed by the hydrophilic cavity created by the end of H5 and H6 helices of *a* subunit, and H3 helixes of *b* and *f* subunits. The H^+ pathway starts from *a*His-168 and *a*His-172 on H5 helix and ends at *a*Glu-203 on H6 helix of *a* subunit, which acts as intermediate H^+-donor to *c*Glu-58 of the *c*-ring [128,151]. The Glu-His interactions establish multiple H-bridges in the membrane half-channel that, by changing the carboxylic group pK_a, allow its protonation. On the opposite half-channel on the luminal side viewed from the *intracrista* space, the *a*Glu-145 on H5 (matrix) helix acts as "de-protonator" which restores the original pK_a of *c*Glu-58 allowing the H^+ detachment. The half-channel arises where a conserved *a*Pro-153 bends the *a*H5 helix [151]. Then, the folding is enforced by DAPIT subunit and H4 helix of *a* subunit in order to make the half-channel hydrophilic core conformationally adapt to the *c*-ring shape (Figure 4).

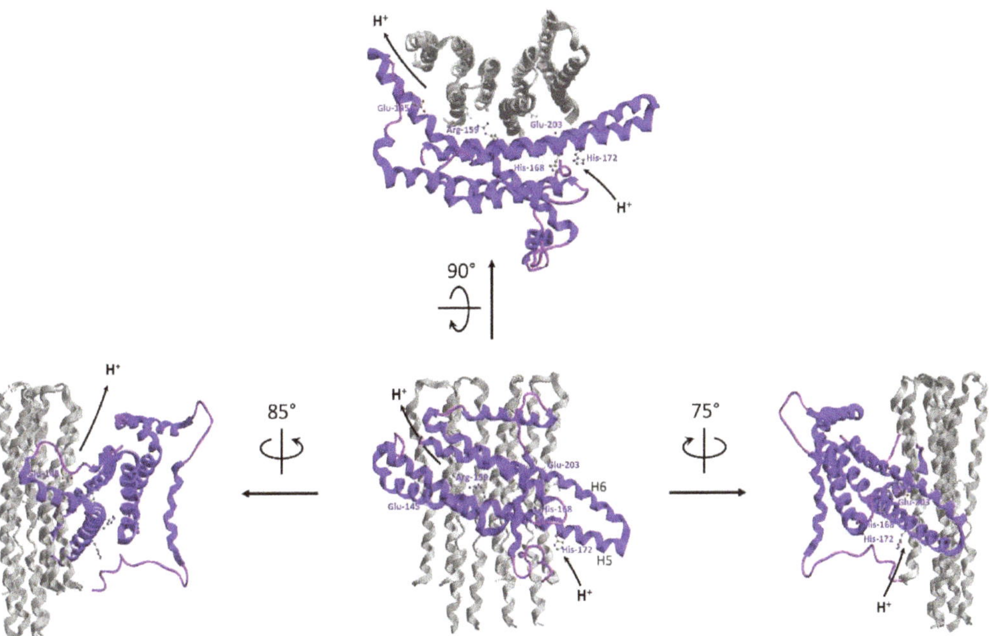

Figure 4. Proton translocation pathway within F_O during ATP synthesis. On the top the H^+ entry and exit into and from the half channels, viewed from the IMS. On the left and right sides, the H^+ outlet and inlet half-channels, respectively, after the rotation angles above the arrows are shown. The horizontal helices H5 and H6 of *a* subunit (violet) and four *c* subunits (gray), drawn as ribbon representations, were obtained from modified PDB ID code: 6TT7. The amino acids of *a* subunit involved in H^+ translocation are drawn as ball and stick models.

During the H^+ translocation through the membrane-embedded F_O domain, a putative participation of hydronium ion (H_3O^+) was also suggested [152]. Accordingly, in prokaryotes and eukaryotes the bell-shaped pH profile of the F_1F_O-ATPase inhibition by dicyclohexyl-carbodiimide (DCCD), which covalently binds to *c*-subunit carboxylates, is not compatible with carboxylate protonation of H^+ binding sites, but it may be explained by H_3O^+ coordination [143,153]. Indeed, on considering the position of molecules able to establish hydrogen bonds in the half-channels, whose core is hydrophilic, H^+ translocation may be coordinated by water molecules by the Grotthuss mechanism, namely by proton jumping from covalent to H-bonds and vice versa [128,154]. During ion translocation, the *c*Glu-58 adopts different conformations, namely the H^+-unlocked anionic form (carboxylate) to face the aqueous environment of the half channels, and the H^+-locked form of the protonated carboxylic group to face the hydrophobic environment of the mtIM. In the locally hydrated luminal half-channel, the *c*Glu-58 carboxylate is oriented in an outward-facing (H^+-unlocked) conformation before protonation and turns to an inward-facing closed (H^+-locked) conformation when protonated. On the opposite half-channel, which faces the matrix, the *c*Glu-58 in the H^+-locked conformation turns to the H^+-unlocked conformation after deprotonation. The *c*-ring key carboxylates embedded in the mtIM are always in the H^+-locked conformation to enter the lipophilic mtIM [155] and make the *c*-ring rotate due to Brownian motion [131]. The rotation direction when viewed from F_O toward F_1 is clockwise during ATP synthesis and counterclockwise during ATP hydrolysis [133,156]. The low pH in the luminal half-channel favors *c*Glu-58 protonation and the consequent locked conformation, pushed by Δp that favors the entry to the mtIM. When an almost entire rotor rotation is completed, the *c*Glu-58 reaches the half-channel on the matrix side where the low H^+ concentration as well as the negative charge of the "H^+-releasing site" on H5 helix of *a* subunit, favors its deprotonation associated with the H^+-unlocked carboxylate form. The positive charge of a crucial arginine, *a*Arg-159, acts as electrostatic barrier between the two half-channels, prevents H^+ short circuiting and attracts the negatively charged *c*Glu-58 carboxylates [157]. Moreover, *a*Arg-159 helps the *c*Glu-58 de-protonation during H^+ translocation and prevents salt bridges between *a* and *c* subunits which would block the rotation. Interestingly, *a*Arg-159 and the H^+-unlocked *c*Glu-58 are too far (about 4.5 Å) and cannot interact [144], so the two half-channels for H^+ entry and exit at the *a*/*c*-ring interface are spatially offset to allow the rotation direction adjust to Δp [135].

During ATP hydrolysis the transition between the three β conformations in F_1 drives the rotations of the γ subunits and of the *c*-ring and increases the Δp by pushing H^+ uphill [134].

The coupling of energy transduction and H^+ translocation is allowed by conformational adaptations of the F_1 and F_O domains [158] that adapt to each other, helped by flexible regions in the peripheral stalk. The latter undergoes torsion during catalysis, but also the *c*-ring shows conformational fluctuations when it contacts the *a* subunit. Therefore, coordinated conformational changes within the enzyme proteins produce and synchronize the torsion.

4.3. The ATP Synthase Supramolecular Arrangement and the Mitochondrial Shape

The mitochondrial F_1F_O-ATPase functions are ensured by the enzyme supermolecular arrangement. By self-assembling in dimers and tetramers, the F_1F_O-ATPase rules most bioenergetic and structural functions in eukaryotic mitochondria [159]. Accordingly, the occurrence of *sms*, which lack in bacteria and chloroplasts, in F_O is linked to the *sms* involvement in supercomplex arrangement and in the mtIM ultrastructure [130]. Consistently, no F_1F_O-ATPase dimers were found in prokaryotes and chloroplasts, while some differences between yeast and mammalian F_1F_O-ATPases occur in the building of contact sites. It seems to have been ascertained that, while the basic subunits allow the overall functionality of the enzyme complex, these *sms* are required to join the monomers and to bend the membrane. The *sms* were differently named according to the taxa: A6L, *e*, *f*, *g*, and DAPIT, which corresponds to the *k* subunit in yeasts, and mammalian 6.8PL, which corresponds to the orthologue *i/j* in yeasts [128]. The membrane subunits are mainly encoded by nuclear genes, apart from *a*, A6L and *c* subunit (identified also as *6*, *8* and *9* subunit) for yeast F_1F_O-ATPase and only *a* and A6L in mammals which are encoded by the mitochondrial DNA. Notably, different *sms* structurally contribute to the F_1F_O-ATPase dimerization in yeasts and in mammals [127,160]. Three different isoforms occur in the mammalian *c* subunit.

F_1F_O-ATPase monomers form dimers arranged in long rows on the mtIM and MICOS (mitochondrial contact site and *cristae* organizing system) complex at *cristae* junctions cooperate to yield the *cristae* morphology in agreement with their opposing effects on membrane curvature [161–163]. Moreover, recent work shows that the F_1F_O-ATPase dimerization factor subunit *e* interacts with the MICOS component Mic10, which stabilizes the enzyme oligomers and possibly temporal orchestration of cristae biogenesis. Moreover, the F_1F_O-ATPase dimerization and the MICOS component Mic60 induce opposite membrane curvature, namely the former promotes a negative curvature and the latter a positive curvature (concave and convex as viewed from the matrix, respectively). Most likely, the interaction between MICOS and dimeric F_1F_O-ATPase, which implies a high mobility of the enzyme monomers in the *crista* membrane, enables spatial and temporal coordination during *crista* biogenesis and dynamics [164].

Interestingly, the dimerization interface in different species shows different subunit composition with no apparent homology [136].

Four different types of F_1F_O-ATPase dimers were described in mitochondria. In mammals (e.g., *Bos taurus, Sus scrofa domesticus,* and *Ovis aries* heart) and yeasts (*Saccharomyces cerevisiae*) V-shaped dimers, named type I dimers, are localized on the rim of the *cristae* with an mtIM convexity of about 90° [137,149]. Type II, III and IV dimers, which differ in the angle between monomers, were only found in algae, protozoa and small invertebrates [148,159].

Mammalian F_1F_O-ATPases can also form a H-shaped tetramer, which in turn consists of two assembled type I dimers that lie antiparallel to each other. The dimers are joined by two IF1 subunits, an endogenous protein inhibitor [127], evolutionary conserved throughout eukaryotes, which only blocks ATP hydrolysis but not ATP synthesis. Most likely, IF1 normally is not essential for cell survival, but can be crucial under pathological or stress conditions [165], as suggested by its overexpression associated to the decrease in OXPHOS in some cancer types [166]. However, IF1 not only prevents ATP dissipation by ATP hydrolysis, but also contributes to mitochondrial morphology [127,167,168]. The IF1 dimer, which becomes active at acidic pHs in the matrix and stems from the dissociation of the tetramer, inhibits ATP hydrolysis when Δp collapses. In the tetramer, IF1 links two F_1 domains of two opposite dimers; it binds to the catalytic interface between the α_{DP} and β_{DP} subunits in loose binding conformation. Therefore, the two monomers of the laterally opposite dimer are connected by the same IF1; these two IF1 proteins make the tetramer steady and block ATP hydrolysis by a ratchet-like action on the rotor. The F_1F_O-ATPase dimers are also maintained in rows by a long-range attractive force that stems from the relief of the overall elastic strain of the mtIM [169]. Other than by interactions with IF1 dimers, the mammalian tetramer is maintained by *e-e* and *g-g* interactions between two opposed and diagonally arranged monomers [170]. Moreover, the monomers are joined side-by-side in a row through DAPIT interaction with *a* subunit at the matrix space and with *g* subunit both at the matrix side and the intra*crista* space. The dimers are linked by interactions of one monomer 6.8PL with *f* and *e* subunits of the other monomer, and vice versa, at the matrix and intra*crista* space, respectively. In addition, the subunits *a-a* in the middle of mtIM, the *f-f* at the matrix and *e-e* at the intra*crista* space establish multiple contacts between each monomer pair [128].

Moreover, the association of two F_O domains depends on *e* and *g* subunits and on a putative conserved dimerization GxxxG motif in the transmembrane α-helices of both subunits [171,172]. The substitution of a glycine residue by leucine into the *e* subunit motif led to the loss of *g* subunit, destabilized the dimer and resulted in an onion-like structure in the *cristae* [171]. The individual α-helix of *e* subunit and the H3 of *g* subunit interact with their respective GxxxG motif and joined to H2 of *b* subunit form the triple transmembrane helix bundle (TTMHB). The TTMHB tilt is driven by a U-turn structure formed by H2 and H3 helices of *b* subunit. This subunit assembly creates a reminiscent BAR-like domain that bends the mtIM [128,173] to form the apex of *cristae,* but *e* and *g* subunits also keep the monomers together. Accordingly, disulfide cross-link experiments showed the interactions between *e-g* or *e-e*/*g-g* increase the stability of F_1F_O-ATPase dimers and oligomers, respectively, in mitochondrial digitonin extracts [172].

In mammals, subunit–subunit interactions produce the F_1F_O-ATPase supermolecular arrangement, in which monomer pairs form dimers and dimer pairs form tetramers, which in turn form long rows of oligomers. This structural arrangement produced by subunit–subunit interactions, also forces the mtIM to be convex at the apex of the *cristae.* The enzyme subunits involved in oligomerization, though not directly involved in catalysis, contribute to the maintenance of *cristae* architecture, which is essential for an efficient respiration by acting as proton sink. A mutation in subunit *k*/DAPIT results in aberrant *cristae* formation, defective F_1F_O-ATPase dimerization and a mild Leigh syndrome phenotype [164]. *Cristae* remodeling and F_1F_O-ATPase dimerization may be also involved in cell differentiation [174].

Interestingly, in some forms of cancer, the overexpression of IF1, which leads to the F_1F_O-ATPase dimerization and *cristae* formation, could contribute to neoplastic degeneration and evasion of apoptosis. The absence of IF1 in Luft's disease, a rare mitochondrial disease, is associated with densely packed mitochondrial *cristae* [175]. Moreover, the acidic phospholipid CL is critical for oligomerization of F_1F_O-ATPase multimeric complexes either by direct interaction with the enzyme or by inducing the proper membrane curvature. CL, acting like a non-bilayer forming phospholipids, interacts with the F_1F_O-ATP synthase and may reduce the free energy of the extreme mtIM curvature to stabilize high-curvature folds [164,176]. From these few examples it seems clear that any change in at least one among the multiple factors which rule the mitochondrial morphology can easily lead to pathology. The F_1F_O-ATPase can work in monomeric and dimeric form; however, the loss of the supramolecular organization of the F_1F_O-ATPase causes aberrant membrane morphology results in respiratory defects.

On these bases CL defects, as in Barth Syndrome (see Section 3.1.3) can affect the supramolecular arrangement of the F_1F_O-ATPase, the mitochondrial shape and, as depicted in the following section, the mitochondrial permeability transition.

During aging, the typical rim shape of the *cristae* disappears and the F_1F_O-ATPase dimers dissociate into monomers. Consistently, the mitochondrial morphology changes, showing a vesicular mtIM, leading to mitochondrial dysfunction and cell death [177]. Even if most results come from in vitro studies on experimental models and should be deepened and confirmed in vivo, recent advances suggest that mitochondrial changes are a driving force, rather than a consequence, of the aging process and neurodegeneration [176]. The connection among the F_1F_O-ATPase supramolecular arrangement, the mitochondrial shape and some human diseases, as pointed out by these few examples, is also made clearer by the enzyme involvement in the mitochondrial permeability transition pore (mPTP), the topic of the following section.

4.4. How the ATP Synthase/Hydrolase Is Involved in the Mitochondrial Permeability Transition Pore

In recent years, the F_1F_O-ATPase has been implicated in the mPTP, a large pore in the mtIM, which permeabilizes the mtIM to ions and other solutes [178–180]. As the membrane becomes permeable, the massive water influx in the matrix results in mitochondrial swelling. For short full openings, the mPTP is reversible, but prolonged openings are irreversible and trigger the release of cyt. *c* and other pre-apoptotic factors which drive the cell to death [181]. The mPTP formation is stimulated by ROS and Ca^{2+} increase, by the binding of the mitochondrial protein cyclophilin D (CyPD) in mitochondria and inhibited by H^+, Mg^{2+}, adenine nucleotides, and cyclosporin A (CsA), which displaces CyPD and desensitizes the mPTP to Ca^{2+} [182]. The mPTP activity mainly appeared as a property of mammalian and yeast mitochondria, where it was intensively investigated. However recently the mPTP was also reported in mussel [183], in sea urchin oocytes [184], in the nematode *Caenorhabditis elegans* [185], in plathelminths [186] and insects [187,188], while it is still controversial in crustaceans [189,190]. Indeed, it is not a "vertebrate invention" to rule cell fate, as initially thought [191].

At present, the problem of the involvement of F_1F_O-ATPase in the mPTP and especially its structural participation in the mechanism of mPTP formation and opening is near to being solved [128]. This quite surprising task of the F_1F_O-ATPase, which has been supposed by different research groups, points out the versatility of this intriguing enzyme complex, which emerges as the playmaker of cell life and cell death. Two main F_1F_O-ATPase sites have been involved in the mPTP formation: the *c*-ring [179,180] or the monomer–monomer interface of the dimer [178].

The hydrolytic function of the F_1F_O-ATPase, namely ATP hydrolysis, can be sustained by metal divalent cations different from the natural cofactor Mg^{2+}, such as Ca^{2+}, which is unable to activate the enzyme complex in the opposite function of ATP synthesis [192–195].

Interestingly, Ca^{2+} rise in the mitochondrial matrix initiates a cascade of events which lead to cell death. So, in recent times Ca^{2+} binding to the F_1F_O-ATPase in replacement of Mg^{2+}, an event which can easily occur in the presence of relatively high Ca^{2+} concentrations in mitochondria, since the enzyme affinity for Ca^{2+} is lower than that for Mg^{2+} [193,196], has been associated with the mPTP opening [192,195,197–199].

Accordingly, the structure of mammalian F_1F_O-ATPase exposed to Ca^{2+} and revealed by cryo-electron microscopy, shows unusual states, not identified when Mg^{2+} acts as cofactor, which can be ascribed to mPTP opening. Indeed, in the new "bent-pull" model, the Sazanov lab highlights the role of *c*-ring and of some *sms* in mPTP formation. Accordingly, when Ca^{2+} replaces Mg^{2+} in the catalytic binding site, the cation insertion in the Ca^{2+}-activated F_1F_O-ATPase, due to the higher steric hindrance, would promote conformational changes of F_1 domain, in turn transmitted by the peripheral stalk [198] to membrane subunits that form the mPTP [198]. Notably, when the Ca^{2+}-activated F_1F_O-ATPase activity is decreased by various inhibitors, the mPTP formation is delayed or even prevented [196,197,200]. The *c*-ring contains two different phospholipids at the two opposite sides of its cavity. At the matrix side, phosphatidyl serine is anchored by ionic interactions to the positive charge of Arg-38 of *c* subunits, while at the intra*crista* side, Lys-71 of *e* subunit coordinates a lyso-phosphatidylserine. Since the phosphatidylserine double acyl-chain does not have enough space around the *c*-ring plug and is linked to the *c*-ring, it rotates together with the rotor. Conversely, the monoacyl chain of lysophosphatidylserine acts as "lubricated" lipid plug. The two lipids inside the *c*-ring are separated by the conserved *c*Val-16 [128]. The F_1F_O-ATPase distortion and tilt induced by Ca^{2+} trigger changes in the conformational states of the Ca^{2+}-activated F_1F_O-ATPase that open the mPTP. Most likely, the signal propagation from F_1 to F_O through the long helix of *b* subunit modifies the TTMHB assembly and changes the position of *e* subunits, which expel the lyso-phosphatidylserine from the central hole inside the *c*-ring and opens the channel at the positive mtIM side [181]. Consequently, the curved *crista* ridges and the F_1F_O-ATPase dimerization are lost [177,199,201]. According to Sazanov's hypothesis, the water molecules inside the *c*-ring destabilize the phosphatidylserine which pushes out the lipid plug and creates a pore through the *c*-ring, while the consequent conformational change detaches F_1 from F_O. Thus, structural/conformational F_1F_O-ATPase changes are involved in the (ir)reversible mechanism of the mPTP, which opens and closes and rules cell fate. The most recent data strongly sustain the hypothesis that these changes mainly involve the *c*-ring, which emerges as the main character in mPTP formation (Figure 5).

However, this matter is still hot and some debate on this topic remains. The possibility that differently sized pores can be formed and coexist in the mtIM satisfactorily combines the conclusions drawn from different experimental approaches. Accordingly, the mtIM could be depolarized by any increase in conductance upon mitochondrial Ca^{2+} overload, due to transmembrane channels and/or transporters [202]. In this scenario, the Ca^{2+}-activated F_1F_OATPase would contribute to the membrane depolarization by inducing the largest mPTP pore. It seems reasonable to think that smaller sub-conductance activities, which contribute to the mPTP can be ascribed to many other mitochondrial channels/transporters [203]. However, the mitochondrial F_1F_O-ATPase remains the most likely candidate as main high-conductance major channel or mPTP by definition, proven to be inhibited by CsA, but not by bongkrekate (BKA), which is known as inhibitor of the adenine nucleotide translocase [202,204].

Moreover, the adenine nucleotide translocase isoforms could form a secondary low-conductance mPTP inhibited by both CsA and BKA [205]. Notably, the mPTP activity can be enhanced by Ca^{2+} and pharmacologically modulated by selective inhibitors of F_1 [196] and F_O domain [206–208]. Newly, purified and functionally active F_1F_O-ATPase monomers [209] and dimers [210] act as a voltage-gated ion channel endowed with mPTP-like properties.

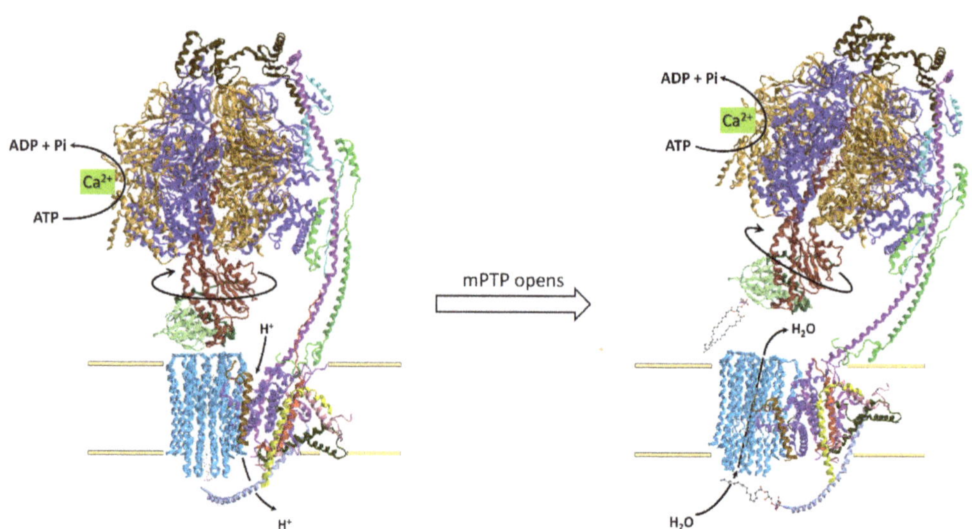

Figure 5. Model of mitochondrial permeability transition pore (mPTP) formation from the Ca^{2+}-activated F_1F_O-ATPase. On the left Ca^{2+} bound to the catalytic sites activates the enzyme by triggering the structural change which opens the mPTP. On the right, the pore forms in the core of the *c*-ring when the lipid plug is pulled out. mPTP opening dissipates the mitochondrial Δp and water entries in the matrix driven by oncotic pressure.

Dysregulated mPTP opening is involved in mitochondrial dysfunctions which feature a variety of diseases such as neurological and cardiovascular disorders, type 1-diabetes [211], cancer [212], inflammatory bone diseases [213] and diseases due to the exposure to contaminants [214]. In general, pathological conditions associated with oxidative stress also involve mPTP dysregulation [182]. Notably, the mPTP has been also involved in lifespan [215] and bone repair [216]. So, the discovery and design of mPTP rulers is at present a great challenge in pharmacology [217–219]. Recently, the use of natural products such as mPTP modulators raised a great interest in ethnomedicine [220].

5. SCs Association Protects from ROS Damage

The supramolecular organization of the respiratory chain has revealed a hitherto unknown link between mitochondrial structure and generation of the so-called ROS. In this section we will provide evidence that there is a biunivocal relation between ROS and SC association, i.e., SC association protects mitochondrial structure from ROS damage, but at the same time the SC association limits ROS generation from the respiratory chain.

ROS, in the past considered exclusively as major factors for cellular toxicity, are recognized today to be important physiological factors involved in cell signaling, by affecting the redox state of signaling proteins [221].

ROS is a collective term including oxygen derivatives, either radical or non-radical, that are oxidizing agents and/or are easily converted into radicals. Transition metals such as iron and copper, when in a free state, have a strong capacity to reduce O_2. Stepwise addition of an electron generates in sequence the superoxide radical anion O_2^-, the peroxide ion (which is a weaker acid and is protonated to hydrogen peroxide H_2O_2), and the hydroxyl radical OH. The hydroxyl radical is extremely reactive with a half-life of less than 1 ns; thus, it reacts close to its site of formation.

5.1. Mitochondrial Sources of ROS

In the cell environment, ROS can stem from exogenous and endogenous sources [222]. Exogenous sources of ROS include UV and visible light, ionizing radiation, drugs and environmental toxins. On the other hand, endogenous sources of ROS embrace enzyme activities such as: xanthine oxidase, cytochrome P-450 enzymes in the endoplasmic reticulum, peroxisomal flavin oxidases and plasma membrane NADPH oxidases. However, the membrane-ETS in the mtIM is probably the major source of ROS. Other mitochondrial enzyme systems can significantly contribute to ROS generation [223], as dihydrolipoamide dehydrogenase (a subunit of the α-ketoglutarate and pyruvate dehydrogenase complexes), monoamine oxidase, mitochondrial nitric oxide synthase and the adaptor protein p66^{Shc}. These systems are not relevant to this review and will not be considered here.

ROS from the Respiratory Chain

In mammalian mitochondria, and particularly in the respiratory chain and related enzymes, at least ten different sites of superoxide/H_2O_2 production have been identified [224]. The FMN and CoQ-binding sites of CI and the Q_O site (at the outer or positive side) of CIII are considered as the most important sites for superoxide production. However, other sites have also been defined (Figure 6).

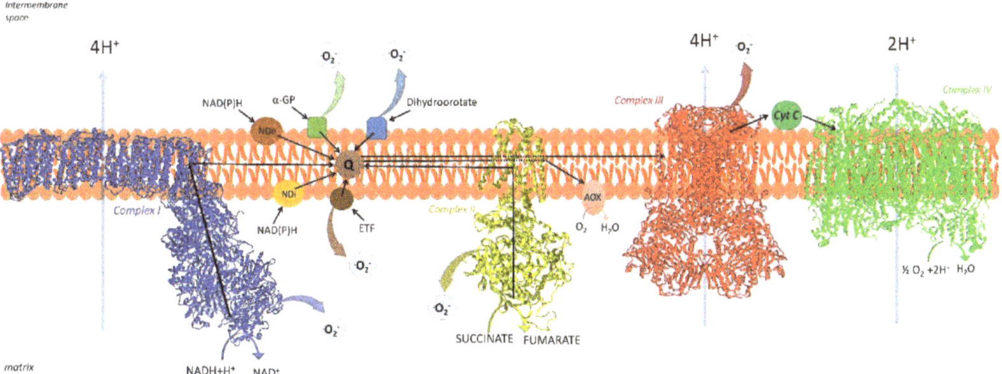

Figure 6. Major sites of superoxide production from the respiratory chain. The arrows represent the sources of superoxide at different sites in relation to the inner mitochondrial membrane. Blue, CI, NADH-ubiquinone oxidoreductase; yellow, CII, succinate-ubiquinone oxidoreductase; red, CIII, ubiquinol-cytochrome *c* oxidoreductase; green, CIV, cytochrome oxidase; NDi and NDe, internal and external alternative NAD(P)H dehydrogenases; AOX, alternative oxidase; αGP, glycerol-3-phosphate; ETF, electron transfer flavoprotein; Q, Coenzyme Q (ubiquinone); Cyt C, cytochrome *c*; (see text for details).

Most of superoxide is generated at the matrix side of the mtIM, as superoxide is detected in SMP, which are inside-out vesicles, namely have an opposite orientation with respect to mitochondria. However, a significant amount of superoxide is formed at the outer face of the mtIM [225]. It is likely that CI releases ROS in the matrix, while CIII releases ROS mostly in the IMS. The superoxide released at the IMS may be directly exported to the cytoplasm through an anion channel related to the Voltage-Dependent Anion Channel, VDAC.

CI is a major source of superoxide production in different mitochondrial types. Several prosthetic groups in CI were proposed to directly reduce oxygen, including FMN [226–228], ubisemiquinone [229,230], and iron sulfur (FeS) cluster N2 [231–233].

Grivennikova and Vinogradov [234] observed both superoxide and hydrogen peroxide generation by CI and attributed superoxide formation to FeS cluster N2, an hydrogen peroxide formation from 2-electron oxidation of fully reduced FMN. Rotenone, a specific inhibitor of CI, enhances ROS formation during forward electron transfer in the respiratory chain [231].

The highest rate of ROS generation in isolated mitochondria occurs during oxidation of succinate [235]. Since this ROS production is inhibited by rotenone [229,236,237], it is commonly attributed to CI during the energy-dependent reverse electron transfer from succinate to NAD^+ through CoQ [238–240]. See also Section 3 for the role of the reduced CoQ pool in reverse electron transfer and ROS-dependent CI destruction (cf. Guáras et al. [120]).

The formation of superoxide in CIII depends on its peculiar mechanism of electron transfer, the so-called Q-cycle [241]. If the lifetime of the semiquinone (Q) at the outer Q_O site, is prolonged, as in the controlled state when the electric potential is high, it reacts with oxygen, thus forming superoxide [242,243]. This conclusion is also reached by the superoxide increase induced by the specific CIII inhibitor, antimycin [244].

More recent studies, however, suggest that the oxygen reductant is the semiquinone formed in the so-called semi-reverse reaction in which cytochrome b_L reduces the fully oxidized quinone [245,246]. Osyczka [247,248] proposes that a metastable radical state, non-reactive with oxygen, safely holds electrons at a local energetic minimum during the oxidation of ubiquinol. This intermediate state is formed by interaction of a radical with a metal cofactor of a catalytic site, presumably the FeS cluster, under physiological conditions.

The other sites in the respiratory chain which may be a source of ROS (e.g., CII, glycerol phosphate dehydrogenase, dihydroorotate dehydrogenase, ETF and ETF dehydrogenase) are reviewed in [249].

5.2. Control of ROS Production

The steady-state concentration of ROS in a cell depends on many factors, including the rate and site of production, their lifetimes and diffusion constants, the interconversions between different ROS, and the removal rate by different antioxidant systems.

The ROS generation by isolated mitochondria accounts for 0.1–0.2% of oxygen consumed, but the rate and extent vary among different tissues and substrates [250,251].

The mitochondrial ROS level is physiologically regulated, as expected for their role as cellular signaling molecules (cf. Section 5.3).

Mitochondrial ROS production increases in State 4 (controlled non-phosphorylating state) when the membrane potential is high and the electron transfer rate decreases [252]. In fact, under this condition the respiratory carriers capable of donating electrons to oxygen are more reduced. For this reason, uncoupling may limit the production of ROS by releasing excessive membrane proton potential. In rat hepatocytes, a futile cycle of H^+ pumping and proton leak may account for 20–25% of respiration [253] and even more in perfused rat muscle. Uncoupling may be achieved by activating proton leak through uncoupling proteins (UCP) [244] (for a different role postulated for uncoupling proteins, cf. Wojtczak et al. [254]). In this way, although a tissue may dissipate a conspicuous part of the energy conserved by its mitochondria, the mitochondrial respiratory chain is maintained under more oxidized conditions, thus preventing the overproduction of damaging species.

UCP2 structure supports a fatty acid cycling mechanism for uncoupling [255]: UCP2 expels fatty acids in their anionic form from the negative matrix, while the protonated acids flow back from the positive exterior by passive diffusion through the bilayer. The resulting effect is H^+ back diffusion to the matrix, so that Δp dissipation decreases superoxide formation. UCP-mediated antioxidant protection and its impairment are expected to play a major role in cell physiology and pathology [256]. Mitochondrial uncoupling could be exploited to treat human diseases, such as obesity, cardiovascular diseases, or neurological disorders [257] (cf. Section 6).

Besides high membrane potential, other reasons may induce a decrease in the rate of electron transfer in CI leading to overproduction of ROS: for example, subunit phosphorylation that inhibits CI activity, enhances its ROS generating capacity [258–260]. Hence, endocrine alterations may affect ROS formation by changing the phosphorylation state of the complex.

Kadenbach et al. [261] proposed a ROS-generating mechanism that depends on regulation of cytochrome *c* oxidase. The cAMP-dependent phosphorylation of subunit I of the oxidase, by increasing its sensitivity to allosteric ATP inhibition, decreases H^+ translocation by the enzyme, with resulting decrease in mitochondrial membrane potential and, consequently, in ROS generation by the respiratory chain. On the contrary, dephosphorylation of CIV under stress conditions would induce increase in membrane potential and a burst of ROS generation by mitochondria. The discovery of a mitochondrial cAMP signalosome [262], modulating the allosteric inhibition of CIV by ATP, underlies a universal mechanism for metabolic regulation in eukaryotes.

In addition, cyt. *c* is regulated by different mechanisms, including post-translational modifications [263,264]. Specific cyt *c* residues may be phosphorylated in a tissue-specific mode. These modifications adjust electron flux and membrane potential to minimize ROS production under normal conditions. Conversely, under pathologic and acute stress conditions, such as ischemia-reperfusion, the dephosphorylation leads to membrane hyperpolarization, excessive ROS generation, and cyt. *c* release.

Other protein modifications in the respiratory complexes can modulate ROS production [265].

5.3. ROS as Signals

It is now established that ROS can act as second messengers [265–274] by modulating the expression of several genes involved in signal transduction. The effects related to ROS signals are different and even opposite, depending on the concentration. ROS may be oncogenic and promote proliferation, invasiveness, angiogenesis and metastasis [275,276], and accordingly ROS generation increases in various cancers. Nevertheless, ROS can also promote anti-tumorigenic signaling and trigger oxidative stress-induced tumor cell death.

Many of the ROS-mediated cellular signals paradoxically protect the cell against oxidative stress, but above a given threshold ROS become harmful and induce oxidative stress [268]. Not all ROS are equally suitable for signal transduction: the OH· radical is too aggressive to participate in catalyzed reactions, whereas O_2^- and H_2O_2 are known as important signaling molecules. Moreover, other reactive species are involved in redox signaling, such as nitric oxide, hydrogen sulfide and oxidized lipids [274].

Because some ROS such as H_2O_2 are oxidants, they modify the redox state of signaling proteins by oxidizing specific sulfhydryl groups [274,277,278] in enzyme-catalyzed reactions.

Several transcription factors contain redox-sensitive cysteine residues at their DNA-binding sites [279,280]. The ensuing formation of disulfide bonds acts as a redox switch to transduce the response of the protein to redox signaling [271,281].

Following the oxidation of signaling proteins by ROS, other post-translational changes (e.g., phosphorylation, acetylation, ubiquitination, and SUMOylation [270]) may occur in the same protein and in other proteins of the signaling cascade.

Nrf2 is an important example of mitochondrial ROS signaling which leads to nuclear gene expression changes [282]; in the presence of ROS, NRf2 is transferred from the cytoplasm to the nucleus, where it binds the DNA antioxidant response element of genes involved in the antioxidant response (e.g., heme oxygenase, NRF-1 and other inducers of mitochondrial biogenesis) [283].

Signaling proteins modified by ROS include phosphoprotein phosphatases (PTPs), Ras, large G-proteins, serine/threonine kinases of the MAPK families, transcription factors as AP-1, NFκB, p53 and others. A different effect is exerted on these proteins: PTPs are inhibited, while nuclear transcription factors are activated [284]. Different combinations of activation or deactivation of these proteins may drive the cell to either survival or death.

Mitochondrial ROS production is involved in preconditioning, by which low doses of a noxious stimulus promote a protective response against future injury [285–287]. Some protective pathways converge on the inhibition of mPTP opening during reperfusion [287].

Mitochondrial ROS are also important signals which rule the inflammation by activating the inflammasome [288,289]; treatment with antioxidants could scavenge excessive ROS and attenuate inflammatory responses by suppressing inflammasome activation [290]. ROS also regulate autophagic processes including mitophagy [291].

5.3.1. Regulation of Mitochondrial ROS in Cell Signaling

Mitochondrial ROS levels depend on the dynamic equilibrium between their rates of production and removal.

Studies on ROS generation under different substrate, inhibitor conditions and different oxygen tensions in rat liver mitochondria lead to the conclusion that only CI may be a significant source of ROS at physiological O_2 concentration [292]. The factors directly associated with respiratory activity (e.g., the redox potential of the $NAD^+/NADH$ couple and the proton-motive force) regulate the reduction in oxygen [268]) but these factors are in turn dependent on other parameters (cf. also Section 5.2).

Moreover, mitochondria enhance ROS generation in response to external stimuli, such as TNFα [293], hypoxia ([294], cf. Section 5.3.2), serum deprivation [295] and oxidative stress itself (ROS-induced ROS release, Zorov et al. [296]). Several proteins, such as p53, p66^{Shc}, the Bcl-2 family and Romo-1 [267] control mitochondrial ROS generation.

An interplay of mitochondria with non-mitochondrial ROS generating systems also occurs, as with the NADPH oxidases (NOX) family [297–299].

On the other hand, control of ROS signaling may be operated by their removal by a series of enzymatic systems [300], including various isoforms of superoxide dismutase, catalyzing the dismutation of superoxide, whereas catalase, glutathione peroxidase and peroxiredoxins remove H_2O_2. In addition, an integrated set of thiol systems in mitochondria prevents oxidative damage [271] and may transduce redox signals.

Peroxiredoxins (Prxs) are small dithiol proteins constituting a major family of peroxidases (six isoforms in mammals). Due to their high sensitivity to oxidation by H_2O_2, Prxs are sensors and transducers of H_2O_2 signaling [301] by transferring their oxidation state to effector proteins. The mitochondrial matrix contains large amounts of Thioredoxin-2 [302,303] and Prx3 [304].

The appropriate mitochondrial redox environment is greatly ascribed to glutathione [305–307] which is abundant and versatile to counteract hydrogen peroxide and lipid hydroperoxides, being a cofactor of glutathione peroxidase or glutathione-*S*-transferase. The mitochondrial GSH:GSSG ratio generally exceeds 100 and is widely used as an indicator of the redox status [308].

Protein glutathionylation is a regulatory mechanism that occurs through thiol disulfide exchange by protein disulfide isomerase [309,310]; the primary sistem catalyzing de-glutathionylation of protein-mixed disulfides with glutathione is glutaredoxin [311]. Mitochondrial proteins are highly susceptible to reversible *S*-glutathionylation, a post-translational modification which is favored by the unique physico-chemical properties of the mitochondrion [312], which contains a variety of *S*-glutathionylation targets playing important physiological roles: if these reactions are defective, pathological consequences are expected to occur.

5.3.2. Hypoxia and ROS Production

Among the many signaling functions, ROS appear as hypoxic signaling molecules. It seems paradoxical that a decrease in the level of the required substrate, O_2, would result in an increase in ROS production. However, there is large evidence of its occurring [313–315].

Indeed, ROS appear to activate the hypoxia-inducible factor HIF-1α. Rho0 cells, which lack mtDNA and thus electron transport, do not show the HIF-1 DNA-binding activity following hypoxia [316]. Therefore, mitochondrial respiration is required for propagation of the hypoxic signal. Exogenous or endogenous H_2O_2 stabilize HIF-1α, thus inducing its activation during normoxia [316], suggesting that HIF stabilization and activation may be induced by ROS.

Many studies demonstrate activation of HIF-1α by ROS [317,318], that are supposed to act by inhibiting the prolyl-4-hydroxylase that addresses the factor to proteolytic digestion. Once stabilized, HIF-1α binds to hypoxia-responsive elements in the DNA and stimulates the expression of a large array of genes [319].

The oxidized cyt *c* level in the IMS seems to be essential to control mitochondrially-derived ROS [320,321]; accordingly, cyt. *c* in the IMS oxidizes superoxide produced by CIII to O_2, thus preventing H_2O_2 production [321,322].

Inhibition of cytochrome *c* oxidase by hypoxia [323] enhances ROS formation by the respiratory chain; most likely because of the increased membrane potential [324] or by inducing a more reduced state of cyt. *c*.

The ROS produced by mitochondrial CIII seem to be critical for hypoxia signaling [294,325–328]. Hernansanz-Augustin et al. [329], however, showed that CI is involved in the superoxide burst under acute hypoxia in endothelial cells.

In addition, it should be pointed out that the plasma membrane NADPH oxidase [313,330] also seems to be involved in the enhanced ROS production during hypoxia.

5.4. ROS and Mitochondrial Quality Control

A major device for physiological control of the cells is the removal of damaged molecules; in particular, oxidized proteins are ubiquitinated and directed to the proteasome, a member of the ATP-dependent AAA+ proteases, where they are completely digested [331,332].

Mitochondrial proteins may particularly suffer from oxidative damage since they are close to the main sites of their generation, so that they must be removed in order to maintain mitochondrial integrity [333,334]. Internal mitochondrial proteins are translocated to the outer mitochondrial membrane, where they are ubiquitinated and then presented to the proteasome for degradation.

Mitochondrial dysfunction due to loss of cellular protein homeostasis (proteostasis) is a hallmark of aging and aging-related degeneration disorders [335], such as Alzheimer's disease and Parkinson's disease. Thus, mitochondrial proteins should be finely tuned. Although mitochondria form a proteasome-exclusive compartment [336], it is the cytosolic ubiquitin–proteasome system that plays a major role in the quality control of mitochondrial proteins.

The ATP-stimulated Lon protease in the mitochondrial matrix is devoted to the selective degradation of oxidized proteins, as for example the selective degradation of the Cox 4-1 subunit during hypoxia [337,338]. Defects in protein degradation have been involved in the age-related accumulation of oxidized proteins [339,340].

Also, entire cellular organelles undergo turnover and are finally directed to autophagy by digestion in the lysosome compartment [341–343].

Non-selective autophagy as well as selective mitophagy are triggered by ROS in response to several stressing signals [291,344,345]. Damaged mitochondria promote ROS generation, and excessive ROS can trigger mitophagy so as to remove impaired mitochondria and reduce ROS levels. Therefore, mitophagy helps to maintain cellular homeostasis under oxidative stress. Low ROS levels can trigger mitophagy in a mitochondrial fission-dependent way, if these levels are insufficient to trigger non-selective autophagy [346]. Therefore, a very specific and selective signaling cascade initiated by ROS has been suggested.

5.5. Supercomplexes Protect Complex I from ROS Damage and Limit ROS Generation

Our group first suggested that SC arrangement represents the missing link between oxidative stress and energy deficiency [96]. We speculated that oxidative stress dissociates the SCs, with loss of electron channeling and causing electron transfer to depend only upon the diffusion-coupled collisional encounters of the free ubiquinone molecules with the partner complexes.

5.5.1. Supercomplex Association Protects from ROS Damage

As predicted by Lenaz and Genova [96], SC dissociation also induces disassembly of CI and CIII and consequent loss of their catalytic activity. Therefore, the alteration of electron transfer may stimulate ROS generation. Here we briefly summarize the experimental evidence supporting this hypothesis.

Analysis of the SCs in patients with an isolated deficiency of single complexes [347] and in cultured cell models harboring cytochrome *b* mutations [348–350] showed that SCs enhance the stability of CI. Genetic mutations causing loss of CIII prevent SC formation and determine a secondary loss of CI, and consequently primary CIII assembly deficiencies appear as joint CIII/I defects. On the other hand, however, the absence of CI does not affect CIII stability.

It was recently suggested that the absence of CIII blocks CI biogenesis by preventing the incorporation of the NADH module rather than decreasing its stability [351].

Most evidence, however, is in line with the idea that misassembly of CIII affects CI stability because of their physical interaction within the SC. Misassembled CIII would prevent SC formation, and a lack of SC would induce an enhanced ROS generation from CI (see Section 5.5.2), with consequent damage to CI itself [350,352] which is vulnerable to oxidative stress both directly and through the lipid peroxidation, particularly of CL [353]. According to Guaras et al. [120], saturation of CoQ oxidation capacity keeps a highly reduced state of CoQ itself and induces reverse electron transport from reduced CoQ to CI; the resulting local superoxide generation oxidizes specific CI proteins, triggering their degradation and the disintegration of the complex.

An ischemia/reperfusion injury enhances ROS generation in mitochondria which promotes the opening of the nonselective mPTP [354]. The mPTP opening further compromises cellular bioenergetics and increases mtROS, resulting in SC disintegration and ultimately leading to cell death.

A reconstitution study [71] showed that lipid peroxidation before formation of CI-CIII proteoliposomes abolishes the SC formation, as determined by flux control analysis. It is likely that the distortion of the lipid bilayer induced by peroxidation provokes the dissociation of the SC.

Mitochondrial ROS also affect CI and CIV activity through CL peroxidation in beef heart SMP [353,355]. CL liposomes exogenously added to mitochondria from aged rats almost completely restored the activity of these enzyme complexes to the values of young control animals [356,357]. Other phospholipid classes, as well as peroxidized CL, could not mimic the CL effect.

5.5.2. Supercomplex Association Limits ROS Generation

Indirect considerations are compatible with the possibility that loss of SC organization may enhance ROS generation by the respiratory chain [12,96,358]. There are two possible reasons to support this hypothesis: a tight assembly of the respiratory complexes within SCs may screen auto-oxidizable groups hindering their reaction with oxygen (cf. [55]), and/or enhanced electron flow from NADH in the chain because of channeling would keep the prosthetic groups in a more oxidized form, thus preventing their interaction with oxygen [250]. On the other hand, on succinate oxidation, the reverse electron flow through CI keeps its centers more reduced favoring production of superoxide.

The molecular structure of the individual complexes and of the SCs does not provide structural evidence of such screening, since the prosthetic groups of CI in the matrix arm do not seem to be in close contact with CIII [359,360]. However, the SC $I_1III_2IV_1$ from bovine heart has a different CI conformation with respect to the free complex, that appears to be bent in such way as to be closer to CIII [359]. This observation agrees with the notion that CI may undergo important conformational changes [361].

A different kind of indirect evidence on the protection exerted by SC organization against ROS generation comes from the observation that high mitochondrial membrane potential supports ROS generation, while uncoupling decreases ROS production [243,249]. Although these effects may have different explanations, they are compatible with the suggestion [87] that high membrane potential induces SC dissociation.

A direct demonstration that loss of SC organization enhances ROS production by CI was obtained in our laboratory [362] using a model system of reconstituted CI/CIII proteoliposomes at high lipid-to-protein ratio (30:1), where formation of the SC I_1III_2 is prevented. In this system, the generation of superoxide was much higher than in a similar system reconstituted at a 1:1 ratio, which is rich in SCs. In addition, we also dissociated the reconstituted proteoliposomes by mild detergent treatment using dodecyl maltoside and observed that the dissociation of CI was accompanied by a three/four-fold increase in ROS generation. An increase in ROS also occurs in mitochondrial membranes after detergent treatment.

A significant finding supporting our in vitro conclusions comes from a study of mitochondrial CI supramolecular structure in neurons and astrocytes. Lopez-Fabuel et al. [363] observed that CI is largely assembled into SCs in neuronal mitochondria, whereas astrocytes have higher content of free CI. The presence of free CI in astrocytes correlates with the severalfold higher ROS production by astrocytes compared with neurons. Thus, regulation of ROS generation by CI assembly into SCs may contribute to the bioenergetic differences between neurons and astrocytes. Nevertheless, ROS levels were found to be controlled by an efficient antioxidant system in astrocytes, to regulate redox signaling [364]. By exerting this control in ROS levels, metabolic functions are finely tuned in both neural cells.

A study of CI activity in the basidiomycete *Ustilago maydis* [365] suggested that the contacts between CI, $CIII_2$ and CIV in the respirasome increase the catalytic efficiency of CI and regulate its activity to prevent ROS production.

The results of our proteoliposome study [362] are supported by many observations in cellular and animal models showing that SC dissociation is concomitant with enhanced ROS production.

Diaz et al. [352] showed enhanced ROS generation in mouse lung fibroblasts lacking the Rieske FeS protein of CIII and hence devoid of the SCs containing CI.

Mouse fibroblasts expressing the activated form of the *k*-ras oncogene and having low levels of high molecular weight SCs also produce higher ROS in comparison with wild type fibroblasts [366]. Hyperglycemia prevents SC formation, while increasing ROS levels in liver mitochondria of streptozotocin-diabetic rats. On the other hand, physical exercise was found to induce the chronic assembly of CIs into SCs in skeletal muscle [367], thus protecting mitochondria against oxidative damage.

Other studies in yeast mutants lacking the SC assembly factor Rcf1 and thus devoid of SCs CIII-CIV [117,368,369], showed enhanced ROS generation. Since the yeast *S. cerevisiae* lacks CI, in this case we may consider the origin of the extra ROS being presumably CIII. Thus, a major function of SC assembly appears to be this limitation of ROS production [370]: in the above case the decreased ROS generation by the Rcf1-stabilized SC III_2-IV_2 may be due to the more efficient electron transfer between CIII and CIV by cyt. *c*.

Several compounds including phospholipids, proteins, and certain chemicals are known to promote or stabilize mitochondrial SCs directly or indirectly [371], thus hindering ROS generation. Overexpression of the CL conversion enzyme ALCAT1 reduced SC formation and promoted ROS production, while preventing upregulation of coupled respiration [75]. These data suggest that the amount of ALCAT1 is critical for coupling mitochondrial respiration and metabolic plasticity.

The CL defect in Barth syndrome, a cardio-skeletal myopathy with neutropenia characterized by respiratory chain dysfunction, results in destabilization of the SCs, with higher levels of superoxide production in lymphoblasts from patients, compared to control cells [67,68]. Analogous results were obtained un studies on CL-lacking yeast mutants [372,373]. Moreover, Chen et al. [368] observed that yeast mutants, which cannot synthesize CL, exhibit increased protein carbonylation, an indicator of ROS.

Stein et al. [374] described a highly conserved 56-amino-acid microprotein named mitoregulin (Mtln). Mtln localizes to the mtIM where it binds CL and influences protein complex assembly. In cultured cells, Mtln overexpression increases mitochondrial SCs and mitochondrial activity while decreasing mitochondrial ROS.

The subunit composition of CI is crucial for SC formation and control of ROS production. Hou et al. [375] demonstrated that CI subunit AB1 (NDUFAB1), also known as mitochondrial acyl carrier protein, coordinates the assembly of respiratory CI, CII, and CIII, and SCs and is a crucial regulator of mitochondrial energy and ROS metabolism.

Jang and Javadov [376] observed that the pharmacological inhibition of CI and CII stimulated disruption of the respirasome accompanied by reduced ATP formation and increased ROS production. Overall, these studies provide biochemical evidence that the CI activity, and the NDUFA11 subunit are important for assembly and stability of the respirasome. The SDHC subunit of CII is not involved in the respirasome however the complex may play a regulatory role in respirasome formation.

Ramirez-Camacho et al. [377] found that mitochondria from reperfused hearts treated with *N*-acetyl-cysteine reduced oxidative stress and maintained SCs assemblies containing CI, CIII, CIV and the adapter protein SCAF1. The associations of the mitochondrial respiratory chain components into SCs could have pathophysiological relevance in metabolic diseases, as supramolecular arrangements, by sustaining a high electron transport rate, might prevent ROS generation [378].

6. Physiological and Pathological Implications

As we have pointed out in the Introduction, mitochondria are deeply involved in pathological processes. Mitochondrial dysfunction is often present in pathological states: at the end of this review, we intend to demonstrate that mitochondrial dysfunction is causative in the pathogenesis of most systemic diseases.

What are the signs showing mitochondrial dysfunction? There are many of them that are more or less specific. Among these we have changes of electron transfer, such as uncoupling, i.e., the occurrence of electron transfer not accompanied by ATP synthesis, changes of mitochondrial shape and number, alterations in quality control, changes in supramolecular structure (SCs) accompanied by decreased respiration and enhanced ROS generation, and finally opening of the mPTP and cell death.

We first briefly examine OXPHOS uncoupling. As for ROS generation cf. Section 5, for quality control see Section 5.4, and for the nature and role of the mPTP cf. Section 4.4. A possible role of SC disassembly as the initial pathogenetic event in many complex diseases will be described in Section 6.3.

6.1. Uncoupling: When Proton Movement and ATP Synthesis Are Disjointed

The term uncoupling in bioenergetics indicates that the H^+ pumping activity of respiratory complexes which forms the Δp is not exploited to synthesize ATP from ADP and Pi by F_1F_O-ATPase. Uncoupling is often associated with another term, mitochondrial dysfunction, which suggests that mitochondria do not work properly and are unable to fully perform their bioenergetic task. Uncoupling can have variable extent. We can distinguish between mitochondrial uncoupling, due to the downhill H^+ flux through pathways outside the F_1F_O-ATPase, and intrinsic F-ATPase uncoupling, which means that the enzyme complex itself cannot match ATP synthesis to H^+ channeling, due to molecular defects. While the mitochondrial uncoupling has a recognized physiological role, even if many mechanisms remain to be clarified [257], as far as we are aware the F_1F_O-ATPase uncoupling only represents a basic biochemical symptom of pathologies [379].

6.1.1. Mitochondrial Uncoupling Due to Dissipative Pathways

The mitochondrial uncoupling can be defined as the detachment of H^+ movement across the mtIM from the F_1F_O-ATPase activity [134]. Accordingly, the mitochondrial membrane potential generation is not used to build ATP. Even if it seems surprising, normally OXPHOS is not completely coupled, and the coupling extent depends on multiple factors. The mitochondrial uncoupling can be detected directly as a decrease in Δp or indirectly as a decrease in the phosphorylation efficiency, namely in the ADP/O ratio and or in the respiratory control ratio (RCR), which corresponds to the ratios of State 3 (ADP-stimulated) and State 4 (basal) respiratory activities. State 4 respiration, after subtraction of the non-mitochondrial oxygen consumption detected upon respiratory chain inhibition, mirrors the oxygen consumption of the so-called proton leak. The latter consists in the futile cycle of H^+ that from the IMS flow downhill in the mitochondrial matrix through pathways independent of the F_1F_O-ATPase [380]. The proton leak extent can be modified by protein complexes, exogenous compounds or permeability changes due to changes in the mtIM composition. Even if, at first, the mitochondrial uncoupling was only associated with mitochondrial dysfunction, at present, it emerges as key ruler of biological processes, as suggested by the occurrence of endogenous natural uncouplers, even if they can also have pathological meaning [257]. Accordingly, dissipative pathways normally occur in variable amounts in different cell types. In mammals, the most known physiological uncoupling occurs in the brown adipose tissue, where an integral membrane-bound protein (UCP-1) channels H^+ and dissipates the electrochemical gradient resulting in thermogenesis. Other uncoupling proteins (UCP-2 and UCP-3) rule insulin secretion in pancreatic β cells and fatty acid metabolism in muscle, brown adipose tissue and heart, respectively [381]. Additionally, the four-membered family of adenine nucleotide translocators (ANTs), which catalyzes the ATP/ADP exchange across the mitochondrial membrane, possesses uncoupling properties [382].

Lastly, mitochondrial uncoupling can be also produced by exogenous chemicals, protonophores, namely lipophilic weak acids which can cross the mtIM, or non-protonophores, able to activate latent proton leaks.

A severe mitochondrial uncoupling leads to ATP depletion and eventually cell death, while mild mitochondrial uncoupling can have positive effects. Mitochondrial uncoupling can play opposite roles: it can promote cell death but also helps to protect cells against cell death, according to the cell type, the uncoupler and the uncoupling extent [257]. Uncoupling should prevent an excessive $\Delta\Psi$ rise which would block electron transfer along the respiratory chain. The decrease in $\Delta\Psi$ and consequently in Δp also decreases ROS generation and vice versa, an increased ROS generation is known to decrease proton leak. On the other hand, the ATP synthase inhibition leads to electron accumulation in the respiratory complexes and ROS overproduction, which leads to oxidative stress and mitochondrial dysfunction. Endogenous mitochondrial uncoupling could prevent excessive ROS production. Accordingly, mitochondrial uncouplers have been tried to treat diseases featured by oxidative stress such as diabetes, obesity, cardiovascular diseases, neurodegenerative and aging-related diseases [134].

6.1.2. Intrinsic ATP Synthase Uncoupling Due by Amino Acid Changes in the *a* Subunit

The term F-ATPsynthase/ase uncoupling refers to any condition that inhibits the coupling between the catalytic activity carried out by the hydrophilic portion and H^+ translocation by the transmembrane portion F_O [383]. Frequently, it stems from structural changes that modify the H^+ pathway of the F_1F_O-ATPase across the mtIM. The tight relationship between structure and function of mitochondrial complexes, described in the previous sections, means that any change in the primary sequence, due even to a mutation which changes a single amino acid can dramatically modify the protein properties, especially when amino acid substitutions involve crucial enzyme domains such as those in the *a* subunit where some amino acids are essential to build the H^+ route. The mitochondrial F_1F_O-ATPase is a good example of how point mutations in some protein sectors result in severe diseases. Mutations in the nuclear genes that encode F_1F_O-ATPase subunits are rare and associated with severe diseases nearly incompatible with life. The most known mutations, associated with diseases whose severity is related to mitochondrial heteroplasmy [384], are localized in the mtDNA, which encodes *a* and A6L subunits of the F_O domain. In mammals, the mtDNA shows a higher mutational rate than nuclear DNA [385].

The most frequent mutations in the F_1F_O-ATPase associated with human pathologies occur in the mitochondrial *ATP6* gene, which encodes the *a* subunit. The structural arrangement of this subunit which, by positioning specific amino acids and exploiting the chemical properties of their side chains, forms the half-channels for H^+ flow within the mtIM, remained enigmatic for years, thus making it difficult to envisage the link between altered molecular function and pathology. However, recent studies, which highlighted the H^+ route, provided a satisfactory explanation on how point mutations are associated with mitochondrial dysfunctions and depicted a link between the bioenergetic defect and the syndrome. Up to now, several point mutations have been described [379]. The most severe mutation is the m.T8993>G transversion, namely the substitution of thymine by guanine, which results in the replacement of the hydrophobic leucine by the positively charged arginine, namely a missense mutation (*a*Leu156Arg) [386]. This molecular change is related to pathologies known as Neuropathy, Ataxia and Retinitis Pigmentosa (NARP) or Maternally Inherited Leigh Syndrome (MILS). These diseases are both associated with the same molecular defect, but exhibit various degrees of severity and are differently classified depending on the heteroplasmy degree [384]. The different chemical nature of the two amino acid side chains can satisfactorily explain the bioenergetic defect: since the inserted arginine is close to the crucial electrostatic barrier of *a*Arg-159, the two positive guanidine groups are close to each other to hamper both the H^+ flux across the mtIM and ATP synthesis [140]. The consequence is a severe bioenergetic failure. Accordingly, the F_1F_O-ATPase becomes unable to pump H^+ in the IMS and re-energize the mtIM, even if the two sectors F_1 and F_O are still structurally and functionally joined, as proven by the observation that the enzyme complex remains sensitive to the selective inhibitor oligomycin, a clear evidence of the coupling of the two domains [387]. Similarly, the m.T9176>G transversion in the mitochondrial ATP6 gene that changes a conserved leucine into arginine (*a*Leu220Arg) on position 220 of *a* subunit [379] is associated with NARP and MILS diseases. In addition, in this case, as the *a*Leu-220 is close to the essential *a*Arg-159, this transversion changes the situation and makes two Arg residues occur in close positions, thus destabilizing the *a* subunit due to steric hindrance and electrostatic repulsions. Accordingly, the two vicinal Arg would act as a positively charged barrier, which prevents H^+ translocation across the mtIM and decreases ATP synthesis and CIV respiration. Moreover, since ATP hydrolysis becomes uncoupled to H^+ transport, as proven by the oligomycin insensitivity, the membrane potential cannot be restored by the F_1F_O-ATPase which cannot pump H^+ [388]. Since these two transversions, which cause substitution of the hydrophobic side chain of Leu by a basic and positively charged chain of Arg, deeply alter the protein microenvironment and the H^+ pathway, they cause the bioenergetic failure which constitute the biochemical basis of these severe diseases.

The m.T8993>C transition, which yields *a*Leu156Pro substitution [389], results in a less severe disease, as a result of an increased ROS production. In this case, the functionality is somehow preserved, since the *c*-ring can still slowly rotate, allowing the coupling of H^+ flux to a low ATP synthesis [390]. It is most likely that the insertion of Pro which replaces Leu modifies the protein secondary structure. Accordingly, the Pro five-membered ring may cause a kink in the helices [391] which could slow down H^+ transfer.

The de novo transition (m.G8969>A) in mtDNA which encodes the *ATP6* gene [392] has been recently associated with a rare mitochondriopathy, defined Myopathy, Lactic Acidosis, and Sideroblastic Anemia (MLASA) [392]. The consequent missense mutation Ser148Asn in *a* subunit [393] is localized at one helix turn from the *a*Glu-145 which acts as "H^+ transfer group" in the half-channel which opens in the mitochondrial matrix [151]. The *a*Asn-145 bears a positive charge which makes ionic bond with *a*Glu-145, thus blocking H^+ translocation which requires the –COOH deprotonation of *c*Glu-59 [393].

To sum up, the mutations in *a* subunit, which stepwise addresses H^+ and allow H^+ movement, block or hamper the torque generation in F_O, which is essential for ATP synthesis by F_1.

As far as we are aware, mutations in A6L subunit leading to pathologies are much less frequent than those in *a* subunit.

6.2. Supercomplexes and ROS Signaling

The role of mitochondrial ROS in cell signaling has been the subject of excellent reviews (cf. [272,274,394–396]). Here we deal with a possible role of the supramolecular organization of the respiratory chain on ROS signaling.

With ROS being involved in cell signaling, it is expected that their generation is subjected to tight control.

The control of mitochondrial ROS levels depends upon the balance between their rate of generation and of removal, as already considered in Section 5. The steady-state concentrations of the redox species responsible for electron leaking and ROS production are governed by a series of nuclear-encoded protein factors [267] and by the forces directly associated with respiratory activity, which are the redox potential of the NAD^+/NADH couple and the Δp [268]. Hoffman and Brookes [292] have investigated the ROS generation by rat liver mitochondria under different substrate and inhibitor conditions and different oxygen tensions, in order to determine the O_2 affinity of the different O_2-reacting sites: from such data, the apparent K_m for O_2 was lowest for CI during forward flow, followed by CI backflow, CIII Q_O site, and highest for ETF dehydrogenase. They conclude that at physiological O_2 concentration, only CI may be a significant source of ROS.

We first speculated [96] and then obtained experimental demonstration [71] that dissociation of SC I_1III_2 occurs upon ROS addition. Therefore, the facilitated electron channeling in the CoQ region is lost. This condition makes electron-transfer necessarily dependent upon random diffusion of the free ubiquinone molecules and collisions with the partner complexes and may elicit further ROS generation.

In fact, SC disorganization eventually leads to destabilization of CI, decreases NAD-linked respiration and ATP synthesis and increases superoxide production by CI (cf. also Section 5 for experimental evidence).

The study of Guaras et al. [120] pinpoints another aspect of SCs in relation to ROS formation. Hyper-reduction of the CoQ pool by $ETFH_2$ oxidation during extensive fatty acid β-oxidation induces reverse electron transfer with a rise in ROS production by CI. Thus, shifting metabolic fuels from NADH-dependent to $FADH_2$-dependent substrates may adjust ROS generation by way of the specific supramolecular assembly of the respiratory complexes involved [119,397].

6.3. *Supercomplexes in Pathology and Aging*

In order to understand the mechanism by which a mitochondrial dysfunction can lead to failure of physiological functions, it is paradigmatic to discuss the mechanisms responsible for aging of cells and organisms. The mitochondrial theory of aging [398–400] is based on a series of assumptions linked by a causal relationship [401]: (a) mitochondrial ROS generation is the major cause of aging. (b) Mitochondrial ROS damage mitochondrial biomolecules, especially inducing mtDNA somatic mutations, mainly affecting post-mitotic tissues. (c) The mtDNA mutations alter the structure of mtDNA-encoded proteins: since these are subunits of the major OXPHOS complexes, they imply a decreased oxidative phosphorylation, which leads to bioenergetic failure, metabolic derangement and cell death.

A further consequence of the decreased electron transfer during aging is further ROS generation, so that mtDNA damage and ROS production become involved in a vicious circle [398,399], not considered by the original theory.

Even if each of these assertions is sustained by experimental evidence, there are still several controversies and the cause–effect relationships among these events are still partially obscure. The complexity of biochemical, genetic, and regulatory systems and their inter-relationships are still not fully understood and often a single linear cause to effect relationship cannot be established [402]. Barja [403] however proposes that aging is the result of several effectors, i.e., mtROS production, lipid unsaturation, autophagy, mitochondrial DNA repair and putative other events such as apoptosis, proteostasis, or telomere shortening, already considered by different classic theories of aging. The aging regulating system gathers the theories of aging, previously considered as independent and assembles them into a single unified theory.

Wallace [1] hypothesized that mitochondrial dysfunction plays a central role in a wide range of age-related disorders and cancer types. He proposes that the delayed-onset and gradual onset of the age-related diseases is due to the accumulation of somatic mutations in the mtDNAs of post-mitotic tissues. The tissue-specific pathological phenotype may depend on the different roles and energy requirements of the various tissues. The individual varied predisposition to degenerative and cancer diseases may result from the interaction of present environment (dietary caloric intake) and ancient genetic factors (mitochondrial genetic polymorphisms). On these bases, the mitochondria make a direct connection between our environment and our genes.

Inherited and/or epigenomic variation of the mitochondrial genome determines our initial energetic capacity, but the age-related accumulation of somatic mtDNA mutations decreases the energetic capacity leading to disease [4], thus providing a unified pathophysiological and genetic mechanism for neurodegenerative diseases such as Alzheimer and Parkinson disease, metabolic diseases such as diabetes and obesity, autoimmune diseases, aging, and cancer.

Such an integrated model for the genetics and pathophysiology of complex diseases, aging, and cancer [4] is summarized in Figure 7.

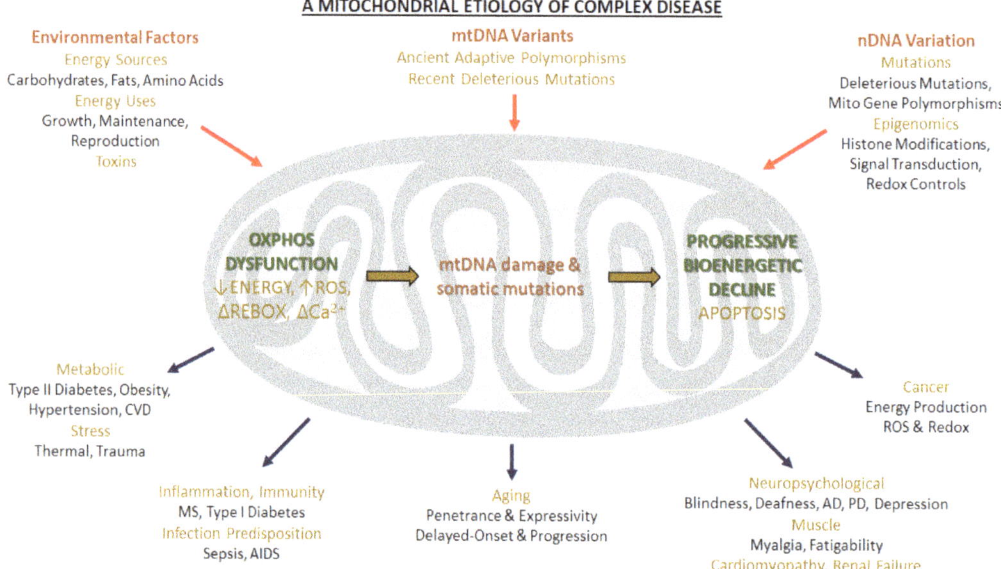

Figure 7. Integrated mitochondrial pattern which shows the genetic and phenotypic features of the "complex" human diseases. The model integrates the genetic and pathophysiological relationships of multi-factor diseases, aging, and cancer (from Wallace [4]). See text for explanations.

Nuclear DNA variations including epigenetic changes, mtDNA variations, including ancient polymorphisms, and environmental influences including diet and calories, all impact the mitochondrial OXPHOS system. The primary defect is the reduction in energy conservation by OXPHOS, that in turn perturbs the mitochondrial biogenesis and enhances ROS production. ROS induce the progressive increase in mtDNA somatic mutations, so that the alteration of proteins in the OXPHOS complexes cause further decline of the mitochondrial function. If the mitochondrial energy production decreases below a specific tissue threshold, the bioenergetic failure can trigger cascade events which lead to cell death by apoptosis or necrosis. The clinical phenotype would result from the reduced energy storage in the tissues that require the most energy, such as brain, heart, muscle, and kidney. The symptom number and severity in these organs mirror the extent and type of the mitochondrial defect. The altered functions in mitochondria are also crucial to determine cancer initiation, promotion, and metastasis [404].

We first proposed SC disassembly as the missing link between oxidative stress and energy failure [96,405]. We proposed that an initial oxidative stress, not necessarily originating from mitochondria, and due to different possible reasons, causes dissociation of SCs. The bioenergetic consequence is the loss of electron channeling leading to random diffusion of ubiquinone, which is less efficient. A further consequence of SC disassembly would be CI dissociation with loss of electron transfer and/or proton translocation; the consequently altered electron transfer may result in further ROS generation. The possible severe metabolic and physiological consequences of SC dissociation are reported in the scheme in Figure 8 [223].

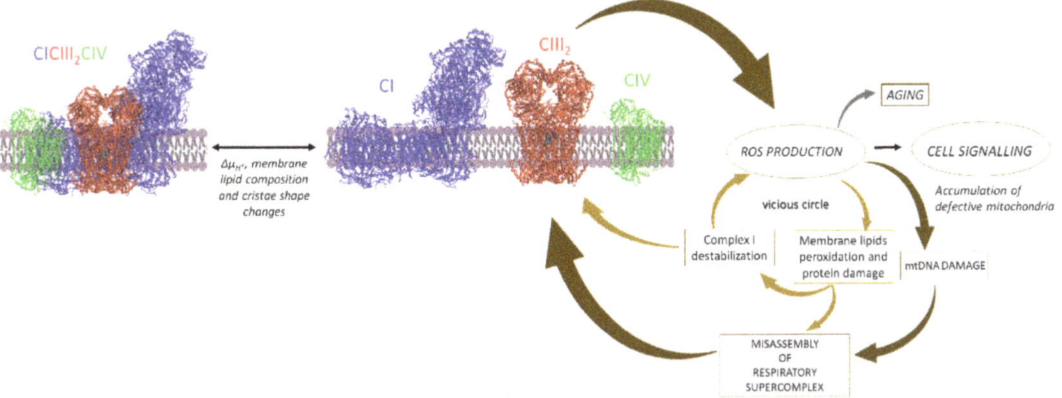

Figure 8. Scheme illustrating how the side chain (SC) organization loss may be involved in a vicious circle of oxidative stress and energy failure. Reactive oxygen species (ROS) production by CI increases due to SC disassembly. Other mitochondrial events, such as membrane phospholipid peroxidation, mtDNA damage and subsequent misassembly of the respiratory complexes with further loss of SC organization, may occur due to the increased oxidative stress, thus fueling and maintaining the vicious circle. In a dose-dependent way, ROS can also operate as molecular signals from mitochondria to the nucleus. See text for explanations.

This model poses SC dissociation and ROS generation in a biunivocal situation. On the one hand, ROS cause dissociation of SCs, but on the other hand, SC dissociation enhances ROS generation. Clearly, these events must be tightly controlled, otherwise they might trigger a vicious circle of ROS generation. We have discussed above the hypothetic vicious circle of ROS generation and mitochondrial failure at the basis of aging and aging-related diseases. The original mitochondrial theory of aging did not imply the existence of a vicious circle [398], nevertheless the evidence that ROS generation increases in aging is overwhelming [403].

Alterations (dissociation) of the SCs often accompany pathological changes, however no direct proof exists that SC dissociation is the direct cause of the pathology.

Even so, circumstantial evidence points to a vicious circle (cf. Genova et al. [405] for review).

The major mitochondrial alterations accompanying aging and several pathologies occur at the level of CI [406,407], in line with the knowledge that this enzyme exerts the main control on respiration. Flux control analysis of respiration in coupled liver mitochondria [408] showed that the control exerted by CI is low in young rats, and becomes very high in the old ones, suggesting that aging induces CI structural changes that eventually affect the whole OXPHOS.

Analysis of the occurrence of respiratory SCs (cf. Section 3.1) in relation to age suggested that the SC destabilization plays a key role in the development of the aging-phenotype [409,410]. In mitochondria of rat cortex, a striking age-associated decline (−40%) in the CI-containing SCs, especially the SC I_1III_2 (−58%) was reported by Frenzel et al. [411].

The so-called mutator mice are a remarkable model of premature aging [407]; they have a defect of the proof-reading function of mtDNA polymerase-γ and hence exhibit multiple mtDNA mutations that result in decreased respiration and altered assembly of respiratory complexes [412]. In these mice, the steady-state level and activity of CI is strongly lowered, possibly as a secondary effect of a decreased assembly of CIV on the stability of CI, which disrupts the SC organization [349].

The observations collected in this review locate SC assembly/disassembly in a physiological signaling network that can easily undergo alterations, leading to dramatic conditions if ROS generation becomes out of control.

We may envisage SC association/dissociation as a physiological phenomenon, as depicted by the plasticity model [64], modulated by a variety of stimuli such as the mitochondrial membrane potential and protein post-translational changes in the respiratory complexes; thus, changes in ROS production modulate the signaling pathways started by ROS. These changes are reversible and must be tightly controlled.

We propose that the primary event responsible for aging and age-related pathologies is the structural damage induced by ROS in mitochondria, as originally stated by the so-called mitochondrial theory of aging [398].

Distinct events may constitute a consecutive series which leads to pathological changes. Progressive ROS-induced damage to the mitochondrial membrane lipids and proteins was found; mtDNA mutations, although present, may not necessarily be an early phenomenon of aging development.

The ROS level may be ruled by the nutrition state and the activity of the mTOR and insulin/IGF pathways. ROS may alter mitochondrial protein structure either directly or by means of peroxidation of CL, whose damage prevents SC association [71]; in turn SC dissociation would lead to further increase in ROS generation. If maintained at low levels, ROS may induce retrograde signals which promote the induction of compensatory mechanisms that try to counteract ROS generation and the consequent damage risk (see Section 5). However, high ROS levels may cause further damage and make the signaling pathways lose their coordination (see Section 5.3). At a later stage, mtDNA mutations would make the whole process irreversible and determine the final aging phenotype.

Author Contributions: Writing—original draft preparation, S.N., A.P., G.L.; writing—review and editing, F.T., V.V., C.A., G.T.; supervision, A.P., G.L.; funding acquisition, S.N. All authors have read and agreed to the published version of the manuscript.

Funding: This research was funded by CARISBO Foundation, grant number 2019.0534 to S.N.

Institutional Review Board Statement: Not applicable.

Informed Consent Statement: Not applicable.

Conflicts of Interest: The authors declare no conflict of interest.

References

1. Wallace, D.C. A Mitochondrial Paradigm of Metabolic and Degenerative Diseases, Aging, and Cancer: A Dawn for Evolutionary Medicine. *Annu. Rev. Genet.* **2005**, *39*, 359–407. [CrossRef] [PubMed]
2. Wallace, D.C. Mitochondrial Genetic Medicine. *Nat. Genet.* **2018**, *50*, 1642–1649. [CrossRef] [PubMed]
3. Picard, M.; Wallace, D.C.; Burelle, Y. The Rise of Mitochondria in Medicine. *Mitochondrion* **2016**, *30*, 105–116. [CrossRef] [PubMed]
4. Wallace, D.C. Bioenergetic Origins of Complexity and Disease. *Cold Spring Harb. Symp. Quant. Biol.* **2011**, *76*, 1–16. [CrossRef]
5. DiMauro, S.; Hirano, M.; Schon, E.A. Approaches to the Treatment of Mitochondrial Diseases. *Muscle Nerve* **2006**, *34*, 265–283. [CrossRef] [PubMed]
6. Reddy, P.H. Mitochondrial Medicine for Aging and Neurodegenerative Diseases. *Neuromol. Med.* **2008**, *10*, 291–315. [CrossRef] [PubMed]
7. Zorov, D.B.; Isaev, N.K.; Plotnikov, E.Y.; Silachev, D.N.; Zorova, L.D.; Pevzner, I.B.; Morosanova, M.A.; Jankauskas, S.S.; Zorov, S.D.; Babenko, V.A. Perspectives of Mitochondrial Medicine. *Biochemistry* **2013**, *78*, 979–990. [CrossRef]
8. Luft, R. The Development of Mitochondrial Medicine. *Proc. Natl. Acad. Sci. USA* **1994**, *91*, 8731–8738. [CrossRef] [PubMed]
9. Barcelos, I.; Shadiack, E.; Ganetzky, R.D.; Falk, M.J. Mitochondrial Medicine Therapies: Rationale, Evidence, and Dosing Guidelines. *Curr. Opin. Pediatr.* **2020**, *32*, 707–718. [CrossRef] [PubMed]

10. Mitchell, P. Coupling of Phosphorylation to Electron and Hydrogen Transfer by a Chemi-Osmotic Type of Mechanism. *Nature* **1961**, *191*, 144–148. [CrossRef] [PubMed]
11. Nicholls, D.G.; Ferguson, S.J. 1-Chemiosmotic Energy Transduction. In *Bioenergetics*, 4th ed.; Academic Press: Boston, MA, USA, 2013; pp. 3–12, ISBN 978-0-12-388425-1.
12. Lenaz, G.; Genova, M.L. Structure and Organization of Mitochondrial Respiratory Complexes: A New Understanding of an Old Subject. *Antioxid. Redox Signal.* **2010**, *12*, 961–1008. [CrossRef] [PubMed]
13. Hatefi, Y.; Haavik, A.G.; Fowler, L.R.; Griffiths, D.E. Studies on the Electron Transfer System. XLII. Reconstitution of the Electron Transfer System. *J. Biol. Chem.* **1962**, *237*, 2661–2669. [CrossRef]
14. Green, D.E.; Tzagoloff, A. The Mitochondrial Electron Transfer Chain. *Arch. Biochem. Biophys.* **1966**, *116*, 293–304. [CrossRef]
15. Houstěk, J.; Cannon, B.; Lindberg, O. Gylcerol-3-Phosphate Shuttle and Its Function in Intermediary Metabolism of Hamster Brown-Adipose Tissue. *Eur. J. Biochem.* **1975**, *54*, 11–18. [CrossRef] [PubMed]
16. Beckmann, J.D.; Frerman, F.E. Reaction of Electron-Transfer Flavoprotein with Electron-Transfer Flavoprotein-Ubiquinone Oxidoreductase. *Biochemistry* **1985**, *24*, 3922–3925. [CrossRef]
17. Evans, D.R.; Guy, H.I. Mammalian Pyrimidine Biosynthesis: Fresh Insights into an Ancient Pathway. *J. Biol. Chem.* **2004**, *279*, 33035–33038. [CrossRef]
18. Salvi, F.; Gadda, G. Human Choline Dehydrogenase: Medical Promises and Biochemical Challenges. *Arch. Biochem. Biophys.* **2013**, *537*, 243–252. [CrossRef]
19. Poole, L.B. The Basics of Thiols and Cysteines in Redox Biology and Chemistry. *Free Radic. Biol. Med.* **2015**, *80*, 148–157. [CrossRef] [PubMed]
20. Sousa, J.S.; D'Imprima, E.; Vonck, J. Mitochondrial Respiratory Chain Complexes. *Subcell. Biochem.* **2018**, *87*, 167–227. [CrossRef]
21. Parey, K.; Wirth, C.; Vonck, J.; Zickermann, V. Respiratory Complex I—Structure, Mechanism and Evolution. *Curr. Opin. Struct. Biol.* **2020**, *63*, 1–9. [CrossRef]
22. Bezawork-Geleta, A.; Rohlena, J.; Dong, L.; Pacak, K.; Neuzil, J. Mitochondrial Complex II: At the Crossroads. *Trends Biochem. Sci.* **2017**, *42*, 312–325. [CrossRef]
23. Xia, D.; Esser, L.; Tang, W.-K.; Zhou, F.; Zhou, Y.; Yu, L.; Yu, C.-A. Structural Analysis of Cytochrome Bc1 Complexes: Implications to the Mechanism of Function. *Biochim. Biophys. Acta* **2013**, *1827*, 1278–1294. [CrossRef]
24. Zong, S.; Wu, M.; Gu, J.; Liu, T.; Guo, R.; Yang, M. Structure of the Intact 14-Subunit Human Cytochrome *c* Oxidase. *Cell Res.* **2018**, *28*, 1026–1034. [CrossRef] [PubMed]
25. Kröger, A.; Klingenberg, M. The Kinetics of the Redox Reactions of Ubiquinone Related to the Electron-Transport Activity in the Respiratory Chain. *Eur. J. Biochem.* **1973**, *34*, 358–368. [CrossRef]
26. Kröger, A.; Klingenberg, M. Further Evidence for the Pool Function of Ubiquinone as Derived from the Inhibition of the Electron Transport by Antimycin. *Eur. J. Biochem.* **1973**, *39*, 313–323. [CrossRef] [PubMed]
27. Hackenbrock, C.R.; Chazotte, B.; Gupte, S.S. The Random Collision Model and a Critical Assessment of Diffusion and Collision in Mitochondrial Electron Transport. *J. Bioenergy Biomembr.* **1986**, *18*, 331–368. [CrossRef] [PubMed]
28. Javadov, S.; Jang, S.; Chapa-Dubocq, X.R.; Khuchua, Z.; Camara, A.K. Mitochondrial Respiratory Supercomplexes in Mammalian Cells: Structural versus Functional Role. *J. Mol. Med.* **2020**. [CrossRef]
29. Chance, B.; Williams, G.R. The Respiratory Chain and Oxidative Phosphorylation. *Adv. Enzymol. Relat. Subj. Biochem.* **1956**, *17*, 65–134. [CrossRef]
30. Hatefi, Y.; Haavik, A.G.; Griffiths, D.E. Studies on the Electron Transfer System. XL. Preparation and Properties of Mitochondrial DPNH-Coenzyme Q Reductase. *J. Biol. Chem.* **1962**, *237*, 1676–1680. [CrossRef]
31. Singer, S.J.; Nicolson, G.L. The Fluid Mosaic Model of the Structure of Cell Membranes. *Science* **1972**, *175*, 720–731. [CrossRef]
32. Ozawa, T.; Papa, S. *Bioenergetics: Structure and Function of Energy Transducing Systems*; Springer: Berlin/Heidelberg, Germany, 1987, ISBN 978-3-540-18007-4.
33. Hochman, J.; Ferguson-Miller, S.; Schindler, M. Mobility in the Mitochondrial Electron Transport Chain. *Biochemistry* **1985**, *24*, 2509–2516. [CrossRef]
34. Schägger, H.; Pfeiffer, K. Supercomplexes in the Respiratory Chains of Yeast and Mammalian Mitochondria. *EMBO J.* **2000**, *19*, 1777–1783. [CrossRef]
35. Wittig, I.; Schägger, H. Advantages and Limitations of Clear-Native PAGE. *Proteomics* **2005**, *5*, 4338–4346. [CrossRef]
36. Lenaz, G.; Tioli, G.; Falasca, A.I.; Genova, M.L. Complex i Function in Mitochondrial Supercomplexes. *Biochim. Biophys. Acta Bioenergy* **2016**, *1857*, 991–1000. [CrossRef]
37. Acin-Perez, R.; Enriquez, J.A. The Function of the Respiratory Supercomplexes: The Plasticity Model. *Biochim. Biophys. Acta* **2014**, *1837*, 444–450. [CrossRef] [PubMed]
38. Genova, M.L.; Lenaz, G. A Critical Appraisal of the Role of Respiratory Supercomplexes in Mitochondria. *Biol. Chem.* **2013**, *394*, 631–639. [CrossRef] [PubMed]
39. Milenkovic, D.; Blaza, J.N.; Larsson, N.-G.; Hirst, J. The Enigma of the Respiratory Chain Supercomplex. *Cell Metab.* **2017**, *25*, 765–776. [CrossRef] [PubMed]
40. den Brave, F.; Becker, T. Supercomplex Formation Boosts Respiration. *EMBO Rep.* **2020**, e51830. [CrossRef]
41. Vonck, J.; Schäfer, E. Supramolecular Organization of Protein Complexes in the Mitochondrial Inner Membrane. *Biochim. Biophys. Acta* **2009**, *1793*, 117–124. [CrossRef] [PubMed]

42. Genova, M.L.; Lenaz, G. Functional Role of Mitochondrial Respiratory Supercomplexes. *Biochim. Biophys. Acta* **2014**, *1837*, 427–443. [CrossRef]
43. Schägger, H.; Pfeiffer, K. The Ratio of Oxidative Phosphorylation Complexes I-V in Bovine Heart Mitochondria and the Composition of Respiratory Chain Supercomplexes. *J. Biol. Chem.* **2001**, *276*, 37861–37867. [CrossRef] [PubMed]
44. Lenaz, G.; Genova, M.L. *Respiratory Cytochrome Supercomplexes: Cytochrome Complexes: Evolution, Structures, Energy Transduction, and Signaling*; Advances in Photosynthesis and Respiration; Springer: Dordrecht, The Netherlands, 2016; ISBN 978-94-017-7479-6.
45. Wang, Y.; Mohsen, A.-W.; Mihalik, S.J.; Goetzman, E.S.; Vockley, J. Evidence for Physical Association of Mitochondrial Fatty Acid Oxidation and Oxidative Phosphorylation Complexes. *J. Biol. Chem.* **2010**, *285*, 29834–29841. [CrossRef] [PubMed]
46. Nsiah-Sefaa, A.; McKenzie, M. Combined Defects in Oxidative Phosphorylation and Fatty Acid β-Oxidation in Mitochondrial Disease. *Biosci. Rep.* **2016**, *36*. [CrossRef] [PubMed]
47. Lenaz, G.; Tioli, G.; Falasca, A.I.; Genova, M.L. Coenzyme Q and Respiratory Supercomplexes: Physiological and Pathological Implications. *Rend. Fis. Acc. Lincei* **2018**, *29*, 383–395. [CrossRef]
48. Fang, J.; Uchiumi, T.; Yagi, M.; Matsumoto, S.; Amamoto, R.; Takazaki, S.; Yamaza, H.; Nonaka, K.; Kang, D. Dihydro-Orotate Dehydrogenase Is Physically Associated with the Respiratory Complex and Its Loss Leads to Mitochondrial Dysfunction. *Biosci. Rep.* **2013**, *33*, e00021. [CrossRef]
49. Hildebrandt, T.M. Modulation of Sulfide Oxidation and Toxicity in Rat Mitochondria by Dehydroascorbic Acid. *Biochim. Biophys. Acta* **2011**, *1807*, 1206–1213. [CrossRef] [PubMed]
50. Ovádi, J. Physiological Significance of Metabolic Channelling. *J. Theor. Biol.* **1991**, *152*, 1–22. [CrossRef]
51. Gupte, S.S.; Hackenbrock, C.R. The Role of Cytochrome c Diffusion in Mitochondrial Electron Transport. *J. Biol. Chem.* **1988**, *263*, 5248–5253. [CrossRef]
52. Althoff, T.; Mills, D.J.; Popot, J.-L.; Kühlbrandt, W. Arrangement of Electron Transport Chain Components in Bovine Mitochondrial Supercomplex I1III2IV1. *EMBO J.* **2011**, *30*, 4652–4664. [CrossRef]
53. Dudkina, N.V.; Kudryashev, M.; Stahlberg, H.; Boekema, E.J. Interaction of Complexes I, III, and IV within the Bovine Respirasome by Single Particle Cryoelectron Tomography. *Proc. Natl. Acad. Sci. USA* **2011**, *108*, 15196–15200. [CrossRef]
54. Letts, J.A.; Fiedorczuk, K.; Sazanov, L.A. The Architecture of Respiratory Supercomplexes. *Nature* **2016**, *537*, 644–648. [CrossRef]
55. Letts, J.A.; Sazanov, L.A. Clarifying the Supercomplex: The Higher-Order Organization of the Mitochondrial Electron Transport Chain. *Nat. Struct. Mol. Biol.* **2017**, *24*, 800–808. [CrossRef]
56. Wu, M.; Gu, J.; Guo, R.; Huang, Y.; Yang, M. Structure of Mammalian Respiratory Supercomplex I1III2IV1. *Cell* **2016**, *167*, 1598–1609.e10. [CrossRef]
57. Sousa, J.S.; Mills, D.J.; Vonck, J.; Kühlbrandt, W. Functional Asymmetry and Electron Flow in the Bovine Respirasome. *eLife* **2016**, *5*. [CrossRef]
58. Mileykovskaya, E.; Dowhan, W. Cardiolipin-Dependent Formation of Mitochondrial Respiratory Supercomplexes. *Chem. Phys. Lipids* **2014**, *179*, 42–48. [CrossRef]
59. Mitchell, P. The Protonmotive Q Cycle: A General Formulation. *FEBS Lett.* **1975**, *59*, 137–139. [CrossRef]
60. Sarewicz, M.; Osyczka, A. Electronic Connection between the Quinone and Cytochrome C Redox Pools and Its Role in Regulation of Mitochondrial Electron Transport and Redox Signaling. *Physiol. Rev.* **2015**, *95*, 219–243. [CrossRef] [PubMed]
61. Covian, R.; Zwicker, K.; Rotsaert, F.A.; Trumpower, B.L. Asymmetric and Redox-Specific Binding of Quinone and Quinol at Center N of the Dimeric Yeast Cytochrome Bc1 Complex. Consequences for Semiquinone Stabilization. *J. Biol. Chem.* **2007**, *282*, 24198–24208. [CrossRef]
62. Letts, J.A.; Fiedorczuk, K.; Degliesposti, G.; Skehel, M.; Sazanov, L.A. Structures of Respiratory Supercomplex I+III2 Reveal Functional and Conformational Crosstalk. *Mol. Cell* **2019**, *75*, 1131–1146.e6. [CrossRef]
63. Bianchi, C.; Genova, M.L.; Parenti Castelli, G.; Lenaz, G. The Mitochondrial Respiratory Chain Is Partially Organized in a Supercomplex Assembly: Kinetic Evidence Using Flux Control Analysis. *J. Biol. Chem.* **2004**, *279*, 36562–36569. [CrossRef] [PubMed]
64. Acín-Pérez, R.; Fernández-Silva, P.; Peleato, M.L.; Pérez-Martos, A.; Enriquez, J.A. Respiratory Active Mitochondrial Supercomplexes. *Mol. Cell* **2008**, *32*, 529–539. [CrossRef] [PubMed]
65. Enríquez, J.A. Supramolecular Organization of Respiratory Complexes. *Annu. Rev. Physiol.* **2016**, *78*, 533–561. [CrossRef] [PubMed]
66. Schlame, M. Protein Crowding in the Inner Mitochondrial Membrane. *Biochim. Biophys. Acta Bioenergy* **2021**, *1862*, 148305. [CrossRef] [PubMed]
67. McKenzie, M.; Lazarou, M.; Thorburn, D.R.; Ryan, M.T. Mitochondrial Respiratory Chain Supercomplexes Are Destabilized in Barth Syndrome Patients. *J. Mol. Biol.* **2006**, *361*, 462–469. [CrossRef] [PubMed]
68. Gonzalvez, F.; D'Aurelio, M.; Boutant, M.; Moustapha, A.; Puech, J.-P.; Landes, T.; Arnauné-Pelloquin, L.; Vial, G.; Taleux, N.; Slomianny, C.; et al. Barth Syndrome: Cellular Compensation of Mitochondrial Dysfunction and Apoptosis Inhibition Due to Changes in Cardiolipin Remodeling Linked to Tafazzin (TAZ) Gene Mutation. *Biochim. Biophys. Acta* **2013**, *1832*, 1194–1206. [CrossRef] [PubMed]
69. Desmurs, M.; Foti, M.; Raemy, E.; Vaz, F.M.; Martinou, J.-C.; Bairoch, A.; Lane, L. C11orf83, a Mitochondrial Cardiolipin-Binding Protein Involved in Bc1 Complex Assembly and Supercomplex Stabilization. *Mol. Cell. Biol.* **2015**, *35*, 1139–1156. [CrossRef] [PubMed]
70. Baker, C.D.; Basu Ball, W.; Pryce, E.N.; Gohil, V.M. Specific Requirements of Nonbilayer Phospholipids in Mitochondrial Respiratory Chain Function and Formation. *Mol. Biol. Cell* **2016**, *27*, 2161–2171. [CrossRef]

71. Genova, M.L.; Baracca, A.; Biondi, A.; Casalena, G.; Faccioli, M.; Falasca, A.I.; Formiggini, G.; Sgarbi, G.; Solaini, G.; Lenaz, G. Is Supercomplex Organization of the Respiratory Chain Required for Optimal Electron Transfer Activity? *Biochim. Biophys. Acta* **2008**, *1777*, 740–746. [CrossRef] [PubMed]
72. Paradies, G.; Paradies, V.; Ruggiero, F.M.; Petrosillo, G. Role of Cardiolipin in Mitochondrial Function and Dynamics in Health and Disease: Molecular and Pharmacological Aspects. *Cells* **2019**, *8*, 728. [CrossRef] [PubMed]
73. Paradies, G.; Petrosillo, G.; Paradies, V.; Ruggiero, F.M. Oxidative Stress, Mitochondrial Bioenergetics, and Cardiolipin in Aging. *Free Radic. Biol. Med.* **2010**, *48*, 1286–1295. [CrossRef]
74. Wasmus, C.; Dudek, J. Metabolic Alterations Caused by Defective Cardiolipin Remodeling in Inherited Cardiomyopathies. *Life* **2020**, *10*, 277. [CrossRef] [PubMed]
75. Rieger, B.; Krajčová, A.; Duwe, P.; Busch, K.B. ALCAT1 Overexpression Affects Supercomplex Formation and Increases ROS in Respiring Mitochondria. *Oxid. Med. Cell Longev.* **2019**, *2019*, 9186469. [CrossRef] [PubMed]
76. Kasahara, T.; Kubota-Sakashita, M.; Nagatsuka, Y.; Hirabayashi, Y.; Hanasaka, T.; Tohyama, K.; Kato, T. Cardiolipin Is Essential for Early Embryonic Viability and Mitochondrial Integrity of Neurons in Mammals. *FASEB J.* **2020**, *34*, 1465–1480. [CrossRef]
77. Ragan, C.I.; Heron, C. The Interaction between Mitochondrial NADH-Ubiquinone Oxidoreductase and Ubiquinol-Cytochrome c Oxidoreductase. Evidence for Stoicheiometric Association. *Biochem. J.* **1978**, *174*, 783–790. [CrossRef] [PubMed]
78. Heron, C.; Ragan, C.I.; Trumpower, B.L. The Interaction between Mitochondrial NADH-Ubiquinone Oxidoreductase and Ubiquinol-Cytochrome c Oxidoreductase. Restoration of Ubiquinone-Pool Behaviour. *Biochem. J.* **1978**, *174*, 791–800. [CrossRef]
79. Lenaz, G.; Fato, R.; Di Bernardo, S.; Jarreta, D.; Costa, A.; Genova, M.L.; Parenti Castelli, G. Localization and Mobility of Coenzyme Q in Lipid Bilayers and Membranes. *Biofactors* **1999**, *9*, 87–93. [CrossRef] [PubMed]
80. García-Poyatos, C.; Cogliati, S.; Calvo, E.; Hernansanz-Agustín, P.; Lagarrigue, S.; Magni, R.; Botos, M.; Langa, X.; Amati, F.; Vázquez, J.; et al. Scaf1 Promotes Respiratory Supercomplexes and Metabolic Efficiency in Zebrafish. *EMBO Rep.* **2020**, *21*, e50287. [CrossRef]
81. Solsona-Vilarrasa, E.; Fucho, R.; Torres, S.; Nuñez, S.; Nuño-Lámbarri, N.; Enrich, C.; García-Ruiz, C.; Fernández-Checa, J.C. Cholesterol Enrichment in Liver Mitochondria Impairs Oxidative Phosphorylation and Disrupts the Assembly of Respiratory Supercomplexes. *Redox Biol.* **2019**, *24*, 101214. [CrossRef]
82. Tomková, V.; Sandoval-Acuña, C.; Torrealba, N.; Truksa, J. Mitochondrial Fragmentation, Elevated Mitochondrial Superoxide and Respiratory Supercomplexes Disassembly Is Connected with the Tamoxifen-Resistant Phenotype of Breast Cancer Cells. *Free Radic. Biol. Med.* **2019**, *143*, 510–521. [CrossRef]
83. Balsa, E.; Soustek, M.S.; Thomas, A.; Cogliati, S.; García-Poyatos, C.; Martín-García, E.; Jedrychowski, M.; Gygi, S.P.; Enriquez, J.A.; Puigserver, P. ER and Nutrient Stress Promote Assembly of Respiratory Chain Supercomplexes through the PERK-EIF2α Axis. *Mol. Cell* **2019**, *74*, 877–890.e6. [CrossRef]
84. Kholodenko, B.N.; Westerhoff, H.V. Metabolic Channelling and Control of the Flux. *FEBS Lett.* **1993**, *320*, 71–74. [CrossRef]
85. Blaza, J.N.; Serreli, R.; Jones, A.J.Y.; Mohammed, K.; Hirst, J. Kinetic Evidence against Partitioning of the Ubiquinone Pool and the Catalytic Relevance of Respiratory-Chain Supercomplexes. *Proc. Natl. Acad. Sci. USA* **2014**, *111*, 15735–15740. [CrossRef]
86. Fato, R.; Estornell, E.; Di Bernardo, S.; Pallotti, F.; Parenti Castelli, G.; Lenaz, G. Steady-State Kinetics of the Reduction of Coenzyme Q Analogs by Complex I (NADH:Ubiquinone Oxidoreductase) in Bovine Heart Mitochondria and Submitochondrial Particles. *Biochemistry* **1996**, *35*, 2705–2716. [CrossRef]
87. Quarato, G.; Piccoli, C.; Scrima, R.; Capitanio, N. Variation of Flux Control Coefficient of Cytochrome *c* Oxidase and of the Other Respiratory Chain Complexes at Different Values of Protonmotive Force Occurs by a Threshold Mechanism. *Biochim. Biophys. Acta* **2011**, *1807*, 1114–1124. [CrossRef] [PubMed]
88. Kaambre, T.; Chekulayev, V.; Shevchuk, I.; Karu-Varikmaa, M.; Timohhina, N.; Tepp, K.; Bogovskaja, J.; Kütner, R.; Valvere, V.; Saks, V. Metabolic Control Analysis of Cellular Respiration in Situ in Intraoperational Samples of Human Breast Cancer. *J. Bioenergy Biomembr.* **2012**, *44*, 539–558. [CrossRef] [PubMed]
89. Kaambre, T.; Chekulayev, V.; Shevchuk, I.; Tepp, K.; Timohhina, N.; Varikmaa, M.; Bagur, R.; Klepinin, A.; Anmann, T.; Koit, A.; et al. Metabolic Control Analysis of Respiration in Human Cancer Tissue. *Front. Physiol.* **2013**, *4*, 151. [CrossRef] [PubMed]
90. Gutman, M.; Silman, N. Mutual Inhibition between NADH Oxidase and Succinoxidase Activities in Respiring Submitochondrial Particles. *FEBS Lett.* **1972**, *26*, 207–210. [CrossRef]
91. Gutman, M.; Kearney, E.B.; Singer, T.P. Control of Succinate Dehydrogenase in Mitochondria. *Biochemistry* **1971**, *10*, 4763–4770. [CrossRef]
92. Gutman, M. Kinetic Analysis of Electron Flux through the Quinones in the Mitochondrial System. In *Coenzyme Q*; Lenaz, G., Ed.; Wiley: Hoboken, NJ, USA, 1985.
93. Ragan, C.I.; Cottingham, I.R. The Kinetics of Quinone Pools in Electron Transport. *Biochim. Biophys. Acta* **1985**, *811*, 13–31. [CrossRef]
94. Jørgensen, B.M.; Rasmussen, H.N.; Rasmussen, U.F. Ubiquinone Reduction Pattern in Pigeon Heart Mitochondria. Identification of Three Distinct Ubiquinone Pools. *Biochem. J.* **1985**, *229*, 621–629. [CrossRef]
95. Lapuente-Brun, E.; Moreno-Loshuertos, R.; Acín-Pérez, R.; Latorre-Pellicer, A.; Colás, C.; Balsa, E.; Perales-Clemente, E.; Quirós, P.M.; Calvo, E.; Rodríguez-Hernández, M.A.; et al. Supercomplex Assembly Determines Electron Flux in the Mitochondrial Electron Transport Chain. *Science* **2013**, *340*, 1567–1570. [CrossRef]

96. Lenaz, G.; Genova, M.L. Kinetics of Integrated Electron Transfer in the Mitochondrial Respiratory Chain: Random Collisions vs. Solid State Electron Channeling. *Am. J. Physiol. Cell Physiol.* **2007**, *292*, C1221–C1239. [CrossRef]
97. Fedor, J.G.; Hirst, J. Mitochondrial Supercomplexes Do Not Enhance Catalysis by Quinone Channeling. *Cell Metab.* **2018**, *28*, 525–531.e4. [CrossRef]
98. Hirst, J. Open Questions: Respiratory Chain Supercomplexes-Why Are They There and What Do They Do? *BMC Biol.* **2018**, *16*, 111. [CrossRef]
99. Szibor, M.; Gainutdinov, T.; Fernandez-Vizarra, E.; Dufour, E.; Gizatullina, Z.; Debska-Vielhaber, G.; Heidler, J.; Wittig, I.; Viscomi, C.; Gellerich, F.; et al. Bioenergetic Consequences from Xenotopic Expression of a Tunicate AOX in Mouse Mitochondria: Switch from RET and ROS to FET. *Biochim. Biophys. Acta Bioenergy* **2020**, *1861*, 148137. [CrossRef]
100. Calvo, E.; Cogliati, S.; Hernansanz-Agustín, P.; Loureiro-López, M.; Guarás, A.; Casuso, R.A.; García-Marqués, F.; Acín-Pérez, R.; Martí-Mateos, Y.; Silla-Castro, J.C.; et al. Functional Role of Respiratory Supercomplexes in Mice: SCAF1 Relevance and Segmentation of the Qpool. *Sci. Adv.* **2020**, *6*, eaba7509. [CrossRef]
101. Estornell, E.; Fato, R.; Castelluccio, C.; Cavazzoni, M.; Parenti Castelli, G.; Lenaz, G. Saturation Kinetics of Coenzyme Q in NADH and Succinate Oxidation in Beef Heart Mitochondria. *FEBS Lett.* **1992**, *311*, 107–109. [CrossRef]
102. Schneider, H.; Lemasters, J.J.; Hackenbrock, C.R. Lateral Diffusion of Ubiquinone during Electron Transfer in Phospholipid- and Ubiquinone-Enriched Mitochondrial Membranes. *J. Biol. Chem.* **1982**, *257*, 10789–10793. [CrossRef]
103. Rosca, M.G.; Vazquez, E.J.; Kerner, J.; Parland, W.; Chandler, M.P.; Stanley, W.; Sabbah, H.N.; Hoppel, C.L. Cardiac Mitochondria in Heart Failure: Decrease in Respirasomes and Oxidative Phosphorylation. *Cardiovasc. Res.* **2008**, *80*, 30–39. [CrossRef]
104. Van Raam, B.J.; Sluiter, W.; de Wit, E.; Roos, D.; Verhoeven, A.J.; Kuijpers, T.W. Mitochondrial Membrane Potential in Human Neutrophils Is Maintained by Complex III Activity in the Absence of Supercomplex Organisation. *PLoS ONE* **2008**, *3*, e2013. [CrossRef]
105. Jones, A.J.Y.; Blaza, J.N.; Bridges, H.R.; May, B.; Moore, A.L.; Hirst, J. A Self-Assembled Respiratory Chain That Catalyzes NADH Oxidation by Ubiquinone-10 Cycling between Complex I and the Alternative Oxidase. *Angew. Chem. Int. Ed. Engl.* **2016**, *55*, 728–731. [CrossRef] [PubMed]
106. Stoner, C.D. Steady-State Kinetics of the Overall Oxidative Phosphorylation Reaction in Heart Mitochondria. Determination of the Coupling Relationships between the Respiratory Reactions and Miscellaneous Observations Concerning Rate-Limiting Steps. *J. Bioenergy Biomembr.* **1984**, *16*, 115–141. [CrossRef] [PubMed]
107. Rauchová, H.; Fato, R.; Drahota, Z.; Lenaz, G. Steady-State Kinetics of Reduction of Coenzyme Q Analogs by Glycerol-3-Phosphate Dehydrogenase in Brown Adipose Tissue Mitochondria. *Arch. Biochem. Biophys.* **1997**, *344*, 235–241. [CrossRef]
108. Mráček, T.; Holzerová, E.; Drahota, Z.; Kovářová, N.; Vrbacký, M.; Ješina, P.; Houštěk, J. ROS Generation and Multiple Forms of Mammalian Mitochondrial Glycerol-3-Phosphate Dehydrogenase. *Biochim. Biophys. Acta* **2014**, *1837*, 98–111. [CrossRef] [PubMed]
109. Nesci, S.; Lenaz, G. The Mitochondrial Energy Conversion Involves Cytochrome c Diffusion into the Respiratory Supercomplexes. *Biochim. Biophys. Acta Bioenergy* **2021**, *1862*, 148394. [CrossRef] [PubMed]
110. Boumans, H.; Grivell, L.A.; Berden, J.A. The Respiratory Chain in Yeast Behaves as a Single Functional Unit. *J. Biol. Chem.* **1998**, *273*, 4872–4877. [CrossRef]
111. Heinemeyer, J.; Braun, H.-P.; Boekema, E.J.; Kouril, R. A Structural Model of the Cytochrome C Reductase/Oxidase Supercomplex from Yeast Mitochondria. *J. Biol. Chem.* **2007**, *282*, 12240–12248. [CrossRef] [PubMed]
112. Mileykovskaya, E.; Penczek, P.A.; Fang, J.; Mallampalli, V.K.P.S.; Sparagna, G.C.; Dowhan, W. Arrangement of the Respiratory Chain Complexes in Saccharomyces Cerevisiae Supercomplex III2IV2 Revealed by Single Particle Cryo-Electron Microscopy. *J. Biol. Chem.* **2012**, *287*, 23095–23103. [CrossRef] [PubMed]
113. Trouillard, M.; Meunier, B.; Rappaport, F. Questioning the Functional Relevance of Mitochondrial Supercomplexes by Time-Resolved Analysis of the Respiratory Chain. *Proc. Natl. Acad. Sci. USA* **2011**, *108*, E1027–E1034. [CrossRef] [PubMed]
114. Rydström Lundin, C.; von Ballmoos, C.; Ott, M.; Ädelroth, P.; Brzezinski, P. Regulatory Role of the Respiratory Supercomplex Factors in Saccharomyces Cerevisiae. *Proc. Natl. Acad. Sci. USA* **2016**, *113*, E4476–E4485. [CrossRef]
115. Stuchebrukhov, A.; Schäfer, J.; Berg, J.; Brzezinski, P. Kinetic Advantage of Forming Respiratory Supercomplexes. *Biochim. Biophys. Acta Bioenergy* **2020**, *1861*, 148193. [CrossRef] [PubMed]
116. Berndtsson, J.; Aufschnaiter, A.; Rathore, S.; Marin-Buera, L.; Dawitz, H.; Diessl, J.; Kohler, V.; Barrientos, A.; Büttner, S.; Fontanesi, F.; et al. Respiratory Supercomplexes Enhance Electron Transport by Decreasing Cytochrome c Diffusion Distance. *EMBO Rep.* **2020**, e51015. [CrossRef]
117. Vukotic, M.; Oeljeklaus, S.; Wiese, S.; Vögtle, F.N.; Meisinger, C.; Meyer, H.E.; Zieseniss, A.; Katschinski, D.M.; Jans, D.C.; Jakobs, S.; et al. Rcf1 Mediates Cytochrome Oxidase Assembly and Respirasome Formation, Revealing Heterogeneity of the Enzyme Complex. *Cell Metab.* **2012**, *15*, 336–347. [CrossRef]
118. Moreno-Loshuertos, R.; Enríquez, J.A. Respiratory Supercomplexes and the Functional Segmentation of the CoQ Pool. *Free Radic. Biol. Med.* **2016**, *100*, 5–13. [CrossRef] [PubMed]
119. Speijer, D. Oxygen Radicals Shaping Evolution: Why Fatty Acid Catabolism Leads to Peroxisomes While Neurons Do without It: $FADH_2$/NADH Flux Ratios Determining Mitochondrial Radical Formation Were Crucial for the Eukaryotic Invention of Peroxisomes and Catabolic Tissue Differentiation. *Bioessays* **2011**, *33*, 88–94. [CrossRef]

120. Guarás, A.; Perales-Clemente, E.; Calvo, E.; Acín-Pérez, R.; Loureiro-Lopez, M.; Pujol, C.; Martínez-Carrascoso, I.; Nuñez, E.; García-Marqués, F.; Rodríguez-Hernández, M.A.; et al. The CoQH2/CoQ Ratio Serves as a Sensor of Respiratory Chain Efficiency. *Cell Rep.* **2016**, *15*, 197–209. [CrossRef] [PubMed]
121. Schertl, P.; Braun, H.-P. Respiratory Electron Transfer Pathways in Plant Mitochondria. *Front. Plant Sci.* **2014**, *5*, 163. [CrossRef]
122. Muench, S.P.; Trinick, J.; Harrison, M.A. Structural Divergence of the Rotary ATPases. *Q. Rev. Biophys.* **2011**, *44*, 311–356. [CrossRef]
123. Okuno, D.; Iino, R.; Noji, H. Rotation and Structure of FoF1-ATP Synthase. *J. Biochem.* **2011**, *149*, 655–664. [CrossRef]
124. Nesci, S.; Trombetti, F.; Ventrella, V.; Pagliarani, A. The a Subunit Asymmetry Dictates the Two Opposite Rotation Directions in the Synthesis and Hydrolysis of ATP by the Mitochondrial ATP Synthase. *Med. Hypotheses* **2015**, *84*, 53–57. [CrossRef]
125. Boyer, P.D. The ATP Synthase—A Splendid Molecular Machine. *Annu. Rev. Biochem.* **1997**, *66*, 717–749. [CrossRef]
126. Yoshida, M.; Muneyuki, E.; Hisabori, T. ATP Synthase—A Marvellous Rotary Engine of the Cell. *Nat. Rev. Mol. Cell Biol.* **2001**, *2*, 669–677. [CrossRef] [PubMed]
127. Gu, J.; Zhang, L.; Zong, S.; Guo, R.; Liu, T.; Yi, J.; Wang, P.; Zhuo, W.; Yang, M. Cryo-EM Structure of the Mammalian ATP Synthase Tetramer Bound with Inhibitory Protein IF1. *Science* **2019**, *364*, 1068–1075. [CrossRef]
128. Pinke, G.; Zhou, L.; Sazanov, L.A. Cryo-EM Structure of the Entire Mammalian F-Type ATP Synthase. *Nat. Struct. Mol. Biol.* **2020**, *27*, 1077–1085. [CrossRef]
129. Nesci, S.; Trombetti, F.; Algieri, C.; Pagliarani, A. A Therapeutic Role for the F1FO-ATP Synthase. *SLAS Discov.* **2019**, *24*, 893–903. [CrossRef] [PubMed]
130. Nesci, S.; Pagliarani, A.; Algieri, C.; Trombetti, F. Mitochondrial F-Type ATP Synthase: Multiple Enzyme Functions Revealed by the Membrane-Embedded FO Structure. *Crit. Rev. Biochem. Mol. Biol.* **2020**, 1–13. [CrossRef]
131. Junge, W.; Lill, H.; Engelbrecht, S. ATP Synthase: An Electrochemical Transducer with Rotatory Mechanics. *Trends Biochem. Sci.* **1997**, *22*, 420–423. [CrossRef]
132. Junge, W.; Sielaff, H.; Engelbrecht, S. Torque Generation and Elastic Power Transmission in the Rotary F(O)F(1)-ATPase. *Nature* **2009**, *459*, 364–370. [CrossRef] [PubMed]
133. Nesci, S.; Trombetti, F.; Ventrella, V.; Pagliarani, A. Opposite Rotation Directions in the Synthesis and Hydrolysis of ATP by the ATP Synthase: Hints from a Subunit Asymmetry. *J. Membr. Biol.* **2015**, *248*, 163–169. [CrossRef]
134. Lippe, G.; Coluccino, G.; Zancani, M.; Baratta, W.; Crusiz, P. Mitochondrial F-ATP Synthase and Its Transition into an Energy-Dissipating Molecular Machine. *Oxid. Med. Cell Longev.* **2019**, *2019*, 8743257. [CrossRef] [PubMed]
135. Symersky, J.; Osowski, D.; Walters, D.E.; Mueller, D.M. Oligomycin Frames a Common Drug-Binding Site in the ATP Synthase. *Proc. Natl. Acad. Sci. USA* **2012**, *109*, 13961–13965. [CrossRef] [PubMed]
136. Kühlbrandt, W. Structure and Mechanisms of F-Type ATP Synthases. *Annu. Rev. Biochem.* **2019**, *88*, 515–549. [CrossRef]
137. Hahn, A.; Parey, K.; Bublitz, M.; Mills, D.J.; Zickermann, V.; Vonck, J.; Kühlbrandt, W.; Meier, T. Structure of a Complete ATP Synthase Dimer Reveals the Molecular Basis of Inner Mitochondrial Membrane Morphology. *Mol. Cell* **2016**, *63*, 445–456. [CrossRef]
138. Boyer, P.D. Catalytic Site Occupancy during ATP Synthase Catalysis. *FEBS Lett.* **2002**, *512*, 29–32. [CrossRef]
139. Murataliev, M.B.; Boyer, P.D. The Mechanism of Stimulation of MgATPase Activity of Chloroplast F1-ATPase by Non-Catalytic Adenine-Nucleotide Binding. Acceleration of the ATP-Dependent Release of Inhibitory ADP from a Catalytic Site. *Eur. J. Biochem.* **1992**, *209*, 681–687. [CrossRef] [PubMed]
140. Xu, T.; Pagadala, V.; Mueller, D.M. Understanding Structure, Function, and Mutations in the Mitochondrial ATP Synthase. *Microb. Cell* **2015**, *2*, 105–125. [CrossRef] [PubMed]
141. Pogoryelov, D.; Klyszejko, A.L.; Krasnoselska, G.O.; Heller, E.-M.; Leone, V.; Langer, J.D.; Vonck, J.; Müller, D.J.; Faraldo-Gómez, J.D.; Meier, T. Engineering Rotor Ring Stoichiometries in the ATP Synthase. *Proc. Natl. Acad. Sci. USA* **2012**, *109*, E1599–E1608. [CrossRef] [PubMed]
142. Von Ballmoos, C.; Cook, G.M.; Dimroth, P. Unique Rotary ATP Synthase and Its Biological Diversity. *Annu. Rev. Biophys.* **2008**, *37*, 43–64. [CrossRef] [PubMed]
143. Nesci, S.; Ventrella, V.; Trombetti, F.; Pirini, M.; Pagliarani, A. Mussel and Mammalian ATP Synthase Share the Same Bioenergetic Cost of ATP. *J. Bioenergy Biomembr.* **2013**, *45*, 289–300. [CrossRef]
144. Hahn, A.; Vonck, J.; Mills, D.J.; Meier, T.; Kühlbrandt, W. Structure, Mechanism, and Regulation of the Chloroplast ATP Synthase. *Science* **2018**, *360*, eaat4318. [CrossRef] [PubMed]
145. Guo, H.; Suzuki, T.; Rubinstein, J.L. Structure of a Bacterial ATP Synthase. *eLife* **2019**, *8*. [CrossRef]
146. Allegretti, M.; Klusch, N.; Mills, D.J.; Vonck, J.; Kühlbrandt, W.; Davies, K.M. Horizontal Membrane-Intrinsic α-Helices in the Stator a-Subunit of an F-Type ATP Synthase. *Nature* **2015**, *521*, 237–240. [CrossRef]
147. Kühlbrandt, W.; Davies, K.M. Rotary ATPases: A New Twist to an Ancient Machine. *Trends Biochem. Sci.* **2016**, *41*, 106–116. [CrossRef] [PubMed]
148. Klusch, N.; Murphy, B.J.; Mills, D.J.; Yildiz, Ö.; Kühlbrandt, W. Structural Basis of Proton Translocation and Force Generation in Mitochondrial ATP Synthase. *eLife* **2017**, *6*, e33274. [CrossRef] [PubMed]
149. Strauss, M.; Hofhaus, G.; Schröder, R.R.; Kühlbrandt, W. Dimer Ribbons of ATP Synthase Shape the Inner Mitochondrial Membrane. *EMBO J.* **2008**, *27*, 1154–1160. [CrossRef]

150. Wolf, D.M.; Segawa, M.; Kondadi, A.K.; Anand, R.; Bailey, S.T.; Reichert, A.S.; van der Bliek, A.M.; Shackelford, D.B.; Liesa, M.; Shirihai, O.S. Individual Cristae within the Same Mitochondrion Display Different Membrane Potentials and Are Functionally Independent. *EMBO J.* **2019**, *38*, e101056. [CrossRef]
151. Srivastava, A.P.; Luo, M.; Zhou, W.; Symersky, J.; Bai, D.; Chambers, M.G.; Faraldo-Gómez, J.D.; Liao, M.; Mueller, D.M. High-Resolution Cryo-EM Analysis of the Yeast ATP Synthase in a Lipid Membrane. *Science* **2018**, *360*, eaas9699. [CrossRef] [PubMed]
152. Boyer, P.D. Bioenergetic Coupling to Protonmotive Force: Should We Be Considering Hydronium Ion Coordination and Not Group Protonation? *Trends Biochem. Sci.* **1988**, *13*, 5–7. [CrossRef]
153. Von Ballmoos, C.; Dimroth, P. Two Distinct Proton Binding Sites in the ATP Synthase Family. *Biochemistry* **2007**, *46*, 11800–11809. [CrossRef] [PubMed]
154. Spikes, T.E.; Montgomery, M.G.; Walker, J.E. Structure of the Dimeric ATP Synthase from Bovine Mitochondria. *Proc. Natl. Acad. Sci. USA* **2020**, *117*, 23519–23526. [CrossRef] [PubMed]
155. Pogoryelov, D.; Krah, A.; Langer, J.D.; Yildiz, Ö.; Faraldo-Gómez, J.D.; Meier, T. Microscopic Rotary Mechanism of Ion Translocation in the F(o) Complex of ATP Synthases. *Nat. Chem. Biol.* **2010**, *6*, 891–899. [CrossRef] [PubMed]
156. Vinogradov, A.D. New Perspective on the Reversibility of ATP Synthesis and Hydrolysis by Fo·F1-ATP Synthase (Hydrolase). *Biochem. Moscow* **2019**, *84*, 1247–1255. [CrossRef]
157. Mitome, N.; Ono, S.; Sato, H.; Suzuki, T.; Sone, N.; Yoshida, M. Essential Arginine Residue of the F(o)-a Subunit in F(o)F(1)-ATP Synthase Has a Role to Prevent the Proton Shortcut without c-Ring Rotation in the F(o) Proton Channel. *Biochem. J.* **2010**, *430*, 171–177. [CrossRef] [PubMed]
158. Murphy, B.J.; Klusch, N.; Langer, J.; Mills, D.J.; Yildiz, Ö.; Kühlbrandt, W. Rotary Substates of Mitochondrial ATP Synthase Reveal the Basis of Flexible F1-Fo Coupling. *Science* **2019**, *364*. [CrossRef]
159. Blum, T.B.; Hahn, A.; Meier, T.; Davies, K.M.; Kühlbrandt, W. Dimers of Mitochondrial ATP Synthase Induce Membrane Curvature and Self-Assemble into Rows. *Proc. Natl. Acad. Sci. USA* **2019**. [CrossRef]
160. Arnold, I.; Pfeiffer, K.; Neupert, W.; Stuart, R.A.; Schägger, H. Yeast Mitochondrial F1F0-ATP Synthase Exists as a Dimer: Identification of Three Dimer-Specific Subunits. *EMBO J.* **1998**, *17*, 7170–7178. [CrossRef]
161. Davies, K.M.; Anselmi, C.; Wittig, I.; Faraldo-Gómez, J.D.; Kühlbrandt, W. Structure of the Yeast F1Fo-ATP Synthase Dimer and Its Role in Shaping the Mitochondrial Cristae. *Proc. Natl. Acad. Sci. USA* **2012**, *109*, 13602–13607. [CrossRef] [PubMed]
162. Eydt, K.; Davies, K.M.; Behrendt, C.; Wittig, I.; Reichert, A.S. Cristae Architecture Is Determined by an Interplay of the MICOS Complex and the F1FO ATP Synthase via Mic27 and Mic10. *Microb. Cell* **2017**, *4*, 259–272. [CrossRef] [PubMed]
163. Rampelt, H.; van der Laan, M. The Yin & Yang of Mitochondrial Architecture—Interplay of MICOS and F1Fo-ATP Synthase in Cristae Formation. *Microb. Cell* **2017**, *4*, 236–239. [CrossRef]
164. Colina-Tenorio, L.; Horten, P.; Pfanner, N.; Rampelt, H. Shaping the Mitochondrial Inner Membrane in Health and Disease. *J. Intern. Med.* **2020**, *287*, 645–664. [CrossRef]
165. Faccenda, D.; Tan, C.H.; Seraphim, A.; Duchen, M.R.; Campanella, M. IF1 Limits the Apoptotic-Signalling Cascade by Preventing Mitochondrial Remodelling. *Cell Death Differ.* **2013**, *20*, 686–697. [CrossRef]
166. Esparza-Moltó, P.B.; Cuezva, J.M. The Role of Mitochondrial H+-ATP Synthase in Cancer. *Front. Oncol.* **2018**, *8*, 53. [CrossRef] [PubMed]
167. García, J.J.; Morales-Ríos, E.; Cortés-Hernandez, P.; Rodríguez-Zavala, J.S. The Inhibitor Protein (IF1) Promotes Dimerization of the Mitochondrial F1F0-ATP Synthase. *Biochemistry* **2006**, *45*, 12695–12703. [CrossRef]
168. Nakamura, J.; Fujikawa, M.; Yoshida, M. IF1, a Natural Inhibitor of Mitochondrial ATP Synthase, Is Not Essential for the Normal Growth and Breeding of Mice. *Biosci. Rep.* **2013**, *33*. [CrossRef] [PubMed]
169. Anselmi, C.; Davies, K.M.; Faraldo-Gómez, J.D. Mitochondrial ATP Synthase Dimers Spontaneously Associate Due to a Long-Range Membrane-Induced Force. *J. Gen. Physiol.* **2018**, *150*, 763–770. [CrossRef] [PubMed]
170. Nesci, S.; Pagliarani, A. Emerging Roles for the Mitochondrial ATP Synthase Supercomplexes. *Trends Biochem. Sci.* **2019**, *44*, 821–823. [CrossRef]
171. Arselin, G.; Giraud, M.-F.; Dautant, A.; Vaillier, J.; Brèthes, D.; Coulary-Salin, B.; Schaeffer, J.; Velours, J. The GxxxG Motif of the Transmembrane Domain of Subunit e Is Involved in the Dimerization/Oligomerization of the Yeast ATP Synthase Complex in the Mitochondrial Membrane. *Eur. J. Biochem.* **2003**, *270*, 1875–1884. [CrossRef]
172. Bustos, D.M.; Velours, J. The Modification of the Conserved GXXXG Motif of the Membrane-Spanning Segment of Subunit g Destabilizes the Supramolecular Species of Yeast ATP Synthase. *J. Biol. Chem.* **2005**, *280*, 29004–29010. [CrossRef]
173. Guo, H.; Bueler, S.A.; Rubinstein, J.L. Atomic Model for the Dimeric FO Region of Mitochondrial ATP Synthase. *Science* **2017**, *358*, 936–940. [CrossRef] [PubMed]
174. Teixeira, F.K.; Sanchez, C.G.; Hurd, T.R.; Seifert, J.R.K.; Czech, B.; Preall, J.B.; Hannon, G.J.; Lehmann, R. ATP Synthase Promotes Germ Cell Differentiation Independent of Oxidative Phosphorylation. *Nat. Cell. Biol.* **2015**, *17*, 689–696. [CrossRef]
175. Faccenda, D.; Campanella, M. Molecular Regulation of the Mitochondrial F(1)F(o)-ATPsynthase: Physiological and Pathological Significance of the Inhibitory Factor 1 (IF(1)). *Int. J. Cell. Biol.* **2012**, *2012*, 367934. [CrossRef]
176. Mnatsakanyan, N.; Jonas, E.A. The New Role of F1Fo ATP Synthase in Mitochondria-Mediated Neurodegeneration and Neuroprotection. *Exp. Neurol.* **2020**, *332*, 113400. [CrossRef] [PubMed]
177. Daum, B.; Walter, A.; Horst, A.; Osiewacz, H.D.; Kühlbrandt, W. Age-Dependent Dissociation of ATP Synthase Dimers and Loss of Inner-Membrane Cristae in Mitochondria. *Proc. Natl. Acad. Sci. USA* **2013**, *110*, 15301–15306. [CrossRef]

178. Giorgio, V.; von Stockum, S.; Antoniel, M.; Fabbro, A.; Fogolari, F.; Forte, M.; Glick, G.D.; Petronilli, V.; Zoratti, M.; Szabó, I.; et al. Dimers of Mitochondrial ATP Synthase Form the Permeability Transition Pore. *Proc. Natl. Acad. Sci. USA* **2013**, *110*, 5887–5892. [CrossRef] [PubMed]
179. Bonora, M.; Bononi, A.; De Marchi, E.; Giorgi, C.; Lebiedzinska, M.; Marchi, S.; Patergnani, S.; Rimessi, A.; Suski, J.M.; Wojtala, A.; et al. Role of the c Subunit of the FO ATP Synthase in Mitochondrial Permeability Transition. *Cell Cycle* **2013**, *12*, 674–683. [CrossRef]
180. Alavian, K.N.; Beutner, G.; Lazrove, E.; Sacchetti, S.; Park, H.-A.; Licznerski, P.; Li, H.; Nabili, P.; Hockensmith, K.; Graham, M.; et al. An Uncoupling Channel within the C-Subunit Ring of the F1FO ATP Synthase Is the Mitochondrial Permeability Transition Pore. *Proc. Natl. Acad. Sci. USA* **2014**, *111*, 10580–10585. [CrossRef]
181. Mnatsakanyan, N.; Jonas, E.A. ATP Synthase C-Subunit Ring as the Channel of Mitochondrial Permeability Transition: Regulator of Metabolism in Development and Degeneration. *J. Mol. Cell. Cardiol.* **2020**, *144*, 109–118. [CrossRef]
182. Bernardi, P.; Rasola, A.; Forte, M.; Lippe, G. The Mitochondrial Permeability Transition Pore: Channel Formation by F-ATP Synthase, Integration in Signal Transduction, and Role in Pathophysiology. *Physiol. Rev.* **2015**, *95*, 1111–1155. [CrossRef] [PubMed]
183. Algieri, C.; Nesci, S.; Trombetti, F.; Fabbri, M.; Ventrella, V.; Pagliarani, A. Mitochondrial F1FO-ATPase and Permeability Transition Pore Response to Sulfide in the Midgut Gland of Mytilus Galloprovincialis. *Biochimie* **2021**, *180*, 222–228. [CrossRef]
184. Torrezan-Nitao, E.; Figueiredo, R.C.B.Q.; Marques-Santos, L.F. Mitochondrial Permeability Transition Pore in Sea Urchin Female Gametes. *Mech. Dev.* **2018**, *154*, 208–218. [CrossRef]
185. Liu, Y.; Zhi, D.; Li, M.; Liu, D.; Wang, X.; Wu, Z.; Zhang, Z.; Fei, D.; Li, Y.; Zhu, H.; et al. Shengmai Formula Suppressed Over-Activated Ras/MAPK Pathway in C. Elegans by Opening Mitochondrial Permeability Transition Pore via Regulating Cyclophilin D. *Sci. Rep.* **2016**, *6*, 38934. [CrossRef] [PubMed]
186. Yuan, Z.; Zhang, J.; Tu, C.; Wang, Z.; Xin, W. The Protective Effect of Blueberry Anthocyanins against Perfluorooctanoic Acid-Induced Disturbance in Planarian (Dugesia Japonica). *Ecotoxicol. Environ. Saf.* **2016**, *127*, 170–174. [CrossRef]
187. Ren, X.; Zhang, L.; Zhang, Y.; Mao, L.; Jiang, H. Mitochondria Response to Camptothecin and Hydroxycamptothecine-Induced Apoptosis in Spodoptera Exigua Cells. *Pestic. Biochem. Physiol.* **2017**, *140*, 97–104. [CrossRef] [PubMed]
188. Pan, C.; Hu, Y.-F.; Song, J.; Yi, H.-S.; Wang, L.; Yang, Y.-Y.; Wang, Y.-P.; Zhang, M.; Pan, M.-H.; Lu, C. Effects of 10-Hydroxycamptothecin on Intrinsic Mitochondrial Pathway in Silkworm BmN-SWU1 Cells. *Pestic. Biochem. Physiol.* **2016**, *127*, 15–20. [CrossRef]
189. Konràd, C.; Kiss, G.; Töröcsik, B.; Lábár, J.L.; Gerencser, A.A.; Mándi, M.; Adam-Vizi, V.; Chinopoulos, C. A Distinct Sequence in the Adenine Nucleotide Translocase from Artemia Franciscana Embryos Is Associated with Insensitivity to Bongkrekate and Atypical Effects of Adenine Nucleotides on Ca2+ Uptake and Sequestration. *FEBS J.* **2011**, *278*, 822–836. [CrossRef]
190. Konrad, C.; Kiss, G.; Torocsik, B.; Adam-Vizi, V.; Chinopoulos, C. Absence of Ca2+-Induced Mitochondrial Permeability Transition but Presence of Bongkrekate-Sensitive Nucleotide Exchange in C. Crangon and P. Serratus. *PLoS ONE* **2012**, *7*, e39839. [CrossRef] [PubMed]
191. Menze, M.A.; Fortner, G.; Nag, S.; Hand, S.C. Mechanisms of Apoptosis in Crustacea: What Conditions Induce versus Suppress Cell Death? *Apoptosis* **2010**, *15*, 293–312. [CrossRef]
192. Nesci, S.; Pagliarani, A. Incoming News on the F-Type ATPase Structure and Functions in Mammalian Mitochondria. *BBA Adv.* **2021**, *1*, 100001. [CrossRef]
193. Nesci, S.; Trombetti, F.; Ventrella, V.; Pirini, M.; Pagliarani, A. Kinetic Properties of the Mitochondrial F1FO-ATPase Activity Elicited by Ca(2+) in Replacement of Mg(2+). *Biochimie* **2017**, *140*, 73–81. [CrossRef] [PubMed]
194. Papageorgiou, S.; Melandri, A.B.; Solaini, G. Relevance of Divalent Cations to ATP-Driven Proton Pumping in Beef Heart Mitochondrial F0F1-ATPase. *J. Bioenergy Biomembr.* **1998**, *30*, 533–541. [CrossRef] [PubMed]
195. Nesci, S.; Pagliarani, A. Ca2+ as Cofactor of the Mitochondrial H+ -Translocating F1 FO -ATP(Hydrol)Ase. *Proteins* **2020**. [CrossRef]
196. Algieri, C.; Trombetti, F.; Pagliarani, A.; Ventrella, V.; Bernardini, C.; Fabbri, M.; Forni, M.; Nesci, S. Mitochondrial Ca2+-Activated F1 FO -ATPase Hydrolyzes ATP and Promotes the Permeability Transition Pore. *Ann. N. Y. Acad. Sci.* **2019**, *1457*, 142–157. [CrossRef]
197. Algieri, V.; Algieri, C.; Maiuolo, L.; De Nino, A.; Pagliarani, A.; Tallarida, M.A.; Trombetti, F.; Nesci, S. 1,5-Disubstituted-1,2,3-Triazoles as Inhibitors of the Mitochondrial Ca2+ -Activated F1 FO -ATP(Hydrol)Ase and the Permeability Transition Pore. *Ann. N. Y. Acad. Sci.* **2020**. [CrossRef]
198. Giorgio, V.; Burchell, V.; Schiavone, M.; Bassot, C.; Minervini, G.; Petronilli, V.; Argenton, F.; Forte, M.; Tosatto, S.; Lippe, G.; et al. Ca(2+) Binding to F-ATP Synthase β Subunit Triggers the Mitochondrial Permeability Transition. *EMBO Rep.* **2017**, *18*, 1065–1076. [CrossRef] [PubMed]
199. Nesci, S. Mitochondrial Permeability Transition, F_1F_O-ATPase and Calcium: An Enigmatic Triangle. *EMBO Rep.* **2017**, *18*, 1265–1267. [CrossRef] [PubMed]
200. Nesci, S.; Algieri, C.; Trombetti, F.; Ventrella, V.; Fabbri, M.; Pagliarani, A. Sulfide Affects the Mitochondrial Respiration, the Ca2+-Activated F1FO-ATPase Activity and the Permeability Transition Pore but Does Not Change the Mg2+-Activated F1FO-ATPase Activity in Swine Heart Mitochondria. *Pharmacol. Res.* **2021**, *166*, 105495. [CrossRef] [PubMed]
201. Ader, N.R.; Hoffmann, P.C.; Ganeva, I.; Borgeaud, A.C.; Wang, C.; Youle, R.J.; Kukulski, W. Molecular and Topological Reorganizations in Mitochondrial Architecture Interplay during Bax-Mediated Steps of Apoptosis. *eLife* **2019**, *8*. [CrossRef]

202. Carraro, M.; Carrer, A.; Urbani, A.; Bernardi, P. Molecular Nature and Regulation of the Mitochondrial Permeability Transition Pore(s), Drug Target(s) in Cardioprotection. *J. Mol. Cell. Cardiol.* **2020**, *144*, 76–86. [CrossRef] [PubMed]
203. Szabo, I.; Zoratti, M. Mitochondrial Channels: Ion Fluxes and More. *Physiol. Rev.* **2014**, *94*, 519–608. [CrossRef] [PubMed]
204. Neginskaya, M.A.; Solesio, M.E.; Berezhnaya, E.V.; Amodeo, G.F.; Mnatsakanyan, N.; Jonas, E.A.; Pavlov, E.V. ATP Synthase C-Subunit-Deficient Mitochondria Have a Small Cyclosporine A-Sensitive Channel, but Lack the Permeability Transition Pore. *Cell Rep.* **2019**, *26*, 11–17.e2. [CrossRef] [PubMed]
205. Karch, J.; Bround, M.J.; Khalil, H.; Sargent, M.A.; Latchman, N.; Terada, N.; Peixoto, P.M.; Molkentin, J.D. Inhibition of Mitochondrial Permeability Transition by Deletion of the ANT Family and CypD. *Sci. Adv.* **2019**, *5*, eaaw4597. [CrossRef]
206. Bonora, M.; Morganti, C.; Morciano, G.; Pedriali, G.; Lebiedzinska-Arciszewska, M.; Aquila, G.; Giorgi, C.; Rizzo, P.; Campo, G.; Ferrari, R.; et al. Mitochondrial Permeability Transition Involves Dissociation of F1FO ATP Synthase Dimers and C-Ring Conformation. *EMBO Rep.* **2017**, *18*, 1077–1089. [CrossRef]
207. Morciano, G.; Preti, D.; Pedriali, G.; Aquila, G.; Missiroli, S.; Fantinati, A.; Caroccia, N.; Pacifico, S.; Bonora, M.; Talarico, A.; et al. Discovery of Novel 1,3,8-Triazaspiro[4.5]Decane Derivatives That Target the c Subunit of F1/FO-Adenosine Triphosphate (ATP) Synthase for the Treatment of Reperfusion Damage in Myocardial Infarction. *J. Med. Chem.* **2018**, *61*, 7131–7143. [CrossRef] [PubMed]
208. Algieri, C.; Trombetti, F.; Pagliarani, A.; Ventrella, V.; Nesci, S. Phenylglyoxal Inhibition of the Mitochondrial F1FO-ATPase Activated by Mg2+ or by Ca2+ Provides Clues on the Mitochondrial Permeability Transition Pore. *Arch. Biochem. Biophys.* **2020**, *681*, 108258. [CrossRef] [PubMed]
209. Mnatsakanyan, N.; Llaguno, M.C.; Yang, Y.; Yan, Y.; Weber, J.; Sigworth, F.J.; Jonas, E.A. A Mitochondrial Megachannel Resides in Monomeric F1FO ATP Synthase. *Nat. Commun.* **2019**, *10*, 5823. [CrossRef]
210. Urbani, A.; Giorgio, V.; Carrer, A.; Franchin, C.; Arrigoni, G.; Jiko, C.; Abe, K.; Maeda, S.; Shinzawa-Itoh, K.; Bogers, J.F.M.; et al. Purified F-ATP Synthase Forms a Ca2+-Dependent High-Conductance Channel Matching the Mitochondrial Permeability Transition Pore. *Nat. Commun.* **2019**, *10*, 4341. [CrossRef] [PubMed]
211. Papu John, A.S.; Kundu, S.; Pushpakumar, S.; Amin, M.; Tyagi, S.C.; Sen, U. Hydrogen Sulfide Inhibits Ca2+-Induced Mitochondrial Permeability Transition Pore Opening in Type-1 Diabetes. *Am. J. Physiol. Endocrinol. Metab.* **2019**, *317*, E269–E283. [CrossRef] [PubMed]
212. Bonora, M.; Pinton, P. The Mitochondrial Permeability Transition Pore and Cancer: Molecular Mechanisms Involved in Cell Death. *Front. Oncol.* **2014**, *4*, 302. [CrossRef] [PubMed]
213. Gan, X.; Zhang, L.; Liu, B.; Zhu, Z.; He, Y.; Chen, J.; Zhu, J.; Yu, H. CypD-MPTP Axis Regulates Mitochondrial Functions Contributing to Osteogenic Dysfunction of MC3T3-E1 Cells in Inflammation. *J. Physiol. Biochem.* **2018**, *74*, 395–402. [CrossRef]
214. Visalli, G.; Facciolà, A.; Currò, M.; Laganà, P.; La Fauci, V.; Iannazzo, D.; Pistone, A.; Di Pietro, A. Mitochondrial Impairment Induced by Sub-Chronic Exposure to Multi-Walled Carbon Nanotubes. *Int. J. Environ. Res. Public Health* **2019**, *16*, 792. [CrossRef] [PubMed]
215. Zhou, B.; Kreuzer, J.; Kumsta, C.; Wu, L.; Kamer, K.J.; Cedillo, L.; Zhang, Y.; Li, S.; Kacergis, M.C.; Webster, C.M.; et al. Mitochondrial Permeability Uncouples Elevated Autophagy and Lifespan Extension. *Cell* **2019**, *177*, 299–314.e16. [CrossRef]
216. Shares, B.H.; Smith, C.O.; Sheu, T.-J.; Sautchuk, R.; Schilling, K.; Shum, L.C.; Paine, A.; Huber, A.; Gira, E.; Brown, E.; et al. Inhibition of the Mitochondrial Permeability Transition Improves Bone Fracture Repair. *Bone* **2020**, *137*, 115391. [CrossRef]
217. Nesci, S. The Mitochondrial Permeability Transition Pore in Cell Death: A Promising Drug Binding Bioarchitecture. *Med. Res. Rev.* **2020**, *40*, 811–817. [CrossRef]
218. Briston, T.; Selwood, D.L.; Szabadkai, G.; Duchen, M.R. Mitochondrial Permeability Transition: A Molecular Lesion with Multiple Drug Targets. *Trends Pharmacol. Sci.* **2019**, *40*, 50–70. [CrossRef]
219. Cui, Y.; Pan, M.; Ma, J.; Song, X.; Cao, W.; Zhang, P. Recent Progress in the Use of Mitochondrial Membrane Permeability Transition Pore in Mitochondrial Dysfunction-Related Disease Therapies. *Mol. Cell Biochem.* **2020**. [CrossRef]
220. Li, Y.; Sun, J.; Wu, R.; Bai, J.; Hou, Y.; Zeng, Y.; Zhang, Y.; Wang, X.; Wang, Z.; Meng, X. Mitochondrial MPTP: A Novel Target of Ethnomedicine for Stroke Treatment by Apoptosis Inhibition. *Front. Pharmacol.* **2020**, *11*, 352. [CrossRef] [PubMed]
221. Halliwell, B.; Gutteridge, J.M.C. *Free Radicals in Biology and Medicine*; Oxford University Press: Oxford, UK, 2015, ISBN 978-0-19-180213-3.
222. Lenaz, G. Mitochondria and Reactive Oxygen Species. Which Role in Physiology and Pathology? *Adv. Exp. Med. Biol.* **2012**, *942*, 93–136. [CrossRef]
223. Lenaz, G.; Strocchi, P. Reactive Oxygen Species in the Induction of Toxicity. In *General, Applied and Systems Toxicology*; American Cancer Society: Atlanta, GA, USA, 2011, ISBN 978-0-470-74430-7.
224. Quinlan, C.L.; Perevoshchikova, I.V.; Hey-Mogensen, M.; Orr, A.L.; Brand, M.D. Sites of Reactive Oxygen Species Generation by Mitochondria Oxidizing Different Substrates. *Redox Biol.* **2013**, *1*, 304–312. [CrossRef]
225. St-Pierre, J.; Buckingham, J.A.; Roebuck, S.J.; Brand, M.D. Topology of Superoxide Production from Different Sites in the Mitochondrial Electron Transport Chain. *J. Biol. Chem.* **2002**, *277*, 44784–44790. [CrossRef]
226. Galkin, A.; Brandt, U. Superoxide Radical Formation by Pure Complex I (NADH:Ubiquinone Oxidoreductase) from Yarrowia Lipolytica. *J. Biol. Chem.* **2005**, *280*, 30129–30135. [CrossRef] [PubMed]
227. Kussmaul, L.; Hirst, J. The Mechanism of Superoxide Production by NADH:Ubiquinone Oxidoreductase (Complex I) from Bovine Heart Mitochondria. *Proc. Natl. Acad. Sci. USA* **2006**, *103*, 7607–7612. [CrossRef]

228. Esterházy, D.; King, M.S.; Yakovlev, G.; Hirst, J. Production of Reactive Oxygen Species by Complex I (NADH:Ubiquinone Oxidoreductase) from Escherichia Coli and Comparison to the Enzyme from Mitochondria. *Biochemistry* **2008**, *47*, 3964–3971. [CrossRef]
229. Lambert, A.J.; Brand, M.D. Inhibitors of the Quinone-Binding Site Allow Rapid Superoxide Production from Mitochondrial NADH:Ubiquinone Oxidoreductase (Complex I). *J. Biol. Chem.* **2004**, *279*, 39414–39420. [CrossRef]
230. Lambert, A.J.; Buckingham, J.A.; Boysen, H.M.; Brand, M.D. Diphenyleneiodonium Acutely Inhibits Reactive Oxygen Species Production by Mitochondrial Complex I during Reverse, but Not Forward Electron Transport. *Biochim. Biophys. Acta* **2008**, *1777*, 397–403. [CrossRef]
231. Genova, M.L.; Ventura, B.; Giuliano, G.; Bovina, C.; Formiggini, G.; Parenti Castelli, G.; Lenaz, G. The Site of Production of Superoxide Radical in Mitochondrial Complex I Is Not a Bound Ubisemiquinone but Presumably Iron-Sulfur Cluster N2. *FEBS Lett.* **2001**, *505*, 364–368. [CrossRef]
232. Fato, R.; Bergamini, C.; Bortolus, M.; Maniero, A.L.; Leoni, S.; Ohnishi, T.; Lenaz, G. Differential Effects of Mitochondrial Complex I Inhibitors on Production of Reactive Oxygen Species. *Biochim. Biophys. Acta* **2009**, *1787*, 384–392. [CrossRef] [PubMed]
233. Lenaz, G.; Fato, R.; Genova, M.L.; Bergamini, C.; Bianchi, C.; Biondi, A. Mitochondrial Complex I: Structural and Functional Aspects. *Biochim. Biophys. Acta* **2006**, *1757*, 1406–1420. [CrossRef]
234. Grivennikova, V.G.; Vinogradov, A.D. Mitochondrial Production of Reactive Oxygen Species. *Biochem. Mosc.* **2013**, *78*, 1490–1511. [CrossRef]
235. Ralph, S.J.; Moreno-Sánchez, R.; Neuzil, J.; Rodríguez-Enríquez, S. Inhibitors of Succinate: Quinone Reductase/Complex II Regulate Production of Mitochondrial Reactive Oxygen Species and Protect Normal Cells from Ischemic Damage but Induce Specific Cancer Cell Death. *Pharm. Res.* **2011**, *28*, 2695–2730. [CrossRef]
236. Vinogradov, A.D.; Grivennikova, V.G. Generation of Superoxide-Radical by the NADH:Ubiquinone Oxidoreductase of Heart Mitochondria. *Biochemistry* **2005**, *70*, 120–127. [CrossRef]
237. Ohnishi, S.T.; Ohnishi, T.; Muranaka, S.; Fujita, H.; Kimura, H.; Uemura, K.; Yoshida, K.; Utsumi, K. A Possible Site of Superoxide Generation in the Complex I Segment of Rat Heart Mitochondria. *J. Bioenergy Biomembr.* **2005**, *37*, 1–15. [CrossRef] [PubMed]
238. Korshunov, S.S.; Skulachev, V.P.; Starkov, A.A. High Protonic Potential Actuates a Mechanism of Production of Reactive Oxygen Species in Mitochondria. *FEBS Lett.* **1997**, *416*, 15–18. [CrossRef]
239. Kushnareva, Y.; Murphy, A.N.; Andreyev, A. Complex I-Mediated Reactive Oxygen Species Generation: Modulation by Cytochrome c and NAD(P)+ Oxidation-Reduction State. *Biochem. J.* **2002**, *368*, 545–553. [CrossRef] [PubMed]
240. Starkov, A.A.; Fiskum, G. Regulation of Brain Mitochondrial H2O2 Production by Membrane Potential and NAD(P)H Redox State. *J. Neurochem.* **2003**, *86*, 1101–1107. [CrossRef] [PubMed]
241. Crofts, A.R. The Cytochrome Bc1 Complex: Function in the Context of Structure. *Annu. Rev. Physiol.* **2004**, *66*, 689–733. [CrossRef]
242. Muller, F.L.; Roberts, A.G.; Bowman, M.K.; Kramer, D.M. Architecture of the Qo Site of the Cytochrome Bc1 Complex Probed by Superoxide Production. *Biochemistry* **2003**, *42*, 6493–6499. [CrossRef]
243. Jezek, P.; Hlavatá, L. Mitochondria in Homeostasis of Reactive Oxygen Species in Cell, Tissues, and Organism. *Int. J. Biochem. Cell. Biol.* **2005**, *37*, 2478–2503. [CrossRef]
244. Casteilla, L.; Rigoulet, M.; Pénicaud, L. Mitochondrial ROS Metabolism: Modulation by Uncoupling Proteins. *IUBMB Life* **2001**, *52*, 181–188. [CrossRef] [PubMed]
245. Dröse, S.; Brandt, U. The Mechanism of Mitochondrial Superoxide Production by the Cytochrome Bc1 Complex. *J. Biol. Chem.* **2008**, *283*, 21649–21654. [CrossRef]
246. Sarewicz, M.; Borek, A.; Cieluch, E.; Swierczek, M.; Osyczka, A. Discrimination between Two Possible Reaction Sequences That Create Potential Risk of Generation of Deleterious Radicals by Cytochrome Bc_1. Implications for the Mechanism of Superoxide Production. *Biochim. Biophys. Acta* **2010**, *1797*, 1820–1827. [CrossRef] [PubMed]
247. Sarewicz, M.; Bujnowicz, Ł.; Bhaduri, S.; Singh, S.K.; Cramer, W.A.; Osyczka, A. Metastable Radical State, Nonreactive with Oxygen, Is Inherent to Catalysis by Respiratory and Photosynthetic Cytochromes Bc1/B6f. *Proc. Natl. Acad. Sci. USA* **2017**, *114*, 1323–1328. [CrossRef] [PubMed]
248. Bujnowicz, Ł.; Borek, A.; Kuleta, P.; Sarewicz, M.; Osyczka, A. Suppression of Superoxide Production by a Spin-Spin Coupling between Semiquinone and the Rieske Cluster in Cytochrome Bc1. *FEBS Lett.* **2019**, *593*, 3–12. [CrossRef] [PubMed]
249. Lenaz, G. Role of Mitochondria in the Generation of Reactive Oxygen Species. In *Handbook on Reactive Oxygen Species (ROS)*; Suzuki, M., Yamamoto, S., Eds.; Nova Biomedical: Waltham, MA, USA, 2014.
250. Panov, A.; Dikalov, S.; Shalbuyeva, N.; Hemendinger, R.; Greenamyre, J.T.; Rosenfeld, J. Species- and Tissue-Specific Relationships between Mitochondrial Permeability Transition and Generation of ROS in Brain and Liver Mitochondria of Rats and Mice. *Am. J. Physiol. Cell Physiol.* **2007**, *292*, C708–C718. [CrossRef] [PubMed]
251. Tahara, E.B.; Navarete, F.D.T.; Kowaltowski, A.J. Tissue-, Substrate-, and Site-Specific Characteristics of Mitochondrial Reactive Oxygen Species Generation. *Free Radic. Biol. Med.* **2009**, *46*, 1283–1297. [CrossRef] [PubMed]
252. Skulachev, V.P. Role of Uncoupled and Non-Coupled Oxidations in Maintenance of Safely Low Levels of Oxygen and Its One-Electron Reductants. *Q. Rev. Biophys.* **1996**, *29*, 169–202. [CrossRef]
253. Brand, M.D. Uncoupling to Survive? The Role of Mitochondrial Inefficiency in Ageing. *Exp. Gerontol.* **2000**, *35*, 811–820. [CrossRef]
254. Wojtczak, L.; Lebiedzińska, M.; Suski, J.M.; Więckowski, M.R.; Schönfeld, P. Inhibition by Purine Nucleotides of the Release of Reactive Oxygen Species from Muscle Mitochondria: Indication for a Function of Uncoupling Proteins as Superoxide Anion Transporters. *Biochem. Biophys. Res. Commun.* **2011**, *407*, 772–776. [CrossRef] [PubMed]

255. Ježek, P.; Holendová, B.; Garlid, K.D.; Jabůrek, M. Mitochondrial Uncoupling Proteins: Subtle Regulators of Cellular Redox Signaling. *Antioxid. Redox Signal.* **2018**, *29*, 667–714. [CrossRef]
256. Zhao, R.-Z.; Jiang, S.; Zhang, L.; Yu, Z.-B. Mitochondrial Electron Transport Chain, ROS Generation and Uncoupling (Review). *Int. J. Mol. Med.* **2019**, *44*, 3–15. [CrossRef]
257. Demine, S.; Renard, P.; Arnould, T. Mitochondrial Uncoupling: A Key Controller of Biological Processes in Physiology and Diseases. *Cells* **2019**, *8*, 795. [CrossRef]
258. Raha, S.; Myint, A.T.; Johnstone, L.; Robinson, B.H. Control of Oxygen Free Radical Formation from Mitochondrial Complex I: Roles for Protein Kinase A and Pyruvate Dehydrogenase Kinase. *Free Radic. Biol. Med.* **2002**, *32*, 421–430. [CrossRef]
259. Maj, M.C.; Raha, S.; Myint, T.; Robinson, B.H. Regulation of NADH/CoQ Oxidoreductase: Do Phosphorylation Events Affect Activity? *Protein J.* **2004**, *23*, 25–32. [CrossRef]
260. Scacco, S.; Petruzzella, V.; Bertini, E.; Luso, A.; Papa, F.; Bellomo, F.; Signorile, A.; Torraco, A.; Papa, S. Mutations in Structural Genes of Complex I Associated with Neurological Diseases. *Ital. J. Biochem.* **2006**, *55*, 254–262. [PubMed]
261. Kadenbach, B.; Ramzan, R.; Vogt, S. High Efficiency versus Maximal Performance—The Cause of Oxidative Stress in Eukaryotes: A Hypothesis. *Mitochondrion* **2013**, *13*, 1–6. [CrossRef] [PubMed]
262. Hess, K.C.; Liu, J.; Manfredi, G.; Mühlschlegel, F.A.; Buck, J.; Levin, L.R.; Barrientos, A. A Mitochondrial CO2-Adenylyl Cyclase-CAMP Signalosome Controls Yeast Normoxic Cytochrome *c* Oxidase Activity. *FASEB J.* **2014**, *28*, 4369–4380. [CrossRef]
263. Kalpage, H.A.; Bazylianska, V.; Recanati, M.A.; Fite, A.; Liu, J.; Wan, J.; Mantena, N.; Malek, M.H.; Podgorski, I.; Heath, E.I.; et al. Tissue-Specific Regulation of Cytochrome *c* by Post-Translational Modifications: Respiration, the Mitochondrial Membrane Potential, ROS, and Apoptosis. *FASEB J.* **2019**, *33*, 1540–1553. [CrossRef]
264. Kalpage, H.A.; Vaishnav, A.; Liu, J.; Varughese, A.; Wan, J.; Turner, A.A.; Ji, Q.; Zurek, M.P.; Kapralov, A.A.; Kagan, V.E.; et al. Serine-47 Phosphorylation of Cytochrome *c* in the Mammalian Brain Regulates Cytochrome *c* Oxidase and Caspase-3 Activity. *FASEB J.* **2019**, *33*, 13503–13514. [CrossRef] [PubMed]
265. Handy, D.E.; Loscalzo, J. Redox Regulation of Mitochondrial Function. *Antioxid. Redox Signal.* **2012**, *16*, 1323–1367. [CrossRef]
266. Janssen-Heininger, Y.M.W.; Mossman, B.T.; Heintz, N.H.; Forman, H.J.; Kalyanaraman, B.; Finkel, T.; Stamler, J.S.; Rhee, S.G.; van der Vliet, A. Redox-Based Regulation of Signal Transduction: Principles, Pitfalls, and Promises. *Free Radic. Biol. Med.* **2008**, *45*, 1–17. [CrossRef]
267. Bae, Y.S.; Oh, H.; Rhee, S.G.; Yoo, Y.D. Regulation of Reactive Oxygen Species Generation in Cell Signaling. *Mol. Cell.* **2011**, *32*, 491–509. [CrossRef] [PubMed]
268. Rigoulet, M.; Yoboue, E.D.; Devin, A. Mitochondrial ROS Generation and Its Regulation: Mechanisms Involved in H(2)O(2) Signaling. *Antioxid. Redox Signal.* **2011**, *14*, 459–468. [CrossRef]
269. Gough, D.R.; Cotter, T.G. Hydrogen Peroxide: A Jekyll and Hyde Signalling Molecule. *Cell Death Dis.* **2011**, *2*, e213. [CrossRef] [PubMed]
270. Wang, Y.; Yang, J.; Yi, J. Redox Sensing by Proteins: Oxidative Modifications on Cysteines and the Consequent Events. *Antioxid. Redox Signal.* **2012**, *16*, 649–657. [CrossRef]
271. Murphy, M.P. Mitochondrial Thiols in Antioxidant Protection and Redox Signaling: Distinct Roles for Glutathionylation and Other Thiol Modifications. *Antioxid. Redox Signal.* **2012**, *16*, 476–495. [CrossRef]
272. Reczek, C.R.; Chandel, N.S. ROS-Dependent Signal Transduction. *Curr. Opin. Cell. Biol.* **2015**, *33*, 8–13. [CrossRef]
273. Schieber, M.; Chandel, N.S. ROS Function in Redox Signaling and Oxidative Stress. *Curr. Biol.* **2014**, *24*, R453–R462. [CrossRef]
274. Sies, H.; Jones, D.P. Reactive Oxygen Species (ROS) as Pleiotropic Physiological Signalling Agents. *Nat. Rev. Mol. Cell. Biol.* **2020**, *21*, 363–383. [CrossRef]
275. Wu, W.-S. The Signaling Mechanism of ROS in Tumor Progression. *Cancer Metast. Rev.* **2006**, *25*, 695–705. [CrossRef]
276. Moloney, J.N.; Cotter, T.G. ROS Signalling in the Biology of Cancer. *Semin. Cell Dev. Biol.* **2018**, *80*, 50–64. [CrossRef]
277. D'Autréaux, B.; Toledano, M.B. ROS as Signalling Molecules: Mechanisms That Generate Specificity in ROS Homeostasis. *Nat. Rev. Mol. Cell. Biol.* **2007**, *8*, 813–824. [CrossRef]
278. Bak, D.W.; Weerapana, E. Cysteine-Mediated Redox Signalling in the Mitochondria. *Mol. BioSyst.* **2015**, *11*, 678–697. [CrossRef]
279. Haddad, J.J. Antioxidant and Prooxidant Mechanisms in the Regulation of Redox(y)-Sensitive Transcription Factors. *Cell Signal* **2002**, *14*, 879–897. [CrossRef]
280. Trachootham, D.; Lu, W.; Ogasawara, M.A.; Nilsa, R.-D.V.; Huang, P. Redox Regulation of Cell Survival. *Antioxid. Redox Signal.* **2008**, *10*, 1343–1374. [CrossRef] [PubMed]
281. Murphy, M.P. How Mitochondria Produce Reactive Oxygen Species. *Biochem. J.* **2009**, *417*, 1–13. [CrossRef]
282. Whelan, S.P.; Zuckerbraun, B.S. Mitochondrial Signaling: Forwards, Backwards, and in Between. *Oxid. Med. Cell Longev.* **2013**, *2013*, 351613. [CrossRef]
283. Itoh, K.; Wakabayashi, N.; Katoh, Y.; Ishii, T.; Igarashi, K.; Engel, J.D.; Yamamoto, M. Keap1 Represses Nuclear Activation of Antioxidant Responsive Elements by Nrf2 through Binding to the Amino-Terminal Neh2 Domain. *Genes Dev.* **1999**, *13*, 76–86. [CrossRef]
284. Valko, M.; Rhodes, C.J.; Moncol, J.; Izakovic, M.; Mazur, M. Free Radicals, Metals and Antioxidants in Oxidative Stress-Induced Cancer. *Chem. Biol. Interact.* **2006**, *160*, 1–40. [CrossRef]
285. Semenza, G.L. Oxygen Sensing, Homeostasis, and Disease. *N. Engl. J. Med.* **2011**, *365*, 537–547. [CrossRef]

286. Chen, R.; Lai, U.H.; Zhu, L.; Singh, A.; Ahmed, M.; Forsyth, N.R. Reactive Oxygen Species Formation in the Brain at Different Oxygen Levels: The Role of Hypoxia Inducible Factors. *Front. Cell Dev. Biol.* **2018**, *6*. [CrossRef]
287. Cadenas, S. ROS and Redox Signaling in Myocardial Ischemia-Reperfusion Injury and Cardioprotection. *Free Radic. Biol. Med.* **2018**, *117*, 76–89. [CrossRef]
288. Finkel, T. Signal Transduction by Mitochondrial Oxidants. *J. Biol. Chem.* **2012**, *287*, 4434–4440. [CrossRef] [PubMed]
289. Pelletier, M.; Lepow, T.S.; Billingham, L.K.; Murphy, M.P.; Siegel, R.M. New Tricks from an Old Dog: Mitochondrial Redox Signaling in Cellular Inflammation. *Semin. Immunol.* **2012**, *24*, 384–392. [CrossRef] [PubMed]
290. Sho, T.; Xu, J. Role and Mechanism of ROS Scavengers in Alleviating NLRP3-Mediated Inflammation. *Biotechnol. Appl. Biochem.* **2019**, *66*, 4–13. [CrossRef]
291. Scherz-Shouval, R.; Elazar, Z. Regulation of Autophagy by ROS: Physiology and Pathology. *Trends Biochem. Sci.* **2011**, *36*, 30–38. [CrossRef]
292. Hoffman, D.L.; Brookes, P.S. Oxygen Sensitivity of Mitochondrial Reactive Oxygen Species Generation Depends on Metabolic Conditions. *J. Biol. Chem.* **2009**, *284*, 16236–16245. [CrossRef]
293. Kim, J.J.; Lee, S.B.; Park, J.K.; Yoo, Y.D. TNF-Alpha-Induced ROS Production Triggering Apoptosis Is Directly Linked to Romo1 and Bcl-X(L). *Cell Death Differ* **2010**, *17*, 1420–1434. [CrossRef] [PubMed]
294. Ns, C.; Ds, M.; Ce, F.; Tm, W.; Ja, M.; Am, R.; Pt, S. Reactive Oxygen Species Generated at Mitochondrial Complex III Stabilize Hypoxia-Inducible Factor-1alpha during Hypoxia: A Mechanism of O2 Sensing. *J. Biol. Chem.* **2000**, *275*, 25130–25138. [CrossRef]
295. Lee, S.B.; Kim, J.J.; Kim, T.W.; Kim, B.S.; Lee, M.-S.; Yoo, Y.D. Serum Deprivation-Induced Reactive Oxygen Species Production Is Mediated by Romo1. *Apoptosis* **2010**, *15*, 204–218. [CrossRef] [PubMed]
296. Zorov, D.B.; Juhaszova, M.; Sollott, S.J. Mitochondrial ROS-Induced ROS Release: An Update and Review. *Biochim. Biophys. Acta* **2006**, *1757*, 509–517. [CrossRef] [PubMed]
297. Dikalov, S. Crosstalk between Mitochondria and NADPH Oxidases. *Free Radic. Biol. Med.* **2011**, *51*, 1289–1301. [CrossRef]
298. Veith, C.; Boots, A.W.; Idris, M.; van Schooten, F.-J.; van der Vliet, A. Redox Imbalance in Idiopathic Pulmonary Fibrosis: A Role for Oxidant Cross-Talk Between NADPH Oxidase Enzymes and Mitochondria. *Antioxid. Redox Signal.* **2019**, *31*, 1092–1115. [CrossRef] [PubMed]
299. Koju, N.; Taleb, A.; Zhou, J.; Lv, G.; Yang, J.; Cao, X.; Lei, H.; Ding, Q. Pharmacological Strategies to Lower Crosstalk between Nicotinamide Adenine Dinucleotide Phosphate (NADPH) Oxidase and Mitochondria. *Biomed. Pharmacother.* **2019**, *111*, 1478–1498. [CrossRef]
300. Fourquet, S.; Huang, M.-E.; D'Autreaux, B.; Toledano, M.B. The Dual Functions of Thiol-Based Peroxidases in H2O2 Scavenging and Signaling. *Antioxid. Redox Signal.* **2008**, *10*, 1565–1576. [CrossRef] [PubMed]
301. Rhee, S.G.; Kil, I.S. Multiple Functions and Regulation of Mammalian Peroxiredoxins. *Annu. Rev. Biochem.* **2017**, *86*, 749–775. [CrossRef]
302. Rhee, S.G.; Chae, H.Z.; Kim, K. Peroxiredoxins: A Historical Overview and Speculative Preview of Novel Mechanisms and Emerging Concepts in Cell Signaling. *Free Radic. Biol. Med.* **2005**, *38*, 1543–1552. [CrossRef]
303. Lillig, C.H.; Holmgren, A. Thioredoxin and Related Molecules—From Biology to Health and Disease. *Antioxid. Redox Signal.* **2007**, *9*, 25–47. [CrossRef] [PubMed]
304. Cox, A.G.; Winterbourn, C.C.; Hampton, M.B. Mitochondrial Peroxiredoxin Involvement in Antioxidant Defence and Redox Signalling. *Biochem. J.* **2009**, *425*, 313–325. [CrossRef] [PubMed]
305. Calabrese, V.; Scapagnini, G.; Ravagna, A.; Colombrita, C.; Spadaro, F.; Butterfield, D.A.; Giuffrida Stella, A.M. Increased Expression of Heat Shock Proteins in Rat Brain during Aging: Relationship with Mitochondrial Function and Glutathione Redox State. *Mech. Ageing Dev.* **2004**, *125*, 325–335. [CrossRef] [PubMed]
306. Calabrese, V.; Cornelius, C.; Leso, V.; Trovato-Salinaro, A.; Ventimiglia, B.; Cavallaro, M.; Scuto, M.; Rizza, S.; Zanoli, L.; Neri, S.; et al. Oxidative Stress, Glutathione Status, Sirtuin and Cellular Stress Response in Type 2 Diabetes. *Biochim. Biophys. Acta Mol. Basis Dis.* **2012**, *1822*, 729–736. [CrossRef]
307. Calabrese, G.; Morgan, B.; Riemer, J. Mitochondrial Glutathione: Regulation and Functions. *Antioxid. Redox Signal.* **2017**, *27*, 1162–1177. [CrossRef] [PubMed]
308. Jones, D.P. Redox Potential of GSH/GSSG Couple: Assay and Biological Significance. *Methods Enzymol.* **2002**, *348*, 93–112. [CrossRef] [PubMed]
309. Yin, F.; Sancheti, H.; Cadenas, E. Mitochondrial Thiols in the Regulation of Cell Death Pathways. *Antioxid. Redox Signal.* **2012**, *17*, 1714–1727. [CrossRef]
310. Dalle-Donne, I.; Colombo, G.; Gagliano, N.; Colombo, R.; Giustarini, D.; Rossi, R.; Milzani, A. S-Glutathiolation in Life and Death Decisions of the Cell. *Free Radic. Res.* **2011**, *45*, 3–15. [CrossRef] [PubMed]
311. Allen, E.M.G.; Mieyal, J.J. Protein-Thiol Oxidation and Cell Death: Regulatory Role of Glutaredoxins. *Antioxid. Redox Signal.* **2012**, *17*, 1748–1763. [CrossRef]
312. Young, A.; Gill, R.; Mailloux, R.J. Protein S-Glutathionylation: The Linchpin for the Transmission of Regulatory Information on Redox Buffering Capacity in Mitochondria. *Chem. Biol. Interact.* **2019**, *299*, 151–162. [CrossRef]
313. Smith, K.A.; Waypa, G.B.; Schumacker, P.T. Redox Signaling during Hypoxia in Mammalian Cells. *Redox Biol.* **2017**, *13*, 228–234. [CrossRef]

314. Bell, E.L.; Klimova, T.A.; Eisenbart, J.; Schumacker, P.T.; Chandel, N.S. Mitochondrial Reactive Oxygen Species Trigger Hypoxia-Inducible Factor-Dependent Extension of the Replicative Life Span during Hypoxia. *Mol. Cell. Biol.* **2007**, *27*, 5737–5745. [CrossRef]
315. Grishko, V.; Solomon, M.; Breit, J.F.; Killilea, D.W.; Ledoux, S.P.; Wilson, G.L.; Gillespie, M.N. Hypoxia Promotes Oxidative Base Modifications in the Pulmonary Artery Endothelial Cell VEGF Gene. *FASEB J.* **2001**, *15*, 1267–1269. [CrossRef] [PubMed]
316. Chandel, N.S.; Maltepe, E.; Goldwasser, E.; Mathieu, C.E.; Simon, M.C.; Schumacker, P.T. Mitochondrial Reactive Oxygen Species Trigger Hypoxia-Induced Transcription. *Proc. Natl. Acad. Sci. USA* **1998**, *95*, 11715–11720. [CrossRef] [PubMed]
317. Kietzmann, T.; Görlach, A. Reactive Oxygen Species in the Control of Hypoxia-Inducible Factor-Mediated Gene Expression. *Semin. Cell Dev. Biol.* **2005**, *16*, 474–486. [CrossRef]
318. Movafagh, S.; Crook, S.; Vo, K. Regulation of Hypoxia-Inducible Factor-1a by Reactive Oxygen Species: New Developments in an Old Debate. *J. Cell Biochem.* **2015**, *116*, 696–703. [CrossRef] [PubMed]
319. Kaluz, S.; Kaluzová, M.; Stanbridge, E.J. Rational Design of Minimal Hypoxia-Inducible Enhancers. *Biochem. Biophys. Res. Commun.* **2008**, *370*, 613–618. [CrossRef] [PubMed]
320. Cai, J.; Jones, D.P. Superoxide in Apoptosis. Mitochondrial Generation Triggered by Cytochrome c Loss. *J. Biol. Chem.* **1998**, *273*, 11401–11404. [CrossRef] [PubMed]
321. Pasdois, P.; Parker, J.E.; Griffiths, E.J.; Halestrap, A.P. The Role of Oxidized Cytochrome c in Regulating Mitochondrial Reactive Oxygen Species Production and Its Perturbation in Ischaemia. *Biochem. J.* **2011**, *436*, 493–505. [CrossRef]
322. Chen, Q.; Moghaddas, S.; Hoppel, C.L.; Lesnefsky, E.J. Ischemic Defects in the Electron Transport Chain Increase the Production of Reactive Oxygen Species from Isolated Rat Heart Mitochondria. *Am. J. Physiol. Cell Physiol.* **2008**, *294*, C460–C466. [CrossRef]
323. Chandel, N.; Budinger, G.R.; Kemp, R.A.; Schumacker, P.T. Inhibition of Cytochrome-c Oxidase Activity during Prolonged Hypoxia. *Am. J. Physiol.* **1995**, *268*, L918–L925. [CrossRef]
324. Zuckerbraun, B.S.; Chin, B.Y.; Bilban, M.; d'Avila, J.C.; de Rao, J.; Billiar, T.R.; Otterbein, L.E. Carbon Monoxide Signals via Inhibition of Cytochrome *c* Oxidase and Generation of Mitochondrial Reactive Oxygen Species. *FASEB J.* **2007**, *21*, 1099–1106. [CrossRef]
325. Mansfield, K.D.; Guzy, R.D.; Pan, Y.; Young, R.M.; Cash, T.P.; Schumacker, P.T.; Simon, M.C. Mitochondrial Dysfunction Resulting from Loss of Cytochrome c Impairs Cellular Oxygen Sensing and Hypoxic HIF-Alpha Activation. *Cell Metab.* **2005**, *1*, 393–399. [CrossRef] [PubMed]
326. Guzy, R.D.; Hoyos, B.; Robin, E.; Chen, H.; Liu, L.; Mansfield, K.D.; Simon, M.C.; Hammerling, U.; Schumacker, P.T. Mitochondrial Complex III Is Required for Hypoxia-Induced ROS Production and Cellular Oxygen Sensing. *Cell Metab.* **2005**, *1*, 401–408. [CrossRef]
327. Guzy, R.D.; Schumacker, P.T. Oxygen Sensing by Mitochondria at Complex III: The Paradox of Increased Reactive Oxygen Species during Hypoxia. *Exp. Physiol.* **2006**, *91*, 807–819. [CrossRef]
328. Korde, A.S.; Yadav, V.R.; Zheng, Y.-M.; Wang, Y.-X. Primary Role of Mitochondrial Rieske Iron-Sulfur Protein in Hypoxic ROS Production in Pulmonary Artery Myocytes. *Free Radic. Biol. Med.* **2011**, *50*, 945–952. [CrossRef]
329. Hernansanz-Agustín, P.; Ramos, E.; Navarro, E.; Parada, E.; Sánchez-López, N.; Peláez-Aguado, L.; Cabrera-García, J.D.; Tello, D.; Buendia, I.; Marina, A.; et al. Mitochondrial Complex I Deactivation Is Related to Superoxide Production in Acute Hypoxia. *Redox Biol.* **2017**, *12*, 1040–1051. [CrossRef]
330. Mittal, M.; Gu, X.Q.; Pak, O.; Pamenter, M.E.; Haag, D.; Fuchs, D.B.; Schermuly, R.T.; Ghofrani, H.A.; Brandes, R.P.; Seeger, W.; et al. Hypoxia Induces Kv Channel Current Inhibition by Increased NADPH Oxidase-Derived Reactive Oxygen Species. *Free Radic. Biol. Med.* **2012**, *52*, 1033–1042. [CrossRef] [PubMed]
331. Sauer, R.T.; Baker, T.A. AAA+ Proteases: ATP-Fueled Machines of Protein Destruction. *Annu. Rev. Biochem.* **2011**, *80*, 587–612. [CrossRef]
332. Tomko, R.J.; Hochstrasser, M. Molecular Architecture and Assembly of the Eukaryotic Proteasome. *Annu. Rev. Biochem.* **2013**, *82*, 415–445. [CrossRef] [PubMed]
333. Luce, K.; Weil, A.C.; Osiewacz, H.D. Mitochondrial Protein Quality Control Systems in Aging and Disease. *Adv. Exp. Med. Biol.* **2010**, *694*, 108–125. [CrossRef]
334. Taylor, E.B.; Rutter, J. Mitochondrial Quality Control by the Ubiquitin-Proteasome System. *Biochem. Soc. Trans.* **2011**, *39*, 1509–1513. [CrossRef]
335. Bragoszewski, P.; Turek, M.; Chacinska, A. Control of Mitochondrial Biogenesis and Function by the Ubiquitin-Proteasome System. *Open Biol.* **2017**, *7*. [CrossRef]
336. Yu, T.; Zhang, Y.; Li, P.-F. Mitochondrial Ubiquitin Ligase in Cardiovascular Disorders. *Adv. Exp. Med. Biol.* **2017**, *982*, 327–333. [CrossRef]
337. Fukuda, R.; Zhang, H.; Kim, J.; Shimoda, L.; Dang, C.V.; Semenza, G.L. HIF-1 Regulates Cytochrome Oxidase Subunits to Optimize Efficiency of Respiration in Hypoxic Cells. *Cell* **2007**, *129*, 111–122. [CrossRef]
338. Venkatesh, S.; Lee, J.; Singh, K.; Lee, I.; Suzuki, C.K. Multitasking in the Mitochondrion by the ATP-Dependent Lon Protease. *Biochim. Biophys. Acta* **2012**, *1823*, 56–66. [CrossRef] [PubMed]
339. Bulteau, A.-L.; Szweda, L.I.; Friguet, B. Mitochondrial Protein Oxidation and Degradation in Response to Oxidative Stress and Aging. *Exp. Gerontol.* **2006**, *41*, 653–657. [CrossRef]

340. Ugarte, N.; Petropoulos, I.; Friguet, B. Oxidized Mitochondrial Protein Degradation and Repair in Aging and Oxidative Stress. *Antioxid. Redox Signal.* **2010**, *13*, 539–549. [CrossRef]
341. Levine, B.; Klionsky, D.J. Development by Self-Digestion: Molecular Mechanisms and Biological Functions of Autophagy. *Dev. Cell* **2004**, *6*, 463–477. [CrossRef]
342. Bergamini, E. Autophagy: A Cell Rep.air Mechanism That Retards Ageing and Age-Associated Diseases and Can Be Intensified Pharmacologically. *Mol. Asp. Med.* **2006**, *27*, 403–410. [CrossRef]
343. Chen, Y.; Gibson, S.B. Is Mitochondrial Generation of Reactive Oxygen Species a Trigger for Autophagy? *Autophagy* **2008**, *4*, 246–248. [CrossRef] [PubMed]
344. Lee, J.; Giordano, S.; Zhang, J. Autophagy, Mitochondria and Oxidative Stress: Cross-Talk and Redox Signalling. *Biochem. J.* **2012**, *441*, 523–540. [CrossRef] [PubMed]
345. Fan, P.; Xie, X.-H.; Chen, C.-H.; Peng, X.; Zhang, P.; Yang, C.; Wang, Y.-T. Molecular Regulation Mechanisms and Interactions Between Reactive Oxygen Species and Mitophagy. *DNA Cell Biol.* **2019**, *38*, 10–22. [CrossRef] [PubMed]
346. Frank, M.; Duvezin-Caubet, S.; Koob, S.; Occhipinti, A.; Jagasia, R.; Petcherski, A.; Ruonala, M.O.; Priault, M.; Salin, B.; Reichert, A.S. Mitophagy Is Triggered by Mild Oxidative Stress in a Mitochondrial Fission Dependent Manner. *Biochim. Biophys. Acta* **2012**, *1823*, 2297–2310. [CrossRef]
347. Schägger, H.; de Coo, R.; Bauer, M.F.; Hofmann, S.; Godinot, C.; Brandt, U. Significance of Respirasomes for the Assembly/Stability of Human Respiratory Chain Complex I. *J. Biol. Chem.* **2004**, *279*, 36349–36353. [CrossRef]
348. Acín-Pérez, R.; Bayona-Bafaluy, M.P.; Fernández-Silva, P.; Moreno-Loshuertos, R.; Pérez-Martos, A.; Bruno, C.; Moraes, C.T.; Enríquez, J.A. Respiratory Complex III Is Required to Maintain Complex I in Mammalian Mitochondria. *Mol. Cell* **2004**, *13*, 805–815. [CrossRef]
349. D'Aurelio, M.; Gajewski, C.D.; Lenaz, G.; Manfredi, G. Respiratory Chain Supercomplexes Set the Threshold for Respiration Defects in Human MtDNA Mutant Cybrids. *Hum. Mol. Genet.* **2006**, *15*, 2157–2169. [CrossRef] [PubMed]
350. Diaz, F.; Garcia, S.; Padgett, K.R.; Moraes, C.T. A Defect in the Mitochondrial Complex III, but Not Complex IV, Triggers Early ROS-Dependent Damage in Defined Brain Regions. *Hum. Mol. Genet.* **2012**, *21*, 5066–5077. [CrossRef] [PubMed]
351. Protasoni, M.; Pérez-Pérez, R.; Lobo-Jarne, T.; Harbour, M.E.; Ding, S.; Peñas, A.; Diaz, F.; Moraes, C.T.; Fearnley, I.M.; Zeviani, M.; et al. Respiratory Supercomplexes Act as a Platform for Complex III-Mediated Maturation of Human Mitochondrial Complexes I and IV. *EMBO J.* **2020**, *39*, e102817. [CrossRef] [PubMed]
352. Diaz, F.; Enríquez, J.A.; Moraes, C.T. Cells Lacking Rieske Iron-Sulfur Protein Have a Reactive Oxygen Species-Associated Decrease in Respiratory Complexes I and IV. *Mol. Cell Biol.* **2012**, *32*, 415–429. [CrossRef] [PubMed]
353. Paradies, G.; Petrosillo, G.; Pistolese, M.; Ruggiero, F.M. Reactive Oxygen Species Affect Mitochondrial Electron Transport Complex I Activity through Oxidative Cardiolipin Damage. *Gene* **2002**, *286*, 135–141. [CrossRef]
354. Jang, S.; Lewis, T.S.; Powers, C.; Khuchua, Z.; Baines, C.P.; Wipf, P.; Javadov, S. Elucidating Mitochondrial Electron Transport Chain Supercomplexes in the Heart During Ischemia-Reperfusion. *Antioxid. Redox Signal.* **2017**, *27*, 57–69. [CrossRef]
355. Paradies, G.; Petrosillo, G.; Pistolese, M.; Ruggiero, F.M. Reactive Oxygen Species Generated by the Mitochondrial Respiratory Chain Affect the Complex III Activity via Cardiolipin Peroxidation in Beef-Heart Submitochondrial Particles. *Mitochondrion* **2001**, *1*, 151–159. [CrossRef]
356. Petrosillo, G.; Matera, M.; Casanova, G.; Ruggiero, F.M.; Paradies, G. Mitochondrial Dysfunction in Rat Brain with Aging Involvement of Complex I, Reactive Oxygen Species and Cardiolipin. *Neurochem. Int.* **2008**, *53*, 126–131. [CrossRef]
357. Petrosillo, G.; Matera, M.; Moro, N.; Ruggiero, F.M.; Paradies, G. Mitochondrial Complex I Dysfunction in Rat Heart with Aging: Critical Role of Reactive Oxygen Species and Cardiolipin. *Free Radic. Biol. Med.* **2009**, *46*, 88–94. [CrossRef]
358. Dencher, N.A.; Frenzel, M.; Reifschneider, N.H.; Sugawa, M.; Krause, F. Proteome Alterations in Rat Mitochondria Caused by Aging. *Ann. N. Y. Acad. Sci.* **2007**, *1100*, 291–298. [CrossRef]
359. Schäfer, E.; Dencher, N.A.; Vonck, J.; Parcej, D.N. Three-Dimensional Structure of the Respiratory Chain Supercomplex I1III2IV1 from Bovine Heart Mitochondria. *Biochemistry* **2007**, *46*, 12579–12585. [CrossRef] [PubMed]
360. Peters, K.; Dudkina, N.V.; Jänsch, L.; Braun, H.-P.; Boekema, E.J. A Structural Investigation of Complex I and I+III2 Supercomplex from Zea Mays at 11-13 A Resolution: Assignment of the Carbonic Anhydrase Domain and Evidence for Structural Heterogeneity within Complex I. *Biochim. Biophys. Acta* **2008**, *1777*, 84–93. [CrossRef]
361. Radermacher, M.; Ruiz, T.; Clason, T.; Benjamin, S.; Brandt, U.; Zickermann, V. The Three-Dimensional Structure of Complex I from Yarrowia Lipolytica: A Highly Dynamic Enzyme. *J. Struct. Biol.* **2006**, *154*, 269–279. [CrossRef] [PubMed]
362. Maranzana, E.; Barbero, G.; Falasca, A.I.; Lenaz, G.; Genova, M.L. Mitochondrial Respiratory Supercomplex Association Limits Production of Reactive Oxygen Species from Complex i. *Antioxid. Redox Signal.* **2013**, *19*, 1469–1480. [CrossRef]
363. Lopez-Fabuel, I.; Le Douce, J.; Logan, A.; James, A.M.; Bonvento, G.; Murphy, M.P.; Almeida, A.; Bolaños, J.P. Complex I Assembly into Supercomplexes Determines Differential Mitochondrial ROS Production in Neurons and Astrocytes. *Proc. Natl. Acad. Sci. USA* **2016**, *113*, 13063–13068. [CrossRef] [PubMed]
364. Vicente-Gutiérrez, C.; Jiménez-Blasco, D.; Quintana-Cabrera, R. Intertwined ROS and Metabolic Signaling at the Neuron-Astrocyte Interface. *Neurochem. Res.* **2020**. [CrossRef]
365. Reyes-Galindo, M.; Suarez, R.; Esparza-Perusquía, M.; de Lira-Sánchez, J.; Pardo, J.P.; Martínez, F.; Flores-Herrera, O. Mitochondrial Respirasome Works as a Single Unit and the Cross-Talk between Complexes I, III2 and IV Stimulates NADH Dehydrogenase Activity. *Biochim. Biophys. Acta Bioenergy* **2019**, *1860*, 618–627. [CrossRef]

366. Lenaz, G.; Baracca, A.; Barbero, G.; Bergamini, C.; Dalmonte, M.E.; Del Sole, M.; Faccioli, M.; Falasca, A.; Fato, R.; Genova, M.L.; et al. Mitochondrial Respiratory Chain Super-Complex I-III in Physiology and Pathology. *Biochim. Biophys. Acta* **2010**, *1797*, 633–640. [CrossRef]
367. Huertas, J.R.; Al Fazazi, S.; Hidalgo-Gutierrez, A.; López, L.C.; Casuso, R.A. Antioxidant Effect of Exercise: Exploring the Role of the Mitochondrial Complex I Superassembly. *Redox Biol.* **2017**, *13*, 477–481. [CrossRef] [PubMed]
368. Chen, Y.-C.; Taylor, E.B.; Dephoure, N.; Heo, J.-M.; Tonhato, A.; Papandreou, I.; Nath, N.; Denko, N.C.; Gygi, S.P.; Rutter, J. Identification of a Protein Mediating Respiratory Supercomplex Stability. *Cell Metab.* **2012**, *15*, 348–360. [CrossRef] [PubMed]
369. Strogolova, V.; Furness, A.; Robb-McGrath, M.; Garlich, J.; Stuart, R.A. Rcf1 and Rcf2, Members of the Hypoxia-Induced Gene 1 Protein Family, Are Critical Components of the Mitochondrial Cytochrome Bc1-Cytochrome c Oxidase Supercomplex. *Mol. Cell Biol.* **2012**, *32*, 1363–1373. [CrossRef]
370. Winge, D.R. Sealing the Mitochondrial Respirasome. *Mol. Cell Biol.* **2012**, *32*, 2647–2652. [CrossRef]
371. Azuma, K.; Ikeda, K.; Inoue, S. Functional Mechanisms of Mitochondrial Respiratory Chain Supercomplex Assembly Factors and Their Involvement in Muscle Quality. *Int. J. Mol. Sci.* **2020**, *21*, 3182. [CrossRef]
372. Pfeiffer, K.; Gohil, V.; Stuart, R.A.; Hunte, C.; Brandt, U.; Greenberg, M.L.; Schägger, H. Cardiolipin Stabilizes Respiratory Chain Supercomplexes. *J. Biol. Chem.* **2003**, *278*, 52873–52880. [CrossRef] [PubMed]
373. Zhang, M.; Mileykovskaya, E.; Dowhan, W. Gluing the Respiratory Chain Together. Cardiolipin Is Required for Supercomplex Formation in the Inner Mitochondrial Membrane. *J. Biol. Chem.* **2002**, *277*, 43553–43556. [CrossRef] [PubMed]
374. Stein, C.S.; Jadiya, P.; Zhang, X.; McLendon, J.M.; Abouassaly, G.M.; Witmer, N.H.; Anderson, E.J.; Elrod, J.W.; Boudreau, R.L. Mitoregulin: A LncRNA-Encoded Microprotein That Supports Mitochondrial Supercomplexes and Respiratory Efficiency. *Cell Rep.* **2018**, *23*, 3710–3720.e8. [CrossRef]
375. Hou, T.; Zhang, R.; Jian, C.; Ding, W.; Wang, Y.; Ling, S.; Ma, Q.; Hu, X.; Cheng, H.; Wang, X. NDUFAB1 Confers Cardio-Protection by Enhancing Mitochondrial Bioenergetics through Coordination of Respiratory Complex and Supercomplex Assembly. *Cell Res.* **2019**, *29*, 754–766. [CrossRef] [PubMed]
376. Jang, S.; Javadov, S. Elucidating the Contribution of ETC Complexes I and II to the Respirasome Formation in Cardiac Mitochondria. *Sci. Rep.* **2018**, *8*, 17732. [CrossRef] [PubMed]
377. Ramírez-Camacho, I.; Correa, F.; El Hafidi, M.; Silva-Palacios, A.; Ostolga-Chavarría, M.; Esparza-Perusquía, M.; Olvera-Sánchez, S.; Flores-Herrera, O.; Zazueta, C. Cardioprotective Strategies Preserve the Stability of Respiratory Chain Supercomplexes and Reduce Oxidative Stress in Reperfused Ischemic Hearts. *Free Radic. Biol. Med.* **2018**, *129*, 407–417. [CrossRef]
378. Ramírez-Camacho, I.; García-Niño, W.R.; Flores-García, M.; Pedraza-Chaverri, J.; Zazueta, C. Alteration of Mitochondrial Supercomplexes Assembly in Metabolic Diseases. *Biochim. Biophys. Acta Mol. Basis Dis.* **2020**, *1866*, 165935. [CrossRef] [PubMed]
379. Dautant, A.; Meier, T.; Hahn, A.; Tribouillard-Tanvier, D.; di Rago, J.-P.; Kucharczyk, R. ATP Synthase Diseases of Mitochondrial Genetic Origin. *Front. Physiol.* **2018**, *9*, 329. [CrossRef]
380. Nicholls, D.G.; Ferguson, S.J. 4-The Chemiosmotic Proton Circuit in Isolated Organelles: Theory and Practice. In *Bioenergetics*, 4th ed.; Academic Press: Boston, MA, USA, 2013; pp. 53–87, ISBN 978-0-12-388425-1.
381. Nicholls, D.G. The Physiological Regulation of Uncoupling Proteins. *Biochim. Biophys. Acta* **2006**, *1757*, 459–466. [CrossRef]
382. Bertholet, A.M.; Chouchani, E.T.; Kazak, L.; Angelin, A.; Fedorenko, A.; Long, J.Z.; Vidoni, S.; Garrity, R.; Cho, J.; Terada, N.; et al. H+ Transport Is an Integral Function of the Mitochondrial ADP/ATP Carrier. *Nature* **2019**, *571*, 515–520. [CrossRef] [PubMed]
383. Feniouk, B.A.; Mulkidjanian, A.Y.; Junge, W. Proton Slip in the ATP Synthase of Rhodobacter Capsulatus: Induction, Proton Conduction, and Nucleotide Dependence. *Biochim. Biophys. Acta* **2005**, *1706*, 184–194. [CrossRef] [PubMed]
384. Uziel, G.; Moroni, I.; Lamantea, E.; Fratta, G.M.; Ciceri, E.; Carrara, F.; Zeviani, M. Mitochondrial Disease Associated with the T8993G Mutation of the Mitochondrial ATPase 6 Gene: A Clinical, Biochemical, and Molecular Study in Six Families. *J. Neurol. Neurosurg. Psychiatry* **1997**, *63*, 16–22. [CrossRef] [PubMed]
385. Jonckheere, A.I.; Smeitink, J.A.M.; Rodenburg, R.J.T. Mitochondrial ATP Synthase: Architecture, Function and Pathology. *J. Inherit. Metab. Dis.* **2012**, *35*, 211–225. [CrossRef]
386. Trounce, I.; Neill, S.; Wallace, D.C. Cytoplasmic Transfer of the MtDNA Nt 8993 T–>G (ATP6) Point Mutation Associated with Leigh Syndrome into MtDNA-Less Cells Demonstrates Cosegregation with a Decrease in State III Respiration and ADP/O Ratio. *Proc. Natl. Acad. Sci. USA* **1994**, *91*, 8334–8338. [CrossRef]
387. Sgarbi, G.; Baracca, A.; Lenaz, G.; Valentino, L.M.; Carelli, V.; Solaini, G. Inefficient Coupling between Proton Transport and ATP Synthesis May Be the Pathogenic Mechanism for NARP and Leigh Syndrome Resulting from the T8993G Mutation in MtDNA. *Biochem. J.* **2006**, *395*, 493–500. [CrossRef] [PubMed]
388. Kucharczyk, R.; Dautant, A.; Godard, F.; Tribouillard-Tanvier, D.; di Rago, J.-P. Functional Investigation of an Universally Conserved Leucine Residue in Subunit a of ATP Synthase Targeted by the Pathogenic m.9176 T>G Mutation. *Biochim. Biophys. Acta Bioenergy* **2019**, *1860*, 52–59. [CrossRef]
389. Kucharczyk, R.; Rak, M.; di Rago, J.-P. Biochemical Consequences in Yeast of the Human Mitochondrial DNA 8993T>C Mutation in the ATPase6 Gene Found in NARP/MILS Patients. *Biochim. Biophys. Acta* **2009**, *1793*, 817–824. [CrossRef]
390. Baracca, A.; Sgarbi, G.; Mattiazzi, M.; Casalena, G.; Pagnotta, E.; Valentino, M.L.; Moggio, M.; Lenaz, G.; Carelli, V.; Solaini, G. Biochemical Phenotypes Associated with the Mitochondrial ATP6 Gene Mutations at Nt8993. *Biochim. Biophys. Acta* **2007**, *1767*, 913–919. [CrossRef]

391. Schmidt, T.; Situ, A.J.; Ulmer, T.S. Structural and Thermodynamic Basis of Proline-Induced Transmembrane Complex Stabilization. *Sci. Rep.* **2016**, *6*, 29809. [CrossRef] [PubMed]
392. Burrage, L.C.; Tang, S.; Wang, J.; Donti, T.R.; Walkiewicz, M.; Luchak, J.M.; Chen, L.-C.; Schmitt, E.S.; Niu, Z.; Erana, R.; et al. Mitochondrial Myopathy, Lactic Acidosis, and Sideroblastic Anemia (MLASA) plus Associated with a Novel de Novo Mutation (m.8969G>A) in the Mitochondrial Encoded ATP6 Gene. *Mol. Genet. Metab.* **2014**, *113*, 207–212. [CrossRef] [PubMed]
393. Skoczeń, N.; Dautant, A.; Binko, K.; Godard, F.; Bouhier, M.; Su, X.; Lasserre, J.-P.; Giraud, M.-F.; Tribouillard-Tanvier, D.; Chen, H.; et al. Molecular Basis of Diseases Caused by the MtDNA Mutation m.8969G>A in the Subunit a of ATP Synthase. *Biochim. Biophys. Acta* **2018**, *1859*, 602–611. [CrossRef] [PubMed]
394. Shadel, G.S.; Horvath, T.L. Mitochondrial ROS Signaling in Organismal Homeostasis. *Cell* **2015**, *163*, 560–569. [CrossRef] [PubMed]
395. Brand, M.D. Mitochondrial Generation of Superoxide and Hydrogen Peroxide as the Source of Mitochondrial Redox Signaling. *Free Radic. Biol. Med.* **2016**, *100*, 14–31. [CrossRef] [PubMed]
396. Angelova, P.R.; Abramov, A.Y. Role of Mitochondrial ROS in the Brain: From Physiology to Neurodegeneration. *FEBS Lett.* **2018**, *592*, 692–702. [CrossRef]
397. Speijer, D. Being Right on Q: Shaping Eukaryotic Evolution. *Biochem. J.* **2016**, *473*, 4103–4127. [CrossRef]
398. Miquel, J.; Economos, A.C.; Fleming, J.; Johnson, J.E. Mitochondrial Role in Cell Aging. *Exp. Gerontol.* **1980**, *15*, 575–591. [CrossRef]
399. Linnane, A.W.; Marzuki, S.; Ozawa, T.; Tanaka, M. Mitochondrial DNA Mutations as an Important Contributor to Ageing and Degenerative Diseases. *Lancet* **1989**, *1*, 642–645. [CrossRef]
400. Barja, G. Updating the Mitochondrial Free Radical Theory of Aging: An Integrated View, Key Aspects, and Confounding Concepts. *Antioxid. Redox Signal.* **2013**, *19*, 1420–1445. [CrossRef]
401. Lenaz, G. Role of Mitochondria in Oxidative Stress and Ageing. *Biochim. Biophys. Acta* **1998**, *1366*, 53–67. [CrossRef]
402. Webb, M.; Sideris, D.P. Intimate Relations-Mitochondria and Ageing. *Int. J. Mol. Sci.* **2020**, *21*, 7580. [CrossRef]
403. Barja, G. Towards a Unified Mechanistic Theory of Aging. *Exp. Gerontol.* **2019**, *124*, 110627. [CrossRef]
404. Wallace, D.C. Mitochondria and Cancer. *Nat. Rev. Cancer* **2012**, *12*, 685–698. [CrossRef]
405. Genova, M.L.; Lenaz, G. The Interplay Between Respiratory Supercomplexes and ROS in Aging. *Antioxid. Redox Signal.* **2015**, *23*, 208–238. [CrossRef]
406. Barrientos, A.; Moraes, C.T. Titrating the Effects of Mitochondrial Complex I Impairment in the Cell Physiology. *J. Biol. Chem.* **1999**, *274*, 16188–16197. [CrossRef]
407. Lenaz, G.; D'Aurelio, M.; Merlo Pich, M.; Genova, M.L.; Ventura, B.; Bovina, C.; Formiggini, G.; Parenti Castelli, G. Mitochondrial Bioenergetics in Aging. *Biochim. Biophys. Acta* **2000**, *1459*, 397–404. [CrossRef]
408. Ventura, B.; Genova, M.L.; Bovina, C.; Formiggini, G.; Lenaz, G. Control of Oxidative Phosphorylation by Complex I in Rat Liver Mitochondria: Implications for Aging. *Biochim. Biophys. Acta* **2002**, *1553*, 249–260. [CrossRef]
409. Gómez, L.A.; Monette, J.S.; Chavez, J.D.; Maier, C.S.; Hagen, T.M. Supercomplexes of the Mitochondrial Electron Transport Chain Decline in the Aging Rat Heart. *Arch. Biochem. Biophys.* **2009**, *490*, 30–35. [CrossRef] [PubMed]
410. Gómez, L.A.; Hagen, T.M. Age-Related Decline in Mitochondrial Bioenergetics: Does Supercomplex Destabilization Determine Lower Oxidative Capacity and Higher Superoxide Production? *Semin. Cell Dev. Biol.* **2012**, *23*, 758–767. [CrossRef]
411. Frenzel, M.; Rommelspacher, H.; Sugawa, M.D.; Dencher, N.A. Ageing Alters the Supramolecular Architecture of OxPhos Complexes in Rat Brain Cortex. *Exp. Gerontol.* **2010**, *45*, 563–572. [CrossRef] [PubMed]
412. Edgar, D.; Shabalina, I.; Camara, Y.; Wredenberg, A.; Calvaruso, M.A.; Nijtmans, L.; Nedergaard, J.; Cannon, B.; Larsson, N.-G.; Trifunovic, A. Random Point Mutations with Major Effects on Protein-Coding Genes Are the Driving Force behind Premature Aging in MtDNA Mutator Mice. *Cell Metab.* **2009**, *10*, 131–138. [CrossRef] [PubMed]

Review

Role of Mitochondria in Viral Infections

Srikanth Elesela [1,*] and Nicholas W. Lukacs [2]

1 Department of Pathology, Michigan Medicine, Ann Arbor, MI 48109, USA
2 Mary H. Weiser Food Allergy Center, Michigan Medicine, Ann Arbor, MI 48109, USA; nlukacs@med.umich.edu
* Correspondence: srelesel@med.umich.edu

Abstract: Viral diseases account for an increasing proportion of deaths worldwide. Viruses maneuver host cell machinery in an attempt to subvert the intracellular environment favorable for their replication. The mitochondrial network is highly susceptible to physiological and environmental insults, including viral infections. Viruses affect mitochondrial functions and impact mitochondrial metabolism, and innate immune signaling. Resurgence of host-virus interactions in recent literature emphasizes the key role of mitochondria and host metabolism on viral life processes. Mitochondrial dysfunction leads to damage of mitochondria that generate toxic compounds, importantly mitochondrial DNA, inducing systemic toxicity, leading to damage of multiple organs in the body. Mitochondrial dynamics and mitophagy are essential for the maintenance of mitochondrial quality control and homeostasis. Therefore, metabolic antagonists may be essential to gain a better understanding of viral diseases and develop effective antiviral therapeutics. This review briefly discusses how viruses exploit mitochondrial dynamics for virus proliferation and induce associated diseases.

Keywords: mitochondria; mitochondrial dynamics; viral infections; MAVS; RIG-I; MDA5; innate immune response; SARS CoV-2; RSV; influenza

Citation: Elesela, S.; Lukacs, N.W. Role of Mitochondria in Viral Infections. *Life* **2021**, *11*, 232. https://doi.org/10.3390/life11030232

Academic Editor: Giorgio Lenaz

Received: 9 February 2021
Accepted: 10 March 2021
Published: 11 March 2021

1. Introduction

Mitochondria are intracellular organelles that are considered as the powerhouse of the cell. They comprise an outer membrane, an inner membrane and a matrix. The highly complex metabolic process of conversion of carbohydrates and fatty acids to adenosine triphosphate (ATP) occurs in mitochondria. During cellular stress, mitochondria rapidly increase energy production [1]. Mitochondria have their own genomic DNA (mitochondrial DNA, mtDNA) and can replicate by using their own transcriptional machinery. [2,3]. Recent advances in the role of mitochondrial dysfunction in causing human diseases has led to an increasing number of studies targeting mitochondrial proteins, metabolic processes and subsequent signaling pathways for drug discovery. Mitochondria can sense inflammation, infection, and/or environmental insults through structural changes in mitochondrial membranes, and protein expression, resulting in dysfunction [4–7]. Mitochondrial dysfunction also affects metabolism, calcium regulation, airway contractility in lungs, gene and protein housekeeping, oxidative stress, cell proliferation and apoptosis. Dysfunctional mitochondria alter homeostatic cellular processes including aging and senescence [8], as well as airway diseases [9,10]. Thus, understanding how mitochondrial dynamics affect various disease conditions would open new avenues that enhance the development of novel therapeutics targeting dysfunctional mitochondria.

Viruses are obligate parasites that completely depend on host cell machinery for their replication and proliferation. They hijack host cell metabolism and cause substantial alterations in cellular and physiological functions [11]. The role of mitochondrial dynamics in viral infections is still emerging, but unequivocally depict mitochondria as a key for cellular metabolism and innate immunity, a promising avenue for further molecular investigations in viral pathogenesis. During viral infections, mitochondria are directly targeted by viral proteins or influenced by physiological alterations such as oxidative stress, hypoxia, endoplasmic reticulum stress (ER stress) and dysregulated calcium homeostasis [11,12]. A plausible metabolic link between mitochondria and influenza A, and with herpes viruses was shown in the 1950s by Ackerman and colleagues [13,14]. Recent studies showed how hepatitis B virus (HBV) [15,16] and hepatitis C virus (HCV) adopt the changes in mitochondrial dynamics for persistent infection [17,18]. Our laboratory has also shown that respiratory syncytial virus (RSV) infection affects mitochondrial function, leading to an altered immune response in lungs [19,20]. More investigations on the role of mitochondrial dynamics in viral pathogenesis will enhance our understanding of host–virus interactions, leading to the design and development of new antiviral strategies.

In this review, we discuss the importance of mitochondrial bioenergetics during specific viral infections and the impact on virus-induced mitochondrial dysfunction, leading to changes in innate immune responses. While there are common pathways that are affected during viral infections, such as inflammasome activation, we have separated individual viruses, as the observed changes in each virus may be distinct.

2. Mitochondrial Dynamics

Mitochondrial biogenesis is a complex process that involves coordination of both nuclear and mitochondrial genes to ensure precise function of proteins of the mitochondrial electron transport chain. Mitochondria are highly dynamic, but they cannot be generated de novo. Every mitochondrion consists of a porous outer membrane; the intermembranous space; an inner membrane where the electron transport chain (ETC) occurs; and a mitochondrial matrix, the main site for metabolic pathways such as the TCA cycle and fatty acid oxidation (FAO). Human mtDNA is a double-stranded, circular DNA molecule consisting of 16,569 base pairs [21]. The mitochondrial genome is comprised of 37 genes that encode 13 polypeptides (ETC essential genes), 2 rRNA genes (12S and 16S rRNA), and 22 tRNA genes required for mitochondrial protein synthesis [22]. The remaining mitochondrial proteins are encoded by nuclear genes and approximately 1500 nuclear encoded proteins are involved in regulating mitochondrial functions in humans [23,24].

Mitochondrial homeostasis is maintained predominantly by mitochondrial dynamics and mitophagy [25], as depicted in Figure 1. Mitochondria form a tubular network that continuously changes by fission and fusion [26], and both of these processes are regulated by large guanosine triphosphatases (GTPases) [27]. Fission and fusion are continuous processes, and dysfunctional or impaired mitochondria are eliminated by a tightly regulated process known as mitophagy [28]. Fission has been implicated in the correction of mutations in mtDNA copies [29], while fusion is involved in a swift exchange and equilibration of matrix metabolites and recycling of the partially impaired mitochondrion to a fully healthy mitochondrial network [25]. The irreversibly damaged mitochondria are selectively eliminated by mitophagy [28]. Mitochondrial fusion and mitophagy appear to have the same function yet are in fact distinctly complementary, and simultaneously play a critical role in mitochondrial homeostasis [25,26]. Mitochondrial membrane potential ($\Delta\Psi m$) plays an important role in the process of identifying and segregating impaired mitochondria [30,31]. During a cellular cycle, the mitochondria follow a "kiss-and-run pattern" that promotes fusion for a very brief period (seconds), followed by a cellular change that shifts to fission [32]. Thus, the fission and fusion and mitophagy machinery work together to preclude circulation of impaired or dysfunctional mitochondria from the healthy pool of mitochondria. Quality and function of mitochondria are determined by the precise balance between incessant fission and fusion, and the subsequent processes that are induced manifest an enormous impact on the consequences of immune response during viral infections.

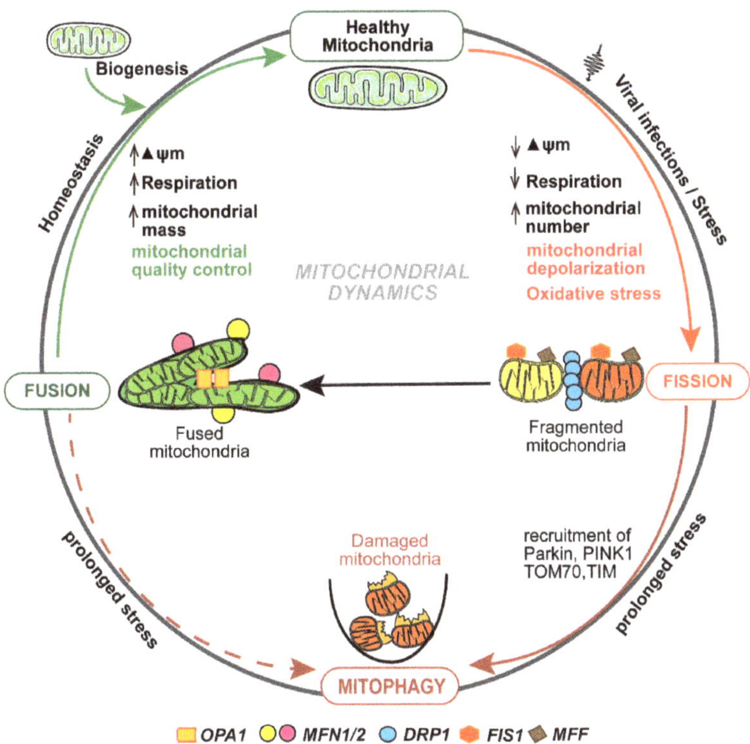

Figure 1. Mitochondrial dynamics: Mitochondrial fission and fusion are tightly regulated and continuous processes to maintain mitochondrial homeostasis. Fission is regulated by Drp1, Fis1, and Mff. Fusion is regulated by Opa1 and Mfn 1 and 2. Viral infections maneuver mitochondrial dynamics and alter mitochondrial membrane potential ($\Delta\Psi$m), mtDNA function, and respiration rate. Interruption in any of these functions/pathways results in the accumulation of dysfunctional mitochondria that are eliminated by mitophagy.

Viruses cause changes in mitochondrial function to promote viral translation and assembly. One theory is that virus–mitochondria interactions hamper mitochondria-associated antiviral signaling mechanisms [33]. In hepatitis viruses, mitochondrial antiviral signaling (MAVS) occurs through mitofusins that interact with MAVS to initiate effective antiviral immunity [34]. Blocking of mitofusins in cells resulted in the loss of mitochondrial membrane potential ($\Delta\Psi$m), leading to defective antiviral immune responses, suggesting that mitochondrial integrity is essential for antiviral innate immunity. Similarly, increased $\Delta\Psi$m induces apoptosis, while decreased $\Delta\Psi$m prevents apoptosis, and viruses, such as human cytomegalovirus (HCMV), decrease $\Delta\Psi$m to prevent cell death and promote their replication [35]. HCMV-encoded RNA (β2.7), localized in the mitochondria, interact with electron transport chain complex I and inhibit cells undergoing apoptosis [36], rapidly downregulating mitochondrial activity and enhancing viral replication. Virus-induced changes in mitochondrial integrity also result in an enhanced TCA cycle and further upregulation of lipid biosynthesis essential for viral envelopment, enlargement of the nucleus and of vesicular bodies of the infected cells [37,38]. During HCV infection, cells expressing HCV polyprotein have shown enhanced glycolytic function mediated by HIF-1α stabilization, with subsequently lower mitochondrial function, even in the presence of cellular oxygen, leading to increased cellular ATP content [39]. Increases in ATP levels have been reported in HCMV and herpes simplex-1 virus (HSV-1) infections [40]. Viral infections also induce ROS that control replication by altering mitochondrial function. In HCV infection, accumulation of defective mitochondria leads to oxidative stress and cell death [41]. Elevated ROS generation in the cells induces MAVS downstream, IRF3 and NFκB, to inhibit viral replication, linking a protective immune response with the virus infection. On the other hand, mitophagy decreases ROS production by removing dysfunctional mitochondria to control exacerbating immune responses [42]. The role of mitophagy in controlling viral replication is yet to be established. Nevertheless, it is interesting to note that mitophagy protects the cell from vulnerable cellular metabolic states. These viral effects are likely different depending on the virus itself and how it infects, replicates, and modifies innate immune cells.

3. Mitochondrial Metabolism and Innate Immune Responses

Mitochondria originated from symbiotic bacteria but co-evolved with their host as most of the mitochondrial proteins are encoded by the nucleus. However, the mitochondrial genome encodes proteins critical for respiration. Mitochondria play a central role in cellular metabolism as key pathways such as TCA, FAO, oxidative phosphorylation (OXPHOS), calcium buffering, and heme biosynthesis occur in mitochondria [43]. It is well established that ATP is generated through oxidative phosphorylation [2]. Communication between the nucleus, mitochondria, and the cytosol is essential for the maintenance of proper mitochondrial function and cellular homeostasis [44]. Mitochondrial dysfunction has serious physiological consequences that led to pathogenesis of many neurodegenerative disorders, cancer, inflammation, metabolic syndrome, cardiac dysfunctions, and viral diseases [45,46]. Several studies have shown the role of mitochondria in the activation of the NLRP3 and NLRP6 inflammasomes, microbial- and host-derived metabolites, and metabolism that effects subsequent immune responses [47,48]. During infection, the activation of pattern recognition receptors (PRR) sends signals to mitochondria, which then shift the metabolic switch from oxidative phosphorylation to glycolysis in order to equip cells to effectively combat the pathogens [49], making mitochondria a primary target during microbial-triggered PRR activation. Several studies have shown that innate immune cells under various stimuli trigger unique metabolic signatures necessary for subsequent immune function [45,50].

Enzymes involved in metabolism are being extensively investigated due to similarities with immune regulators. Methylcrotonyl-CoA carboxylase 1 has been shown to be associated with TRAF6 and enhances MAVS signaling in order to induce antiviral type I IFN (interferon) secretion [51]. Type 1 IFNs were also shown to induce FAO and oxidative phosphorylation [52,53]. Metabolic intermediates of the TCA cycle such as succinate, fumarate and citrate are associated with various processes that are coupled with inflammatory pathways in both innate and adaptive immune cells. The preference of metabolic pathways for immune cells depends on several factors including cell type, differentiation state, activation conditions and the cellular microenvironment [45]. Macrophages stimulated with LPS and IFN prefer glycolysis, but when stimulated with IL-4, macrophages prefer OXPHOS and FAO to meet the energy demands of the cell [54]. Dendritic cells prefer glycolysis once they are infected and activated through PRR [44]. It is interesting to note that resting T lymphocytes and memory T lymphocytes rely on OXPHOS, but the proliferating T lymphocytes prefer glycolysis through the upregulation of glucose transporter glut-1 [55]. Neutrophils engage in glycolysis, including the release of neutrophil extracellular traps (NETs) with increased glut-1 ex-pression and glycolytic function [56]. Activated B lymphocytes undergo metabolic reprogramming as per bioenergetic and biosynthetic demands. Plasma cells are unique in that they take up more glucose and glutamine to potentiate both glycolysis and mitochondrial OXPHOS, essential for promoting cell survival [57,58]. The innate immune response has a critical role in both the detection and regulation of infectious insults. The recognition of the insult by PRR triggers specific innate immune cells and its respective receptor and ligand, also resulting in a swift reaction to additional immune cells of the disease. This early innate immune signaling is central to recognition of infecting viruses, stimulating recruitment of additional immune cells to the site, activating the specific adaptive immune response, and inducing the production of molecules necessary to combat infection for elimination of the infectious agent as well as repair of damaged tissues [59,60].

Mitochondrial structure and function can affect innate immune responses. The most direct effect of mitochondria on immune response is due to mitochondrial damage while it can also occur as a result of normal mitochondrial physiology and function. The innate immune system specifically recognizes pathogen-associated molecular patterns (PAMPs) and damage-associated molecular patterns (DAMPs) as alarmins in order to trigger the appropriate immune response [59]. The release of alarmins by mitochondria is due to cellular stress and loss of homeostasis. The DAMPs released by mitochondria include unmethylated CpG mtDNA [61], ROS [42], cardiolipin [62], and n-formyl peptides (n-fp) [63]. The exact mechanism by which these mitochondrial alarmins are released is still unknown. However, numerous studies have shown that it is mainly due to loss of mitochondrial membrane integrity. Importantly, these molecules are recognized by discrete receptors and trigger specific inflammatory pathways that restore normal cellular function.

4. Mitochondrial Antiviral Signaling (MAVS)

Viral genomes usually replicate in the host cell cytoplasm, where they are not recognized by TLRs such as TLR3, TLR7 or TLR8 due to TLR localization to endosomes [64]. However, RNA viruses can still be sensed in the cytosol through RIG-I-like receptors (RLRs) such as Retinoic acid-Inducible Gene-I (RIG-I), Melanoma Differentiation-Associated gene-5 (MDA5) and Laboratory of Genetics and Physiology 2 (LGP2). While RIG-I and MDA-5 are prototypical PRRs, LGP2 is a regulator of RIG-I and MDA5 signaling [65]. Both ss-RNA and dsRNA are the known ligands of RIG-I and MDA5 [66]. RIG-I can also sense RNA polymers generated by the RNA polymerase III (pol III) from DNA templates, indirectly detecting dsDNA from intracellular pathogens [67]. RIG-I and MDA5 are cytosolic helicases with ATPase activity and consist of a regulatory C-terminal domain that binds to viral RNA, but the N-terminal domain comprises two tandem CARD domains (caspase activation and recruitment domains) [68]. The ATPase activity of these helicases is critical for translocation along dsRNA and in order to expose the CARDs that are masked by the C-terminal domain [69]. Once the specific 5′-triphosphate RNA structures are recognized, the E3 ubiquitin ligases TRIM25 and RIPLET enhance lysine 63-linked polyubiquitination of RIG-I, releasing CARDs from regulatory domain repression [70]. This conformational change leads to an essential interaction between the two CARD domains of RIG-I or MDA5 with the CARD domain of mitochondrial antiviral signaling protein (MAVS; also known as CARDIF, IPS-1 or VISA) [71,72]. MAVS, localized in the mitochondrial outer membrane, acts as a central signaling molecule in the RLR signaling pathway by linking upstream viral RNA recognition to downstream signal activation.

MAVS is required to localize to the mitochondria to exert its function, indicating that the mitochondrial environment is essential for signal transduction [71]. MAVS-deficient mice failed to induce type I IFN production and specific immune response against poly(I:C) suggests an essential role of MAVS in antiviral innate immunity [73,74]. RIG-I- and MDA5-mediated immune recognition and MAVS interaction are shown in Figure 2. Interaction between RIG-I or MDA5 with MAVS recruits a complex interactome to transduce the immune signaling. The MAVS-interacting proteins involved in antiviral response are TRAF3, TRAF5, IKKi/IKKε (IKKi), NEMO, DDX3, WDR5 IRF3, IRF7, and STING. The proteins that are involved in inflammatory responses are NLRC5, NLRX1, TRAF2, TRAF5, TRAF6, TAK1, and IKKα/β [75,76]. MAVS also interacts with mitochondrial proteins such as Mfn1, Mfn2, Tom70, and VDAC1; proteins involved in cell death (TRADD, FADD, RIP1) or autophagy (Atg5-Atg12); and with kinases (IKKi, PLK1, c-Abl, c-Src) or E3 ubiquitin ligases (PCBP2/AIP4, RNF5 and RNF125) that promote MAVS post-translational modifications [76,77]. Many of these proteins are indispensable and play a critical role in the canonical RLR pathway that is central to antiviral innate immune responses. Once activated, MAVS forms a signaling platform and recruits TNF receptor-associated factor (TRAF) 3 and TRAF6, inducing type I IFN [78] and inflammatory responses [79], respectively.

TRAF3−/− cells have shown an impaired type I IFN response against viral infections [80]. TRAF3, along with NF-κB modulator protein NEMO [76], TRAF family member-associated NF-κB activator (TANK) [81] and NAK-associated protein 1 (NAP1) [82], regulates the activity of two noncanonical IKK-related kinases, TANK-binding kinase 1 (TBK1) and inducible IκB kinase (IKKi). The phosphorylation of interferon regulatory factors (IRFs), IRF3 and IRF7, by TBK1 and IKKi leads to the induction of type I IFN genes and a set of IFN-inducible genes that bind to IFN-stimulated response elements (ISREs) in the nucleus [83]. MAVS activates IRF3 through the ubiquitin-binding domains of NEMO, while NEMO itself activates TBK1 [84] through TRAF3 [85]. FAS-associated death domain-containing protein (FADD) was also found in a complex with MAVS that activates NF-κB downstream of MAVS through the FADD/caspase-8-dependent pathway [86]. TRADD, a tumor necrosis factor receptor (TNFRI) adaptor protein is recruited to MAVS and induces the activation of IRF3 and NF-κB by initiating a complex formation with TRAF3, TANK, FADD and RIP1 [87]. RIG-I-mediated activation of NF-κB requires MAVS and a complex of CARD9 and Bcl-10 adaptor proteins [88]. RIG-I also binds to the adaptor protein ASC and stimulates caspase-1-dependent inflammasome activation by a mechanism independent of MAVS, which suggests that RIG-I activates the inflammasome in response to certain RNA viruses [88–90].

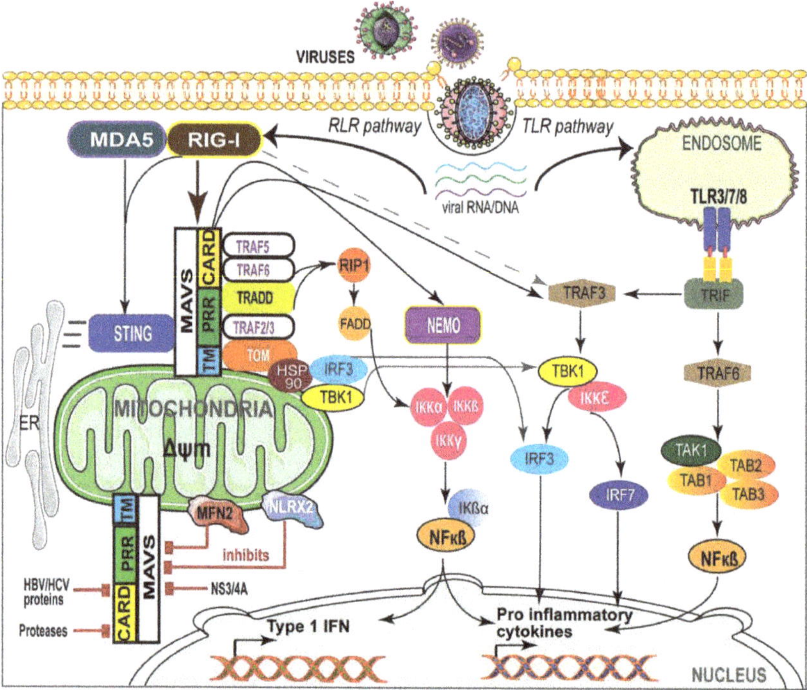

Figure 2. RIG-I/MDA-5 and MAVS interaction in viral disease. The cytosolic viral RNA/DNA is recognized by the RLR and/or TLR pathways. RIG-I-like receptors (RLRs) and MDA-5 activate MAVS through CARD and recruit signaling molecules to induce canonical nuclear factor-κB (NF-κB). NF-κB translocates into the nucleus and initiates pro-inflammatory cytokine gene expression. MAVS activates the stimulator of interferon genes (STING) and further mediates the activation of TANK-binding kinase 1 (TBK1) which phosphorylates interferon regulatory factor (IRF) signaling factors IRF-3 and IRF-7. IRF-3 then translocates into the nucleus and induces type I interferon (IFN) genes. NS3-4A, mitofusin 2 (MFN2), and NLR family member X1 (NLRX1) inhibit MAVS by preventing the formation of the MAVS–IKKi signaling complex. Hepatitis B virus (HBV) X protein promotes polyubiquitin conjugation of MAVS. ER—endoplasmic reticulum; MAM—mitochondria-associated membrane.

5. Viral Infections and Mitochondrial Biogenesis

Viruses impede mitochondrial biogenesis, causing alterations in mitochondrial function in order to promote viral translation and assembly. One theory is that virus–mitochondria interactions hamper mitochondria-associated antiviral signaling mechanisms [33,91]. The regulation of mitochondrial dynamics in order to cause physiological perturbations in the cellular environment due to viral infections makes mitochondrial dynamics a primary target. Innate immune responses against viral infections led to type I Interferon (IFN-α/β) and other proinflammatory cytokines and chemokine responses. The specific molecules involved in mitochondrial biogenesis are peroxisome proliferator-activated receptor-γ coactivator (PGC)-1α, the main regulator of mitochondrial biogenesis [92,93]; PTEN-induced putative kinase 1 (PINK1) [94] that activates protein synthesis in damaged mitochondria; and the ligand-activated transcription factor aryl hydrocarbon receptor that functions to protect the cell from oxidative stress [95]. The silent information regulator-1 (SIRT1) activates the PGC1α-mediated transcription of nuclear and mitochondrial genes encoding for proteins during mitochondria proliferation, oxidative phosphorylation and energy production [96]. SIRT3, on the other hand, stimulates the proteins important for oxidative phosphorylation, the tricarboxylic acid cycle and fatty acid oxidation, and indirectly, PGC-1α and AMPK. SIRT1 deacetylates histone and numerous non-histone proteins during transcription, including PGC-1α [96]. Viruses have developed discrete strategies to regulate MAVS signaling by regulating mitochondria biogenesis, and thus regulating early innate immune responses.

5.1. SARS-CoV-2

A novel severe acute respiratory syndrome-related coronavirus (SARS-CoV-2) has recently emerged as a serious pathogen that causes high morbidity and substantial mortality. It is causing a global pandemic and worldwide social and economic disruption. Patients with severe SARS-CoV-2 infection develop dyspnea that can rapidly manifest as acute respiratory distress syndrome, leading to death [97–100]. SARS-CoV-2 is a single-stranded positive-sense RNA virus which encodes over 28 proteins, including 4 structural proteins (spike, membrane, envelope, and nucleocapsid), 16 non-structural proteins (NSP1–NSP16), and 8 auxiliary proteins (ORF3a, ORF3b, ORF6, ORF7a, ORF7b, ORF8, ORF9b and ORF14) [101,102]. The pathophysiology of SARS-CoV-2 infection shows exaggerated inflammatory responses, causing severe damage to the airways [103]. During SARS-CoV-2 infection in lungs, monocytes and macrophages are recruited to the site of infection and release cytokines and activate T and B cells. An impaired immune response during this process leads to chronic lung pathology. COVID-19 patients have shown dysregulated type I IFN response. However, the mechanisms by which SARS-CoV-2 evades host immunity have not been fully understood. SARS-CoV-2 M protein has been identified as a factor that interacts with MAVS to inhibit RLR-mediated induction of the host's type I IFN response [104]. The M protein suppressed RIG-I-, MDA5- and MAVS-mediated signaling but did not show any effect on their downstream components TBK1 or p65. The authors have shown that the M protein directly interacts with MAVS and impairs viral RNA-induced MAVS through the downstream components TRAF3, TBK1, and IRF3 [104]. Screening of SARS-CoV-2 has identified several proteins including M, N, ORF3a, ORF6, and NSP (non-structural protein) family proteins as potential candidates that downregulate IFNβ responses [105]. Mitochondrial dysfunction-mediated reduced oxygen sensing, and mitochondrial oxidative stress-mediated platelet dysfunction and coagulation pathways have been reported in SARS-CoV-2 infection [106,107]. SARS-CoV-2 main protease Mpro (nsp5) impairs both the virus-induced type I IFN production and the induction of downstream antiviral interferon-stimulated genes (ISGs) [108]. Another protein, Orf9b, localizes to mitochondria, binds to TOM70, an adaptor protein of the mitochondrial outer membrane, and suppresses the antiviral type I IFN response [109,110]. However, the molecular consequences of Orf9b binding to TOM70 are not yet clear.

5.2. *Respiratory Syncytial Virus*

Human respiratory syncytial virus (RSV) of the Paramyxoviridae family is a single-stranded, negative-sense RNA virus that causes serious respiratory complications especially in infants and the older adults worldwide [111,112]. Quantitative proteomic analysis of RSV-infected cells has identified several nuclear-encoded mitochondrial proteins which include OMM complex subunits, respiratory complex I proteins, VDAC protein (voltage-dependent anion channel), and prohibitin (PHB) that play a critical role in the regulation of mitochondrial structure, function and biogenesis [113,114]. Hu et al. have shown for the first time that RSV infection hijacks host mitochondria, maneuvering for its replication and causing mitochondrial redistribution towards the perinuclear region of the microtubule organizing center [115]. This redistribution is a dynein-dependent mode of transport that causes perturbances in mitochondrial membrane polarization, leading to decreased mitochondrial membrane potential and significantly elevated levels of ROS [116]. Blocking dynein or the microtubule function resulted in a significant inhibition of RSV effect on mitochondrial function. In another study, deletion of a mitochondrial biogenesis factor, clustered mitochondria homolog (CLUH), resulted in enhanced mitochondrial ROS production during RSV infection [116]. The mitochondrial ROS scavenger MitoQ has been shown to remarkably reduce viral proliferation and restore mitochondrial function during RSV infection, suggesting that RSV-induced mitochondrial ROS contributes to sustained viral infection [115]. Similarly, our group has shown that SIRT1 is necessary to promote dendritic cell activation and autophagy during RSV infection, and the absence of SIRT1 led to exacerbated pathology [19]. In another study, we have also shown that mitochondrial function regulates RSV-induced innate immune response, leading to instruction of adaptive immune responses through SIRT1 [20]. The central role of acetyl coA carboxylase (ACC1) that activates acetyl CoA requires regulation by SIRT1 (via AMPK) in order to control the fatty acid synthesis pathway that leads to dysregulated innate cytokine responses. The inhibition of ACC1 has allowed the SIRT1-deficient dendritic cells to manifest a more appropriate innate and acquired immune response. The inhibition of ACC1 with a specific inhibitor led to correction of the altered metabolic state and resulted in the stabilization of the altered innate and acquired immune responses driven by RSV in DC and altered the pathologic responses in the lung [20]. However, the molecular mechanisms involving RIG-I/MDA5 and MAVS in RSV infection are yet to be explored.

5.3. *Influenza Virus*

Influenza virus is a respiratory pathogen that causes contagious respiratory illness known as influenza or flu, which accounts for millions of deaths worldwide. The three main types of influenza virus that cause disease in humans are A, B, and C, which are classified based on antigenic differences in matrix and nucleoproteins [116]. Once infected, the influenza virus is recognized by various PRRs such as TLRs, RIG-I, NLRP3, and cGAS pathways. The influenza virus replicates in the nucleus but how RIG-I signaling is activated during this is not very clear. However, the NS1 protein has been shown to suppress type 1 IFN responses by directly interrupting RIG-I signaling [117,118]. The nucleotide-binding oligomerization domain-containing protein 2 (NOD2) and receptor interacting protein kinase 2 (RIPK2) promotes ULK1 phosphorylation and induces mitophagy that protected mice from viral immunopathology in influenza A virus infection [119,120]. Defective mitophagy along with segregation of dysfunctional mitochondria and subsequent inflammasome activation was observed in RIPK2-depleted cells. Increased mitochondrial dynamics have been shown to downregulate IL-18 secretion and inflammasome activation [119,120]. Another protein, PB1-F2, disrupts mitochondrial membrane potential, binds to MAVS and downregulates innate immune responses, especially type 1 IFNs, and NLRP3 activation [121,122].

5.4. Hepatitis Viruses

Hepatitis C virus (HCV) is a positive-strand RNA virus of family Flaviviridae. During infection, HCV proteins localize to mitochondrial membranes, induce ER stress and cause depletion of ER calcium stores, leading to mitochondrial dysfunction [123,124]. The non-structural protein 5A (NS5A) of HCV inhibits electron transport chain enzyme complex I activity to promote mitochondrial calcium uptake, mitochondrial permeability transition, and ROS production [17,125]. NS3/4a protease, on the other hand, cleaves MAVS and facilitates immune evasion [126]. Mitochondrial damage during HCV infection inhibits FAO and enhances lipogenesis [127]. HCV induces translocation of Drp1 by phosphorylating it at S616 and promotes mitochondrial dynamics, subverts MAVS and increases IFN responses [17]. HCV infection induces the recruitment of Parkin and PINK1 and enhances the removal of accumulated impaired mitochondria in a Parkin-dependent manner [17]. Several studies indicate that HCV-induced regulation of mitochondrial dynamics favors viral persistence and illuminate how viruses exploit mitochondrial dynamics, leading to exacerbated pathology.

Hepatitis B virus (HBV) belongs to the family Hepadnaviridae and its genome consists of a partially double-stranded circular DNA that replicates via an RNA intermediate. HBx, a regulatory protein of HBV, is associated with VDAC, localizes to mitochondrial membranes and affects the membrane potential, inducing remarkably high levels of calcium and ROS, leading to mitochondrial dysfunction [128]. This HBx-regulated calcium signaling and ROS activate STAT3, NF-kB and NFAT [129]. Like HCV, HBV also induced Drp1 phosphorylation at S616 to promote mitochondrial dynamics and Parkin-mediated mitophagy [15]. Inhibition of Parkin during HBV infection increased the release of cytochrome C, activation of caspase-3, and cleaving of PARP (poly ADP-ribose polymerase), resulting in an enhanced apoptosis [130,131]. During infection, RIG-I in the cytosol detects HBV dsRNA in the cytosol [132], binds through its C-terminal RNA helicase domain and activates IKKi and TBK1 by CARD, which is at the N-terminal. MAVS then links RIG-I to IKKi and TBK1 activation. The role of MAVS/IPS-1 is essential for induction of IFN by cytosolic DNA [132,133].

5.5. Measles

Measles virus consists of a negative-sense RNA genome that causes highly contagious respiratory sickness including pneumonia, seizures, brain damage, and even death. The attenuated measles virus of the Edmonston strain (MV-Edm) activates p62-mediated mitophagy in non-small-cell lung cancer (NSCLC) cells by disrupting MAVS and resulting in the significant inhibition of type I IFN responses [134]. It utilizes apoptosis to sustain viral propagation and replication [135]. Defects in autophagy resulted in decreased viral titers and MV-Edm induced cell death in NSCLC cells. When p62 expression was silenced, it led to the restoration of mitochondrial mass in MV-Edm-infected cells and inhibition of mitophagy. Therefore, it appears that MV usurps mitophagy to mitigate the RIG-I/MAVS-mediated innate immune signaling pathways [134].

6. Concluding Remarks

Despite the exhilarating scientific advances in recent times that have identified many important metabolic pathways that might be targets in order to enhance immune responses, there continue to be new and exciting questions. Various aspects of mitochondrial dynamics, including mitophagy, have been of great interest, with recent studies showing that viruses circumvent host innate immune responses through altering mitochondrial functions. Viral infections induce metabolic re-programming, resulting in discrete bioenergetic phenotypes, strategically utilizing them for viral propagation and replication. Viruses exploit RIG-I-MDA5-MAVS antiviral signaling pathways and aim at disrupting mitochondrial membrane potential, mitochondrial-associated proteins and mitochondrial dynamics that essentially impede virus-induced type I IFN responses. Undeniably, there exists an intimate association between mitochondrial dynamics and viral infections. However, more comprehensive mechanistic studies and their significance to chronic pathology are necessary in understanding complex viral life cycle processes. A deeper understanding of tightly regulated mitochondrial functions such as bioenergetics, innate antiviral immunity, apoptosis, and inter-organelle cross-talk needs to be extensively investigated to analyze their effect on viral infections. The ultimate goal of identifying mechanisms, which may differ with individual viruses, may provide important information for targeted therapeutic interventions to redirect the immune response toward a less pathogenic response.

Author Contributions: S.E. and N.W.L. have conceptualized and written the manuscript. All authors have read and agreed to the published version of the manuscript.

Funding: Authors acknowledge the support from NIH/NIAID grant funding (RO1HL144858 and RO1AI036302) and the Mary H. Weiser Food Allergy Center at the University of Michigan.

Conflicts of Interest: The authors declare no conflict of interest.

References

1. Eisner, V.; Picard, M.; Hajnóczky, G. Mitochondrial dynamics in adaptive and maladaptive cellular stress responses. *Nat. Cell Biol.* **2018**, *20*, 755–765. [CrossRef] [PubMed]
2. Ernster, L.; Schatz, G. Mitochondria: A historical review. *J. Cell Biol.* **1981**, *91*, 227s–255s. [CrossRef]
3. Xia, M.; Zhang, Y.; Jin, K.; Lu, Z.; Zeng, Z.; Xiong, W. Communication between mitochondria and other organelles: A brand-new perspective on mitochondria in cancer. *Cell Biosci.* **2019**, *9*, 27. [CrossRef]
4. Tschopp, J. Mitochondria: Sovereign of inflammation? *Eur. J. Immunol.* **2011**, *41*, 1196–1202. [CrossRef] [PubMed]
5. Osellame, L.D.; Blacker, T.S.; Duchen, M.R. Cellular and molecular mechanisms of mitochondrial function. *Best Pract. Res. Clin. Endocrinol. Metab.* **2012**, *26*, 711–723. [CrossRef] [PubMed]
6. Murphy, M.P.; Hartley, R.C. Mitochondria as a therapeutic target for common pathologies. *Nat. Rev. Drug Discov.* **2018**, *17*, 865–886. [CrossRef] [PubMed]
7. Weissig, V. Drug Development for the Therapy of Mitochondrial Diseases. *Trends Mol. Med.* **2020**, *26*, 40–57. [CrossRef] [PubMed]
8. Kauppila, T.E.S.; Kauppila, J.H.K.; Larsson, N.G. Mammalian Mitochondria and Aging: An Update. *Cell Metab.* **2017**, *25*, 57–71. [CrossRef] [PubMed]
9. Vernochet, C.; Kahn, C.R. Mitochondria, obesity and aging. *Aging* **2012**, *4*, 859–860. [CrossRef]
10. Pan, S.; Conaway, S., Jr.; Deshpande, D.A. Mitochondrial regulation of airway smooth muscle functions in health and pulmonary diseases. *Arch. Biochem. Biophys.* **2019**, *663*, 109–119. [CrossRef] [PubMed]
11. Thaker, S.K.; Ch'ng, J.; Christofk, H.R. Viral hijacking of cellular metabolism. *BMC Biol.* **2019**, *17*, 59. [CrossRef] [PubMed]
12. Asha, K.; Sharma-Walia, N. Virus and tumor microenvironment induced ER stress and unfolded protein response: From complexity to therapeutics. *Oncotarget* **2018**, *9*, 31920–31936. [CrossRef] [PubMed]
13. Ackermann, W.W.; Johnson, R.B. Some energy relations in a host-virus system. *J. Exp. Med.* **1953**, *97*, 315–322. [CrossRef]
14. Ackermann, W.W.; Klernschmidt, E. Concerning the relation of the Krebs cycle to virus propagation. *J. Biol. Chem.* **1951**, *189*, 421–428. [CrossRef]
15. Kim, S.J.; Khan, M.; Quan, J.; Till, A.; Subramani, S.; Siddiqui, A. Hepatitis B virus disrupts mitochondrial dynamics: Induces fission and mitophagy to attenuate apoptosis. *PLoS Pathog.* **2013**, *9*, e1003722. [CrossRef] [PubMed]
16. Hossain, M.G.; Akter, S.; Ohsaki, E.; Ueda, K. Impact of the Interaction of Hepatitis B Virus with Mitochondria and Associated Proteins. *Viruses* **2020**, *12*, 175. [CrossRef]
17. Kim, S.J.; Syed, G.H.; Khan, M.; Chiu, W.W.; Sohail, M.A.; Gish, R.G.; Siddiqui, A. Hepatitis C virus triggers mitochondrial fission and attenuates apoptosis to promote viral persistence. *Proc. Natl. Acad. Sci. USA* **2014**, *111*, 6413–6418. [CrossRef] [PubMed]
18. Javed, F.; Manzoor, S. HCV non-structural NS4A protein of genotype 3a induces mitochondria mediated death by activating Bax and the caspase cascade. *Microb. Pathog.* **2018**, *124*, 346–355. [CrossRef]
19. Owczarczyk, A.B.; Schaller, M.A.; Reed, M.; Rasky, A.J.; Lombard, D.B.; Lukacs, N.W. Sirtuin 1 Regulates Dendritic Cell Activation and Autophagy during Respiratory Syncytial Virus-Induced Immune Responses. *J. Immunol.* **2015**, *195*, 1637–1646. [CrossRef]
20. Elesela, S.; Morris, S.B.; Narayanan, S.; Kumar, S.; Lombard, D.B.; Lukacs, N.W. Sirtuin 1 regulates mitochondrial function and immune homeostasis in respiratory syncytial virus infected dendritic cells. *PLoS Pathog.* **2020**, *16*, e1008319. [CrossRef] [PubMed]
21. Anderson, S.; Bankier, A.T.; Barrell, B.G.; de Bruijn, M.H.; Coulson, A.R.; Drouin, J.; Eperon, I.C.; Nierlich, D.P.; Roe, B.A.; Sanger, F.; et al. Sequence and organization of the human mitochondrial genome. *Nature* **1981**, *290*, 457–465. [CrossRef] [PubMed]
22. Kolesnikov, A.A. The Mitochondrial Genome. The Nucleoid. *Biochemistry* **2016**, *81*, 1057–1065. [CrossRef] [PubMed]
23. Saneto, R.P. Genetics of Mitochondrial Disease. *Adv. Genet.* **2017**, *98*, 63–116. [CrossRef]
24. Cheong, A.; Archambault, D.; Degani, R.; Iverson, E.; Tremblay, K.D.; Mager, J. Nuclear-encoded mitochondrial ribosomal proteins are required to initiate gastrulation. *Development* **2020**, *147*, dev188714. [CrossRef]
25. Yoo, S.M.; Jung, Y.K. A Molecular Approach to Mitophagy and Mitochondrial Dynamics. *Mol. Cells* **2018**, *41*, 18–26. [CrossRef] [PubMed]

26. Meyer, J.N.; Leuthner, T.C.; Luz, A.L. Mitochondrial fusion, fission, and mitochondrial toxicity. *Toxicology* **2017**, *391*, 42–53. [CrossRef] [PubMed]
27. Feng, S.T.; Wang, Z.Z.; Yuan, Y.H.; Wang, X.L.; Sun, H.M.; Chen, N.H.; Zhang, Y. Dynamin-related protein 1: A protein critical for mitochondrial fission, mitophagy, and neuronal death in Parkinson's disease. *Pharmacol. Res.* **2020**, *151*, 104553. [CrossRef]
28. Dombi, E.; Mortiboys, H.; Poulton, J. Modulating Mitophagy in Mitochondrial Disease. *Curr. Med. Chem.* **2018**, *25*, 5597–5612. [CrossRef] [PubMed]
29. Otten, A.B.C.; Sallevelt, S.; Carling, P.J.; Dreesen, J.; Drüsedau, M.; Spierts, S.; Paulussen, A.D.C.; de Die-Smulders, C.E.M.; Herbert, M.; Chinnery, P.F.; et al. Mutation-specific effects in germline transmission of pathogenic mtDNA variants. *Hum. Reprod.* **2018**, *33*, 1331–1341. [CrossRef] [PubMed]
30. Mattenberger, Y.; James, D.I.; Martinou, J.C. Fusion of mitochondria in mammalian cells is dependent on the mitochondrial inner membrane potential and independent of microtubules or actin. *FEBS Lett.* **2003**, *538*, 53–59. [CrossRef]
31. Zorova, L.D.; Popkov, V.A.; Plotnikov, E.Y.; Silachev, D.N.; Pevzner, I.B.; Jankauskas, S.S.; Babenko, V.A.; Zorov, S.D.; Balakireva, A.V.; Juhaszova, M.; et al. Mitochondrial membrane potential. *Anal. Biochem.* **2018**, *552*, 50–59. [CrossRef] [PubMed]
32. Lee, H.; Yoon, Y. Mitochondrial fission and fusion. *Biochem. Soc. Trans.* **2016**, *44*, 1725–1735. [CrossRef] [PubMed]
33. Refolo, G.; Vescovo, T.; Piacentini, M.; Fimia, G.M.; Ciccosanti, F. Mitochondrial Interactome: A Focus on Antiviral Signaling Pathways. *Front. Cell Dev. Biol.* **2020**, *8*, 8. [CrossRef]
34. Koshiba, T.; Yasukawa, K.; Yanagi, Y.; Kawabata, S. Mitochondrial membrane potential is required for MAVS-mediated antiviral signaling. *Sci. Signal.* **2011**, *4*, ra7. [CrossRef] [PubMed]
35. Anand, S.K.; Tikoo, S.K. Viruses as modulators of mitochondrial functions. *Adv. Virol.* **2013**, *2013*, 738794. [CrossRef] [PubMed]
36. Combs, J.A.; Norton, E.B.; Saifudeen, Z.R.; Bentrup, K.H.Z.; Katakam, P.V.; Morris, C.A.; Myers, L.; Kaur, A.; Sullivan, D.E.; Zwezdaryk, K.J. Human Cytomegalovirus Alters Host Cell Mitochondrial Function during Acute Infection. *J. Virol.* **2020**, *94*. [CrossRef]
37. Yu, Y.; Clippinger, A.J.; Alwine, J.C. Viral effects on metabolism: Changes in glucose and glutamine utilization during human cytomegalovirus infection. *Trends Microbiol.* **2011**, *19*, 360–367. [CrossRef] [PubMed]
38. Xi, Y.; Harwood, S.; Wise, L.M.; Purdy, J.G. Human Cytomegalovirus pUL37x1 Is Important for Remodeling of Host Lipid Metabolism. *J. Virol.* **2019**, *93*, e00843-19. [CrossRef] [PubMed]
39. Yogev, O.; Lagos, D.; Enver, T.; Boshoff, C. Kaposi's sarcoma herpesvirus microRNAs induce metabolic transformation of infected cells. *PLoS Pathog.* **2014**, *10*, e1004400. [CrossRef] [PubMed]
40. Alves, V.S.; Leite-Aguiar, R.; Silva, J.P.D.; Coutinho-Silva, R.; Savio, L.E.B. Purinergic signaling in infectious diseases of the central nervous system. *Brain. Behav. Immun.* **2020**, *89*, 480–490. [CrossRef] [PubMed]
41. Zhang, L.; Qin, Y.; Chen, M. Viral strategies for triggering and manipulating mitophagy. *Autophagy* **2018**, *14*, 1665–1673. [CrossRef]
42. West, A.P.; Shadel, G.S.; Ghosh, S. Mitochondria in innate immune responses. *Nat. Rev. Immunol.* **2011**, *11*, 389–402. [CrossRef] [PubMed]
43. Mishra, P.; Chan, D.C. Metabolic regulation of mitochondrial dynamics. *J. Cell Biol.* **2016**, *212*, 379–387. [CrossRef] [PubMed]
44. O'Neill, L.A.; Pearce, E.J. Immunometabolism governs dendritic cell and macrophage function. *J. Exp. Med.* **2016**, *213*, 15–23. [CrossRef] [PubMed]
45. O'Neill, L.A.; Kishton, R.J.; Rathmell, J. A guide to immunometabolism for immunologists. *Nat. Rev. Immunol.* **2016**, *16*, 553–565. [CrossRef]
46. Chen, Y.; Zhou, Z.; Min, W. Mitochondria, Oxidative Stress and Innate Immunity. *Front. Physiol.* **2018**, *9*, 1487. [CrossRef] [PubMed]
47. Yu, J.W.; Lee, M.S. Mitochondria and the NLRP3 inflammasome: Physiological and pathological relevance. *Arch. Pharm. Res.* **2016**, *39*, 1503–1518. [CrossRef]
48. Próchnicki, T.; Latz, E. Inflammasomes on the Crossroads of Innate Immune Recognition and Metabolic Control. *Cell Metab.* **2017**, *26*, 71–93. [CrossRef] [PubMed]
49. Banoth, B.; Cassel, S.L. Mitochondria in innate immune signaling. *Transl. Res.* **2018**, *202*, 52–68. [CrossRef]
50. Lachmandas, E.; Boutens, L.; Ratter, J.M.; Hijmans, A.; Hooiveld, G.J.; Joosten, L.A.; Rodenburg, R.J.; Fransen, J.A.; Houtkooper, R.H.; van Crevel, R.; et al. Microbial stimulation of different Toll-like receptor signalling pathways induces diverse metabolic programmes in human monocytes. *Nat. Microbiol.* **2016**, *2*, 16246. [CrossRef]
51. Cao, Z.; Xia, Z.; Zhou, Y.; Yang, X.; Hao, H.; Peng, N.; Liu, S.; Zhu, Y. Methylcrotonoyl-CoA carboxylase 1 potentiates RLR-induced NF-κB signaling by targeting MAVS complex. *Sci. Rep.* **2016**, *6*, 33557. [CrossRef] [PubMed]
52. Truong, N.T.T.; Lydic, T.A.; Bazil, J.N.; Suryadevara, A.; Olson, L.K. Regulation of lipid metabolism in pancreatic beta cells by interferon gamma: A link to anti-viral function. *Cytokine* **2020**, *133*, 155147. [CrossRef] [PubMed]
53. Fritsch, S.D.; Weichhart, T. Effects of Interferons and Viruses on Metabolism. *Front. Immunol.* **2016**, *7*, 630. [CrossRef] [PubMed]
54. Rodríguez-Prados, J.C.; Través, P.G.; Cuenca, J.; Rico, D.; Aragonés, J.; Martín-Sanz, P.; Cascante, M.; Boscá, L. Substrate fate in activated macrophages: A comparison between innate, classic, and alternative activation. *J. Immunol.* **2010**, *185*, 605–614. [CrossRef]
55. Pearce, E.L.; Pearce, E.J. Metabolic pathways in immune cell activation and quiescence. *Immunity* **2013**, *38*, 633–643. [CrossRef] [PubMed]

56. Rodríguez-Espinosa, O.; Rojas-Espinosa, O.; Moreno-Altamirano, M.M.; López-Villegas, E.O.; Sánchez-García, F.J. Metabolic requirements for neutrophil extracellular traps formation. *Immunology* **2015**, *145*, 213–224. [CrossRef] [PubMed]
57. Waters, L.R.; Ahsan, F.M.; Wolf, D.M.; Shirihai, O.; Teitell, M.A. Initial B Cell Activation Induces Metabolic Reprogramming and Mitochondrial Remodeling. *iScience* **2018**, *5*, 99–109. [CrossRef]
58. Choi, S.-C.; Morel, L. Immune metabolism regulation of the germinal center response. *Exp. Mol. Med.* **2020**, *52*, 348–355. [CrossRef] [PubMed]
59. Iwasaki, A.; Medzhitov, R. Control of adaptive immunity by the innate immune system. *Nat. Immunol.* **2015**, *16*, 343–353. [CrossRef]
60. Antonelli, M.; Kushner, I. It's time to redefine inflammation. *FASEB J.* **2017**, *31*, 1787–1791. [CrossRef]
61. Moretton, A.; Morel, F.; Macao, B.; Lachaume, P.; Ishak, L.; Lefebvre, M.; Garreau-Balandier, I.; Vernet, P.; Falkenberg, M.; Farge, G. Selective mitochondrial DNA degradation following double-strand breaks. *PLoS ONE* **2017**, *12*, e0176795. [CrossRef] [PubMed]
62. Shen, Z.; Ye, C.; McCain, K.; Greenberg, M.L. The Role of Cardiolipin in Cardiovascular Health. *Biomed. Res. Int.* **2015**, *2015*, 891707. [CrossRef] [PubMed]
63. Schiffmann, E.; Corcoran, B.A.; Wahl, S.M. N-formylmethionyl peptides as chemoattractants for leucocytes. *Proc. Natl. Acad. Sci. USA* **1975**, *72*, 1059–1062. [CrossRef]
64. Lester, S.N.; Li, K. Toll-like receptors in antiviral innate immunity. *J. Mol. Biol.* **2014**, *426*, 1246–1264. [CrossRef]
65. Hei, L.; Zhong, J. Laboratory of genetics and physiology 2 (LGP2) plays an essential role in hepatitis C virus infection-induced interferon responses. *Hepatology* **2017**, *65*, 1478–1491. [CrossRef] [PubMed]
66. Tal, M.C.; Iwasaki, A. Autophagy and innate recognition systems. *Curr. Top. Microbiol. Immunol.* **2009**, *335*, 107–121. [CrossRef] [PubMed]
67. Chiu, Y.H.; Macmillan, J.B.; Chen, Z.J. RNA polymerase III detects cytosolic DNA and induces type I interferons through the RIG-I pathway. *Cell* **2009**, *138*, 576–591. [CrossRef] [PubMed]
68. Kowalinski, E.; Lunardi, T.; McCarthy, A.A.; Louber, J.; Brunel, J.; Grigorov, B.; Gerlier, D.; Cusack, S. Structural basis for the activation of innate immune pattern-recognition receptor RIG-I by viral RNA. *Cell* **2011**, *147*, 423–435. [CrossRef]
69. Louber, J.; Brunel, J.; Uchikawa, E.; Cusack, S.; Gerlier, D. Kinetic discrimination of self/non-self RNA by the ATPase activity of RIG-I and MDA5. *BMC Biol.* **2015**, *13*, 54. [CrossRef] [PubMed]
70. Hayman, T.J.; Hsu, A.C.; Kolesnik, T.B.; Dagley, L.F.; Willemsen, J.; Tate, M.D.; Baker, P.J.; Kershaw, N.J.; Kedzierski, L.; Webb, A.I.; et al. RIPLET, and not TRIM25, is required for endogenous RIG-I-dependent antiviral responses. *Immunol. Cell Biol.* **2019**, *97*, 840–852. [CrossRef]
71. Seth, R.B.; Sun, L.; Ea, C.K.; Chen, Z.J. Identification and characterization of MAVS, a mitochondrial antiviral signaling protein that activates NF-kappaB and IRF 3. *Cell* **2005**, *122*, 669–682. [CrossRef]
72. Park, Y.J.; Oanh, N.T.K.; Heo, J.; Kim, S.G.; Lee, H.S.; Lee, H.; Lee, J.H.; Kang, H.C.; Lim, W.; Yoo, Y.S.; et al. Dual targeting of RIG-I and MAVS by MARCH5 mitochondria ubiquitin ligase in innate immunity. *Cell. Signal.* **2020**, *67*, 109520. [CrossRef]
73. Kumar, H.; Kawai, T.; Kato, H.; Sato, S.; Takahashi, K.; Coban, C.; Yamamoto, M.; Uematsu, S.; Ishii, K.J.; Takeuchi, O.; et al. Essential role of IPS-1 in innate immune responses against RNA viruses. *J. Exp. Med.* **2006**, *203*, 1795–1803. [CrossRef]
74. Sun, Q.; Sun, L.; Liu, H.H.; Chen, X.; Seth, R.B.; Forman, J.; Chen, Z.J. The specific and essential role of MAVS in antiviral innate immune responses. *Immunity* **2006**, *24*, 633–642. [CrossRef]
75. Belgnaoui, S.M.; Paz, S.; Samuel, S.; Goulet, M.L.; Sun, Q.; Kikkert, M.; Iwai, K.; Dikic, I.; Hiscott, J.; Lin, R. Linear ubiquitination of NEMO negatively regulates the interferon antiviral response through disruption of the MAVS-TRAF3 complex. *Cell Host Microbe* **2012**, *12*, 211–222. [CrossRef]
76. Belgnaoui, S.M.; Paz, S.; Hiscott, J. Orchestrating the interferon antiviral response through the mitochondrial antiviral signaling (MAVS) adapter. *Curr. Opin. Immunol.* **2011**, *23*, 564–572. [CrossRef] [PubMed]
77. Lubin, A.; Zhang, L.; Chen, H.; White, V.M.; Gong, F. A human XPC protein interactome—A resource. *Int. J. Mol. Sci.* **2013**, *15*, 141–158. [CrossRef] [PubMed]
78. Zhu, W.; Li, J.; Zhang, R.; Cai, Y.; Wang, C.; Qi, S.; Chen, S.; Liang, X.; Qi, N.; Hou, F. TRAF3IP3 mediates the recruitment of TRAF3 to MAVS for antiviral innate immunity. *EMBO J.* **2019**, *38*, e102075. [CrossRef] [PubMed]
79. Zhang, X.; Zhu, C.; Wang, T.; Jiang, H.; Ren, Y.; Zhang, Q.; Wu, K.; Liu, F.; Liu, Y.; Wu, J. GP73 represses host innate immune response to promote virus replication by facilitating MAVS and TRAF6 degradation. *PLoS Pathog.* **2017**, *13*, e1006321. [CrossRef] [PubMed]
80. Mao, H.T.; Wang, Y.; Cai, J.; Meng, J.L.; Zhou, Y.; Pan, Y.; Qian, X.P.; Zhang, Y.; Zhang, J. HACE1 Negatively Regulates Virus-Triggered Type I IFN Signaling by Impeding the Formation of the MAVS-TRAF3 Complex. *Viruses* **2016**, *8*, 146. [CrossRef] [PubMed]
81. Lin, Y.C.; Huang, D.Y.; Chu, C.L.; Lin, Y.L.; Lin, W.W. The tyrosine kinase Syk differentially regulates Toll-like receptor signaling downstream of the adaptor molecules TRAF6 and TRAF3. *Sci. Signal.* **2013**, *6*, ra71. [CrossRef]
82. Helgason, E.; Phung, Q.T.; Dueber, E.C. Recent insights into the complexity of Tank-binding kinase 1 signaling networks: The emerging role of cellular localization in the activation and substrate specificity of TBK1. *FEBS Lett.* **2013**, *587*, 1230–1237. [CrossRef] [PubMed]
83. Hayakari, R.; Matsumiya, T.; Xing, F.; Yoshida, H.; Hayakari, M.; Imaizumi, T. Critical Role of IRF-3 in the Direct Regulation of dsRNA-Induced Retinoic Acid-Inducible Gene-I (RIG-I) Expression. *PLoS ONE* **2016**, *11*, e0163520. [CrossRef] [PubMed]

84. Dalrymple, N.A.; Cimica, V.; Mackow, E.R. Dengue Virus NS Proteins Inhibit RIG-I/MAVS Signaling by Blocking TBK1/IRF3 Phosphorylation: Dengue Virus Serotype 1 NS4A Is a Unique Interferon-Regulating Virulence Determinant. *mBio* **2015**, *6*, e00553-15. [CrossRef] [PubMed]
85. Fang, P.; Fang, L.; Xia, S.; Ren, J.; Zhang, J.; Bai, D.; Zhou, Y.; Peng, G.; Zhao, S.; Xiao, S. Porcine Deltacoronavirus Accessory Protein NS7a Antagonizes IFN-β Production by Competing With TRAF3 and IRF3 for Binding to IKKε. *Front. Cell Infect. Microbiol.* **2020**, *10*, 257. [CrossRef] [PubMed]
86. El Maadidi, S.; Faletti, L.; Berg, B.; Wenzl, C.; Wieland, K.; Chen, Z.J.; Maurer, U.; Borner, C. A novel mitochondrial MAVS/Caspase-8 platform links RNA virus-induced innate antiviral signaling to Bax/Bak-independent apoptosis. *J. Immunol.* **2014**, *192*, 1171–1183. [CrossRef]
87. Wang, H.; Cebotaru, L.; Lee, H.W.; Yang, Q.; Pollard, B.S.; Pollard, H.B.; Guggino, W.B. CFTR Controls the Activity of NF-κB by Enhancing the Degradation of TRADD. *Cell Physiol. Biochem.* **2016**, *40*, 1063–1078. [CrossRef] [PubMed]
88. Poeck, H.; Bscheider, M.; Gross, O.; Finger, K.; Roth, S.; Rebsamen, M.; Hannesschläger, N.; Schlee, M.; Rothenfusser, S.; Barchet, W.; et al. Recognition of RNA virus by RIG-I results in activation of CARD9 and inflammasome signaling for interleukin 1 beta production. *Nat. Immunol.* **2010**, *11*, 63–69. [CrossRef]
89. Kell, A.M.; Gale, M., Jr. RIG-I in RNA virus recognition. *Virology* **2015**, *479–480*, 110–121. [CrossRef] [PubMed]
90. Zhi, X.; Zhang, Y.; Sun, S.; Zhang, Z.; Dong, H.; Luo, X.; Wei, Y.; Lu, Z.; Dou, Y.; Wu, R.; et al. NLRP3 inflammasome activation by Foot-and-mouth disease virus infection mainly induced by viral RNA and non-structural protein 2B. *RNA Biol.* **2020**, *17*, 335–349. [CrossRef] [PubMed]
91. Burtscher, J.; Cappellano, G.; Omori, A.; Koshiba, T.; Millet, G.P. Mitochondria: In the Cross Fire of SARS-CoV-2 and Immunity. *iScience* **2020**, *23*, 101631. [CrossRef] [PubMed]
92. Untereiner, A.A.; Fu, M.; Módis, K.; Wang, R.; Ju, Y.; Wu, L. Stimulatory effect of CSE-generated H2S on hepatic mitochondrial biogenesis and the underlying mechanisms. *Nitric Oxide* **2016**, *58*, 67–76. [CrossRef]
93. Wenz, T. Regulation of mitochondrial biogenesis and PGC-1α under cellular stress. *Mitochondrion* **2013**, *13*, 134–142. [CrossRef]
94. Zhang, Y.; Xu, H. Translational regulation of mitochondrial biogenesis. *Biochem. Soc. Trans.* **2016**, *44*, 1717–1724. [CrossRef] [PubMed]
95. Wang, X.; Li, S.; Liu, L.; Jian, Z.; Cui, T.; Yang, Y.; Guo, S.; Yi, X.; Wang, G.; Li, C.; et al. Role of the aryl hydrocarbon receptor signaling pathway in promoting mitochondrial biogenesis against oxidative damage in human melanocytes. *J. Dermatol. Sci.* **2019**, *96*, 33–41. [CrossRef] [PubMed]
96. Lagouge, M.; Argmann, C.; Gerhart-Hines, Z.; Meziane, H.; Lerin, C.; Daussin, F.; Messadeq, N.; Milne, J.; Lambert, P.; Elliott, P.; et al. Resveratrol improves mitochondrial function and protects against metabolic disease by activating SIRT1 and PGC-1alpha. *Cell* **2006**, *127*, 1109–1122. [CrossRef] [PubMed]
97. Tay, M.Z.; Poh, C.M.; Rénia, L.; MacAry, P.A.; Ng, L.F.P. The trinity of COVID-19: Immunity, inflammation and intervention. *Nat. Rev. Immunol.* **2020**, *20*, 363–374. [CrossRef]
98. Huang, C.; Wang, Y.; Li, X.; Ren, L.; Zhao, J.; Hu, Y.; Zhang, L.; Fan, G.; Xu, J.; Gu, X.; et al. Clinical features of patients infected with 2019 novel coronavirus in Wuhan, China. *Lancet* **2020**, *395*, 497–506. [CrossRef]
99. Wang, Z.; Yang, B.; Li, Q.; Wen, L.; Zhang, R. Clinical Features of 69 Cases with Coronavirus Disease 2019 in Wuhan, China. *Clin. Infect Dis.* **2020**, *71*, 769–777. [CrossRef]
100. Wu, F.; Zhao, S.; Yu, B.; Chen, Y.M.; Wang, W.; Song, Z.G.; Hu, Y.; Tao, Z.W.; Tian, J.H.; Pei, Y.Y.; et al. A new coronavirus associated with human respiratory disease in China. *Nature* **2020**, *579*, 265–269. [CrossRef] [PubMed]
101. Gordon, D.E.; Jang, G.M.; Bouhaddou, M.; Xu, J.; Obernier, K.; O'Meara, M.J.; Guo, J.Z.; Swaney, D.L.; Tummino, T.A.; Hüttenhain, R.; et al. A SARS-CoV-2-Human Protein-Protein Interaction Map Reveals Drug Targets and Potential Drug-Repurposing. *bioRxiv* 2020. [CrossRef]
102. Wu, A.; Peng, Y.; Huang, B.; Ding, X.; Wang, X.; Niu, P.; Meng, J.; Zhu, Z.; Zhang, Z.; Wang, J.; et al. Genome Composition and Divergence of the Novel Coronavirus (2019-nCoV) Originating in China. *Cell Host Microbe* **2020**, *27*, 325–328. [CrossRef]
103. Polidoro, R.B.; Hagan, R.S.; de Santis Santiago, R.; Schmidt, N.W. Overview: Systemic Inflammatory Response Derived from Lung Injury Caused by SARS-CoV-2 Infection Explains Severe Outcomes in COVID-19. *Front. Immunol.* **2020**, *11*, 1626. [CrossRef] [PubMed]
104. Fu, Y.Z.; Wang, S.Y.; Zheng, Z.Q.; Yi, H.; Li, W.W.; Xu, Z.S.; Wang, Y.Y. SARS-CoV-2 membrane glycoprotein M antagonizes the MAVS-mediated innate antiviral response. *Cell. Mol. Immunol.* **2020**, *18*, 613–620. [CrossRef] [PubMed]
105. Lei, X.; Dong, X.; Ma, R.; Wang, W.; Xiao, X.; Tian, Z.; Wang, C.; Wang, Y.; Li, L.; Ren, L.; et al. Activation and evasion of type I interferon responses by SARS-CoV-2. *Nat. Commun.* **2020**, *11*, 3810. [CrossRef] [PubMed]
106. Archer, S.L.; Sharp, W.W.; Weir, E.K. Differentiating COVID-19 Pneumonia From Acute Respiratory Distress Syndrome and High Altitude Pulmonary Edema: Therapeutic Implications. *Circulation* **2020**, *142*, 101–104. [CrossRef] [PubMed]
107. Saleh, J.; Peyssonnaux, C.; Singh, K.K.; Edeas, M. Mitochondria and microbiota dysfunction in COVID-19 pathogenesis. *Mitochondrion* **2020**, *54*, 1–7. [CrossRef] [PubMed]
108. Wu, Y.; Ma, L.; Zhuang, Z.; Cai, S.; Zhao, Z.; Zhou, L.; Zhang, J.; Wang, P.H.; Zhao, J.; Cui, J. Main protease of SARS-CoV-2 serves as a bifunctional molecule in restricting type I interferon antiviral signaling. *Signal Transduct. Target. Ther.* **2020**, *5*, 221. [CrossRef] [PubMed]

109. Jiang, H.W.; Zhang, H.N.; Meng, Q.F.; Xie, J.; Li, Y.; Chen, H.; Zheng, Y.X.; Wang, X.N.; Qi, H.; Zhang, J.; et al. SARS-CoV-2 Orf9b suppresses type I interferon responses by targeting TOM70. *Cell Mol. Immunol.* **2020**, *17*, 998–1000. [CrossRef] [PubMed]
110. Kreimendahl, S.; Rassow, J. The Mitochondrial Outer Membrane Protein Tom70-Mediator in Protein Traffic, Membrane Contact Sites and Innate Immunity. *Int. J. Mol. Sci.* **2020**, *21*, 7262. [CrossRef]
111. Fonseca, W.; Lukacs, N.W.; Ptaschinski, C. Factors Affecting the Immunity to Respiratory Syncytial Virus: From Epigenetics to Microbiome. *Front. Immunol.* **2018**, *9*, 226. [CrossRef] [PubMed]
112. Ptaschinski, C.; Lukacs, N.W. Early Life Respiratory Syncytial Virus Infection and Asthmatic Responses. *Immunol. Allergy Clin. N. Am.* **2019**, *39*, 309–319. [CrossRef]
113. Munday, D.C.; Howell, G.; Barr, J.N.; Hiscox, J.A. Proteomic analysis of mitochondria in respiratory epithelial cells infected with human respiratory syncytial virus and functional implications for virus and cell biology. *J. Pharm. Pharmacol.* **2015**, *67*, 300–318. [CrossRef] [PubMed]
114. Kipper, S.; Hamad, S.; Caly, L.; Avrahami, D.; Bacharach, E.; Jans, D.A.; Gerber, D.; Bajorek, M. New host factors important for respiratory syncytial virus (RSV) replication revealed by a novel microfluidics screen for interactors of matrix (M) protein. *Mol. Cell Proteom.* **2015**, *14*, 532–543. [CrossRef]
115. Hu, M.; Schulze, K.E.; Ghildyal, R.; Henstridge, D.C.; Kolanowski, J.L.; New, E.J.; Hong, Y.; Hsu, A.C.; Hansbro, P.M.; Wark, P.A.; et al. Respiratory syncytial virus co-opts host mitochondrial function to favour infectious virus production. *eLife* **2019**, *8*, e42448. [CrossRef] [PubMed]
116. Hayashida, H.; Toh, H.; Kikuno, R.; Miyata, T. Evolution of influenza virus genes. *Mol. Biol. Evol.* **1985**, *2*, 289–303. [CrossRef]
117. DeDiego, M.L.; Nogales, A.; Lambert-Emo, K.; Martinez-Sobrido, L.; Topham, D.J. NS1 Protein Mutation I64T Affects Interferon Responses and Virulence of Circulating H3N2 Human Influenza A Viruses. *J. Virol.* **2016**, *90*, 9693–9711. [CrossRef] [PubMed]
118. Nogales, A.; Martinez-Sobrido, L.; Topham, D.J.; DeDiego, M.L. Modulation of Innate Immune Responses by the Influenza A NS1 and PA-X Proteins. *Viruses* **2018**, *10*, 708. [CrossRef] [PubMed]
119. Lupfer, C.; Thomas, P.G.; Anand, P.K.; Vogel, P.; Milasta, S.; Martinez, J.; Huang, G.; Green, M.; Kundu, M.; Chi, H.; et al. Receptor interacting protein kinase 2-mediated mitophagy regulates inflammasome activation during virus infection. *Nat. Immunol.* **2013**, *14*, 480–488. [CrossRef] [PubMed]
120. Eng, V.V.; Wemyss, M.A.; Pearson, J.S. The diverse roles of RIP kinases in host-pathogen interactions. *Semin. Cell Dev. Biol.* **2020**, *109*, 125–143. [CrossRef] [PubMed]
121. Pasricha, G.; Mukherjee, S.; Chakrabarti, A.K. Apoptotic and Early Innate Immune Responses to PB1-F2 Protein of Influenza A Viruses Belonging to Different Subtypes in Human Lung Epithelial A549 Cells. *Adv. Virol.* **2018**, *2018*, 5057184. [CrossRef]
122. Cheung, P.H.; Lee, T.T.; Kew, C.; Chen, H.; Yuen, K.Y.; Chan, C.P.; Jin, D.Y. Virus subtype-specific suppression of MAVS aggregation and activation by PB1-F2 protein of influenza A (H7N9) virus. *PLoS Pathog.* **2020**, *16*, e1008611. [CrossRef] [PubMed]
123. Scrima, R.; Piccoli, C.; Moradpour, D.; Capitanio, N. Targeting Endoplasmic Reticulum and/or Mitochondrial Ca(2+) Fluxes as Therapeutic Strategy for HCV Infection. *Front. Chem.* **2018**, *6*, 73. [CrossRef]
124. Horner, S.M.; Liu, H.M.; Park, H.S.; Briley, J.; Gale, M., Jr. Mitochondrial-associated endoplasmic reticulum membranes (MAM) form innate immune synapses and are targeted by hepatitis C virus. *Proc. Natl. Acad. Sci. USA* **2011**, *108*, 14590–14595. [CrossRef]
125. Jassey, A.; Liu, C.H.; Changou, C.A.; Richardson, C.D.; Hsu, H.Y.; Lin, L.T. Hepatitis C Virus Non-Structural Protein 5A (NS5A) Disrupts Mitochondrial Dynamics and Induces Mitophagy. *Cells* **2019**, *8*, 290. [CrossRef] [PubMed]
126. Wong, M.T.; Chen, S.S. Emerging roles of interferon-stimulated genes in the innate immune response to hepatitis C virus infection. *Cell Mol. Immunol.* **2016**, *13*, 11–35. [CrossRef] [PubMed]
127. Douglas, D.N.; Pu, C.H.; Lewis, J.T.; Bhat, R.; Anwar-Mohamed, A.; Logan, M.; Lund, G.; Addison, W.R.; Lehner, R.; Kneteman, N.M. Oxidative Stress Attenuates Lipid Synthesis and Increases Mitochondrial Fatty Acid Oxidation in Hepatoma Cells Infected with Hepatitis C Virus. *J. Biol. Chem.* **2016**, *291*, 1974–1990. [CrossRef]
128. Rahmani, Z.; Huh, K.W.; Lasher, R.; Siddiqui, A. Hepatitis B virus X protein colocalizes to mitochondria with a human voltage-dependent anion channel, HVDAC3, and alters its transmembrane potential. *J. Virol.* **2000**, *74*, 2840–2846. [CrossRef] [PubMed]
129. Xie, W.H.; Ding, J.; Xie, X.X.; Yang, X.H.; Wu, X.F.; Chen, Z.X.; Guo, Q.L.; Gao, W.Y.; Wang, X.Z.; Li, D. Hepatitis B virus X protein promotes liver cell pyroptosis under oxidative stress through NLRP3 inflammasome activation. *Inflamm. Res.* **2020**, *69*, 683–696. [CrossRef] [PubMed]
130. Khan, M.; Syed, G.H.; Kim, S.J.; Siddiqui, A. Hepatitis B Virus-Induced Parkin-Dependent Recruitment of Linear Ubiquitin Assembly Complex (LUBAC) to Mitochondria and Attenuation of Innate Immunity. *PLoS Pathog.* **2016**, *12*, e1005693. [CrossRef] [PubMed]
131. Tornesello, M.L.; Buonaguro, L.; Izzo, F.; Buonaguro, F.M. Molecular alterations in hepatocellular carcinoma associated with hepatitis B and hepatitis C infections. *Oncotarget* **2016**, *7*, 25087–25102. [CrossRef] [PubMed]
132. Liu, Y.; Olagnier, D.; Lin, R. Host and Viral Modulation of RIG-I-Mediated Antiviral Immunity. *Front. Immunol.* **2016**, *7*, 662. [CrossRef] [PubMed]
133. Megahed, F.A.K.; Zhou, X.; Sun, P. The Interactions between HBV and the Innate Immunity of Hepatocytes. *Viruses* **2020**, *12*, 285. [CrossRef] [PubMed]

134. Xia, M.; Gonzalez, P.; Li, C.; Meng, G.; Jiang, A.; Wang, H.; Gao, Q.; Debatin, K.M.; Beltinger, C.; Wei, J. Mitophagy enhances oncolytic measles virus replication by mitigating DDX58/RIG-I-like receptor signaling. *J. Virol.* **2014**, *88*, 5152–5164. [CrossRef]
135. Richetta, C.; Grégoire, I.P.; Verlhac, P.; Azocar, O.; Baguet, J.; Flacher, M.; Tangy, F.; Rabourdin-Combe, C.; Faure, M. Sustained autophagy contributes to measles virus infectivity. *PLoS Pathog.* **2013**, *9*, e1003599. [CrossRef] [PubMed]

 life

Review

Mitochondrial Functionality in Inflammatory Pathology-Modulatory Role of Physical Activity

Rafael A. Casuso * and Jesús R. Huertas *

Department of Physiology, Institute of Nutrition and Food Technology, University of Granada, 18100 Granada, Spain
* Correspondence: racasuso@ugr.es (R.A.C.); jhuertas@ugr.es (J.R.H.); Tel.: +34-958-241-000 (ext. 20318) (R.A.C. & J.R.H.)

Abstract: The incidence and severity of metabolic diseases can be reduced by introducing healthy lifestyle habits including moderate exercise. A common observation in age-related metabolic diseases is an increment in systemic inflammation (the so-called inflammaging) where mitochondrial reactive oxygen species (ROS) production may have a key role. Exercise prevents these metabolic pathologies, at least in part, due to its ability to alter immunometabolism, e.g., reducing systemic inflammation and by improving immune cell metabolism. Here, we review how exercise regulates immunometabolism within contracting muscles. In fact, we discuss how circulating and resident macrophages alter their function due to mitochondrial signaling, and we propose how these effects can be triggered within skeletal muscle in response to exercise. Finally, we also describe how exercise-induced mitochondrial adaptations can help to fight against virus infection. Moreover, the fact that moderate exercise increases circulating immune cells must be taken into account by public health agencies, as it may help prevent virus spread. This is of interest in order to face not only acute respiratory-related coronavirus (SARS-CoV) responsible for the COVID-19 pandemic but also for future virus infection challenges.

Keywords: exercise; mitochondria; immune system; metabolic disease; COVID-19

Citation: Casuso, R.A.; Huertas, J.R. Mitochondrial Functionality in Inflammatory Pathology-Modulatory Role of Physical Activity. *Life* **2021**, *11*, 61. https://doi.org/10.3390/life11010061

Received: 21 December 2020
Accepted: 13 January 2021
Published: 15 January 2021

1. Introduction

Immunometabolism is a recently proposed term which highlight the close relationship between systemic and cellular metabolism and the immune system [1]. On one hand, immunometabolism studies show how the metabolism of immune cells ultimately regulates their function [2]. On the other hand, they also cover the chronic activation of the immune system that is observed in several pathologies such as obesity, Type 2 Diabetes, cardiovascular diseases and cancer [1,2]. However, these two events should not be viewed as independent processes. In fact, mitochondrial signaling from bone marrow cells stimulates macrophage polarization and their infiltration into tissues thereby controlling diet-inducing obesity [3]. In fact, mitochondrial metabolism has a key role in regulating both innate immune cell function during infection and metabolic disease development (for detailed reviews, see [4,5]).

In the present review, we describe how exercise regulates immune function, reduces systemic inflammation and ultimately results in improving age-related metabolic diseases. We also highlight a close relationship between the mitochondrial function of immune cells and their functional infiltration into tissues thereby altering systemic metabolism. Finally, we propose a scenario where exercise improves skeletal muscle macrophage–mitochondria crosstalk, which will likely enhance skeletal muscle metabolism and prevent age-related skeletal muscle dysfunction.

2. Exercise Alters Systemic Inflammatory State: Role of Exercise Intensity

The immune system responds to exercise following a hormetic curve. Notably, an immunological threshold has been recently proposed for exercise intensity (up to 60% of oxygen uptake) and duration (up to 60 min) [6]. In this scenario of moderate exercise intensity and/or duration (i.e., moderate exercise), an acute exercise bout increase the exchange of leukocytes between circulation and tissues; in addition, anti-inflammatory cytokines are released and tissue macrophages show antipathogen activity [7,8]. Moreover, the fact that natural killer cells and CD38+ T lymphocytes enrich the blood compartment [7,9] has been used to state that acute moderate exercise may protect from subsequent infection [6]. Therefore, this kind of exercise exerts a number of physiological responses that may act as a barrier for acute virus infection and maybe its spread.

In contrast, high-volume vigorous exercise (i.e., vigorous exercise) is associated with up to six-fold upper respiratory tract infections in the weeks following it. This effect has been associated with increments in circulating lymphocytes and neutrophils [10] probably due to a decreased levels of NK activity as well as T cell and macrophage function [11,12]. Importantly, this effect not only reflects a high metabolic stress but can also be triggered by skeletal muscle damage leading to pro-inflammatory cytokine release [13] and a strong innate immune response [14,15]. Notably, Casuso et al. asked highly trained subjects in both swimming and running to perform 8 bouts of 30 s at maximal intensity (~30 min) one day by swimming and one day by running. It was reported that circulating levels of interleukin 6 (IL-6) raised 2 h after running but not after swimming. Moreover, this effect was not due to muscle damage nor to metabolic stress (i.e., cortisol release), and a general anti-inflammatory response was observed after both exercise types [16]. This suggests that (i) immune response to exercise may differ between exercise modes and/or due to the contracting muscle and that (ii) extremely intense exercise does not perturb the immune response in trained subjects, at least if there is no excessive duration.

The chronic application of moderate-intensity exercise bouts (i.e., training) is known to decrease illness and lower systemic inflammation [6,17]. It is less clear how regular vigorous exercise affects the immune system. Nevertheless, it is tempting to suggest that the acute depression of the immune system described above would follow a homeostatic response thereby showing an enhanced effect in the long term. What is clear is that regular moderate exercise can prevent systemic inflammation.

3. Exercise Impact Chronic Inflammation in Age and Age-Related Metabolic Diseases

Many chronic diseases are associated with persistent low-grade systemic inflammation. For instance, immune cells can infiltrate in adipose tissue thereby promoting a pro-inflammatory environment leading to insulin resistance and Type 2 Diabetes [18,19]. Moreover, a similar polarization and infiltration of the immune cells has been observed in cardiovascular diseases and in some cancers [20–22]. Franceschi et al. (2018) proposed that aging and ageing-related metabolic diseases should be studied as a whole as they converge in systemic inflammation (i.e., inflammaging) and both directly and indirectly influence each other. In fact, they propose that metabolic inflammation driven, for example, by nutrient excess and inflammaging probably share the same molecular pathways [23]. Although this remains to be empirically demonstrated, it seems evident that age and age-related metabolic diseases show increased levels of pro-inflammatory cytokines [24]. This is likely due to a decrease in the body´s ability to repair damaged cells or tissues, thus leading to accumulated damage over time. In fact, a chronic activation of the innate immune system underlies the inflammaging process where macrophages have a key role [25].

In this scenario, twenty years ago, Pedersen´s laboratory first described that contracting muscle release IL-6 into the circulation indeed altered the function of other parts of the body [26]. For instance, muscle-derived IL-6 is known to reduce systemic inflammation by promoting cortisol and IL-10 release [27,28]; this is of importance because IL-10 blocks the transcription of the inflammatory cytokine tumor necrosis factor apha (TNF-α) [29]. It is now known that skeletal muscle secretes hundreds of peptides (i.e., myokines), most of them with unknown effects. While the main studied one remains IL-6, other myokines have been described to affect bone, liver, adipose tissue, brain and muscle function, among others (see Sverinsen and Pedersen [30] for a detailed updated review). Nevertheless, it is important to note that IL-6 has a double-edged effect depending on the secreting cell. In contrast to exercise, chronic sedentarism polarizes macrophages, thereby releasing pro-inflammatory cytokines such as IL-6, IL-1b and TNFα [31].

Exercise has been recommended as a therapy for a number of metabolic diseases such as obesity, type 2 diabetes as well as for cardiovascular diseases and cancer [32]. Notably, most of these diseases are directly or indirectly affected by contracting muscle-derived IL-6. For instance, IL-6 receptor blockade impedes exercise-induced visceral and cardiac adipose tissue reduction [33,34]. Additionally, IL-6 regulates glucose uptake by promoting glucose transporter 4 (GLUT4) translocation to the sarcolemma [35].

Moreover, sarcopenia (i.e., age-related muscle and/or force decline) has also been associated with increments in circulating levels of IL-6 and TNF-α [36]. Studies in mice showed that genetic loss of IL-6 impair skeletal muscle hypertrophy [37]. While these observations seem to contrast, it can be explained because, as noted above, chronic muscle disuse promotes IL-6 release from macrophages, thereby inducing a pro-inflammatory environment.

Nevertheless, few studies have addressed macrophage infiltration in response to exercise. Walton et al. [38] reported that 14 weeks of resistance training increases anti-inflammatory macrophage content within aged human skeletal muscle. Moreover, studies analyzing cycling training show that anti-inflammatory macrophages may have a key role in stimulating muscle growth [39]. More recently, Jensen et al. [40] analyzed the time course of macrophage infiltration and polarization following acute resistance exercise. They found that within the range of 4 to 7 days after exercise, older males show an increase in anti-inflammatory macrophages, while older females show an increase in both anti- and pro-inflammatory macrophages [40]. Notably, while young females show unchanged macrophage polarization and infiltration, 5 days following exercise, there is an increase in total macrophage and anti-inflammatory cytokines in older compared to young females [40]. Taken together, these data suggest that macrophage polarization influences circulating pro- and anti-inflammatory cytokines which may directly or indirectly impact age-related muscle waste. Moreover, there is a need to unravel how resistance exercise-induced alterations in macrophage polarization influence sarcopenia and the potential role (if any) of sexual dimorphism.

4. The Mitochondria–Macrophage Connection Regulates Metabolism

Mitochondria are double membrane organelles that support cellular function including metabolism and signaling. The inner mitochondrial membrane contains the electron transport chain (ETC), a multiprotein complex system that pumps protons from the mitochondrial matrix into the intermembrane; complex V can use the stored potential energy to generate ATP. This highlights the importance of mitochondria in integrating fuel metabolism to cellular ATP production. However, mitochondria are now widely recognized as biosynthetic hubs such as for nucleotide synthesis, fatty acid and cholesterol synthesis, amino acid synthesis, and glucose and heme synthesis, but they also orchestrate waste management [41]. Therefore, mitochondria are contemporarily viewed as a multifaceted organelle. Of particular relevance is the fact that mitochondrial reactive oxygen species (ROS) production acts as a signaling molecule that ultimately results (if produced in excess) in the increased inflammation observed in obesity and obesity-related diseases [42]. In this regard, excessive ROS production by ETC complex I was observed when it was not assembled with complex III and/or complex IV forming supercomplexes [43,44]. In agreement, skeletal muscle of diabetic subjects showed a decrease in the content of supercomplexes [45]. Therefore, ETC supramolecular organization may contribute to regulating metabolic diseases by controlling ROS production. However, the precise role of mitochondrial ROS production in health and disease is yet to be elucidated. For instance, it seems that mitochondrial

ROS production via reverse electron transport improves lifespan in model organisms [46]. This may suggest that "when" and "where" mitochondrial ROS are produced are relevant factors in order to switch from beneficial to adverse adaptations.

As stated above, immunometabolism can refer either to (i) how immune cells function depending on their own metabolism or to (ii) how chronic systemic inflammation influences the development and progression of metabolic diseases. However, these two apparently different processes might be influenced by each other. For instance, infiltration of pro-inflammatory macrophages into adipose tissue drives metabolic dysfunction through the expression of TNF-a, IL-1B and IL-6 [47]. Notably, excessive ROS production leads to Fgr activation in macrophages which is associated with increased mitochondrial complex II activity and complex I degradation leading to pro-inflammatory macrophage polarization [48,49]. Recently, Acín-Perez et al. [3] effectively demonstrated that mice lacking Fgr are protected against high fat diet-induced obesity and insulin resistance. Furthermore, transplanting bone marrow cells from mice overexpressing mitochondrial catalase protects mice from high fat diet-induced obesity and changes tissue macrophage towards an anti-inflammatory polarization. Therefore these data not only suggest that bone-marrow-derived cell metabolism might influence obesity and insulin resistance but also that prevention of excessive mitochondrial H_2O_2 production may be an interesting target for the prevention and treatment of metabolic disease. Moreover, it also opens the possibility that local mitochondrial ROS production has a key role in regulating systemic inflammation and metabolic disease. In this regard, it is important to note that moderate exercise training gradually decreases ROS production within contracting muscle [50]. This ultimately maintains ROS production at physiological levels, thus avoiding pathological ROS production. In addition, we recently reported that moderate exercise training changes the supramolecular organization of the ETC, which prevents excessive mitochondrial ROS production and systemic lipid peroxidation [51]. These findings open new interesting hypotheses such as whether exercise training alters mitochondrial function from circulating and/or skeletal muscle resident macrophages, thereby improving overall health (Figure 1).

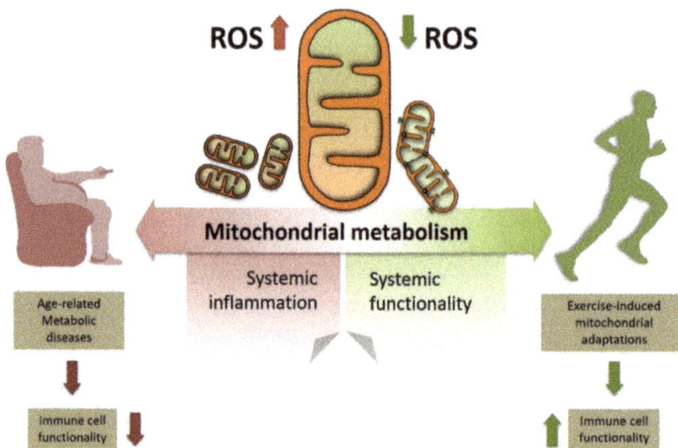

Figure 1. Exercise improves systemic inflammation by enhancing mitochondrial metabolism. This effect likely involves decreased mitochondrial reactive oxygen species (ROS) production below pathological levels which may ultimately affect immune cell function.

Additionally, these observations could also anticipate that exercise-induced skeletal muscle mitochondrial metabolic adaptations result in enhancing resident macrophage function. In this regard, a second recent discovery shows that cardiomyocytes release subcellular particles called exophers which mainly transport defective mitochondria. These exophers are captured and eliminated by cardiac macrophages, thus maintaining cardiomyocyte homeostasis [52]. Notably, the authors identified Mertk as the macrophage phagocytic receptor of exophers [52]. This study has a profound impact not only on understanding the metabolic regulation of cardiac cells, but it also may help to unravel how skeletal muscle responds to stress. Runyan et al. [53] analyzed skeletal muscle macrophages from old and young mice during recovery from influenza A infection-induced pneumonia. They found that macrophages from old mice had a reduced phagocytic function which coincided with reduced Merk expression, thus leading to impaired muscle recovery. Moreover, when they knocked down Mertk in young mice, they found that muscle recovery was also impaired [53]. Cardiomyocytes have evolved to allow continuous beating by maintaining mitochondrial homeostasis through ejecting dysfunctional mitochondrial portions [52]. Given the similar structure between skeletal muscle and cardiac cells and that Mertk might have an impact on skeletal muscle recovery following infection, it would be important to study whether highly trained skeletal muscle may also have such a macrophage function as mitochondrial waste management. In fact, highly trained skeletal muscle shows a highly specialized mitochondrial network which facilitates energy diffusion through skeletal muscle cells [54]. Notably, the capacity to reach this adaptation is almost intact in aged human skeletal muscle [55]. This is of importance as muscle regenerative capacity is impaired in aged skeletal muscle [56]. Therefore, aged skeletal muscle cells show a low renewal rate but almost intact mitochondrial function (at least in response to exercise) which somewhat mimics the physiology of cardiomyocytes [57]. These recent studies open a new line of investigation of age-related skeletal muscle dysfunction.

Finally, it is important to highlight another key finding that might have an important implication in understanding how skeletal muscle mitochondria connect immunity and metabolism in mammals. There is growing interest in elucidating how contracting muscles use lipid droplets (LD) in order to sustain exercise metabolism and/or for unknown purposes. For instance, during high-volume, high-intensity exercise (i.e., 57 min and 11 mmol/L blood lactate), LD within myofibrils but not those LD located close to the sarcolemma are used [58]. This is surprising as this kind of exercise mostly relies on oxidative metabolism mainly through fatty acid and glucose oxidation [59]. LDs have important implications in cellular innate defense and, in hepatocytes, this effect seems to be dependent upon LD–mitochondria binding [60]. In fact, pathogen-associated molecular pattern lipopolysaccharide (LPS) stimulation-induced innate immune response was prevented when perilipin-5 (PLN5) was overexpressed [60]. Notably, both in sedentary and trained skeletal muscle, moderate exercise seems to decrease LD–PLN5 association [61]. As PLN5 is involved in mitochondria–LD tethering, these observations could suggest that moderate exercise liberates LD from mitochondria, thus impeding any competition between mitochondria and bacteria for LD tethering. This further supports the observation that acute moderate exercise prevents subsequent infection in this case within skeletal muscle. Furthermore, during innate immune response, LD increase their size [62]. This may also reveal different effects of LDs within skeletal muscle fibers, for instance, type 2 fibers have lower LD density, but they seems to have higher size than those LD form type 1 fibers in the leg skeletal muscle [58]. Although the current evidence is scarce, it would be interesting to test whether exercise impacts innate immune function by altering LD–mitochondria contact.

5. Mitochondria at the Crossroad to Viral Infection

Although it is not the primary aim of the present review, given the actual situation due to the COVID-19 pandemic, we would like to discuss how skeletal muscle mitochondrial fitness may help to face virus infection. It has been reported in a cohort of 249 subjects (59 ± 12 years old) that maximal exercise capacity was independently and inversely associated with the likelihood for hospitalization due to COVID-19 [63]. This is in accordance with the fact that obesity is a risk factor for increasing severity due to virus infection and that obesity impairs antibody response to the influenza vaccine [64]. Therefore, general fitness is a protective factor to virus infection severity in general and COVID-19 disease in particular.

Damage-associated molecular patterns are the pathway by which innate immune cells recognize pathogens leading to specific immune response. A growing body of evidence highlights the key role of mitochondria, in addition to the ROS-mediated mechanisms, in innate immunity. The reader is referred to some excellent new reviews on the topic [65,66], because we will not describe in detail these molecular mechanisms. In brief, when viruses reach the cytoplasm, RNAs perceived as foreign activate the aggregation of an outer mitochondrial membrane protein called mitochondrial antiviral-signaling protein (MAVs). MAV aggregation induces the expression factor kappa light chain enhancer of activated B cell (NF-kB) and type I interferons via interferon regulatory factors (IRFs) in the nucleus, thereby leading to antiviral defense [65,66].

However, this pathway alters mitochondrial metabolism and morphology. It has been reported that mitochondrial dynamics (i.e., balance between mitochondrial fusion and fission) and mitophagy seem to be important to the regulation of immune response to virus infection [67]. In fact, some viruses are able to alter the innate immune system in cells by increasing DRP1 s616 phosphorylation [68], which ultimately results in mitochondrial fragmentation and dysfunction [66]. Additionally, peripheral mononuclear cells showed altered mitochondrial metabolism by SARS-CoV-2 in patients with COVID-19 [69]. Moreover, the authors proposed that disease severity was positively associated with the degree of mitochondrial dysfunction in peripheral mononuclear cells [69].

These observations are of great interest in order to describe how exercise can hamper virus severity and function. For instance, sedentary aged skeletal muscle shows an enlarged and dysfunctional mitochondrial network due to impaired fission and mitophagy [55]. This may contribute to the increased risk of older subjects to show a higher severity COVID-19 disease. In this regard, moderate exercise is able to improve skeletal muscle mitochondrial function and dynamics [54,70] even in aged skeletal muscle [55]. In particular, only 12 weeks of moderate exercise can decrease skeletal muscle DRP1 s616 phosphorylation, thereby enhancing mitochondrial network function in older subjects [71]. Furthermore, short-term exercise training improves the mitochondrial function and dynamics of peripheral blood mononuclear cells even in previously trained subjects [72]. Therefore, it is tempting to suggest that moderate exercise inducing skeletal muscle and systemic mitochondrial adaptations may lower the degree of virus severity and complications related to COVID-19.

6. Conclusions

Several age-associated pathologies show increased systemic inflammatory markers. These metabolic diseases are worsened if a sedentary state is maintained for a long period, and in this process, the pro-inflammatory polarization of tissue resident macrophages seems to have a key role. In contrast, exercise stimulates a raising number of known myokines which target several tissues and protect them against metabolic diseases. In this scenario, we suggest that acute exercise might polarize circulating macrophages through limiting mitochondrial ROS production, thereby reducing tissue metabolic dysregulation. Moreover, we propose that both skeletal muscle macrophages and LDs may have an important role in the improvement of innate immunity observed in response to training in a process fine-tuned by mitochondria (Figure 2). Finally, we would like to highlight that acute moderate exercise training likely reduces virus infection through increasing circulating lymphocytes and leukocytes and by improving mitochondrial function. This must be acknowledged by public health agencies in order to face the current COVID-19 disease and future virus-related pandemics.

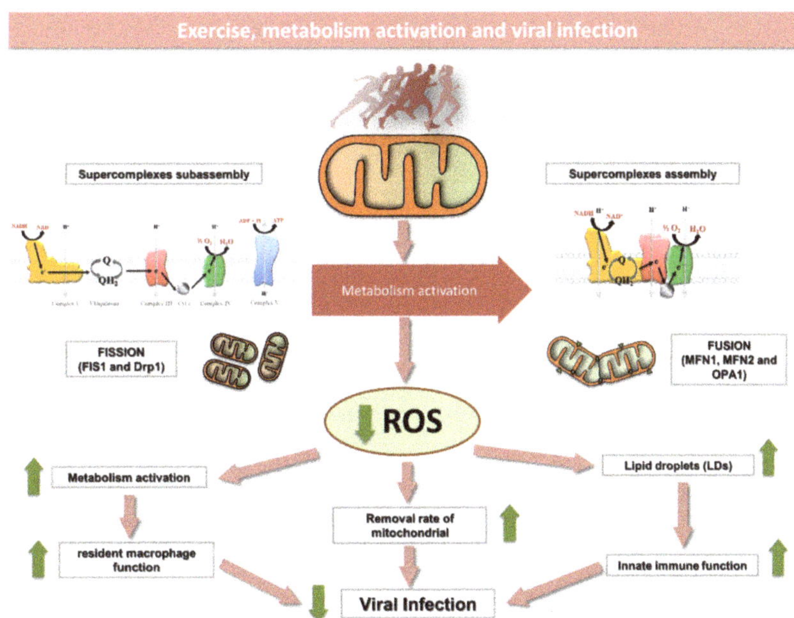

Figure 2. Exercise induces a number of structural and metabolic changes within skeletal muscle mitochondria that may affect immune function. Exercise enhances mitochondrial function by (1) inducing the assembly of respiratory complexes into supercomplexes; (2) promoting an enlargement of functional mitochondrial network; (3) increasing mitochondrial turnover and (4) reducing mitochondrial ROS production. The immunological role of skeletal muscle lipid droplets and resident macrophages in response to exercise is yet to be elucidated. Mechanistic studies, however, suggest that LDs separated from mitochondria can improve innate immune function in a mechanism likely mimicked by exercise. In addition, resident macrophages are known to improve mitochondrial function through eliminating dysfunctional parts of the mitochondrial network. As some viruses alter the innate immune system in cells by inducing excessive mitochondrial fragmentation, exercise may help to face virus infection by maintaining an enlarged and fully functional mitochondrial network. ROS, reactive oxygen species; DRP, dynamin-like protein; FIS1, fission, mitochondrial 1; MFN, mitofusin; OPA1, optic atrophy 1.

Author Contributions: R.A.C. conceived the manuscript; R.A.C. and J.R.H., writing and editing. All authors have read and agreed to the published version of the manuscript.

Funding: This research received no external funding.

Data Availability Statement: No new data were created or analyzed in this study. Data sharing is not applicable to this article.

Acknowledgments: The authors would like to thank to all members of the laboratory.

Conflicts of Interest: The authors declare no conflict of interest.

References

1. Mathis, D.; Shoelson, S.E. Immunometabolism: An emerging frontier. *Nat. Rev. Immunol.* **2011**, *11*, 81–83. [CrossRef] [PubMed]
2. Bay, M.L.; Pedersen, B.K. Muscle-Organ Crosstalk: Focus on Immunometabolism. *Front. Physiol.* **2020**, *11*, 567881. [CrossRef] [PubMed]
3. Acín-Pérez, R.; Iborra, S.; Martí-Mateos, Y.; Cook, E.C.L.; Conde-Garrosa, R.; Petcherski, A.; Muñoz, M.M.; Martínez de Mena, R.; Krishnan, K.C.; Jiménez, C.; et al. Fgr kinase is required for proinflammatory macrophage activation during diet-induced obesity. *Nat. Metab.* **2020**, *2*, 974–988. [CrossRef] [PubMed]

4. Escoll, P.; Platon, L.; Buchrieser, C. Roles of Mitochondrial Respiratory Complexes during Infection. *Immunometabolism* **2019**, *1*, e190011. [CrossRef]
5. Diaz-Vegas, A.; Sanchez-Aguilera, P.; Krycer, J.R.; Morales, P.E.; Monsalves-Alvarez, M.; Cifuentes, M.; Rothermel, B.A.; Lavandero, S. Is Mitochondrial Dysfunction a Common Root of Noncommunicable Chronic Diseases? *Endocr. Rev.* **2020**, *41*, 491–517. [CrossRef]
6. Nieman, D.C.; Wentz, L.M. The compelling link between physical activity and the body's defense system. *J. Sport Health Sci.* **2019**, *8*, 201–217. [CrossRef]
7. Bigley, A.B.; Rezvani, K.; Chew, C.; Sekine, T.; Pistillo, M.; Crucian, B.; Bollard, C.M.; Simpson, R.J. Acute exercise preferentially redeploys NK-cells with a highly-differentiated phenotype and augments cytotoxicity against lymphoma and multiple myeloma target cells. *Brain Behav. Immun.* **2014**, *39*, 160–171. [CrossRef]
8. Lavoy, E.C.P.; Bollard, C.M.; Hanley, P.J.; Blaney, J.W.; O'Connor, D.P.; Bosch, J.A.; Simpson, R.J. A single bout of dynamic exercise enhances the expansion of MAGE-A4 and PRAME-specific cytotoxic T-cells from healthy adults. *Exerc. Immunol. Rev.* **2015**, *21*, 144–153.
9. Campbell, J.P.; Riddell, N.E.; Burns, V.E.; Turner, M.; van Zanten, J.J.C.S.V.; Drayson, M.T.; Bosch, J.A. Acute exercise mobilises CD8+ T lymphocytes exhibiting an effector-memory phenotype. *Brain Behav. Immun.* **2009**, *23*, 767–775. [CrossRef]
10. Gleeson, M. Immune function in sport and exercise. *J. Appl. Physiol.* **2007**, *103*, 693–699. [CrossRef]
11. Siedlik, J.A.; Benedict, S.H.; Landes, E.J.; Weir, J.P.; Vardiman, J.P.; Gallagher, P.M. Acute bouts of exercise induce a suppressive effect on lymphocyte proliferation in human subjects: A meta-analysis. *Brain Behav. Immun.* **2016**, *56*, 343–351. [CrossRef] [PubMed]
12. Peake, J.M.; Neubauer, O.; Gatta, P.A.D.; Nosaka, K. Muscle damage and inflammation during recovery from exercise. *J. Appl. Physiol.* **2017**, *122*, 559–570. [CrossRef] [PubMed]
13. van de Vyver, M.; Myburgh, K.H. Variable inflammation and intramuscular STAT3 phosphorylation and myeloperoxidase levels after downhill running. *Scand. J. Med. Sci. Sports* **2014**, *24*, e360–e371. [CrossRef] [PubMed]
14. Nehlsen-Cannarella, S.L.; Fagoaga, O.R.; Nieman, D.C.; Henson, D.A.; Butterworth, D.E.; Schmitt, R.L.; Bailey, E.M.; Warren, B.J.; Utter, A.; Davis, J.M. Carbohydrate and the cytokine response to 2.5 h of running. *J. Appl. Physiol.* **1997**, *82*, 1662–1667. [CrossRef] [PubMed]
15. Markworth, J.F.; Maddipati, K.R.; Cameron-Smith, D. Emerging roles of pro-resolving lipid mediators in immunological and adaptive responses to exercise-induced muscle injury. *Exerc. Immunol. Rev.* **2016**, *22*, 110–134.
16. Casuso, R.A.; Aragon-Vela, J.; Huertas, J.R.; Ruiz-Ariza, A.; Martínez-Lopez, E.J. Comparison of the inflammatory and stress response between sprint interval swimming and running. *Scand. J. Med. Sci. Sports* **2018**, *28*, 1371–1378. [CrossRef]
17. Nieman, D.C. Is infection risk linked to exercise workload? *Med. Sci. Sports Exerc.* **2000**, *32*, S406–S411. [CrossRef]
18. Hotamisligil, G.S. Inflammation and metabolic disorders. *Nature* **2006**, *444*, 860–867. [CrossRef]
19. Handschin, C.; Spiegelman, B.M. The role of exercise and PGC1α in inflammation and chronic disease. *Nature* **2008**, *454*, 463–469. [CrossRef]
20. Haffner, S.M. The metabolic syndrome: Inflammation, diabetes mellitus, and cardiovascular disease. *Am. J. Cardiol.* **2006**, *97*, 3–11. [CrossRef]
21. Matter, C.M.; Handschin, C. RANTES (regulated on activation, normal T cell expressed and secreted), inflammation, obesity, and the metabolic syndrome. *Circulation* **2007**, *115*, 946–948. [CrossRef] [PubMed]
22. Lin, W.W.; Karin, M. A cytokine-mediated link between innate immunity, inflammation, and cancer. *J. Clin. Investig.* **2007**, *117*, 1175–1183. [CrossRef] [PubMed]
23. Franceschi, C.; Garagnani, P.; Parini, P.; Giuliani, C.; Santoro, A. Inflammaging: A new immune–metabolic viewpoint for age-related diseases. *Nat. Rev. Endocrinol.* **2018**, *14*, 576–590. [CrossRef] [PubMed]
24. Fulop, T.; Larbi, A.; Dupuis, G.; Page, A.L.; Frost, E.H.; Cohen, A.A.; Witkowski, J.M.; Franceschi, C. Immunosenescence and inflamm-aging as two sides of the same coin: Friends or Foes? *Front. Immunol.* **2018**, *8*, 1960. [CrossRef] [PubMed]
25. Franceschi, C.; Bonafè, M.; Valensin, S.; Olivieri, F.; De Luca, M.; Ottaviani, E.; De Benedictis, G. Inflamm-aging. An evolutionary perspective on immunosenescence. *Ann. N. Y. Acad. Sci.* **2000**, *908*, 244–254. [CrossRef]
26. Steensberg, A.; Van Hall, G.; Osada, T.; Sacchetti, M.; Saltin, B.; Pedersen, B.K. Production of interleukin-6 in contracting human skeletal muscles can account for the exercise-induced increase in plasma interleukin-6. *J. Physiol.* **2000**, *529*, 237–242. [CrossRef]
27. Steensberg, A.; Fischer, C.P.; Keller, C.; Møller, K.; Pedersen, B.K. IL-6 enhances plasma IL-1ra, IL-10, and cortisol in humans. *Am. J. Physiol. Endocrinol. Metab.* **2003**, *285*, E433–E437. [CrossRef]
28. Pedersen, B.K.; Febbraio, M.A. Muscle as an endocrine organ: Focus on muscle-derived interleukin-6. *Physiol. Rev.* **2008**, *88*, 1379–1406. [CrossRef]
29. Petersen, A.M.W.; Pedersen, B.K. The anti-inflammatory effect of exercise. *J. Appl. Physiol.* **2005**, *98*, 1154–1162. [CrossRef]
30. Severinsen, M.C.K.; Pedersen, B.K. Muscle-Organ Crosstalk: The Emerging Roles of Myokines. *Endocr. Rev.* **2020**, *41*, 594–609. [CrossRef]
31. Pedersen, B.K. Muscles and their myokines. *J. Exp. Biol.* **2011**, *214*, 337–346. [CrossRef] [PubMed]
32. Pedersen, B.K.; Saltin, B. Exercise as medicine Evidence for prescribing exercise as therapy in 26 different chronic diseases. *Scand. J. Med. Sci. Sports* **2015**, *25*, 1–72. [CrossRef] [PubMed]

33. Wedell-Neergaard, A.S.; Lang Lehrskov, L.; Christensen, R.H.; Legaard, G.E.; Dorph, E.; Larsen, M.K.; Launbo, N.; Fagerlind, S.R.; Seide, S.K.; Nymand, S.; et al. Exercise-Induced Changes in Visceral Adipose Tissue Mass Are Regulated by IL-6 Signaling: A Randomized Controlled Trial. *Cell Metab.* **2019**, *29*, 844–855. [CrossRef] [PubMed]
34. Christensen, R.H.; Lehrskov, L.L.; Wedell-Neergaard, A.S.; Legaard, G.E.; Ried-Larsen, M.; Karstoft, K.; Krogh-Madsen, R.; Pedersen, B.K.; Ellingsgaard, H.; Rosenmeier, J.B. Aerobic exercise induces cardiac fat loss and alters cardiac muscle mass through an interleukin-6 receptor-dependent mechanism. *Circulation* **2019**, *140*, 1684–1686. [CrossRef]
35. Carey, A.L.; Steinberg, G.R.; Macaulay, S.L.; Thomas, W.G.; Holmes, A.G.; Ramm, G.; Prelovsek, O.; Hohnen-Behrens, C.; Watt, M.J.; James, D.E.; et al. Interleukin-6 increases insulin-stimulated glucose disposal in humans and glucose uptake and fatty acid oxidation in vitro via AMP-activated protein kinase. *Diabetes* **2006**, *55*, 2688–2697. [CrossRef]
36. Tuttle, C.S.L.; Thang, L.A.N.; Maier, A.B. Markers of inflammation and their association with muscle strength and mass: A systematic review and meta-analysis. *Ageing Res. Rev.* **2020**, *64*, 101185. [CrossRef]
37. Serrano, A.L.; Baeza-Raja, B.; Perdiguero, E.; Jardí, M.; Muñoz-Cánoves, P. Interleukin-6 Is an Essential Regulator of Satellite Cell-Mediated Skeletal Muscle Hypertrophy. *Cell Metab.* **2008**, *7*, 33–44. [CrossRef]
38. Walton, R.G.; Dungan, C.M.; Long, D.E.; Tuggle, S.C.; Kosmac, K.; Peck, B.D.; Bush, H.M.; Villasante Tezanos, A.G.; McGwin, G.; Windham, S.T.; et al. Metformin blunts muscle hypertrophy in response to progressive resistance exercise training in older adults: A randomized, double-blind, placebo-controlled, multicenter trial: The MASTERS trial. *Aging Cell* **2019**, *18*, e13039. [CrossRef]
39. Walton, R.G.; Kosmac, K.; Mula, J.; Fry, C.S.; Peck, B.D.; Groshong, J.S.; Finlin, B.S.; Zhu, B.; Kern, P.A.; Peterson, C.A. Human skeletal muscle macrophages increase following cycle training and are associated with adaptations that may facilitate growth. *Sci. Rep.* **2019**, *9*, 1–14. [CrossRef]
40. Jensen, S.M.; Bechshøft, C.J.L.; Heisterberg, M.F.; Schjerling, P.; Andersen, J.L.; Kjaer, M.; Mackey, A.L. Macrophage Subpopulations and the Acute Inflammatory Response of Elderly Human Skeletal Muscle to Physiological Resistance Exercise. *Front. Physiol.* **2020**, *11*, 811. [CrossRef]
41. Spinelli, J.B.; Haigis, M.C. The multifaceted contributions of mitochondria to cellular metabolism. *Nat. Cell Biol.* **2018**, *20*, 745–754. [CrossRef] [PubMed]
42. de Mello, A.H.; Costa, A.B.; Engel, J.D.G.; Rezin, G.T. Mitochondrial dysfunction in obesity. *Life Sci.* **2018**, *192*, 26–32. [CrossRef] [PubMed]
43. Maranzana, E.; Barbero, G.; Falasca, A.I.; Lenaz, G.; Genova, M.L. Mitochondrial respiratory supercomplex association limits production of reactive oxygen species from complex i. *Antioxid. Redox Signal.* **2013**, *19*, 1469–1480. [CrossRef] [PubMed]
44. Enríquez, J.A. Supramolecular Organization of Respiratory Complexes. *Annu. Rev. Physiol.* **2016**, *78*, 533–561. [CrossRef] [PubMed]
45. Antoun, G.; McMurray, F.; Thrush, A.B.; Patten, D.A.; Peixoto, A.C.; Slack, R.S.; McPherson, R.; Dent, R.; Harper, M.E. Impaired mitochondrial oxidative phosphorylation and supercomplex assembly in rectus abdominis muscle of diabetic obese individuals. *Diabetologia* **2015**, *58*, 2861–2866. [CrossRef] [PubMed]
46. Scialò, F.; Sriram, A.; Fernández-Ayala, D.; Gubina, N.; Lõhmus, M.; Nelson, G.; Logan, A.; Cooper, H.M.; Navas, P.; Enríquez, J.A.; et al. Mitochondrial ROS Produced via Reverse Electron Transport Extend Animal Lifespan. *Cell Metab.* **2016**, *23*, 725–734. [CrossRef]
47. Curat, C.A.; Miranville, A.; Sengenès, C.; Diehl, M.; Tonus, C.; Busse, R.; Bouloumié, A. From Blood Monocytes to Adipose Tissue-Resident Macrophages: Induction of Diapedesis by Human Mature Adipocytes. *Diabetes* **2004**, *53*, 1285–1292. [CrossRef]
48. Garaude, J.; Acín-Pérez, R.; Martínez-Cano, S.; Enamorado, M.; Ugolini, M.; Nistal-Villán, E.; Hervás-Stubbs, S.; Pelegrín, P.; Sander, L.E.; Enríquez, J.A.; et al. Mitochondrial respiratory-chain adaptations in macrophages contribute to antibacterial host defense. *Nat. Immunol.* **2016**, *17*, 1037–1045. [CrossRef]
49. Acín-Pérez, R.; Carrascoso, I.; Baixauli, F.; Roche-Molina, M.; Latorre-Pellicer, A.; Fernández-Silva, P.; Mittelbrunn, M.; Sanchez-Madrid, F.; Pérez-Martos, A.; Lowell, C.A.; et al. ROS-triggered phosphorylation of complex II by Fgr kinase regulates cellular adaptation to fuel use. *Cell Metab.* **2014**, *19*, 1020–1033. [CrossRef]
50. Brooks, S.V.; Vasilaki, A.; Larkin, L.M.; McArdle, A.; Jackson, M.J. Repeated bouts of aerobic exercise lead to reductions in skeletal muscle free radical generation and nuclear factor κB activation. *J. Physiol.* **2008**, *586*, 3979–3990. [CrossRef]
51. Huertas, J.R.; Al Fazazi, S.; Hidalgo-Gutierrez, A.; López, L.C.; Casuso, R.A. Antioxidant effect of exercise: Exploring the role of the mitochondrial complex I superassembly. *Redox Biol.* **2017**, *13*, 477–481. [CrossRef] [PubMed]
52. Nicolás-Ávila, J.A.; Lechuga-Vieco, A.V.; Esteban-Martínez, L.; Sánchez-Díaz, M.; Díaz-García, E.; Santiago, D.J.; Rubio-Ponce, A.; Li, J.L.Y.; Balachander, A.; Quintana, J.A.; et al. A Network of Macrophages Supports Mitochondrial Homeostasis in the Heart. *Cell* **2020**, *183*, 94–109.e23. [CrossRef] [PubMed]
53. Runyan, C.E.; Welch, L.C.; Lecuona, E.; Shigemura, M.; Amarelle, L.; Abdala-Valencia, H.; Joshi, N.; Lu, Z.; Nam, K.; Markov, N.S.; et al. Impaired phagocytic function in CX3CR1+ tissue-resident skeletal muscle macrophages prevents muscle recovery after influenza A virus-induced pneumonia in old mice. *Aging Cell* **2020**, 19. [CrossRef]
54. Huertas, J.R.; Ruiz-Ojeda, F.J.; Plaza-Díaz, J.; Nordsborg, N.B.; Martín-Albo, J.; Rueda-Robles, A.; Casuso, R.A. Human muscular mitochondrial fusion in athletes during exercise. *FASEB J.* **2019**, *33*, 12087–12098. [CrossRef]
55. Casuso, R.A.; Huertas, J.R. The emerging role of skeletal muscle mitochondrial dynamics in exercise and ageing. *Ageing Res. Rev.* **2020**, *58*, 101025. [CrossRef]

56. Yamakawa, H.; Kusumoto, D.; Hashimoto, H.; Yuasa, S. Stem cell aging in skeletal muscle regeneration and disease. *Int. J. Mol. Sci.* **2020**, *21*, 1830. [CrossRef]
57. Bergmann, O.; Bhardwaj, R.D.; Bernard, S.; Zdunek, S.; Barnabé-Heide, F.; Walsh, S.; Zupicich, J.; Alkass, K.; Buchholz, B.A.; Druid, H.; et al. Evidence for cardiomyocyte renewal in humans. *Science* **2009**, *324*, 98–102. [CrossRef]
58. Koh, H.C.E.; Nielsen, J.; Saltin, B.; Holmberg, H.C.; Ørtenblad, N. Pronounced limb and fibre type differences in subcellular lipid droplet content and distribution in elite skiers before and after exhaustive exercise. *J. Physiol.* **2017**, *595*, 5781–5795. [CrossRef]
59. Huertas, J.R.; Casuso, R.A.; Agustín, P.H.; Cogliati, S. Stay Fit, Stay Young: Mitochondria in Movement: The Role of Exercise in the New Mitochondrial Paradigm. *Oxid. Med. Cell. Longev.* **2019**, *2019*, 1–18. [CrossRef]
60. Bosch, M.; Sánchez-Álvarez, M.; Fajardo, A.; Kapetanovic, R.; Steiner, B.; Dutra, F.; Moreira, L.; López, J.A.; Campo, R.; Marí, M.; et al. Mammalian lipid droplets are innate immune hubs integrating cell metabolism and host defense. *Science* **2020**, *370*, eaay8085. [CrossRef]
61. Shepherd, S.O.; Cocks, M.; Tipton, K.D.; Ranasinghe, A.M.; Barker, T.A.; Burniston, J.G.; Wagenmakers, A.J.M.; Shaw, C.S. Sprint interval and traditional endurance training increase net intramuscular triglyceride breakdown and expression of perilipin 2 and 5. *J. Physiol.* **2013**, *591*, 657–675. [CrossRef] [PubMed]
62. Green, D.R. Immiscible immunity. *Science* **2020**, *370*, 294–295. [CrossRef] [PubMed]
63. Brawner, C.A.; Ehrman, J.K.; Bole, S.; Kerrigan, D.J.; Parikh, S.S.; Lewis, B.K.; Gindi, R.M.; Keteyian, C.; Abdul-Nour, K.; Keteyian, S.J. Maximal Exercise Capacity is Inversely Related to Hospitalization Secondary to Coronavirus Disease 2019. *Mayo Clin. Proc.* **2020**. [CrossRef]
64. Nieman, D.C. Coronavirus disease-2019: A tocsin to our aging, unfit, corpulent, and immunodeficient society. *J. Sport Health Sci.* **2020**, *9*, 293–301. [CrossRef]
65. Burtscher, J.; Cappellano, G.; Omori, A.; Koshiba, T.; Millet, G.P. Mitochondria: In the Cross Fire of SARS-CoV-2 and Immunity. *iScience* **2020**, *23*, 101631. [CrossRef]
66. Tan, J.X.; Finkel, T. Mitochondria as intracellular signaling platforms in health and disease. *J. Cell Biol.* **2020**, 219. [CrossRef]
67. Gkikas, I.; Palikaras, K.; Tavernarakis, N. The role of mitophagy in innate immunity. *Front. Immunol.* **2018**, *9*, 1283. [CrossRef]
68. Kim, S.J.; Khan, M.; Quan, J.; Till, A.; Subramani, S.; Siddiqui, A. Hepatitis B Virus Disrupts Mitochondrial Dynamics: Induces Fission and Mitophagy to Attenuate Apoptosis. *PLoS Pathog.* **2013**, *9*, e1003722. [CrossRef]
69. Ajaz, S.; McPhail, M.J.; Singh, K.K.; Mujib, S.; Trovato, F.M.; Napoli, S.; Agarwal, K. Mitochondrial metabolic manipulation by SARS-CoV-2 in peripheral blood mononuclear cells of COVID-19 patients. *Am. J. Physiol. Cell Physiol.* **2020**. [CrossRef]
70. Granata, C.; Jamnick, N.A.; Bishop, D.J. Training-Induced Changes in Mitochondrial Content and Respiratory Function in Human Skeletal Muscle. *Sport. Med.* **2018**, *48*, 1809–1828. [CrossRef]
71. Fealy, C.E.; Mulya, A.; Lai, N.; Kirwan, J.P. Exercise training decreases activation of the mitochondrial fission protein dynamin-related protein-1 in insulin-resistant human skeletal muscle. *J. Appl. Physiol.* **2014**, *117*, 239–245. [CrossRef] [PubMed]
72. Busquets-Cortés, C.; Capó, X.; Martorell, M.; Tur, J.A.; Sureda, A.; Pons, A. Training and acute exercise modulates mitochondrial dynamics in football players' blood mononuclear cells. *Eur. J. Appl. Physiol.* **2017**, *117*, 1977–1987. [CrossRef] [PubMed]

Article

Mitochondrial Dysfunction in Pancreatic Alpha and Beta Cells Associated with Type 2 Diabetes Mellitus

Vladimir Grubelnik [1], Jan Zmazek [2], Rene Markovič [1,2], Marko Gosak [2,3] and Marko Marhl [2,3,4,*]

[1] Faculty of Electrical Engineering and Computer Science, University of Maribor, SI-2000 Maribor, Slovenia; vlado.grubelnik@um.si (V.G.); rene.markovic@um.si (R.M.)

[2] Faculty of Natural Sciences and Mathematics, University of Maribor, SI-2000 Maribor, Slovenia; jan.zmazek@um.si (J.Z.); marko.gosak@um.si (M.G.)

[3] Faculty of Medicine, University of Maribor, SI-2000 Maribor, Slovenia

[4] Faculty of Education, University of Maribor, SI-2000 Maribor, Slovenia

* Correspondence: marko.marhl@um.si

Received: 23 October 2020; Accepted: 10 December 2020; Published: 14 December 2020

Abstract: Type 2 diabetes mellitus is a complex multifactorial disease of epidemic proportions. It involves genetic and lifestyle factors that lead to dysregulations in hormone secretion and metabolic homeostasis. Accumulating evidence indicates that altered mitochondrial structure, function, and particularly bioenergetics of cells in different tissues have a central role in the pathogenesis of type 2 diabetes mellitus. In the present study, we explore how mitochondrial dysfunction impairs the coupling between metabolism and exocytosis in the pancreatic alpha and beta cells. We demonstrate that reduced mitochondrial ATP production is linked with the observed defects in insulin and glucagon secretion by utilizing computational modeling approach. Specifically, a 30–40% reduction in alpha cells' mitochondrial function leads to a pathological shift of glucagon secretion, characterized by oversecretion at high glucose concentrations and insufficient secretion in hypoglycemia. In beta cells, the impaired mitochondrial energy metabolism is accompanied by reduced insulin secretion at all glucose levels, but the differences, compared to a normal beta cell, are the most pronounced in hyperglycemia. These findings improve our understanding of metabolic pathways and mitochondrial bioenergetics in the pathology of type 2 diabetes mellitus and might help drive the development of innovative therapies to treat various metabolic diseases.

Keywords: pancreatic endocrine cells; mathematical model; mitochondrial dysfunction; cellular bioenergetics; diabetes; glucagon; insulin

1. Introduction

Mitochondrial bioenergetics is a critical element of our life. A decline in mitochondrial function is the root cause of many inborn metabolic destructions, e.g., the Leigh Syndrome [1], and mitochondrial dysfunction has been related to several age-related diseases, e.g., type 2 diabetes mellitus (T2DM) [2–4], neurodegenerative [5,6], cancer [7], cardiovascular [8], and others (recently reviewed in [9]). The interrelation, particularly the causality between mitochondrial dysfunction and age-related diseases is complex, a matter of permanent discussion. Multifactorial aging processes impact mitochondrial functioning, and the loss of intrinsic mitochondrial quality and dynamics affect several other processes in the body [10]. The complexity of the interrelationship between the mitochondrial dysfunction and the diseases requires intensive research, and it is promising that the accumulation of knowledge about mitochondria and its role in pathophysiology is enormous, representing a "phase shift" in the last two decades [11].

The pathogenesis of T2DM is rather complicated and not completely understood. The link between mitochondrial dysfunction and T2DM appears to be particularly relevant for understanding the pathophysiology of T2DM. Alterations in mitochondrial structure have been found in the muscles of patients with T2DM [12]. Gene expression analyses in skeletal muscles from healthy and diabetic subjects have shown that a decreased PGC1 expression can lead to metabolic disturbances characteristic for insulin resistance and T2DM [13]. Moreover, it has been reported that myotubes with mitochondrial dysfunction exhibited insulin resistance, oxidative stress, and inflammation with impaired insulin signaling activities [14]. In subjects with T2DM, the mitochondrial DNA (mtDNA) density is mainly depleted; however, in early phases of T2DM, a compensatory mechanism can appear that increases mitochondrial gene expression even though the mtDNA copy number is reduced [15]. This compensation may be responsible for the striking contrast between the severity of mtDNA depletion and the clinical manifestation of T2DM, such as the late onset and the slowly progressive nature of the disease. Assessing mitochondrial dysfunction via the mtDNA density requires adjustments for different tissues, as the content of mtDNA varies among tissues, being relatively higher in the liver than in muscle or other tissues, for example [16].

In pancreatic tissue, a strong association between mitochondrial dysfunction and the impaired secretory response of beta cells to glucose has been established [2,15,17–20]. Disturbances in insulin secretion are related to functional and molecular alterations of beta cells, rather than a decrease in beta cell mass [21]. In T2DM, both the morphology and function of beta cells is altered. Transmission electron microscopy investigations show that mitochondria in beta cells are swollen and have disordered cristae [18–20]. Functionally, the hyperpolarization of the mitochondria inner-membrane potential is diminished in beta cells, partially due to UCP-2 overexpression, which causes a reduced glucose-stimulated ATP/ADP ratio, and consequently an impaired insulin secretion [18,19]. Measurements, quantifying the mitochondrial dysfunction in insulin-resistant subjects, point to approximately 30–40% decrease in mitochondrial oxidative and phosphorylation activity [16,22,23]. In beta cells, mitochondrial dysfunction has also been linked, at least partially, to the lipotoxicity-mediated suppression of glucose-stimulated insulin secretion [24]. Studying the association between diabetes and mitochondrial metabolism in pancreatic beta cells has shown that hyperglycemia markedly reduces mitochondrial metabolism and ATP synthesis [3]. In addition to the intra-mitochondrial defects, Rutter et al. [25] have pointed out the impact of a defective interconnected mitochondrial network on disturbances in insulin secretion. Moreover, they have suggested that altered mitochondrial metabolism may also impair intracellular communication and coherent multicellular activity [25]. Recently, it has been shown that in the Western-diet-induced T2DM in mice, besides the beta cells, pancreatic alpha cells undergo considerable mitochondrial alterations, affecting the T2DM-related shift in glucagon secretion [26]. The growing awareness concerning T2DM-associated mitochondrial dysfunction in both alpha and beta cells might considerably contribute to a better understanding of T2DM and may point toward new therapies for treating metabolic diseases.

Mitochondrial bioenergetics is essential for the regulation of both glucagon and insulin secretion. In our previous study, we coupled glycolysis and mitochondrial metabolism with the hormone secretion in pancreatic alpha and beta cells and presented turning on/off secretory mechanism in both cells, guided by the energy-driven metabolic switch [27]. The model includes the metabolism of both glucose and free fatty acids (FFA). At low glucose concentration, the FFA oxidation (FFAO) in mitochondria is responsible for physiologically required glucagon secretion from alpha cells. At high glucose levels, glucagon secretion is primarily switched off by the glycolytic ATP production. In beta cells, during hypoglycemia, the ATP production is low. When glucose is elevated, the mitochondrial ATP production is the primary energy source required for adequate insulin secretion. This model, linking cell metabolism with the hormone secretion, provides a useful platform for studying pathophysiology in insulin and glucagon secretion.

In the present study, we extend our previous computational models and implement the models to analyze the impact of mitochondrial dysfunction on ATP production and the consequent hormonal disorders in pancreatic alpha and beta cells. To some extent, the mitochondrial dysfunction in alpha cells has been studied previously [26]. Here, we provide a more comprehensive comparative study of both insulin and glucagon secretion in dependence on the grade of mitochondrial dysfunction. The study improves our understanding of how a decline in mitochondrial bioenergetics is linked with the typical hormonal disorders in T2DM: a decreased insulin secretion, and in-/decreased glucagon secretion in hyper-/hypoglycemia, respectively.

2. Model

We designed computational models incorporating intracellular mechanisms from metabolic processes to hormone secretion in the pancreatic alpha and beta cells. Specifically, models for both cell types include a detailed description of metabolite and oxygen uptake, mitochondrial bioenergetic pathways, and ATP production, as well as the ATP/ADP ratio-mediated KATP-channel activity, which in turn orchestrates the hormone secretion. Particularities of individual segments and parameters of both models were adjusted and specified to match experimental data. A more detailed description of both models is given in Section 2.1. Furthermore, we incorporate in both cellular models a T2DM-associated mitochondrial dysfunction, which leads to a reduced net flux of ATP from mitochondria and impaired hormone secretion patterns. How the effect of altered mitochondrial bioenergetics is integrated into the models is described in Section 2.2.

2.1. *Bioenergetics and ATP-Driven Switch for Hormone Secretion*

Hormone secretion in alpha and beta cells depends crucially on the metabolic pathways in the cells. Unique properties of alpha and beta cells' metabolism enable glucose-sensing mechanisms, necessary for the modulation of glucose homeostasis. Because metabolic pathways facilitate the intracellular production of high-energy molecules ATP, the ATP concentration plays an essential role in the stimulus-secretion coupling. In the herein presented model, we divide the metabolic pathways into the glycolytic part (which can produce ATP anaerobically, in the absence of oxygen) and the mitochondrial metabolism (producing ATP at the electron transport chain (ETC) in the presence of oxygen), see Figure 1. The glycolytic and mitochondrial model components are coupled by glycolysis-produced pyruvate and NADH molecules. Whereas the ATP is produced directly by glycolysis (ADP molecules are phosphorylated during the payoff phase of glycolysis), the ATP production in mitochondria requires the ETC, driven by the oxidation of NADH and $FADH_2$ molecules, acting as the electron donors of the redox reaction. Oxygen, which is consumed by this process, serves as the final electron receptor. Therefore, besides the influx of critical metabolites glucose and FFA, the model also implements the fluxes of oxygen and the intermediate fluxes of the electron donors. However, to maximally simplify the computational model, only the most abundant electron donor, namely NADH, is included, whereas the flux of $FADH_2$ is considered as a part of the NADH flux.

In the process of glycolysis, each 6-carbon glucose is converted into two 3-carbon pyruvate molecules. During the first reaction of glycolysis, glucose is converted to glucose-6-phosphate (G6P). The velocity of the reaction is glucose-dependent due to the expression of low-affinity, high-Km enzyme glucokinase (hexokinase IV), in both alpha and beta cells. Consequently, the established glycolytic flux is glucose-dependent, which is crucial for the downstream energy-sensing mechanisms. Overall, glycolysis yields two molecules of anaerobically-produced ATP per one glucose molecule entering the glycolytic pathway. For the quantitative description of the glycolytic pathway, we use the theoretical framework described in Ref. [27].

On the other hand, aerobic ATP production is driven by the electrochemical proton gradient between intermembrane space and mitochondrial matrix. The proton gradient is established by the ETC, a series of complexes that transfer electrons from electron donors (reducing equivalents) to electron receptors, terminating with molecular oxygen being the final electron receptor. In the model, the reducing equivalents are the products of glucose oxidation and FFAO pathways. During glycolysis, two NAD^+ molecules are reduced to NADH in addition to the direct (anaerobic) phosphorylation of two ADP molecules to ATP. The NADH shuttle systems transport the glycolysis-derived NADH molecules through the mitochondrial membrane, where they enter the redox reactions of the ETC (see Figure 2). The majority of reducing equivalents, originating from the glucose oxidation pathway is, however, produced by the oxidation of pyruvate, the end-product of the glycolytic pathway. The flux of pyruvate passes the mitochondrial membrane and is converted to acetyl-CoA, which enters the tricarboxylic acid cycle. Pyruvate dehydrogenase (PDH) and tricarboxylic acid cycle reactions yield 4 molecules of NADH, 1 $FADH_2$, and 1 GTP, which is energetically equal to 5 molecules of NADH (for an extended explanation, see [27,28]). Following the oxidation, reducing equivalents enter the ETC, enabling aerobic ATP production.

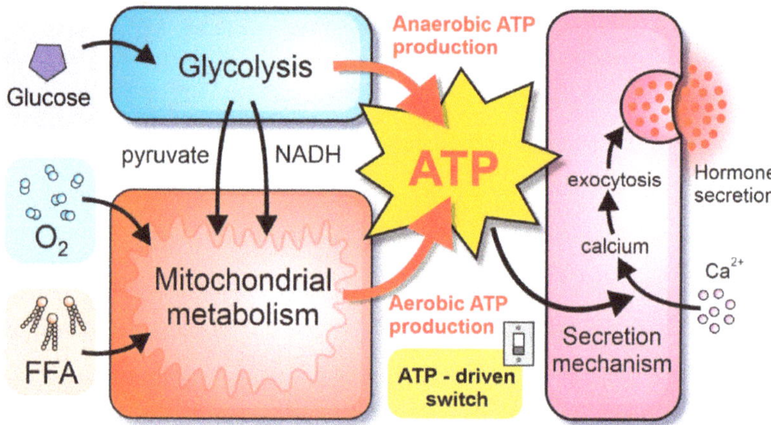

Figure 1. Schematic presentation of the model, coupling metabolic pathways in pancreatic alpha and beta cells with the hormone secretion. Intracellular ATP concentration drives the ATP-driven switch of hormone secretion, modulating the calcium influx and consequent calcium-dependent exocytosis of hormone granules. The ATP concentration is governed by the ATP production rates. The anaerobic ATP is yielded by direct phosphorylation of ADP during glycolysis. Conversely, the production of reducing equivalents from glycolysis, pyruvate oxidation, and FFAO drives the ETC (consuming molecular oxygen O_2), resulting in aerobic ATP production.

The second metabolic pathway involved in the aerobic ATP production is the FFAO pathway. Rather than the rate of FFA entry into the FFAO reactions, the input to the model is the net rate of oxygen consumption, for which the experimental data are more widely available. In the model approximation, the net rate of oxygen consumption is fully utilized by the ETC for the oxidation of glucose- and FFA-derived reducing equivalents. Consequently, the net oxygen consumption rate can be split into oxygen consumption due to glucose oxidation and oxygen consumption due to FFAO. The net aerobic ATP production due to FFAO is proportional to the oxygen consumption due to FFAO, which is calculated as the difference between the net oxygen consumption and oxygen consumption due to glucose oxidation. For the quantitative description of the metabolic pathways in mitochondria, we utilized our previous computational models for the alpha cell [26,27] and the beta cells [27].

The rates of anaerobic and aerobic ATP production are denoted in Figure 2 by $J_{ATP,anaerobic}$ and $J_{ATP,aerobic}$. Glycolysis and mitochondrial respiration modulate the intracellular ATP concentration by affecting the K_{ATP} channels. Since the density of K_{ATP} channels is comparable in alpha and beta cells [29], the ATP-dependent activity of K_{ATP} channels is also similar. The increased ATP concentration during hyperglycemia lowers the K_{ATP} channel conductance, depolarizing cell's membrane potential. In the beta cell, depolarization causes increased electrical activity, which results in higher calcium influx through voltage-dependent calcium channels, VDCC, [30] and higher insulin secretion. On the other hand, depolarization in alpha cells is followed by higher action potential (AP) frequency but lower amplitude, resulting in lower calcium influx [31] and glucagon secretion.

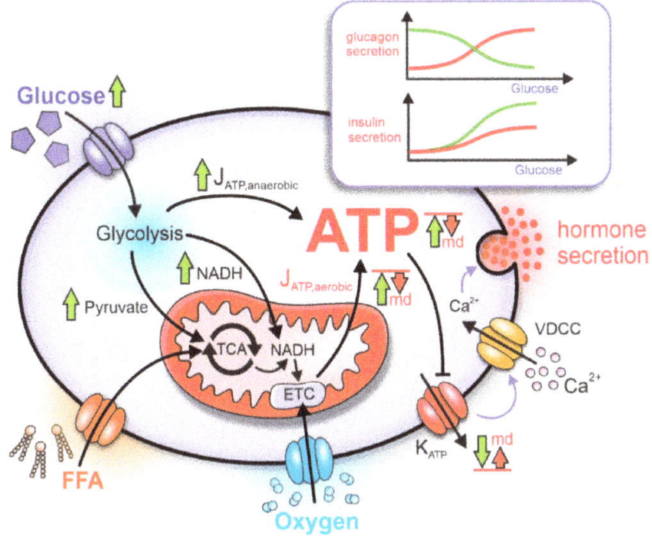

Figure 2. Impaired mitochondrial bioenergetics causes dysregulations in insulin and glucagon secretion. The increased glucose uptake during high plasma glucose concentration causes the acceleration of glycolytic and mitochondrial metabolism. Due to mitochondrial dysfunction (md), the aerobic ATP production rate ($J_{ATP,aerobic}$) and subsequent absolute ATP concentration are reduced, causing insufficient decrease of K_{ATP} conductance and impaired effects on the VDCC. As the glucose dependency of glucagon and insulin secretion is severely impacted. Green arrows represent a response to the rise in glucose under normal physiological conditions, and red arrows represent a lack of response due to mitochondrial dysfunction.

Our description of the alpha and beta cell's electrical activity is based on previous computational endeavors. Specifically, for the coupling of alpha cell's K_{ATP}-channel conductance with glucagon secretion, we used model proposed by Grubelnik et al. [26], with the same parameter values. The model of glucagon granule exocytosis is founded on the theoretical framework proposed by Montefusco et al. [32]. For the beta cell, we coupled K_{ATP}-channel conductance and insulin secretion as described in the model by Grubelnik et al. [27], which relies on the fundamental work of Pedersen et al. [33].

2.2. Mitochondrial Dysfunction and Pathophysiological Hormone Secretion

The separate modeling of aerobic and anaerobic pathways producing ATP in alpha and beta cells allows studying the effect of mitochondrial dysfunction on glucagon and insulin secretion. In cells with dysfunctional mitochondria, the net rate of aerobic ATP production is reduced, causing an insufficient lowering of K_{ATP}-channel conductance. As a result, the calcium influx through VDCC during hyperglycemia is inadequately increased in the beta cell and decreased in the alpha cell [29–31]. Dysregulation of ATP concentration due to the mitochondrial dysfunction severely impacts hormone granule exocytosis [34,35]. The qualitative effects of the impaired mitochondrial bioenergetics are schematically presented in Figure 2. In the alpha cell, the glucagon secretion typically peaks at low glucose concentrations. However, due to impaired mitochondrial function, the peak is shifted towards higher glucose concentrations. Conversely, in beta cells, the insulin secretion rates monotonically increase with increasing glucose concentrations and the anomalous mitochondrial functioning causes the reduction of insulin secretion across the whole glucose concentration interval. The mechanisms behind the dysregulated hormone secretion in both cell types are modeled as described in more detail in the continuation.

The intracellular ATP concentration is in the steady-state approximation governed by the ATP production and ATP hydrolysis rates. The reduced net flux of ATP from mitochondria is modeled by multiplying the mitochondrial ATP flux $J_{\mathrm{ATP,aerobic}}$, as described by Grubelnik et al. [27], by an additional coefficient of mitochondrial dysfunction, k_{md}:

$$J_{\mathrm{ATP}} = J_{\mathrm{ATP,anaerobic}} + (1 - k_{\mathrm{md}}) J_{\mathrm{ATP,aerobic}}, \tag{1}$$

where k_{md} is defined as the degree of mitochondrial dysfunction. k_{md} lies in the range between 0 and 1, where 0 represents a normal mitochondrial function, and 1 signifies a complete mitochondrial breakdown. According to the model by Grubelnik et al. [27], the rate of ATP hydrolysis, J_{ATPase}, obeys the Michaelis–Menten kinetic, which is modeled by:

$$J_{\mathrm{ATPase}} = J_{\mathrm{max,ATPase}} \frac{[\mathrm{ATP}]}{K_{\mathrm{m,ATPase}} + [\mathrm{ATP}]} (1 - k_{\mathrm{ATPase,r}} k_{\mathrm{md}}), \tag{2}$$

where [ATP] is the ATP concentration, $J_{\mathrm{max,ATPase}}$ is the maximal reaction rate (132 μM/s for the alpha cell, and 178 μM/s for the beta cell), $K_{\mathrm{m,ATPase}}$ is the Michaelis-Menten constant ($K_{\mathrm{m,ATPase}}$ = 2000 μM), and $k_{\mathrm{ATPase,r}}$ is a scaling constant for the mitochondrial-dysfunction-dependent decrease in ATP hydrolysis. The $k_{\mathrm{ATPase,r}}$ takes on a value of 0.6. Considering Equations (1) and (2), the ATP concentration is in the steady state given by:

$$[\mathrm{ATP}] = \frac{K_{\mathrm{m,ATPase}} J_{\mathrm{ATP}}}{J_{\mathrm{max,ATPase}} (1 - k_{\mathrm{ATPase,r}} k_{\mathrm{md}}) - J_{\mathrm{ATP}}}, \tag{3}$$

The model Equations (1)–(3) enable us to quantify the impact of mitochondrial dysfunction on glucagon and insulin secretion in pancreatic alpha and beta cells, respectively, as we describe in more detail in the next section.

3. Results

Herein we present how our computational model (Appendix A) behaves under normal physiological conditions and how its output is affected by the mitochondrial dysfunction. We first focus on the mitochondrial ATP production and the so-called energy-driven metabolic switch, the main determinants for glucagon and insulin secretion in alpha and beta cells, respectively. Then, we further explore how pathologically altered ATP generation affects the hormonal secretion process.

3.1. Mitochondrial Function Affects ATP Production

In models for the alpha as well as for the beta cell, the ATP production depends on glycolysis and mitochondrial metabolism and is driven by glucose and FFA oxidation. Altered mitochondrial metabolism has therefore a decisive effect on the drop in intracellular ATP concentration and consequently on the secretion of the two pancreatic hormones. In Figure 3, we show how mitochondrial ATP synthesis in both cell types depends on glucose concentration. In both panels the blue lines correspond to the cell's normal physiological function without mitochondrial dysfunction. A comparison of ATP production with regard to the cell type reveals that in both alpha and beta cells the ATP production monotonically increases with increasing glucose concentrations. However, the differences provoked by different stimulation levels are much more pronounced in the beta cell. Namely, in alpha cells at low glucose concentration, the ATP production is high compared to beta cells due to the oxidation of FFAs in mitochondria. As the concentration of glucose increases, part of FFAO in mitochondria is replaced by glucose oxidation (Figure 3a). At the same time, glucose is also metabolized by glycolysis in the cytoplasm, which is the key element that rises the ATP concentration at higher glucose levels in the alpha cell [27]. In contrast, in the beta cell the ATP production is low during hypoglycemia (Figure 3b), whereas in hyperglycemia it reaches approximately the same value as it is observed in the alpha cell. Because the production of ATP in beta cells is predominantly mediated by the mitochondrial glucose oxidation [2,36,37], this allows for very significant increase in ATP concentration when glucose levels are increased.

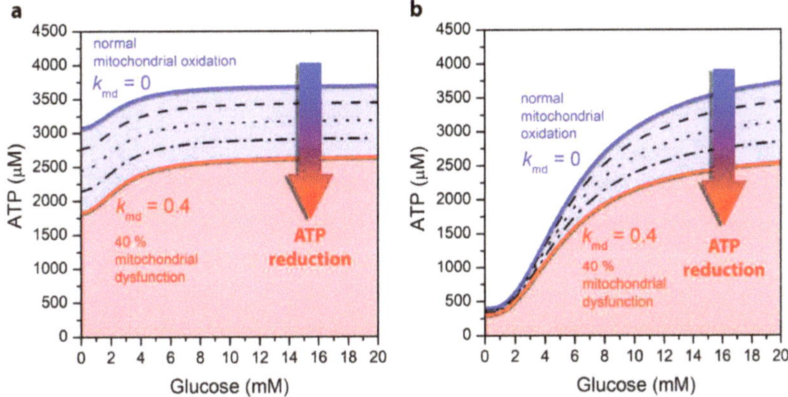

Figure 3. Glucose-dependent ATP concentration levels under normal mitochondrial conditions (blue line) and under different levels of mitochondrial dysfunction: 10% (dashed line), 20% (dotted line), 30% dash-dotted line, and 40% (red line), for the pancreatic alpha (**a**) and beta (**b**) cell.

Results in Figure 3 also demonstrate the change in cytosolic ATP concentration due to a defective mitochondrial metabolism associated with the pathogenesis of T2DM. It is known that the mitochondrial function is decreased by 30–40% in T2DM [16,22,23]. The ATP concentration due to mitochondrial dysfunction, which is determined in Equation (1), specifically by the k_{md} parameter, is shown by black lines, with each line illustrating an additional decrease in mitochondrial activity by 10%. The red line illustrates a maximum of 40% mitochondrial dysfunction. It can be observed that mitochondrial dysfunction results in a drop in ATP concentration in both alpha and beta cells. In alpha cells, the decline is approximately the same for all glucose concentrations and reflects a direct decrease in mitochondrial ATP production, which is weakly modulated by glucose. This drop is due to a direct decrease in mitochondrial ATP production, which is approximately the same at low and high glucose. Namely, mitochondrial ATP production in alpha cells is close to saturated even at low glucose levels due to FFAO. The increase in cytosolic ATP concentration with increasing glucose is due to anaerobic glycolytic ATP production, which is not affected by the mitochondrial dysfunction. In contrast to alpha cells, in beta cells, a decrease in ATP production is pronounced mainly in the hyperglycemic state, which occurs primarily at the expense of increased glucose oxidation in mitochondria, which is perturbed by mitochondrial dysfunction.

3.2. Impaired Glucagon and Insulin Secretion in the Pathogenesis of T2DM

In the previous section we explored how the glucose concentration-dependent ATP production is affected by the mitochondrial dysfunction. Here we further investigate how T2DM-associated mitochondrial perturbations impair the secretion of glucagon and insulin secretion in the pancreatic alpha (Figure 4a) and beta cells (Figure 4b).

Under normal physiological circumstances, an increase in ATP concentration at high glucose concentration in alpha cells leads to a decrease in the conductivity of K_{ATP} channels and to cell depolarization [38]. As a result, the Ca^{2+} inflow as well as the glucagon secretion rate is lower (Figure 4a, blue line). Impaired mitochondrial ATP synthesis lowers the intracellular ATP concentration and the ATP concentrations required for glucagon secretion are reached at higher glucose values. It turns out that the mitochondrial dysfunction causes an increased glucagon secretion in hyperglycemia, whereas glucagon secretion decreases in the hypoglycemic region, as the ATP production is not sufficient to fulfil the conditions for elevated glucagon secretion. Apparently, the T2DM-associated pathological state with mitochondrial dysfunction shifts glucagon secretion to higher glucose concentrations (Figure 4a). A detailed explanation of the cellular processes underlying the switch in secretion pattern can be found in our previous work [26]. It should be noted that here, in contrast to the previously published model [26], the glycolysis is modeled on the basis of the stationary state of ATP concentration (see Model). Despite this difference in the models, both lead to qualitatively the same results.

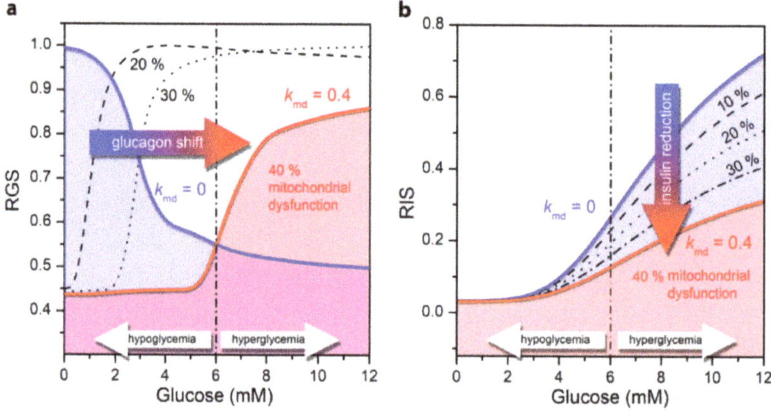

Figure 4. Relative glucagon and insulin secretion as a function of glucose concentration for normal cell function (blue line) and for the pathological state of T2DM under different levels of mitochondrial dysfunction: 10–40%. (**a**) Relative secretion of glucagon (RGS) in the alpha cell; (**b**) Relative insulin secretion (RIS) in the beta cell.

In beta cells under normal physiological conditions, higher glucose concentrations and the corresponding increase in ATP concentration enhances electrical activity. This in turn leads to a higher calcium influx, and finally to higher insulin secretion (Figure 4b). An impaired mitochondrial ATP production lowers the insulin secretion rates, which corresponds to the decrease in the intracellular ATP level. The differences in secretion profiles are mostly pronounced in hyperglycemic conditions. Abnormalities in mitochondrial function thus lead to lower glucose-stimulated insulin secretion in beta cells, which is a hallmark of T2DM.

4. Discussion

Using computational modeling approaches, we analyzed how impaired mitochondrial bioenergetics affects the hormone secretion patterns in pancreatic alpha and beta cells. The model predictions show that a reduced net flux of ATP from mitochondria can importantly contribute to dysregulations in the K_{ATP}-channel conductance, calcium influx through VDCC, and consequently the pathology in glucagon and insulin secretion. We have shown that mitochondrial-dysfunction-induced anomalies of hormone secretion in alpha and beta cells can be explained on the basis of very similar ATP-driven switching mechanisms. Even though the dysregulations of both glucagon and insulin secretion may result from the same pathological changes in mitochondria, they are reflected in a completely different matter in terms of glucose-dependency. Furthermore, we have shown that mitochondria play a key role in ensuring adequate ATP levels during the process of regulating hormone secretion, although their role in terms of ATP-driven switch in alpha and beta cells is different. In alpha cells, mitochondrial FFAO provides high levels of ATP, which is not significantly increased at elevated glucose levels. The modest rise in ATP concentration coincides with published measurements showing an approximately 15% increase in ATP in the hyperglycemic range [39]. Mitochondrial dysfunction thereby causes a drop in ATP (Figure 3a) and dysregulation of glucagon (Figure 4a) at both low and high glucose levels. In contrast to alpha cells, in beta cells, mitochondria provide a decisive increase in ATP due to oxidation of glucose in mitochondria (Figure 3a), which in the case of mitochondrial dysfunction leads to insulin dysregulation, especially in hyperglycemia (Figure 4b). In alpha cells, mitochondrial dysfunction causes a shift in glucagon secretion towards higher glucose concentration. The physiological values of glucagon in hypoglycemia are markedly reduced, whereas unphysiologically high glucagon levels are observed in hyperglycemia, which is a hallmark of T2DM [38,40]. In beta cells, insulin secretion is reduced over the broad range of glucose concentrations. The model results resemble the pathology of T2DM, showing that insulin secretion is indeed markedly reduced in patients with T2DM [19,41–44]. Supported by the experimental evidence that in T2DM the mitochondrial bioenergetic capacity is reduced by 30–40% [16,22,23], the model predictions considering mitochondrial dysfunction in the same range (30–40%) also matches quantitatively well the pathophysiology of glucagon and insulin secretion.

For beta cells, the results showing that the mitochondrial ATP production plays the most critical role in glucose-mediated insulin secretion are well in agreement with previous studies [2,36,45,46]. In 2019, Haythorne et al. [3] have reported that a markedly reduction in mitochondrial glucose metabolism is closely related to the pathogenesis of T2DM in mouse. This suggests that the primary cause of the insufficient insulin secretion in T2DM is the impaired mitochondrial bioenergetics, which is in contradiction to an old paradigm that gave more prominence to the beta-cell loss [3,47–49].

In alpha cells, the model-predicted role of anaerobic glycolysis, representing the ATP-driven switch for glucagon secretion, agrees with previous experimental observations showing that inhibition of glycolysis suppresses glucose-induced glucagon secretion [50]. In this study by Olsen et al. [50], it has also been demonstrated that the inhibition of glycolysis does not influence basal glucagon secretion. In our model, these experimental findings correspond with the FFAO in intact mitochondria, providing sufficient levels of ATP for keeping the K_{ATP} channels nearly closed and providing the glucagon secretion. Therefore, normal physiological conditions are characterized by intact mitochondria that efficiently use FFA as the main fuel. Moreover, FFA are crucial in hypoglycemic conditions for effective glucagon secretion at low glucose levels [51]. At high glucose concentrations, however, the model predictions show that glucose metabolism, particularly the anaerobic glycolytic ATP production, is the main contributor to the energy-driven switch turning off the glucagon secretion at the threshold level, which have been shown previously with a similar model [27].

It should be noted that amino acids, besides glucose and FFA studied here, also play a role in insulin and glucagon secretion. In alpha cells, an elevation in circulating amino acids causes glucagon secretion [52,53], and glucagon regulates amino acid metabolism at a systemic level [54,55]. It becomes even more evident that the liver–alpha-cell axis plays a crucial role in this regulation [55,56]. The contribution of amino acids to the energy-production in alpha cells is less relevant. In particular, in hypoglycemia under physiological conditions, alpha cells preferentially consume FFA for ATP production, in a similar way as amino acids do not fuel ATP production in hepatocytes, but instead the hepatic FFA oxidation is enhanced to supply the energy required to sustain gluconeogenesis [57]. Furthermore, amino acids were also found to affect beta cell function and insulin secretion. Amino acids such as glutamine, alanine, arginine, and others are known to cause increments in insulin secretion via activating mitochondrial metabolism similar to the downstream glucose metabolism in beta cells [58]. Therefore, the amino-acid-stimulation of insulin secretion could be in essence mathematically modeled in the same way as presented here for glucose stimulation. However, the physiological mechanisms by which amino acids confer their regulatory effects are multifactorial and involve mitochondrial metabolism [58,59] as well as other pathways such as electrical stimulation by electrogenic transport into the cells [60]. Therefore, amino acids add another layer of complexity to the islet cell metabolism and hormone secretion dynamics, and they represent a viable route for further investigation, including computational modeling approaches. Understanding the mechanisms by which amino acids along with other nutrients regulate islet hormone secretion has the potential to assess novel targets for future diabetes therapies [61].

Several studies have investigated the role of signaling pathways in alpha and beta cells, particularly the cAMP-PKA signaling pathway [62–66], the AMPK-signal-driven K_{ATP}-channel-density regulation [67], and several other signaling pathways. For alpha cells, many of them were recently reviewed by Onyango [68]. In contrast to beta cells, where the intracellular calcium concentration is undoubtedly the main factor which drives insulin secretion, the role of calcium in alpha cells is less well defined [69]. Furthermore, it has been recently shown that fixing cAMP at high levels prevents the glucose-induced inhibition of glucagon secretion [66]. Therefore, compared to the beta cell, the intrinsic glucose-induced pathway of hormone secretion is probably more complex in alpha cells, involving a calcium-independent inhibitory pathway, paracrine signaling, and juxtracrine factors [69]. In further studies, these aspects should also be included. In this vein, mathematical modeling represents a promising approach to investigate how these additional pathways contribute to cellular function and to bioenergetic processes studied here.

The computational models are a valuable repertoire for improving of our understanding of how cellular bioenergetics impacts hormone secretion in pancreatic alpha and beta cells. In addition to the qualitative holistic understanding of the mechanisms, the model also provides quantitative insights into the relative impacts of particular sub-processes. The model predictions point to the crucial role of mitochondria, revealing how specific pathways in relation to impaired mitochondrial bioenergetics lead to the T2DM-related dysregulations of glucagon and insulin secretion. If the reduced insulin secretion in patients with T2DM is due to impaired mitochondrial bioenergetics, it would be promising to reanimate or even replace the pathologically exhaustive mitochondria. Very recently, Rackham et al. [70] reported for the first time that insulin secretion can be restored by mitochondria transfer into insulin-secreting beta cells of diabetic mouse and human islets. The mitochondria were transferred by mesenchymal stromal cells. In the experiments, it has been observed that the mesenchymal stromal cells-derived mitochondria have reached approx. 30% of beta cells, which was sufficient to reinduce a functional phenotype in the intact islets [70]. This might indicate a new epoch in clinical treatments of T2DM. For the first time, it has been shown that exactly the transfer of functional mitochondria is an important mechanism underlying the beneficial effects of the mesenchymal-stromal-cell therapies. This might even have a broader impact on a better understanding of mitochondrial bioenergetics' role in other diseases. These therapies, and the corresponding mitochondrial transfer, are already recognized in treating injuries in the lung [71] and other tissues [72], vascular diseases [73], neurodegenerative diseases [74,75], and also COVID-19 [76,77].

Moreover, in addition to the intracellular processes that govern the function of alpha and beta cells, the pancreatic islets of Langerhans are multicellular micro-organs. Endocrine islet cells interact with each other, and this is another key player in hormonal rhythm control [69,78–83]. Beta cells are electrically coupled through gap-junctions, ensuring that the otherwise heterogeneous cell population works in synchrony [84–86]. In contrast, alpha cell populations' activity is not synchronized, which collaborates the idea that these cells function individually within the islet to secrete glucagon rather than act as a syncytium [46]. However, it is known that secretion of glucagon is also influenced by local paracrine signals, mediated by beta and delta cells [69,87–89].

From the viewpoint of the results presented herein, the question arises of how mitochondrial dysfunction manifests itself not only at the level of individual endocrine cells but also in their socio-cellular context. It has been postulated that genetic or environmental factors, such as glucotoxicity, lipotoxicity, or inflammation, impair the beta-cell network dynamics and, hence, insulin secretion via a defective mitochondrial function [25]. A multifaceted heterogeneity characterizes beta cells, and they form subgroups with distinct metabolic properties [90–92]. For addressing the spatio-temporal complexity of the beta cell populations, network analyzes have been applied, which have revealed a high level of heterogeneity in beta cell connectivity [93–95]. Specifically, a subset of specialized and well-connected cells were identified. These so-called hub cells are not only believed to substantially affect the collective cellular activity, but were also found to have the highest rate of energy consumption [96], they are metabolically highly active, exhibit hyperpolarized mitochondria [94,97]. These highly connected cells might be strongly affected by aging [98] or T2DM pathogenesis, when the secretory demand in beta cells increases [99]. We might speculate that increasing energetic needs affect mostly the hub cells, which operate at a higher basal level of energy consumption. Moreover, it is known that realistic biological networks are in general very susceptible to hub removal [100]. This implies that impaired mitochondrial bioenergetics perturbs significantly functional beta cell networks and collective beta cell activity as well. The activity of alpha cells, under normal as well as under pathophysiological conditions, and particularly in relation to their operation in the multicellular environment, is much less explored. Some results have been provided; however, the authors emphasize tremendous heterogeneity of the alpha-cell populations [101,102]. Accordingly, in a recent computational study, it has been emphasized that considering the cell-to-cell variability is necessary when investigating how the intrinsic and paracrine factors affect the glucose-dependent glucagon release [103].

Considering both intra- and inter-cellular processes, we encourage future theoretical and experimental studies to incorporate modes of coupling among islet cells, as they represent an important consideration underlying their function. This would enable a more holistic assessment of the complex system of the pancreatic islets and the exploration of how impaired mitochondrial energetics contributes to the pathological endocrine function.

Author Contributions: The study was conceptualized by M.M. and V.G. The models were implemented by V.G., R.M., and J.Z. The evaluation of single- and multi-cellular models was done by M.G. All authors have participated in writing the manuscript. In particular, M.M. wrote the main part of the Introduction, J.Z., M.G., and M.M. mostly participated in the model description, V.G., R.M., and M.G. in the Results, and M.M. and M.G. in the Discussion. The visualization, involving all figures and the graphical abstract, has been provided by V.G., M.M. supervised the study. All authors have read and agreed to the published version of the manuscript.

Funding: This research was supported by Javna Agencija za Raziskovalno Dejavnost RS, grand numbers P1-0055, P3-0396, N3-0133, J3-9289.

Conflicts of Interest: The authors declare no conflict of interest. The funders had no role in the design of the study; in the collection, analysis, or interpretation of data; in the writing of the manuscript; or in the decision to publish the results.

Appendix A

The source code for the computational model is available at the Figshare open access repository: https://doi.org/10.6084/m9.figshare.13132880.

References

1. Legault, J.T.; Strittmatter, L.; Tardif, J.; Sharma, R.; Tremblay-Vaillancourt, V.; Aubut, C.; Boucher, G.; Clish, C.B.; Cyr, D.; Daneault, C.; et al. A Metabolic Signature of Mitochondrial Dysfunction Revealed through a Monogenic Form of Leigh Syndrome. *Cell Rep.* **2015**, *13*, 981–989. [CrossRef]
2. Maechler, P.; Wollheim, C.B. Mitochondrial function in normal and diabetic β-cells. *Nature* **2001**, *414*, 807–812. [CrossRef] [PubMed]
3. Haythorne, E.; Rohm, M.; van de Bunt, M.; Brereton, M.F.; Tarasov, A.I.; Blacker, T.S.; Sachse, G.; dos Santos, M.S.; Exposito, R.T.; Davis, S.; et al. Diabetes causes marked inhibition of mitochondrial metabolism in pancreatic β-cells. *Nat. Commun.* **2019**, *10*. [CrossRef] [PubMed]
4. Madreiter-Sokolowski, C.T.; Malli, R.M.; Graier, W.F. Mitochondrial–Endoplasmic Reticulum Interplay: A Lifelong On–Off Relationship? *Contact* **2019**, *2*. [CrossRef] [PubMed]
5. Kim, Y.; Zheng, X.; Ansari, Z.; Bunnell, M.C.; Herdy, J.R.; Traxler, L.; Lee, H.; Paquola, A.C.M.; Blithikioti, C.; Ku, M.; et al. Mitochondrial Aging Defects Emerge in Directly Reprogrammed Human Neurons due to Their Metabolic Profile. *Cell Rep.* **2018**, *23*, 2550–2558. [CrossRef] [PubMed]
6. Cabral-Costa, J.V.; Kowaltowski, A.J. Neurological disorders and mitochondria. *Mol. Asp. Med.* **2020**, *71*, 100826. [CrossRef]

7. Porporato, P.E.; Filigheddu, N.; Pedro, J.M.B.S.; Kroemer, G.; Galluzzi, L. Mitochondrial metabolism and cancer. *Cell Res.* **2018**, *28*, 265–280. [CrossRef]
8. Groschner, L.N.; Waldeck-Weiermair, M.; Malli, R.; Graier, W.F. Endothelial mitochondria-less respiration, more integration. *Pflug. Arch. Eur. J. Physiol.* **2012**, *464*, 63–76. [CrossRef]
9. Bozi, L.H.M.; Campos, J.C.; Zambelli, V.O.; Ferreira, N.D.; Ferreira, J.C.B. Mitochondrially-targeted treatment strategies. *Mol. Asp. Med.* **2020**, *71*, 100836. [CrossRef]
10. Madreiter-Sokolowski, C.T.; Sokolowski, A.A.; Waldeck-Weiermair, M.; Malli, R.; Graier, W.F. Targeting mitochondria to counteract age-related cellular dysfunction. *Genes* **2018**, *9*, 165. [CrossRef]
11. Kowaltowski, A.J.; Oliveira, M.F. Mitochondria: New developments in pathophysiology. *Mol. Asp. Med.* **2020**, *71*. [CrossRef] [PubMed]
12. Kelley, D.E.; He, J.; Menshikova, E.V.; Ritov, V.B. Dysfunction of mitochondria in human skeletal muscle in type 2 diabetes. *Diabetes* **2002**, *51*, 2944–2950. [CrossRef]
13. Patti, M.E.; Butte, A.J.; Crunkhorn, S.; Cusi, K.; Berria, R.; Kashyap, S.; Miyazaki, Y.; Kohane, I.; Costello, M.; Saccone, R.; et al. Coordinated reduction of genes of oxidative metabolism in humans with insulin resistance and diabetes: Potential role of PGC1 and NRF1. *Proc. Natl. Acad. Sci. USA* **2003**, *100*, 8466–8471. [CrossRef] [PubMed]
14. Abu Bakar, M.H.; Sarmidi, M.R. Association of cultured myotubes and fasting plasma metabolite profiles with mitochondrial dysfunction in type 2 diabetes subjects. *Mol. Biosyst.* **2017**, *13*, 1838–1853. [CrossRef] [PubMed]
15. Lee, H.K. Mitochondria in Diabetes Mellitus. In *Mitochondria in Health and Disease*; Berdanier, C.D., Ed.; CRC Press: Boca Raton, FL, USA, 2005; pp. 377–454.
16. Patti, M.E.; Corvera, S. The role of mitochondria in the pathogenesis of type 2 diabetes. *Endocr. Rev.* **2010**, *31*, 364–395. [CrossRef] [PubMed]
17. Nunemaker, C.S.; Zhang, M.; Satin, L.S. Insulin feedback alters mitochondrial activity through an ATP-sensitive K+ channel-dependent pathway in mouse islets and β-cells. *Diabetes* **2004**, *53*, 1765–1772. [CrossRef] [PubMed]
18. Anello, M.; Lupi, R.; Spampinato, D.; Piro, S.; Masini, M.; Boggi, U.; Del Prato, S.; Rabuazzo, A.M.; Purrello, F.; Marchetti, P. Functional and morphological alterations of mitochondria in pancreatic beta cells from type 2 diabetic patients. *Diabetologia* **2005**, *48*, 282–289. [CrossRef]
19. Lu, H.; Koshkin, V.; Allister, E.M.; Gyulkhandanyan, A.V.; Wheeler, M.B. Molecular and metabolic evidence for mitochondrial defects associated with β-cell dysfunction in a mouse model of type 2 diabetes. *Diabetes* **2010**, *59*, 448–459. [CrossRef]
20. Sivitz, W.I.; Yorek, M.A. Mitochondrial dysfunction in diabetes: From molecular mechanisms to functional significance and therapeutic opportunities. *Antioxid. Redox Signal.* **2010**, *12*, 537–577. [CrossRef]
21. Del Guerra, S.; Lupi, R.; Marselli, L.; Masini, M.; Bugliani, M.; Sbrana, S.; Torri, S.; Pollera, M.; Boggi, U.; Mosca, F.; et al. Functional and molecular defects of pancreatic islets in human type 2 diabetes. *Diabetes* **2005**, *54*, 727–735. [CrossRef]
22. Petersen, K.F.; Befroy, D.; Dufour, S.; Dziura, J.; Ariyan, C.; Rothman, D.L.; DiPietro, L.; Cline, G.W.; Shulman, G.I. Mitochondrial dysfunction in the elderly: Possible role in insulin resistance. *Science* **2003**, *300*, 1140–1142. [CrossRef]
23. Petersen, K.F.; Dufour, S.; Befroy, D.; Garcia, R.; Shulman, G.I. Impaired Mitochondrial Activity in the Insulin-Resistant Offspring of Patients with Type 2 Diabetes. *N. Engl. J. Med.* **2004**, *350*, 664–671. [CrossRef]
24. Assali, E.A.; Shlomo, D.; Zeng, J.; Taddeo, E.P.; Trudeau, K.M.; Erion, K.A.; Colby, A.H.; Grinstaff, M.W.; Liesa, M.; Las, G.; et al. Nanoparticle-mediated lysosomal reacidification restores mitochondrial turnover and function in β cells under lipotoxicity. *FASEB J.* **2019**, *33*, 4154–4165. [CrossRef] [PubMed]
25. Rutter, G.A.; Georgiadou, E.; Martinez-Sanchez, A.; Pullen, T.J. Metabolic and functional specialisations of the pancreatic beta cell: Gene disallowance, mitochondrial metabolism and intercellular connectivity. *Diabetologia* **2020**, *63*, 1990–1998. [CrossRef] [PubMed]
26. Grubelnik, V.; Markovič, R.; Lipovšek, S.; Leitinger, G.; Gosak, M.; Dolenšek, J.; Valladolid-Acebes, I.; Berggren, P.O.; Stožer, A.; Perc, M.; et al. Modelling of dysregulated glucagon secretion in type 2 diabetes by considering mitochondrial alterations in pancreatic α-cells. *R. Soc. Open Sci.* **2020**, *7*. [CrossRef] [PubMed]
27. Grubelnik, V.; Zmazek, J.; Markovič, R.; Gosak, M.; Marhl, M. Modelling of energy-driven switch for glucagon and insulin secretion. *J. Theor. Biol.* **2020**, *493*. [CrossRef]

28. Wilson, D.F.; Cember, A.T.J.; Matschinsky, F.M. The thermodynamic basis of glucose-stimulated insulin release: A model of the core mechanism. *Physiol. Rep.* **2017**, *5*, 1–13. [CrossRef]
29. Rorsman, P.; Braun, M.; Zhang, Q. Regulation of calcium in pancreatic α- and β-cells in health and disease. *Cell Calcium* **2012**, *51*, 300–308. [CrossRef]
30. Rorsman, P.; Ashcroft, F.M. Pancreatic β-cell electrical activity and insulin secretion: Of mice and men. *Physiol. Rev.* **2018**, *98*, 117–214. [CrossRef]
31. Ramracheya, R.; Ward, C.; Shigeto, M.; Walker, J.N.; Amisten, S.; Zhang, Q.; Johnson, P.R.; Rorsman, P.; Braun, M. Membrane potential-dependent inactivation of voltage-gated ion channels in α-cells inhibits glucagon secretion from human islets. *Diabetes* **2010**, *59*, 2198–2208. [CrossRef]
32. Montefusco, F.; Pedersen, M.G. Mathematical modelling of local calcium and regulated exocytosis during inhibition and stimulation of glucagon secretion from pancreatic alpha-cells. *J. Physiol.* **2015**, *593*, 4519–4530. [CrossRef] [PubMed]
33. Pedersen, M.G.; Bertram, R.; Sherman, A. Intra- and inter-islet synchronization of metabolically driven insulin secretion. *Biophys. J.* **2005**, *89*, 107–119. [CrossRef] [PubMed]
34. Doliba, N.M.; Qin, W.; Najafi, H.; Liu, C.; Buettger, C.W.; Sotiris, J.; Collins, H.W.; Li, C.; Stanley, C.A.; Wilson, D.F.; et al. Glucokinase activation repairs defective bioenergetics of islets of Langerhans isolated from type 2 diabetics. *Am. J. Physiol. Endocrinol. Metab.* **2012**, *302*, E87–E102. [CrossRef]
35. Knudsen, J.G.; Hamilton, A.; Ramracheya, R.; Tarasov, A.I.; Brereton, M.; Haythorne, E.; Chibalina, M.V.; Spégel, P.; Mulder, H.; Zhang, Q.; et al. Dysregulation of Glucagon Secretion by Hyperglycemia-Induced Sodium-Dependent Reduction of ATP Production. *Cell Metab.* **2019**, *29*, 430–442. [CrossRef] [PubMed]
36. Kennedy, E.D.; Wollheim, C.B. Role of mitochondrial calcium in metabolism-secretion coupling in nutrient-stimulated insulin release. *Diabetes Metab.* **1998**, *24*, 15–24. [PubMed]
37. Quesada, I.; Todorova, M.G.; Alonso-Magdalena, P.; Beltrá, M.; Carneiro, E.M.; Martin, F.; Nadal, A.; Soria, B. Glucose induces opposite intracellular Ca2+ concentration oscillatory patterns in identified {alpha}- and β-cells within intact human islets of Langerhans. *Diabetes* **2006**, *55*, 2463–2469. [CrossRef]
38. Zhang, Q.; Ramracheya, R.; Lahmann, C.; Tarasov, A.; Bengtsson, M.; Braha, O.; Braun, M.; Brereton, M.; Collins, S.; Galvanovskis, J.; et al. Role of KATPchannels in glucose-regulated glucagon secretion and impaired counterregulation in type 2 diabetes. *Cell Metab.* **2013**, *18*, 871–882. [CrossRef]
39. Li, J.; Yu, Q.; Ahooghalandari, P.; Gribble, F.M.; Reimann, F.; Tengholm, A.; Gylfe, E. Submembrane ATP and Ca2+ kinetics in α-cells: Unexpected signaling for glucagon secretion. *FASEB J.* **2015**, *29*, 3379–3388. [CrossRef]
40. Kusminski, C.M.; Chen, S.; Ye, R.; Sun, K.; Wang, Q.A.; Spurgin, S.B.; Sanders, P.E.; Brozinick, J.T.; Geldenhuys, W.J.; Li, W.H.; et al. MitoNEET-Parkin effects in pancreatic α- and β-cells, cellular survival, and intrainsular cross talk. *Diabetes* **2016**, *65*, 1534–1555. [CrossRef]
41. Hosker, J.P.; Rudenski, A.S.; Burnett, M.A.; Matthews, D.R.; Turner, R.C. Similar reduction of first- and second-phase B-cell responses at three different glucose levels in type II diabetes and the effect of gliclazide therapy. *Metabolism* **1989**, *38*, 767–772. [CrossRef]
42. Silva, J.P.; Köhler, M.; Graff, C.; Oldfors, A.; Magnuson, M.A.; Berggren, P.O.; Larsson, N.G. Impaired insulin secretion and β-cell loss in tissue-specific knockout mice with mitochondrial diabetes. *Nat. Genet.* **2000**, *26*, 336–340. [CrossRef] [PubMed]
43. Santulli, G.; Lombardi, A.; Sorriento, D.; Anastasio, A.; Del Giudice, C.; Formisano, P.; Béguinot, F.; Trimarco, B.; Miele, C.; Iaccarino, G. Age-related impairment in insulin release: The essential role of β 2-adrenergic receptor. *Diabetes* **2012**, *61*, 692–701. [CrossRef] [PubMed]
44. Grespan, E.; Giorgino, T.; Arslanian, S.; Natali, A.; Ferrannini, E.; Mari, A. Defective amplifying pathway of β-cell secretory response to glucose in type 2 diabetes: Integrated modeling of in vitro and in vivo evidence. *Diabetes* **2018**, *67*, 496–506. [CrossRef] [PubMed]
45. Kennedy, E.D.; Maechler, P.; Wollheim, C.B. Effects of depletion of mitochondrial DNA in metabolism secretion coupling in INS-1 cells. *Diabetes* **1998**, *47*, 374–380. [CrossRef] [PubMed]
46. Quesada, I.; Todorova, M.G.; Soria, B. Different metabolic responses in α-, β-, and δ-cells of the islet of langerhans monitored by redox confocal microscopy. *Biophys. J.* **2006**, *90*, 2641–2650. [CrossRef] [PubMed]
47. Cantley, J.; Ashcroft, F.M. Q & A: Insulin secretion and type 2 diabetes: Why do β-cells fail? *BMC Biol.* **2015**, *13*, 1–7. [CrossRef]

48. Rahier, J.; Guiot, Y.; Goebbels, R.M.; Sempoux, C.; Henquin, J.C. Pancreatic β-cell mass in European subjects with type 2 diabetes. *Diabetes Obes. Metab.* **2008**, *10*, 32–42. [CrossRef]
49. Weir, G.C.; Bonner-Weir, S. Islet β cell mass in diabetes and how it relates to function, birth, and death. *Ann. N. Y. Acad. Sci.* **2013**, *1281*, 92–105. [CrossRef]
50. Olsen, H.L.; Theander, S.; Bokvist, K.; Buschard, K.; Wollheim, C.B.; Gromada, J. Glucose stimulates glucagon release in single rat α-cells by mechanisms that mirror the stimulus-secretion coupling in β-cells. *Endocrinology* **2005**, *146*, 4861–4870. [CrossRef]
51. Briant, L.J.B.; Dodd, M.S.; Chibalina, M.V.; Rorsman, N.J.G.; Johnson, P.R.V.; Carmeliet, P.; Rorsman, P.; Knudsen, J.G. CPT1a-Dependent Long-Chain Fatty Acid Oxidation Contributes to Maintaining Glucagon Secretion from Pancreatic Islets. *Cell Rep.* **2018**, *23*, 3300–3311. [CrossRef]
52. Janah, L.; Kjeldsen, S.; Galsgaard, K.D.; Winther-Sørensen, M.; Stojanovska, E.; Pedersen, J.; Knop, F.K.; Holst, J.J.; Albrechtsen, N.J.W. Glucagon Receptor Signaling and Glucagon Resistance. *Int. J. Mol. Sci.* **2019**, *20*, 3314. [CrossRef] [PubMed]
53. Unger, R.H.; Ohneda, A.; Aguilar-Parada, E.; Eisentraut, A.M. The role of aminogenic glucagon secretion in blood glucose homeostasis. *J. Clin. Investig.* **1969**, *48*, 810–822. [CrossRef] [PubMed]
54. Basco, D.; Zhang, Q.; Salehi, A.; Tarasov, A.; Dolci, W.; Herrera, P.; Spiliotis, I.; Berney, X.; Tarussio, D.; Rorsman, P.; et al. A-Cell Glucokinase Suppresses Glucose-Regulated Glucagon Secretion. *Nat. Commun.* **2018**, *9*. [CrossRef] [PubMed]
55. Holst, J.J.; Albrechtsen, N.J.W.; Pedersen, J.; Knop, F.K. Glucagon and Amino Acids Are Linked in a Mutual Feedback Cycle: The Liver-α-Cell Axis. *Diabetes* **2017**, *66*, 235–240. [CrossRef] [PubMed]
56. Dean, E.D. A primary role for α-cells as amino acid sensors. *Diabetes* **2020**, *69*, 542–549. [CrossRef] [PubMed]
57. Dean, E.D.; Li, M.; Prasad, N.; Wisniewski, S.N.; Von Deylen, A.; Spaeth, J.; Maddison, L.; Botros, A.; Sedgeman, L.R.; Bozadjieva, N.; et al. Interrupted Glucagon Signaling Reveals Hepatic α Cell Axis and Role for L-Glutamine in α Cell Proliferation. *Cell Metab.* **2017**, *25*, 1362–1373. [CrossRef]
58. Newsholme, P.; Brennan, L.; Bender, K. Amino acid metabolism, β-cell function, and diabetes. *Diabetes* **2006**, *55*, 39–47. [CrossRef]
59. Zhang, T.; Li, C. Mechanisms of amino acid-stimulated insulin secretion in congenital hyperinsulinism. *Acta Biochim. Biophys. Sin.* **2013**, *45*, 36–43. [CrossRef]
60. Smith, P.A.; Sakura, H.; Coles, B.; Gummerson, N.; Proks, P.; Ashcroft, F.M. Electrogenic arginine transport mediates stimulus-secretion coupling in mouse pancreatic beta-cells. *J. Physiol.* **1997**, *499*, 625–635. [CrossRef]
61. Newsholme, P.; Krause, M. Nutritional Regulation of Insulin Secretion: Implications for Diabetes. *Clin. Biochem. Rev.* **2012**, *33*, 35–47.
62. Tengholm, A. Cyclic AMP dynamics in the pancreatic β-cell. *Ups. J. Med. Sci.* **2012**, *117*, 355–369. [CrossRef] [PubMed]
63. Seino, S. Cell signalling in insulin secretion: The molecular targets of ATP, cAMP and sulfonylurea. *Diabetologia* **2012**, *55*, 2096–2108. [CrossRef] [PubMed]
64. Tian, G.; Sandler, S.; Gylfe, E.; Tengholm, A. Glucose- and hormone-induced cAMP oscillations in α- and β-cells within intact pancreatic islets. *Diabetes* **2011**, *60*, 1535–1543. [CrossRef] [PubMed]
65. Tengholm, A.; Gylfe, E. cAMP signalling in insulin and glucagon secretion. *Diabetes Obes. Metab.* **2017**, *19*, 42–53. [CrossRef] [PubMed]
66. Yu, Q.; Shuai, H.; Ahooghalandari, P.; Gylfe, E.; Tengholm, A. Glucose controls glucagon secretion by directly modulating cAMP in alpha cells. *Diabetologia* **2019**, *62*, 1212–1224. [CrossRef] [PubMed]
67. Han, Y.E.; Chun, J.N.; Kwon, M.J.; Ji, Y.S.; Jeong, M.H.; Kim, H.H.; Park, S.H.; Rah, J.C.; Kang, J.S.; Lee, S.H.; et al. Endocytosis of KATP Channels Drives Glucose-Stimulated Excitation of Pancreatic β Cells. *Cell Rep.* **2018**, *22*, 471–481. [CrossRef]
68. Onyango, A.N. Mechanisms of the Regulation and Dysregulation of Glucagon Secretion. *Oxid. Med. Cell. Longev.* **2020**, *2020*. [CrossRef]
69. Hughes, J.W.; Ustione, A.; Lavagnino, Z.; Piston, D.W. Regulation of islet glucagon secretion: Beyond calcium. *Diabetes Obes. Metab.* **2018**, *20*, 127–136. [CrossRef]
70. Rackham, C.L.; Hubber, E.L.; Czajka, A.; Malik, A.N.; King, A.J.F.; Jones, P.M. Optimizing beta cell function through mesenchymal stromal cell-mediated mitochondria transfer. *Stem Cells* **2020**, *38*, 574–584. [CrossRef]

71. Islam, M.N.; Das, S.R.; Emin, M.T.; Wei, M.; Sun, L.; Westphalen, K.; Rowlands, D.J.; Quadri, S.K.; Bhattacharya, S.; Bhattacharya, J. Mitochondrial transfer from bone-marrow-derived stromal cells to pulmonary alveoli protects against acute lung injury. *Nat. Med.* **2012**, *18*, 759–765. [CrossRef]
72. Wang, J.; Li, H.; Yao, Y.; Zhao, T.; Chen, Y.Y.; Shen, Y.L.; Wang, L.L.; Zhu, Y. Stem cell-derived mitochondria transplantation: A novel strategy and the challenges for the treatment of tissue injury. *Stem Cell Res. Ther.* **2018**, *9*, 1–10. [CrossRef]
73. Liu, K.; Ji, K.; Guo, L.; Wu, W.; Lu, H.; Shan, P.; Yan, C. Mesenchymal stem cells rescue injured endothelial cells in an in vitro ischemia-reperfusion model via tunneling nanotube like structure-mediated mitochondrial transfer. *Microvasc. Res.* **2014**, *92*, 10–18. [CrossRef] [PubMed]
74. Vasic, V.; Barth, K.; Schmidt, M.H.H. Neurodegeneration and neuro-regeneration—Alzheimer's disease and stem cell therapy. *Int. J. Mol. Sci.* **2019**, *20*, 4272. [CrossRef] [PubMed]
75. Raoof, R.; van der Vlist, M.; Willemen, H.L.D.M.; Prado, J.; Versteeg, S.; Vos, M.; Lockhorst, R.; Pasterkamp, J.; Khoury-Hanold, W.; Meyaard, L.; et al. Macrophages transfer mitochondria to sensory neurons to resolve inflammatory pain. *BioRxiv* **2020**. [CrossRef]
76. Leng, Z.; Zhu, R.; Hou, W.; Feng, Y.; Yang, Y.; Han, Q.; Shan, G.; Meng, F.; Du, D.; Wang, S.; et al. Transplantation of ACE2-Mesenchymal stem cells improves the outcome of patients with covid-19 pneumonia. *Aging Dis.* **2020**, *11*, 216–228. [CrossRef] [PubMed]
77. Thorlund, K.; Dron, L.; Park, J.; Hsu, G.; Forrest, J.I.; Mills, E.J. A real-time dashboard of clinical trials for COVID-19. *Lancet Digit. Health* **2020**, *2*, e286–e287. [CrossRef]
78. Bavamian, S.; Klee, P.; Britan, A.; Populaire, C.; Caille, D.; Cancela, J.; Charollais, A.; Meda, P. Islet-cell-to-cell communication as basis for normal insulin secretion. *Diabetes Obes. Metab.* **2007**, *9*, 118–132. [CrossRef]
79. Jain, R.; Lammert, E. Cell-cell interactions in the endocrine pancreas. *Diabetes Obes. Metab.* **2009**, *11*, 159–167. [CrossRef]
80. Briant, L.J.B.; Reinbothe, T.M.; Spiliotis, I.; Miranda, C.; Rodriguez, B.; Rorsman, P. Δ-Cells and B-Cells Are Electrically Coupled and Regulate A-Cell Activity Via Somatostatin. *J. Physiol.* **2018**, *596*, 197–215. [CrossRef]
81. Salem, V.; Silva, L.D.; Suba, K.; Georgiadou, E.; Gharavy, S.N.M.; Akhtar, N.; Martin-Alonso, A.; Gaboriau, D.C.A.; Rothery, S.M.; Stylianides, T.; et al. Leader β-cells coordinate Ca2+ dynamics across pancreatic islets in vivo. *Nat. Metab.* **2019**, *1*, 615–629. [CrossRef]
82. Zavala, E.; Wedgwood, K.C.A.; Voliotis, M.; Tabak, J.; Spiga, F.; Lightman, S.L.; Tsaneva-Atanasova, K. Mathematical Modelling of Endocrine Systems. *Trends Endocrinol. Metab.* **2019**, *30*, 244–257. [CrossRef] [PubMed]
83. Klemen, M.S.; Dolenšek, J.; Rupnik, M.S.; Stožer, A. The triggering pathway to insulin secretion: Functional similarities and differences between the human and the mouse β cells and their translational relevance. *Islets* **2017**, *9*, 109–139. [CrossRef] [PubMed]
84. Bosco, D.; Haefliger, J.A.; Meda, P. Connexins: Key mediators of endocrine function. *Physiol. Rev.* **2011**, *91*, 1393–1445. [CrossRef] [PubMed]
85. Benninger, R.K.P.; Piston, D.W. Cellular communication and heterogeneity in pancreatic islet insulin secretion dynamics. *Trends Endocrinol. Metab.* **2014**, *25*, 399–406. [CrossRef] [PubMed]
86. Šterk, M.; Dolenšek, J.; Bombek, L.K.; Markovič, R.; Zakelšek, D.; Perc, M.; Pohorec, V.; Stožer, A.; Gosak, M. Assessing the origin and velocity of Ca2+ waves in three-dimensional tissue: Insights from a mathematical model and confocal imaging in mouse pancreas tissue slices. *Commun. Nonlinear Sci. Numer. Simul.* **2021**, *93*, 105495. [CrossRef]
87. Hauge-Evans, A.C.; King, A.J.; Fairhall, K.; Persaud, S.J.; Jones, P.M. A role for islet somatostatin in mediating sympathetic regulation of glucagon secretion. *Islets* **2010**, *2*, 341–344. [CrossRef]
88. Vergari, E.; Knudsen, J.G.; Ramracheya, R.; Salehi, A.; Zhang, Q.; Adam, J.; Asterholm, I.W.; Benrick, A.; Briant, L.J.B.; Chibalina, M.V.; et al. Insulin inhibits glucagon release by SGLT2-induced stimulation of somatostatin secretion. *Nat. Commun.* **2019**, *10*, 1–11. [CrossRef]
89. Kellard, J.A.; Rorsman, N.J.G.; Hill, T.G.; Armour, S.L.; van de Bunt, M.; Rorsman, P.; Knudsen, J.G.; Briant, L.J.B. Reduced somatostatin signalling leads to hypersecretion of glucagon in mice fed a high-fat diet. *Mol. Metab.* **2020**, *40*, 101021. [CrossRef]
90. Xavier, G.D.S.; Rutter, G.A. Metabolic and Functional Heterogeneity in Pancreatic β Cells. *J. Mol. Biol.* **2020**, *432*, 1395–1406. [CrossRef]

91. Dwulet, J.A.M.; Ludin, N.W.F.; Piscopio, R.A.; Schleicher, W.E.; Moua, O.; Westacott, M.J.; Benninger, R.K.P. How Heterogeneity in Glucokinase and Gap-Junction Coupling Determines the Islet [Ca2+] Response. *Biophys. J.* **2019**, *117*, 2188–2203. [CrossRef]
92. Stožer, A.; Markovič, R.; Dolenšek, J.; Perc, M.; Marhl, M.; Rupnik, M.S.; Gosak, M. Heterogeneity and delayed activation as hallmarks of self-organization and criticality in excitable tissue. *Front. Physiol.* **2019**, *10*, 1–19. [CrossRef] [PubMed]
93. Stožer, A.; Gosak, M.; Dolenšek, J.; Perc, M.; Marhl, M.; Rupnik, M.S.; Korošak, D. Functional Connectivity in Islets of Langerhans from Mouse Pancreas Tissue Slices. *PLoS Comput. Biol.* **2013**, *9*. [CrossRef] [PubMed]
94. Johnston, N.R.; Mitchell, R.K.; Haythorne, E.; Pessoa, M.P.; Semplici, F.; Ferrer, J.; Piemonti, L.; Marchetti, P.; Bugliani, M.; Bosco, D.; et al. Beta Cell Hubs Dictate Pancreatic Islet Responses to Glucose. *Cell Metab.* **2016**, *24*, 389–401. [CrossRef] [PubMed]
95. Gosak, M.; Markovič, R.; Dolenšek, J.; Rupnik, M.S.; Marhl, M.; Stožer, A.; Perc, M. Network science of biological systems at different scales: A review. *Phys. Life Rev.* **2018**, *24*, 118–135. [CrossRef] [PubMed]
96. Gosak, M.; Stožer, A.; Markovič, R.; Dolenšek, J.; Marhl, M.; Rupnik, M.S.; Perc, M. The relationship between node degree and dissipation rate in networks of diffusively coupled oscillators and its significance for pancreatic beta cells. *Chaos* **2015**, *25*. [CrossRef] [PubMed]
97. Benninger, R.K.P.; Hodson, D.J. New understanding of β-cell heterogeneity and in situ islet function. *Diabetes* **2018**, *67*, 537–547. [CrossRef]
98. Westacott, M.J.; Farnsworth, N.L.; Clair, J.R.S.; Poffenberger, G.; Heintz, A.; Ludin, N.W.; Hart, N.J.; Powers, A.C.; Benninger, R.K.P. Age-dependent decline in the coordinated [Ca2+] and insulin secretory dynamics in human pancreatic islets. *Diabetes* **2017**, *66*, 2436–2445. [CrossRef]
99. Kahn, S.E.; Hull, R.L.; Utzschneider, K.M. Mechanisms linking obesity to insulin resistance and type 2 diabetes. *Nature* **2006**, *444*, 840–846. [CrossRef]
100. Estrada, E. Network robustness to targeted attacks. the interplay of expansibility and degree distribution. *Eur. Phys. J. B* **2006**, *52*, 563–574. [CrossRef]
101. Huang, Y.C.; Rupnik, M.; Gaisano, H.Y. Unperturbed islet α-cell function examined in mouse pancreas tissue slices. *J. Physiol.* **2011**, *589*, 395–408. [CrossRef]
102. Briant, L.J.B.; Zhang, Q.; Vergari, E.; Kellard, J.A.; Rodriguez, B.; Ashcroft, F.M.; Rorsman, P. Functional identification of islet cell types by electrophysiological fingerprinting. *J. R. Soc. Interface* **2017**, *14*. [CrossRef] [PubMed]
103. Montefusco, F.; Cortese, G.; Pedersen, M.G. Heterogeneous alpha-cell population modeling of glucose-induced inhibition of electrical activity. *J. Theor. Biol.* **2020**, *485*. [CrossRef] [PubMed]

 life

Review

Human Mitochondrial Pathologies of the Respiratory Chain and ATP Synthase: Contributions from Studies of *Saccharomyces cerevisiae*

Leticia V. R. Franco [1,2,*], Luca Bremner [1] and Mario H. Barros [2]

[1] Department of Biological Sciences, Columbia University, New York, NY 10027, USA; lib2109@columbia.edu
[2] Department of Microbiology, Institute of Biomedical Sciences, Universidade de Sao Paulo, Sao Paulo 05508-900, Brazil; mariohb@usp.br
* Correspondence: lv2404@columbia.edu

Received: 27 October 2020; Accepted: 19 November 2020; Published: 23 November 2020

Abstract: The ease with which the unicellular yeast *Saccharomyces cerevisiae* can be manipulated genetically and biochemically has established this organism as a good model for the study of human mitochondrial diseases. The combined use of biochemical and molecular genetic tools has been instrumental in elucidating the functions of numerous yeast nuclear gene products with human homologs that affect a large number of metabolic and biological processes, including those housed in mitochondria. These include structural and catalytic subunits of enzymes and protein factors that impinge on the biogenesis of the respiratory chain. This article will review what is currently known about the genetics and clinical phenotypes of mitochondrial diseases of the respiratory chain and ATP synthase, with special emphasis on the contribution of information gained from *pet* mutants with mutations in nuclear genes that impair mitochondrial respiration. Our intent is to provide the yeast mitochondrial specialist with basic knowledge of human mitochondrial pathologies and the human specialist with information on how genes that directly and indirectly affect respiration were identified and characterized in yeast.

Keywords: mitochondrial diseases; respiratory chain; yeast; *Saccharomyces cerevisiae*; *pet* mutants

---

1. Introduction

Mitochondria are dynamic organelles that supply most of the ATP needed to sustain the different energy-demanding activities of eukaryotic cells. Their ATP generating pathway consists of the oxphos complexes—four hetero-oligomeric complexes that make up the electron transfer chain plus the ATP synthase. Some complexes of this pathway are genetic hybrids composed of both mitochondrial and nuclear gene products. Most of the organelle, however, consists of proteins that are encoded by nuclear genes, their mRNAs translated on cytoplasmic ribosomes and the protein products transported to their proper internal membrane and soluble compartments. It is estimated that the nuclear proteome of *Saccharomyces cerevisiae* dedicated to the maintenance of respiratory competent mitochondria consists of at least 900 proteins [1]. Much of our information about the functions of this class of nuclear genes has been learned from studies of yeast. The identity of proteins localized in mitochondria and the phenotypic consequences of null mutations in their genes has come from large-scale proteomic studies. Information about the functions of nuclear gene products at the molecular level, however, has been gathered from a large body of previously known information of protein structure and function and from more recent in-depth biochemical analyses of yeast nuclear *pet* mutants. A search of the current Saccharomyces Genome Database (SGD) [2] indicates that the functions of some 170 proteins of the mitochondrial proteome essential for respiration and/or ATP synthesis are still not known.

2. Yeast, a Model for Studying Mitochondrial Function and Biogenesis

The last 50 years have witnessed unparalleled technical advances in deciphering the genetic compositions of whole genomes, so much so that whole new specialties have been born with the goal of developing tools for analyzing and dealing with this wealth of data in almost every major area of biological research. Of course, genes are only a starting point for the more interesting question of what their protein products do. This is one of the central questions of proteomics, which strives to develop methods for the simultaneous analysis of the entire complement of proteins in organisms, tissues and cells. Although the field as it stands today is highly successful in many important areas, such as ascertaining subcellular protein localization, their transient, as well as stable, physical interactions and patterns of expression during cell division, development, and diseased states, an understanding of their molecular functions and the specific cellular process they participate in continues to depend on slugfest genetics and biochemistry on a single or a small number of genes.

Respiratory deficient *pet* mutants of *S. cerevisiae*, particularly those obtained in Alexander Tzagoloff's laboratory [3], have been helpful in identifying a number of nuclear gene products essential for maintaining structurally and functionally competent mitochondria. The genes represented by about two thirds of the 215 complementation groups in such collections [3] have been characterized and their functions deduced.

One of the unexpected finding to have emerged from the functional analyses of *pet* mutants is the large extent to which expression of mitochondrial genes depends on mRNA-specific factors encoded in nuclear DNA. Also unexpected are the many accessory proteins that function in translation and assembly of the respiratory and ATP synthase complexes. For the most part this class of mitochondrial proteins target translation of specific mitochondrial mRNAs and maturation and biogenesis of their encoded proteins. For example, some three dozen proteins that are not constituents of cytochrome oxidase are currently known to be required specifically for the assembly of this single respiratory complex. Foreseeably, still unrecognized assembly factors may be discovered with further biochemical and genetic screens of uncharacterized *pet* mutants.

Mutations in human mitochondrial genes for subunit polypeptides of NADH-coenzyme Q reductase, cytochrome oxidase (COX, Complex IV), coenzyme Q-cytochrome *c* reductase (bc1 complex, Complex III) and ATP synthase (Complex V), were the first to be identified [4]. Subsequently, mutations presenting different clinical phenotypes were reported in nuclear genes that code for protein subunits of the ATP synthase and factors that function as chaperones during its assembly [5–7] and enzymes of biosynthetic pathways for heme *a* [8] and coenzyme Q [9]. A more complete blueprint of the regulatory proteins and chaperone factors that contribute to the biogenesis of respiratory competent mitochondria will uncover new chaperones and regulatory factors, some of which will undoubtedly have human homologs.

Human cells are more complex than yeast cells; and the same can be stated about the mitochondria of these two organisms. The higher the organization and the complexity, the more the consequences of one given deficiency differ. In many tested mutations, the phenotypes observed in humans are more deleterious for cell survival than in the yeast counterpart, which is not only true because yeast can ferment but also because of the variable energy demand of a complex organism with different tissues. As an example, the deletion of *MRX10* in yeast did not impair its respiratory capacity but mutations in the human counterpart led to respiratory impairment even in cells with low energetic demand such as fibroblasts [10]. In other circumstances, due to the need of proper protein-protein interactions, or just because of evolutionary divergence, the possibility of heterologous complementation is lost. For instance, yeast *shy1* mutants are not complemented by the human homolog *SURF1*, even with chimeric versions of the gene [11]. However, when the human genes do not complement the respective yeast mutant, it is still possible to evaluate the pathogenicity of a given mutation by constructing an allele with the corresponding change in the yeast gene.

3. Strategy for Determining the Function of Unknown Mitochondrial Proteins

A useful initial step for identifying the biochemical lesion of *pet* mutants is to assign them to one of three broad phenotypic classes based on the spectral properties of their mitochondria. This substantially reduces the number of subsequent assays. For example, a strain showing a normal complement of mitochondrial cytochromes and respiratory chain complexes can be excluded from harboring a mutation in a gene that affects mitochondrial translation, as both COX and the bc1 complex contain subunit polypeptides (cytochromes a, a_3, and cytochrome b) that are translated on mitochondrial ribosomes [12]. By the same token, a selective loss of cytochrome *a* generally signals a mutation in a gene required either for:

(1) Expression of one of the mitochondrial gene products of the COX complex;
(2) Assembly of this complex; or
(3) Maturation of the heme active centers of the enzyme.

A similar argument can be made for mutants lacking cytochrome *b*, except that a safe assumption is that the lesion affects some aspect of bc1 biogenesis.

Finally, mutations in genes that are directly or indirectly required for the maintenance of mitochondrial DNA (mtDNA), undergo a large deletion or complete loss of the genome resulting in a population of cytoplasmic petites (ϱ- and ϱ^0 mutants). Particularly prevalent in this class are mitochondrial protein synthesis mutants, (e.g. aminoacyl tRNA synthases and ribosomal mutants) [13] and mutants with defective ATP synthase [14,15]. Both mitochondrial translation and ATP synthase mutants display the absence of "*a*" and "*b*" type cytochromes for the reasons indicated above. An outline of the screens useful in identifying different classes of *pet* mutants is shown in Figure 1.

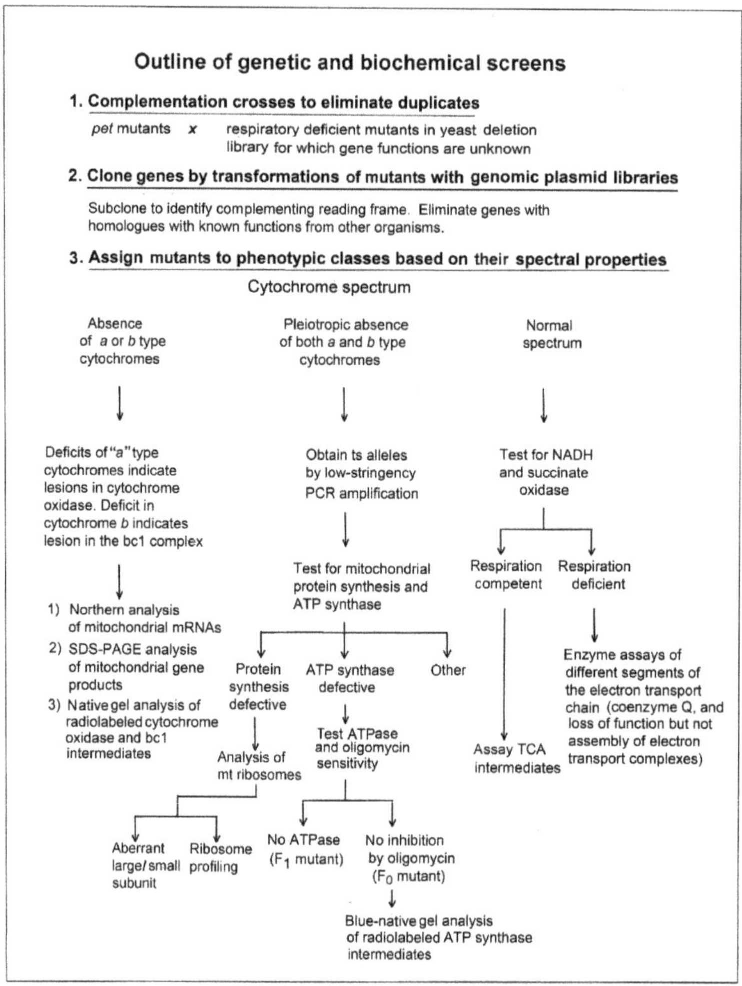

Figure 1. Genetic and biochemical screening of *pet* mutants. The initial complementation tests are done to identify *pet* mutants with counterparts in the knockout strain collection. This eliminates the need to clone and sequence genes already annotated.

4. Pathological Mutations in Respiratory Chain and ATP Synthase Human Genes with *S. cerevisiae* Homologs

In this section, the reader will find information on complexes II, III and IV, ATP synthase, Coenzyme Q and cytochrome *c*. All genes with pathogenic variants encoding subunits of the respiratory complexes and ATP synthase will be described. Regarding assembly factors, we have included only the ones that have yeast homologs.

4.1. Complex II

Succinate-coenzyme Q oxidoreductase, or Complex II, is a membrane-bound enzyme that functions in both the TCA cycle and the electron transfer chain. In the TCA cycle, it catalyzes the oxidation of succinate to fumarate using coenzyme Q as the electron acceptor without accompanying ATP synthesis [16]. The reduced coenzyme Q formed in this reaction is then reoxidized by Complex III, a reaction coupled to the translocation of protons across the mitochondrial inner membrane and ATP synthesis. Fungal and mammalian Complex II is embedded in the mitochondrial inner membrane, with a large portion protruding into the matrix. It is composed of four protein subunits, including the flavoprotein succinate dehydrogenase with covalently bound FAD and the iron sulfur subunit. Both of these catalytic subunits are peripheral proteins facing the matrix side of the inner membrane [17,18]. All four subunits of Complex II are encoded in the nucleus. The two catalytic subunits of Complex II are encoded by *SDHA* and *SDHB* in humans and by *SDH1* and *SDH2* in yeast. The other two subunits are integral membrane proteins that form a dimer that houses a single heme *b* group of cytochrome b_{560} and the two coenzyme Q binding sites of the complex. These two membrane anchors of the catalytic sector are encoded by human *SDHC* and *SDGD* and yeast *SDH3* and *SDH4*.

In yeast, Sdh3 is a bifunctional protein that is also a subunit of the TIM22 protein translocase complex responsible for transporting and integrating members of the substrate exchange carrier family into the inner membrane [19]. Electrons released during the oxidation of succinate first reduce the FAD cofactor of SDHA and are then sequentially transferred to three iron-sulfur clusters in SDHB before reacting with coenzyme Q [18,20–22]. *S. cerevisiae sdh1–4* mutants are respiratory deficient and display a severe growth defect on non-fermentable carbon sources such as glycerol and ethanol. The function of the cytochrome b_{560} is not fully understood, but it is thought to shuttle electrons between the two ubiquinone binding sites [23].

4.1.1. Mutations in Complex II Catalytic and Structural Subunits

Patients with lactic acidosis resulting from reduced succinate dehydrogenase activity have been linked to mutations in all four gene products of human Complex II. Although patients with deficiencies in the respiratory chain complexes, including Complex II, had been reported earlier [24], *SDHA* was the first instance of a nuclear encoded protein of the electron transfer chain with a mutation shown to cause a respiratory defect [25]. In that study, two siblings were homozygous for an R554T substitution in *SDHA* which resulted in Leigh syndrome, a severe neurological disorder that affects the central nervous system first described by Denis Leigh in 1951 [26]. The attribution of the phenotype to the mutation in *SDHA* was confirmed when the homologous mutation in the yeast flavoprotein was shown to have a deleterious effect on Complex II activity [25]. In the past 20 years, other *SDHA* mutations have been reported in patients presenting different clinical phenotypes (Table 1).

Interestingly, the same homozygous G555E substitution was identified in patients with distinct phenotypes: Leigh syndrome [27] and neonatal isolated cardiomyopathy [28]. This mutation was also found in a baby that died at five months of age following a respiratory infection before developing other phenotypes [29]. There is evidence that the G555E mutation prevents an adequate interaction between SDHA and SDHB [29]. This is supported by earlier studies on yeast Complex II assembly involving chimeric human/yeast genes [30].

More recently, germline mutations in *SDHA* were found in three patients with persistent polyclonal B cell lymphocytosis. In contrast to the other cases mentioned, the mutations resulted in a substantial increase of Complex II activity, leading to fumarate accumulation which engaged the KEAP1–Nrf2 system to drive the expression of inflammatory cytokines [31]. Mitochondrial pathologies have also been ascribed to mutations in *SDHB* and *SDHD* (Table 1).

4.1.2. Mutations in Complex II Assembly Factors

Respiratory deficiency is also elicited by mutations in accessory proteins that are required for assembly but are not constituents of Complex II [32,33]. Four such assembly factors have been identified for the human complex: *SDHAF1-4*, with yeast homologs *SDH6*, *SDH5*, *SDH7*, and *SDH8*, respectively [32,34–36]. Mutations in *SDHAF1* (yeast *SDH6*), that codes for an assembly factor of Complex II, result in infantile leukoencephalopathy and have been reported in five patients, some sharing substitutions at the same residues [34,37].

4.1.3. Complex II and Paragangliomas

Mutations in *SDHA*, *SDHB*, *SDHC*, *SDHD*, *SDHAF2*, and *SDHAF3* have been identified in an increasing number of neoplasms, mainly paragangliomas (rare neoplasms of the autonomic nervous system) and gastrointestinal stromal tumors. A discussion of these mutations is beyond the scope of this article but the topic has been extensively reviewed by others [38–43]. Variants in these genes have been associated with a high probability of developing cancer. To improve the identification of putative cancer-inducing human genetic variants, 22 known human variants of *SDHA* were examined in yeast [44]. Complementation tests of the homologous mutations in yeast *sdh1* identified 16 variants that affected the growth of yeast on the non-fermentable carbon source glycerol. The corresponding 16 human variants were proposed to be putative cancer inducing amino acid substitutions.

Table 1. Pathologies resulting from mutations in genes encoding Complex II subunits and assembly factors, and their yeast homologs.

Human Gene	Yeast Gene	Clinical Phenotype [1]	Mutation	Confirmation in Yeast	Reference
SDHA	*SDH1*	Leigh syndrome	homozygous R554W	yes	[25]
			compound heterozygous A524V, M1L	no	[45]
			compound heterozygous W119*, A83V	no	[46]
			homozygous G555E	no	[27]
		late-onset optic atrophy, ataxia, myopathy	heterozygous R408C	no	[47]
		neonatal isolated cardiomyopathy	homozygous G555E	no	[28]
		undefined [2]	homozygous G555E	no	[29]
		encephalopathy	compound heterozygous—stop codons at residues 56 and 81	no	[48]
		cardiomyopathy, leukodystrophy	compound heterozygous T508I, S509L	no	[49]
		optic atrophy, progressive, polyneuropathy, cardiomyopathy	heterozygous R451C	no	[50]
SDHB	*SDH2*	hypotonia, leukodystrophy	homozygous D48V	yes	[49]
		leukoencephalopathy	compound heterozygous D48V, R230H; homozygous L257V [3]	no	[51]
SDHD	*SDH4*	early progressive encephalomyopathy	compound heterozygous E69K, *164Lext*3	no	[52]
SDHAF1	*SDH6*	leukoencephalopathy	homozygous G57R; homozygous R55P [3]	yes	[34]
			homozygous R55P; homozygous Q8*; homozygous G57E [3]	no	[37]

[1] Paragangliomas were not included in the table. [2] The patient died at five months of age following a respiratory infection before developing other phenotypes. [3] In different patients.

4.2. Complex III

Complex III or the *bc1* complex is an integral inner membrane homodimeric complex of the mitochondrial inner membrane that catalyzes the oxidation of reduced coenzyme Q and reduction of cytochrome *c*, a reaction coupled to the translocation of protons from the matrix to the inter-membrane space [53]. Both human and yeast Complex III contain three catalytic subunits: a mitochondrially-encoded cytochrome *b*, with two non-covalently bound heme *b* containing redox centers corresponding to cytochromes b_H and b_L; cytochrome c_1, with a covalently linked heme *b*;

and the Rieske iron-sulfur protein [53–56]. In addition to the three catalytic subunits, Complex III contains seven other subunits, four of which are essential for the assembly and stability of the complex but do not participate in either electron transfer or proton translocation (Table 2). Like cytochrome c_1 and the Rieske iron sulfur protein, all the non-catalytic subunits are products of nuclear genes.

Assembly of Complex III also depends on nuclear encoded chaperones and on factors that regulate translation and assembly of cytochrome *b*. Most of the currently known Complex III assembly factors that were first described in yeast are conserved in humans (Table 2). Among yeast factors that have human homologs associated with diseases, there are Cbp6 and Cbp4. Cbp6, together with Cbp3, forms a complex with nascent apocytochrome *b* [57] for subsequent addition of heme to form the redox center at the cytochrome b_L site, followed by stabilization of the partially mature protein by Cbp4 [58] and further hemylation of the cytochrome b_H site [59]. Mitochondrial pathologies have also been reported in patients with mutations in human homologs of two other factors, Bcs1 and Mzm1, both needed for maturation and insertion of the Rieske iron sulfur protein into the complex [60–62].

Most laboratory strains of *S. cerevisiae* have a mitochondrial cytochrome *b* gene (*COB*) containing group I and II introns that are post-transcriptionally removed [63,64]. Some of the group II introns contain reading frames that code for factors, termed maturases, that function in splicing their own introns [65]. Splicing of the terminal group I intron is aided by a protein factor encoded by a nuclear gene [66]. In addition to these splicing factors, expression of *COB* depends on other factors that stabilize and activate translation of the mRNA [67,68]. Due to the absence of introns in the human cytochrome *b* gene and of 5′- non-coding sequences in the human mRNA, none of the yeast RNA splicing factors and translational activators have human homologs.

Complex III disorders are relatively rare but, like mutations in the other respiratory complexes, they present a wide spectrum of phenotypes. Complex III deficiency can be caused by mutations in the mitochondrially-encoded cytochrome *b*, in nuclear genes coding for catalytic and structural subunits and in ancillary proteins that function in assembly of the complex.

Table 2. Yeast complex III subunits and their human homologs. The table also shows Complex III assembly factors that are associated with diseases.

	Yeast Gene	Genome	*pet* Mutant	Human Gene	Function
Enzyme Subunits	*COB*	mitochondrial	N/A	*MT-CYB*	catalysis
	CYT1	nuclear	yes	*CYC1*	catalysis
	RIP1	nuclear	yes	*UQCRFS1*	catalysis
	COR1	nuclear	yes	*UQCRC1*	structure
	COR2	nuclear	yes	*UQCRC2*	structure
	QCR6	nuclear	no	*UQCRH*	structure
	QCR8	nuclear	yes	*UQCRQ*	structure
	QCR7	nuclear	yes	*UQCRB*	structure
	QCR9	nuclear	no	*UQCR10*	structure
	QCR10	nuclear	no	*UQCR11*	structure
Assembly Factors	*CBP3*	nuclear	yes	*UQCC1*	translation/assembly of cyt. b
	CBP6	nuclear	yes	*UQCC2*	translation/assembly of cyt. b
	CBP4	nuclear	yes	*UQCC3*	assembly of cyt. b
	MZM1	nuclear	no	*LYRM7*	assembly of Rieske protein
	BCS1	nuclear	yes	*BCS1L*	assembly of Rieske protein

4.2.1. Mutations in Complex III Catalytic Subunits

Most of the known Complex III associated pathologies result from mutations in *MT-CYB*, the human mitochondrial cytochrome *b* gene, that have so far been described in 49 different positions of the genome [69]. Typical phenotypes include MELAS (mitochondrial myopathy, encephalopathy, lactic acidosis, and stroke-like episodes), LHON (Leber's Hereditary Optic Neuropathy), hearing loss and in some cases less severe phenotypes expressed in exercise intolerance (reviewed in [70]). While some cytochrome *b* mutations are maternally inherited, others are heteroplasmic and are present mainly in muscle tissue, suggesting that they arise de novo after differentiation of the primary germ layers. Additionally, some LHON patients with a mutation in one of the mitochondrially encoded Complex I genes have a second mutation in cytochrome *b*, which exacerbates the severity of the pathology [71–73]. Mutations in the nuclear *UQCR4* (yeast *CYC1*) and *UQCRFS1* (yeast *RIP1*) genes are much rarer and were more recently identified (Table 3).

4.2.2. Mutations in Complex III Structural Subunits

A 4 bp deletion in *UQCRB* resulting in a change in the last seven amino acids and an addition of a stretch of 14 amino acids at the C-terminal end of the protein was the first described case of a Complex III deficiency resulting from mutations in a nuclear encoded subunit of the complex [74]. Interestingly, earlier studies on yeast Complex III showed that deletions in the C-terminal helical domain of the *QCR7* homolog resulted in reduced levels of the subunit and of cytochrome *b*, Rip1, and Qcr8 [75]. Taken together, these studies show the importance of the helical domain of *UQCRB* in the maintenance or assembly of human and yeast Complex III.

The discovery of the deletions in *UQCRB* was followed by the identification of a pathogenic variant of *UQCRQ* [76]. A pathogenic mutation in *UQCRC2* was also identified: three patients from a Mexican consanguineous family with neonatal onset of hypoglycemia and lactic acidosis were found to have a homozygous mutation leading to a R183W substitution in the Core 2 subunit. The patients also showed hyperammonemia, high urine organic acids and elevated plasma hydroxyl fatty acids, suggesting that the *UQCRC2* mutation may elicit secondary effects on TCA and urea cycles and beta-oxidation. Modeling of the crystal structure of bovine Complex III predicted that the R183W mutation would disrupt the hydrophobic interface of the UQCRC2 homodimer, leading to Complex III destabilization. This was supported by an 80% decrease of the complex in the patients, even though Complex III activity was only marginally affected [77]. More recently, the same homozygous mutation with similar symptoms was described in a French child [78]. Furthermore, the expression of *UQCRC2* is upregulated in multiple human tumors, whereas its suppression inhibits cancer cells and induces senescence [79].

4.2.3. Mutations in Complex III Assembly Factors

Together with cytochrome *b*, most Complex III pathologies are derived from mutations in the nuclear *BCS1L* gene. Its yeast homolog *BCS1* is an AAA protease that was shown to be required for the expression/maturation of Rip1 [60,61]. Later studies indicated that Bcs1 promotes one of the steps in the translocation of Rip1 necessary for incorporation of the iron-sulfur center [80]. Bcs1 exists as a heptamer with a contractile central cavity that participates in the translocation of the folded Rip1. Additionally, Bcs1 is associated with the complex III assembly module and its dissociation ends the maturation process [81].

Mutations in *BCS1L* comprise a wide spectrum of pathologies, including: GRACILE syndrome (growth retardation, aminoaciduria, cholestasis, iron overload, lactic acidosis, early death) [82–88]; Björstand syndrome, characterized by hearing loss and *pili torti* [89–94]; encephalopathy [95–98]; lactic acidosis, liver dysfunction and tubulopathy [99–102]; muscle weakness, focal motor seizures and optic atrophy [103], among others. Besides low steady state levels of Complex III, cells from patients with *BSC1L* mutations have impaired mitochondrial import of the protein as evidenced by its accumulation in the cytosol [91]. An adult harboring a R69C missense mutation was diagnosed with aminoaciduria, seizures, bilateral sensorineural deafness, and learning difficulties. Yeast complementation studies corroborate that the R69C mutation impairs the respiratory capacity of the cell [104].

Assembly of Rip1 is thought to be the last step in the biogenesis of Complex III [81]. In addition to Bcs1, this assembly step was found to require the product of the yeast nuclear *MZM1* gene [105].

The first case of Complex III deficiency caused by a mutation in *LYRM7*, the human homolog of *MZM1*, was reported in 2013 [106]. The equivalent mutation in yeast resulted in decreased oxygen consumption as a result of reduced steady state levels of Rieske protein and Complex III. Since then, Complex III deficiency caused by *LYRM7* has been identified in patients presenting with leukoencephalopathy [107–109] and liver dysfunction [110].

Mutations have also been reported in recent years in the human *UQCC2* and *UQCC3* genes that code for protein homologs of yeast complex III assembly factors Cbp6 and Cbp4. Complex III deficiency was found in a patient with a homozygous mutation in *UQCC2* [111]. This study demonstrated that the biochemical phenotype produced by the *UQCC2* mutation is similar to that reported in yeast [57], as cytochrome *b* synthesis and stability was decreased in the patient's fibroblasts [111]. A homozygous mutation in *UQCC2* leading to a Complex III deficiency was also reported in a consanguineous baby presenting neonatal encephalomyopathy. This mutation resulted in a secondary deficiency of Complex I [112]. The authors proposed that assembled Complex III is required for the stability or assembly of complexes I and IV, which may be related to supercomplex formation. Interestingly, a recent study [113] showed that the ND1 subunit of Complex I co-immunoprecipitated with newly synthesized UQCRFS1 of Complex III in mammalian mitochondria, indicating a possible coordination of the assembly of the two complexes.

Table 3. Pathologies resulting from mutations in genes encoding Complex III subunits and assembly factors, and their yeast homologs.

Human Gene	Yeast Gene	Clinical Phenotype	Mutation	Confirmation in Yeast	Reference
MT-CYB	*COB*	MELAS, LHON, hearing loss, exercise intolerance	many		[70] [1]
UQCR4	*CYT1*	ketoacidosis and insulin-responsive hyperglycemia	homozygous L215F; homozygous T96C [2]	yes	[115]
UQCRFS1	*RIP1*	cardiomyopathy and alopecia totalis	homozygous V72_T81del10; heterozygous V14D; heterozygous R204* [2]	no	[116]
UQCRC2	*COR2*	neonatal onset of hypoglycemia	homozygous R183W	no	[77,78]
UQCRB	*QCR7*	hypoglycemia	homozygous 4 bp deletion at nucleotides 338–341	yes [75]	[74]
UQCRQ	*QCR8*	severe psychomotor retardation, dystonia, athetosis, ataxia, dementia	homozygous S45F	no	[76]
UQCC2	*CBP6*	intrauterine growth retardation, renal tubular dysfunction	homozygous c.214-3C>G [3]	no	[111]
		neonatal encephalomyopathy	homozygous R8P and L10F [4]	no	[112]
UQCC3	*CBP4*	hypoglycemia, hypotonia, delayed development	homozygous V20E	no	[114]
BCS1L	*BCS1*	GRACILE syndrome, Björstand syndrome, encephalopathy, muscle weakness	many		see main text

Table 3. *Cont.*

Human Gene	Yeast Gene	Clinical Phenotype	Mutation	Confirmation in Yeast	Reference
LYRM7	*MZM1*	early onset severe encephalopathy	homozygous D25N	yes	[106]
		leukoencephalopathy	homozygous c.243_244 + 2del [3]	no	[107]
			several	yes	[108]
			homozygous 4bp deletion c.[243_244 + 2delGAGT] [3]	no	[109]
		liver dysfunction	homozygous R18Dfs*12	no	[110]

[1] Review describing many mutations. [2] In different patients. [3] Mutation causing a splicing defect. [4] In the same patient.

During assembly of yeast Complex III, Cbp4 is recruited by the Cbp3–Cbp6-cytochrome *b* ternary complex following release of the latter from the mitoribosome [57]. Wanschers et al. [114] described a homozygous mutation in *UQCC3*, the human homolog of *CBP4*, in a patient diagnosed with isolated Complex III deficiency. Cultured fibroblasts from the patient were partially deficient in cytochrome *b* and had no detectable UQCC3 protein. The authors concluded that UQCC3 functions in Complex III assembly downstream of UQCC1 and UQCC2, as the absence of UQCC3 did not affect the levels of UQCC1 and UQCC2 [114]. These observations are consistent with the above mentioned sequential interaction of Cbp4 with the Cbp3-Cbp6-cytochrome *b* complex during synthesis and assembly of cytochrome *b* [57].

4.3. Complex IV

Complex IV or cytochrome oxidase (COX) is an integral mitochondrial inner membrane protein complex that catalyzes the oxidation of cytochrome *c* and the reduction of molecular oxygen to water. This reaction is coupled to the translocation of protons from the matrix to the inter-membrane space [117]. In both human and yeast mitochondria, the COX catalytic core is composed of three proteins encoded in the mitochondrial DNA. They are Cox1, Cox2, and Cox3 in yeast, and MTCO1, MTCO2, and MTCO3 in humans. All the redox centers of COX are located in the Cox1 and Cox2 subunits. Yeast but not human Cox2 is synthesized with a cleavable N-terminal presequence that is required for correct insertion of the protein into the membrane [118]. Cox3 does not contain redox centers. It is thought to stabilize the catalytic core and to enhance the uptake of protons from the mitochondrial matrix [119]. One of the redox centers of Cox1, corresponding to cytochrome *a*, contains heme *a*. The second center, corresponding to cytochrome a_3, consists of a binuclear heme a-Cu_B. The third center, located on Cox2, is the binuclear Cu_A. In addition to the catalytic core, COX is composed of several other structural subunits, all encoded in nuclear DNA (Table 4). Recently NDUFA4/COXFA4, a subunit previously thought to be part of Complex I, has been shown to be a subunit of COX [120].

COX catalyzes the consecutive transfer of 4 electrons from cytochrome *c* to a molecule of oxygen bound to the heme *a* of cytochrome a_3 with the formation of water. Each electron of cytochrome *c* first reduces the Cu_A center of Cox2, from which it is then transferred to the heme *a* of cytochrome *a* and finally to the binuclear heme *a* –Cu center of cytochrome a_3 [121,122].

The heme *a* differs from heme *b* at two positions of the porphyrin ring. While heme *b* is present in hemoglobin and most heme containing enzymes, heme *a* appears only in COX. The biosynthesis of heme *a* is initiated by the addition of a farnesyl group to the C-2 position of the heme *b* porphyrin ring by farnesyl transferase, encoded by *COX10* [123]. The resulting heme *o* is then converted to heme *a* by the oxidation of a methyl to a formyl group on C-8 of the porphyrin ring in a reaction that requires Cox15, mitochondrial ferredoxin, and ferredoxin reductase [124,125]. Some other factors, such as Shy1 (human SURF1) and Pet117, have been shown to be involved in the hemylation of Cox1 [126,127]. Additionally, several proteins with human homologs, including Cox17, Sco1, Sco2, Cox11, Cox19, Cox23, Cox16, and Cmc1 have been implicated in the trafficking of copper and maturation of the Cu_A and heme a-Cu_B centers, as reviewed elsewhere [128].

Pathological mutations have been reported in both mitochondrial and nuclear encoded COX subunits, as well as in proteins involved in the biogenesis of the complex. As of today, mutations in more than 20 genes can lead to COX deficiency with a broad spectrum of clinical phenotypes (Table 5).

Furthermore, COX interacts with other complexes of the electron transport chain in entities called supercomplexes or respirasomes that are thought to provide a kinetic advantage by allowing for a more efficient transfer of electrons between the respiratory complexes and their intermediary carriers cytochrome *c* and coenzyme Q [129,130]. In humans, the respirasome is composed of Complex I, III, and IV in variable stoichiometry, while in yeast it is composed of Complex III and IV in strict 2:2 and 2:1 ratios [61,131,132]. Mutations resulting in COX deficiency affect respirasome biogenesis, which could ultimately lead to complex secondary phenotypes, as discussed elsewhere [133].

Table 4. Yeast complex IV subunits and their human homologs. The table also shows Complex IV assembly factors that are associated with diseases.

	Yeast Gene	Genome	*pet* Mutant	Human Gene	Function
Enzyme Subunits	*COX1*	mitochondrial	N/A	*MTCO1*	catalysis
	COX2	mitochondrial	N/A	*MTCO2*	catalysis
	COX3	mitochondrial	N/A	*MTCO3*	structure/catalysis
	COX5a	nuclear	yes	*COX4*	structure
	COX5b [1]	nuclear	no	-	structure
	COX6	nuclear	yes	*COX5A*	structure
	COX4	nuclear	yes	*COX5B*	structure
	COX13	nuclear	no	*COX6A*	structure
	COX12	nuclear	yes	*COX6B*	structure
	COX9	nuclear	yes	*COX6C*	structure
	COX7	nuclear	yes	*COX7A*	structure
	-	nuclear	-	*COX7B*	structure
	COX8	nuclear	no	*COX7C*	structure
	-	nuclear	-	*COX8*	structure
	-	nuclear	-	*NDUFA4/COXFA4*	structure
Assembly Factors	*COX20*	nuclear	yes	*COX20/FAM36A*	membrane insertion of Cox2
	COX14	nuclear	yes	*COX14*	regulation of *COX1* expression, maintenance of monomeric Cox1
	PET117	nuclear	yes	*PET117*	couples synthesis of heme *a* to COX assembly
	PET191	nuclear	yes	*COA5*	required for COX assembly
	PET100	nuclear	yes	*PET100*	required for COX assembly
	COA6	nuclear	no	*COA6*	maturation of the Cu_A site
	COX25/COA3	nuclear	yes	*COA3*	translational regulation of *COX1* mRNA
	COX15	nuclear	yes	*COX15*	conversion of heme *o* to heme *a*
	COX10	nuclear	yes	*COX10*	farnesylation of heme *b*
	SCO1	nuclear	yes	*SCO1*	maturation of the Cu_A site
	SCO2	nuclear	no	*SCO2*	maturation of the Cu_A site
	SHY1	nuclear	yes	*SURF1*	hemylation of Cox1

[1] Yeast subunit 5b is a paralog of subunit 5a and under standard conditions of growth is present at low concentrations [134].

4.3.1. Mutations in Complex IV Catalytic Subunits

Generally, when compared to nuclear structural subunits and factors, mitochondrial COX genes are associated with milder and late onset clinical phenotypes [135]. As of today, there are 42 pathogenic mutations reported for MTCO1, 26 for MTCO2, and 24 for MTCO3 [69]. The phenotypes associated with mutations in these subunits are briefly summarized in the paragraphs bellow and are cited from the MITOMAP [69].

The most frequent homoplasmic pathogenic mutations in *MTCO1* are associated with prostate cancer, LHON, SNHL (sensorineural hearing loss) and DEAF (maternally-inherited deafness). The most frequent diseases caused by homoplasmic variants are dilated cardiomyopathy and maternally inherited epilepsy and ataxia. Clinical phenotypes associated with heteroplasmic variants include epilepsy partialis continua, Leigh syndrome, asthenozoospermic infertility, MELAS, myoglobinuria, motor neuron disease, Rhabdomyolysis and acquired idiopathic sideroblastic anemia. Additionally, both homoplasmic and heteroplasmic variants can lead to exercise intolerance.

Homoplasmic pathogenic *MTCO2* variants are mostly associated with progressive encephalomyopathy, possible susceptibility to hypertrophic cardiomyopathy (HCM), SNHL, DEAF, and LHON. For example, there are 147 sequences containing the m.7859G>A substitution that causes progressive encephalomyopathy. Less frequent mutations can cause Alpers-Huttenlocher-like, Asthenozoospermia, developmental delay, ataxia, seizure, hypotonia, hepatic failure, myopathy, MELAS, cerebellar and pyramidal syndrome with cognitive impairment, pseudoexfoliation glaucoma, multisystem disorder, Rhabdomyolysis, biliary atresia, and MIDD (maternally-inherited diabetes and deafness).

The most frequent pathogenic *MTCO3* variants are homoplasmic and lead to LHON. Other clinical phenotypes associated with mutations in *MTCO3* include Alzheimer's disease, MELAS, Leigh syndrome, cardiomyopathy, exercise intolerance, myoglobinuria, myopathy, asthenozoospermia, failure to thrive, cognitive impairment, optic atrophy, encephalopathy, rhabdomyolysis, and sporadic bilateral optic neuropathy.

4.3.2. Mutations in Complex IV Structural Subunits

In the past two decades, pathogenic mutations that result in COX deficiency have been identified in the structural subunits *COX4*, *COX5A*, *COX6A*, *COX6B*, *COX7B*, *COX8*, and *NDUFA4* (Table 5). With the exception of *COX7B*, patients with described mutations in structural COX subunits are born to consanguineous parents and therefore carry homozygous mutant alleles. Clinical phenotypes include, among others, Leigh or Leigh-like syndrome, encephalopathy, myopathy, and anemia.

Interestingly, *COX7B* is located in the X chromosome and is the only X-linked subunit of COX. Different heterozygous mutations in *COX7B* have been described in patients presenting with microphthalmia with linear skin lesions (MLS), a neurocutaneous X-linked dominant male-lethal disorder [136]. In addition to *COX7B*, MLS has also been associated with mutations in *HCCS*, the holocytochrome *c*-type synthase, and in *NDUFB11*, a subunit of Complex I. Somatic mosaicism and the degree of X chromosome inactivation in different tissues could explain the variability of additional clinical phenotypes that accompany MLS, such as developmental delay, abnormalities of the central nervous system, short stature, cardiac defects, and several ocular anomalies [137]. Indeed, the majority of MLS patients have severe skewing of X chromosome inactivation, probably because during embryonic development, respiratory competent cells multiply faster and outgrow cells harboring mutations in these structural genes [137].

4.3.3. Mutations in Complex IV Assembly Factors

SURF1, the human homolog of yeast *SHY1*, has been implicated in the maturation of the heme *a* centers of Complex IV [138]. Mutations in this gene are the most frequent cause of Leigh syndrome stemming from COX deficiency [139]. The first cases of Leigh syndrome caused by *SURF1* mutations were described in 1998 [140,141]. Since then, many other cases have been reported. A systematic review by Wedatilake et al. [139] lists 43 records describing 129 cases of Leigh syndrome with SURF1 deficiency caused by 83 different mutations. The authors also performed a study that included about 50 patients, in which the most frequently occurring mutation was L105* (16 homozygous and 11 compound heterozygous) and no specific correlation of genotype to phenotype was established [139]. Besides Leigh syndrome, there are also reports of *SURF1* mutations associated with Charcot–Marie–Tooth disease [142].

As already mentioned, Cox10 and Cox15 are required for the conversion of heme *b* to the heme *a* of the two redox centers in Cox1 (MTCO1). In *S. cerevisiae*, *cox10* and *cox15* mutants have no Complex IV activity [143,144]. Yeast *cox10* mutants can be functionally complemented by the human homolog of *COX10* [145]. Pathogenic mutations in *COX10* and *COX15* are associated with Leigh syndrome, cardiomyopathy, and encephalopathy, among others (Table 6), and, typically, such patients have an early fatal outcome due to respiratory failure. However, a single adult patient, a 37-year old woman, was identified with isolated COX deficiency associated with a relatively mild clinical phenotype (myopathy, demyelinating neuropathy, premature ovarian failure, short stature, hearing loss, pigmentary maculopathy, and renal tubular dysfunction) due to compound heterozygous mutations resulting in D336V and R339W substitutions in *COX10* [146]. Surprisingly, no COX was detected in blue native gels on mitochondria extracted from the patient's muscle cells. The mutations were introduced into yeast both individually and in combination, all resulting in the loss of the respiratory capacity, which supported the pathogenicity of the mutations [146].

Table 5. Pathologies resulting from mutations in genes encoding Complex IV subunits and assembly factors, and their yeast homologs.

Human Gene	Yeast Gene	Clinical Phenotype	Mutation	Confirmation in Yeast	Reference
MTCO1	*COX1*	prostate cancer, LHON, SNHL, DEAF	many		See main text
MTCO2	*COX2*	progressive encephalomyopathy, HCM, SNHL, DEAF, LHON	many		See main text
MTCO3	*COX3*	LHON, Alzheimer's disease	many		See main text
COX4	*COX5a*	pancreatic insufficiency, dyserythropoeitic anemia, calvarial hyperostosis	homozygous E138K	no	[155]
		short stature, poor weight gain, mild dysmorphic features with highly suspected Fanconi anemia	homozygous K101N	no	[156]
COX5A	*COX6*	early-onset pulmonary arterial hypertension, failure to thrive	homozygous R107C	no	[157]
COX6a	*COX13*	axonal or mixed form of Charcot-Marie-Tooth disease	homozygous c.247−10_247−6delCACTC [3]	no	[158,159]
COX6b	*COX12*	infantile encephalomyopathy, myopathy, growth retardation	homozygous R19H	yes	[160]
		encephalomyopathy, hydrocephalus, HCM	homozygous R20C	no	[161]
		cystic leukodystrophy	homozygous c.241A>C	no	[162]
COX7b		microphthalmia with linear skin lesions	heterozygous c.196delC; heterozygous c.41-2A>G [3]; heterozygous Q19* [2]	no	[136]
COX8		Leigh-like syndrome, epilepsy	homozygous c.115-1G>C [3]	no	[163]
NDUFA4		Leigh syndrome	homozygous c.42+1G>C [3]	no	[164]

Table 5. *Cont.*

Human Gene	Yeast Gene	Clinical Phenotype	Mutation	Confirmation in Yeast	Reference
FAM36A	*COX20*	dystonia and ataxia	homozygous T52P	no	[165,166]
		dysarthria, ataxia, sensory neuropathy	compound heterozygous K14R, G114S, c.157+3G>C [3]	no	[167]
		axonal neuropathy, static encephalopathy	compound heterozygous L14R, W74C	no	[168]
COX14	*COX14*	fatal neonatal lactic acidosis, dysmorphic features	homozygous M19I	no	[169]
PET117	*PET117*	neurodevelopmental regression, medulla oblongata lesions	homozygous c.172C>T (stop codon at residue 58)	no	[170]
COA5	*PET191*	fatal neonatal cardiomyopathy	homozygous A53P	no	[171]
PET100	*PET100*	Leigh syndrome	homozygous c.3G>C (p.Met1?)	no	[172]
		fatal infantile lactic acidosis	homozygous Q48*	no	[173]
COA6	*COA6*	HCM	homozygous W66R	no	[174]
			compound heterozygous W59C, E87*	yes [175]	[162]
COA3	*COX25*	neuropathy, exercise intolerance, obesity, short stature	compound heterozygous L67Pfs*21, Y72C	no	[176]
COX15	*COX15*	Leigh syndrome	homozygous L139V	no	[177]
			homozygous R217W	no	[178]
			compound heterozygous S152*, S344P	no	[147]
		infantile cardioencephalopathy	compound heterozygous S151*, R217W	no	[179]
		early-onset fatal HCM	compound heterozygous R217W, c.C447-3G [3]	no	[180]
COX10	*COX10*	severe muscle weakness, hypotonia, ataxia, ptosis, pyramidal syndrome, status epilepticus	homozygous N204K	yes	[8]
		Leigh-like disease	homozygous T>C in the ATG start codon	no	[181]
		anemia, sensorineural deafness, fatal infantile HCM	compound heterozygous T196K, P225L	no	[182]
		Leigh syndrome, anemia	compound heterozygous D336V, D336G	no	[182]
		myopathy, demyelinating neuropathy, premature ovarian failure, short stature, hearing loss	compound heterozygous D336V, R339W	yes	[146]
		Leigh syndrome, anemia	homozygous P225L	no	[183]
		hypotony, sideroblastic anemia, progressive encephalopathy	compound heterozygous M344V, L424Pfs	no	[183]
		developmental delay, short stature	compound heterozygous R228H, deletion disrupting the last 2 exons	no	[184]

Table 5. *Cont.*

Human Gene	Yeast Gene	Clinical Phenotype	Mutation	Confirmation in Yeast	Reference
SCO1	*SCO1*	encephalopathy, hepatopathy, hypotonia, or cardiac involvement	homozygous G106del	no	[185]
		fatal infantile encephalopathy	compound heterozygous M294V, V93*	no	[186]
		early onset HCM, encephalopathy, hypotonia, hepatopathy	homozygous G132S	no	[187]
		neonatal-onset hepatic failure, encephalopathy	compound heterozygous P174L, ΔGA nt 363–364	no	[8]
SCO2	*SCO2*	cardioencephalomyopathy, Leigh syndrome, high myopia, Charcot-Marie-Tooth disease [2]	many		[151,152, 154] [1]
SURF1	*SHY1*	Leigh syndrome, Charcot-Marie-Tooth disease [2]	many		[139,142] [1;]

[1] Review describing many mutations. [2] In different patients. [3] Splicing mutation.

There is also a single case of a long surviving Leigh syndrome patient resulting from compound heterozygous mutations leading to S152* and S344P substitutions in Cox15. The 16 year old patient presented 42% and 22% of residual COX activity in skeletal muscle cells and fibroblasts, respectively. A normal amount of assembled COX holoenzyme was present in cultured fibroblasts, which could account for the slower clinical progression of the disease [147].

Human SCO1 and SCO2 are copper proteins involved in the metalation of MTCO2 [148]. Although both proteins have homologs in yeast, only Sco1 is needed for metalation of Cox2 [149]. A role of Sco1 in assembly of COX is supported by its presence in a Cox2 assembly intermediate [150]. Pathological mutations in *SCO1* and *SCO2* have been described to result in cardioencephalomyopathy. Gurgel-Giannetti et al. [151] reviewed about 40 patients with mutations in *SCO2*, the majority presenting with cardioencephalomyopathy while two patients suffered from Leigh syndrome. With the exception of one patient that had a homozygous G193S substitution, all other patients presented the E140K substitution which was found either in homozygosis or in association with a second mutation [151]. Furthermore, mutations in *SCO2* have also been associated with Charcot–Marie–Tooth disease [152] and with high degree myopia [153,154].

Table 6. Human ATP Synthase subunits and theirs yeast homologs.

ATP Synthase Sector	Human Subunit	Human Gene	Genome in Human	Yeast Subunit	Yeast Gene	Genome in Yeast	*pet* Mutant
F_1 catalytic barrel	α	*ATP5F1A*	nuclear	α	*ATP1*	nuclear	yes
	β	*ATP5F1B*	nuclear	β	*ATP2*	nuclear	yes
F_1 central stalk	γ	*ATP5F1C*	nuclear	γ	*ATP3*	nuclear	yes
	ε	*ATP5F1E*	nuclear	ε	*ATP15*	nuclear	yes
	δ	*ATP5F1D*	nuclear	δ	*ATP16*	nuclear	yes
Peripheral stalk	b	*ATP5PB*	nuclear	b	*ATP4*	nuclear	yes
	d	*ATP5PD*	nuclear	d	*ATP7*	nuclear	yes
	OSCP	*ATP5PO*	nuclear	OSCP	*ATP5*	nuclear	yes
	F_6	*ATP5PF*	nuclear	h	*ATP14*	nuclear	yes

Table 6. *Cont.*

ATP Synthase Sector	Human Subunit	Human Gene	Genome in Human	Yeast Subunit	Yeast Gene	Genome in Yeast	*pet* Mutant
F_o rotor	Atp6	*MT-ATP6*	mitochondrial	Atp6	*ATP6*	mitochondrial	N/A
	Atp8	*MT-ATP8*	mitochondrial	Atp8	*ATP8*	mitochondrial	N/A
	c1	*ATP5MC1*	nuclear	Atp9	*ATP9*	mitochondrial	N/A
	c2	*ATP5MC2*	nuclear				
	c3	*ATP5MC3*	nuclear				
F_o supernumerary	f	*ATP5MF*	nuclear	f	*ATP17*	nuclear	yes
	6.8 PL	*ATP5MPL*	nuclear	i/j	*ATP18*	nuclear	no
	DAPIT	*ATP5MD*	nuclear	k	*ATP19*	nuclear	no
	g	*ATP5MG*	nuclear	g	*ATP20*	nuclear	no
	e	*ATP5ME*	nuclear	e	*ATP21*	nuclear	no

Additionally, in the past 10 years, pathogenic mutations have been identified in COX assembly factors, products of *FAM36A*, *COX14*, *PET117*, *COA5*, *PET100*, *COA6*, and *COA3* with a variety of clinical phenotypes (Table 5). The pathologies were found in infants with either homozygous or compound heterozygous mutations.

4.4. ATP Synthase

The F_1F_o-ATP synthase or Complex V is a large multimeric protein complex located in the inner membrane of mitochondria. Its principal function is to phosphorylate ADP to ATP using the energy of the proton gradient formed during oxidation of NADH and succinate [188]. This means of making ATP, historically referred to as oxidative phosphorylation, is reversible, allowing the energy released when the ATP synthase hydrolyzes ATP to be stored as a proton gradient capable of driving other energy-demanding chemical, transport, and physical processes. The mitochondrial ATP synthase consists of three distinct structural components: the F_1 ATPase, the peripheral stalk, and a membrane-embedded unit referred to as F_o [189]. In both mammalian and yeast cells, the F_1 ATPase is composed of five distinct subunits: three α and three β subunits, that form a hexameric barrel structure, and the monomeric γ, δ, and ε subunits that constitute the central stalk [190]. The peripheral stalk is composed of four nuclear encoded subunits: b, d, OSCP, and h (yeast) or F6 (human). The membrane-embedded domain is comprised of eight subunits, three of which (Atp6, Atp8, and Atp9) are encoded in the mitochondrial genome of *S. cerevisiae* [191], but only two (MT-ATP6 and MT-ATP8) in mammalian mitochondria [192]. Atp9 is present in multiple copies in a ring that rotates during catalysis. The number of Atp9 molecules per ring differs depending on the organism. The yeast and human rings consist of ten and eight subunits of Atp9, respectively [193–195].

4.4.1. Mutations in ATP Synthase Mitochondrially Encoded Subunits

The first mitochondrial disease caused by an ATP synthase dysfunction was found in patients with mutations in *MT-ATP6*, a mitochondrial gene encoding a key component of the F_o proton channel [196]. The clinical phenotypes of these patients depend largely on the relative levels of heteroplasmy. For instance, patients harboring an m.T8993G substitution with a 70–90% mutation load often exhibit 'milder' syndromes, such as NARP (neurogenic muscle weakness, ataxia, and retinitis pigmentosa) or FBSN (familial bilateral striated necrosis) [196]. Patients with the same m.T8993G substitution, but with a mutation load exceeding 90–95%, present with a more severe neurological disorder: maternally inherited Leigh syndrome characterized by fatal infantile encephalopathy [196].

The m.T8993G pathogenic mutation in *MT-ATP6* was the first reported mitochondrial disease associated with an ATP synthase defect [197]. Since then, more than 40 pathogenic variants have been reported [69]. Four of these, m.T8993G, m.T8993C, m.T9185C, and m.T9185C, constitute most of all known cases, while the remaining variants mostly appear as isolated cases [198]. The clinical presentations of these patients are heterogeneous, with Leigh syndrome and NARP being the most frequently reported (Table 7). Recurrent phenotypes also include Charcot–Marie–Tooth peripheral neuropathy, spinocerebellar ataxia, and familiar upper motor neuron disease. Isolated cases of MLASA (mitochondrial lactic acidosis and sideroblastic anemia), HCM, primary lactic acidosis, 3-methylglutaconic aciduria and optic neuropathy have also been reported [198].

Due to the heteroplasmy accompanying mitochondrial deficiencies caused by *MT-ATP6* mutations, the pathogenic mechanisms underlying these disorders have been challenging to elucidate. In this respect, homoplasmic *S. cerevisiae* clones carrying the pathogenic *ATP6* mutations have served as a good model for characterizing the etiology of *MT-ATP6* diseases. The first modeling in yeast of a pathogenic mutation in *ATP6* was done by Rak et al. [199]. These authors have introduced in yeast the equivalent of the T8993G mutation responsible for NARP. Although the ATP synthase was correctly assembled and present at 80% of wild-type levels, the yeast mutant showed poor respiratory growth and mitochondrial ATP synthesis was only 10% of that of the wild-type. The mutant also had lower steady state levels of COX, suggesting a co-regulation of ATP synthase activity and COX expression [199]. Additionally, there is more evidence that the biogenesis of these two enzymes may be coupled. It was recently showed that Atco, an assembly intermediate composed of COX and ATP synthase subunits, namely Cox6 and Atp9, is a precursor and the sole Atp9 source for ATP synthase assembly [200].

Another study showed that a Leigh syndrome causing m.T9191C mutation with an L242P substitution in *ATP6* reduced both assembly of the yeast ATP synthase and the efficiency of ATP synthesis by 90% [201]. Based on the biochemical phenotypes of the mutant and of revertants with amino acid substitutions at this position, the proline was proposed to disrupt the terminal α-helix causing a displacement of other neighboring helices involved in proton transfer at the Atp6 and Atp9 interface [202]. Additionally, the conformational change induced by the mutations promoted proteolysis of Atp6, thereby accounting for the reduction of assembled ATP synthase. Revertants with replacements of the helix-disrupting proline by threonine or serine completely restored the assembly defect, but only partially the efficiency of proton translocation. These observations, combined with modeling of the interface between the two subunits, suggested that the suppressor mutations did not completely compensate for the displacement of residues involved in release of protons from the Atp9 ring and their transfer to a neighboring aspartic acid residue in Atp6 [202]. These findings are consistent with previous models of the disease, suggesting that certain pathogenic *ATP6* mutations do not completely block the proton translocation mechanisms of the F_o domain, but rather disrupt proton transport during rotation of the ring. As the conformational changes in the F_1 moiety are induced by rotation of the Atp9 ring, the uncoupling leads to the inability of ATP synthase to harness the energy normally released from the translocation of protons for ATP synthesis [203].

The different biochemical anomalies discerned in the numerous *MT-ATP6* pathogenic variants suggest several pathophysiological mechanisms responsible for diseases associated with defects in the ATP synthase [198]. These include decreased holoenzyme assembly, destabilization of the proton pore resulting in mitochondrial membrane potential buildup, impairment of the proton pump leading to decreased membrane potential, reduced ATP synthesis and abnormal sensitivity to the ATP synthase inhibitor oligomycin [198].

A smaller subset of mitochondrial mutations causing ATP synthase deficiencies has been ascribed to the second mitochondrial gene, *MT-ATP8*, encoding a subunit of F_o. The first pathogenic mutation in *MT-ATP8* was identified in a 16-year-old patient with apical HCM and neuropathy, with a marked reduction in ATP synthase activity in fibroblasts and muscle tissue [204]. Since then, several other patients have been identified with pathogenic *MT-ATP8* mutations. One of the patients was a seven-month-old infant diagnosed with tetralogy of Fallot, the most common type of congenital heart defect characterized by ventricular septal defect, pulmonary stenosis, right ventricular hypertrophy and aortic dextroposition [205].

MT-ATP6 and *MT-ATP8* genes overlap 46 nucleotides that span 16 codons [206]. On the MITOMAP [69] are presently listed four known pathogenic mutations in this region that contribute to an amino acid change in both genes, and three other variants that have a hypothesized deleterious effect. An ATP synthase deficiency has been associated with an m.C8561G mutation in two siblings with cerebral ataxia and loss of neuromuscular function. The mutation, resulting in P12R and P66A changes in *MT-ATP6* and *MT-ATP8*, respectively, when expressed in homoplasmic myoblasts, reduced cellular ATP by 15% but had only a marginal effect on steady state concentrations of the ATP synthase [207]. Even though there was no obvious decrease in the steady-state concentration of ATP synthase, a substantial increase of an uncharacterized assembly intermediate was noted. Another study reported an m.C8561T mutation with other amino acid substitutions in a patient with a similar biochemical but more severe clinical presentation [208]. Neither studies examined the effect of the singular or combined *ATP8* and *ATP6* mutations on the yeast ATP synthase.

4.4.2. Mutations in ATP Synthase Nuclear Structural Subunits

At present, only a small number of mitochondrial pathologies have been attributed to mutations in nuclear genes coding for subunits of the ATP synthase and for chaperones that function in assembly of the enzyme. Several studies found that patients with lactic acidosis, persisting 3-methylglutaconic aciduria, cardiomyopathy, and early death were correlated with severe deficiencies in ATP synthase, but the genetic lesions in these studies were not identified [209–211]. *ATP5F1E* and *ATP5F1D*, the human homologs of the yeast *ATP15* and *ATP16*, code for the ε and δ subunits, respectively, of the central stalk of F_1. A homozygous missense mutation in *ATP5F1E* was the first reported instance of a patient with an ATP synthase defect stemming from a mutation in a nuclear encoded subunit of the enzyme. The mutation caused mild mental retardation and the patient developed peripheral neuropathy [6]. The mitochondrial synthesis of ATP, the oligomycin ATPase activity and the ATP synthase content were reduced by 60–70%. The residual enzyme with the mutated ε subunit had normal activity, indicating that the primary effect of the mutation was on assembly of the synthase. Interestingly, this mutation, when introduced in the ε subunit of yeast ATP synthase, did not elicit any detectable effect on either the activity of assembly of the enzyme [212]. This suggested that the mutation in the human ε subunit, perhaps because of a weaker physical interaction with the c ring, exerted a deleterious effect on assembly of the human but not of the yeast enzyme [212]. In a more recent report, two patients with homozygous mutations in *ATP5F1D* were shown to suffer from metabolic disorders in one case and from encephalopathy in another [213]. *ATP5F1A*, coding for human α subunit of F_1, is the third nuclear ATP synthase gene identified in two patients presenting fatal neonatal encephalopathy with intractable seizures [5]. Interestingly, reduced levels of cellular ATP5F1 were also shown to correlate significantly with earlier-onset prostate cancer [214]. Indeed, transitioning from oxidative phosphorylation to anaerobic glycolysis for energy production occurs in many types of tumors, and could explain the pathophysiology of the disease.

Table 7. Pathologies resulting from mutations in genes encoding ATP synthase subunits and assembly factors, and their yeast homologs.

Human Gene	Yeast Gene	Clinical Phenotype	Mutation	Confirmation in Yeast	Reference
MT-ATP6	*ATP6*	NARP, FBSN, Leigh syndrome, Charcot-Marie-Tooth disorder, HCM, MLASA	many		[198] [1]
MT-ATP8	*ATP8*	apical HCM, neuropathy	homoplasmic W55*	no	[204]
		tetralogy of Fallot	homoplasmic G9804A, C8481T, heteroplamic T7501C [3]	no	[205]

Table 7. *Cont.*

Human Gene	Yeast Gene	Clinical Phenotype	Mutation	Confirmation in Yeast	Reference
ATP5F1D	*ATP16*	metabolic disorders, encephalopathy	homozygous P82L; homozygous V106G [2]	no	[213]
ATP5F1E	*ATP15*	mild mental retardation, developed peripheral neuropathy	homozygous Y12C	no	[6]
ATP5F1A	*ATP1*	fatal neonatal encephalopathy	heterozygous R329C	no	[5]
ATPAF2	*ATP12*	microcephaly etc	homozygous W94R	yes	[7,211]

[1] Review describing many mutations. [2] In different patients. [3] In the same patient.

None of the patients with mutations in the nuclear ATP synthase genes discussed in this and the next sections exhibit the clinical NARP and Leigh syndrome phenotypes characteristic of patients with mutations in the two mitochondrially-encoded genes of the enzyme.

4.4.3. Nuclear ATP Synthase Assembly Gene Mutations

A substantial number of nuclear gene products of *S. cerevisiae* are known to regulate and chaperone different steps of ATP synthase assembly [215]. At present, however, the only known human regulatory factors with identified mutations in a small number of patients are TMEM70, a protein essential for ATP synthase assembly and ATPAF2, the homolog of the yeast Atp12 chaperone that interacts with the α subunit of F_1 during assembly of this ATP synthase module [216]. TMEM70 does not have a yeast homolog. The phenotypes associated with mutations in TMEM70 and the role of its product in ATP synthase assembly have been reviewed elsewhere [217,218].

A study in which two patients, ascertained to have nuclear mutations that affected mitochondrial ATP synthase assembly, were screened by sequencing human homologs of yeast genes previously shown to affect F_1 biogenesis led to the identification in one patient of a mutation in *ATPAF2*. This patient, diagnosed with lactic acidosis, glutaconic aciduria, encephalomyopathy and a range of different developmental problems, was found to have a homozygous W94R amino acid substitution that resulted in severe deficits of ATP synthase in the heart, liver and to a lesser degree in skeletal muscle, resulting in death at 14 months of age [219]. The deleterious effect of the W94R substitution in a highly conserved region of *ATPAF2* on ATP synthase assembly was confirmed by Meulemans et al. [7], who showed that the wild-type, but not the mutant human gene, restored the ATP synthase activity of a yeast *atp12* mutant. The *atp12* mutant also failed to be complemented by the yeast *ATP12* harboring the equivalent W102R mutation expressed from a low copy yeast CEN, but not from a high-copy plasmid. The rescue by the W102R, however, depended on the presence of wild-type *FMC1*, a gene implicated in regulating the activity of yeast Atp12 [220].

4.5. Coenzyme Q

Coenzyme Q (ubiquinone, CoQ or CoQ_{10} in humans) is a lipophilic redox molecule found in virtually all eukaryotic organisms and most bacteria. It is composed of a quinone ring connected to a polyisoprenoid side chain of variable length. Coenzyme Q serves several crucial functions in mitochondria, including transfer of electrons from Complexes I and II to Complex III, acting as an essential cofactor in the uncoupler protein mediated transfer of protons to the matrix, prevention of lipid peroxidation, biosynthesis of uridine, beta-oxidation of fatty acids, and binding to and regulating the permeability transition of the voltage-dependent anion channel [221].

The CoQ biosynthetic pathway in eukaryotes has been studied in yeast *coq* mutants arrested at different steps of the pathway [222,223]. Like human cells, yeast relies on de novo synthesis of CoQ; and any deficiency in the biosynthetic pathway results in a growth arrest on media containing non-fermentable carbon sources [3,224]. Most enzymes of the CoQ biosynthetic pathway are organized in a multi-subunit complex known as the CoQ synthome [222,225]. The CoQ synthome is spatially linked to the endoplasmic reticulum–mitochondria contact sites, providing optimal CoQ production with an efficient intracellular distribution as well as minimizing the escape of toxic intermediates [226,227]. Expression of functional CoQ in yeast depends on at least 14 nuclear gene products (Coq1-Coq11, Yah1, Arh1, and Hfd1), all located in mitochondria [227].

CoQ biosynthesis is achieved by three separate and highly conserved pathways:

(1) Synthesis of the quinone ring from 4- hydroxybenzoate (4HB), derived from tyrosine [228], or from p-aminobenzoic acid (pABA), in yeast but not in humans [225,227,229]. The early steps of 4HB formation are still to be determined but deamination of tyrosine starts with Aro8 or Aro9 catalyzed transamination [230]. The formation of the final intermediate 4-hydroxybenzaldehyde (4 HBz) is catalyzed by the aldehyde dehydrogenase Hfd1 [230,231].

(2) Synthesis of isopentenyl pyrophosphate (IPP) and dimethylally pyrophosphate (DMAPP), catalyzed by Coq1 in yeast [232] and PDSS1 and PDSS2 in humans.

(3) Prenylation of parahydroxybenzoate by the polisoprenyl transferase Coq2 [233] and further modifications of the benzoquinone ring by hydroxylases and methyl transferases. The isoprenoid side chain is important for proper CoQ localization at the mid-plane of phospholipid bilayers. Yeast coenzyme Q contains 6 isopentenyl units (CoQ_6) while in humans the major coenzyme Q isoform contains 10 isopentenyl units (CoQ_{10}).

The benzoquinone head group is modified by hydroxylations catalyzed by Coq6 and Coq7 [234–236], methylation of the resultant hydroxyls by the Coq3 methyl transferase [237], methylation of the ring by Coq5 [238] and a decarboxylation step catalyzed by a still unidentified enzyme of this pathway. Other gene products linked to coenzyme Q biosynthesis and utilization include Coq4 and Coq9, that have been assigned a role in assembly and stability of the CoQ synthome, and Coq8, a member of a protein family that includes kinases and ATP-dependent ligases. Coq8 has been implicated in the phosphorylation state of Coq3, Coq5 and Coq7 [239–241]. The steady-state concentrations of Coq4, 6, 7, and 9 are markedly decreased in a *coq8* null mutant, as a result of which assembly of the CoQ synthome is abrogated [242]. The *coq8* null mutant, however, contains a complex of Coq6 and Coq7 thought to be an early intermediate of the synthome [243]. Coq10 is a low molecular weight member of the START protein family that binds coenzyme Q. Although Coq10 is required for respiration, its synthesis is only partially affected in log but not in stationary phase yeast cells, suggesting that its function is related to the delivery of coenzyme Q from the synthome located at the endoplasmic-mitochondrial contact sites to the regions of the inner membrane containing the respiratory chain complexes [226,244]. The respiratory deficiency of *coq10* mutants is partially rescued in a *coq11* mutant, which codes for a component of the synthome that has been proposed to down- regulate the synthesis of coenzyme Q [244]. Coq11 (human NDUFA9), a separate protein in yeast, is a component of the CoQ synthome [245] and appears to regulate its formation and stability [244]. CoQ yeast and human genes are listed in Table 8.

Table 8. Coenzyme Q genes required for functional expression of CoQ and their yeast homologs.

Yeast Gene	Human Gene
COQ1	*PDSS1*
COQ1	*PDSS2*
COQ2	*COQ2*
COQ3	*COQ3*
COQ4	*COQ4*
COQ5	*COQ5*
COQ6	*COQ6*

Table 8. *Cont.*

Yeast Gene	Human Gene
COQ7/CAT5	*COQ7*
COQ8/ABC1	*COQ8A*
COQ8/ABC2	*COQ8B*
COQ9	*COQ9*
COQ10	*COQ10A, COQ10B*
COQ11	*NDUFA9*

Mutations in COQ Genes

CoQ_{10} deficiency, a biochemical lesion first described over three decades ago by Ogasahara et al. [246], is subdivided into primary CoQ_{10} deficiency, when caused by a pathogenic mutation in one of the genes required for the coenzyme's biosynthesis, and secondary CoQ_{10} deficiency, when the mutated gene is not directly related to the biosynthetic pathway [247]. Of the two, the latter has been reported more frequently in patients, with phenotypes including mitochondrial myopathies, mitochondrial DNA depletion syndrome and multiple acyl-CoA dehydrogenase deficiency (MADD) [248]. However, the pathogenic mechanisms linking these disorders with the observed CoQ_{10} deficiency have yet to be elucidated.

Primary CoQ_{10} deficiency is far rarer [248]. Emmanuele et al. [247] first classified the clinical manifestations of primary CoQ_{10} deficiency into five distinct phenotypes: encephalomyopathy, isolated myopathy, nephropathy, infantile multisystemic disease, and cerebellar ataxia. However, it has been argued that this subdivision should be updated, as new cases have been discovered with novel mutations presenting a wide range of other clinical phenotypes, as well as different combinations of the previously described symptoms [248]. To date, mutations in ten genes have been associated with primary CoQ_{10} deficiency (Table 9).

With the exception of Coq3, patients have been reported with mutations in all other components of the CoQ multi-subunit complex [249,250]. These patients can be treated with CoQ_{10} supplementation with partial success. Early treatment based on early diagnosis is critical for the best outcome [251]. Because of its poor solubility, CoQ_{10} is only administrated in oral formulations, despite its destitute bioavailability [252,253]. Similarly, uptake of CoQ6 in yeast *coq* mutants is inefficient [254,255].

Due to the striking homology between human and yeast COQ genes [227], studies of CoQ proteins in *S. cerevisiae* may provide insight into human homologs, leading to the identification of residues critical for protein function and, therefore, with higher pathogenic potential [256]. Indeed, yeast *coq3*, *coq8*, *coq9*, and *coq10* mutants are complemented by the human counterparts [239,257–259], while yeast *coq5* null mutants are complemented by human *COQ5* combined with overexpression of *COQ8* [260]. Furthermore, studies of pathogenic mutations in *COQ* genes have been validated in yeast *coq1* [261], *coq2* [261,262], *coq4* [263], *coq6* [264], *coq8* [241], and *coq9* mutants [265].

PDSS1 and *PDSS2*, both human homologs of yeast *COQ1*, encode two proteins that form a heterotetramer that catalyzes the elongation of the isoprenoid side chain. *PDSS1* does not complement the yeast *coq1* null mutant [261]. Mutations in both of these genes lead to infantile multisystemic disease, a heterogeneous disorder characterized by psychomotor regression, encephalopathy, optic atrophy, retinopathy, hearing loss, renal dysfunction and heart valvulopathy [247,261]. Mutations in *PDSS2* have been associated with additional phenotypes, including Leigh syndrome, steroid resistant nephrotic syndrome (SRNS)—an atypical manifestation for other mitochondrial disorders but quite common for CoQ_{10} deficiencies; and hepatocellular carcinoma [247,266,267]. It has been shown that the downregulation of *PDSS2* can induce a shift from aerobic metabolism to anaerobic glycolysis, as well as increased chromosomal instability—a possible pathogenic mechanism for hepatocellular carcinoma [267].

Pathogenic mutations in *COQ2*, encoding a parahydroxybenzoate-polyprenyltransferase that catalyzes the addition of the isoprenoid chain to the benzoquinone ring, were the first to be associated with primary CoQ_{10} deficiency [268]. Pathogenic mutations in human *COQ2* have been confirmed in yeast by complementation studies of the yeast *coq2* null mutant [261,262]. Clinical manifestations of these mutations include isolated SRNS, SRNS with encephalomyopathy resembling MELAS, fatal infantile multisystemic disease, and late-onset multiple-system atrophy and retinopathy [247,248].

The first pathogenic mutation in *COQ4*, required for the stability of the CoQ synthome, was found as a haploinsufficiency, with a phenotype similar to that of a heterozygous yeast mutant [263]. The patient presented facial dysmorphism and muscle hypotonia, which improved significantly with CoQ_{10} supplementation [263]. Since then, a total of 19 patients, all infants, have been identified with mutations in *COQ4*, with clinical phenotypes that included cerebellar atrophy, lactic acidosis, seizures, muscle weakness, cardiomyopathy, ataxia, and Leigh syndrome [269,270]. None of the patients suffered from nephropathy, typically found in primary CoQ_{10} deficiency.

To date, only three cases of pathogenic mutations in the methyltransferase encoded by *COQ5*, have been recorded. The affected individuals were three female siblings presenting with non-progressive cerebellar ataxia, dysarthria, and mild to moderate cognitive disability [250]. Two of the three siblings exhibited myoclonic jerks and generalized tonic-clonic seizures in adolescence and early 20s. Next-generation sequencing identified a tandem duplication of the last four exons of *COQ5*, while biochemical studies showed a 33% reduction of CoQ_{10} in skeletal muscle—sufficient to sustain a basal rate but insufficient to reach maximal efficiency of respiration [250].

Mutations in *COQ6*, encoding an enzyme involved in hydroxylation and deamination reactions during CoQ biosynthesis, have been primarily associated with SRNS [264], characterized by significant proteinuria with resulting hypoalbuminemia and edema and presenting with focal segmental glomerulosclerosis [248]. The pathogenicity of the first six *COQ6* mutations in human patients was confirmed by complementation studies using the yeast *coq6* null mutant [264]. Three patients harboring a *COQ6* mutation also suffered from infantile multisystemic disease [247]. Based on in vitro and in vivo studies of renal podocyte cell lines, it has been hypothesized that the pathogenicity of *COQ6* mutations relates to respiratory chain deficiency, ROS generation, disruption of podocyte cytoskeleton and induction of cellular apoptosis, ultimately resulting in SRNS [271].

Three cases of primary CoQ_{10} deficiency caused by mutations in *COQ7*, responsible for the penultimate step of CoQ biosynthesis, have been reported in two children with similar phenotypes of spasticity, sensorineural hearing loss and muscle hypotonia; and a third more severe case of fatal mitochondrial encephalo-myo-nephro-cardiopathy, persistent lactic acidosis, and basal ganglia lesions [272–274]. In one of the patients, treatment with the unnatural biosynthesis precursor 2,4-dihydroxybenzoate (DHB), a hydroxylated variant of the native 4-hydroxybenzoic acid (4-HB) normally modified by COQ7, increased CoQ_{10} levels and partially restored mitochondrial respiration [273].

The human genes *ADCK3* and *ADCK4*, also known as *COQ8A* and *COQ8B*, are both homologs of yeast *COQ8*. Mutations in *COQ8A* have mostly been associated with autosomal recessive progressive cerebellar ataxia (ARCA), often accompanied by childhood onset cerebellar atrophy, with and without seizures, and exercise intolerance [241,248,275]. The pathogenic nature of the *ADCK3* mutations was corroborated using the yeast counterpart system [241]. However, isolated cases of psychiatric disorders, seizures, migraines, and dysarthria have been reported [248]. Studies of fibroblast cell lines isolated from ARCA patients showed an increased sensitivity to oxidative stress induced by hydrogen peroxide, high levels of oxidative stress and changes in mitochondrial homeostasis as a result of loss-of-function mutations in *COQ8A* [275]. Interestingly, concomitant with the upregulation of ROS production, increased respiratory supercomplex stability and basal respiratory rate were observed, suggesting that the loss of ADCK3 could result in a compensatory elevation of respirasome formation [275]. The relationship between the two human paralogs *COQ8A* and *COQ8B* is unclear. However, pathogenic mutations in the two genes lead to completely different clinical phenotypes.

Indeed, all patients with *COQ8B* mutations suffered from SNRS, with only a single case of neurological involvement reported [248,276].

Table 9. Pathologies resulting from mutations in Coenzyme Q genes and their yeast homologs.

Human Gene	Yeast Gene	Clinical Phenotype	Mutation	Confirmation in Yeast	Reference
PDSS1	*COQ1*	infantile multisystemic disease	homozygous D308E	yes	[261]
PDSS2	*COQ1*	infantile multisystemic disease, SRNS, LS, hepatocellular carcinoma	compound heterozygous Q322*, S382L	yes	[262]
COQ2	*COQ2*	SRNS, SRNS with encephalomyopathy resembling MELAS, fatal infantile multisystemic disease	homozygous Y297C	yes	[268]
COQ4	*COQ4*	LS, cerebellar atrophy, lactic acidosis, etc.	many		[263,269,270] [1]
COQ5	*COQ5*	cerebellar ataxia, dysarthria, myoclonic jerks	biallelic 9590 bp duplication	no	[250]
COQ6	*COQ6*	SRNS, infantile multisystemic disease	many		[264] [1]
COQ7	*COQ7*	spasticity, sensorineural hearing loss etc.	homozygous V141E; homozygous L111P; compound heterozygous R107W, K200Ifs*56 [2]	no	[272–274]
COQ8A	*COQ8*	ARCA, seizures, dystonia, spasticity	many		[248] [1]
COQ8B	*COQ8*	SRNS	many		[248,276] [1]
COQ9	*COQ9*	infantile multisystemic disease, Leigh-like Syndrome, microcephaly	many		See main text

[1] Review describing many mutations. [2] In different patients.

Primary CoQ_{10} deficiency caused by mutations in *COQ9* is extremely rare, with only seven cases from four families having been reported. The *COQ9* gene product binds to the polyprenyl tail of CoQ intermediates with high specificity, allowing the modification of the benzene ring by Coq7 and other components of the CoQ synthome [259,277]. The reported cases include an infant suffering from lethal lactic acidosis, seizures, cerebral atrophy, HCM, and renal dysfunction [248,265], a boy diagnosed with neonatal Leigh-like syndrome who died at 18 days of age from cardio-respiratory failure [278], four siblings with an unknown and ultimately lethal condition characterized by dilated cardiomyopathy, anemia, abnormal appearing kidney, and suspected Leigh syndrome [279]; and a nine-month old girl presenting with microcephaly, truncal hypotonia, and dysmorphic features [280].

It has been shown that the pathogenicity of primary and secondary CoQ_{10} deficiencies is linked to the impairment of electron transfer to Complex III, ultimately resulting in decreased mitochondrial respiration and ATP synthesis [248,281]. It has also been noted that the neurological presentation of CoQ_{10} deficiency is likely associated with oxidative damage and caspase-independent apoptotic cell death in the brain, as a result of mitochondrial impairment [281]. While renal dysfunction is not uncharacteristic of mitochondrial deficiencies, the glomerular involvement, as opposed to the tubular damage seen in other mitochondrial cytopathies, is perplexing and could be a result of impaired CoQ_{10} antioxidant function [248]. Equally puzzling is the remarkable diversity of clinical phenotypes resulting from mutations in different COQ genes. Based on in vivo studies of mouse models harboring *COQ9* mutations, Luna-Sánchez et al. [281] have hypothesized that a key factor in determining the degree of severity and particular clinical phenotypes of CoQ_{10} deficiencies is the stability of the CoQ synthome. Lastly, decreased CoQ_{10} levels impair the activity of sulfide:quinone oxidoreductase, an enzyme involved in the catabolism of H_2S [282]. While at physiological conditions H_2S serves as an electron donor to the mitochondrial respiratory chain, at elevated levels it inhibits Complex IV activity, resulting in reduced cellular respiration. Over-physiological levels of H_2S could therefore participate in the pathogenic mechanism of CoQ_{10} deficiency [282].

4.6. *Cytochrome c*

Cytochrome *c*, a low-molecular weight heme protein loosely tethered to the inner mitochondrial membrane, is an important component of the respiratory chain that accepts electrons from Complex III and transfers them to Complex IV. Cytochrome *c* also functions as an initiator of apoptosis. In humans, cytochrome *c* is encoded by *CYCS* and in yeast by two isoforms *CYC1* and *CYC7*, the first one being predominant. To date, four mutations in *CYCS* have been associated with non-syndromic and mild thrombocytopenia, a rare autosomal dominant disorder characterized by low platelet levels in the blood but no other hematological findings. These mutations include a G41S substitution found in a New Zealander family [283], a Y48H mutation in an Italian family [284], an A52V mutation in a British family [285], and a K101del deletion in a 64-year-old woman from a Japanese family with multiple cases of Hemophilia A [286]. Unlike other non-syndromic thrombocytopenias, the morphology of circulating platelets in these patients was normal [287]. Some of these mutations, recreated using yeast model systems, led to reduced cytochrome *c* expression, reduced respiratory rate, and increased apoptotic rate [284,286]. However, the link between loss-of-function mutations in *CYSC* and abnormal platelet formation has yet to be elucidated.

The addition of the heme moiety to cytochrome *c* is catalyzed by a cytochrome *c* type heme lyase encoded by the *HCCS* gene, the human homolog of yeast *CYC3*. HCCS is able to complement the yeast Cyc3 deficiency [288]. Mutations in *HCCS* have been associated with MLS, an X-linked, male-lethal disorder described in the Complex III section [289]. It has been hypothesized that the characteristic phenotype of MLS patients is a consequence of the activation of non-canonical, caspase-9 mediated cell death in the brain and eyes as a result of HCCS impairment [290].

5. Human Pathologies Resulting from Mutations in Genes with No Homologs in *S. cerevisiae*

5.1. *Complex I*

While *S. cerevisie* mitochondria contain three different single subunit NADH dehydrogenases in the inner membrane [291], the human counterpart forms a large hetero-oligomeric complex (Complex I) comprised of 38 nuclear-encoded and seven mitochondrially encoded subunits. As such, the genes for the numerous structural proteins and assembly factors related to Complex I have no yeast homologs. The majority of Complex I genes have been associated with mitochondrial disorders. Complex I deficiency is the most common biochemical lesion identified in childhood-onset mitochondrial diseases, accounting for approximately 30% of all cases [292]. The clinical presentations of these disorders are varied, including Leigh syndrome, fatal infantile lactic acidosis, HCM, exercise intolerance, LHON, leukoencephalopathy, and MELAS [292]. Although divergent, yeast NADH-dehydrogenase Ndi1 has been tested as a candidate for gene therapy for human Complex I deficiencies, with success in preventing ischemia-reperfusion injury in transgenic rats [293,294].

Other yeasts, such the obligate aerobic *Yarrowia lipolytica*, however, have Complex I and have been used to study and model diseases of this complex [295,296]. In one study, modeling mutations in *Y. lipolityca* made it possible to determine the most pathogenic mutation among two mutated alleles. The compound heterozygous Y53C and Y308C substitutions in *NDUFS2* were found in siblings presenting with non-syndromic LHON-like disease but normal levels of Complex I [297]. The parents, each carrying one of the mutations, presented no clinical phenotype. Equivalent substitutions (H57C and Y311C) were introduced independently in the *Y. lipolytica* homolog *NUCM*, resulting in normal levels of Complex I in mutant H57C and no detectable Complex I in mutant Y311C. The authors concluded that the normal levels of Complex I in the patient's fibroblast arose from the expression of the Y53C allele and that the non-syndromic LHON-like phenotype was possibly caused by an instability of Complex I containing this mutation [297]. In another study, 13 out of 16 single amino acid substitutions of NDUFV1 were confirmed as pathogenic using *Y. lipolytica* [298]. This yeast is also useful in the evaluation of pathogenic mutations in Complex I assembly factors, such as NUBPL (nucleotide binding protein-like), which were recreated in the homologous Ind1 protein of the yeast model [299]. Recently, the cryo-EM structure of *Y. lipolytica* Complex I has been solved [300,301] and can provide helpful information for those modeling Complex I diseases in this yeast.

5.2. Assembly Factors

Pathogenic mutations have also been described for some human genes with no yeast homologs encoding assembly factors of oxphos complexes, such as *TCC19* (Complex III), *TACO1* (Complex IV), and *TMEM70* (ATP synthase). They are beyond the scope of this review.

6. Concluding Remarks

Mammalian mitochondrial DNA codes for 13 proteins, all of which are subunits of the respiratory complexes and ATP synthase. Most of the organelle, however, is encoded in the nuclear genome, by what, in yeast, is estimated to be at least 900 genes or 15% of the genome. It stands to reason that a large number of human genetic disorders will stem from mutations that affect mitochondrial function. In the past, studies of yeast *pet* mutants have had a direct impact on understanding the genetic and mechanistic basis of human mitochondrial diseases. However, about 20% of the eukaryotic proteome has not yet been characterized, even in well-studied model organisms such as *S. cerevisiae* [302]. A substantial number of proteins still lacking an ascribed function are encoded by genes that affect mitochondrial respiration. In this context, yeast is still a powerful platform for discovering the function of uncharacterized mitochondrial proteins, as well as to gain a better understanding of the underlying molecular consequences of pathogenic mutations that could prove to be promising targets for mitochondrial disease therapies.

Author Contributions: Conceptualization: L.V.R.F.; writing: L.V.R.F., L.B., and M.H.B. All authors have read and agreed to the published version of the manuscript.

Funding: L.V.R.F. was supported by FAPESP Post-Doctoral Fellowship 2019/02799-2.

Acknowledgments: The authors would like to express gratitude for the guidance of Alexander Tzagoloff in the preparation of this manuscript.

Conflicts of Interest: The authors declare no conflict of interest.

References

1. Morgenstern, M.; Stiller, S.B.; Lübbert, P.; Peikert, C.D.; Dannenmaier, S.; Drepper, F.; Weill, U.; Höß, P.; Feuerstein, R.; Gebert, M.; et al. Definition of a High-Confidence Mitochondrial Proteome at Quantitative Scale. *Cell Rep.* **2017**, *19*, 2836–2852. [CrossRef] [PubMed]
2. Saccharomyces Genome Database | SGD. Available online: https://www.yeastgenome.org/ (accessed on 23 October 2020).
3. Tzagoloff, A.; Dieckmann, C.L. PET genes of *Saccharomyces cerevisiae*. *Microbiol. Rev.* **1990**, *54*, 211–225. [CrossRef] [PubMed]
4. DiMauro, S. Mitochondrial diseases. *Biochim. Biophys. Acta* **2004**, *1658*, 80–88. [CrossRef] [PubMed]
5. Jonckheere, A.I.; Renkema, G.H.; Bras, M.; van den Heuvel, L.P.; Hoischen, A.; Gilissen, C.; Nabuurs, S.B.; Huynen, M.A.; de Vries, M.C.; Smeitink, J.A.M.; et al. A complex V ATP5A1 defect causes fatal neonatal mitochondrial encephalopathy. *Brain J. Neurol.* **2013**, *136*, 1544–1554. [CrossRef]
6. Mayr, J.A.; Havlícková, V.; Zimmermann, F.; Magler, I.; Kaplanová, V.; Jesina, P.; Pecinová, A.; Nusková, H.; Koch, J.; Sperl, W.; et al. Mitochondrial ATP synthase deficiency due to a mutation in the ATP5E gene for the F1 epsilon subunit. *Hum. Mol. Genet.* **2010**, *19*, 3430–3439. [CrossRef]
7. Meulemans, A.; Seneca, S.; Pribyl, T.; Smet, J.; Alderweirldt, V.; Waeytens, A.; Lissens, W.; Van Coster, R.; De Meirleir, L.; di Rago, J.-P.; et al. Defining the pathogenesis of the human Atp12p W94R mutation using a *Saccharomyces cerevisiae* yeast model. *J. Biol. Chem.* **2010**, *285*, 4099–4109. [CrossRef]

8. Valnot, I.; von Kleist-Retzow, J.C.; Barrientos, A.; Gorbatyuk, M.; Taanman, J.W.; Mehaye, B.; Rustin, P.; Tzagoloff, A.; Munnich, A.; Rötig, A. A mutation in the human heme A:farnesyltransferase gene (COX10) causes cytochrome c oxidase deficiency. *Hum. Mol. Genet.* **2000**, *9*, 1245–1249. [CrossRef]
9. Quinzii, C.M.; Kattah, A.G.; Naini, A.; Akman, H.O.; Mootha, V.K.; DiMauro, S.; Hirano, M. Coenzyme Q deficiency and cerebellar ataxia associated with an aprataxin mutation. *Neurology* **2005**, *64*, 539–541. [CrossRef]
10. Garcia-Diaz, B.; Barros, M.H.; Sanna-Cherchi, S.; Emmanuele, V.; Akman, H.O.; Ferreiro-Barros, C.C.; Horvath, R.; Tadesse, S.; El Gharaby, N.; DiMauro, S.; et al. Infantile encephaloneuromyopathy and defective mitochondrial translation are due to a homozygous RMND1 mutation. *Am. J. Hum. Genet.* **2012**, *91*, 729–736. [CrossRef]
11. Barrientos, A. Yeast models of human mitochondrial diseases. *IUBMB Life* **2003**, *55*, 83–95. [CrossRef]
12. Foury, F.; Roganti, T.; Lecrenier, N.; Purnelle, B. The complete sequence of the mitochondrial genome of *Saccharomyces cerevisiae*. *FEBS Lett.* **1998**, *440*, 325–331. [CrossRef]
13. Myers, A.M.; Pape, L.K.; Tzagoloff, A. Mitochondrial protein synthesis is required for maintenance of intact mitochondrial genomes in *Saccharomyces cerevisiae*. *EMBO J.* **1985**, *4*, 2087–2092. [CrossRef] [PubMed]
14. Rak, M.; Tetaud, E.; Godard, F.; Sagot, I.; Salin, B.; Duvezin-Caubet, S.; Slonimski, P.P.; Rytka, J.; di Rago, J.-P. Yeast cells lacking the mitochondrial gene encoding the ATP synthase subunit 6 exhibit a selective loss of complex IV and unusual mitochondrial morphology. *J. Biol. Chem.* **2007**, *282*, 10853–10864. [CrossRef] [PubMed]
15. Rak, M.; Tzagoloff, A. F1-dependent translation of mitochondrially encoded Atp6p and Atp8p subunits of yeast ATP synthase. *Proc. Natl. Acad. Sci. USA* **2009**, *106*, 18509–18514. [CrossRef] [PubMed]
16. Doeg, K.A.; Krueger, S.; Ziegler, D.M. Studies on the electron transfer system. 29. The isolation and properties of a succinic dehydrogenase-cytochrome b complex from beef-heart mitochondria. *Biochim. Biophys. Acta* **1960**, *41*, 491–497. [CrossRef]
17. Capaldi, R.A.; Sweetland, J.; Merli, A. Polypeptides in the succinate-coenzyme Q reductase segment of the respiratory chain. *Biochemistry* **1977**, *16*, 5707–5710. [CrossRef]
18. Lemire, B.D.; Oyedotun, K.S. The *Saccharomyces cerevisiae* mitochondrial succinate:ubiquinone oxidoreductase. *Biochim. Biophys. Acta* **2002**, *1553*, 102–116. [CrossRef]
19. Gebert, N.; Gebert, M.; Oeljeklaus, S.; von der Malsburg, K.; Stroud, D.A.; Kulawiak, B.; Wirth, C.; Zahedi, R.P.; Dolezal, P.; Wiese, S.; et al. Dual function of Sdh3 in the respiratory chain and TIM22 protein translocase of the mitochondrial inner membrane. *Mol. Cell* **2011**, *44*, 811–818. [CrossRef]
20. Baginsky, M.L.; Hatefi, Y. Reconstitution of succinate-coenzyme Q reductase (complex II) and succinate oxidase activities by a highly purified, reactivated succinate dehydrogenase. *J. Biol. Chem.* **1969**, *244*, 5313–5319.
21. Davis, K.A.; Hatefi, Y. Resolution and reconstitution of complex II (succinate-ubiquinone reductase) by salts. *Arch. Biochem. Biophys.* **1972**, *149*, 505–512. [CrossRef]
22. Sun, F.; Huo, X.; Zhai, Y.; Wang, A.; Xu, J.; Su, D.; Bartlam, M.; Rao, Z. Crystal structure of mitochondrial respiratory membrane protein complex II. *Cell* **2005**, *121*, 1043–1057. [CrossRef]
23. Kim, H.J.; Jeong, M.-Y.; Na, U.; Winge, D.R. Flavinylation and assembly of succinate dehydrogenase are dependent on the C-terminal tail of the flavoprotein subunit. *J. Biol. Chem.* **2012**, *287*, 40670–40679. [CrossRef] [PubMed]
24. Rivner, M.H.; Shamsnia, M.; Swift, T.R.; Trefz, J.; Roesel, R.A.; Carter, A.L.; Yanamura, W.; Hommes, F.A. Kearns-Sayre syndrome and complex II deficiency. *Neurology* **1989**, *39*, 693–696. [CrossRef] [PubMed]
25. Bourgeron, T.; Rustin, P.; Chretien, D.; Birch-Machin, M.; Bourgeois, M.; Viegas-Péquignot, E.; Munnich, A.; Rötig, A. Mutation of a nuclear succinate dehydrogenase gene results in mitochondrial respiratory chain deficiency. *Nat. Genet.* **1995**, *11*, 144–149. [CrossRef]
26. Leigh, D. Subacute Necrotizing Encephalomyelopathy in an Infant. *J. Neurol. Neurosurg. Psychiatry* **1951**, *14*, 216–221. [CrossRef]
27. Pagnamenta, A.T.; Hargreaves, I.P.; Duncan, A.J.; Taanman, J.-W.; Heales, S.J.; Land, J.M.; Bitner-Glindzicz, M.; Leonard, J.V.; Rahman, S. Phenotypic variability of mitochondrial disease caused by a nuclear mutation in complex II. *Mol. Genet. Metab.* **2006**, *89*, 214–221. [CrossRef]
28. Levitas, A.; Muhammad, E.; Harel, G.; Saada, A.; Caspi, V.C.; Manor, E.; Beck, J.C.; Sheffield, V.; Parvari, R. Familial neonatal isolated cardiomyopathy caused by a mutation in the flavoprotein subunit of succinate dehydrogenase. *Eur. J. Hum. Genet. EJHG* **2010**, *18*, 1160–1165. [CrossRef]

29. Van Coster, R.; Seneca, S.; Smet, J.; Van Hecke, R.; Gerlo, E.; Devreese, B.; Van Beeumen, J.; Leroy, J.G.; De Meirleir, L.; Lissens, W. Homozygous Gly555Glu mutation in the nuclear-encoded 70 kDa flavoprotein gene causes instability of the respiratory chain complex II. *Am. J. Med. Genet. Part A* **2003**, *120*, 13–18. [CrossRef]
30. Saghbini, M.; Broomfield, P.L.; Scheffler, I.E. Studies on the assembly of complex II in yeast mitochondria using chimeric human/yeast genes for the iron-sulfur protein subunit. *Biochemistry* **1994**, *33*, 159–165. [CrossRef]
31. Burgener, A.-V.; Bantug, G.R.; Meyer, B.J.; Higgins, R.; Ghosh, A.; Bignucolo, O.; Ma, E.H.; Loeliger, J.; Unterstab, G.; Geigges, M.; et al. SDHA gain-of-function engages inflammatory mitochondrial retrograde signaling via KEAP1-Nrf2. *Nat. Immunol.* **2019**, *20*, 1311–1321. [CrossRef]
32. Van Vranken, J.G.; Bricker, D.K.; Dephoure, N.; Gygi, S.P.; Cox, J.E.; Thummel, C.S.; Rutter, J. SDHAF4 promotes mitochondrial succinate dehydrogenase activity and prevents neurodegeneration. *Cell Metab.* **2014**, *20*, 241–252. [CrossRef] [PubMed]
33. Moosavi, B.; Berry, E.A.; Zhu, X.-L.; Yang, W.-C.; Yang, G.-F. The assembly of succinate dehydrogenase: A key enzyme in bioenergetics. *Cell. Mol. Life Sci. CMLS* **2019**, *76*, 4023–4042. [CrossRef] [PubMed]
34. Ghezzi, D.; Goffrini, P.; Uziel, G.; Horvath, R.; Klopstock, T.; Lochmüller, H.; D'Adamo, P.; Gasparini, P.; Strom, T.M.; Prokisch, H.; et al. SDHAF1, encoding a LYR complex-II specific assembly factor, is mutated in SDH-defective infantile leukoencephalopathy. *Nat. Genet.* **2009**, *41*, 654–656. [CrossRef] [PubMed]
35. Hao, H.-X.; Khalimonchuk, O.; Schraders, M.; Dephoure, N.; Bayley, J.-P.; Kunst, H.; Devilee, P.; Cremers, C.W.R.J.; Schiffman, J.D.; Bentz, B.G.; et al. SDH5, a gene required for flavination of succinate dehydrogenase, is mutated in paraganglioma. *Science* **2009**, *325*, 1139–1142. [CrossRef] [PubMed]
36. Na, U.; Yu, W.; Cox, J.; Bricker, D.K.; Brockmann, K.; Rutter, J.; Thummel, C.S.; Winge, D.R. The LYR factors SDHAF1 and SDHAF3 mediate maturation of the iron-sulfur subunit of succinate dehydrogenase. *Cell Metab.* **2014**, *20*, 253–266. [CrossRef]
37. Ohlenbusch, A.; Edvardson, S.; Skorpen, J.; Bjornstad, A.; Saada, A.; Elpeleg, O.; Gärtner, J.; Brockmann, K. Leukoencephalopathy with accumulated succinate is indicative of SDHAF1 related complex II deficiency. *Orphanet J. Rare Dis.* **2012**, *7*, 69. [CrossRef]
38. Rustin, P.; Rötig, A. Inborn errors of complex II–unusual human mitochondrial diseases. *Biochim. Biophys. Acta* **2002**, *1553*, 117–122. [CrossRef]
39. Rutter, J.; Winge, D.R.; Schiffman, J.D. Succinate dehydrogenase—Assembly, regulation and role in human disease. *Mitochondrion* **2010**, *10*, 393–401. [CrossRef]
40. Baysal, B.E. Mitochondrial complex II and genomic imprinting in inheritance of paraganglioma tumors. *Biochim. Biophys. Acta* **2013**, *1827*, 573–577. [CrossRef]
41. Miettinen, M.; Lasota, J. Succinate dehydrogenase deficient gastrointestinal stromal tumors (GISTs)—A review. *Int. J. Biochem. Cell Biol.* **2014**, *53*, 514–519. [CrossRef]
42. Lussey-Lepoutre, C.; Buffet, A.; Gimenez-Roqueplo, A.-P.; Favier, J. Mitochondrial Deficiencies in the Predisposition to Paraganglioma. *Metabolites* **2017**, *7*, 17. [CrossRef] [PubMed]
43. Nazar, E.; Khatami, F.; Saffar, H.; Tavangar, S.M. The Emerging Role of Succinate Dehyrogenase Genes (SDHx) in Tumorigenesis. *Int. J. Hematol. Oncol. Stem Cell Res.* **2019**, *13*, 72–82. [CrossRef] [PubMed]
44. Bannon, A.E.; Kent, J.; Forquer, I.; Town, A.; Klug, L.R.; McCann, K.; Beadling, C.; Harismendy, O.; Sicklick, J.K.; Corless, C.; et al. Biochemical, Molecular, and Clinical Characterization of Succinate Dehydrogenase Subunit A Variants of Unknown Significance. *Clin. Cancer Res. Off. J. Am. Assoc. Cancer Res.* **2017**, *23*, 6733–6743. [CrossRef] [PubMed]
45. Parfait, B.; Chretien, D.; Rötig, A.; Marsac, C.; Munnich, A.; Rustin, P. Compound heterozygous mutations in the flavoprotein gene of the respiratory chain complex II in a patient with Leigh syndrome. *Hum. Genet.* **2000**, *106*, 236–243. [CrossRef] [PubMed]
46. Horváth, R.; Abicht, A.; Holinski-Feder, E.; Laner, A.; Gempel, K.; Prokisch, H.; Lochmüller, H.; Klopstock, T.; Jaksch, M. Leigh syndrome caused by mutations in the flavoprotein (Fp) subunit of succinate dehydrogenase (SDHA). *J. Neurol. Neurosurg. Psychiatry* **2006**, *77*, 74–76. [CrossRef] [PubMed]
47. Birch-Machin, M.A.; Taylor, R.W.; Cochran, B.; Ackrell, B.A.; Turnbull, D.M. Late-onset optic atrophy, ataxia, and myopathy associated with a mutation of a complex II gene. *Ann. Neurol.* **2000**, *48*, 330–335. [CrossRef]

48. Ma, Y.-Y.; Wu, T.-F.; Liu, Y.-P.; Wang, Q.; Li, X.-Y.; Ding, Y.; Song, J.-Q.; Shi, X.-Y.; Zhang, W.-N.; Zhao, M.; et al. Two compound frame-shift mutations in succinate dehydrogenase gene of a Chinese boy with encephalopathy. *Brain Dev.* **2014**, *36*, 394–398. [CrossRef]
49. Alston, C.L.; Davison, J.E.; Meloni, F.; van der Westhuizen, F.H.; He, L.; Hornig-Do, H.-T.; Peet, A.C.; Gissen, P.; Goffrini, P.; Ferrero, I.; et al. Recessive germline SDHA and SDHB mutations causing leukodystrophy and isolated mitochondrial complex II deficiency. *J. Med. Genet.* **2012**, *49*, 569–577. [CrossRef]
50. Courage, C.; Jackson, C.B.; Hahn, D.; Euro, L.; Nuoffer, J.-M.; Gallati, S.; Schaller, A. SDHA mutation with dominant transmission results in complex II deficiency with ocular, cardiac, and neurologic involvement. *Am. J. Med. Genet. Part A* **2017**, *173*, 225–230. [CrossRef]
51. Grønborg, S.; Darin, N.; Miranda, M.J.; Damgaard, B.; Cayuela, J.A.; Oldfors, A.; Kollberg, G.; Hansen, T.V.O.; Ravn, K.; Wibrand, F.; et al. Leukoencephalopathy due to Complex II Deficiency and Bi-Allelic SDHB Mutations: Further Cases and Implications for Genetic Counselling. *JIMD Rep.* **2017**, *33*, 69–77. [CrossRef]
52. Jackson, C.B.; Nuoffer, J.-M.; Hahn, D.; Prokisch, H.; Haberberger, B.; Gautschi, M.; Häberli, A.; Gallati, S.; Schaller, A. Mutations in SDHD lead to autosomal recessive encephalomyopathy and isolated mitochondrial complex II deficiency. *J. Med. Genet.* **2014**, *51*, 170–175. [CrossRef] [PubMed]
53. Hatefi, Y.; Haavik, A.G.; Griffiths, D.E. Studies on the electron transfer system. XLI. Reduced coenzyme Q (QH2)-cytochrome c reductase. *J. Biol. Chem.* **1962**, *237*, 1681–1685. [PubMed]
54. Beckmann, J.D.; Ljungdahl, P.O.; Lopez, J.L.; Trumpower, B.L. Isolation and characterization of the nuclear gene encoding the Rieske iron-sulfur protein (RIP1) from *Saccharomyces cerevisiae*. *J. Biol. Chem.* **1987**, *262*, 8901–8909. [PubMed]
55. Hunte, C.; Koepke, J.; Lange, C.; Rossmanith, T.; Michel, H. Structure at 2.3 A resolution of the cytochrome bc(1) complex from the yeast *Saccharomyces cerevisiae* co-crystallized with an antibody Fv fragment. *Structure* **2000**, *8*, 669–684. [CrossRef]
56. Trumpower, B.L. Cytochrome bc1 complexes of microorganisms. *Microbiol. Rev.* **1990**, *54*, 101–129. [CrossRef] [PubMed]
57. Gruschke, S.; Kehrein, K.; Römpler, K.; Gröne, K.; Israel, L.; Imhof, A.; Herrmann, J.M.; Ott, M. Cbp3-Cbp6 interacts with the yeast mitochondrial ribosomal tunnel exit and promotes cytochrome b synthesis and assembly. *J. Cell Biol.* **2011**, *193*, 1101–1114. [CrossRef]
58. Crivellone, M.D. Characterization of CBP4, a new gene essential for the expression of ubiquinol-cytochrome c reductase in *Saccharomyces cerevisiae*. *J. Biol. Chem.* **1994**, *269*, 21284–21292.
59. Hildenbeutel, M.; Hegg, E.L.; Stephan, K.; Gruschke, S.; Meunier, B.; Ott, M. Assembly factors monitor sequential hemylation of cytochrome b to regulate mitochondrial translation. *J. Cell Biol.* **2014**, *205*, 511–524. [CrossRef]
60. Nobrega, F.G.; Nobrega, M.P.; Tzagoloff, A. BCS1, a novel gene required for the expression of functional Rieske iron-sulfur protein in *Saccharomyces cerevisiae*. *EMBO J.* **1992**, *11*, 3821–3829. [CrossRef]
61. Cruciat, C.M.; Hell, K.; Fölsch, H.; Neupert, W.; Stuart, R.A. Bcs1p, an AAA-family member, is a chaperone for the assembly of the cytochrome bc(1) complex. *EMBO J.* **1999**, *18*, 5226–5233. [CrossRef]
62. Atkinson, A.; Smith, P.; Fox, J.L.; Cui, T.-Z.; Khalimonchuk, O.; Winge, D.R. The LYR protein Mzm1 functions in the insertion of the Rieske Fe/S protein in yeast mitochondria. *Mol. Cell. Biol.* **2011**, *31*, 3988–3996. [CrossRef] [PubMed]
63. Nobrega, F.G.; Tzagoloff, A. Assembly of the mitochondrial membrane system. DNA sequence and organization of the cytochrome b gene in *Saccharomyces cerevisiae* D273-10B. *J. Biol. Chem.* **1980**, *255*, 9828–9837. [PubMed]
64. Lazowska, J.; Jacq, C.; Slonimski, P.P. Sequence of introns and flanking exons in wild-type and box3 mutants of cytochrome b reveals an interlaced splicing protein coded by an intron. *Cell* **1980**, *22*, 333–348. [CrossRef]
65. Church, G.M.; Slonimski, P.P.; Gilbert, W. Pleiotropic mutations within two yeast mitochondrial cytochrome genes block mRNA processing. *Cell* **1979**, *18*, 1209–1215. [CrossRef]
66. Gampel, A.; Nishikimi, M.; Tzagoloff, A. CBP2 protein promotes in vitro excision of a yeast mitochondrial group I intron. *Mol. Cell. Biol.* **1989**, *9*, 5424–5433. [CrossRef]
67. Rödel, G. Two yeast nuclear genes, CBS1 and CBS2, are required for translation of mitochondrial transcripts bearing the 5′-untranslated COB leader. *Curr. Genet.* **1986**, *11*, 41–45. [CrossRef]

68. Dieckmann, C.L.; Homison, G.; Tzagoloff, A. Assembly of the mitochondrial membrane system. Nucleotide sequence of a yeast nuclear gene (CBP1) involved in 5′ end processing of cytochrome b pre-mRNA. *J. Biol. Chem.* **1984**, *259*, 4732–4738.
69. WebHome < MITOMAP < Foswiki. Available online: https://www.mitomap.org/foswiki/bin/view/MITOMAP/WebHome (accessed on 19 October 2020).
70. Meunier, B.; Fisher, N.; Ransac, S.; Mazat, J.-P.; Brasseur, G. Respiratory complex III dysfunction in humans and the use of yeast as a model organism to study mitochondrial myopathy and associated diseases. *Biochim. Biophys. Acta* **2013**, *1827*, 1346–1361. [CrossRef]
71. Mackey, D.A.; Oostra, R.J.; Rosenberg, T.; Nikoskelainen, E.; Bronte-Stewart, J.; Poulton, J.; Harding, A.E.; Govan, G.; Bolhuis, P.A.; Norby, S. Primary pathogenic mtDNA mutations in multigeneration pedigrees with Leber hereditary optic neuropathy. *Am. J. Hum. Genet.* **1996**, *59*, 481–485.
72. Brown, M.D.; Voljavec, A.S.; Lott, M.T.; Torroni, A.; Yang, C.C.; Wallace, D.C. Mitochondrial DNA complex I and III mutations associated with Leber's hereditary optic neuropathy. *Genetics* **1992**, *130*, 163–173.
73. Johns, D.R.; Neufeld, M.J. Cytochrome b mutations in Leber hereditary optic neuropathy. *Biochem. Biophys. Res. Commun.* **1991**, *181*, 1358–1364. [CrossRef]
74. Haut, S.; Brivet, M.; Touati, G.; Rustin, P.; Lebon, S.; Garcia-Cazorla, A.; Saudubray, J.M.; Boutron, A.; Legrand, A.; Slama, A. A deletion in the human QP-C gene causes a complex III deficiency resulting in hypoglycaemia and lactic acidosis. *Hum. Genet.* **2003**, *113*, 118–122. [CrossRef] [PubMed]
75. Hemrika, W.; De Jong, M.; Berden, J.A.; Grivell, L.A. The C-terminus of the 14-kDa subunit of ubiquinol-cytochrome-c oxidoreductase of the yeast *Saccharomyces cerevisiae* is involved in the assembly of a functional enzyme. *Eur. J. Biochem.* **1994**, *220*, 569–576. [CrossRef] [PubMed]
76. Barel, O.; Shorer, Z.; Flusser, H.; Ofir, R.; Narkis, G.; Finer, G.; Shalev, H.; Nasasra, A.; Saada, A.; Birk, O.S. Mitochondrial complex III deficiency associated with a homozygous mutation in UQCRQ. *Am. J. Hum. Genet.* **2008**, *82*, 1211–1216. [CrossRef]
77. Miyake, N.; Yano, S.; Sakai, C.; Hatakeyama, H.; Matsushima, Y.; Shiina, M.; Watanabe, Y.; Bartley, J.; Abdenur, J.E.; Wang, R.Y.; et al. Mitochondrial complex III deficiency caused by a homozygous UQCRC2 mutation presenting with neonatal-onset recurrent metabolic decompensation. *Hum. Mutat.* **2013**, *34*, 446–452. [CrossRef]
78. Gaignard, P.; Eyer, D.; Lebigot, E.; Oliveira, C.; Therond, P.; Boutron, A.; Slama, A. UQCRC2 mutation in a patient with mitochondrial complex III deficiency causing recurrent liver failure, lactic acidosis and hypoglycemia. *J. Hum. Genet.* **2017**, *62*, 729–731. [CrossRef]
79. Han, Y.; Wu, P.; Wang, Z.; Zhang, Z.; Sun, S.; Liu, J.; Gong, S.; Gao, P.; Iwakuma, T.; Molina-Vila, M.A.; et al. Ubiquinol-cytochrome C reductase core protein II promotes tumorigenesis by facilitating p53 degradation. *EBioMedicine* **2019**, *40*, 92–105. [CrossRef]
80. Wagener, N.; Ackermann, M.; Funes, S.; Neupert, W. A pathway of protein translocation in mitochondria mediated by the AAA-ATPase Bcs1. *Mol. Cell* **2011**, *44*, 191–202. [CrossRef]
81. Ndi, M.; Marin-Buera, L.; Salvatori, R.; Singh, A.P.; Ott, M. Biogenesis of the bc1 Complex of the Mitochondrial Respiratory Chain. *J. Mol. Biol.* **2018**, *430*, 3892–3905. [CrossRef]
82. Akduman, H.; Eminoglu, T.; Okulu, E.; Erdeve, O.; Atasay, B.; Arsan, S. A Neonate Presenting with Gracile Syndrome and Bjornstad Phenotype Associated with Bcs1l Mutation. *Genet. Couns. Geneva Switz.* **2016**, *27*, 509–512.
83. Serdaroğlu, E.; Takcı, Ş.; Kotarsky, H.; Çil, O.; Utine, E.; Yiğit, Ş.; Fellman, V. A Turkish BCS1L mutation causes GRACILE-like disorder. *Turk. J. Pediatr.* **2016**, *58*, 658–661. [CrossRef] [PubMed]
84. Kasapkara, Ç.S.; Tümer, L.; Ezgü, F.S.; Küçükçongar, A.; Hasanoğlu, A. BCS1L gene mutation causing GRACILE syndrome: Case report. *Ren. Fail.* **2014**, *36*, 953–954. [CrossRef] [PubMed]
85. Lynn, A.M.; King, R.I.; Mackay, R.J.; Florkowski, C.M.; Wilson, C.J. BCS1L gene mutation presenting with GRACILE-like syndrome and complex III deficiency. *Ann. Clin. Biochem.* **2012**, *49*, 201–203. [CrossRef] [PubMed]
86. Kotarsky, H.; Karikoski, R.; Mörgelin, M.; Marjavaara, S.; Bergman, P.; Zhang, D.-L.; Smet, J.; van Coster, R.; Fellman, V. Characterization of complex III deficiency and liver dysfunction in GRACILE syndrome caused by a BCS1L mutation. *Mitochondrion* **2010**, *10*, 497–509. [CrossRef]
87. Fellman, V.; Lemmelä, S.; Sajantila, A.; Pihko, H.; Järvelä, I. Screening of BCS1L mutations in severe neonatal disorders suspicious for mitochondrial cause. *J. Hum. Genet.* **2008**, *53*, 554–558. [CrossRef]

88. Visapää, I.; Fellman, V.; Vesa, J.; Dasvarma, A.; Hutton, J.L.; Kumar, V.; Payne, G.S.; Makarow, M.; Van Coster, R.; Taylor, R.W.; et al. GRACILE syndrome, a lethal metabolic disorder with iron overload, is caused by a point mutation in BCS1L. *Am. J. Hum. Genet.* **2002**, *71*, 863–876. [CrossRef]
89. Falco, M.; Franzè, A.; Iossa, S.; De Falco, L.; Gambale, A.; Marciano, E.; Iolascon, A. Novel compound heterozygous mutations in BCS1L gene causing Bjornstad syndrome in two siblings. *Am. J. Med. Genet. Part A* **2017**, *173*, 1348–1352. [CrossRef]
90. Shigematsu, Y.; Hayashi, R.; Yoshida, K.; Shimizu, A.; Kubota, M.; Komori, M.; Shimomura, Y.; Niizeki, H. Novel heterozygous deletion mutation c.821delC in the AAA domain of BCS1L underlies Björnstad syndrome. *J. Dermatol.* **2017**, *44*, e111–e112. [CrossRef]
91. Zhang, J.; Duo, L.; Lin, Z.; Wang, H.; Yin, J.; Cao, X.; Zhao, J.; Dai, L.; Liu, X.; Zhang, J.; et al. Exome sequencing reveals novel BCS1L mutations in siblings with hearing loss and hypotrichosis. *Gene* **2015**, *566*, 84–88. [CrossRef]
92. Yanagishita, T.; Sugiura, K.; Kawamoto, Y.; Ito, K.; Marubashi, Y.; Taguchi, N.; Akiyama, M.; Watanabe, D. A case of Björnstad syndrome caused by novel compound heterozygous mutations in the BCS1L gene. *Br. J. Dermatol.* **2014**, *170*, 970–973. [CrossRef]
93. Siddiqi, S.; Siddiq, S.; Mansoor, A.; Oostrik, J.; Ahmad, N.; Kazmi, S.A.R.; Kremer, H.; Qamar, R.; Schraders, M. Novel mutation in AAA domain of BCS1L causing Bjornstad syndrome. *J. Hum. Genet.* **2013**, *58*, 819–821. [CrossRef] [PubMed]
94. Hinson, J.T.; Fantin, V.R.; Schönberger, J.; Breivik, N.; Siem, G.; McDonough, B.; Sharma, P.; Keogh, I.; Godinho, R.; Santos, F.; et al. Missense mutations in the BCS1L gene as a cause of the Björnstad syndrome. *N. Engl. J. Med.* **2007**, *356*, 809–819. [CrossRef] [PubMed]
95. Tegelberg, S.; Tomašić, N.; Kallijärvi, J.; Purhonen, J.; Elmér, E.; Lindberg, E.; Nord, D.G.; Soller, M.; Lesko, N.; Wedell, A.; et al. Respiratory chain complex III deficiency due to mutated BCS1L: A novel phenotype with encephalomyopathy, partially phenocopied in a Bcs1l mutant mouse model. *Orphanet J. Rare Dis.* **2017**, *12*, 1–14. [CrossRef] [PubMed]
96. Blázquez, A.; Gil-Borlado, M.C.; Morán, M.; Verdú, A.; Cazorla-Calleja, M.R.; Martín, M.A.; Arenas, J.; Ugalde, C. Infantile mitochondrial encephalomyopathy with unusual phenotype caused by a novel BCS1L mutation in an isolated complex III-deficient patient. *Neuromuscul. Disord. NMD* **2009**, *19*, 143–146. [CrossRef]
97. Fernandez-Vizarra, E.; Bugiani, M.; Goffrini, P.; Carrara, F.; Farina, L.; Procopio, E.; Donati, A.; Uziel, G.; Ferrero, I.; Zeviani, M. Impaired complex III assembly associated with BCS1L gene mutations in isolated mitochondrial encephalopathy. *Hum. Mol. Genet.* **2007**, *16*, 1241–1252. [CrossRef]
98. de Lonlay, P.; Valnot, I.; Barrientos, A.; Gorbatyuk, M.; Tzagoloff, A.; Taanman, J.W.; Benayoun, E.; Chrétien, D.; Kadhom, N.; Lombès, A.; et al. A mutant mitochondrial respiratory chain assembly protein causes complex III deficiency in patients with tubulopathy, encephalopathy and liver failure. *Nat. Genet.* **2001**, *29*, 57–60. [CrossRef]
99. De Meirleir, L.; Seneca, S.; Damis, E.; Sepulchre, B.; Hoorens, A.; Gerlo, E.; García Silva, M.T.; Hernandez, E.M.; Lissens, W.; Van Coster, R. Clinical and diagnostic characteristics of complex III deficiency due to mutations in the BCS1L gene. *Am. J. Med. Genet. Part A* **2003**, *121*, 126–131. [CrossRef]
100. Gil-Borlado, M.C.; González-Hoyuela, M.; Blázquez, A.; García-Silva, M.T.; Gabaldón, T.; Manzanares, J.; Vara, J.; Martín, M.A.; Seneca, S.; Arenas, J.; et al. Pathogenic mutations in the 5′ untranslated region of BCS1L mRNA in mitochondrial complex III deficiency. *Mitochondrion* **2009**, *9*, 299–305. [CrossRef]
101. Ramos-Arroyo, M.A.; Hualde, J.; Ayechu, A.; De Meirleir, L.; Seneca, S.; Nadal, N.; Briones, P. Clinical and biochemical spectrum of mitochondrial complex III deficiency caused by mutations in the BCS1L gene. *Clin. Genet.* **2009**, *75*, 585–587. [CrossRef]
102. Ezgu, F.; Senaca, S.; Gunduz, M.; Tumer, L.; Hasanoglu, A.; Tiras, U.; Unsal, R.; Bakkaloglu, S.A. Severe renal tubulopathy in a newborn due to BCS1L gene mutation: Effects of different treatment modalities on the clinical course. *Gene* **2013**, *528*, 364–366. [CrossRef]
103. Tuppen, H.A.L.; Fehmi, J.; Czermin, B.; Goffrini, P.; Meloni, F.; Ferrero, I.; He, L.; Blakely, E.L.; McFarland, R.; Horvath, R.; et al. Long-term survival of neonatal mitochondrial complex III deficiency associated with a novel BCS1L gene mutation. *Mol. Genet. Metab.* **2010**, *100*, 345–348. [CrossRef] [PubMed]

104. Oláhová, M.; Ceccatelli Berti, C.; Collier, J.J.; Alston, C.L.; Jameson, E.; Jones, S.A.; Edwards, N.; He, L.; Chinnery, P.F.; Horvath, R.; et al. Molecular genetic investigations identify new clinical phenotypes associated with BCS1L-related mitochondrial disease. *Hum. Mol. Genet.* **2019**, *28*, 3766–3776. [CrossRef] [PubMed]
105. Cui, T.-Z.; Smith, P.M.; Fox, J.L.; Khalimonchuk, O.; Winge, D.R. Late-stage maturation of the Rieske Fe/S protein: Mzm1 stabilizes Rip1 but does not facilitate its translocation by the AAA ATPase Bcs1. *Mol. Cell. Biol.* **2012**, *32*, 4400–4409. [CrossRef] [PubMed]
106. Invernizzi, F.; Tigano, M.; Dallabona, C.; Donnini, C.; Ferrero, I.; Cremonte, M.; Ghezzi, D.; Lamperti, C.; Zeviani, M. A homozygous mutation in LYRM7/MZM1L associated with early onset encephalopathy, lactic acidosis, and severe reduction of mitochondrial complex III activity. *Hum. Mutat.* **2013**, *34*, 1619–1622. [CrossRef] [PubMed]
107. Zhang, J.; Liu, M.; Zhang, Z.; Zhou, L.; Kong, W.; Jiang, Y.; Wang, J.; Xiao, J.; Wu, Y. Genotypic Spectrum and Natural History of Cavitating Leukoencephalopathies in Childhood. *Pediatr. Neurol.* **2019**, *94*, 38–47. [CrossRef] [PubMed]
108. Dallabona, C.; Abbink, T.E.M.; Carrozzo, R.; Torraco, A.; Legati, A.; van Berkel, C.G.M.; Niceta, M.; Langella, T.; Verrigni, D.; Rizza, T.; et al. LYRM7 mutations cause a multifocal cavitating leukoencephalopathy with distinct MRI appearance. *Brain J. Neurol.* **2016**, *139*, 782–794. [CrossRef] [PubMed]
109. Hempel, M.; Kremer, L.S.; Tsiakas, K.; Alhaddad, B.; Haack, T.B.; Löbel, U.; Feichtinger, R.G.; Sperl, W.; Prokisch, H.; Mayr, J.A.; et al. LYRM7—Associated complex III deficiency: A clinical, molecular genetic, MR tomographic, and biochemical study. *Mitochondrion* **2017**, *37*, 55–61. [CrossRef]
110. Kremer, L.S.; L'hermitte-Stead, C.; Lesimple, P.; Gilleron, M.; Filaut, S.; Jardel, C.; Haack, T.B.; Strom, T.M.; Meitinger, T.; Azzouz, H.; et al. Severe respiratory complex III defect prevents liver adaptation to prolonged fasting. *J. Hepatol.* **2016**, *65*, 377–385. [CrossRef]
111. Tucker, E.J.; Wanschers, B.F.J.; Szklarczyk, R.; Mountford, H.S.; Wijeyeratne, X.W.; van den Brand, M.A.M.; Leenders, A.M.; Rodenburg, R.J.; Reljić, B.; Compton, A.G.; et al. Mutations in the UQCC1-interacting protein, UQCC2, cause human complex III deficiency associated with perturbed cytochrome b protein expression. *PLoS Genet.* **2013**, *9*, e1004034. [CrossRef]
112. Feichtinger, R.G.; Brunner-Krainz, M.; Alhaddad, B.; Wortmann, S.B.; Kovacs-Nagy, R.; Stojakovic, T.; Erwa, W.; Resch, B.; Windischhofer, W.; Verheyen, S.; et al. Combined Respiratory Chain Deficiency and UQCC2 Mutations in Neonatal Encephalomyopathy: Defective Supercomplex Assembly in Complex III Deficiencies. *Oxid. Med. Cell. Longev.* **2017**, *2017*. [CrossRef]
113. Bogenhagen, D.F.; Haley, J.D. Pulse-chase SILAC-based analyses reveal selective oversynthesis and rapid turnover of mitochondrial protein components of respiratory complexes. *J. Biol. Chem.* **2020**, *295*, 2544–2554. [CrossRef] [PubMed]
114. Wanschers, B.F.J.; Szklarczyk, R.; van den Brand, M.A.M.; Jonckheere, A.; Suijskens, J.; Smeets, R.; Rodenburg, R.J.; Stephan, K.; Helland, I.B.; Elkamil, A.; et al. A mutation in the human CBP4 ortholog UQCC3 impairs complex III assembly, activity and cytochrome b stability. *Hum. Mol. Genet.* **2014**, *23*, 6356–6365. [CrossRef] [PubMed]
115. Gaignard, P.; Menezes, M.; Schiff, M.; Bayot, A.; Rak, M.; Ogier de Baulny, H.; Su, C.-H.; Gilleron, M.; Lombes, A.; Abida, H.; et al. Mutations in CYC1, encoding cytochrome c1 subunit of respiratory chain complex III, cause insulin-responsive hyperglycemia. *Am. J. Hum. Genet.* **2013**, *93*, 384–389. [CrossRef] [PubMed]
116. Gusic, M.; Schottmann, G.; Feichtinger, R.G.; Du, C.; Scholz, C.; Wagner, M.; Mayr, J.A.; Lee, C.-Y.; Yépez, V.A.; Lorenz, N.; et al. Bi-Allelic UQCRFS1 Variants Are Associated with Mitochondrial Complex III Deficiency, Cardiomyopathy, and Alopecia Totalis. *Am. J. Hum. Genet.* **2020**, *106*, 102–111. [CrossRef]
117. Wikström, M. Identification of the electron transfers in cytochrome oxidase that are coupled to proton-pumping. *Nature* **1989**, *338*, 776–778. [CrossRef]
118. Hell, K.; Tzagoloff, A.; Neupert, W.; Stuart, R.A. Identification of Cox20p, a novel protein involved in the maturation and assembly of cytochrome oxidase subunit 2. *J. Biol. Chem.* **2000**, *275*, 4571–4578. [CrossRef]
119. Sharma, V.; Enkavi, G.; Vattulainen, I.; Róg, T.; Wikström, M. Proton-coupled electron transfer and the role of water molecules in proton pumping by cytochrome c oxidase. *Proc. Natl. Acad. Sci. USA* **2015**, *112*, 2040–2045. [CrossRef]
120. Pitceathly, R.D.S.; Taanman, J.-W. NDUFA4 (Renamed COXFA4) Is a Cytochrome-c Oxidase Subunit. *Trends Endocrinol. Metab. TEM* **2018**, *29*, 452–454. [CrossRef]

121. Hill, B.C. Modeling the sequence of electron transfer reactions in the single turnover of reduced, mammalian cytochrome c oxidase with oxygen. *J. Biol. Chem.* **1994**, *269*, 2419–2425.
122. Brunori, M.; Giuffrè, A.; Sarti, P. Cytochrome c oxidase, ligands and electrons. *J. Inorg. Biochem.* **2005**, *99*, 324–336. [CrossRef]
123. Tzagoloff, A.; Nobrega, M.; Gorman, N.; Sinclair, P. On the functions of the yeast COX10 and COX11 gene products. *Biochem. Mol. Biol. Int.* **1993**, *31*, 593–598. [PubMed]
124. Barros, M.H.; Carlson, C.G.; Glerum, D.M.; Tzagoloff, A. Involvement of mitochondrial ferredoxin and Cox15p in hydroxylation of heme O. *FEBS Lett.* **2001**, *492*, 133–138. [CrossRef]
125. Barros, M.H.; Nobrega, F.G.; Tzagoloff, A. Mitochondrial ferredoxin is required for heme A synthesis in *Saccharomyces cerevisiae*. *J. Biol. Chem.* **2002**, *277*, 9997–10002. [CrossRef] [PubMed]
126. Bestwick, M.; Jeong, M.-Y.; Khalimonchuk, O.; Kim, H.; Winge, D.R. Analysis of Leigh syndrome mutations in the yeast SURF1 homolog reveals a new member of the cytochrome oxidase assembly factor family. *Mol. Cell. Biol.* **2010**, *30*, 4480–4491. [CrossRef] [PubMed]
127. Taylor, N.G.; Swenson, S.; Harris, N.J.; Germany, E.M.; Fox, J.L.; Khalimonchuk, O. The Assembly Factor Pet117 Couples Heme a Synthase Activity to Cytochrome Oxidase Assembly. *J. Biol. Chem.* **2017**, *292*, 1815–1825. [CrossRef] [PubMed]
128. Barros, M.H.; McStay, G.P. Modular biogenesis of mitochondrial respiratory complexes. *Mitochondrion* **2020**, *50*, 94–114. [CrossRef]
129. Genova, M.L.; Lenaz, G. Functional role of mitochondrial respiratory supercomplexes. *Biochim. Biophys. Acta* **2014**, *1837*, 427–443. [CrossRef]
130. Berndtsson, J.; Aufschnaiter, A.; Rathore, S.; Marin-Buera, L.; Dawitz, H.; Diessl, J.; Kohler, V.; Barrientos, A.; Büttner, S.; Fontanesi, F.; et al. Respiratory supercomplexes enhance electron transport by decreasing cytochrome c diffusion distance. *EMBO Rep.* **2020**, e51015. [CrossRef]
131. Schägger, H.; Pfeiffer, K. Supercomplexes in the respiratory chains of yeast and mammalian mitochondria. *EMBO J.* **2000**, *19*, 1777–1783. [CrossRef]
132. Stroh, A.; Anderka, O.; Pfeiffer, K.; Yagi, T.; Finel, M.; Ludwig, B.; Schägger, H. Assembly of respiratory complexes I, III, and IV into NADH oxidase supercomplex stabilizes complex I in Paracoccus denitrificans. *J. Biol. Chem.* **2004**, *279*, 5000–5007. [CrossRef]
133. Rak, M.; Bénit, P.; Chrétien, D.; Bouchereau, J.; Schiff, M.; El-Khoury, R.; Tzagoloff, A.; Rustin, P. Mitochondrial cytochrome c oxidase deficiency. *Clin. Sci. Lond. Engl. 1979* **2016**, *130*, 393–407. [CrossRef] [PubMed]
134. Trueblood, C.E.; Poyton, R.O. Differential effectiveness of yeast cytochrome c oxidase subunit genes results from differences in expression not function. *Mol. Cell. Biol.* **1987**, *7*, 3520–3526. [CrossRef] [PubMed]
135. DiMauro, S.; Tanji, K.; Schon, E.A. The many clinical faces of cytochrome c oxidase deficiency. *Adv. Exp. Med. Biol.* **2012**, *748*, 341–357. [CrossRef] [PubMed]
136. Indrieri, A.; van Rahden, V.A.; Tiranti, V.; Morleo, M.; Iaconis, D.; Tammaro, R.; D'Amato, I.; Conte, I.; Maystadt, I.; Demuth, S.; et al. Mutations in COX7B cause microphthalmia with linear skin lesions, an unconventional mitochondrial disease. *Am. J. Hum. Genet.* **2012**, *91*, 942–949. [CrossRef]
137. van Rahden, V.A.; Fernandez-Vizarra, E.; Alawi, M.; Brand, K.; Fellmann, F.; Horn, D.; Zeviani, M.; Kutsche, K. Mutations in NDUFB11, encoding a complex I component of the mitochondrial respiratory chain, cause microphthalmia with linear skin defects syndrome. *Am. J. Hum. Genet.* **2015**, *96*, 640–650. [CrossRef]
138. Smith, D.; Gray, J.; Mitchell, L.; Antholine, W.E.; Hosler, J.P. Assembly of cytochrome-c oxidase in the absence of assembly protein Surf1p leads to loss of the active site heme. *J. Biol. Chem.* **2005**, *280*, 17652–17656. [CrossRef]
139. Wedatilake, Y.; Brown, R.M.; McFarland, R.; Yaplito-Lee, J.; Morris, A.A.M.; Champion, M.; Jardine, P.E.; Clarke, A.; Thorburn, D.R.; Taylor, R.W.; et al. SURF1 deficiency: A multi-centre natural history study. *Orphanet J. Rare Dis.* **2013**, *8*. [CrossRef]
140. Zhu, Z.; Yao, J.; Johns, T.; Fu, K.; De Bie, I.; Macmillan, C.; Cuthbert, A.P.; Newbold, R.F.; Wang, J.; Chevrette, M.; et al. SURF1, encoding a factor involved in the biogenesis of cytochrome c oxidase, is mutated in Leigh syndrome. *Nat. Genet.* **1998**, *20*, 337–343. [CrossRef]
141. Tiranti, V.; Hoertnagel, K.; Carrozzo, R.; Galimberti, C.; Munaro, M.; Granatiero, M.; Zelante, L.; Gasparini, P.; Marzella, R.; Rocchi, M.; et al. Mutations of SURF-1 in Leigh disease associated with cytochrome c oxidase deficiency. *Am. J. Hum. Genet.* **1998**, *63*, 1609–1621. [CrossRef]

142. Echaniz-Laguna, A.; Ghezzi, D.; Chassagne, M.; Mayençon, M.; Padet, S.; Melchionda, L.; Rouvet, I.; Lannes, B.; Bozon, D.; Latour, P.; et al. SURF1 deficiency causes demyelinating Charcot-Marie-Tooth disease. *Neurology* **2013**, *81*, 1523–1530. [CrossRef]
143. Nobrega, M.P.; Nobrega, F.G.; Tzagoloff, A. COX10 codes for a protein homologous to the ORF1 product of Paracoccus denitrificans and is required for the synthesis of yeast cytochrome oxidase. *J. Biol. Chem.* **1990**, *265*, 14220–14226. [PubMed]
144. Glerum, D.M.; Muroff, I.; Jin, C.; Tzagoloff, A. COX15 codes for a mitochondrial protein essential for the assembly of yeast cytochrome oxidase. *J. Biol. Chem.* **1997**, *272*, 19088–19094. [CrossRef] [PubMed]
145. Glerum, D.M.; Tzagoloff, A. Isolation of a human cDNA for heme A:farnesyltransferase by functional complementation of a yeast cox10 mutant. *Proc. Natl. Acad. Sci. USA* **1994**, *91*, 8452–8456. [CrossRef] [PubMed]
146. Pitceathly, R.D.S.; Taanman, J.-W.; Rahman, S.; Meunier, B.; Sadowski, M.; Cirak, S.; Hargreaves, I.; Land, J.M.; Nanji, T.; Polke, J.M.; et al. COX10 mutations resulting in complex multisystem mitochondrial disease that remains stable into adulthood. *JAMA Neurol.* **2013**, *70*, 1556–1561. [CrossRef] [PubMed]
147. Bugiani, M.; Tiranti, V.; Farina, L.; Uziel, G.; Zeviani, M. Novel mutations in COX15 in a long surviving Leigh syndrome patient with cytochrome c oxidase deficiency. *J. Med. Genet.* **2005**, *42*, e28. [CrossRef] [PubMed]
148. Rigby, K.; Cobine, P.A.; Khalimonchuk, O.; Winge, D.R. Mapping the Functional Interaction of Sco1 and Cox2 in Cytochrome Oxidase Biogenesis. *J. Biol. Chem.* **2008**, *283*, 15015–15022. [CrossRef]
149. Glerum, D.M.; Shtanko, A.; Tzagoloff, A. SCO1 and SCO2 act as high copy suppressors of a mitochondrial copper recruitment defect in *Saccharomyces cerevisiae*. *J. Biol. Chem.* **1996**, *271*, 20531–20535. [CrossRef]
150. Franco, L.V.R.; Su, C.-H.; McStay, G.P.; Yu, G.J.; Tzagoloff, A. Cox2p of yeast cytochrome oxidase assembles as a stand-alone subunit with the Cox1p and Cox3p modules. *J. Biol. Chem.* **2018**, *293*, 16899–16911. [CrossRef]
151. Gurgel-Giannetti, J.; Oliveira, G.; Brasileiro Filho, G.; Martins, P.; Vainzof, M.; Hirano, M. Mitochondrial cardioencephalomyopathy due to a novel SCO2 mutation in a Brazilian patient: Case report and literature review. *JAMA Neurol.* **2013**, *70*, 258–261. [CrossRef]
152. Rebelo, A.P.; Saade, D.; Pereira, C.V.; Farooq, A.; Huff, T.C.; Abreu, L.; Moraes, C.T.; Mnatsakanova, D.; Mathews, K.; Yang, H.; et al. SCO_2 mutations cause early-onset axonal Charcot-Marie-Tooth disease associated with cellular copper deficiency. *Brain J. Neurol.* **2018**, *141*, 662–672. [CrossRef]
153. Wakazono, T.; Miyake, M.; Yamashiro, K.; Yoshikawa, M.; Yoshimura, N. Association between SCO2 mutation and extreme myopia in Japanese patients. *Jpn. J. Ophthalmol.* **2016**, *60*, 319–325. [CrossRef] [PubMed]
154. Tran-Viet, K.-N.; Powell, C.; Barathi, V.A.; Klemm, T.; Maurer-Stroh, S.; Limviphuvadh, V.; Soler, V.; Ho, C.; Yanovitch, T.; Schneider, G.; et al. Mutations in SCO2 are associated with autosomal-dominant high-grade myopia. *Am. J. Hum. Genet.* **2013**, *92*, 820–826. [CrossRef] [PubMed]
155. Shteyer, E.; Saada, A.; Shaag, A.; Al-Hijawi, F.A.; Kidess, R.; Revel-Vilk, S.; Elpeleg, O. Exocrine pancreatic insufficiency, dyserythropoeitic anemia, and calvarial hyperostosis are caused by a mutation in the COX4I2 gene. *Am. J. Hum. Genet.* **2009**, *84*, 412–417. [CrossRef]
156. Abu-Libdeh, B.; Douiev, L.; Amro, S.; Shahrour, M.; Ta-Shma, A.; Miller, C.; Elpeleg, O.; Saada, A. Mutation in the COX4I1 gene is associated with short stature, poor weight gain and increased chromosomal breaks, simulating Fanconi anemia. *Eur. J. Hum. Genet. EJHG* **2017**, *25*, 1142–1146. [CrossRef] [PubMed]
157. Baertling, F.; Al-Murshedi, F.; Sánchez-Caballero, L.; Al-Senaidi, K.; Joshi, N.P.; Venselaar, H.; van den Brand, M.A.; Nijtmans, L.G.; Rodenburg, R.J. Mutation in mitochondrial complex IV subunit COX5A causes pulmonary arterial hypertension, lactic acidemia, and failure to thrive. *Hum. Mutat.* **2017**, *38*, 692–703. [CrossRef]
158. Tamiya, G.; Makino, S.; Hayashi, M.; Abe, A.; Numakura, C.; Ueki, M.; Tanaka, A.; Ito, C.; Toshimori, K.; Ogawa, N.; et al. A mutation of COX6A1 causes a recessive axonal or mixed form of Charcot-Marie-Tooth disease. *Am. J. Hum. Genet.* **2014**, *95*, 294–300. [CrossRef]
159. Laššuthová, P.; Šafka Brožková, D.; Krůtová, M.; Mazanec, R.; Züchner, S.; Gonzalez, M.A.; Seeman, P. Severe axonal Charcot-Marie-Tooth disease with proximal weakness caused by de novo mutation in the MORC2 gene. *Brain J. Neurol.* **2016**, *139*, e26. [CrossRef]
160. Massa, V.; Fernandez-Vizarra, E.; Alshahwan, S.; Bakhsh, E.; Goffrini, P.; Ferrero, I.; Mereghetti, P.; D'Adamo, P.; Gasparini, P.; Zeviani, M. Severe infantile encephalomyopathy caused by a mutation in COX6B1, a nucleus-encoded subunit of cytochrome c oxidase. *Am. J. Hum. Genet.* **2008**, *82*, 1281–1289. [CrossRef]

161. Abdulhag, U.N.; Soiferman, D.; Schueler-Furman, O.; Miller, C.; Shaag, A.; Elpeleg, O.; Edvardson, S.; Saada, A. Mitochondrial complex IV deficiency, caused by mutated COX6B1, is associated with encephalomyopathy, hydrocephalus and cardiomyopathy. *Eur. J. Hum. Genet. EJHG* **2015**, *23*, 159–164. [CrossRef]
162. Calvo, S.E.; Compton, A.G.; Hershman, S.G.; Lim, S.C.; Lieber, D.S.; Tucker, E.J.; Laskowski, A.; Garone, C.; Liu, S.; Jaffe, D.B.; et al. Molecular diagnosis of infantile mitochondrial disease with targeted next-generation sequencing. *Sci. Transl. Med.* **2012**, *4*, 118ra10. [CrossRef]
163. Hallmann, K.; Kudin, A.P.; Zsurka, G.; Kornblum, C.; Reimann, J.; Stüve, B.; Waltz, S.; Hattingen, E.; Thiele, H.; Nürnberg, P.; et al. Loss of the smallest subunit of cytochrome c oxidase, COX8A, causes Leigh-like syndrome and epilepsy. *Brain J. Neurol.* **2016**, *139*, 338–345. [CrossRef] [PubMed]
164. Pitceathly, R.D.S.; Rahman, S.; Wedatilake, Y.; Polke, J.M.; Cirak, S.; Foley, A.R.; Sailer, A.; Hurles, M.E.; Stalker, J.; Hargreaves, I.; et al. NDUFA4 mutations underlie dysfunction of a cytochrome c oxidase subunit linked to human neurological disease. *Cell Rep.* **2013**, *3*, 1795–1805. [CrossRef] [PubMed]
165. Szklarczyk, R.; Wanschers, B.F.J.; Nijtmans, L.G.; Rodenburg, R.J.; Zschocke, J.; Dikow, N.; van den Brand, M.A.M.; Hendriks-Franssen, M.G.M.; Gilissen, C.; Veltman, J.A.; et al. A mutation in the FAM36A gene, the human ortholog of COX20, impairs cytochrome c oxidase assembly and is associated with ataxia and muscle hypotonia. *Hum. Mol. Genet.* **2013**, *22*, 656–667. [CrossRef] [PubMed]
166. Doss, S.; Lohmann, K.; Seibler, P.; Arns, B.; Klopstock, T.; Zühlke, C.; Freimann, K.; Winkler, S.; Lohnau, T.; Drungowski, M.; et al. Recessive dystonia-ataxia syndrome in a Turkish family caused by a COX20 (FAM36A) mutation. *J. Neurol.* **2014**, *261*, 207–212. [CrossRef] [PubMed]
167. Otero, M.G.; Tiongson, E.; Diaz, F.; Haude, K.; Panzer, K.; Collier, A.; Kim, J.; Adams, D.; Tifft, C.J.; Cui, H.; et al. Novel pathogenic COX20 variants causing dysarthria, ataxia, and sensory neuropathy. *Ann. Clin. Transl. Neurol.* **2019**, *6*, 154–160. [CrossRef]
168. Xu, H.; Ji, T.; Lian, Y.; Wang, S.; Chen, X.; Li, S.; Yin, Y.; Dong, X. Observation of novel COX20 mutations related to autosomal recessive axonal neuropathy and static encephalopathy. *Hum. Genet.* **2019**, *138*, 749–756. [CrossRef]
169. Weraarpachai, W.; Sasarman, F.; Nishimura, T.; Antonicka, H.; Auré, K.; Rötig, A.; Lombès, A.; Shoubridge, E.A. Mutations in C12orf62, a factor that couples COX I synthesis with cytochrome c oxidase assembly, cause fatal neonatal lactic acidosis. *Am. J. Hum. Genet.* **2012**, *90*, 142–151. [CrossRef]
170. Renkema, G.H.; Visser, G.; Baertling, F.; Wintjes, L.T.; Wolters, V.M.; van Montfrans, J.; de Kort, G.A.P.; Nikkels, P.G.J.; van Hasselt, P.M.; van der Crabben, S.N.; et al. Mutated PET117 causes complex IV deficiency and is associated with neurodevelopmental regression and medulla oblongata lesions. *Hum. Genet.* **2017**, *136*, 759–769. [CrossRef]
171. Huigsloot, M.; Nijtmans, L.G.; Szklarczyk, R.; Baars, M.J.H.; van den Brand, M.A.M.; Hendriksfranssen, M.G.M.; van den Heuvel, L.P.; Smeitink, J.A.M.; Huynen, M.A.; Rodenburg, R.J.T. A mutation in C2orf64 causes impaired cytochrome c oxidase assembly and mitochondrial cardiomyopathy. *Am. J. Hum. Genet.* **2011**, *88*, 488–493. [CrossRef]
172. Lim, S.C.; Smith, K.R.; Stroud, D.A.; Compton, A.G.; Tucker, E.J.; Dasvarma, A.; Gandolfo, L.C.; Marum, J.E.; McKenzie, M.; Peters, H.L.; et al. A founder mutation in PET100 causes isolated complex IV deficiency in Lebanese individuals with Leigh syndrome. *Am. J. Hum. Genet.* **2014**, *94*, 209–222. [CrossRef]
173. Oláhová, M.; Haack, T.B.; Alston, C.L.; Houghton, J.A.; He, L.; Morris, A.A.; Brown, G.K.; McFarland, R.; Chrzanowska-Lightowlers, Z.M.; Lightowlers, R.N.; et al. A truncating PET100 variant causing fatal infantile lactic acidosis and isolated cytochrome c oxidase deficiency. *Eur. J. Hum. Genet. EJHG* **2015**, *23*, 935–939. [CrossRef] [PubMed]
174. Baertling, F.; van den Brand, M.A.M.; Hertecant, J.L.; Al-Shamsi, A.; van den Heuvel, L.P.; Distelmaier, F.; Mayatepek, E.; Smeitink, J.A.; Nijtmans, L.G.J.; Rodenburg, R.J.T. Mutations in COA6 cause cytochrome c oxidase deficiency and neonatal hypertrophic cardiomyopathy. *Hum. Mutat.* **2015**, *36*, 34–38. [CrossRef]
175. Ghosh, A.; Trivedi, P.P.; Timbalia, S.A.; Griffin, A.T.; Rahn, J.J.; Chan, S.S.L.; Gohil, V.M. Copper supplementation restores cytochrome c oxidase assembly defect in a mitochondrial disease model of COA6 deficiency. *Hum. Mol. Genet.* **2014**, *23*, 3596–3606. [CrossRef]
176. Ostergaard, E.; Weraarpachai, W.; Ravn, K.; Born, A.P.; Jønson, L.; Duno, M.; Wibrand, F.; Shoubridge, E.A.; Vissing, J. Mutations in COA3 cause isolated complex IV deficiency associated with neuropathy, exercise intolerance, obesity, and short stature. *J. Med. Genet.* **2015**, *52*, 203–207. [CrossRef] [PubMed]

177. Miryounesi, M.; Fardaei, M.; Tabei, S.M.; Ghafouri-Fard, S. Leigh syndrome associated with a novel mutation in the COX15 gene. *J. Pediatr. Endocrinol. Metab. JPEM* **2016**, *29*, 741–744. [CrossRef] [PubMed]
178. Oquendo, C.E.; Antonicka, H.; Shoubridge, E.A.; Reardon, W.; Brown, G.K. Functional and genetic studies demonstrate that mutation in the COX15 gene can cause Leigh syndrome. *J. Med. Genet.* **2004**, *41*, 540–544. [CrossRef] [PubMed]
179. Alfadhel, M.; Lillquist, Y.P.; Waters, P.J.; Sinclair, G.; Struys, E.; McFadden, D.; Hendson, G.; Hyams, L.; Shoffner, J.; Vallance, H.D. Infantile cardioencephalopathy due to a COX15 gene defect: Report and review. *Am. J. Med. Genet. Part A* **2011**, *155*, 840–844. [CrossRef]
180. Antonicka, H.; Mattman, A.; Carlson, C.G.; Glerum, D.M.; Hoffbuhr, K.C.; Leary, S.C.; Kennaway, N.G.; Shoubridge, E.A. Mutations in COX15 produce a defect in the mitochondrial heme biosynthetic pathway, causing early-onset fatal hypertrophic cardiomyopathy. *Am. J. Hum. Genet.* **2003**, *72*, 101–114. [CrossRef]
181. Coenen, M.J.H.; van den Heuvel, L.P.; Ugalde, C.; Ten Brinke, M.; Nijtmans, L.G.J.; Trijbels, F.J.M.; Beblo, S.; Maier, E.M.; Muntau, A.C.; Smeitink, J.A.M. Cytochrome c oxidase biogenesis in a patient with a mutation in COX10 gene. *Ann. Neurol.* **2004**, *56*, 560–564. [CrossRef]
182. Antonicka, H.; Leary, S.C.; Guercin, G.-H.; Agar, J.N.; Horvath, R.; Kennaway, N.G.; Harding, C.O.; Jaksch, M.; Shoubridge, E.A. Mutations in COX10 result in a defect in mitochondrial heme A biosynthesis and account for multiple, early-onset clinical phenotypes associated with isolated COX deficiency. *Hum. Mol. Genet.* **2003**, *12*, 2693–2702. [CrossRef]

183. Pronicka, E.; Piekutowska-Abramczuk, D.; Ciara, E.; Trubicka, J.; Rokicki, D.; Karkucińska-Więckowska, A.; Pajdowska, M.; Jurkiewicz, E.; Halat, P.; Kosińska, J.; et al. New perspective in diagnostics of mitochondrial disorders: Two years' experience with whole-exome sequencing at a national paediatric centre. *J. Transl. Med.* **2016**, *14*. [CrossRef] [PubMed]
184. Kohda, M.; Tokuzawa, Y.; Kishita, Y.; Nyuzuki, H.; Moriyama, Y.; Mizuno, Y.; Hirata, T.; Yatsuka, Y.; Yamashita-Sugahara, Y.; Nakachi, Y.; et al. A Comprehensive Genomic Analysis Reveals the Genetic Landscape of Mitochondrial Respiratory Chain Complex Deficiencies. *PLoS Genet.* **2016**, *12*, e1005679. [CrossRef] [PubMed]
185. Brix, N.; Jensen, J.M.; Pedersen, I.S.; Ernst, A.; Frost, S.; Bogaard, P.; Petersen, M.B.; Bender, L. Mitochondrial Disease Caused by a Novel Homozygous Mutation (Gly106del) in the SCO1 Gene. *Neonatology* **2019**, *116*, 290–294. [CrossRef] [PubMed]
186. Leary, S.C.; Antonicka, H.; Sasarman, F.; Weraarpachai, W.; Cobine, P.A.; Pan, M.; Brown, G.K.; Brown, R.; Majewski, J.; Ha, K.C.H.; et al. Novel mutations in SCO1 as a cause of fatal infantile encephalopathy and lactic acidosis. *Hum. Mutat.* **2013**, *34*, 1366–1370. [CrossRef]
187. Stiburek, L.; Vesela, K.; Hansikova, H.; Hulkova, H.; Zeman, J. Loss of function of Sco1 and its interaction with cytochrome c oxidase. *Am. J. Physiol. Cell Physiol.* **2009**, *296*, C1218–C1226. [CrossRef]
188. Mitchell, P. Coupling of Phosphorylation to Electron and Hydrogen Transfer by a Chemi-Osmotic type of Mechanism. *Nature* **1961**, *191*, 144–148. [CrossRef]
189. Kagawa, Y.; Racker, E. Partial Resolution of the Enzymes Catalyzing Oxidative Phosphorylation VIII. Properties of a factor conferring oligomycin sensitivity on mitochondrial adenosine triphosphatase. *J. Biol. Chem.* **1966**, *241*, 2461–2466.
190. Boyer, P.D. The Atp synthase—A splendid molecular machine. *Annu. Rev. Biochem.* **1997**, *66*, 717–749. [CrossRef]
191. Velours, J.; Arselin, G. The *Saccharomyces cerevisiae* ATP Synthase. *J. Bioenerg. Biomembr. N. Y.* **2000**, *32*, 383–390. [CrossRef]
192. Anderson, S.; Bankier, A.T.; Barrell, B.G.; de Bruijn, M.H.L.; Coulson, A.R.; Drouin, J.; Eperon, I.C.; Nierlich, D.P.; Roe, B.A.; Sanger, F.; et al. Sequence and organization of the human mitochondrial genome. *Nature* **1981**, *290*, 457–465. [CrossRef] [PubMed]
193. Dautant, A.; Velours, J.; Giraud, M.-F. Crystal Structure of the Mg·ADP-inhibited State of the Yeast F1c10-ATP Synthase. *J. Biol. Chem.* **2010**, *285*, 29502–29510. [CrossRef] [PubMed]
194. Symersky, J.; Pagadala, V.; Osowski, D.; Krah, A.; Meier, T.; Faraldo-Gómez, J.D.; Mueller, D.M. Structure of the c10 ring of the yeast mitochondrial ATP synthase in the open conformation. *Nat. Struct. Mol. Biol.* **2012**, *19*, 485–491. [CrossRef] [PubMed]
195. Watt, I.N.; Montgomery, M.G.; Runswick, M.J.; Leslie, A.G.W.; Walker, J.E. Bioenergetic cost of making an adenosine triphosphate molecule in animal mitochondria. *Proc. Natl. Acad. Sci. USA* **2010**, *107*, 16823–16827. [CrossRef]
196. Schon, E.A.; Santra, S.; Pallotti, F.; Girvin, M.E. Pathogenesis of primary defects in mitochondrial ATP synthesis. *Semin. Cell Dev. Biol.* **2001**, *12*, 441–448. [CrossRef] [PubMed]
197. Holt, I.J.; Harding, A.E.; Petty, R.K.; Morgan-Hughes, J.A. A new mitochondrial disease associated with mitochondrial DNA heteroplasmy. *Am. J. Hum. Genet.* **1990**, *46*, 428–433.
198. Ganetzky, R.D.; Stendel, C.; McCormick, E.M.; Zolkipli-Cunningham, Z.; Goldstein, A.C.; Klopstock, T.; Falk, M.J. MT-ATP6 mitochondrial disease variants: Phenotypic and biochemical features analysis in 218 published cases and cohort of 14 new cases. *Hum. Mutat.* **2019**, *40*, 499–515. [CrossRef]
199. Rak, M.; Tetaud, E.; Duvezin-Caubet, S.; Ezkurdia, N.; Bietenhader, M.; Rytka, J.; di Rago, J.-P. A yeast model of the neurogenic ataxia retinitis pigmentosa (NARP) T8993G mutation in the mitochondrial ATP synthase-6 gene. *J. Biol. Chem.* **2007**, *282*, 34039–34047. [CrossRef]
200. Franco, L.V.R.; Su, C.-H.; Burnett, J.; Teixeira, L.S.; Tzagoloff, A. Atco, a yeast mitochondrial complex of Atp9 and Cox6, is an assembly intermediate of the ATP synthase. *PLoS ONE* **2020**, *15*, e0233177. [CrossRef]
201. Kabala, A.M.; Lasserre, J.-P.; Ackerman, S.H.; di Rago, J.-P.; Kucharczyk, R. Defining the impact on yeast ATP synthase of two pathogenic human mitochondrial DNA mutations, T9185C and T9191C. *Biochimie* **2014**, *100*, 200–206. [CrossRef]

202. Su, X.; Dautant, A.; Godard, F.; Bouhier, M.; Zoladek, T.; Kucharczyk, R.; di Rago, J.-P.; Tribouillard-Tanvier, D. Molecular Basis of the Pathogenic Mechanism Induced by the m.9191T>C Mutation in Mitochondrial ATP6 Gene. *Int. J. Mol. Sci.* **2020**, *21*, 5083. [CrossRef]
203. Houštěk, J.; Pícková, A.; Vojtíšková, A.; Mráček, T.; Pecina, P.; Ješina, P. Mitochondrial diseases and genetic defects of ATP synthase. *Biochim. Biophys. Acta BBA Bioenerg.* **2006**, *1757*, 1400–1405. [CrossRef] [PubMed]
204. Jonckheere, A.I.; Hogeveen, M.; Nijtmans, L.; van den Brand, M.; Janssen, A.; Diepstra, H.; van den Brandt, F.; van den Heuvel, B.; Hol, F.; Hofste, T.; et al. A novel mitochondrial ATP8 gene mutation in a patient with apical hypertrophic cardiomyopathy and neuropathy. *BMJ Case Rep.* **2009**, *2009*. [CrossRef] [PubMed]
205. Tansel, T.; Paçal, F.; Ustek, D. A novel ATP8 gene mutation in an infant with tetralogy of Fallot. *Cardiol. Young Camb.* **2014**, *24*, 531–533. [CrossRef]
206. Dautant, A.; Meier, T.; Hahn, A.; Tribouillard-Tanvier, D.; di Rago, J.-P.; Kucharczyk, R. ATP Synthase Diseases of Mitochondrial Genetic Origin. *Front. Physiol.* **2018**, *9*. [CrossRef]
207. Kytövuori, L.; Lipponen, J.; Rusanen, H.; Komulainen, T.; Martikainen, M.H.; Majamaa, K. A novel mutation m.8561C>G in MT-ATP6/8 causing a mitochondrial syndrome with ataxia, peripheral neuropathy, diabetes mellitus, and hypergonadotropic hypogonadism. *J. Neurol.* **2016**, *263*, 2188–2195. [CrossRef]
208. Fragaki, K.; Chaussenot, A.; Serre, V.; Acquaviva, C.; Bannwarth, S.; Rouzier, C.; Chabrol, B.; Paquis-Flucklinger, V. A novel variant m.8561C>T in the overlapping region of MT-ATP6 and MT-ATP8 in a child with early-onset severe neurological signs. *Mol. Genet. Metab. Rep.* **2019**, *21*. [CrossRef]
209. Holme, E.; Greter, J.; Jacobson, C.-E.; Larsson, N.-G.; Lindstedt, S.; Nilsson, K.O.; Oldfors, A.; Tulinius, M. Mitochondrial ATP-Synthase Deficiency in a Child with 3-Methylglutaconic Aciduria. *Pediatr. Res.* **1992**, *32*, 731–736. [CrossRef]
210. Houštek, J.; Klement, P.; Floryk, D.; Antonická, H.; Hermanská, J.; Kalous, M.; Hansíková, H.; Houšt'ková, H.; Chowdhury, S.K.R.; Rosipal, Š.; et al. A Novel Deficiency of Mitochondrial ATPase of Nuclear Origin. *Hum. Mol. Genet.* **1999**, *8*, 1967–1974. [CrossRef]
211. Sperl, W.; Ješina, P.; Zeman, J.; Mayr, J.A.; DeMeirleir, L.; VanCoster, R.; Pícková, A.; Hansíková, H.; Houšt'ková, H.; Krejčík, Z.; et al. Deficiency of mitochondrial ATP synthase of nuclear genetic origin. *Neuromuscul. Disord.* **2006**, *16*, 821–829. [CrossRef]
212. Sardin, E.; Donadello, S.; di Rago, J.-P.; Tetaud, E. Biochemical investigation of a human pathogenic mutation in the nuclear ATP5E gene using yeast as a model. *Front. Genet.* **2015**, *6*. [CrossRef]
213. Oláhová, M.; Yoon, W.H.; Thompson, K.; Jangam, S.; Fernandez, L.; Davidson, J.M.; Kyle, J.E.; Grove, M.E.; Fisk, D.G.; Kohler, J.N.; et al. Biallelic Mutations in ATP5F1D, which Encodes a Subunit of ATP Synthase, Cause a Metabolic Disorder. *Am. J. Hum. Genet.* **2018**, *102*, 494–504. [CrossRef] [PubMed]
214. Feichtinger, R.G.; Schäfer, G.; Seifarth, C.; Mayr, J.A.; Kofler, B.; Klocker, H. Reduced Levels of ATP Synthase Subunit ATP5F1A Correlate with Earlier-Onset Prostate Cancer. *Oxid. Med. Cell. Longev.* **2018**, *2018*. [CrossRef] [PubMed]
215. Franco, L.V.R.; Su, C.H.; Tzagoloff, A. Modular assembly of yeast mitochondrial ATP synthase and cytochrome oxidase. *Biol. Chem.* **2020**, *401*. [CrossRef] [PubMed]
216. Wang, Z.-G.; Sheluho, D.; Gatti, D.L.; Ackerman, S.H. The α-subunit of the mitochondrial F1 ATPase interacts directly with the assembly factor Atp12p. *EMBO J.* **2000**, *19*, 1486–1493. [CrossRef]
217. Kratochvílová, H.; Hejzlarová, K.; Vrbacký, M.; Mráček, T.; Karbanová, V.; Tesařová, M.; Gombitová, A.; Cmarko, D.; Wittig, I.; Zeman, J.; et al. Mitochondrial membrane assembly of TMEM70 protein. *Mitochondrion* **2014**, *15*, 1–9. [CrossRef]
218. Sánchez-Caballero, L.; Elurbe, D.M.; Baertling, F.; Guerrero-Castillo, S.; van den Brand, M.; van Strien, J.; van Dam, T.J.P.; Rodenburg, R.; Brandt, U.; Huynen, M.A.; et al. TMEM70 functions in the assembly of complexes I and V. *Biochim. Biophys. Acta BBA Bioenerg.* **2020**, *1861*. [CrossRef]
219. De Meirleir, L. Respiratory chain complex V deficiency due to a mutation in the assembly gene ATP12. *J. Med. Genet.* **2004**, *41*, 120–124. [CrossRef]
220. Lefebvre-Legendre, L.; Vaillier, J.; Benabdelhak, H.; Velours, J.; Slonimski, P.P.; Rago, J.-P. di Identification of a Nuclear Gene (FMC1) Required for the Assembly/Stability of Yeast Mitochondrial F1-ATPase in Heat Stress Conditions. *J. Biol. Chem.* **2001**, *276*, 6789–6796. [CrossRef]
221. Turunen, M.; Olsson, J.; Dallner, G. Metabolism and function of coenzyme Q. *Biochim. Biophys. Acta BBA Biomembr.* **2004**, *1660*, 171–199. [CrossRef]

222. Tran, U.C.; Clarke, C.F. Endogenous synthesis of coenzyme Q in eukaryotes. *Mitochondrion* **2007**, *7*, S62–S71. [CrossRef]
223. Kawamukai, M. Biosynthesis and bioproduction of coenzyme Q_{10} by yeasts and other organisms. *Biotechnol. Appl. Biochem.* **2009**, *53*, 217–226. [CrossRef] [PubMed]
224. Tzagoloff, A.; Akai, A.; Needleman, R.B. Assembly of the mitochondrial membrane system. Characterization of nuclear mutants of *Saccharomyces cerevisiae* with defects in mitochondrial ATPase and respiratory enzymes. *J. Biol. Chem.* **1975**, *250*, 8228–8235. [PubMed]
225. Marbois, B.; Gin, P.; Gulmezian, M.; Clarke, C.F. The yeast Coq4 polypeptide organizes a mitochondrial protein complex essential for coenzyme Q biosynthesis. *Biochim. Biophys. Acta BBA Mol. Cell Biol. Lipids* **2009**, *1791*, 69–75. [CrossRef] [PubMed]
226. Subramanian, K.; Jochem, A.; Le Vasseur, M.; Lewis, S.; Paulson, B.R.; Reddy, T.R.; Russell, J.D.; Coon, J.J.; Pagliarini, D.J.; Nunnari, J. Coenzyme Q biosynthetic proteins assemble in a substrate-dependent manner into domains at ER–mitochondria contacts. *J. Cell Biol.* **2019**, *218*, 1353–1369. [CrossRef] [PubMed]
227. Awad, A.M.; Bradley, M.C.; Fernández-del-Río, L.; Nag, A.; Tsui, H.S.; Clarke, C.F. Coenzyme Q_{10} deficiencies: Pathways in yeast and humans. *Essays Biochem.* **2018**, *62*, 361–376. [CrossRef]
228. Gloor, U.; Wiss, O. On the biosynthesis of ubiquinone (50). *Arch. Biochem. Biophys.* **1959**, *83*, 216–222. [CrossRef]
229. Pierrel, F.; Hamelin, O.; Douki, T.; Kieffer-Jaquinod, S.; Mühlenhoff, U.; Ozeir, M.; Lill, R.; Fontecave, M. Involvement of Mitochondrial Ferredoxin and Para-Aminobenzoic Acid in Yeast Coenzyme Q Biosynthesis. *Chem. Biol.* **2010**, *17*, 449–459. [CrossRef]
230. Payet, L.-A.; Leroux, M.; Willison, J.C.; Kihara, A.; Pelosi, L.; Pierrel, F. Mechanistic Details of Early Steps in Coenzyme Q Biosynthesis Pathway in Yeast. *Cell Chem. Biol.* **2016**, *23*, 1241–1250. [CrossRef]
231. Stefely, J.A.; Pagliarini, D.J. Biochemistry of Mitochondrial Coenzyme Q Biosynthesis. *Trends Biochem. Sci.* **2017**, *42*, 824–843. [CrossRef]
232. Ashby, M.N.; Edwards, P.A. Elucidation of the deficiency in two yeast coenzyme Q mutants. Characterization of the structural gene encoding hexaprenyl pyrophosphate synthetase. *J. Biol. Chem.* **1990**, *265*, 13157–13164.
233. Ashby, M.N.; Kutsunai, S.Y.; Ackerman, S.; Tzagoloff, A.; Edwards, P.A. COQ2 is a candidate for the structural gene encoding para-hydroxybenzoate: Polyprenyltransferase. *J. Biol. Chem.* **1992**, *267*, 4128–4136.
234. Gin, P.; Hsu, A.Y.; Rothman, S.C.; Jonassen, T.; Lee, P.T.; Tzagoloff, A.; Clarke, C.F. The *Saccharomyces cerevisiae* COQ6 Gene Encodes a Mitochondrial Flavin-dependent Monooxygenase Required for Coenzyme Q Biosynthesis. *J. Biol. Chem.* **2003**, *278*, 25308–25316. [CrossRef]
235. Ozeir, M.; Mühlenhoff, U.; Webert, H.; Lill, R.; Fontecave, M.; Pierrel, F. Coenzyme Q Biosynthesis: Coq6 Is Required for the C5-Hydroxylation Reaction and Substrate Analogs Rescue Coq6 Deficiency. *Chem. Biol.* **2011**, *18*, 1134–1142. [CrossRef]
236. Marbois, B.N.; Clarke, C.F. The COQ7 Gene Encodes a Protein in *Saccharomyces cerevisiae* Necessary for Ubiquinone Biosynthesis. *J. Biol. Chem.* **1996**, *271*, 2995–3004. [CrossRef]
237. Poon, W.W.; Barkovich, R.J.; Hsu, A.Y.; Frankel, A.; Lee, P.T.; Shepherd, J.N.; Myles, D.C.; Clarke, C.F. Yeast and Rat Coq3 and Escherichia coli UbiG Polypeptides Catalyze Both O-Methyltransferase Steps in Coenzyme Q Biosynthesis. *J. Biol. Chem.* **1999**, *274*, 21665–21672. [CrossRef]
238. Barkovich, R.J.; Shtanko, A.; Shepherd, J.A.; Lee, P.T.; Myles, D.C.; Tzagoloff, A.; Clarke, C.F. Characterization of the COQ5 Gene from *Saccharomyces cerevisiae* evidence for a C-methyltransferase in ubiquinone biosynthesis. *J. Biol. Chem.* **1997**, *272*, 9182–9188. [CrossRef] [PubMed]
239. Xie, L.X.; Hsieh, E.J.; Watanabe, S.; Allan, C.M.; Chen, J.Y.; Tran, U.C.; Clarke, C.F. Expression of the human atypical kinase ADCK3 rescues coenzyme Q biosynthesis and phosphorylation of Coq polypeptides in yeast coq8 mutants. *Biochim. Biophys. Acta BBA Mol. Cell Biol. Lipids* **2011**, *1811*, 348–360. [CrossRef] [PubMed]
240. Tauche, A.; Krause-Buchholz, U.; Rödel, G. Ubiquinone biosynthesis in *Saccharomyces cerevisiae*: The molecular organization of O-methylase Coq3p depends on Abc1p/Coq8p. *FEMS Yeast Res.* **2008**, *8*, 1263–1275. [CrossRef] [PubMed]
241. Lagier-Tourenne, C.; Tazir, M.; López, L.C.; Quinzii, C.M.; Assoum, M.; Drouot, N.; Busso, C.; Makri, S.; Ali-Pacha, L.; Benhassine, T.; et al. ADCK3, an Ancestral Kinase, Is Mutated in a Form of Recessive Ataxia Associated with Coenzyme Q_{10} Deficiency. *Am. J. Hum. Genet.* **2008**, *82*, 661–672. [CrossRef]

242. Hsieh, E.J.; Gin, P.; Gulmezian, M.; Tran, U.C.; Saiki, R.; Marbois, B.N.; Clarke, C.F. *Saccharomyces cerevisiae* Coq9 polypeptide is a subunit of the mitochondrial coenzyme Q biosynthetic complex. *Arch. Biochem. Biophys.* **2007**, *463*, 19–26. [CrossRef] [PubMed]
243. Awad, A.M.; Nag, A.; Pham, N.V.B.; Bradley, M.C.; Jabassini, N.; Nathaniel, J.; Clarke, C.F. Intragenic suppressor mutations of the COQ8 protein kinase homolog restore coenzyme Q biosynthesis and function in *Saccharomyces cerevisiae*. *PLoS ONE* **2020**, *15*, e0234192. [CrossRef] [PubMed]
244. Bradley, M.C.; Yang, K.; Fernández-del-Río, L.; Ngo, J.; Ayer, A.; Tsui, H.S.; Novales, N.A.; Stocker, R.; Shirihai, O.S.; Barros, M.H.; et al. COQ11 deletion mitigates respiratory deficiency caused by mutations in the gene encoding the coenzyme Q chaperone protein Coq10. *J. Biol. Chem.* **2020**, *295*, 6023–6042. [CrossRef] [PubMed]
245. Allan, C.M.; Awad, A.M.; Johnson, J.S.; Shirasaki, D.I.; Wang, C.; Blaby-Haas, C.E.; Merchant, S.S.; Loo, J.A.; Clarke, C.F. Identification of Coq11, a New Coenzyme Q Biosynthetic Protein in the CoQ-Synthome in *Saccharomyces cerevisiae*. *J. Biol. Chem.* **2015**, *290*, 7517–7534. [CrossRef]
246. Ogasahara, S.; Engel, A.G.; Frens, D.; Mack, D. Muscle coenzyme Q deficiency in familial mitochondrial encephalomyopathy. *Proc. Natl. Acad. Sci. USA* **1989**, *86*, 2379–2382. [CrossRef] [PubMed]
247. Emmanuele, V.; López, L.C.; Berardo, A.; Naini, A.; Tadesse, S.; Wen, B.; D'Agostino, E.; Solomon, M.; DiMauro, S.; Quinzii, C.; et al. Heterogeneity of Coenzyme Q_{10} Deficiency: Patient Study and Literature Review. *Arch. Neurol.* **2012**, *69*, 978–983. [CrossRef] [PubMed]
248. Desbats, M.A.; Lunardi, G.; Doimo, M.; Trevisson, E.; Salviati, L. Genetic bases and clinical manifestations of coenzyme Q_{10} (CoQ_{10}) deficiency. *J. Inherit. Metab. Dis.* **2015**, *38*, 145–156. [CrossRef] [PubMed]
249. Acosta, M.J.; Vazquez Fonseca, L.; Desbats, M.A.; Cerqua, C.; Zordan, R.; Trevisson, E.; Salviati, L. Coenzyme Q biosynthesis in health and disease. *Biochim. Biophys. Acta BBA Bioenerg.* **2016**, *1857*, 1079–1085. [CrossRef]
250. Malicdan, M.C.V.; Vilboux, T.; Ben-Zeev, B.; Guo, J.; Eliyahu, A.; Pode-Shakked, B.; Dori, A.; Kakani, S.; Chandrasekharappa, S.C.; Ferreira, C.R.; et al. A novel inborn error of the coenzyme Q_{10} biosynthesis pathway: Cerebellar ataxia and static encephalomyopathy due to COQ5 C-methyltransferase deficiency. *Hum. Mutat.* **2018**, *39*, 69–79. [CrossRef]
251. Montini, G.; Malaventura, C.; Salviati, L. Early Coenzyme Q_{10} Supplementation in Primary Coenzyme Q_{10} Deficiency. *N. Engl. J. Med.* **2008**, *26*, 2849–2850. [CrossRef]
252. Quinzii, C.M.; Hirano, M. Coenzyme Q and mitochondrial disease. *Dev. Disabil. Res. Rev.* **2010**, *16*, 183–188. [CrossRef]
253. López, L.C.; Quinzii, C.M.; Area, E.; Naini, A.; Rahman, S.; Schuelke, M.; Salviati, L.; DiMauro, S.; Hirano, M. Treatment of CoQ_{10} Deficient Fibroblasts with Ubiquinone, CoQ Analogs, and Vitamin C: Time- and Compound-Dependent Effects. *PLoS ONE* **2010**, *5*, e11897. [CrossRef] [PubMed]
254. Do, T.Q.; Hsu, A.Y.; Jonassen, T.; Lee, P.T.; Clarke, C.F. A Defect in Coenzyme Q Biosynthesis Is Responsible for the Respiratory Deficiency in *Saccharomyces cerevisiae* abc1Mutants. *J. Biol. Chem.* **2001**, *276*, 18161–18168. [CrossRef] [PubMed]
255. Zampol, M.A.; Busso, C.; Gomes, F.; Ferreira-Junior, J.R.; Tzagoloff, A.; Barros, M.H. Over-expression of COQ10 in *Saccharomyces cerevisiae* inhibits mitochondrial respiration. *Biochem. Biophys. Res. Commun.* **2010**, *402*, 82–87. [CrossRef]
256. Busso, C.; Ferreira-Júnior, J.R.; Paulela, J.A.; Bleicher, L.; Demasi, M.; Barros, M.H. Coq7p relevant residues for protein activity and stability. *Biochimie* **2015**, *119*, 92–102. [CrossRef]
257. Jonassen, T.; Clarke, C.F. Isolation and Functional Expression of Human COQ3, a Gene Encoding a Methyltransferase Required for Ubiquinone Biosynthesis. *J. Biol. Chem.* **2000**, *275*, 12381–12387. [CrossRef]
258. Barros, M.H.; Johnson, A.; Gin, P.; Marbois, B.N.; Clarke, C.F.; Tzagoloff, A. The *Saccharomyces cerevisiae* COQ10 Gene Encodes a START Domain Protein Required for Function of Coenzyme Q in Respiration. *J. Biol. Chem.* **2005**, *280*, 42627–42635. [CrossRef] [PubMed]
259. He, C.H.; Black, D.S.; Allan, C.M.; Meunier, B.; Rahman, S.; Clarke, C.F. Human COQ9 Rescues a coq9 Yeast Mutant by Enhancing Coenzyme Q Biosynthesis from 4-Hydroxybenzoic Acid and Stabilizing the CoQ-Synthome. *Front. Physiol.* **2017**, *8*. [CrossRef] [PubMed]
260. Nguyen, T.P.T.; Casarin, A.; Desbats, M.A.; Doimo, M.; Trevisson, E.; Santos-Ocaña, C.; Navas, P.; Clarke, C.F.; Salviati, L. Molecular characterization of the human COQ5 C-methyltransferase in coenzyme Q_{10} biosynthesis. *Biochim. Biophys. Acta BBA Mol. Cell Biol. Lipids* **2014**, *1841*, 1628–1638. [CrossRef] [PubMed]

261. Mollet, J.; Giurgea, I.; Schlemmer, D.; Dallner, G.; Chretien, D.; Delahodde, A.; Bacq, D.; de Lonlay, P.; Munnich, A.; Rötig, A. *Prenyldiphosphate synthase, subunit 1* (*PDSS1*) and *OH-benzoate polyprenyltransferase* (*COQ2*) mutations in ubiquinone deficiency and oxidative phosphorylation disorders. *J. Clin. Investig.* **2007**, *117*, 765–772. [CrossRef]
262. López-Martín, J.M.; Salviati, L.; Trevisson, E.; Montini, G.; DiMauro, S.; Quinzii, C.; Hirano, M.; Rodriguez-Hernandez, A.; Cordero, M.D.; Sánchez-Alcázar, J.A.; et al. Missense mutation of the COQ2 gene causes defects of bioenergetics and de novo pyrimidine synthesis. *Hum. Mol. Genet.* **2007**, *16*, 1091–1097. [CrossRef]
263. Salviati, L.; Trevisson, E.; Hernandez, M.A.R.; Casarin, A.; Pertegato, V.; Doimo, M.; Cassina, M.; Agosto, C.; Desbats, M.A.; Sartori, G.; et al. Haploinsufficiency of COQ4 causes coenzyme Q_{10} deficiency. *J. Med. Genet.* **2012**, *49*, 187–191. [CrossRef] [PubMed]
264. Heeringa, S.F.; Chernin, G.; Chaki, M.; Zhou, W.; Sloan, A.J.; Ji, Z.; Xie, L.X.; Salviati, L.; Hurd, T.W.; Vega-Warner, V.; et al. *COQ6* mutations in human patients produce nephrotic syndrome with sensorineural deafness. *J. Clin. Investig.* **2011**, *121*, 2013–2024. [CrossRef] [PubMed]
265. Duncan, A.J.; Bitner-Glindzicz, M.; Meunier, B.; Costello, H.; Hargreaves, I.P.; López, L.C.; Hirano, M.; Quinzii, C.M.; Sadowski, M.I.; Hardy, J.; et al. A Nonsense Mutation in COQ9 Causes Autosomal-Recessive Neonatal-Onset Primary Coenzyme Q_{10} Deficiency: A Potentially Treatable Form of Mitochondrial Disease. *Am. J. Hum. Genet.* **2009**, *84*, 558–566. [CrossRef] [PubMed]
266. Lake, N.J.; Compton, A.G.; Rahman, S.; Thorburn, D.R. Leigh syndrome: One disorder, more than 75 monogenic causes. *Ann. Neurol.* **2016**, *79*, 190–203. [CrossRef] [PubMed]
267. Li, Y.; Lin, S.; Li, L.; Tang, Z.; Hu, Y.; Ban, X.; Zeng, T.; Zhou, Y.; Zhu, Y.; Gao, S.; et al. PDSS2 Deficiency Induces Hepatocarcinogenesis by Decreasing Mitochondrial Respiration and Reprogramming Glucose Metabolism. *Cancer Res.* **2018**, *78*, 4471–4481. [CrossRef]
268. Quinzii, C.; Naini, A.; Salviati, L.; Trevisson, E.; Navas, P.; DiMauro, S.; Hirano, M. A Mutation in Para-Hydroxybenzoate-Polyprenyl Transferase (COQ2) Causes Primary Coenzyme Q_{10} Deficiency. *Am. J. Hum. Genet.* **2006**, *78*, 345–349. [CrossRef]
269. Caglayan, A.O.; Gumus, H.; Sandford, E.; Kubisiak, T.L.; Ma, Q.; Ozel, A.B.; Per, H.; Li, J.Z.; Shakkottai, V.G.; Burmeister, M. COQ4 Mutation Leads to Childhood-Onset Ataxia Improved by CoQ_{10} Administration. *Cerebellum Lond. Engl.* **2019**, *18*, 665–669. [CrossRef]
270. Lu, M.; Zhou, Y.; Wang, Z.; Xia, Z.; Ren, J.; Guo, Q. Clinical phenotype, in silico and biomedical analyses, and intervention for an East Asian population-specific c.370G>A (p.G124S) COQ4 mutation in a Chinese family with CoQ_{10} deficiency-associated Leigh syndrome. *J. Hum. Genet.* **2019**, *64*, 297–304. [CrossRef]
271. Song, C.-C.; Hong, Q.; Geng, X.-D.; Wang, X.; Wang, S.-Q.; Cui, S.-Y.; Guo, M.-D.; Li, O.; Cai, G.-Y.; Chen, X.-M.; et al. New Mutation of Coenzyme Q_{10} Monooxygenase 6 Causing Podocyte Injury in a Focal Segmental Glomerulosclerosis Patient. *Chin. Med. J.* **2018**, *131*, 2666–2675. [CrossRef]
272. Freyer, C.; Stranneheim, H.; Naess, K.; Mourier, A.; Felser, A.; Maffezzini, C.; Lesko, N.; Bruhn, H.; Engvall, M.; Wibom, R.; et al. Rescue of primary ubiquinone deficiency due to a novel COQ7 defect using 2,4–dihydroxybensoic acid. *J. Med. Genet.* **2015**, *52*, 779–783. [CrossRef]
273. Wang, Y.; Smith, C.; Parboosingh, J.S.; Khan, A.; Innes, M.; Hekimi, S. Pathogenicity of two COQ7 mutations and responses to 2,4-dihydroxybenzoate bypass treatment. *J. Cell. Mol. Med.* **2017**, *21*, 2329–2343. [CrossRef] [PubMed]
274. Kwong, A.K.-Y.; Chiu, A.T.-G.; Tsang, M.H.-Y.; Lun, K.-S.; Rodenburg, R.J.T.; Smeitink, J.; Chung, B.H.-Y.; Fung, C.-W. A fatal case of COQ7-associated primary coenzyme Q_{10} deficiency. *JIMD Rep.* **2019**, *47*, 23–29. [CrossRef] [PubMed]
275. Cullen, J.K.; Abdul Murad, N.; Yeo, A.; McKenzie, M.; Ward, M.; Chong, K.L.; Schieber, N.L.; Parton, R.G.; Lim, Y.C.; Wolvetang, E.; et al. AarF Domain Containing Kinase 3 (ADCK3) Mutant Cells Display Signs of Oxidative Stress, Defects in Mitochondrial Homeostasis and Lysosomal Accumulation. *PLoS ONE* **2016**, *11*, e0148213. [CrossRef] [PubMed]
276. Song, X.; Fang, X.; Tang, X.; Cao, Q.; Zhai, Y.; Chen, J.; Liu, J.; Zhang, Z.; Xiang, T.; Qian, Y.; et al. COQ8B nephropathy: Early detection and optimal treatment. *Mol. Genet. Genomic Med.* **2020**, *8*, e1360. [CrossRef] [PubMed]

277. Lohman, D.C.; Aydin, D.; Von Bank, H.C.; Smith, R.W.; Linke, V.; Weisenhorn, E.; McDevitt, M.T.; Hutchins, P.; Wilkerson, E.M.; Wancewicz, B.; et al. An Isoprene Lipid-Binding Protein Promotes Eukaryotic Coenzyme Q Biosynthesis. *Mol. Cell* **2019**, *73*, 763–774.e10. [CrossRef]
278. Danhauser, K.; Herebian, D.; Haack, T.B.; Rodenburg, R.J.; Strom, T.M.; Meitinger, T.; Klee, D.; Mayatepek, E.; Prokisch, H.; Distelmaier, F. Fatal neonatal encephalopathy and lactic acidosis caused by a homozygous loss-of-function variant in COQ9. *Eur. J. Hum. Genet. EJHG* **2016**, *24*, 450–454. [CrossRef]
279. Smith, A.C.; Ito, Y.; Ahmed, A.; Schwartzentruber, J.A.; Beaulieu, C.L.; Aberg, E.; Majewski, J.; Bulman, D.E.; Horsting-Wethly, K.; Koning, D.V.; et al. A family segregating lethal neonatal coenzyme Q_{10} deficiency caused by mutations in COQ9. *J. Inherit. Metab. Dis.* **2018**, *41*, 719–729. [CrossRef]
280. Olgac, A.; Öztoprak, Ü.; Kasapkara, Ç.S.; Kılıç, M.; Yüksel, D.; Derinkuyu, E.B.; Taşçı Yıldız, Y.; Ceylaner, S.; Ezgu, F.S. A rare case of primary coenzyme Q_{10} deficiency due to COQ9 mutation. *J. Pediatr. Endocrinol. Metab. JPEM* **2020**, *33*, 165–170. [CrossRef]
281. Luna-Sánchez, M.; Díaz-Casado, E.; Barca, E.; Tejada, M.Á.; Montilla-García, Á.; Cobos, E.J.; Escames, G.; Acuña-Castroviejo, D.; Quinzii, C.M.; López, L.C. The clinical heterogeneity of coenzyme Q_{10} deficiency results from genotypic differences in the Coq9 gene. *EMBO Mol. Med.* **2015**, *7*, 670–687. [CrossRef]
282. Quinzii, C.M.; Luna-Sanchez, M.; Ziosi, M.; Hidalgo-Gutierrez, A.; Kleiner, G.; Lopez, L.C. The Role of Sulfide Oxidation Impairment in the Pathogenesis of Primary CoQ Deficiency. *Front. Physiol.* **2017**, *8*. [CrossRef] [PubMed]
283. Morison, I.M.; Cramer Bordé, E.M.; Cheesman, E.J.; Cheong, P.L.; Holyoake, A.J.; Fichelson, S.; Weeks, R.J.; Lo, A.; Davies, S.M.K.; Wilbanks, S.M.; et al. A mutation of human cytochrome c enhances the intrinsic apoptotic pathway but causes only thrombocytopenia. *Nat. Genet.* **2008**, *40*, 387–389. [CrossRef] [PubMed]
284. De Rocco, D.; Cerqua, C.; Goffrini, P.; Russo, G.; Pastore, A.; Meloni, F.; Nicchia, E.; Moraes, C.T.; Pecci, A.; Salviati, L.; et al. Mutations of cytochrome c identified in patients with thrombocytopenia THC4 affect both apoptosis and cellular bioenergetics. *Biochim. Biophys. Acta BBA Mol. Basis Dis.* **2014**, *1842*, 269–274. [CrossRef] [PubMed]
285. Johnson, B.; Lowe, G.C.; Futterer, J.; Lordkipanidzé, M.; MacDonald, D.; Simpson, M.A.; Sanchez-Guiú, I.; Drake, S.; Bem, D.; Leo, V.; et al. Whole exome sequencing identifies genetic variants in inherited thrombocytopenia with secondary qualitative function defects. *Haematologica* **2016**, *101*, 1170–1179. [CrossRef] [PubMed]
286. Uchiyama, Y.; Yanagisawa, K.; Kunishima, S.; Shiina, M.; Ogawa, Y.; Nakashima, M.; Hirato, J.; Imagawa, E.; Fujita, A.; Hamanaka, K.; et al. A novel CYCS mutation in the α-helix of the CYCS C-terminal domain causes non-syndromic thrombocytopenia. *Clin. Genet.* **2018**, *94*, 548–553. [CrossRef] [PubMed]
287. Ong, L.; Morison, I.M.; Ledgerwood, E.C. Megakaryocytes from CYCS mutation-associated thrombocytopenia release platelets by both proplatelet-dependent and -independent processes. *Br. J. Haematol.* **2017**, *176*, 268–279. [CrossRef]
288. Schwarz, Q.P.; Cox, T.C. Complementation of a yeast CYC3 deficiency identifies an X-linked mammalian activator of apocytochrome c. *Genomics* **2002**, *79*, 51–57. [CrossRef]
289. van Rahden, V.A.; Rau, I.; Fuchs, S.; Kosyna, F.K.; de Almeida, H.L.; Fryssira, H.; Isidor, B.; Jauch, A.; Joubert, M.; Lachmeijer, A.M.A.; et al. Clinical spectrum of females with HCCS mutation: From no clinical signs to a neonatal lethal form of the microphthalmia with linear skin defects (MLS) syndrome. *Orphanet J. Rare Dis.* **2014**, *9*. [CrossRef]
290. Indrieri, A.; Conte, I.; Chesi, G.; Romano, A.; Quartararo, J.; Tatè, R.; Ghezzi, D.; Zeviani, M.; Goffrini, P.; Ferrero, I.; et al. The impairment of HCCS leads to MLS syndrome by activating a non-canonical cell death pathway in the brain and eyes. *EMBO Mol. Med.* **2013**, *5*, 280–293. [CrossRef]
291. Luttik, M.A.; Overkamp, K.M.; Kötter, P.; de Vries, S.; van Dijken, J.P.; Pronk, J.T. The *Saccharomyces cerevisiae* NDE1 and NDE2 genes encode separate mitochondrial NADH dehydrogenases catalyzing the oxidation of cytosolic NADH. *J. Biol. Chem.* **1998**, *273*, 24529–24534. [CrossRef]
292. Fassone, E.; Rahman, S. Complex I deficiency: Clinical features, biochemistry and molecular genetics. *J. Med. Genet.* **2012**, *49*, 578–590. [CrossRef]
293. Forte, M.; Palmerio, S.; Bianchi, F.; Volpe, M.; Rubattu, S. Mitochondrial complex I deficiency and cardiovascular diseases: Current evidence and future directions. *J. Mol. Med. Berl. Ger.* **2019**, *97*, 579–591. [CrossRef] [PubMed]

294. Perry, C.N.; Huang, C.; Liu, W.; Magee, N.; Carreira, R.S.; Gottlieb, R.A. Xenotransplantation of Mitochondrial Electron Transfer Enzyme, Ndi1, in Myocardial Reperfusion Injury. *PLoS ONE* **2011**, *6*, e16288. [CrossRef] [PubMed]
295. Kerscher, S.; Dröse, S.; Zwicker, K.; Zickermann, V.; Brandt, U. Yarrowia lipolytica, a yeast genetic system to study mitochondrial complex I. *Biochim. Biophys. Acta BBA Bioenerg.* **2002**, *1555*, 83–91. [CrossRef]
296. Kerscher, S.; Grgic, L.; Garofano, A.; Brandt, U. Application of the yeast Yarrowia lipolytica as a model to analyse human pathogenic mutations in mitochondrial complex I (NADH:ubiquinone oxidoreductase). *Biochim. Biophys. Acta* **2004**, *1659*, 197–205. [CrossRef] [PubMed]
297. Gerber, S.; Ding, M.G.; Gérard, X.; Zwicker, K.; Zanlonghi, X.; Rio, M.; Serre, V.; Hanein, S.; Munnich, A.; Rotig, A.; et al. Compound heterozygosity for severe and hypomorphic NDUFS2 mutations cause non-syndromic LHON-like optic neuropathy. *J. Med. Genet.* **2017**, *54*, 346–356. [CrossRef]
298. Varghese, F.; Atcheson, E.; Bridges, H.R.; Hirst, J. Characterization of clinically identified mutations in NDUFV1, the flavin-binding subunit of respiratory complex I, using a yeast model system. *Hum. Mol. Genet.* **2015**, *24*, 6350–6360. [CrossRef]
299. Maclean, A.E.; Kimonis, V.E.; Balk, J. Pathogenic mutations in NUBPL affect complex I activity and cold tolerance in the yeast model Yarrowia lipolytica. *Hum. Mol. Genet.* **2018**, *27*, 3697–3709. [CrossRef]
300. Parey, K.; Haapanen, O.; Sharma, V.; Köfeler, H.; Züllig, T.; Prinz, S.; Siegmund, K.; Wittig, I.; Mills, D.J.; Vonck, J.; et al. High-resolution cryo-EM structures of respiratory complex I: Mechanism, assembly, and disease. *Sci. Adv.* **2019**, *5*. [CrossRef]
301. Parey, K.; Brandt, U.; Xie, H.; Mills, D.J.; Siegmund, K.; Vonck, J.; Kühlbrandt, W.; Zickermann, V. Cryo-EM structure of respiratory complex I at work. *eLife* **2018**, *7*. [CrossRef]
302. Wood, V.; Lock, A.; Harris, M.A.; Rutherford, K.; Bähler, J.; Oliver, S.G. Hidden in plain sight: What remains to be discovered in the eukaryotic proteome? *Open Biol.* **2019**, *9*. [CrossRef]

Publisher's Note: MDPI stays neutral with regard to jurisdictional claims in published maps and institutional affiliations.

 life

Review

Analysis of Human Mutations in the Supernumerary Subunits of Complex I

Quynh-Chi L. Dang, Duong H. Phan, Abigail N. Johnson, Mukund Pasapuleti, Hind A. Alkhaldi, Fang Zhang and Steven B. Vik *

Department of Biological Sciences, Southern Methodist University, Dallas, TX 75287, USA; quynhchid@smu.edu (Q.-C.L.D.); lisap@smu.edu (D.H.P.); abbyjohnson@smu.edu (A.N.J.); mpasapuleti@smu.edu (M.P.); halkhaldi@smu.edu (H.A.A.); fangz@smu.edu (F.Z.)

* Correspondence: svik@smu.edu

Received: 30 October 2020; Accepted: 16 November 2020; Published: 20 November 2020

Abstract: Complex I is the largest member of the electron transport chain in human mitochondria. It comprises 45 subunits and requires at least 15 assembly factors. The subunits can be divided into 14 "core" subunits that carry out oxidation–reduction reactions and proton translocation, as well as 31 additional supernumerary (or accessory) subunits whose functions are less well known. Diminished levels of complex I activity are seen in many mitochondrial disease states. This review seeks to tabulate mutations in the supernumerary subunits of humans that appear to cause disease. Mutations in 20 of the supernumerary subunits have been identified. The mutations were analyzed in light of the tertiary and quaternary structure of human complex I (PDB id = 5xtd). Mutations were found that might disrupt the folding of that subunit or that would weaken binding to another subunit. In some cases, it appeared that no protein was made or, at least, could not be detected. A very common outcome is the lack of assembly of complex I when supernumerary subunits are mutated or missing. We suggest that poor assembly is the result of disrupting the large network of subunit interactions that the supernumerary subunits typically engage in.

Keywords: mitochondria; mammalian complex I; NADH dehydrogenase; complex I assembly; complex I structure; complex I deficiency; supernumerary subunits; electron transport chain; mitochondrial dysfunction; Leigh syndrome

---

1. Introduction

Mutations in the genes that encode complex I are responsible for a large fraction of all mitochondrial diseases. For example, 20–30% of cases in childhood mitochondrial disease (MD) are related to complex I dysfunction [1,2]. There seem to be many reasons for this. Complex I genes make up a large component of the mitochondrial and nuclear genome. Complex I is encoded by seven mitochondrial genes (out of 13 total) and 37 nuclear genes. There are also at least 15 complex I assembly factors [3]. Mitochondrial DNA (mtDNA) is especially prone to mutation due to insufficient DNA repair systems. All of the various functions of complex I can be impacted by mutation. Complex I forms supercomplexes with other members of the respiratory chain, such as complex III (cytochrome bc_1) and complex IV (cytochrome *c* oxidase). It is metabolically linked to the citric acid cycle by NADH and to ATP synthesis by proton translocation. Furthermore, it is a major site of superoxide generation in mitochondria. Therefore, mutations that alter or degrade complex I function will typically have wider effects on mitochondrial function.

Complex I is a boot-shaped multi-subunit enzyme embedded in the inner mitochondrial membrane (see Figure 1). Its primary role is to oxidize NADH while reducing ubiquinone and translocating protons across the membrane. One arm extends into the matrix space, and it contains the flavin mononucleotide (FMN) and all of the iron–sulfur (FeS) clusters necessary for electron transfer to ubiquinone. The membrane arm contains subunits that translocate the protons from the matrix space to the intermembrane space (IMS). These core functions are carried out by the fourteen "core" subunits that appear in all known examples of complex I, including bacteria. Seven of the core subunits are membrane-embedded, and these seven are all encoded in mtDNA: ND1, ND2, ND3, ND4, ND4L, ND5, and ND6. The other seven core subunits are found in the matrix arm and contain the FMN and all of the FeS clusters. NDUFV1 contains the flavin and one tetranuclear FeS cluster, N3. NDUFV2 contains one binuclear FeS cluster N1a, which is not on the main electron transfer path. NDUFS1 contains one binuclear FeS cluster, N1b, and two tetranuclear clusters, N4 and N5. NDUFS8 contains two ferredoxin-like clusters, N6a and N6b. NDUFS7 contains the tetranuclear cluster N2, which is proximal to the ubiquinone binding site. For recent reviews of complex I, see [4–7]; for supernumerary subunits, see [8].

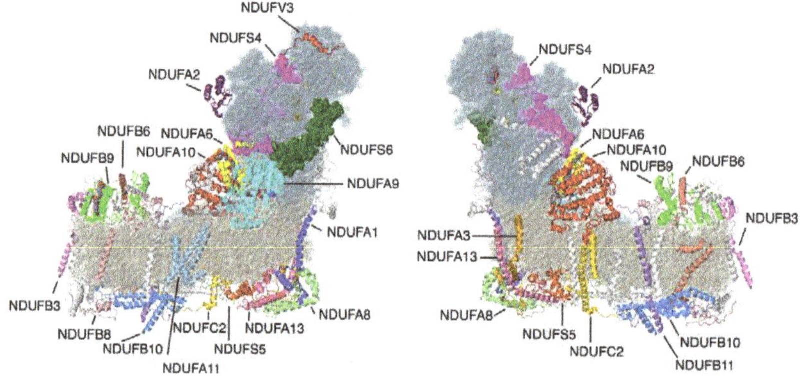

Figure 1. Structure of complex I with supernumerary subunits highlighted. Core subunits in the matrix arm are colored light blue. Core subunits in the membrane arm are colored beige. Supernumerary subunits that are described in this review are colored and labeled. Other supernumerary subunits are white. The two views are rotated 180° relative to each other. The structure is from the Protein Data Bank file 5xtd (PDB id = 5xtd) [9]. All structural images were generated using Jmol (http://www.jmol.org).

The remaining thirty-one subunits (one is found in two copies) are supernumerary (or accessory) subunits, and much less is known about their functions. They are typically much smaller than the core subunits, and they are distributed on all surfaces of complex I. Some cross the membrane, while others are localized to the matrix face or the IMS. The naming of these subunits has generally followed their co-purification with various fractions of complex I: FV for the flavoprotein fraction, FS for the FeS protein fraction, FA for the alpha fraction associated with the matrix arm subunits, and FB for the beta fraction associated with the membrane proteins. The exception is NDUFAB1, the acyl carrier protein, which resembles an enzyme in lipid biosynthesis. This subunit appears to have an essential role apart from complex I, and it is the only protein that appears in two copies.

From an evolutionary point of view, the core subunits can be organized into three modules. The N-module contains NDUFV1, NDUFV2, and NDUFS1, and it is defined by the source of electrons to the complex, the substrate NADH. This module is related to various NAD-linked dehydrogenases. The Q-module contains the remaining peripheral subunits NDUFS2, NDUFS3, NDUFS7, and NDUFS8, as well as two membrane subunits, ND1 and ND3, that contain the remaining FeS clusters and contribute to the ubiquinone binding site. This module is related to various membrane-bound hydrogenases. Finally, the remaining membrane subunits, ND2, ND4, ND4L, ND5, and ND6 compose the P-module (for proton translocating) and are related to subunits of an Na^+/H^+ antiporter, the Mrp complex [10]. This grouping of subunits also corresponds to the assembly pathway. Q-module subunits appear to assemble first, followed by the stepwise addition of the P-module, associated with the three major subunits ND2, ND4, and ND5. ND5 and the N-module enter the complex last.

In this review, 20 supernumerary subunits for which clinical mutations have been identified are described (see Table 1). Some of the mutations are interpreted in light of the structure of the human complex I (Protein Data Bank file 5xtd) or, in some cases, from other species. Some of the mutations are likely to be null mutations in which no protein is made, but the evidence is not always clear for that. Currently, the understanding of the effects of mutations is limited by a lack of knowledge. For example, how do mutations in one subunit affect the stability or import of that subunit? How does the absence of one subunit affect the expression, import, or stability of other subunits? How do mutations affect the assembly of that subunit or of other subunits? Do supernumerary subunits have roles that can be affected by mutation while assembly remains normal?

Table 1. Clinical missense mutations in supernumerary subunits and their surface contacts. IMS: intermembrane space.

Location	Subunit	Mutated Residues	Primary Contacts
N-module	NDUFA2	p. Lys45Thr p. Glu57Ala p. Asp50Asn + others *	NDUFS1
	NDUFV3	p. Arg26Gln p. Lys56Asn p. Gly103Asp p. Glu276Lys	NDUFV1, NDUFV2
Q-module	NDUFS4	p. Trp114Arg p. Asp119His + others *	NDUFS1, NDUFS3, NDUFS8, NDUFV1, NDUFA6, NDUFA9
	NDUFS6	p. Cys115Tyr + others *	NDUFA9, NDUFA12, NDUFS1, NDUFV2, NDUFS2, NDUFS8
	NDUFA9	p. Arg321Pro p. Arg360Cys	NDUFS3, NDUFS7, NDUFS4, NDUFS6, NDUFA6
ND1- module	NDUFA1	p. Gly8Arg p. Arg37Ser p. Gly32Arg p. Pro19Ser	ND1, NDUFA8
	NDUFA3	Deletion of one allele	ND1, ND3, NDUFS8, NDUFA8, NDUFA13
	NDUFA13	p. Arg57His	ND1, ND6, NDUFS2, NDUFS5, NDUFA3, NDUFA8
LYR proteins	NDUFA6	p. Arg64Pro + others *	NDUFAB1, NDUFS1, NDUFS3, NDUSA9
	NDUFB9	p. Leu64Pro p. Arg47Leu	NDUFB1, NDUFB3, NDUFB4, NDUFB5, NDUFB6
IMS proteins	NDUFS5	p. Pro96Ser	ND2, ND4L, NDUFB5, ND6, NDUFB13
	NDUFA8	p. Arg47Cys p. Glu109Lys p. Arg135Gln	NDUFA13, NDUFA3, NDUFA1
	NDUFB10	Glu79 p. Cys107Ser	NDUFB5, NDUFB6, NDUFB11
ND2-module	NDUFA10	p. Gln142Arg p. Leu294Pro p. Gly99Glu + others	ND2, NDUFS2
	NDUFC2	p. His58Leu + others *	ND2, NDUFC1, NDUFB5, NDUFA8

Table 1. *Cont.*

Location	Subunit	Mutated Residues	Primary Contacts
ND4-module	NDUFA11	p. Ala132Pro p. Thr106Ile (isoform two) + others *	ND2, ND4, ND5, NDUFB5
	NDUFB11	p. Arg134Ser p. Glu219Lys + others	ND4, NDUFB10
ND5-module	NDUFB3	p. Trp22Arg + others	NDUFB9, NDUFAB1, NDUFB2, ND5
	NDUFB6	Promoter mutation	NDUFB5, NDUFAB1, NDUFB9, ND5, NDUFB10, NDUFB7
	NDUFB8	p. Pro76Gln p. Cys144Trp p. Tyr62His + others *	ND5, NDUFB4, ND4, NDUFAB1, NDUFB7, NDUFB9, NDUFB10

* Other mutations include nonsense mutations, splicing mutations, start codon mutations, insertions, and deletions. Mutations are defined by the change in the protein (p.) and the amino acid substitution, Lys45Thr, lysine at position 45 changed to threonine. See text for more details.

2. Review of Mutations in the Supernumerary Subunits

2.1. N-Module Subunits

The first subunits to be described are two subunits that primarily have contacts to core subunits of the N-module (see Supplementary File Table S1 for a Table of Subunit Interactions). Overall, they are rather isolated from other supernumerary subunits, as shown in Figure 2. Beyond those similarities, their properties appear divergent.

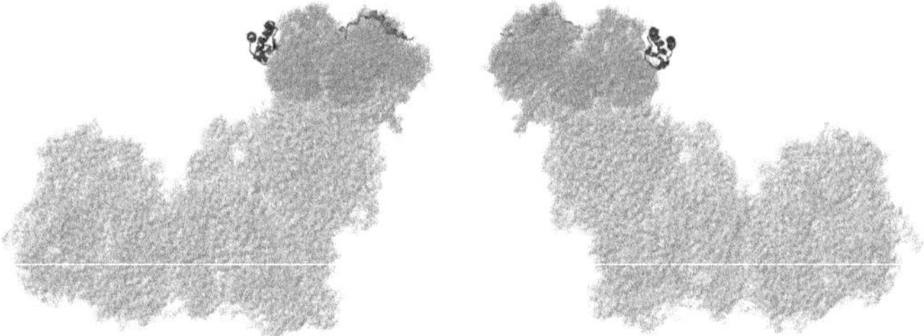

Figure 2. Location of NDUFA2 and NDUFV3. Most of Complex I is colored gray. Core subunits in the N-module arm are colored light blue (NDUFV1, NDUFV2, and NDUFS1). NDUFA2 and NDUFV3 are shown in ribbons, with NDUFA2 colored purple and NDUFV3 colored red. The two views are rotated 180° relative to each other. The structure is from PDB id = 5xtd [9].

2.1.1. NDUFA2

The gene for NDUFA2 is located on chromosome 5. The protein contains 99 amino acids with a mass of about 10.9 kDa before the processing of the amino-terminal methionine. This subunit is bound to the matrix arm in a unique fashion, with exclusive contacts to core subunit NDUFS1. It is bound to C-terminal domains of NDUFS1, and it is at least 40 Å from any of the FeS clusters. It has a compact structure that resembles a thioredoxin fold and includes a flat four-stranded beta-sheet (see Figure 3). Two Cys residues, 24 and 58, are found in the loops of the beta-sheet, but they do not appear to be oxidized in the existing structures of mammalian complex I. Contact with NDUFS1 is primarily with the beta-sheet region, including both Cys residues. In culture, NDUFA2 knockout cell lines have been found to lack complex I activity, and an analysis by blue native (BN) gel electrophoresis showed a limited assembly of the N-module—in particular, the loss of core subunits NDUFS1, NDUFV1, and NDUFV2 and supernumerary subunits NDUFS4, NDUFS6, NDUFA7, and NDUFV3. In a cell line with an NDUFA2 knockout, complex I assembly is slightly affected, with a prominent band in BN gel electrophoresis but apparently missing some subunits, probably the N-module. Complex I activity has been found to be extremely low, consistent with a missing NADH binding site [11].

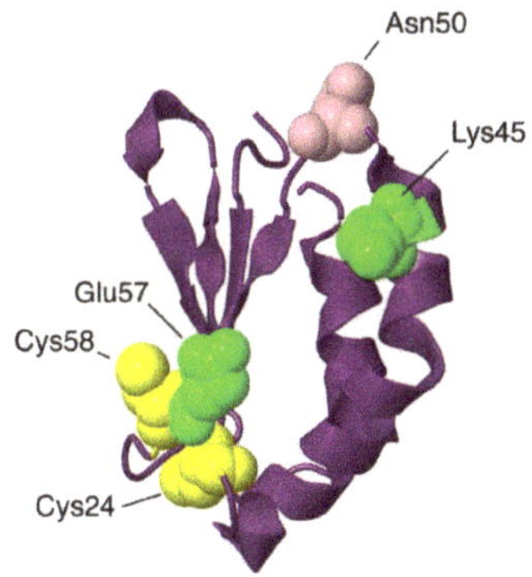

Figure 3. Structural features of NDUFA2 from the N-module. The protein is portrayed in a purple-colored ribbon. Sites of mutations are shown in space-filling views: Lys45Thr and Glu57Ala are mitochondrial disease mutations (colored lime). Asn50Asp has been found in breast cancer patients (colored pink). Two Cys residues are colored yellow.

An individual with Leigh syndrome and hypertrophic cardiomyopathy was discovered to be homozygous with a mutation, c.208 + 5 G > A, in the NDUFA2 gene that caused a reduction in correct splicing [12]. Normally, NDUFA2 is coded by three exons, but in this individual, most of the transcripts lacked exon 2, causing a frameshift. No protein or transcript was found, making it a null mutation. The patient died of cardiovascular arrest at eleven months.

Two individuals with leukoencephalopathy were discovered to have mutations in NDUFA2 [13]. One was homozygous for p.Lys45Thr, and the other had compound heterozygous mutations, p.Lys45Thr and a deletion at Asn76, c.225del, and p.Asn76Metfs*4, thus causing a frameshift and a stop codon after 4 codons. The first was found to have complex I deficiency and lacked an assembled complex I in BN gel electrophoresis. This individual also had a systemic deficiency in carnitine due to a homozygous mutation in SLC22A5. She developed focal epilepsy at six years, was wheelchair-bound at nine years, and was last evaluated at 12 years. The second patient had a similar presentation, including abnormal white matter in the brain. She also had movement difficulties and was last evaluated at age four. Lys45 is found at the interface with NDUFS1 and has close contacts with several NDUFS1 residues, including Gly376, Asp380, and Ser672. It is likely that the binding of NDUFA2 to NDUFS1 is destabilized.

In 2020 [14], another individual was described with a homozygous mutation in NDUFA2, p.Glu57Ala. This residue is at the interface with NDUFS1 and makes a possible ion pair with Arg382 of NDUFS1. It also makes nonbonding contacts with Gly661, Ala662, Asn663, Tyr664, Leu381, Arg382, and Ser383, suggesting a likely disruption of binding. This individual had abnormal white matter in the brain, microcephaly, seizures, and movement disorders. She was last evaluated at four years of age.

In a screen of breast cancer patients [15], a homozygous p.Asp50Asn mutation in NDUFA2 was detected in one individual. This residue is not at the interface with NDUFS1, and so it perhaps does not disrupt the assembly of the enzyme. The significance of this finding is uncertain.

2.1.2. NDUFV3

The gene for NDUFV3 is found on chromosome 21. This was the first subunit of human complex I known to have multiple isoforms [16–18], which can be found by alternative splicing. It is not an essential protein, as shown by knockout studies in cell culture [11]. The loss of this gene does not prevent the assembly of complex I, as seen in BN gels, and cells retain the ability to grow in a galactose medium, indicative of oxidative phosphorylation activity. Exogenous NDUFV3 has been seen to exchange with fully assembled complex I, suggesting that it can be incorporated last into the enzyme. The more common isoform has 108 amino acids (74 after the cleavage of the transit peptide) and a mass of 11.8 kDa, while the second isoform has 473 amino acids and a mass of about 51 kDa. Isoform two has an additional exon between the second and the last exons of the shorter form. The additional exon contains a serine-rich region that was found to be phosphorylated [19]. NDUFV3 is not present in fungal complex I from *Yarrowia lipolytica* [7].

NDUFV3 was originally identified in the flavoprotein fraction of bovine heart complex I and now can be seen to have primary contacts with NDUFV1 and NDUFV2 in the human enzyme. It also has limited contact with NDUFS1 and NDUFS4. In the human model of complex I (PDB id = 5xtd) [9], only 33 amino acids are visible, corresponding to residues 74–106. Residues 82–5 are alpha-helical. The longer isoform has an insertion at residue 56 and retains the same C-terminus as isoform one. Therefore, both isoforms can likely bind in the same fashion to other complex I subunits. Both isoforms have been seen in a variety of tissues, but isoform two is more common in cultured cells and in brain tissue [16]. Others have found that isoform one is more prevalent in bovine and murine complex I of heart tissue, as well as that increased Km values for NADH consumption by complex I correlate with an increased extent of the short form in heart tissue [18]. In a cross-linking study of native mouse heart mitochondria [20], both isoforms were identified in cross-links with a surface peptide of malate dehydrogenase, found near the NAD^+ binding site. The binding of malate dehydrogenase and several other enzymes to porcine complex I was demonstrated in 1984 [21]. Other cross-links identified with NDUFV3 were with NDUFV1, NDUFV2, Cox7A1, and Atp5a1—the latter two supporting the proximity of the complexes of oxidative phosphorylation.

Sequence variations have been discovered in NDUFV3 among individuals with decreased levels of complex I activity, but none have been shown to be causative for disease. In one such cohort, four individuals with mutations in NDUFV3 were found [22]. The first individual had p.Arg26Gln, which would occur in the mitochondrial transit peptide and thus might affect import. A second mutation, was found in the same individual in the mitochondrial DNA polymerase G gene POLG, p.Gly11Asp. The second individual had the mutation p.Lys56Asn, which occurs at the splice junction and might affect expression, but this individual also carried a mutation in AMACR, an enzyme associated with branched-chain fatty acid metabolism, p.Val185Ala. The third individual had the mutation p.Gly103Asp, which would occur in both isoforms, in the region that binds to other subunits. Gly103 contacts Trp166 of NDUFS4 in a non-bonding interaction, and it is adjacent to several charged residues. Arg104 of NDUFV3 is near Arg169 of NDUFS4 and Asp426 of core subunit NDUFS1. Glu105 of NDUFV3 and Glu72 of NDUFV2 are nearby. Ser106 is also in this region, and it was found to be phosphorylated in a screen of human cancer cells [23]. Therefore the p.Gly103Asp mutation appears likely to be disruptive to the binding of NDUFV3. The fourth individual had the mutation p.Glu276Lys, which is found only in the long form, near the sites of phosphorylation. This individual also carried a p.Gly154Ser mutation in the core subunit NDUFS8, which is a buried residue, and that substitution is more likely to be deleterious.

2.2. *Q-Module Subunits*

NDUFS4, NDUFS6, and NDUFA9 are found in the Q-module, as shown in Figure 4. They interact with each other and with many other subunits in the matrix arm. They appear to have roles in assembling and stabilizing the N- and Q-modules.

2.2.1. NDUFS4

The gene for NDUFS4 is found on chromosome 5. NDUFS4 is a protein that was initially found in the iron–sulfur protein fraction of bovine heart complex I. It is synthesized as 175 amino acids, and after the cleavage of the mitochondrial transit peptide, the mature protein is 133 amino acids with a mass of about 15.3 kDa. As shown in Figure 1, it is located in the matrix arm of complex I and contacts core subunits NDUFS1, NDUFS3, NDUFS8, and NDUFV1, as well as supernumerary subunits NDUFA6 and NDUFA9. It fits between two lobes of core subunit NDUFS1. It can be understood as a globular protein with a mixed alpha/beta architecture and three extensions that reach towards the chain of FeS clusters (see Figure 5). The C-terminus, in particular, is within about 9 Å of FeS clusters N3 and N1b, found in NDUFV1 and NDUFS1, respectively. A loop ending in Met87 is within 10 Å of FeS clusters N4 and N5 of NDUFS1, and a second loop including Met112 is pointing toward the last three FeS clusters—N6a, N6b, and N2—but is at least 13 Å away. NDUFS4 has been found to be phosphorylated at residue Ser173, near the C-terminus [24]. It has been suggested that this improves the net import of the subunit by inhibiting its return to the cytoplasm. In a knockout cell line of cultured human cells, the loss of NDUFS4 had a mild effect on assembly of complex I, causing it to migrate as a smaller-than-normal-sized complex in BN gel electrophoresis, probably lacking the N-module [11].

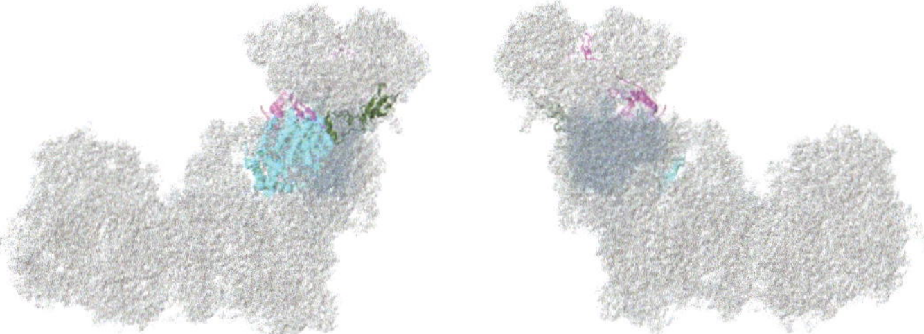

Figure 4. Location of NDUFS4, NDUFS6, and NDUFA9. Most of Complex I is colored gray. Core subunits in the Q-module are colored light blue (NDUFS2, NDUFS3, NDUFS7, and NDUFS8). NDUFS4, NDUFS6, and NDUFA9 are shown in ribbons, with NDUFS4 colored magenta, NDUFS6 colored green, and NDUFA9 colored cyan. The two views are rotated 180° relative to each other. The structure is from PDB id = 5xtd [9].

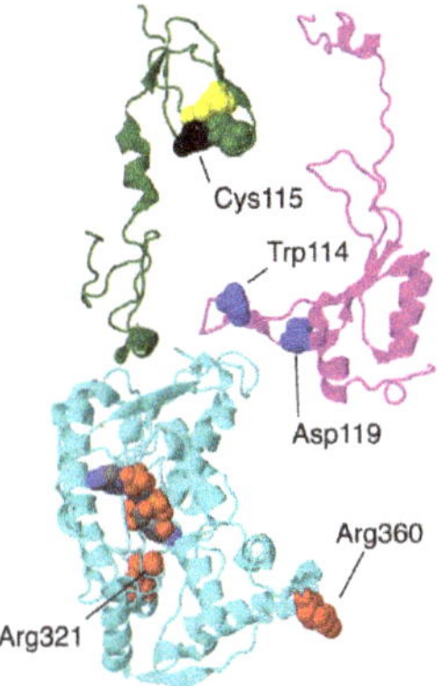

Figure 5. Structural features of 3 interacting subunits from the Q-module. The proteins are portrayed in ribbons. NDUFS4 is colored magenta. The sites of two mutations in this subunit, Trp114Arg and Asp119His, are shown in space-filling and are colored blue. NDUFS6 is colored green. The zinc bind residues are shown in space-filling, with Cys colored yellow and His colored green. The site of the mutation, Cys115Tyr is colored black. NDUFA9 is colored cyan, and its bound ligand $NADP^+$ is shown in space-filling with CPK colors (e.g., carbon gray, nitrogen blue, oxygen red, phosphorus orange). The sites of 2 mutations, Arg321Pro and Arg360Cys, are shown in space-filling and are colored red.

Three human mutations in NDUFS4 from Leigh syndrome patients were examined biochemically [25]: a duplication of AAGTC at position 466–470 of the coding sequence, a single G deletion at position 289 (in Trp97), and a nonsense mutation, c.44 G > A, p.Trp15X, in the first exon of the gene. Nonsense-mediated decay apparently eliminated the transcript in the second case. In the third case, c.44 G > A, it was later determined that three alternative splice variants were produced [26]. In all three cases, the mutations were homozygous, and it was demonstrated that little or no protein was made. An analysis of these null mutants from cultured fibroblasts showed that complex I did not assemble by BN gel electrophoresis and that little or no activity could be measured [25].

Two clinical missense mutations have been identified in this subunit, p.Trp114Arg [27] and p.Asp119His [28], both resulting in Leigh syndrome. Several frameshift mutations have been identified, including the homozygous c.221delC (p.Thr74Ifs*17) and the compound heterozygous mutations c.462delA (p.Lys154Nfs*34) and c.99-1 G > A) (p.Ser34Ifs*4), which also appear to cause Leigh syndrome [22]. Both missense mutations are found in the loop nearest to the N2/N6ab FeS clusters. Perhaps more importantly, this extension of NDFS4 is part of a junction of three subunits, with NDUFA9, the $NADP^+$ binding protein, and an extension of core subunit NDUFS3. The individual with the p.Trp114Arg mutation was found to be homozygous for this allele, while both parents were found to be heterozygous [27]. This individual was diagnosed with myocarditis, respiratory failure, delirium, and basal ganglia abnormalities in the brain as seen by magnetic resonance imaging (MRI), and she was found to have complex I deficiency. Trp114 is a conserved residue that has close contact with Gln228 of NDUFS3 and is near several residues of core subunit NDUFS8. In the human structure (PDB id = 5xtd), the glutamine oxygen from the sidechain is pointing into the tryptophan ring, but it seems that the amino group would more logically assume that position. In conclusion, it seems likely that the p.Trp114Arg mutation would be disruptive and might impact the transfer of electrons through complex I.

The individual with the p.Asp119His mutation was compound heterozygous, with a second allele containing a Lys154 frame shift mutation, and he died in his third year [28]. While these mutations were found in a heterozygous situation (mother, father, and brother), there was no clinical presentation, suggesting that both mutations contributed to declining health. In a different study, the transcript for the frameshift allele was shown to exist, indicating that nonsense-mediated decay did not occur and suggesting that the protein was made. Asp119 is primarily contacted by other residues in NDUFS4, including an ion pairing with Arg75. The closest contact with subunit NDUFA9 is residue Lys45 at a distance of over 5 Å. Therefore, it is likely that the Asp119His substitution would be deleterious. If we assume that the frameshift mutation would produce a protein that is truncated at residue 154, that would remove a small but highly conserved domain that normally interacts with NDUFV3, and with core subunits NDUFV1, NDUFV2, and NDUFS1, in the vicinity of the FeS clusters N3 and N1b. Even this loss of about 20 amino acids would appear to be disruptive, and it seems possible that the truncated protein might be found at lower-than-normal levels due to instability.

2.2.2. NDUFS6

The gene for NDUFS6 is located in chromosome 5 and contains four exons. The protein has 124 amino acids before the N-terminal 28 are cleaved upon entry to the mitochondrion. The original mass is 13.7 kDa. The protein is bound to the matrix arm and has two domains connected by a short alpha-helix, residues 67–75 (see Figure 5). The N-terminal domain, residues 29–66, is closer to the membrane and primarily contacts subunits NDUFA9 and NDUFA12. The C-terminal domain, residues 76–124, contains several short beta-strands and is in contact with core subunits NDUFS1, NDUFV2, NDUFS2, and NDUFS8. A zinc ion is bound by three cysteines and one histidine that are part of a $C–X_8–H–X_7–C–X_2–C$ motif with Cys87, Cys112, Cys115, and His96. The significance of the bound zinc is not clear, but this subunit is found not only in the fungal *Y. lipolytica* enzyme [29,30] but also in some bacterial enzymes (e.g., *Paracoccus denitrificans*) [31]. In cell culture, the knockout of NDUFS6 results in a mild assembly defect in which complexes of somewhat smaller size with a reduced oxidative phosphorylation capacity are seen [11]. In *Y. lipolytica*, the knockout of the NDUFS6 gene results in a complex lacking NDUFA12 and the N4 FeS cluster. Complementation with constructs lacking one of the zinc binding residues failed to regenerate the wild-type activity, suggesting a role in the incorporation of this FeS cluster [29].

The first mutations associated with NDUFS6 were described in 2004 [32]. Three individuals presented with lethal neonatal mitochondrial complex I deficiency and died within days of birth. Two were siblings with homozygous mutations, c.186 + 2 T > A, which affected splicing, leading to 26 bp of intronic RNA incorporated at the exon 2/3 junction, causing a frameshift and a protein predicted to be 71 amino acids. A small amount (3%) of normally spliced transcript was also found. The third individual was also homozygous, with a deletion of 4.75 kb that removed exons 3 and 4. This individual died five days after birth.

A mutation in one of the residues that bind zinc, p.Cys115Tyr, was found in two different families [33]. The affected individuals, two from each family, presented with fatal neonatal lactic acidemia and died within a few days of birth. All were homozygous for this allele.

Two additional mutations in NDUFS6 were identified, first in 2017 [34] and then again in 2019 [35] when a more extensive analysis was done. Both were characterized as having Leigh syndrome, and the individuals were compound heterozygous, carrying both mutant alleles, c.309 + 5 G > A and c.343 C > A (p.Cys115Arg). The individual analyzed in 2019 lived for eleven months. It was determined that the splicing error led to a transcript missing exon 3, but that a small amount of normally spliced form was present. Complex I assembly was very limited, as indicated by BN gel electrophoresis. The mutation of Cys115 to Arg or Tyr could disrupt the folding of the protein and its binding to the matrix arm, which could have led to the loss of the FeS cluster N4, as was shown to happen in the fungal *Y. lipolytica* enzyme [29].

2.2.3. NDUFA9

The gene for NDUFA9 is located on chromosome 12. It is encoded as a protein of 377 amino acids, about 42.5 kDa, but the N-terminal 35 amino acids are cleaved upon entry to the mitochondrion. The protein contains two domains, one being a Rossmann fold with an $NADP^+$ bound to the C-terminal ends of the parallel beta-strands of the central sheet. A second domain, which contacts the $NADP^+$, is largely alpha-helical. The second domain is largely C-terminal, and the Rossmann fold is largely N-terminal, except for two helices have been domain-swapped (see Figure 5). This protein is related to a family of short chain dehydrogenases, as first identified by Fearnley and Walker [36]. It is also one of the proteins that undergoes conformational changes in the active/deactive transition, as seen in mouse (PDB id = 6g2j, 6g72) [37] and sheep open and closed structures [38]. NDUFA9 is found at the junction of the Q-module and the membrane arm. It primarily has contact with core subunits NDUFS3 and NDUFS7, as well as with NDUFS4, NDUFS6, and NDUFA6. It also has minor contact with the core subunits ND1, ND3, ND6, NDUFS1, and NDUFS8. There is no evidence that this subunit exhibits any enzyme activity.

A human cell line was established with a knockout of NDUFA9 using transcription activator-like effector nucleases (TALEN) technology [39]. A clear assembly defect was discovered in which an 880 kDa complex was transiently seen, but a more stable 600 kDa complex was later found. The latter complex had the membrane arm subunits and some of the Q-module subunits of the matrix arm. In a more comprehensive study of all supernumerary subunits [11], the same group later established that levels of N-module subunits decreased the most in cell lines with the NDUFA9 knockout. The offered interpretation was that NDUFA9 is important for stabilizing the binding of the N-module to the Q-module during assembly. The complex seen in cells with the NDUFA9 knockout was smaller than those seen when NDUFS4 or NDUFS6 were knocked out, indicating that the Q-module was also destabilized.

Two missense mutations have been discovered in patients. The first was p.Arg321Pro, a homozygous mutation found in a boy of consanguineous parents [40]. He had complications in respiration after birth, with lactic acidosis and vision and hearing loss. He died after one month of respiratory failure, and it was classified as Leigh syndrome. An analysis of cultured fibroblasts showed very low levels of NDUFA9, and several other subunits, as well as essentially no assembled complex I in BN gel electrophoresis. Arg321 is found in the $NADP^+$ binding pocket near the nicotinamide end but not in contact with it. It does not contact any other subunits, as it is completely buried. Therefore, it is likely that the substitution to Pro would disrupt the folding of the subunit and render it unable to properly bind to the assembling complex I.

The second discovered mutation also involved a homozygous Arg mutation, p.Arg360Cys [41]. This patient first exhibited symptoms of dystonia at age seven, leading to a loss of speech and becoming wheelchair-bound. His conditions stabilized in adulthood. A biochemical analysis did not reveal any metabolic disease, but an MRI of his brain revealed atrophy consistent with Leigh syndrome. An analysis of fibroblast mitochondria revealed reduced levels of complexes I and IV by BN gel electrophoresis, as well as reduced levels of several complex I subunits by immunoblotting. Complex I activity was 17–61% of control samples. Complementation with wild-type NDUFA9 confirmed that this mutation was causative for the reduced complex I activity. Arg360 is near the C-terminus of the subunit and is sandwiched between Trp361 of NDUFA9 and Tyr78 of ND6, a core subunit. A similar arrangement is seen in the mouse structure of the active conformation [37], except that Glu80 of ND6 is also involved. This junction between NDUFA9 and ND6 appears to be disrupted in the deactive conformation, as shown in the mouse structure [37] in which the C-terminus of NDUFA9 is not visible and is perhaps disordered. A similar situation has been seen in the higher resolution structures from sheep [38] in which the loops between NDUFA9 and ND6 are formed in the closed structures (PDB id = 6zko and 6zkc), while these loops, including Arg360 are disordered in the open structures (PDB id = 6zkd, 6zke, 6zkf, 6zkp, and 6zkr). Therefore, it is possible that even if the p.Arg360Cys subunit can assemble, it might be less able to undergo conformational changes during turnover, and so activity would decrease.

2.3. ND1-Module Subunits

Three membrane subunits from the "heel" of complex I are described next: NDUFA1, NDUFA3, and NDUFA13, as shown in Figure 6. These subunits are part of the ND1-module for assembly. Each crosses the membrane once. NDUFA3 interacts with the transmembrane domain of NDUFA13, while NDUFA1 interacts with NDUFA13 in the IMS.

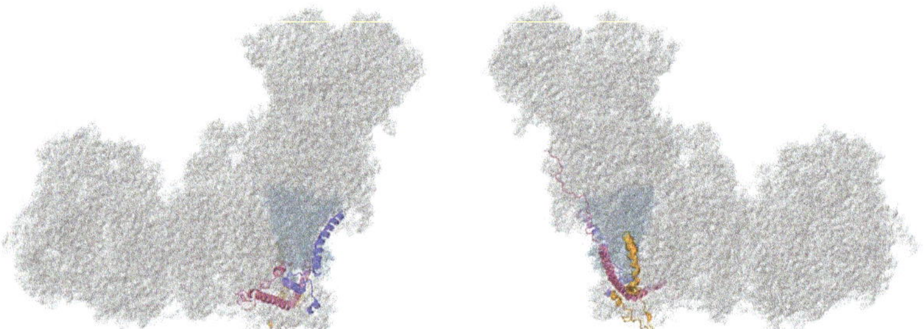

Figure 6. Location of NDUFA1, NDUFA3, and NDUFA13. Most of Complex I is colored gray. Core subunits in the ND1-modules are colored light blue (ND1). NDUFA1, NDUFA3, and NDUFA13 are shown in ribbons, with NDUFA1 colored blue, NDUFA3 colored orange, and NDUFA13 colored pink. The two views are rotated 180° relative to each other. The structure is from PDB id = 5xtd [9].

2.3.1. NDUFA1

The NDUFA1 gene is located on the X chromosome. The NDUFA1 subunit, also known as MWFE, contains 70 amino acids and has a mass of 8.0 kDa. It is a single-pass transmembrane protein that lies at the junction of the membrane and matrix arms of complex I. The N-terminus lies at the matrix surface. The protein consists of two separate domains: residues 1–31 form an alpha-helix that is situated in a groove between the first and seventh transmembrane helices of core subunit ND1. The C-terminal domain lies in the IMS and contains an alpha-helix (residues 42–56) and a short 3–10 helix (residues 65–70). This domain primarily contacts NDUFA8 (see Figure 7). A Ser residue is phosphorylated at position 55, but the significance of this modification remains unknown [42]. In a knockout strain of cultured human cells, the loss of NDUFA1 resulted in a reduced expression of complex I, and it migrated as a smaller-than-normal-sized complex in BN gel electrophoresis [11]. NDUFA1 mainly contacts core subunit ND1 and supernumerary subunit NDUFA8, but it also weakly contacts core subunit ND6 and supernumerary subunits NDUFS5, NDUFS8, and NDUFA13.

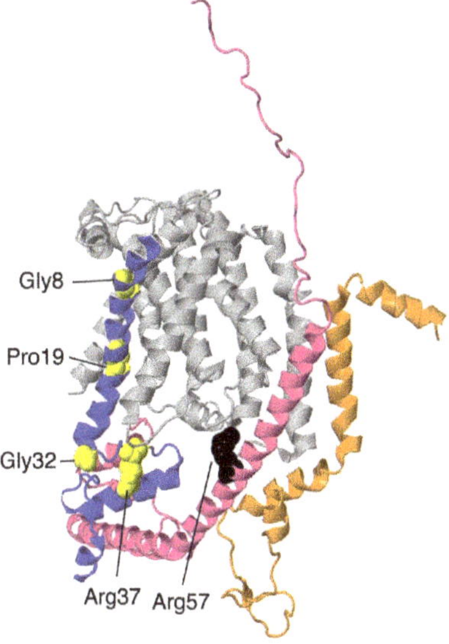

Figure 7. Structural features of ND1-module subunits. The proteins are portrayed in ribbons. ND1 is in the background and colored light gray. NDUFA1 is colored blue with the sites of 4 mutations, Gly8Arg, Pro19Ser, Gly32Arg, and Arg37Ser, shown in space-filling and colored yellow. NDUFA3 is shown in orange. No point mutations have been discovered yet. NDUFA13 is colored pink. The site of one mutation, Arg57His, is shown in space-filling and colored black.

Four clinical mutations have been identified in this subunit: p.Gly8Arg, p.Arg37Ser, p.Gly32Arg, and p.Pro19Ser, and so would be hemizygous in males. The p.Gly8Arg mutation was identified in two half-brothers diagnosed with Leigh syndrome who both died in infancy [43]. The brothers had different biological fathers but shared a mother who carried the p.Gly8Arg mutation. The older brother developed psychomotor retardation at nine months and generalized hypotonia and choreoathetosis. At 19 months of age, brainstem lesions were observed, and the patient died from cardiorespiratory failure. The younger brother presented with axial hypotonia, vertical rolling nystagmus, choreoathetosis, and bilateral lesions. He developed respiratory insufficiency at 13 months of age and died at 14 months of age from cardiorespiratory arrest. The family pedigree shows that three maternal uncles had died of an unknown disease. The severity of the disease is reflected in the structural importance of the Gly8 residue on NDUFA1. Gly8 lies in the conserved hydrophobic N-terminal region of NDUFA1. Gly8 is in loose contact with residues Thr23, Lys26, and Leu43 in core subunit ND1. The substitution of the glycine for an arginine would be disruptive to interactions in this region.

The p.Arg37Ser mutation was identified in a boy who was diagnosed with generalized hypotonia, myoclonic epilepsy, and cerebellar atrophy [43]. His clinical evolution stabilized, and he was still living at 10 years old. His mother was a heterozygous carrier for the p.Arg37Ser mutation. BN gel electrophoresis demonstrated a low level of complex I formation, and muscle cells showed 15–30% of the normal level of complex I activity. The loss of a positive charge may disrupt the local structure or affect the phosphorylation of the subunit at Ser55. Arg37 contacts two residues from supernumerary subunit NDUFA8: it forms a hydrogen bond with Ser22 and is in contact with Gln92. It is also located near Asp90 and Gly93 from NDUFA8. Furthermore, according to mouse active (PDB id-6g2j) and deactive (PDB id = 6g72) structures [37], the residue may engage in different intermolecular interactions depending on the state of complex I, although this was not seen when comparing open and closed structures from the ovine enzyme [38].

The p.Gly32Arg mutation has been well-studied. Three separate clinical studies have been performed on this mutation. The first study [44] focused on two male patients who were maternal cousins from the same healthy non-consanguineous family. One patient experienced deterioration of motor and verbal skills at age four and an unsteady gait, retinitis pigmentosa, and cerebellar atrophy at age seven. His cousin developed an ataxia and proximal muscle weakness at age five and myoclonic seizures and bilateral sensorineural hearing loss at age 10. Both probands had mothers who were heterozygous carriers. Both men had lived into their thirties when the study was conducted. A six-year-old boy was also found to carry the p.Gly32Arg mutation [45]. Though the age of onset was not recorded, he experienced episodic neuroregression and encephalopathy but was seizure-free. His family pedigree was unknown. This finding is in concordance with the fact that the p.Gly32Arg mutation is not as deleterious as the p.Gly8Arg mutation. In 2011, a female patient with a heterozygous p.Gly32Arg mutation was identified [46]. She did not show signs of deficiency until 11 months of age, when she developed an X-linked respiratory chain deficiency in skeletal muscle tissue. She developed frequent upper airway infections and experienced somnolence and muscle hypotonia during these illnesses. However, at age five, she showed nearly normal psychomotor development. Though there was only a 25% expression of the normal allele in her skeletal muscle tissue, she showed a relatively mild clinical phenotype. It is hypothesized that X-inactivation due to selection advantage may have favored the expression of the normal allele. Neither her mother nor her father had the p.Gly32Arg mutation, suggesting that the mutation arose spontaneously. The fact that all four patients carrying the p.Gly32Arg mutation survived into adulthood may be attributed to the fact that Gly32 in NDUFA1 lies in the membrane near the cytoplasmic side and is not located at the interfaces of NDUFA1 and other subunits.

In 2014, a patient diagnosed with Leigh syndrome and mitochondrial respiratory chain disorder (MRCD) was discovered to have the p.Pro19Ser mutation [47]. The age of onset was five years old, but before 10 months of age, the boy had already experienced hypotonia, nystagmus, generalized epilepsy, and high blood lactate and pyruvate levels. Pro19 in NDUFA1 is a highly conserved in vertebrates and lies in the N-terminal membrane spanning helix (Pro7–Arg28). The residue contacts Leu9, Pro12, and Met91 of core subunit ND1 in the membrane region of the bilayer, and the substitution of the polar serine for proline could be deleterious.

2.3.2. NDUFA3

The NDUFA3 gene is located on chromosome 19. The NDUFA3 protein is a 9.3 kDa protein that consists of 83 amino acids. It is a single-pass protein located in the inner mitochondrial membrane. It lies in the junction between the membrane and matrix arms of complex I and contains a kinked transmembrane alpha-helix domain that spans residues 17–35 and 37–49. During processing, the initiator methionine is cleaved, and Ala at position 2 is N-acetylated by analogy with the bovine enzyme [48]. NDUFA3 contacts core subunits ND1, ND3, and NDUFS8. Additionally, it contacts supernumerary subunits NDUFA8 and NDUFA13 (see Figure 7). In a knockout strain of cultured human cells, the loss of NDUFA3 resulted in a reduced expression of complex I and migrated as a smaller-than-normal-sized complex in BN gel electrophoresis [11]. An earlier study demonstrated the role of NDUFA3 in the assembly of the Q-module of complex I [49].

Though no clinical mutations in NDUFA3 have been identified yet, it has been shown that the full deletion of the NDUFA3 gene, along with other genes on chromosome 19, results in retinitis pigmentosa (RP), a disorder caused by photoreceptor cell degeneration. It is characterized by vision loss following night-blindness. In 2006, a family was identified as carrying a 30 kb deletion in chromosome 19 [50]. The family members afflicted with RP contained at least one copy of chromosome 19 that lacked a region containing genes NDUFA3, TFPT, PRP31, and the OSCAR promoter. The ages of onset of the RP symptoms varied from 3 to 30 years of age. Surprisingly, the patients were not afflicted by any disease other than RP. From this study, the absence of symptoms relating to complex I deficiencies suggests that only one copy of the gene for NDUFA3 may be sufficient for complex I activity.

Similarly, in 2011, another family was discovered to harbor a 112 kb deletion that encompassed the PRP31 gene and five of its upstream genes: TFPT, OSCAR, NDUFA3, TARM-1, and VSTM-1 [51]. Only one family member, a 33-year-old female, was diagnosed with RP. Once again, the RP patient showed no signs of complex I deficiency, and the authors suggested that one copy of NDUFA3 was sufficient for complex I activity.

2.3.3. NDUFA13

The NDUFA13 gene is located on chromosome 19. NDUFA13 is a unique complex I subunit because it functions as a cell-death regulatory protein, GRIM-19, as well. It has been suggested that GRIM-19 regulates STAT3, a signal transducer and transcription activator, and is involved in interferon-β- and retinoic acid-induced cancer cell death [52]. For this reason, although NDUFA13 is mainly found in the inner mitochondrial membrane as a single-pass membrane protein, it can also be translocated to the nucleus to serve its apoptotic purpose. NDUFA13 is a 16 kDa protein synthesized as a 144-residue polypeptide, but the initiator methionine is removed during processing. It has a transmembrane alpha helix that spans residues 30–51 (see Figure 7). NDUFA13 has a longer alternate isoform that exists as a 222-residue polypeptide. NDUFA13 shares a large contact surface area with core subunits ND1, ND6, and NDUFS2, NDUFS5, NDUFA3, and NDUFA8. It weakly contacts ND3, NDUFS8, and supernumerary subunits NDUFA1 and NDUFA7. In a knockout line of cultured human cells, the loss of NDUFA13 resulted in a reduced expression of complex I, and it migrated as a smaller-than-normal-sized complex in BN gel electrophoresis [11]. Furthermore, when homozygous GRIM-19-deficient mice were generated, embryonic lethality resulted [53].

At least two diseases have been associated with mutations in NDUFA13. In 2015, a pair of sisters were discovered to harbor the first pathogenic germinal mutation identified in the short isoform one of NDUFA13, p.Arg57His (c.170 G > A in exon 2) [54]. Both sisters were homozygous for the mutation and presented with early onset hypotonia, dyskinesia, auditory neuropathy, and severe optic neuropathy. The patients were siblings who had been born from a consanguineous marriage. Their parents and elder sister were heterozygous for the p.Arg57His mutation and asymptomatic. The sisters were still living in 2015, the elder sister being 12 years old at the time. Though the sisters experienced early-onset neurological symptoms, the progression of neurological symptoms has been slow. Arg57 is a highly conserved residue across species. Upon biochemical analysis, NDUFA13 expression was reduced by a mean of 70% in the two sisters. Additionally, NDUFA9 and NDUFB8 expression was reduced by 95% and 90%, respectively. Due to the absence of intermediary complex I structures in the BN gel electrophoresis, the p.Arg57His mutation was concluded to cause major instability in NDUFA13 and prevent complex I assembly. It was suggested that the varied clinical symptoms in these sisters could be attributed to the dual function of NDUFA13/GRIM-19.

Somatic mutations in the long isoform two of NDUFA13 have been implicated in Hurthle cell thyroid tumors. Isoform two of NDUFA13 has 222 amino acids instead of 144 amino acids like isoform one. In 2005, four unrelated patients with Hurthle cell thyroid carcinomas were found to harbor four different heterozygous missense mutations that affected the N-terminal sequence of NDUFA13 [55]. Because the patients' transcripts arose from an alternate reading frame and coded for isoform two, a structural analysis of these mutations cannot be performed.

2.4. LYR Family Subunits

Two subunits from the LYR family are described next: NDUFA6 and NDUFB9, as shown in Figure 8. These proteins have the LYR sequence motif, which helps them bind to one of the two acyl carrier proteins, NDUFAB1. They also bind the substrate 4′-phosphopantethiene for the acyl carrier proteins. They are found on the matrix side of the membrane arm.

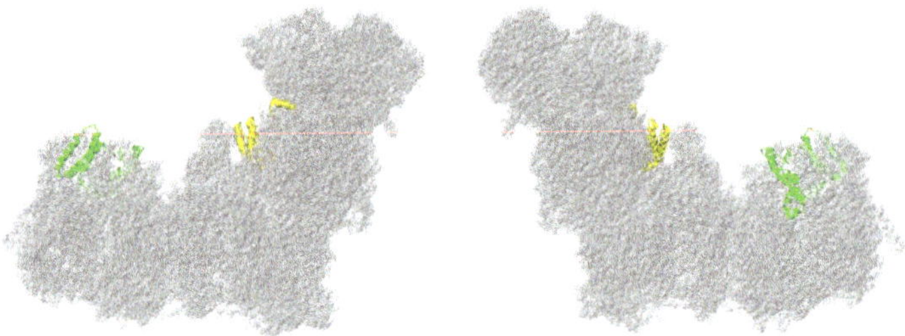

Figure 8. Location of NDUFA6 and NDUFB9. Most of Complex I is colored gray. They are shown in ribbons, with NDUFA6 colored yellow and NDUFB9 colored lime. The two views are rotated 180° relative to each other. The structure is from PDB id = 5xtd [9].

2.4.1. NDUFA6

The gene for NDUFA6 is located on chromosome 22. The encoded protein has 128 amino acids, with a mass of about 15.1 kDa. This subunit is found on the matrix arm. Its tertiary structure resembles a four-helix bundle in which the fourth helix is pulled away from the other three (see Figure 9a). Helices 1 and 2 interact with one of the two NDUFAB1 subunits, which are also known as the acyl carrier proteins and presumed to be involved in fatty acid biosynthesis. The short C-terminal helix interacts with core subunit NDUFS1, while core subunit NDUFS3 and supernumerary subunit NDUFA9 interact with both the N-terminal helices and the extended C-terminus. Thus, NDUFA6 bridges the N- and Q-modules. NDUFA6 is a member of the LYR family [56], whose members are known to interact with large mitochondrial complexes. Leu35 of the LYR motif fits inside the bundle of three helices, while Tyr36 and Arg37 contact NDUFAB1. In knockout cell lines, the assembly of full-size complex I was diminished, as seen in BN gel electrophoresis, and the levels of all subunits associated with the N-module of the peripheral arm were decreased by two-fold [11].

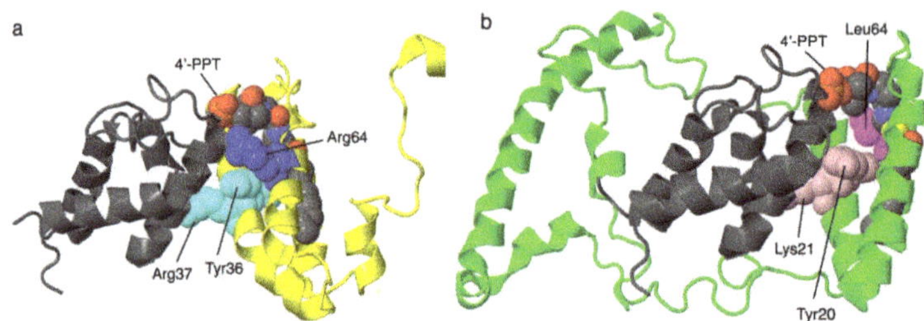

Figure 9. Structural features of the LYR proteins, NDUFA6 and NDUFB9. (**a**) NDUFA6 (colored yellow) is shown with its partner NDUFAB1 (colored gray), known as one of the two acyl carrier proteins (ACPs). NDUFA6 binds the 4′-phosphopantethiene (4′-PPT) analog substrate for the ACP. Tyr36 and Arg37 of the LYR motif are shown in space-filling and colored cyan. The site of the Arg64Pro mutation is shown in space-filling and colored blue. (**b**) NDUFB9 (colored lime) is shown with its partner NDUFAB1 (colored gray), the other one of the two ACPs. NDUFB9 binds the 4′-PPT analog substrate for the ACP. Tyr20 and Lys21 of the LYR motif are shown in space-filling and colored pink. The site of the Arg64Pro mutation is shown in space-filling and colored magenta.

Mutations in NDUFA6 have been described for four individuals, though only recently [57]. The most severe of the illnesses occurred in an individual with compound heterozygous mutations: p.Arg64Pro and c.265 G > T, a nonsense mutation at Glu89. An analysis of tissue demonstrated almost no fully assembled complex I and almost no in-gel activity in BN gel electrophoresis. This individual died after about two days. Tissue from two other individuals was analyzed and showed low levels of assembled complex I with activity in BN gel electrophoresis. Residue Arg64 is ion-paired with Asp111 of NDUFAB1 and is near Asp92 of NDUFAB1 and Tyr36 of the LYR motif of NDUFA6. The adjacent Ser112 of NDUFAB1 is part of the catalytic site of the acyl carrier enzyme and is bound by a substrate in the human structure (PDB id = 5xtd) [9].

The second individual was homozygous for a two-nucleotide deletion at Glu111, c.331_332del, resulting in 35 altered amino acids before a stop codon appeared [57]. This transcript was found at normal levels, thus indicating the lack of nonsense-mediated decay, presumably because it occurs in the last of three exons near the normal termination codon. If this transcript was translated, the protein would lack the C-terminal alpha-helix that interacts with core subunit NDUFS1. The third individual was homozygous for a c.3 G > A, substitution in the start codon. It was suggested that a transcript starting from an upstream Met codon, as predicted by another isoform, might have permitted some expression of a functional protein. Both individuals showed low levels of fully assembled complex I with activity, and their outcomes were only somewhat better than the first, with abnormal white matter in the brain, a loss of movement and vision, and seizures. The fourth individual was compound heterozygous with two frameshift mutations: p.Met104Cysfs*35 and p.Leu119Tyrfs*20. This child died in infancy, and no biochemical analyses were performed.

2.4.2. NDUFB9

The gene for NDUFB9 is located on chromosome 8. It is encoded as a protein of 179 amino acids, with a mass of about 21.8 kDa. The N-terminal Met is likely removed, and other possible modifications include the acetylation of Ala2 [58] and the phosphorylation of Ser85 [59]. NDUFB9 is a peripheral protein that is bound to the matrix side of core subunit ND5, with significant contacts to NDUFAB1, one of the acyl carrier proteins (see Figure 9b), and to NDUFB3, NDUFB4, NDUFB5, and NDUFB6. The protein is U-shaped, with each arm consisting of a three-helix bundle and pointing away from ND5. NDUFB9, along with NDUFA6, is a member of the LYR protein family, characterized by a Leu–Tyr–Arg motif near the N-terminus. In the case of NDUFB9, the motif is Leu19–Tyr20–Lys21, and it is found in the N-terminal alpha-helix. The acyl carrier protein NDUFAB1 fits between the two helical domains and binds to Tyr20 and Lys21 of this motif. In a cell line in which NDUFB9 was knocked out, complex I failed to assemble [11].

In 2012, a large-scale mutation-screening identified two patients who carried NDUFB9 mutations [60]. One patient, who presented with lactic acidemia and muscular hypotonia, was homozygous for the missense p.Leu64Pro. This residue is found in the third helix and packs against the first helix in the N-terminal three-helix bundle. Complex I activity was about 39% of the normal activity. After lentiviral rescue with a wild-type NDUFB9 gene, both complex I activity and levels of other complex I subunits were restored to normal levels. A second individual was found to carry the heterozygous mutation Arg47Leu, but this mutation was not found in a sibling with similar symptoms and could not be rescued by the complementation with a wild-type NDUFB9 gene. Therefore, it was concluded that illness was due to another gene.

2.5. *Subunits of the Twin C–X_9–C Family in the IMS*

Complex I has four proteins that are members of the twin C–X_9–C family and three with identified mutations that are described next: NDUFS5, NDUFA8, and NDUFB10, as shown in Figure 10. All are found in the IMS. They are substrates of the IMS oxidoreductase CHCHD4, also known as hMia40, which is essential for the export of FeS clusters from the mitochondrial matrix [61].

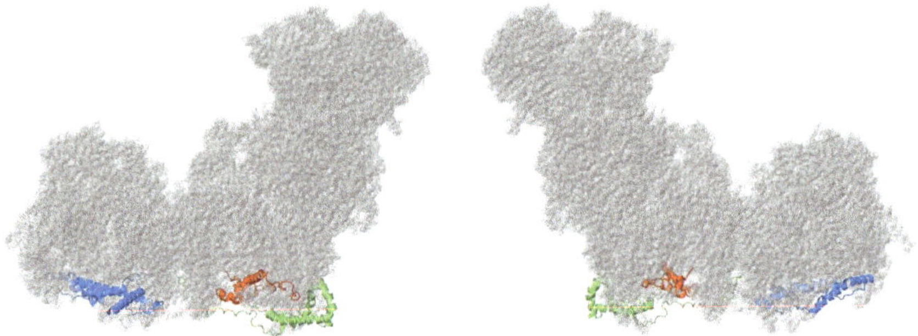

Figure 10. Location of NDUFS5, NDUFA8, and NDUFB10. Most of Complex I is colored gray. NDUFS5, NDUFA8, and NDUFB10 are shown as ribbons, with NDUFS5 colored red, NDUFA8 colored green, and NDUFB10 colored blue. They are found facing the intermembrane space. The two views are rotated 180° relative to each other. The structure is from PDB id = 5xtd [9].

2.5.1. NDUFS5

The gene for NDUFS5 is found on chromosome 1. The protein contains 106 amino acids, with a mass of 12.5 kDa. In cultured human cell lines, the knockout of NDUFS5 completely eliminates the assembly of complex I [11]. Though this protein was initially identified with the iron–sulfur protein fraction of bovine complex I, it is now known to be localized to the IMS and does not contact any proteins with FeS clusters. NDUFS5 is bound to the surface of the membrane and lacks transmembrane helices. It has a limited tertiary structure and includes a helix–coil–helix with two C–X_9–C sequence motifs that form two disulfides between the two alpha-helices at residue pairs 33–66 and 43–56. Another alpha-helix is formed by residues 70–86, but the rest of the protein has little regular secondary structure. The N-terminal domain (2–30) primarily contacts core subunits ND2 and ND4L, and supernumerary subunit NDUFB5, while the C-terminal domain (70–106) primarily contacts core subunit ND6, and NDUFB13.

One missense mutation in NDUFS5 has been reported in a compound heterozygous individual, p.Pro96Ser [22]. This individual has a second mutation in core subunit NDUFS8, p.Arg2Cys, of unknown consequence. These mutations were identified in a cohort of 103 patients with complex I deficiency. Pro96 in NDUFS5 is highly conserved among vertebrates and appears as the third Pro in an unusual sequence motif of Pro-Pro-Pro-His-His near the C-terminus. In human complex I (Video S1), Pro96 contacts Thr129 of NDUFA13. There is a nearby ion pair between Lys101 of NDUFS5 and Glu89 of NDUFA13. Somewhat different interactions can be seen in the mouse complex I in both active and deactive conformations [37]. Therefore, it seems likely that this missense mutation would disrupt local structure, but it is not clear that it would be deleterious to complex I assembly. The mutation in NDUFS8, p. Arg2Cys, should also be considered, as it might be defective, for example, in translation initiation.

2.5.2. NDUFA8

The gene for NDUFA8 is located on chromosome 9. The protein has 172 amino acids with a mass of about 20 kDa, and the initial methionine is cleaved. NDUFA8 is found in the IMS and has large contact surfaces with supernumerary subunits NDUFA13, NDUFA3, and NDUFA1, and it is near core subunit ND1. It has two pairs of alpha helices, each having two pairs of Cys residues that likely form disulfides (see Figure 11). They have characteristic spacings of either 9 or 11 amino acids between the cysteine residues. In the human structure (PDB id = 5xtd), not all disulfides are shown as formed, but it is likely that they do form in the IMS: C36–C66, C46–C56, C78–C110, and C88–C100. In the mouse structures (PDB id = 6g2j and 6g72), they are shown as disulfides [37]. The N-terminal residues 2–22 are extended and interact with NDUFA13, NDUFS5, and NDUFA1. The C-terminal region extends from residues 113 to 172 without a regular secondary structure. It makes an interesting junction with

NDUFC2 and NDUFB5 in a conserved multi-centered ion pair including Arg166 of NDUFA8, Asp86 of NDUFC2, and Glu153 of NDUFB5.

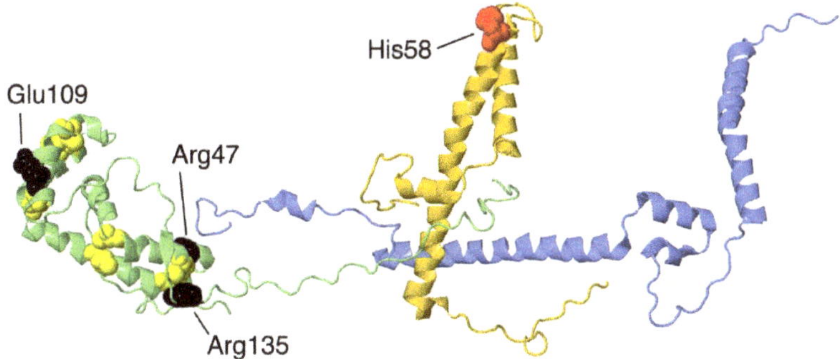

Figure 11. Structural features of NDUFA8, a member of the twin C–X_9–C. The proteins are portrayed in ribbons. NDUFA8 is colored green, NDUFC2 is colored gold, and NDUFB5 is colored blue. In NDUFA8, the eight Cys residues that form disulfide bonds are shown in space-filling and colored yellow. The sites of three mutations, Arg47Cys, Glu109Lys, and Arg135Gln, are shown in space-filling and colored black. NDUFB5 makes contact near Arg47, and both NDUFB5 and NDUFA8 contact NDUFC2. The site of one mutation in NDUFC2, His58Leu, is shown in space-filling and colored red.

At least three missense mutations have been described in NDUFA8. The p.Arg47Cys mutation was described in a patient that was homozygous, having received a mutant allele from each parent, who were heterozygous and asymptomatic [62]. This individual presented with developmental delay and epilepsy, and they were bed-ridden by age 26. Fibroblasts showed reduced levels of NDUFA8, as well as other complex I subunits, and reduced complex I activity. The mutation creates an additional Cys residue next to Cys46, which possibly leads to incorrect disulfide bond formation. In addition, Arg47 has multiple interactions within the NDUFA8 subunit, as well as with NDUFA13 and NDUFB5, which would be disrupted in the mutant p.Arg47Cys.

The second individual was compound heterozygous with p.Glu109Lys in NDUFA8 and p.Ala224Val in core subunit NDUFS2 [63]. The transcript for the mutant NDUFA8 subunit was not found, and so the mutation might have affected its stability. Ala224 in the core subunit NDUFS2 is highly conserved and is found near Cys 153 and 160 of a FeS cluster, and so this substitution might also have been deleterious. The infant showed neonatal hypotonia and died at two months.

The third individual was found in a screen of patients with reduced complex I levels [22]. Two potential deleterious mutations were discovered. First was a heterozygous mutation p.Arg135Gln in NDUFA8, and second was a homozygous mutation p.Leu229Pro in C20orf, a spindle assembly factor for microtubles. Since the individual showed decreased levels of complex I activity, the NDUFA8 mutation was implicated, but it was not clear whether the spindle assembly factor had any impact on complex I levels. Arg135 is highly conserved, and it makes an ion pair with Glu59, also a conserved amino acid of NDUFA8. Glu59 is near the disulfide formed by Cys46 and Cys56. This ion pair is exposed on the IMS and does not contact any other subunits.

2.5.3. NDUFB10

The gene for NDUFB10 is located on chromosome 16. The NDUFB10 protein has 172 amino acids and a mass of 20.8 kDa. It lacks a mitochondrial import sequence. It is one of four IMS proteins found in complex I, and all have four cysteine residues that form two disulfide bonds. In NDUFB10, two cysteine residues, Cys78 and Cys107, are found near the ends of a long alpha-helix (M77–E111). These cysteine residues form disulfides with nearby cysteines, connected to the long helix by loops of 6–11 residues. Cys78 forms its disulfide with Cys71, and Cys107 forms its disulfide with Cys119. In addition, there is another cysteine, Cys145, conserved among vertebrates, and of unknown significance. The import of NDUFB10 is dependent upon its oxidation by CHCHD4/Mia40 and its formation of correct disulfide bonds [64]. NDUFB10 also contains three additional alpha-helices, two C-terminal ones and one N-terminal one. These extensions allow it to contact many subunits in the vicinity of ND4 and ND5. It primarily contacts NDUFB5, NDUFB6, and NDUFB11, with lesser contact to ND4, ND5, and NDUFC2. In an analysis of cultured cells, the knockout of NDUFB10 resulted in a loss of complex I assembly and, especially, the ND4/5 modules.

One individual with mutations in NDUFB10 was studied [64]. She was compound heterozygous with a paternally inherited termination codon at Glu70 of NDUFB10 and a maternally inherited p.Cys107Ser substitution. This infant presented with lactic acidosis and cardiomyopathy and survived for only about one day. An analysis revealed reduced levels of complex I activity, reduced levels of in-gel assays of complex I in BN gels, and the presence of the Cys107Ser protein in the cytosol. The transcript of the nonsense allele was not found by RT-PCR. Cys107 is not near any other subunits, the closest being NDUFB5, NDUFB6, and NDUFB11. Given the healthy status of the parents, it appears that the Cys107S mutant does not have a dominant negative phenotype. Rather, it primarily does not enter the mitochondrion, and if another allele is available, complex I can assemble without difficulty.

2.6. *Subunits of the ND2-Module*

The next two subunits to be described are members of the ND2-module: NDUFA10 and NDUFC2, as shown in Figure 12. NDUFA10 is a globular protein bound on the matrix side, while NDUFC2 is a membrane protein. These two subunits have contacts on the matrix side of the membrane.

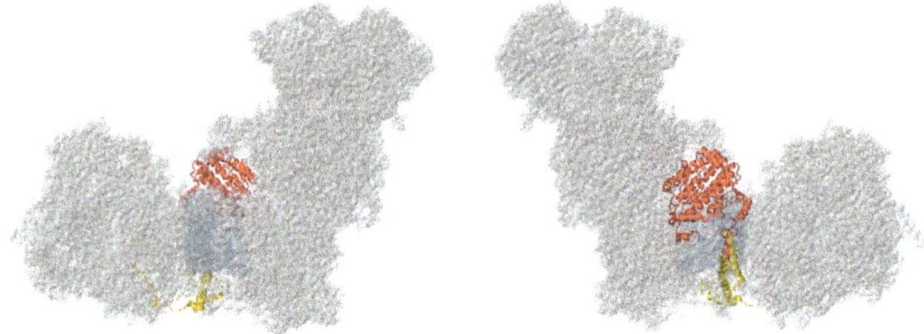

Figure 12. Location of NDUFA10 and NDUFC2. Most of Complex I is colored gray. Core subunits in the ND2-module are colored light blue (ND2). NDUFA10 and NDUFC2 are shown in ribbons, with NDUFA10 colored red and NDUFC2 colored gold. The two views are rotated 180° relative to each other. The structure is from PDB id = 5xtd [9].

2.6.1. NDUFA10

The gene for NDUFA10 is found on chromosome 2. The encoded protein has 355 amino acids and a mass of about 40.8 kDa, with 35 N-terminal amino acids cleaved upon entry to the mitochondrion. Structurally, NDUFA10 has been identified as a member of the deoxynucleoside kinase (dNK) family [8], with the dNK domain (also known as the PF01712 family in the Pfam database), although it is unlikely to be an active enzyme. The protein is compact with a four-stranded parallel beta-sheet surrounded by numerous alpha-helices (see Figure 13). The human structure (PDB id = 5xtd) [9] did not reveal any ligands, but adenosine nucleotides have been found at the C-terminal ends of the beta-sheet in a pocket surrounded by alpha-helices in the mouse (PDB id-6g2j) [37] and sheep [65] structures. NDUFA10 is found on the matrix side of the membrane and primarily interacts with core subunits ND2 and NDUFS2. These interactions with the extended N-terminus of NDUFS2, and the matrix-side peripheral helices of ND2 are all centered over the C-terminal sector of ND2, which contains the broken helix that is part of the proton translocation pathway. NDUFA10 also makes limited contact with the N-terminal regions of NDUFC1, NDUFC2, and NDUFB11 on the matrix side.

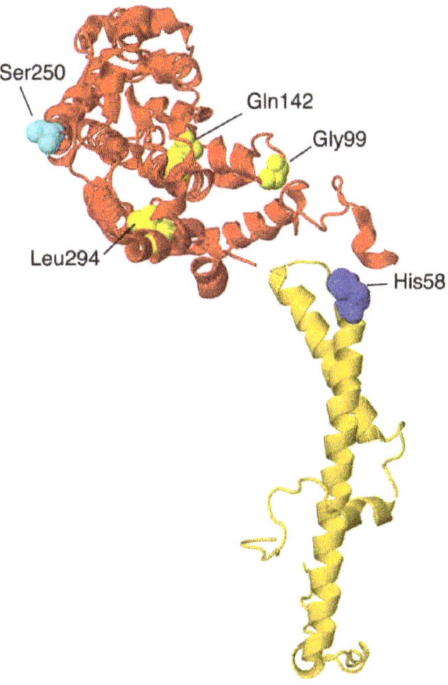

Figure 13. Structural features of two subunits from the ND2-module. NDUFA10, colored red, and NDUFC2, colored gold, are portrayed in ribbons. The site of one mutation, His48Leu in NDUFC2, is shown in space-filling and colored blue. The sites of three mutations in NDUFA10—Gly99Glu, Gln142Arg, and Leu294Pro—are shown in space-filling and colored yellow. The site of phosphorylation by PINK1, Ser250, is shown in space-filling and colored cyan. NDUFA10 likely binds an adenosine nucleoside, not shown.

In knockout human cell line BN gel electrophoresis, only faint bands of complexes containing primarily membrane subunits are seen [11]. Perhaps because of its contacts with a Q-module, core subunit NDUFS2, and a core membrane subunit ND2, along with subunits at the interfaces of ND2/ND4 (NDUFC1 and NDUFC2) and of ND4/ND5 (NDUFB11), it appears to be a key subunit in assembly. Residue Ser250, which is found on the matrix surface, was reported to be phosphorylated by the kinase PINK1 [66], a kinase that is known to be imported into the mitochondrial matrix space [66,67]. This is a conserved amino acid, and phosphorylation appears to be important for complex I activity.

Three reports of clinical mutations in NDUFA10 have been described. The first patient had compound heterozygous mutations including one allele in which the start codon was changed to GTG, and a second in which p.Gln142Arg occurred [68]. This individual showed developmental problems at 10 months and was eventually diagnosed with Leigh syndrome. It is likely that the first mutation would significantly impair the translation unless an alternative start codon were available. The second mutation is found in the interior of the protein, and so the Arg likely disrupts the packing of the protein, especially because of the positive charge.

The second patient was a boy with Leigh syndrome who, in a screen, was determined to have compound heterozygous mutations: p.Leu294Pro and c.383_384insTAA (p.Ser218delinslS) [27]. These were found to be inherited from the father and mother, respectively. The latter mutation would likely lead to a degraded mRNA or protein. The former mutation p.Leu294Pro resides on an alpha-helix at an interior location. The substitution of Leu by Pro might be disruptive, as this alpha-helix contacts two distinct regions of ND2. The immunoblotting of fibroblasts showed reduced levels of NDUFA10.

The third patient was initially diagnosed as having a nonlethal infantile mitochondrial disorder, and the mutation was discovered in a screen [69]. This individual developed brain lesions and was considered to have Leigh syndrome. His parents were third cousins, and DNA sequencing revealed him to be homozygous for p.Gly99Glu, while both parents were heterozygous for the same mutation. A deltoid muscle biopsy showed significant reduction of complex I activity (about 40–70% of normal levels). This amino acid is packed against the N-terminal amino acids of NDUFC1, and so the introduction of the negatively-charged Glu could disrupt the assembly of the membrane arm.

2.6.2. NDUFC2

The gene for NDUFC2 is located on chromosome 11. The NDUFC2 protein is 119 amino acids, with a mass of 14.2 kDa. It is a double-pass transmembrane protein that contacts core subunit ND2, with both termini on the IMS side of the membrane. From the N-terminus, the protein enters the membrane in a non-helical stretch from Ser 19 to Arg29, and then residues 29–47 are alpha-helical. The second crossing of the membrane is alpha-helical from residues 56–97, extending into the IMS. NDUFC2 also contacts NDUFC1, NDUFB5, and NDUFA8 (see Figure 11), and it is part of the ND2 assembly module. NDUFC1 has a special role in sealing NDUFC2 from the lipid bilayer, with its single transmembrane helix parallel to the long helix of NDUFC2 and very limited contact to any other subunits in the membrane. The extension of the long alpha-helix into the IMS, residues 84–97, is the region that contacts NDUFA8 and NDUFB5 (see Figure 11). On the matrix side it contacts NDUFA10 (see Figure 13).

Mutations of NDUFC2 in three patients from two families were reported in 2020 [70]. The patients presented with symptoms of Leigh syndrome and reduced complex I activity but had different outcomes. In both families, the parents were healthy consanguineous first cousins and were heterozygous for the mutations. In one family, the mutation was a deletion of 22 base pairs near the C-terminus at residue His116, p.His116_Arg119delins21. This would cause a frame shift and elimination of the normal stop codon at position 120. It was not clear how long the new reading frame would be. The transcript of this gene was identified at low levels, but the immunoblotting of fibroblast samples was negative, suggesting the degradation of the protein or a lack of expression. Little or no complex I was seen in native gels. Complexome profiling found evidence of Q-module assembly and ND4-module assembly but little else of complex I subunits. The daughter had no seizures and survived until at least age six, while her brother passed away at three years of age.

In the second family, the mutation was p.His58Leu. This patient had normal transcript levels but only slightly more evidence of complex I assembly. His58 is found on the matrix side of the membrane and starts the long helix that crosses the membrane. It sits between Val44 of NDUFC1 and Trp353 of NDUFA10 and is very near the phosphate group of a bound lipid. In the mouse structure (PDB id = 6g2j), the corresponding His59 is even more tightly packed between NDUFC1 and NDUFA10 residues, suggesting that a Leu substitution could be deleterious. This child had impaired growth and seizures and passed away at eight months.

NDUFC2 has a unique position in complex I. It is embedded in the membrane near the ND2-ND4 junction but contacts two subunits that are embedded at distant sites, with long extension to NDUFC2: NDUFA8 is found at the "heel" of the complex I boot, in the IMS, while NDUFB5 is found at the junction of ND4-ND5. Interactions in this network of subunits appear to be essential for complex I assembly.

2.7. Subunits Form the ND4-Module

Two subunits from the ND4-module are described next: NDUFA11 and NDUFB11, as shown in Figure 14. Both are membrane proteins but are found on alternate sides of ND4 and do not contact each other.

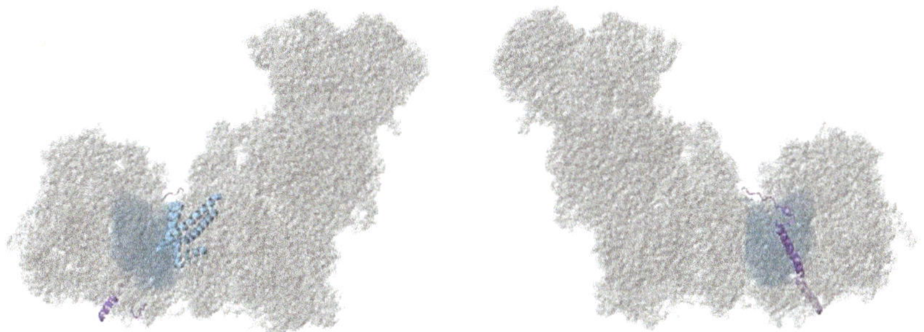

Figure 14. Location of NDUFA11 and NDUFB11. Most of Complex I is colored gray. Core subunits in the ND4-module are colored light blue (ND4). NDUFA11 and NDUFB11are shown in ribbons, NDUFA11 colored blue and NDUFB11 colored purple. The two views are rotated 180° relative to each other. The structure is from PDB id = 5xtd [9].

2.7.1. NDUFA11

The NDUFA11 gene is located on chromosome 19. NDUFA11 is a 14.7 kDa protein with four transmembrane helices and little exposure outside the membrane. NDUFA11 has two isoforms. The first isoform contains 140 amino acids after the cleavage of the initiator methionine and is N-acetylated at Ser2. It consists of five alpha helices: a short helix is found on the IMS side, followed by membrane spanning helices 3–11, 17–45, 48–82, 87–106, and 107–137. NDUFA11 contacts the core subunits ND2, ND4, and ND5, as well as the supernumerary subunit NDUFB5. The second isoform is rare and contains 228 amino acids. In a previous study, it was found that the knockdown of NDUFA11 by RNA interference (siRNA) in human cell culture led to partially assembled subcomplexes visualized by blue native gels [71]. It was concluded that NDUFA11 acts like an assembly factor in complex I. Furthermore, it was shown that, in a knockout strain of cultured human cells, the loss of NDUFA11 resulted in no expression of complex I in BN gel electrophoresis [11].

The first identified NDUFA11 clinical mutation, a G to A mutation at the exon-intron junction (exon 1-IVS1) donor splice site, c.99 + 5 G > A, was identified in 2008 [72]. Six patients from three unrelated families were products of consanguineous marriages and were found to be homozygous for the mutation. They were clinically affected with either fatal infantile lactic acidemia or encephalocardiomyopathy. Though heterozygotes were identified in two families, all family members besides the patients were healthy. The patient from Family A and the two patients from Family B had similar clinical presentations. They all developed severe metabolic acidosis and hyperlactatemia within 10–24 h of age. All three patients from Families A and B died from acidosis within 6–40 days of age. In Family C, three patients were homozygous for this splice mutation. They experienced slow psychomotor development, hypertrophy of myocardial walls, acidosis, and generalized brain atrophy. Two out of the three patients from Family C died at 18 months and four years of age, but the third patient from this family lived until at least six months of age. The presence of both wild-type and mutant mRNA transcripts in the patients' fibroblasts indicates that the varied clinical presentation may have been caused by variable splicing that produced variable mutant/wild-type transcript ratios. They suggested that normal splicing would have produced a protein that was responsible for the detected activity, while alternative splicing would have yielded a nonfunctional protein.

In 2019, a two more NDUFA11 clinical mutations, p.Ala132Pro and p.Thr106Ile, were found in isoform two of NDUFA11 [73]. These mutations were found simultaneously in a compound heterozygote patient with mitochondrial myopathy. The patient developed late-onset symptoms of a neuromuscular disorder, bilateral hearing loss, saccadic eye movements, and proximal leg weakness. The good health of the patient's offspring and the absence of neuromuscular disease in the patient's family history suggested that these two mutations were spontaneous. In contrast to isoform one, NDUFA11 isoform two is rare and appears in skeletal muscle. Thus, a mutation in isoform two may only lead to mild muscular impairment. The structure of this rare isoform of NDUFA11 has not yet been determined, so we cannot confirm that the mutation causes a structural disruption within this subunit.

2.7.2. NDUFB11

Similar to NDUFA1, the NDUFB11 gene is located on the X chromosome. It has three exons. It is synthesized as a 153 amino acid protein with a mass of about 17.3 kDa, but its first 29 amino acids serve as a transit peptide that is cleaved off during processing. NDUFB11 is a single-pass membrane protein with the N-terminus on the matrix side and one helix that spans residues 80–107, followed by another helix into the IMS of residues 115–132 (see Figure 15). Its N- and C-termini are extended and point in the same direction, like a letter C. It has a large contact surface with core subunit ND4 and NDUFB10, but it also weakly contacts core subunit NDUFS2, NDUFA10, NDUFB5, and NDUFC2. In a knockout line of cultured human cells, the loss of NDUFB11 resulted in almost no assembly of complex I [11]. Furthermore, NDUFB11 short-hairpin RNA (shRNA) knockdown in HeLa cells on complex I resulted in decreased expression of subunit NDUFB8, the failure of the membrane arm and the holocomplex to assemble, and decreased oxygen consumption. Moreover, NDUFB11 reduction was associated with decreased cell growth and increased apoptosis. For this reason, cell death caused by NDUFB11 mutations may be responsible for embryonic lethality in males and developmental defects in female patients [74].

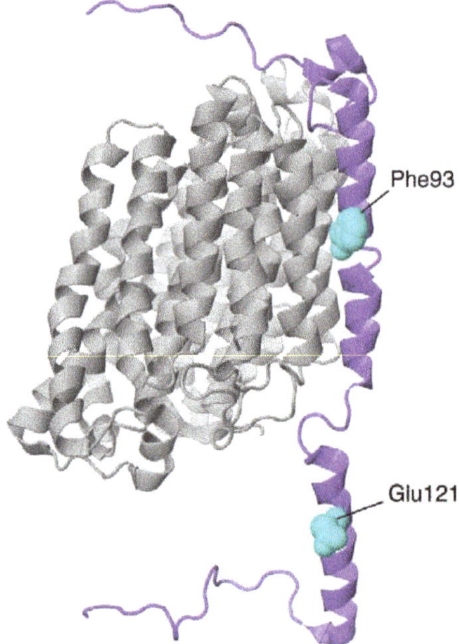

Figure 15. Structural features of NDUFB11 of the ND4-module. The proteins are shown as ribbons, with NDUFB11 colored purple and ND4 colored light gray. The sites of two mutations are shown in space-filling, Phe93 (a deletion) and Glu121Lys (in the IMS).

Two studies conducted in 2015 found that nonsense mutations in NDUFB11 are associated with histiocytoid cardiomyopathy (histiocytoid CM) and microphthalmia with linear skin defects syndrome. Clinical characteristics of histiocytoid CM, an arrhythmogenic disorder, are incessant ventricular tachycardia, cardiomegaly, and sudden death within the first two years of life. The study did not expand on specific patient symptoms or survival. In the first study, two unrelated female patients with histiocytoid CM were identified to harbor de novo nonsense mutations in NDUFA11 [74]. The first patient had a mutation that changed Tyr108 to a stop codon (c. 324 T > G). The second patient had a mutation that changed Trp85 to a stop codon (c. 255 G > A). Both mutated residues are located in exon 2 of NDUFB11. The authors concluded that these de novo mutations result in a dominant haploinsufficient phenotype, which contrasts with the Mendelian recessive inheritance pattern of many complex I deficiencies.

In the second study, two female patients were discovered with microphthalmia and linear skin defects syndrome (MLS) [75]. MLS is an X-linked disease found exclusively in females and is embryonically lethal in males. Both patients presented with linear skin defects but not microphthalmia. Patient 1 was heterozygous for a de novo nonsense mutation, c.262 C > T in exon 2, that changed Arg88 to a premature stop codon. Besides MLS and histiocytoid CM, she experienced additional symptoms such as axial hypotonia, failure to thrive, oncocytic metaplasia, and histiocytoid cardiomyopathy. She died at six months of age from cardiac arrest. This mutation was also identified in a later study [76]. Patient 2 was heterozygous for an inherited one base-pair deletion, c.402delG in exon 3, that caused a p.Arg134Ser mutation and changed Val136 to a premature stop codon. She was observed to have corpus callosum agenesis and dilated lateral ventricles during the fetal stage and experienced seizures, cardiomyopathy, myopia, nystagmus, severe psychomotor developmental delay, and muscular hypotonia after birth. She was still living at seven years of age. Her mother was a healthy carrier of the mutation. Though the authors of the previous study [74] concluded that one copy of wild-type NDUFB11 was not sufficient for normal cell function, the existence of a heterozygous, healthy carrier of a NDUFB11 deletion contradicts this conclusion. We conclude that the variable outcomes for females might be due to mosaicism; phenotypic differences may depend on which tissues receive the wild-type NDUFB11 due to X-inactivation. After Patient 2's birth, her mother was pregnant with another fetus who had the same frameshift mutation. The fetus had to be aborted due to severe intrauterine growth retardation. Only wild-type NDUFB11 transcripts were found in the fibroblasts and leukocytes of Patient 1 and Patient 2, indicating that the two mutations resulted in no transcript or expression of NDUFB11.

Another NDUFB11 mutation was found in 2016 [27]. A male patient who harbored a hemizygous de novo mutation (c.361 G > A, p.Glu121Lys) presented with lethal infantile mitochondrial disorder (LIMD), heart and respiratory failure, and complex I deficiency. There was no NDUFB11 expression in the patient's fibroblasts, and he died 55 h after birth. Glu121 is a highly conserved residue. It lies on the intermembrane space side of complex I and sits between Arg124 and Tyr117 of the same subunit and near His50 of NDUFB10.

Additionally in 2016 [77], another mutation in NDUFB11 was discovered in five males from four families in a screen of patients with congenital sideroblastic anemia. The mutation was a three-nucleotide deletion, c.276_278del, p.F93del, occurring in three consecutive phenylalanine codons (TTC). These three Phe occur in the membrane helix between ND4 and NDUFB5 (see Figure 15). It is possible that the deletion could be partially accommodated by the chain of three consecutive Phe residues. The subjects ranged from 2 to 76 years of age, and many had associated symptoms of short stature, congenital optic atrophy, myopathy, and lactic acidosis.

2.8. Subunits from the ND5-Module

Finally, three subunits are next described from the ND5-module: NDUFB3, NDUFB6, and NDUFB8, as shown in Figure 16. These subunits do not interact directly with each other, but they serve to enclose the distal end of complex I on all sides. In particular, NDUFB8 clamps the lateral helix of core subunit ND5.

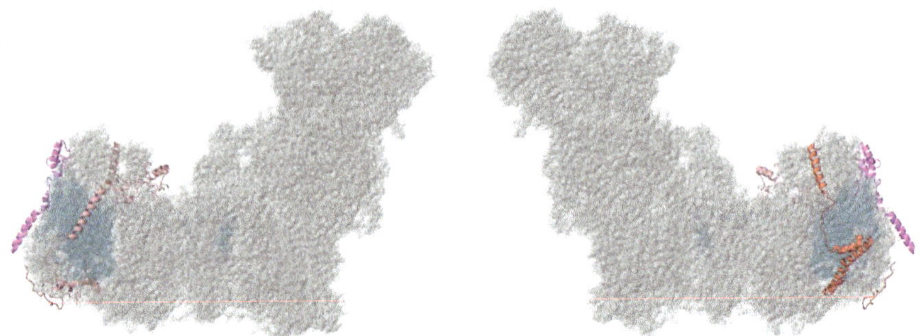

Figure 16. Location of NDUFB3, NDUFB6, and NDUFB8. Most of Complex I is colored gray. Core subunits in the ND5-module are colored light blue (ND5). NDUFB3, NDUFB6, and NDUFB8 are shown in ribbons, with NDUFB3 colored violet, NDUFB6 colored salmon, and NDUFB8 colored pink. The two views are rotated 180° relative to each other. The structure is from PDB id = 5xtd [9].

2.8.1. NDUFB3

The gene for NDUFB3 is found on chromosome 2. The encoded protein has 99 amino acids with a mass of about 11.4 kDa. The N-terminal Met appears to be cleaved. NDUFB3 is found at the distal end of complex I near core subunit ND5. It has a small N-terminal domain composed of three alpha-helices on the matrix side, and the C-terminus forms an alpha-helix, residues 62–89, that crosses the membrane. NDUFB3 contacts NDUFB9 and NDUFAB1 on the matrix side, as well as NDUFB2 and core subunit ND5, primarily in the membrane region. In knockout cell lines, the absence of NDUFB3 was found to result in the near total loss of complex I assembly and a reduction in the level of subunits from both N- and ND5 assembly modules [11].

Two mutations have been identified in the gene for NDUFB3 among over 10 individuals. In a report from [78], p.Trp22Arg appeared as a homozygous mutation in a girl who exhibited intrauterine growth retardation and premature birth at 31 weeks. She died at four months of age. Very low levels of complex I activity (<15%) were measured in fibroblasts, and this phenotype was rescued by lentivector complementation. A second individual was identified with compound heterozygous mutations: p.Trp22Arg and p.Gly70X [79]. The level of complex I activity was <25% of normal, and the individual exhibited muscular hypotonia, developmental delay, and lactic acidosis. In fibroblasts, complex I activity could be recovered by the ectopic expression of wild-type NDUFB3, though not either mutant form. The health status of this individual was not reported. In contrast, two more recent reports have identified the p.Trp22Arg mutation in older and seemingly healthier individuals. Ten children, up to 10 years of age, were identified as homozygous carriers of p.Trp22Arg in NDUFB3 [80]. These individuals with short statures and prominent foreheads ranged from 10 months to 10 years in age, and their complex I activity levels ranged from 25 to 35% of normal levels. Various levels of assembled complex I were seen from these individuals using blue native gel electrophoresis. More recently, a 32-year-old previously diagnosed with non-alcoholic steatohepatitis was discovered to be homozygous p.Trp22Arg [81]. This individual had reduced levels of complex I activity (<30% of normal) and suffered from oculomotor dysfunction with optic nerve anomalies, episodes of lactic acidosis during surgical interventions, and progressive fatigue.

The latter two studies emphasized that the context of these two mutations in NDUFB3 appears to be significant. The level of complex I activity, and its assembly, is likely on the borderline between tolerable and deleterious. Trp22 is found in a conserved region of NDUFB3 on the matrix side of the membrane. The NH side chain of Trp is in position to H-bond to the OH side chain of Tyr85 of NDUFAB1. It also makes nonbonding interactions with Arg45 and Gln46 of NDUFB2. Therefore, the mutation to the positively charged Arg is likely to be disruptive. The second mutation, p.Gly80X, occurs in the last exon and so is not likely to trigger nonsense-mediated decay. This residue occurs in the transmembrane region, and the stop codon would truncate the protein and eliminate the transmembrane helix. Gly70 is flanked by Lys69 and Lys72, which probably demarcates the matrix end of the lipid bilayer. Thus, it seems possible that the truncated NDUFB3 would retain some ability to assemble and provide normal interactions on the matrix side.

2.8.2. NDUFB6

The NDUFB6 gene is located on chromosome 9. NDUFB6 is a single-pass membrane protein with its N-terminus found on the matrix side. It is a 15 kDa protein with 127 amino acids, with residues 6–26 forming an alpha helix on the matrix side and residues 56–92 forming an alpha-helix that crosses the membrane. Its initiator methionine is removed during processing. Furthermore, it contains an N-acetylated threonine at residue 2 [58] and a N6-acetylated lysine at residue 24 by analogy with mouse [82]. NDUFB6 contacts NDUFB5, NDUFAB1, and NDUFB9 on the matrix side, core subunit ND5 in the membrane, and supernumerary subunits NDUFB7 and NDUFB10 in the IMS. In a knockout strain of cultured human cells, the loss of NDUFB6 resulted in no assembly of complex I [11].

It has been established that insulin-resistance and type 2 diabetes are associated with reduced NDUFB6 expression [83]. To date, only one polymorphism has been found in NDUFB6 [84]. This polymorphism, rs629566 (A/G or G/G), is located in the promoter region of the NDUFB6 gene. It changes the DNA sequence at position 544 from CA to CG. This polymorphism introduces a fourth potential methylation site in the promoter of the NDUFB6 gene. The introduction of a fourth methylation site resulted in increased DNA methylation at the NDUFB6 gene promoter and decreased NDUFB6 mRNA expression in the muscle of elderly patients. As predicted, elderly patients with the rs629566 G/G genotype had lower NDUFB6 mRNA expression than elderly patients with the A/G genotype. Surprisingly, young patients with the A/G polymorphism had the same level of NDUFB6 mRNA expression as young patients with the wild-type A/A gene. Moreover, young patients with the G/G genotype actually had the highest levels of NDUFB6 mRNA. Thus, it was concluded that age significantly affects NDUFB6 DNA methylation and influences NDUFB6 expression.

Furthermore, it has been suggested that NDUFB6 is a possible tumor suppressor of metastatic clear cell renal cell carcinoma (ccRNC) [85]. In an analysis of 50 primary ccRNC samples, copy number alteration at the 9p24.3–p13.3 region (the chromosomal region in which NDUFB6 gene is located) was found to strongly correlated with poor ccRNC prognosis. Patients missing this chromosomal region had higher cancer recurrence rates, higher metastasis rates, and lower survival rates. It was determined that NDUFB6 expression was downregulated in primary ccRNC samples due to gene copy number loss. More focused experimentation revealed that when NDUFB6-encoded lentiviruses were transduced into a ccRNC cell line, cell proliferation was suppressed. Consequently, it was observed that siRNA knockdown of NDUFB6 led to cell proliferation with the same cell line. The study concluded that loss of region 9p24.1–p13.3 results in NDUFB6 downregulation, which causes cell proliferation in metastatic ccRNC tumors.

2.8.3. NDUFB8

The NDUFB8 gene is located on chromosome 10. The NDUFB8 subunit is initially synthesized with 186 residues and a mass of about 22 kDa, but the first 28 residues serve as a transit peptide and are cleaved off during processing. NDUFB8 is a single-pass protein with the C-terminus in the IMS and the N-terminus in the matrix (see Figure 17). It is located near the end of complex I, and its single transmembrane span lies near ND5. The helix residues (126–151) are not conserved, but the small domains on each side of the helix are much more conserved. On the matrix side NDUFB8 mainly contacts core subunit ND5 and NDUFB4, but it also contacts core subunit ND4 and NDUFB9. On the IMS side, NDUFB8 contacts NDUFB7 and NDUFB10. In a knockout strain of cultured human cells, the loss of NDUFB8 resulted in no assembly of complex I [11].

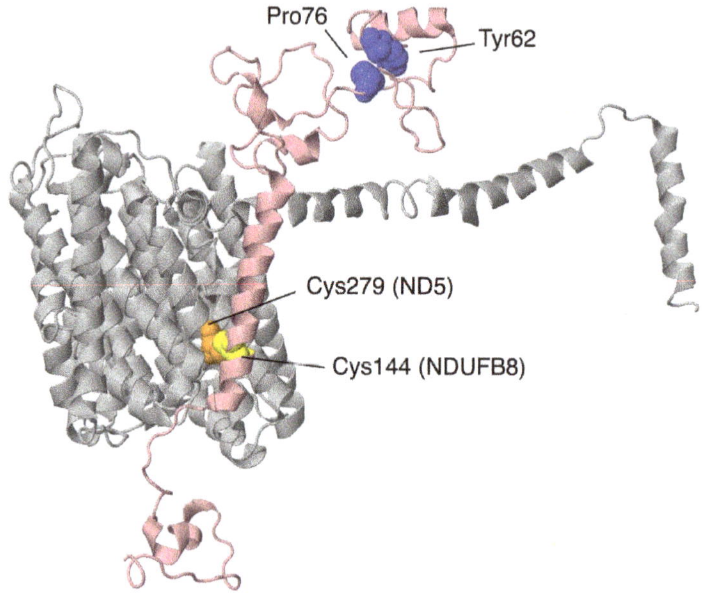

Figure 17. Structural features of NDUFB8 from the ND5-module. The sites of three mutations in NDUFB8 are shown in space-filling. NDUFB8 is colored pink, while the sites of Pro76Gln and Tyr62His are shown in blue on the matrix side, and Cys144Trp is shown in yellow. The Cys279 of ND5 is shown in orange, and it appears to form a disulfide with Cys144 of NDUFB8.

In 2018, two unrelated compound heterozygous patients with biallelic NDUFB8 mutations were identified to have Leigh-like encephalomyopathy [86]. At three months of age, Patient 1 showed symptoms of failure to thrive, muscle hypotonia, and elevated lactate levels. Patient 2 began to show these symptoms at six months of age. Similar to patients with Leigh syndrome, both patients showed symmetrical basal ganglia and capsula interna lesions. Hypertrophy of the left cardiac ventricle was observed in Patient 1, and he died at 15 months of age. Patient 2 was still alive at six years of age.

Patient 1 harbored a p.Pro76Gln mutation and a p.Cys144Trp mutation. Pro76 is located within a highly conserved region. The residue lies on the matrix side above the lateral helix of core subunit ND5, and it is near core subunit ND4. Cys144 is found within the membrane domain. Though it is weakly conserved, it appears to form a disulfide with Cys279 of chain ND5. Surprisingly, it was discovered that the p.Cys144Trp mutation generated a new exonic splicing silencer (ESS). The Human Splicing Finder algorithm was used to identify the ESS, and Sanger sequencing revealed that exon 4 (Met105–Val156) was skipped in the NDUFB8 transcript, thus confirming the splice defect. Nevertheless, 14% of total cDNA from Patient 1 contained the p.Cys144Trp missense mutation, indicating that while skipping exon 4 is favored, normal splicing also occurs at a lower frequency. Upon the analysis of respiratory chain activity in Patient 1's muscle tissue and BN gel electrophoresis in a muscle biopsy sample, isolated complex I deficiency was confirmed.

Patient 2 carried a p.Tyr62His mutation and a loss-of-function deletion Glu63Aspfs*35 (c. 189delA). Tyr62 is completely surrounded by mostly conserved residues of NDUFB8 and is located in the same conserved region as Pro76; the two residues are only 4–5 Å distant from each other. Furthermore, the side chain hydroxyl group of Tyr62 forms a hydrogen bond with the side chain and backbone oxygen atoms of Asp74 in NDUFB8. Thus, although His is a fairly conserved substitution for Tyr, His substitution would likely be disruptive at this residue. Complex I deficiency in Patient 2 was observed through the BN gel electrophoresis of muscle homogenate. The complementation of both mutant cell lines was tested using the lentiviral expression of wild-type NDUFB8. Cells from Patient 1 showed normal levels of complex I activity by in-gel assays after BN gel electrophoresis, and normal levels of the NDUFB8 subunit by immunoblotting. Cells from Patient 2 did not grow well enough for such analysis, but both cell lines showed complementation in assays using microrespirometry and flow cytometry.

3. Conclusions

This review has discussed 20 of the 30 supernumerary subunits of human complex I. Missense and various other point mutations have been described for 17 of the subunits. One subunit, NDUFV3, has several known amino acids substitutions, but there is no evidence that they are causative for disease. All of the mutations, the diseases they are associated with, and their effects on assembly of complex I are summarized in Table 2. The gene of one subunit NDUFA3 was completely deleted, along with several neighboring genes. This provided evidence for haplosufficiency, since no mitochondrial disease characteristics occurred in those individuals. The gene for another subunit NDUFB6 had a mutation in the promoter, resulting in a change in the expression level.

Table 2. Summary of mutations with associated diseases and effects on assembly.

Gene	Mutation (DNA)	Mutation (Protein)	Diagnosis	Assembly [1]	Reference
NDUFA2	c.A170 C	p.Glu57Ala	Microcephaly Leukoencephalopathy	NT	[14]
	1. c.134 A > C 2. c.225del	1. p.Lys45Thr 2. p.Asn76Metfs*4 [2]	Leukoencephalopathy	NT	[13]
	c.134 A > C	p.Lys45Thr	Leukoencephalopathy	• Smaller complex	[13]
	c.208 + 5 G > A	mRNA not detected	Leigh syndrome, hypertrophic cardiomyopathy	•	[12]
NDUFV3	1. c.77 G>A 2. POLG: c.32 G > A	1. p.Arg26Gln 2. POLG: p.Gly11Asp	Complex I deficiency	NT	[22]
	1. c.168 A > C 2. POLG: c.2492 A > G 3. ND1 m3946 A > G	1. pLys56Asn 2. POLG: p.Tyr831Cys 3. ND1: p.Glu214Lys	Complex I deficiency	NT	[22]
	c.308 G > A	p.Gly103Asp	Complex I deficiency	NT	[22]
	1. c.826 G > A 2. NDUFS8:c.460 G > A	1. p.Glu276Lys 2. NDUFS8: p.Gly154Ser	Complex I deficiency	NT	[22]
	1. c.168 A > C 2. AMACR: c.554 T > C	1. pLys56Asn 2. AMACR: p.Val185Ala	Complex I deficiency	NT	[22]
NDUFS4	AAGTC at 466–470 duplication	frameshift	Leigh-like disease	•	[25]
	c.G289*del	p.Tyr97* [2]	Leigh-like disease	•	[25]
	c.G44 A	Splicing variants, no protein detected	Leigh-like disease	•	[25]
	1. c.355 G > C 2. c.462delA	1. p.Asp119His 2. p.Lys154Asnfs*34	Leigh syndrome	Smaller size	[28]
	1. c.99-1 G > A .2. c.462delA	1. p.Ser34Ilefs*4 2. p.Lys154Asnfs*34	Leigh syndrome	NT	[22]
	c.340 T > C	p.Trp114Arg	Leigh syndrome	NT	[27]
	c.221delC	p.Thr74Ilefs*17	Leigh syndrome	NT	[22]

Table 2. *Cont.*

Gene	Mutation (DNA)	Mutation (Protein)	Diagnosis	Assembly [1]	Reference
NDUFS6	c.344 G > A	p.Cys115Tyr	Neonatal lactic acidemia	NT	[33]
	c.186+2 T > A	Affected splice site, eventually leading to premature termination	Lethal infantile mitochondrial disease	NT	[32]
	1. c.343 C > A 2. c.309 + 5 G > A	1. p.Cys115Arg 2. Loss of exon 3, but some normally spliced	Leigh syndrome	•	[34,35]
NDUFA9	c.962 G > C	p.Arg321Pro	Leigh syndrome	•	[40]
	c.1078 C > T	p.Arg360Cys	Leigh syndrome	••	[87]
NDUFA1	c.22 G > C	p.Gly8Arg	Leigh syndrome	••	[43]
	c.55 C > T	p.Pro19Ser	Leigh syndrome	•	[47]
	c.111 G > C	p.Gly32Arg	Mitochondrial encephalopathy	••	[44,46,88]
	c.251 G > C	p.Arg37Ser	Myoclonic epilepsy	••	[43]
NDUFA13	c.170 G > A	p.Arg57His	Mitochondrial encephalopathy	•	[54]
NDUFA6	1. c.191 G > C 2. c.265 G > T	1. p.Arg64Pro 2. p.Glu89*	Auditory and optic neuropathy	• Lacking N- and Q-modules	[57]
	c.331_332del	p.Glu111Serfs*35	Mitochondrial-related infantile death	• Lacking N- and Q-modules	[57]
	c.3 G > A	p.Met1Ile	Brain disorder	• Lacking N- and Q-modules	[57]
	1. c.309del 2. c.355del	1. p.Met104Cysfs*35 2. p.Leu119Tyrfs*20	Leukoencephalopathy	NT	[57]
NDUFB9	c.140 G > T	p.Arg47Leu. (heterozygous)	Complex 1 deficiency	•••	[60]
	c.191 T > C	p.Leu64Pro	Complex 1 deficiency	•	[60]
NDUFS5	1. c.286 C > T 2. NDUFS8:c.4 C > T	1. p.Pro96Ser 2. NDUFS8: p.Arg2Cys	Complex I deficiency	NT	[22]
NDUFA8	1. c.404 G > A 2. C20orf: c.686 T > C	1. p.Arg135Gln 2. C20orf: p.Leu229Pro	Complex I deficiency	NT	[22]
	c.139 C > T	p.Arg47Cys	Microcephaly and epilepsy	•	[62]
	1. c.325 G > A 2. NDUFS2: c.671 C > T	1. p.Glu109Lys, mRNA not found 2. p.Ala224Val	Neonatal hypotonia and epilepsy	NT	[63]
NDUFB10	1. c.319 T > C 2. c.206_207insT	1. p.Cys107Ser 2. p.Glu70X	Fatal lactic acidosis, cardiomyopathy	•	[64]
NDUFA10	1. c.1 A > G 2. c.425 A > G	1. p.Met1Val 2. p.Gln142Arg	Leigh syndrome	•	[68]
	1. c.891 T > C 2. c.383_384insTAA	1. p.Leu294Pro 2. p.Ser218delinslS	Leigh syndrome	NT	[27]
	c.296 G > A	p.Gly99Glu	Leigh syndrome	NT	[69]
NDUFC2	c.346_*7del	p.His116_Arg119delins21	Leigh syndrome	•	[70]
	c.173 A > T	p.His58Leu	Leigh syndrome	•	[70]

Table 2. *Cont.*

Gene	Mutation (DNA)	Mutation (Protein)	Diagnosis	Assembly [1]	Reference
NDUFA11	c.99 C + 5 G > A	G to A mutation at exon 1-IVS1 splice junction	Encephalocardiomyopathy and fatal infantile lactic acidemia	NT	[72]
	1. c.317 C > T 2. c.394 G > C	1. p.Thr106Ile (isoform two) 2. p.Ala132Pro	Neuromuscular disorder	NT	[73]
NDUFB11	c.324 T > G	p.Tyr108*	Histiocytoid cardiomyopathy, and microphthalmia	NT	[74]
	c. 255 G > A	p.Trp85*	Histiocytoid cardiomyopathy, and microphthalmia	NT	[74]
	c. 262 C > T	p.Arg88*	Histiocytoid cardiomyopathy, and micropthlamia	NT	[75]
	c.402delG	p.Arg134Serfs*2	Microphthalmia and cardiomyopathy	NT	[75]
	c.361 G > A	p.Glu121Lys	Lethal infantile mitochondrial disorder	•	[27]
	c.276_278del	p.F93del	Congenital optic atrophy and myopathy	NT	[77]
NDUFB3	1. c.64 T > C 2. c.208 G > T	1. p.Trp22Arg 2. p.Gly70X	Muscular hypotonia and lactic acidosis	•	[79]
	1. c.64 T > C	p.Trp22Arg	Muscular hypotonia and lactic acidosis	••	[78,81,89]
NDUFB8	1. c.227 C > A 2. c.432 C > G	1. p.Pro76Gln 2. p.Cys114Trp	Leigh-like disease	•	[86]
	1. c.184 C > G 2. c.189delA	1. p.Tyr62His 2. p.Glu63Aspfs*35	Leigh-like disease	••	[86]

[1] Assembly: ••• normal or near normal; •• intermediate level; • little or no assembly. NT: not tested. When 2 mutations are listed, they are two alleles of the same gene, or if indicated, one is a second gene. See the text for more details. [2] p.Asn76Metfs*4 is a frameshift mutation at codon 76 that converts Asn (AAT) to Met (ATG) with a stop codon appearing 4 codons downstream. p.Tyr97* is a nonsense mutation at Tyr76 to TAG.

Among the 17 subunits that were found to have missense or other point mutations, nearly all were homozygous or compound heterozygous and therefore lacked a normal allele. This supports the likely general trend of haplosufficiency. One exception might be in NDUFB11, which is found on the X chromosome. We suggest that mosaicism in female patients due to X-inactivation might explain variable outcomes from heterozygous null mutations. To resolve such questions, it is important to extensively sequence the genome to identify other possible mutations related to the disease state. Another discrepancy was found to occur in NDUFB3, in which one girl with one homozygous mutation died in infancy [78]; meanwhile, several boys with the same mutation, in the hemizygous form, survived into adolescence and adulthood [80]. Deep sequencing might be the answer, again to try to identify other mutations.

Among the missense mutations, many are found at subunit interfaces. Such mutations are predicted to negatively impact assembly, at least local assembly. Others are found at interior sites of the proteins, and it can be predicted that protein folding will be disrupted and lead to a nonfunctional state, an effect similar to that of a gene knockout. The results of the knockouts of each of the supernumerary subunits in cell culture [11] have demonstrated that most are essential for complex I assembly. The same general trend can be seen with missense mutations in the supernumerary subunits. Essential subunits might include those that serve to stabilize the structure of complex I or that might act as chaperones in the assembly process. In recent years, analyses of complex I assembly have increased, but there have not been a great number of mutants analyzed at sufficient depth. The network of interactions among supernumerary subunits, as well as with core subunits, suggests that the instability of complex I might reflect a cooperative assembly and binding of the subunits. The loss of one supernumerary subunit could impact multiple other subunits such that assembly is not completed. Even if complex I is able to assemble, its stability might be marginal, eventually leading to a loss of function.

In conclusion, the increasingly high-resolution structures of complex I from human and other species have helped to interpret the possible roles of the supernumerary subunits and the loss of function due to mutation. Unbiased DNA sequencing will be necessary to find all of the mutations that might exist in various disease states, as illustrated by the mutations discovered in NDUFB11 to be associated with congenital sideroblastic anemia [77]. In this way, other rare diseases might become better understood and treated.

Supplementary Materials: The following are available online at http://www.mdpi.com/2075-1729/10/11/296/s1, Table S1: Interactions of Human Complex I Subunits, Video S1: Structure of Human Complex I.

Author Contributions: Conceptualization, S.B.V.; Investigation, Q.-C.L.D., D.H.P., A.N.J., M.P., H.A.A., F.Z. and S.B.V.; Writing—Original Draft Preparation, Q.-C.L.D., D.H.P. and S.B.V.; Writing—Review & Editing, Q.-C.L.D., D.H.P., A.N.J., M.P., H.A.A., F.Z. and S.B.V.; Supervision, S.B.V.; Funding Acquisition, S.B.V. All authors have read and agreed to the published version of the manuscript.

Funding: This research was funded by the National Institutes of Health, USA, grant number 1R15GM126507, and the American Heart Association, grant number 17AIREA33661165.

Acknowledgments: Q.-C.L.D., D.H.P., A.N.J. and M.P. acknowledge support of the Engaged Learning and Hamilton Scholars programs at Southern Methodist University.

Conflicts of Interest: The authors declare no conflict of interest. The funders had no role in the design of the study; in the collection, analyses, or interpretation of data; in the writing of the manuscript, or in the decision to publish the results.

References

1. Fassone, E.; Rahman, S. Complex I Deficiency: Clinical Features, Biochemistry and Molecular Genetics. *J. Med. Genet.* **2012**, *49*, 578–590. [CrossRef] [PubMed]
2. Mayr, J.A.; Haack, T.B.; Freisinger, P.; Karall, D.; Makowski, C.; Koch, J.; Feichtinger, R.G.; Zimmermann, F.A.; Rolinski, B.; Ahting, U.; et al. Spectrum of Combined Respiratory Chain Defects. *J. Inherit. Metab. Dis.* **2015**, *38*, 629–640. [CrossRef] [PubMed]
3. Formosa, L.E.; Dibley, M.G.; Stroud, D.A.; Ryan, M.T. Building a Complex Complex: Assembly of Mitochondrial Respiratory Chain Complex I. *Semin. Cell Dev. Biol.* **2018**, *76*, 154–162. [CrossRef] [PubMed]
4. Agip, A.-N.A.; Blaza, J.N.; Fedor, J.G.; Hirst, J. Mammalian Respiratory Complex I through the Lens of Cryo-EM. *Annu. Rev. Biophys.* **2019**, *48*, 165–184. [CrossRef]
5. Fiedorczuk, K.; Sazanov, L.A. Mammalian Mitochondrial Complex I Structure and Disease-Causing Mutations. *Trends Cell Biol.* **2018**, *28*, 835–867. [CrossRef]
6. Parey, K.; Wirth, C.; Vonck, J.; Zickermann, V. Respiratory Complex I—Structure, Mechanism and Evolution. *Curr. Opin. Struct. Biol.* **2020**, *63*, 1–9. [CrossRef]
7. Wirth, C.; Brandt, U.; Hunte, C.; Zickermann, V. Structure and Function of Mitochondrial Complex I. *Biochim. Biophys. Acta* **2016**, *1857*, 902–914. [CrossRef]
8. Elurbe, D.M.; Huynen, M.A. The Origin of the Supernumerary Subunits and Assembly Factors of Complex I: A Treasure Trove of Pathway Evolution. *Biochim. Biophys. Acta* **2016**, *1857*, 971–979. [CrossRef]
9. Gu, J.; Wu, M.; Guo, R.; Yan, K.; Lei, J.; Gao, N.; Yang, M. The Architecture of the Mammalian Respirasome. *Nat. Cell Biol.* **2016**, *537*, 639–643. [CrossRef]
10. Ito, M.; Morino, M.; Krulwich, T.A. Mrp Antiporters Have Important Roles in Diverse Bacteria and Archaea. *Front. Microbiol.* **2017**, *8*, 2325. [CrossRef]
11. Stroud, D.A.; Surgenor, E.E.; Formosa, L.E.; Reljic, B.; Frazier, A.E.; Dibley, M.G.; Osellame, L.D.; Stait, T.; Beilharz, T.H.; Thorburn, D.R.; et al. Accessory Subunits Are Integral for Assembly and Function of Human Mitochondrial Complex I. *Nat. Cell Biol.* **2016**, *538*, 123–126. [CrossRef] [PubMed]

12. Hoefs, S.J.; Dieteren, C.E.; Distelmaier, F.; Janssen, R.J.; Epplen, A.; Swarts, H.G.; Forkink, M.; Rodenburg, R.J.T.; Nijtmans, L.G.; Willems, P.H.; et al. NDUFA2 Complex I Mutation Leads to Leigh Disease. *Am. J. Hum. Genet.* **2008**, *82*, 1306–1315. [CrossRef] [PubMed]
13. Perrier, S.; Gauquelin, L.; Tétreault, M.; Tran, L.; Webb, N.; Srour, M.; Mitchell, J.; Brunel-Guitton, C.; Majewski, J.; Long, V.; et al. Recessive Mutations in NDUFA2 Cause Mitochondrial Leukoencephalopathy. *Clin. Genet.* **2018**, *93*, 396–400. [CrossRef] [PubMed]
14. Alagia, M.; Cappuccio, G.; Torella, A.; D'Amico, A.; Mazio, F.; Romano, A.; Fecarotta, S.; Casari, G.; Nigro, V.; TUDP; et al. Cavitating and Tigroid-Like Leukoencephalopathy in a Case of NDUFA2-Related Disorder. *JIMD Rep.* **2020**, *52*, 11–16. [CrossRef] [PubMed]
15. Sjöblom, T.; Jones, S.; Wood, L.D.; Parsons, D.W.; Lin, J.; Barber, T.D.; Mandelker, D.; Leary, R.J.; Ptak, J.; Silliman, N.; et al. The Consensus Coding Sequences of Human Breast and Colorectal Cancers. *Science* **2006**, *314*, 268–274. [CrossRef]
16. Bridges, H.R.; Mohammed, K.; Harbour, M.E.; Hirst, J. Subunit NDUFV3 is Present in Two Distinct Isoforms in Mammalian Complex I. *Biochim. Biophys. Acta* **2017**, *1858*, 197–207. [CrossRef]
17. Dibley, M.G.; Formosa, L.E.; Lyu, B.; Reljic, B.; McGann, D.; Muellner-Wong, L.; Kraus, F.; Sharpe, A.J.; Stroud, D.A.; Ryan, M.T. The Mitochondrial Acyl-carrier Protein Interaction Network Highlights Important Roles for LYRM Family Members in Complex I and Mitoribosome Assembly. *Mol. Cell. Proteom.* **2020**, *19*, 65–77. [CrossRef]
18. Guerrero-Castillo, S.; Cabrera-Orefice, A.; Huynen, M.A.; Arnold, S. Identification and Evolutionary Analysis of Tissue-Specific Isoforms of Mitochondrial Complex I Subunit NDUFV3. *Biochim. Biophys. Acta* **2017**, *1858*, 208–217. [CrossRef]
19. Dephoure, N.; Zhou, C.; Villén, J.; Beausoleil, S.A.; Bakalarski, C.E.; Elledge, S.J.; Gygi, S.P. A Quantitative Atlas of Mitotic Phosphorylation. *Proc. Natl. Acad. Sci. USA* **2008**, *105*, 10762–10767. [CrossRef]
20. Liu, F.; Lössl, P.; Rabbitts, B.M.; Balaban, R.S.; Heck, A.J.R. The Interactome of Intact Mitochondria by Cross-Linking Mass Spectrometry Provides Evidence for Coexisting Respiratory Supercomplexes. *Mol. Cell. Proteom.* **2018**, *17*, 216–232. [CrossRef]
21. Sumegi, B.; Srere, P.A. Complex I Binds Several Mitochondrial NAD-Coupled Dehydrogenases. *J. Biol. Chem.* **1984**, *259*, 5040–5045.
22. Calvo, S.E.; Tucker, E.J.; Compton, A.G.; Kirby, D.M.; Crawford, G.; Burtt, N.P.; Rivas, M.; Guiducci, C.; Bruno, D.L.; Goldberger, O.A.; et al. High-Throughput, Pooled Sequencing Identifies Mutations in NUBPL and FOXRED1 in Human Complex I Deficiency. *Nat. Genet.* **2010**, *42*, 851–858. [CrossRef] [PubMed]
23. Zhou, H.; Di Palma, S.; Preisinger, C.; Peng, M.; Polat, A.N.; Heck, A.J.R.; Mohammed, S. Toward a Comprehensive Characterization of a Human Cancer Cell Phosphoproteome. *J. Proteome Res.* **2013**, *12*, 260–271. [CrossRef] [PubMed]
24. De Rasmo, D.; Palmisano, G.; Scacco, S.; Technikova-Dobrova, Z.; Panelli, D.; Cocco, T.; Sardanelli, A.M.; Gnoni, A.; Micelli, L.; Trani, A.; et al. Phosphorylation Pattern of the NDUFS4 Subunit of Complex I of the Mammalian Respiratory Chain. *Mitochondrion* **2010**, *10*, 464–471. [CrossRef]
25. Scacco, S.; Petruzzella, V.; Budde, S.; Vergari, R.; Tamborra, R.; Panelli, D.; Heuvel, L.P.V.D.; Smeitink, J.A.; Papa, S. Pathological Mutations of the Human NDUFS4 Gene of the 18-kDa (AQDQ) Subunit of Complex I Affect the Expression of the Protein and the Assembly and Function of the Complex. *J. Biol. Chem.* **2003**, *278*, 44161–44167. [CrossRef]
26. Petruzzella, V.; Panelli, D.; Torraco, A.; Stella, A.; Papa, S. Mutations in theNDUFS4 Gene of Mitochondrial Complex I Alter Stability of the Splice Variants. *FEBS Lett.* **2005**, *579*, 3770–3776. [CrossRef]
27. Kohda, M.; Tokuzawa, Y.; Kishita, Y.; Nyuzuki, H.; Moriyama, Y.; Mizuno, Y.; Hirata, T.; Yatsuka, Y.; Yamashita-Sugahara, Y.; Nakachi, Y.; et al. A Comprehensive Genomic Analysis Reveals the Genetic Landscape of Mitochondrial Respiratory Chain Complex Deficiencies. *PLoS Genet.* **2016**, *12*, e1005679. [CrossRef]
28. Leshinsky-Silver, E.; Lebre, A.-S.; Minai, L.; Saada, A.; Steffann, J.; Cohen, S.; Rötig, A.; Munnich, A.; Lev, D.; Lerman-Sagie, T. NDUFS4 Mutations Cause Leigh Syndrome With Predominant Brainstem Involvement. *Mol. Genet. Metab.* **2009**, *97*, 185–189. [CrossRef]
29. Kmita, K.; Wirth, C.; Warnau, J.; Guerrero-Castillo, S.; Hunte, C.; Hummer, G.; Kaila, V.R.I.; Zwicker, K.; Brandt, U.; Zickermann, V. Accessory NUMM (NDUFS6) Subunit Harbors a Zn-Binding Site and Is Essential for Biogenesis of Mitochondrial Complex I. *Proc. Natl. Acad. Sci. USA* **2015**, *112*, 5685–5690. [CrossRef]

30. Kmita, K.; Zickermann, V. Accessory Subunits of Mitochondrial Complex I. *Biochem. Soc. Trans.* **2013**, *41*, 1272–1279. [CrossRef]
31. Yip, C.-Y.; Harbour, M.E.; Jayawardena, K.; Fearnley, I.M.; Sazanov, L.A. Evolution of Respiratory Complex I. *J. Biol. Chem.* **2010**, *286*, 5023–5033. [CrossRef] [PubMed]
32. Kirby, D.M.; Salemi, R.; Sugiana, C.; Ohtake, A.; Parry, L.; Bell, K.M.; Kirk, E.P.; Boneh, A.; Taylor, R.W.; Dahl, H.-H.M.; et al. NDUFS6 Mutations Are a Novel Cause of Lethal Neonatal Mitochondrial Complex I Deficiency. *J. Clin. Investig.* **2004**, *114*, 837–845. [CrossRef] [PubMed]
33. Spiegel, R.; Shaag, A.; Mandel, H.; Reich, D.S.; Penyakov, M.; Hujeirat, Y.; Saada, A.; Elpeleg, O.; Shalev, S.A. Mutated NDUFS6 is the Cause of Fatal Neonatal Lactic Acidemia in Caucasus Jews. *Eur. J. Hum. Genet.* **2009**, *17*, 1200–1203. [CrossRef] [PubMed]
34. Ogawa, E.; Shimura, M.; Fushimi, T.; Tajika, M.; Ichimoto, K.; Matsunaga, A.; Tsuruoka, T.; Ishige, M.; Fuchigami, T.; Yamazaki, T.; et al. Clinical Validity of Biochemical and Molecular Analysis in Diagnosing Leigh Syndrome: A Study of 106 Japanese Patients. *J. Inherit. Metab. Dis.* **2017**, *40*, 685–693. [CrossRef]
35. Rouzier, C.; Chaussenot, A.; Fragaki, K.; Serre, V.; Ait-El-Mkadem, S.; Richelme, C.; Paquis-Flucklinger, V.; Bannwarth, S. NDUFS6 related Leigh Syndrome: A Case Report and Review of the Literature. *J. Hum. Genet.* **2019**, *64*, 637–645. [CrossRef]
36. Fearnley, I.M.; Walker, J.E. Conservation of Sequences of Subunits of Mitochondrial Complex I and Their Relationships with Other Proteins. *Biochim. Biophys. Acta* **1992**, *1140*, 105–134. [CrossRef]
37. Agip, A.-N.A.; Blaza, J.N.; Bridges, H.R.; Viscomi, C.; Rawson, S.; Muench, S.P.; Hirst, J. Cryo-EM Structures of Complex I From Mouse Heart Mitochondria in Two Biochemically Defined States. *Nat. Struct. Mol. Biol.* **2018**, *25*, 548–556. [CrossRef]
38. Kampjut, D.; Sazanov, L.A. Structure and Mechanism of Mitochondrial Proton-Translocating Transhydrogenase. *Nature* **2019**, *573*, 291–295. [CrossRef]
39. Stroud, D.A.; Formosa, L.E.; Wijeyeratne, X.W.; Nguyen, T.N.; Ryan, M.T. Gene Knockout Using Transcription Activator-like Effector Nucleases (TALENs) Reveals That Human NDUFA9 Protein Is Essential for Stabilizing the Junction between Membrane and Matrix Arms of Complex I. *J. Biol. Chem.* **2012**, *288*, 1685–1690. [CrossRef]
40. Bosch, B.J.V.D.; Gerards, M.; Sluiter, W.; Stegmann, A.P.; Jongen, E.L.C.; Hellebrekers, D.M.; Oegema, R.; Lambrichs, E.H.; Prokisch, H.; Danhauser, K.; et al. Defective NDUFA9 as a Novel Cause of Neonatally Fatal Complex I Disease. *J. Med. Genet.* **2011**, *49*, 10–15. [CrossRef]
41. Baertling, F.; Sánchez-Caballero, L.; Brand, M.V.D.; Fung, C.-W.; Chan, S.-S.; Wong, V.-N.; Hellebrekers, D.; De Coo, I.; Smeitink, J.; Rodenburg, R.; et al. NDUFA9 Point Mutations Cause a Variable Mitochondrial Complex I Assembly Defect. *Clin. Genet.* **2018**, *93*, 111–118. [CrossRef] [PubMed]
42. Chen, R.; Fearnley, I.M.; Peak-Chew, S.Y.; Walker, J.E. The Phosphorylation of Subunits of Complex I from Bovine Heart Mitochondria. *J. Biol. Chem.* **2004**, *279*, 26036–26045. [CrossRef] [PubMed]
43. Fernandez-Moreira, D.; Ugalde, C.; Smeets, R.; Rodenburg, R.J.T.; Lopez-Laso, E.; Ruiz-Falco, M.L.; Briones, P.; Martin, M.A.; Smeitink, J.A.M.; Arenas, J. X-Linked NDUFA1 Gene Mutations Associated With Mitochondrial Encephalomyopathy. *Ann. Neurol.* **2007**, *61*, 73–83. [CrossRef] [PubMed]
44. Potluri, P.; Davila, A.; Ruiz-Pesini, E.; Mishmar, D.; O'Hearn, S.; Hancock, S.; Simon, M.; Scheffler, I.E.; Wallace, D.C.; Procaccio, V. A Novel NDUFA1 Mutation Leads to a Progressive Mitochondrial Complex I-Specific Neurodegenerative Disease. *Mol. Genet. Metab.* **2009**, *96*, 189–195. [CrossRef] [PubMed]
45. Bindu, P.S.; Sonam, K.; Chiplunkar, S.; Govindaraj, P.; Nagappa, M.; Vekhande, C.C.; Aravinda, H.R.; Ponmalar, J.J.; Mahadevan, A.; Gayathri, N.; et al. Mitochondrial Leukoencephalopathies: A Border Zone Between Acquired and Inherited White Matter Disorders in Children? *Mult. Scler. Relat. Disord.* **2018**, *20*, 84–92. [CrossRef]
46. Mayr, J.A.; Bodamer, O.; Haack, T.B.; Zimmermann, F.A.; Madignier, F.; Prokisch, H.; Rauscher, C.; Koch, J.; Sperl, W. Heterozygous Mutation in the X Chromosomal NDUFA1 Gene in a Girl With Complex I Deficiency. *Mol. Genet. Metab.* **2011**, *103*, 358–361. [CrossRef]
47. Uehara, N.; Mori, M.; Tokuzawa, Y.; Mizuno, Y.; Tamaru, S.; Kohda, M.; Moriyama, Y.; Nakachi, Y.; Matoba, N.; Sakai, T.; et al. New MT-ND6 and NDUFA1 Mutations in Mitochondrial Respiratory Chain Disorders. *Ann. Clin. Transl. Neurol.* **2014**, *1*, 361–369. [CrossRef]

48. Fearnley, I.M.; Skehel, J.M.; Walker, J.E. Electrospray Ionization Mass Spectrometric Analysis of Subunits of NADH: Ubiquinone Oxidoreductase (Complex I) From Bovine Heart Mitochondria. *Biochem. Soc. Trans.* **1994**, *22*, 551–555. [CrossRef]
49. Rak, M.; Rustin, P. Supernumerary Subunits NDUFA3, NDUFA5 and NDUFA12 Are Required for the Formation of the Extramembrane Arm of Human Mitochondrial Complex I. *FEBS Lett.* **2014**, *588*, 1832–1838. [CrossRef]
50. Abu-Safieh, L.; Vithana, E.N.; Mantel, I.; Holder, G.E.; Pelosini, L.; Bird, A.C.; Bhattacharya, S.S. A Large Deletion in the adRP Gene PRPF31: Evidence That Haploinsufficiency Is the Cause of Disease. *Mol. Vis.* **2006**, *12*, 384–388.
51. Rose, A.M.; Mukhopadhyay, R.; Webster, A.R.; Bhattacharya, S.S.; Waseem, N.H. A 112 kb Deletion in Chromosome 19q13.42 Leads to Retinitis Pigmentosa. *Investig. Ophthalmol. Vis. Sci.* **2011**, *52*, 6597–6603. [CrossRef] [PubMed]
52. Lufei, C.; Ma, J.; Huang, G.; Zhang, T.; Novotny-Diermayr, V.; Ong, C.T.; Cao, X. GRIM-19, a Death-Regulatory Gene Product, Suppresses Stat3 Activity via Functional Interaction. *EMBO J.* **2003**, *22*, 1325–1335. [CrossRef] [PubMed]
53. Huang, G.; Lu, H.; Hao, A.; Ng, D.C.H.; Ponniah, S.; Guo, K.; Lufei, C.; Zeng, Q.; Cao, X. GRIM-19, a Cell Death Regulatory Protein, Is Essential for Assembly and Function of Mitochondrial Complex I. *Mol. Cell. Biol.* **2004**, *24*, 8447–8456. [CrossRef] [PubMed]
54. Angebault, C.; Charif, M.; Guegen, N.; Piro-Megy, C.; De Camaret, B.M.; Procaccio, V.; Guichet, P.-O.; Hebrard, M.; Manes, G.; Leboucq, N.; et al. Mutation in NDUFA13/GRIM19 Leads to Early Onset Hypotonia, Dyskinesia and Sensorial Deficiencies, and Mitochondrial Complex I Instability. *Hum. Mol. Genet.* **2015**, *24*, 3948–3955. [CrossRef]
55. Maximo, V.; Botelho, T.; Capela, J.P.; Soares, P.C.; Lima, J.A.C.; Taveira, A.G.; Amaro, T.; Barbosa, A.P.; Preto, A.; Harach, H.R.; et al. Somatic and Germline Mutation in GRIM-19, a Dual Function Gene Involved in Mitochondrial Metabolism and Cell Death, Is Linked to Mitochondrion-Rich (Hürthle Cell) Tumours of the Thyroid. *Br. J. Cancer* **2005**, *92*, 1892–1898. [CrossRef]
56. Angerer, H.; Radermacher, M.; Mańkowska, M.; Steger, M.; Zwicker, K.; Heide, H.; Wittig, I.; Brandt, U.; Zickermann, V. The LYR Protein Subunit NB4M/NDUFA6 of Mitochondrial Complex I Anchors an Acyl Carrier Protein and Is Essential for Catalytic Activity. *Proc. Natl. Acad. Sci. USA* **2014**, *111*, 5207–5212. [CrossRef]
57. Alston, C.L.; Heidler, J.; Dibley, M.; Kremer, L.S.; Taylor, L.S.; Fratter, C.; French, C.E.; Glasgow, R.I.; Feichtinger, R.G.; Delon, I.; et al. Bi-allelic Mutations in NDUFA6 Establish Its Role in Early-Onset Isolated Mitochondrial Complex I Deficiency. *Am. J. Hum. Genet.* **2018**, *103*, 592–601. [CrossRef]
58. Jacome, A.S.V.; Rabilloud, T.; Schaeffer-Reiss, C.; Rompais, M.; Ayoub, D.; Lane, L.; Bairoch, A.; Van Dorsselaer, A.; Carapito, C. N- Terminome Analysis of the Human Mitochondrial Proteome. *Proteomics* **2015**, *15*, 2519–2524. [CrossRef]
59. Daub, H.; Olsen, J.V.; Bairlein, M.; Gnad, F.; Oppermann, F.S.; Körner, R.; Greff, Z.; Kéri, G.; Stemmann, O.; Mann, M. Kinase-Selective Enrichment Enables Quantitative Phosphoproteomics of the Kinome across the Cell Cycle. *Mol. Cell* **2008**, *31*, 438–448. [CrossRef]
60. Haack, T.B.; Madignier, F.; Herzer, M.; Lamantea, E.; Danhauser, K.; Invernizzi, F.; Koch, J.; Freitag, M.; Drost, R.; Hillier, I.; et al. Mutation Screening of 75 Candidate Genes in 152 Complex I Deficiency Cases Identifies Pathogenic Variants in 16 Genes Including NDUFB9. *J. Med Genet.* **2012**, *49*, 83–89. [CrossRef]
61. Murari, A.; Thiriveedi, V.R.; Mohammad, F.; Vengaldas, V.; Gorla, M.; Tammineni, P.; Krishnamoorthy, T.; Sepuri, N.B.V. Human Mitochondrial MIA40 (CHCHD4) is a Component of the Fe–S Cluster Export Machinery. *Biochem. J.* **2015**, *471*, 231–241. [CrossRef] [PubMed]
62. Yatsuka, Y.; Kishita, Y.; Formosa, L.E.; Shimura, M.; Nozaki, F.; Fujii, T.; Nitta, K.R.; Ohtake, A.; Murayama, K.; Ryan, M.T.; et al. A Homozygous Variant in NDUFA8 is Associated with Developmental Delay, Microcephaly, and Epilepsy Due to Mitochondrial Complex I Deficiency. *Clin. Genet.* **2020**, *98*, 155–165. [CrossRef] [PubMed]
63. Bugiani, M.; Invernizzi, F.; Alberio, S.; Briem, E.; Lamantea, E.; Carrara, F.; Moroni, I.; Farina, L.; Spada, M.; Donati, M.; et al. Clinical and Molecular Findings in Children with Complex I Deficiency. *Biochim. Biophys. Acta* **2004**, *1659*, 136–147. [CrossRef]

64. Friederich, M.W.; Erdogan, A.J.; Coughlin, C.R.; Elos, M.T.; Jiang, H.; O'Rourke, C.P.; Lovell, M.A.; Wartchow, E.; Gowan, K.; Chatfield, K.C.; et al. Mutations in the Accessory Subunit Ndufb10 Result in Isolated Complex I Deficiency and Illustrate the Critical Role of Intermembrane Space Import for Complex I Holoenzyme Assembly. *Hum. Mol. Genet.* **2016**, *26*, 702–716. [CrossRef]
65. Kampjut, D.; Sazanov, L.A. The Coupling Mechanism of Mammalian Respiratory Complex I. *Science* **2020**, eabc4209. [CrossRef]
66. Morais, V.A.; Haddad, D.; Craessaerts, K.; De Bock, P.-J.; Swerts, J.; Vilain, S.; Aerts, L.; Overbergh, L.; Grünewald, A.; Seibler, P.; et al. PINK1 Loss-of-Function Mutations Affect Mitochondrial Complex I Activity via NdufA10 Ubiquinone Uncoupling. *Science* **2014**, *344*, 203–207. [CrossRef] [PubMed]
67. Beinlich, F.R.M.; Drees, C.; Piehler, J.; Busch, K.B. Shuttling of PINK1 between Mitochondrial Microcompartments Resolved by Triple-Color Superresolution Microscopy. *ACS Chem. Biol.* **2015**, *10*, 1970–1976. [CrossRef]
68. Hoefs, S.J.G.; Van Spronsen, F.J.; Lenssen, E.W.H.; Nijtmans, L.G.; Rodenburg, R.J.T.; Smeitink, J.A.M.; Heuvel, L.V.D. NDUFA10 Mutations Cause Complex I Deficiency in a Patient with Leigh Disease. *Eur. J. Hum. Genet.* **2010**, *19*, 270–274. [CrossRef]
69. Minoia, F.; Bertamino, M.; Picco, P.; Severino, M.; Rossi, A.; Fiorillo, C.; Minetti, C.; Nesti, C.; Santorelli, F.M.; Di Rocco, M.; et al. Widening the Heterogeneity of Leigh Syndrome: Clinical, Biochemical, and Neuroradiologic Features in a Patient Harboring a NDUFA10 Mutation. *JIMD Rep.* **2017**, *37*, 37–43. [CrossRef]
70. Alahmad, A.; Nasca, A.; Heidler, J.; Thompson, K.; Oláhová, M.; Legati, A.; Lamantea, E.; Meisterknecht, J.; Spagnolo, M.; He, L.; et al. Bi-Allelic Pathogenic Variants in NDUFC2 Cause Early-Onset Leigh Syndrome and Stalled Biogenesis of Complex I. *EMBO Mol. Med.* **2020**, e12619. [CrossRef]
71. Andrews, B.; Carroll, J.; Ding, S.; Fearnley, I.M.; Walker, J.E. Assembly Factors for the Membrane Arm of Human Complex I. *Proc. Natl. Acad. Sci. USA* **2013**, *110*, 18934–18939. [CrossRef] [PubMed]
72. Berger, I.; Hershkovitz, E.; Shaag, A.; Edvardson, S.; Saada, A.; Elpeleg, O. Mitochondrial Complex I Deficiency Caused by a Deleterious NDUFA11 Mutation. *Ann. Neurol.* **2008**, *63*, 405–408. [CrossRef] [PubMed]
73. Peverelli, L.; Legati, A.; Lamantea, E.; Nasca, A.; Lerario, A.; Galimberti, V.; Ghezzi, D.; Lamperti, C. New Missense Variants of NDUFA11 Associated with Late-Onset Myopathy. *Muscle Nerve* **2019**, *60*, E11–E14. [CrossRef] [PubMed]
74. Shehata, B.M.; Cundiff, C.A.; Lee, K.; Sabharwal, A.; Lalwani, M.K.; Davis, A.K.; Agrawal, V.; Sivasubbu, S.; Iannucci, G.J.; Gibson, G. Exome Sequencing of Patients With Histiocytoid Cardiomyopathy Reveals a de Novo NDUFB11 Mutation That Plays a Role in the Pathogenesis of Histiocytoid Cardiomyopathy. *Am. J. Med. Genet. Part A* **2015**, *167*, 2114–2121. [CrossRef] [PubMed]
75. Van Rahden, V.A.; Fernandez-Vizarra, E.; Alawi, M.; Brand, K.; Fellmann, F.; Horn, D.; Zeviani, M.; Kutsche, K. Mutations in NDUFB11, Encoding a Complex I Component of the Mitochondrial Respiratory Chain, Cause Microphthalmia with Linear Skin Defects Syndrome. *Am. J. Hum. Genet.* **2015**, *96*, 640–650. [CrossRef] [PubMed]
76. Rea, G.; Homfray, T.; Till, J.; Roses-Noguer, F.; Buchan, R.J.; Wilkinson, S.; Wilk, A.; Walsh, R.; John, S.; McKee, S.; et al. Histiocytoid Cardiomyopathy and Microphthalmia With Linear Skin Defects Syndrome: Phenotypes Linked by Truncating Variants in NDUFB11. *Mol. Case Stud.* **2017**, 3, a001271. [CrossRef]
77. Lichtenstein, D.A.; Crispin, A.W.; Sendamarai, A.K.; Campagna, D.R.; Schmitz-Abe, K.; Sousa, C.M.; Kafina, M.D.; Schmidt, P.J.; Niemeyer, C.M.; Porter, J.; et al. A Recurring Mutation in the Respiratory Complex 1 Protein NDUFB11 is Responsible for a Novel Form of X-Linked Sideroblastic Anemia. *Blood* **2016**, *128*, 1913–1917. [CrossRef]
78. Calvo, S.E.; Compton, A.G.; Hershman, S.G.; Lim, S.C.; Lieber, D.S.; Tucker, E.J.; Laskowski, A.; Garone, C.; Liu, S.; Jaffe, D.B.; et al. Molecular Diagnosis of Infantile Mitochondrial Disease with Targeted Next-Generation Sequencing. *Sci. Transl. Med.* **2012**, *4*, 118ra10. [CrossRef]
79. Haack, T.; Haberberger, B.; Frisch, E.-M.; Wieland, T.; Iuso, A.; Gorza, M.; Strecker, V.; Graf, E.; Mayr, J.A.; Herberg, U.; et al. Molecular Diagnosis in Mitochondrial Complex I Deficiency Using Exome Sequencing. *J. Med. Genet.* **2012**, *49*, 277–283. [CrossRef]
80. Alston, C.L.; Howard, C.; Oláhová, M.; Hardy, S.A.; He, L.; Murray, P.G.; O'Sullivan, S.; Doherty, G.; Shield, J.P.H.; Hargreaves, I.P.; et al. A Recurrent Mitochondrial p.Trp22Arg NDUFB3 Variant Causes a

Distinctive Facial Appearance, Short Stature and a Mild Biochemical and Clinical Phenotype. *J. Med. Genet.* **2016**, *53*, 634–641. [CrossRef]

81. Hakim, A.; Zhang, X.; Delisle, A.; Oral, E.A.; Dykas, D.; Drzewiecki, K.; Assis, D.N.; Silveira, M.; Batisti, J.; Jain, D.; et al. Clinical Utility of Genomic Analysis in Adults With Idiopathic Liver Disease. *J. Hepatol.* **2019**, *70*, 1214–1221. [CrossRef] [PubMed]
82. Rardin, M.J.; Newman, J.C.; Held, J.M.; Cusack, M.P.; Sorensen, D.J.; Li, B.; Schilling, B.; Mooney, S.D.; Kahn, C.R.; Verdin, E.; et al. Label-Free Quantitative Proteomics of the Lysine Acetylome in Mitochondria Identifies Substrates of SIRT3 in Metabolic Pathways. *Proc. Natl. Acad. Sci. USA* **2013**, *110*, 6601–6606. [CrossRef] [PubMed]
83. Mootha, V.K.; Lindgren, C.M.; Eriksson, K.F.; Subramanian, A.; Sihag, S.; Lehar, J.; Puigserver, P.; Carlsson, E.; Ridderstrale, M.; Laurila, E.; et al. PGC-1alpha-Responsive Genes Involved in Oxidative Phosphorylation Are Coordinately Downregulated in Human Diabetes. *Nat. Genet.* **2003**, *34*, 267–273. [CrossRef] [PubMed]
84. Ling, C.; Poulsen, P.; Simonsson, S.; Rönn, T.; Holmkvist, J.; Almgren, P.; Hagert, P.; Nilsson, E.; Mabey, A.G.; Nilsson, P.; et al. Genetic and Epigenetic Factors Are Associated With Expression of Respiratory Chain Component NDUFB6 in Human Skeletal Muscle. *J. Clin. Investig.* **2007**, *117*, 3427–3435. [CrossRef] [PubMed]
85. Narimatsu, T.; Matsuura, K.; Nakada, C.; Tsukamoto, Y.; Hijiya, N.; Kai, T.; Inoue, T.; Uchida, T.; Nomura, T.; Chisato, N.; et al. Downregulation of NDUFB 6 due to 9p24.1-p13.3 Loss Is Implicated in Metastatic Clear Cell Renal Cell Carcinoma. *Cancer Med.* **2014**, *4*, 112–124. [CrossRef] [PubMed]
86. Piekutowska-Abramczuk, D.; Assouline, Z.; Mataković, L.; Feichtinger, R.G.; Koňaříková, E.; Jurkiewicz, E.; Stawiński, P.; Gusic, M.; Koller, A.; Pollak, A.; et al. NDUFB8 Mutations Cause Mitochondrial Complex I Deficiency in Individuals with Leigh-like Encephalomyopathy. *Am. J. Hum. Genet.* **2018**, *102*, 460–467. [CrossRef] [PubMed]
87. Baertling, F.; Sánchez-Caballero, L.; Timal, S.; Brand, M.A.V.D.; Ngu, L.H.; Distelmaier, F.; Rodenburg, R.J.; Nijtmans, L. Mutations in Mitochondrial Complex I Assembly Factor NDUFAF3 Cause Leigh Syndrome. *Mol. Genet. Metab.* **2017**, *120*, 243–246. [CrossRef]
88. Bindu, P.S.; Sonam, K.; Govindaraj, P.; Govindaraju, C.; Chiplunkar, S.; Nagappa, M.; Kumar, R.; Vekhande, C.C.; Arvinda, H.R.; Narayanappa, G.; et al. Outcome of Epilepsy in Patients With Mitochondrial Disorders: Phenotype Genotype and Magnetic Resonance Imaging Correlations. *Clin. Neurol. Neurosurg.* **2018**, *164*, 182–189. [CrossRef]
89. Alston, C.L.; Compton, A.G.; Formosa, L.E.; Strecker, V.; Oláhová, M.; Haack, T.B.; Smet, J.; Stouffs, K.; Diakumis, P.; Ciara, E.; et al. Biallelic Mutations in TMEM126B Cause Severe Complex I Deficiency with a Variable Clinical Phenotype. *Am. J. Hum. Genet.* **2016**, *99*, 217–227. [CrossRef]

Publisher's Note: MDPI stays neutral with regard to jurisdictional claims in published maps and institutional affiliations.

life

Review

Metabolic Alterations Caused by Defective Cardiolipin Remodeling in Inherited Cardiomyopathies

Christina Wasmus and Jan Dudek *

Comprehensive Heart Failure Center (CHFC), University Clinic Würzburg, 97078 Würzburg, Germany; Wasmus_C@ukw.de

* Correspondence: Dudek_J@ukw.de; Tel.: +49-931-201-46426; Fax: +49-931-201-646502

Received: 19 October 2020; Accepted: 6 November 2020; Published: 11 November 2020

Abstract: The heart is the most energy-consuming organ in the human body. In heart failure, the homeostasis of energy supply and demand is endangered by an increase in cardiomyocyte workload, or by an insufficiency in energy-providing processes. Energy metabolism is directly associated with mitochondrial redox homeostasis. The production of toxic reactive oxygen species (ROS) may overwhelm mitochondrial and cellular ROS defense mechanisms in case of heart failure. Mitochondria are essential cell organelles and provide 95% of the required energy in the heart. Metabolic remodeling, changes in mitochondrial structure or function, and alterations in mitochondrial calcium signaling diminish mitochondrial energy provision in many forms of cardiomyopathy. The mitochondrial respiratory chain creates a proton gradient across the inner mitochondrial membrane, which couples respiration with oxidative phosphorylation and the preservation of energy in the chemical bonds of ATP. Akin to other mitochondrial enzymes, the respiratory chain is integrated into the inner mitochondrial membrane. The tight association with the mitochondrial phospholipid cardiolipin (CL) ensures its structural integrity and coordinates enzymatic activity. This review focuses on how changes in mitochondrial CL may be associated with heart failure. Dysfunctional CL has been found in diabetic cardiomyopathy, ischemia reperfusion injury and the aging heart. Barth syndrome (BTHS) is caused by an inherited defect in the biosynthesis of cardiolipin. Moreover, a dysfunctional CL pool causes other types of rare inherited cardiomyopathies, such as Sengers syndrome and Dilated Cardiomyopathy with Ataxia (DCMA). Here we review the impact of cardiolipin deficiency on mitochondrial functions in cellular and animal models. We describe the molecular mechanisms concerning mitochondrial dysfunction as an incitement of cardiomyopathy and discuss potential therapeutic strategies.

Keywords: cardiolipin; mitochondria; Barth syndrome; Sengers syndrome; respiratory chain; Dilated Cardiomyopathy with Ataxia; cardiomyopathy

1. Introduction

The adult heart shows the highest metabolic activity of all organs in the human body by consuming 6 kg of ATP every day. It converts chemical energy stored in fatty acids, lactate and glucose into the mechanical energy to pump blood through the body. The heart contains the highest amount of mitochondria of any tissue [1], comprising about 35% of the cardiac myocyte cell volume [2,3]. A total of 95% of the energy demand of the heart is covered by oxidative phosphorylation in the mitochondria. Due to their central role in energy metabolism, a defect in mitochondria endangers the tight homeostasis of energy supply and demand in the heart. An imbalance between available energy and energy demand has been observed in almost all etiologies of heart failure [4]. Mitochondrial dysfunction can have deleterious consequences for the heart physiology and affects many forms of heart disease [5,6]. Dysfunctional mitochondria in skeletal muscle also impact heart failure and are associated with exercise intolerance [7].

Mitochondria are double membrane-surrounded organelles. The inner mitochondrial membrane forms cristae structures, which harbor the respiratory chain and form independent units of oxidative phosphorylation [8]. The respiratory chain consists of four complexes (complex I–IV), that are involved in the electron transport from NADH or $FADH_2$ onto molecular oxygen. Electron transport is coupled with proton export across the inner membrane. The corresponding membrane potential is the driving force for the fifth complex, F_1F_o-ATP synthase, to produce ATP. The ADP/ATP carrier (ANT) ensures the exchange of ATP and ADP across the inner membrane.

The reducing equivalents, NADH or $FADH_2$, are yielded in the mitochondrial Krebs cycle and NADH is additionally yielded in glycolysis. A high energy demand results in elevated levels of ADP, which accelerates ATP production at the F_1F_o-ATP synthase and thus increases the activity of the respiratory chain. To avoid draining of reducing equivalents under conditions of high energy demand, production is increased by a compensatory upregulation of the Krebs cycle. Ca^{2+} plays a key role in coupling energy demanding processes of the myofilaments with mitochondrial metabolism. During excitation, a contraction coupling release of cytosolic Ca^{2+} from the sarcoplasmic reticulum stimulates energy conversion in myofilaments. Additionally, Ca^{2+} is transported from the cytosol into mitochondria by the mitochondrial calcium uniporter (MCU). Mitochondrial Ca^{2+} potently activates several mitochondrial dehydrogenases of the Krebs cycle. This direct coupling of cytosolic and mitochondrial signals allows an immediate activation of Krebs cycle flux under conditions of increased workload. Besides their function in energy conversion, mitochondria participate in multiple metabolic pathways, such as the urea cycle, the metabolism of amino acids and lipids, and the biogenesis of heme and iron sulfur clusters. Mitochondria morphology is highly dynamic and maintained by fission and fusion processes. Mitochondrial dynamics are instrumental for many signaling pathways, such as programmed cell death, calcium signaling or innate immune responses. Impaired mitochondrial dynamics promote mitophagy. This mitochondria-specific form of autophagy maintains mitochondrial function which is particularly important for cardiac homeostasis [9].

Most of the key metabolic enzymes of the mitochondria are embedded in the inner membrane. The phospholipid Cardiolipin (CL) is the characteristic lipid of the inner membrane, playing a pivotal role in most mitochondrial metabolic activities. Here, we will discuss how CL is involved in many essential mitochondrial functions including morphology, metabolism and respiration. Defects in the biosynthesis and remodeling of CL have a strong impact on mitochondrial function and particularly affect tissues with a high energetic contribution of mitochondria, such as the heart and neuronal tissue. Diseases with a direct link to CL biosynthesis and remodeling comprise Sengers disease (OMIM 212350), Barth syndrome (OMIM 302060) and Dilated Cardiomyopathy with Ataxia (DCMA, OMIM 610198). Changes in CL levels are also involved in other cardiac modifications including ischemia/reperfusion injury, diabetic cardiomyopathy and the aging heart.

2. CL Biosynthesis

Mitochondrial membranes are characterized by their high content of membrane proteins and unique phospholipid composition. A recent study found the respiratory chain, the ADP/ATP carrier, phosphatidylethanolamine, phosphatidylcholine and cardiolipin to be the main constituents of the inner membrane [10]. The hallmark lipid is the dimeric phospholipid cardiolipin (CL), which is found almost exclusively in the inner mitochondrial membrane. Four different fatty acids can be bound to CL forming different CL species, leading to a highly diversified CL pool [11]. CL in the colon has short and mostly saturated fatty acids, whereas CL in the brain primarily constitutes long and unsaturated CL species [12]. The heart has an unique CL pool consisting mostly of tetralinoleoyl-CL (CL(18:2)). Interestingly, the highest proportions of oxidized CL were found in heart ($1.8\% \pm 0.7\%$) and skeletal muscle [12]. The diversified CL pool is acquired by reshaping CL acyl composition after its initial biosynthesis. CL is synthesized by enzymes located in the inner mitochondrial membrane.

An important step in the biosynthesis of CL in the inner membrane is catalyzed by PTPMT1, converting phosphatidylglycerol phosphate to phosphatidylglycerol (Figure 1). Deletion of this in mice is embryonically lethal and PTPMT1-deficient Mouse Embryonal Fibroblasts (MEFs) of embryos obtained from intercrosses of PTPMT1+/flox mice reveal the essential role of this enzyme in CL biosynthesis and affect mitochondrial morphology and respiration [13]. CL Synthase (CLS1) catalyzes the subsequent addition of a second molecule of CDP-DAG to form premature CL (Figure 1). The embryonic lethality of CLS1-deficient animals underscores the essential role of CLS1 in CL biosynthesis. Neuron-specific knockout results in neuronal loss and gliosis in the forebrains as a result of defective respiration and morphological abnormalities in these mitochondria [14].

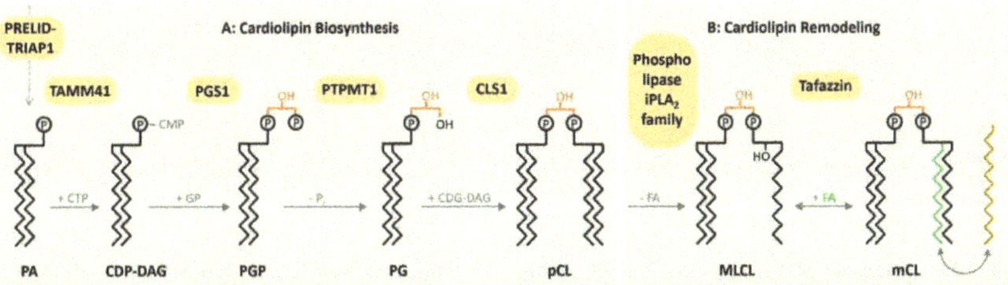

Figure 1. Biosynthesis and remodeling of cardiolipin (CL) in the inner membrane. Phosphatic acid (PA) as a precursor reaches the mitochondrial matrix via the PRELID–TRIAP1 complex. The CDP-DAG synthase TAMM41 activates PA with cytidine triphosphate (CTP), producing CDP-DAG. Subsequently, the phosphatidylglycerol phosphate synthase 1 (PGS1) catalyzes the formation of phosphatidylglycerol phosphate (PGP), followed by the dephosphorylation generating phosphatidylglycerol (PG) via the mitochondrial protein-tyrosine phosphatase 1 (PTPMT1). In a last reaction, Cardiolipin synthase 1 (CLS1) mediates the emergence of premature CL based on one molecule of PG and one CDG-DAG. Following the initial synthesis, CL remodeling ensues. After the removement of fatty acids (FAs) mediated by a member of the phospholipase iPLA2 family, the incorporation of new FAs is proceeded by Tafazzin to form matured CL from Monolysocardiolipn (MLCL). To reach an energetic optimum, the remodeling and the ongoing exchange of FAs is a continual persisting process, culminating in a highly diversified CL pool.

Afterwards, CL remodeling is initiated by deacylation, catalyzed by mitochondrial members of the Ca^{2+} independent phospholipases to form monolysocardiolipin (MLCL). Subsequently, CL is reacetylated by Tafazzin to form mature CL (Figure 1).

The protein Tafazzin is produced by alternative splicing of the TAZ gene. The gene consists of 11 exons and two ATG initiation sites. Transcription of the TAZ gene gives rise to multiple mRNAs by alternative splicing at exons 5–7. A highly hydrophobic segment of 30 residues at the *N*-terminus acts as a membrane anchor. Tafazzin is directed to mitochondria with the help of two independent targeting sequences [15]. The active site of phospholipid-binding has been identified in a 57-amino acid cleft containing positively charged residues [16]. The substrate specificity of Tafazzin was found to be surprisingly low and cannot explain the tissue-specific remodeling of the CL pool [17]. Recent data suggest that Tafazzin's substrate specificity is determined by the thermodynamic properties of lipid domains in the vicinity of the enzyme [17]. Tafazzin exchanges fatty acids in the CL molecule until the molecular species composition with optimal packing conditions and the lowest free energy is established [18]. A recent study from the Schlame lab suggests that the thermodynamic properties of lipid domains are determined by the assembly of respiratory chain complexes. This study proposes that protein crowding in the respiratory chain imposes packing stress on the lipid bilayer, which is relieved by CL remodeling to form tightly packed lipid–protein complexes [19]. Further evidence was provided by an integrative approach of lipidomics and transcriptomics. Oemer et al. found that a transcriptional regulation of CL biosynthesis genes shows no correlation with the tissue-specific CL composition. Interestingly, the only significant compliance with CL content was found for genes of the respiratory chain. This finding supports the idea that the respiratory chain and other protein complexes determine the tissue-specific CL composition of the membrane and hence its properties.

Besides Tafazzin, two other enzymes are involved in CL remodeling. The Coenzyme A-dependent monolysocardiolipin acyltransferase (MLCLAT1) is a splice variant of the HADHA gene encoding for the α-subunit of the human trifunctional enzyme (α-TFP) [20,21]. In a recent study, Miklas et al. show that apart from its function in the β-oxidation cycle, α-TFP plays a role in CL remodeling [22]. As patient-derived HADHA knockout cardiomyocytes do not accumulate MLCL as TAZ mutant cells, it has been concluded that HADHA acts subsequently to TAZ to mature the CL pool. The second enzyme is the ER-MAM resident lysocardiolipin acyltransferase (ALCAT1). Overexpression of ALCAT1 in mouse myoblasts increases the incorporation of docosahexaenoic acid (22:6) into CL, yielding a peroxidation prone form of CL, which has been suggested to reflect a pathogenic CL remodeling [23].

3. Function of CL in Mitochondrial Morphology

Structural abnormalities of mitochondria, including fragmentation, disruption of mitochondrial membranes and loss of the electron-dense matrix have been observed in animal models of heart failure and in patients suffering from hypertrophic cardiomyopathy [24]. In a dog model of chronic heart failure, the mitochondrial ultrastructure changes from tightly packed to disorganized cristae [25]. Mitochondria possess two highly specialized membranes—the outer and the inner mitochondrial membrane. Invaginations in the inner membrane form the cristae structures, which are areas of intensive membrane curvature. CL plays a considerable role in shaping the morphology of mitochondria and prominently locates in the bended regions of the inner membrane [26]. CL adopts a cone-shaped structure due to the high content of unsaturated fatty acids, which is causative for the membrane bending at sites of high CL concentration. In addition, mitochondrial morphology is also shaped by a constant remodeling through the fission and fusion of individual mitochondria [27]. The opposing processes of fusion (merging of two mitochondria) and fission (segregation of two mitochondria) are shown to form a highly dynamic network of mitochondria in the cell. The continuing remodeling by fission and fusion is essential for normal mitochondrial function, downregulation of mitochondrial fusion promotes apoptosis and cardiomyocyte loss [28]. Mitochondrial fragmentation was elucidated to influence heart failure notably [29]. By its role in regulating mitochondrial fission and fusion, CL actively participates in shaping the mitochondrial network. Fission and fusion are regulated by a set of dynamin-related GTPases, whereby CL serves as a modulator of the activity of the fission protein Drp1 and the fusion protein OPA1 [30,31]. Morphological alterations of mitochondria are evident in human patients with defects in CL biosynthesis and remodeling as well as in animal models (Table 1).

Table 1. Role of CL in inherited cardiomyopathies in animal models and human.

Function	Anmial	Human
Mitochondrial morphology	A tafazzin knockdown mouse model of Barth syndrome describes alteration of mitochondrial phospholipid compositions via lipidomics [32]. Morphology alterations as mitochondrial enlargement, concentric layers of cristae or large vacuoles were observed in tafazzin-deficient mice [33].	BTHS patient-derived lymphoblasts (BTHS lymphoblasts) reveal enlarged mitochondria with a lower surface area of cristae and altered morphology [34]. Another study with BTHS lymphoblasts detected giant, partly onion-shaped mitochondria [35].
Oxidative phosphorylation	A BTHS mouse model with an inducible systemic knockdown of tafazzin gene shows a reduced respiration on succinate as well as on pyruvate and malate. Furthermore, respirasome remodeling was detected [36].	BTHS patient-derived induced pluripotent stem cells (BTHS-iPSC) reveal structural remodeling of respiratory chain complexes resulting in decreased mitochondrial respiration [37]. In a second study, mitochondria of BTHS lymphoblasts show a reduced respiratory activity on succinate and ascorbate [34].
Krebs cycle	BTHS mouse model reveals a striking reduction in succinate dehydrogenase activity in cardiac mitochondria [36].	BTHS skin fibroblasts reveal a significant destabilization of 2-oxoglutarate dehydrogenase and branched-chain ketoacid dehydrogenase [38].
Apoptosis	Murine germline TAZ knockout mice model reveals significant increased cardiomyocyte apoptosis and fibrosis occurrence [39].	BTHS lymphoblasts show a requirement of cardiolipin for apoptosis and an apoptotic defect [40].

The formation of cristae structures critically depends on the mitochondrial contact site and cristae organizing system (MICOS). These protein complexes locate at the cristae junctions and seal individual cristae to maintain individual membrane potentials [8]. The MICOS complex is a central component of a large interaction network which includes several complexes in the inner membrane and outer membrane [41]. MIC27 (APOOL) as a MICOS complex constituent directly interacts with CL and this interaction was found to be essential for MIC27 assembly in the MICOS complex [42,43] (Figure 2). MIC27 and the structurally related MIC26 also modulate the cardiolipin remodeling enzyme Tafazzin (see below) [44]. In fibroblasts from Barth syndrome patients, a compensatory increase in MICOS subunits was observed [38,45]. Interestingly, a slightly lower molecular mass was detected for MICOS subunits in Barth syndrome (BTHS) fibroblasts, indicative of structural changes due to CL deficiency. Application of the CL-interacting molecule SS31/Elamipretide normalized deregulated structural proteins, such as Drp1, Mfn2, Opa1 and Mic60 in human heart failure patients [46].

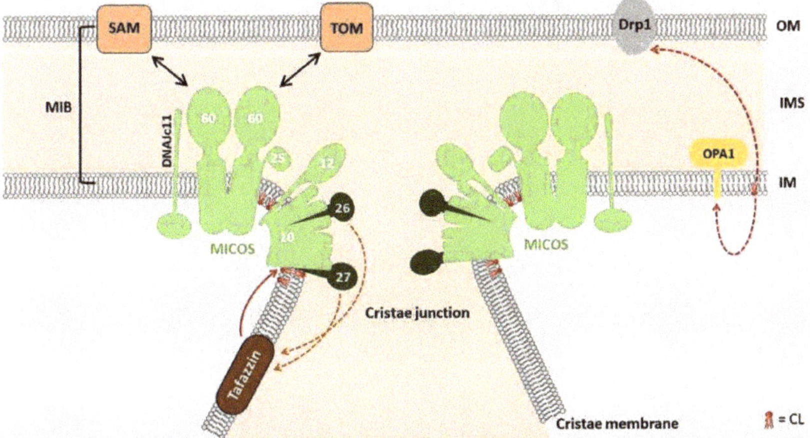

Figure 2. The role of CL in mitochondrial morphology: The mitochondrial contact site and cristae organizing system (MICOS) complex requires CL for optimal structural integrity. MICOS is a protein complex located in the inner mitochondrial membrane, playing an essential role in cristae junction formation. The resulting membrane invaginations harbor the respiratory chain complexes. By interactions with other proteins in the outer mitochondrial membrane such as the Translocase of the Outer membrane (TOM) and the sorting and Assembly Machinery (SAM), the mitochondrial intermembrane space (IMS) bridging complex (MIB) is formed. The MICOS subunits MIC27 and MIC26 influence the regulation of CL levels, while CL is remodeled by Tafazzin [47]. Not only the integration of MIC27 in MICOS, but also the function of other mitochondrial membrane anchored proteins such as the fission and fusion proteins Drp1 and OPA1 is affected by CL [48]. OM, outer membrane; IMS, intermembrane space; IM, inner membrane.

4. Function of CL in Energy Metabolism

In total, 95% of the cardiac energy demand is covered by oxidative phosphorylation of the respiratory chain in the inner membrane, consisting of five complexes (I–V). Four complexes (I–IV) transport electrons from reducing equivalents (NADH, $FADH_2$) to oxygen, forming water. The resulting energy is stored in a membrane potential across the inner membrane and converted to ATP by the last complex in the chain (V). The structure of the respiratory chain has been resolved and specific interaction sites for CL were identified in all respiratory chain complexes [36,49]. Correspondingly, CL is a structural component of the respiratory chain and is essential for its integrity and full enzymatic activity [50–53]. Changes in the CL pool, including the accumulation of MLCL, may interfere with the structure of respiratory chain complexes. MLCL, which strongly accumulates in Barth syndrome, binds to complex IV with a substantially reduced affinity and significantly lowers the enzymatic activity of this complex [54]. The structure of the dimeric ATP synthase from bovine mitochondria recently shed light upon the mechanism of proton uptake in the matrix. Interestingly, this study also described the incidence of CL in the integral membrane subunits of complex V [55]. Furthermore, CL is directly involved in the proton export and required for the organization of the F_1F_O ATPase into highly ordered structures [56,57].

The respiratory chain complexes assemble into higher-ordered structures—the respirasomes. These supercomplexes include complex I, a dimer of complex III, and one or several complex IV units (Figure 3). CL molecules are not an only integral to individual complexes but are also associated with the mitochondrial supercomplex formation. It was found to support the structure of supercomplexes and mediate the interaction with the lipid phase of the inner membrane [58–60]. Due to its role in the structure of the respirasomes, CL deficiency causes a defect in respiratory function and a decrease in membrane potential and in ATP synthesis [36,37,61] (Table 1).

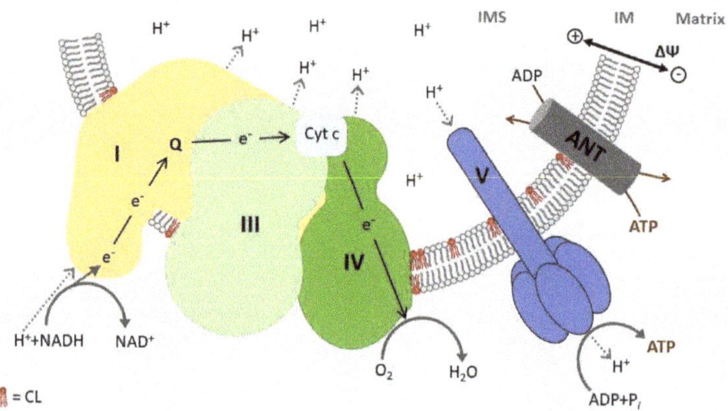

Figure 3. CL is an essential constituent of respirasomes: The mitochondrial respiratory chain assembles into large oligomeric structures called respirasomes. Respirasomes consist of complex I, a dimer of complex III, and several copies of complex IV. CL is required for respirasome formation and is essential for the activity of the complexes. CL molecules partially interacting with membrane protein complexes are shown in red. ANT, ADP/ATP carrier; Cyt c, Cytochrom c; IM inner membrane; IMS, intermembrane space.

In many forms of cardiac disease, reduced CL levels, alterations in the CL pool or Reactive Oxygen Species (ROS)-induced damage of CL have been observed. Given the important structural function of CL, these changes may have direct structural implications for the respiratory chain. Remodeling of the respiratory chain due to changes in CL has been described in aging, ischaemia/reperfusion and heart failure [62–65]. These findings are of particular importance as the architecture of respirasomes prevents ROS production. Structural alterations may induce increased ROS generation at the respirasomes, and increased oxidative stress may be an important contributor to the development of heart failure [66]. Therefore, a large number of studies have focused on preventing mitochondrial ROS in various forms of cardiac disease [67,68]. These studies, although somewhat inconsistent, showed that reducing ROS levels has the potential to ameliorate ROS-mediated cardiac abnormalities [69–71].

5. Function of CL in Intermediate Metabolism

In order to maintain the membrane potential, the inner membrane must be tightly sealed and transport processes across the membrane need to be vigorously controlled. This contrasts with the demands of energy metabolism, which requires an intensive exchange of metabolites between the cytosol and the matrix. This is ensured by the superfamily of carrier proteins in the inner membrane mediating the transport of metabolites across the inner membrane. The most abundant carrier protein is the ADP/ATP carrier (ANT), which exchanges ATP in the matrix with ADP in the cytosol. The recent resolution of the matrix-open state advances our understanding of the molecular mechanism of metabolite cycling, which most likely applies to the whole carrier family [72]. The ADP/ATP carrier possesses not only a tight binding to CL (Figure 3), but CL was also identified in its crystal structure from bovine heart mitochondria [73,74]. In addition, studies of bakers' yeast show that the conformation of the ADP/ATP carrier is controlled by CL acting as a link for dimer formation and the integration of ANT into a complex interaction network with the respiratory chain [75]. A requirement for CL has also been documented for other members of the carrier family including the Phosphate carrier (PiC), the monocarboxylate carrier (MCT1), carnitine/acylcarnitine translocase, pyruvate carrier and the tricarboxylate carrier [76–80]. Carrier proteins are integral proteins of the inner membrane, synthesized in the cytosol and then transported across the outer membrane to become subsequently integrated into the inner membrane. Their incorporation depends on the protein Translocase of Inner Membrane TIM22. TIM22 is a protein complex and is integrated in the inner membrane in a CL-dependent manner. Moreover, one structural component of the TIM22 complex is acylglycerol kinase (AGK). Besides its role in protein translocation, it also has a second function as a kinase in the biosynthesis of phosphatidic acid (PA), which serves as a precursor of CL biosynthesis [81,82].

The protein creatine kinase (CK) catalyzes the reversible conversion of creatine and adenosine triphosphate (ATP) into phosphocreatine (PCr) and adenosine diphosphate (ADP) and is expressed in the heart, skeletal muscle, brain and kidney. Phosphocreatine serves as a buffer for rapid regeneration of ATP in tissue with a high energy demand. Mitochondrial creatine kinase (mtCK) is located in the mitochondrial intermembrane space, where it uses the local ATP concentration to generate phosphocreatine. In heart and skeletal muscle, sarcomeric mtCK catalyzes the reverse reaction to regenerate ATP in close proximity of the site of high energy turnover. In-vitro experiments suggest CL as one of the main binding sites of the creatine kinase to the inner membrane [83,84]. mtCK was also found to form large oligomeric complexes in the intermembrane space [85]. By binding to CL, the creatine kinase was also suggested to mediate CL transfer between the inner and outer membrane [86].

Studies in CL-deficient yeast show reduced levels of acetyl-CoA due to a decreased activity of acetyl-CoA synthetase. Despite a compensatory upregulation of pyruvate dehydrogenase (PDH), the enzymatic activity was not increased, suggesting a defect in the specific PDH activity [75]. A C2C12 myoblast model of BTHS confirmed diminished PDH activity and proposed that an increased level of inhibitory phosphorylation was responsible for the defect. The authors suggested that CL is required to facilitate the binding of the pyruvate dehydrogenase phosphatase to the E2 subunit and reduced binding enhances inhibitory PDH phosphorylation [87]. To compensate for the deficiency in PDH activity, the pyruvate carboxylase is upregulated. A defect in PDH activity was not verified in an induced Pluripotent Stem Cell-derived Cardiomyocyte (iPSC-CM) model of BTHS, which even showed an increased flux of glucose into the Krebs cycle intermediate citrate. Accordingly, the anaplerotic supplementation by carboxylation of pyruvate was reduced in this cell model [88].

CL deficiency can also affect the mitochondrial Krebs cycle (TCA). The α-ketoglutarate dehydrogenase complex was found to be structurally affected in human BTHS patient fibroblasts; however, its enzymatic activity and the metabolic flux of glutamate into the Krebs cycle remained unaffected [38]. In addition, a CL-deficient yeast model showed alterations in the enzymatic activities of aconitase and succinate dehydrogenase [87]. Metabolic flux analyses in iPSC-CM displayed an increased level of the Krebs cycle intermediate citrates and decreased level of fumarate in BTHS, indicative of a reduced turnover of succinate into fumarate [88]. Aconitase and succinate dehydrogenase are strictly dependent on iron sulfur clusters as a cofactor. These data and an increase in mitochondrial iron amounts pointed to a defect in the biogenesis of iron sulfur clusters. CL is required for the correct processing of the mitochondrial protein frataxin, which is an important scaffold protein in the iron sulfur cluster biogenesis [89,90]. Consistent with a defect in the TCA cycle, a recent report showed that anaplerotic pathways are required to ameliorate TCA cycle dysfunction in yeast cells [91]. Other cofactors may also be affected. A recent study assumed that lower levels of the cofactor Coenzyme A contribute to the respiratory deficiency in BTHS [92]. Coenzyme A is an important cofactor for fatty acid metabolism. Studies in yeast, however, described that increased fatty acid oxidation can compensate for the Krebs cycle defects in CL-deficient yeast [91]. Under conditions of a high-fat diet, Cole et al. were able to show intensified accumulation of triacylglycerides in cardiac tissue of BTHS mice compared to control animals. The authors found the upregulated synthesis of fatty acid synthase to be responsible for this effect. Increased triacylglycerides may enhance the susceptibility to lipotoxicity in BTHS under conditions of increased fat uptake [93].

Diseases, which are associated with defects in CL, such as DCMA, Sengers syndrome and Barth syndrome, commonly present with 3-methylglutaconic aciduria (3-MGA) accompanied by increased levels of lactic acid [94]. 3-Methylglutaconic and the related 3-methylglutaric acid are catabolic intermediates of the branched-chain amino acid (BCAA) leucine. Additionally, elevated levels of the intermediate of isoleucine metabolism, 2-ethylhydracrylic acid (2-EHA), occurred in the urine of BTHS patients [95]. A few studies allow for speculations of a defect in the metabolism of the branched-chain amino acids valine, leucine and isoleucine in BTHS. Transcriptome analysis of shTAZ mouse model revealed reduced gene expression of genes involved in BCAA breakdown [32]. A recent analysis of protein complexes in the mitochondria of BTHS patient skin fibroblasts revealed a profound destabilization of the branched-chain ketoacid dehydrogenase complex, which is involved in the initial oxidative degradation step of branched-chain amino acids [38]. Further investigations are required to elucidate the molecular mechanism of the observed secretion of organic acids in BTHS.

6. CL Function in Calcium Homeostasis

Calcium is an important regulator of sarcomere contraction in the heart. In systole, Ca^{2+} influx via L-type Ca^{2+} channels triggers the release of Ca^{2+} from ryanodine receptors (RyRs) in the sarcoplasmic reticulum (SR). Ca^{2+} binding to troponin C induces contraction. During diastole, Ca^{2+} is transported back into the SR by the sarcoplasmic reticulum Ca^{2+}-ATPase (SERCA) or exported across the cell membrane via the Na^+/Ca^{2+} exchanger [96] (Figure 4). By means of the active transport of Ca^{2+} ions, the P-type ATPase SERCA is required for muscle relaxation and the proper regulation of muscle contraction. It also ensures a sufficient Ca^{2+} load in the sarcoplasmic reticulum for systolic contraction. Oxidative stress has been associated with a deregulation of excitation–contraction coupling in the BTHS. Peroxynitrite formed by increased levels of superoxide and nitric oxide (NO) can build adducts with tyrosine residues, which may change the structure or catalytic activity of target proteins [97]. Tyrosine nitrosylation of SERCA in BTHS leads on to a decrease in SERCA activity and results in a decline of SR Ca^{2+} levels. These abnormalities may promote left ventricular diastolic dysfunction. A decrease in SR Ca^{2+} levels has been observed in many forms of heart failure, including dilated cardiomyopathy [98,99].

Calcium is an important regulator of mitochondrial metabolism. Accelerated cardiac workload causes an enhanced demand of ATP. The conversion of ATP to ADP in energy-consuming processes such as contraction increases ADP levels, accelerates respiration and results in an elevated oxidation of reducing equivalents. To compensate for the higher demand for reducing equivalents, Ca^{2+} plays an important role in activating the mitochondrial Krebs cycle [100]. During excitation–contraction coupling, Ca^{2+} is emitted from the sarcoplasmic reticulum and is transmitted into the mitochondria via the Mitochondrial Calcium Uniporter (MCU) (Figure 4). To allow for an efficient calcium transmission, ryanodine receptors (RyRs) in the sarcoplasmic reticulum are located in close proximity to mitochondria, especially next to the mitochondrial calcium uniporter (MCU) in the inner membrane [101]. Ca^{2+} stimulates key dehydrogenases of the Krebs cycle, including pyruvate dehydrogenase, isocitrate dehydrogenase and α-ketoglutarate dehydrogenase, accelerating the regeneration of NADH and $FADH_2$. However, the forced regeneration of reducing equivalents activates the respiratory chain.

Therefore, Ca^{2+} plays a prominent role in adapting mitochondrial metabolism to increased energy demands during accelerated cardiac workload [102]. The pore-forming unit of the mitochondrial calcium uniporter is the protein MCU, which associates EMRE and regulatory subunits MICU1, MICU2, and MCUb into a complex integrated into the inner membrane [103–106] (Figure 4). The association of phospholipids in MCU has been revealed in a recent structural analysis [107]. Moreover, a specific requirement of the MCU for CL was found. Consequently, the ability of MCU to assemble into functional complexes was reduced in cardiac patient samples of BTHS [108].

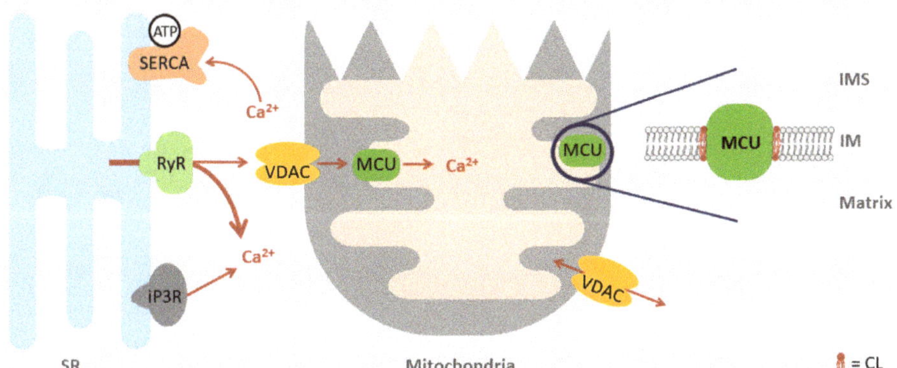

Figure 4. Mitochondrial Calcium Uniporter: Calcium transport from the sarcoplasmic reticulum (SR) is mediated by different proteins—Ryanodine receptors (RyRs) and inositoltriphosphate receptors (iP3Rs) release Ca^{2+} under systolic conditions, while the voltage-dependent anion channel (VDAC) and mitochondrial calcium uniporter (MCU) allow the Ca^{2+} uptake in mitochondria. MCU is embedded in the inner mitochondrial membrane and requires CL for optimal activity. The sarcoplasmic reticulum Ca^{2+}-ATPase (SERCA) controls the Ca^{2+} uptake by the SR. IMS, intermembrane space; IM, inner mitochondrial membrane.

7. Barth Syndrome

BTHS patients have a quite variable clinical presentation of cardiomyopathy ranging from milder cases to severe cases, which require cardiac transplantation. BTHS patients present with a dilated or hypertrophic cardiomyopathy and left ventricular non-compaction [109–112]. The left ventricular ejection fraction (LVEF) was only moderately reduced—50 ± 10% (normal LVEF 50–70%, mild dysfunction LVEF 40–49%, moderate dysfunction LVEF 30–39% and severe dysfunction LVEF <30%) [111,112]. Despite preserved LVEF, many patients show an inability to increase cardiac output during exercise [113]. A very common symptom is skeletal muscle fatigue and exercise intolerance [110,113]. BTHS patients also manifest metabolic abnormalities in the degradation pathway of branched amino acids and show lactic acidosis during exercise as well as an elevated excretion of 3-Methylglutaconic acid (3-MGA) in urine [113,114] (Figure 5). Defects in the immune system include persistent or intermittent neutropenia causing recurrent infections. A growth deficiency during childhood is compensated by a delayed growth spurt after 12–14 years of age in BTHS [111,112,115].

BTHS is frequently associated with a severe exercise intolerance, which makes it difficult to perform activities of daily living and drastically reduces quality of life. Skeletal muscle impairments include harmed functional exercise capacity, diminished extensor strength, and lowered daily activity [116]. Exercise intolerance is thought to be a consequence of cardiac impairment and decreased skeletal muscle oxygen utilization [113]. Changes in CL levels also in the skeletal muscle of Barth syndrome patients cause the destabilization of respirasomes, reduction in respiration, excessive production of reactive oxygen, abnormal mitochondrial morphology and defects in ATP production [115,117,118]. The development of a cardiac-specific knockout model recapitulated cardiomyopathy with reduced fractional shortening, increased hypertrophy and left ventricular dilatation. A constitutive knockout of Taz, however, resulted in very poor survival. Interestingly, the low survival rates were rescued, upon skeletal muscle-specific virus transmitted gene replacement therapy, indicating a particular contribution of skeletal myopathy to the reduced survival of the mice [39].

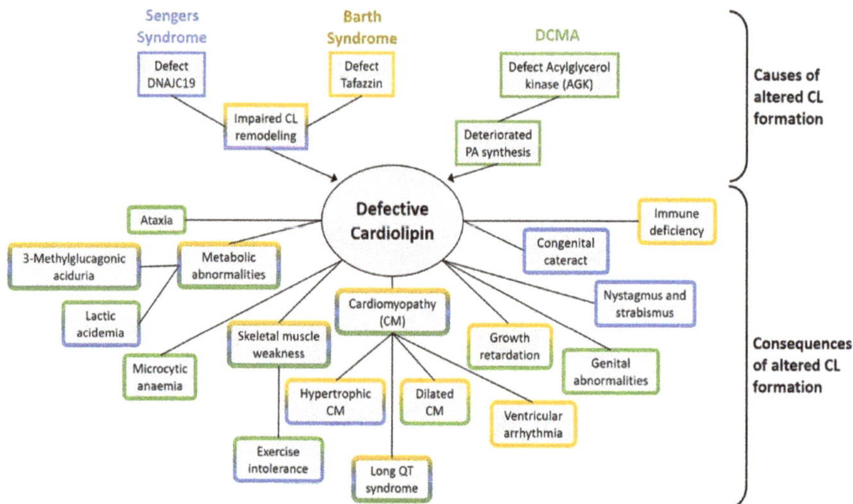

Figure 5. Clinical alterations, caused by a defect in the CL pool: Colors indicate clinical symptoms matching Sengers syndrome (blue), Barth syndrome (yellow) and dilated cardiomyopathy with ataxia (DCMA, green).

Using ^{31}P nuclear magnetic resonance (NMR) spectroscopy, a higher content of glycolytic fibers (type 2, fast-twitch) and a smaller fraction of oxidative fibers (type 1, slow-twitch) has been documented in the skeletal muscle of BTHS patients [119,120]. This finding suggests that BTHS patients rely on glycolytic metabolism to a greater extent than control individuals. Moreover, a higher respiratory exchange ratio during exercise and a greater glucose rate of disposal during a hyperinsulinemic–euglycemic clamp procedure indicate higher glucose usage to compensate for the impaired mitochondrial capacity to generate ATP [113,121]. Interestingly, exercise intolerance has also been described in patients with chronic acquired heart failure [122].

BTHS is caused by mutations in the X chromosomal gene encoding for Tafazzin (TAZ) [123]. Pathogenic mutations in TAZ include frameshift, splice-site, missense, and non-sense mutations [124]. Mutations are associated with a decrease in the amount of functional enzymes, mislocalization of the protein in the cell, protein aggregation or altered macromolecular complex assembly [124,125]. Protein levels and messenger RNA of cardiolipin synthase (CLS), Tafazzin (TAZ) and acyl-CoA:lysocardiolipin acyltransferase-1 (ALCAT-1) are affected in cardiac tissue of BTHS patients. Loss of Tafazzin function causes a significant alteration in the CL pool including a reduction in mature forms of CL and an increase in MLCL, which also serves as a diagnostic marker for BTHS [126,127]. Decreased respiration and reduced activity of single respiratory enzymes have been confirmed in several models of BTHS, including BTHS patient-derived fibroblasts and lymphoblasts as well as cellular and animal models [36,40,128].

BTHS patients have a significantly lower body weight and fat free mass [121]. When measured under resting conditions, the fatty acid oxidation rate related to body mass in BTHS patients was comparable to the control group. However, under exercise conditions the elevated fatty acid oxidation rate was severely blunted in BTHS [121]. In turn, glucose turnover was already increased at rest. Therefore, the metabolic derangements, in particular the inability to upregulate fatty acid metabolism, reflect the exercise inability of BTHS patients [129]. Abnormal blood levels of amino acids were determined in BTHS patient [94,130]—in particular, amino acids, which are associated with anaplerosis in the intermediate metabolism (arginine, ornithine and citrulline), were consistently lower. This strongly supports the hypothesis of alterations in the energy metabolism in BTHS. Increased levels of tyrosine, proline, and asparagine compared to the control were found under starvation conditions [94,130]. Considering that the lower lean and skeletal muscle mass of BTHS patients, the unchanged rates of appearance of the ketogenic amino acid leucine upon starvation is remarkable and led to the hypothesis of a higher proteolysis rate in BTHS patients in order to use amino acids for energy needs [121].

8. Sengers Syndrome

The clinical manifestation of Sengers syndrome is hypertrophic cardiomyopathy and congenital cataracts. Other symptoms including skeletal myopathy, exercise intolerance, lactic acidosis and 3-methylglutaconic aciduria, showing strong similarities to Barth syndrome [131] (Figure 5). Sengers syndrome is caused by mutations in the gene encoding for mitochondrial acylglycerol kinase (AGK) [132]. AGK is a mitochondrial lipid kinase strongly expressed in heart, but also in the skeletal muscle, kidney and brain. This protein can phosphorylate monoacylglycerol to lysophosphatidic acid (LPA) and diacylglycerol to phosphatidic acid (PA), a precursor of CL biosynthesis [132]. AGK has an additional independent function in mitochondrial protein import as a constituent of the mitochondrial carrier translocase TIM22 complex [81,133]. This inner membrane-embedded protein translocase imports proteins of the carrier family from the cytosol. Carrier proteins are a large family of proteins which mediate the transport of metabolites across the inner membrane. A defect in the import and assembly of carrier proteins has been found in AGK-deficient mitochondria [81,133]. In particular, decreased levels of mitochondrial ADP/ATP carrier in heart and muscle tissues was the initial finding in patients of Sengers syndrome [134]. The function in protein transport is independent of AGK kinase activity. It can be assumed that loss of both functions contributes to the pathogenesis in Sengers syndrome.

9. Dilated Cardiomyopathy with Ataxia (DCMA)

Mutations in the gene encoding for the mitochondrial protein DNAJC19 were found to be causative for Dilated Cardiomyopathy with Ataxia (DCMA) [135]. DCMA patients presenting with dilated cardiomyopathy and arrhythmias due to abnormalities in repolarization after a heartbeat (long QT syndrome, LQTS) [136]. Similar to BTHS and Sengers syndrome, 3-methylglutaconic aciduria is commonly described in DCMA (Figure 5). Other symptoms include cerebellar ataxia, growth retardation or genital anomalies in male patients [137]. Patient-derived iPSC-CM models of DCMA have been developed recently [138,139]. The function of the affected gene (DNAJC19) is unknown but it shares sequence similarities with the family of DnaJ proteins, which act as cofactors of Hsp70 chaperones. DNAJC19 binds to the mitochondrial protein prohibitin which is integrated in the inner membrane and oligomerizes into large ring-like structures, which restrict CL into specific membrane domains. The role of prohibitin appears to be to segregate specific membrane domains, which facilitate CL remodeling. The direct interaction of DNAJC19 with prohibitin suggests a participation of DNAJC19 in CL remodeling. In fact, deletion of DNAJC19 in a cell model resulted in changes in the CL species composition [140]. Interestingly, these alterations in the CL pool were not approved in an iPSC-CM model of DCMA [139]. The reason for these inconsistencies have not been resolved yet. A consistent finding, however, is that DNAJC19 deficiency causes highly fragmented and abnormally shaped mitochondria and changes in mitochondrial cristae morphology.

10. Therapeutic Approaches

Improvements in the therapy of BTHS cardiomyopathy and the associated symptoms such as neutropenia and skeletal myopathy have resulted in improved survival of the deadly disease.

Standard therapy addressing the cardiac defects are angiotensin-converting enzyme (ACE) inhibitors or angiotensin receptor blockers, in combination with beta-adrenergic receptor blockers [141]. More severe cases are treated with vasodilators or inotropes, left ventricular assist devices, and/or cardiac transplantation [142–144]. Neutropenia is treated with granulocyte colony-stimulating factor (G-CSF) and complemented with prophylactic antibiotics. Growth hormone (GH) supplementation is used to treat the growth delay and arginine supplementation acts as a complement for arginine depletion [141,145,146].

Gene therapy in inherited diseases allows to directly target affected genes and has been tested in mouse models of Barth syndrome. Adeno-associated virus (AAV)-mediated *TAZ* gene replacement ameliorated cardiac function in *Taz*-KD mice, indicating the reversibility of the clinical phenotype and the feasibility of the approach [147]. Additionally, pharmaceutical intervention in CL biosynthesis has been tested in experimental studies. The mitochondrial enzyme phospholipase A_2 is responsible for the deacylation of premature CL after its initial synthesis. The resulting monolysocardiolipin MLCL is then acetylated to mature CL by Tafazzin [148]. Inhibiting phospholipase A_2 would result in stabilizing the premature CL pool and prevents the accumulation of MLCL. This intervention has been tested in a *Taz* deletion Drosophila model of BTHS. *Taz* deletion causes male sterility in flies, which was prevented by inhibiting phospholipase A_2 [148]. However, as the human homolog of the *Drosophila* phospholipase A_2 has not been identified and its inhibition might have severe side effects, translation into medicine is difficult [149].

Therapeutic approaches are interesting with regard to the targeting of metabolic defects in BTHS. Due to a central role in energy metabolism and mitochondrial bioenergetics, peroxisome proliferator-activated receptors (PPARs) have been considered as potential therapeutic targets. Activation of the PPAR/PGC1α axis using the PPARα agonist bezafibrate has been a successful strategy in various mitochondrial disorders [150]. PPARα activation regulates the transcription of numerous genes involved in mitochondrial energy metabolism and fatty acid oxidation [151,152]. When tested in the BTHS mouse model, bezafibrate prevented the development of systolic dysfunction and improved exercise capacity when combined with voluntary exercise [153].

As the structural change in the respiratory chain in Barth syndrome is associated with a marked increase in ROS emission from mitochondria [36,154,155], an intriguing therapeutic strategy is the use of anitoxidants. The mitochondria-targeted ROS scavenger mitoTEMPO was tested in the human-induced pluripotent stem cell (iPSC)-derived cardiomyocytes of BTHS patients [156,157]. ROS-induced changes in sarcomere assembly and the resulting deficits in contractility undergo an improvement with mitoTEMPO [158]. The translation into a mouse model was tested by expressing the ROS-scavenging catalase in the mitochondrial matrix to target mitochondrial hydrogen peroxide (H_2O_2) emission in the BTHS mouse. This approach efficiently reduced H_2O_2 levels and lipid peroxidation, but did not eradicate cardiac dysfunction or skeletal muscle fatigue in *Taz*-knockdown (KD) mice [159]. The discrepancy of these contradictory results has not been resolved, yet.

Cytochrome c mediates electron transfer from complex III to complex IV in the respiratory chain and is also involved in the peroxidation of cardiolipin (CL), which has been observed in a variety of pathological conditions, including BTHS [160]. Peroxidized CL is associated with energy deficiency and plays a role in the opening of the permeability transition pore (PTP) [161]. The PTP is a large pore in the mitochondrial membranes and consists of the proapoptotic Bcl-2 members Bax/Bak in the outer membrane and the F_1/F_O ATPase in the inner membrane and cyclophilin D in the matrix [162]. PTP opening causes a depletion of the membrane potential and induces apoptosis. The Szeto-Schiller peptide (SS-31 or Elamipretide) is an aromatic-cationic mitochondria-targeting tetrapeptide that penetrates the plasma membrane and localizes to the inner mitochondrial membrane based on its direct interaction with CL. Elamipretide prevents the peroxidase activity of cytochrome c and normalizes the CL pool in models of BTHS and other models of heart failure [163,164]. Elamipretide also improves inner membrane cristae structures and re-establishes mitochondrial respiration and ATP production in models of heart failure including BTHS [165,166]. Unfortunately, a phase II clinical trial (TAZPOWER trial, NCT03098797) treatment with Elamipretide in BTHS patients did not improve exercise capacity. Recently, Elamipretide was also tested in fibroblasts from DCMA patients and was found to rescue mitochondrial fragmentation and increased ROS production [167]. Heart tissue from a rat model of ischemia-reperfusion showed deterioration of mitochondrial complexes I, II, and IV. Application of Elamipretide significantly alleviated the structural changes in respirasomes, improved the fragmentation of mitochondria and enhanced the formation of cristae structures [168]. As changes in the CL pool were also found in the aging heart, Elamipretide was tested in aging mice. Here, it reduced mitochondrial ROS and normalized protein oxidation in old hearts, and even showed beneficial effects for aging-related diastolic defects. Interestingly, expression of the mitochondrial catalase presented similar beneficial effects, which were not further improved by Elamipretide application, indicating normalizing mitochondrial oxidative stress as the main mechanism for Elamipretide in aging hearts [67].

11. Conclusions

Mitochondria play a crucial role in energy metabolism, redox homeostasis and intermediate metabolism, not only having additional anabolic functions but also participating in signaling pathways. Many of these functions are membrane-associated and were shown to be dependent on CL. CL-deficient cells and animal models, including BTHS patient-derived lymphoblasts, *Drosophila*, *C.elegans*, *Trypanosoma* and mice, have been developed [35,169–173]. These model systems have helped to understand the role of CL in mitochondrial biogenesis and in shaping mitochondrial morphology. CL is an integral component of the respiratory chain since CL deficiency causes a decline in respiratory capacity and an increase in ROS production. A defect in MCU uncouples mitochondrial energy metabolism from myocardial Ca^{2+} signaling, which mediates an important response during accelerated workload conditions. Deficiencies in the Krebs cycle require remodeling of intermediate metabolism for compensatory anaplerotic pathways. Finally, defects in the biogenesis of iron sulfur clusters and Coenzyme A might endanger fatty acid oxidation, one of the most prevalent energy sources for the heart. The tight interaction between CL and membrane proteins, such as respirasomes, may also have implications on the half-life of CL itself, although CL presents a longer half-life compared to other phospholipids [118,174,175]. As the slow turnover of CL is dependent on its interaction with the respiratory chain, respiratory chain remodeling as in BTHS may induce a vicious cycle when increased CL turnover may contribute to low CL levels in Barth syndrome [118]. The exact mechanisms of CL degradation remain unclear. The phospholipase HSD10 has been suggested to mediate rapid degradation, particularly of oxidized CL (CL^{OX}) [176]. Patients with mutations in the gene encoding HSD10 suffer from progressive cardiomyopathy and neurodegenerative disease. The contribution of HSD10-mediated CL^{OX} clearance for the maintenance of functional mitochondria has not been studied yet.

Barth syndrome patients suffer from a general metabolic remodeling including changes in serum amino acid levels, lactic acidosis during exercise and elevated urinary excretion of 3-MGA [113,114]. In the future it will be interesting to understand the exact mechanisms behind changes in the metabolism, how they relate to mitochondrial dysfunction and how dysfunctional mitochondria induce compensatory mechanisms such as anaplerotic pathways. Based on the integration of mitochondria in cellular signaling pathways, changes in mitochondrial function might be monitored and trigger a cellular response. ROS may be at the center of this mechanism since alteration in ROS according to CL deficiency has been widely documented. Being a part of a large number of cellular signaling pathways, ROS affect a wide variety of biological processes including responses to hypoxia, apoptosis, autophagy, cell proliferation and differentiation. Alterations in ROS signaling due to CL deficiency cause a defect in key cellular signaling pathways involved in the response to hypoxia [155]. Several signaling pathways are also directly dependent on CL. Kinases of the protein kinase C (PKC) family regulate diverse biological functions such as growth and differentiation and influence multiple physiological processes in the heart, including heart rate, contraction, and relaxation. Members of the PKC family locating to the mitochondria and requiring CL for activation have been described previously [177,178]. As CL predominantly is located in the inner membrane, externalization of CL onto the outer membrane serves as a signaling platform in many signaling events, such as mitophagy and apoptosis. How these changes relate to the clinical picture in heart disease is not understood and remain a challenging task for the future.

Funding: The work in the laboratory of the authors is supported by the Deutsche Forschungsgemeinschaft (DFG; DU1839/2-1), the Bundesministerium für Bildung und Forschung (BMBF) and the Barth Syndrome Foundation.

Conflicts of Interest: The authors declare no conflict of interest.

References

1. Brown, D.A.; Perry, J.B.; Allen, M.E.; Sabbah, H.N.; Stauffer, B.L.; Shaikh, S.R.; Cleland, J.G.; Colucci, W.S.; Butler, J.; Voors, A.A.; et al. Expert consensus document: Mitochondrial function as a therapeutic target in heart failure. *Nat. Rev. Cardiol.* **2017**, *14*, 238–250. [CrossRef] [PubMed]
2. Barth, E.; Stammler, G.; Speiser, B.; Schaper, J. Ultrastructural quantitation of mitochondria and myofilaments in cardiac muscle from 10 different animal species including man. *J. Mol. Cell. Cardiol.* **1992**, *24*, 669–681. [CrossRef]
3. Schaper, J.; Meiser, E.; Stämmler, G. Ultrastructural morphometric analysis of myocardium from dogs, rats, hamsters, mice, and from human hearts. *Circ. Res.* **1985**, *56*, 377–391. [CrossRef] [PubMed]
4. Neubauer, S. The Failing Heart—An Engine Out of Fuel. *N. Engl. J. Med.* **2007**, *356*, 1140–1151. [CrossRef] [PubMed]
5. Dudek, J.; Maack, C. Barth syndrome cardiomyopathy. *Cardiovasc. Res.* **2017**, *113*, 399–410. [CrossRef] [PubMed]
6. Marín-García, J.; Goldenthal, M.J. Mitochondrial centrality in heart failure. *Heart Fail. Rev.* **2008**, *13*, 137–150. [CrossRef] [PubMed]
7. Sabbah, H.N.; Gupta, R.C.; Singh-Gupta, V.; Zhang, K. Effects of elamipretide on skeletal muscle in dogs with experimentally induced heart failure. *ESC Heart Fail.* **2019**, *6*, 328–335. [CrossRef]
8. Wolf, D.M.; Segawa, M.; Kondadi, A.K.; Anand, R.; Bailey, S.T.; Reichert, A.S.; Van Der Bliek, A.M.; Shackelford, D.B.; Liesa, M.; Shirihai, O.S. Individual cristae within the same mitochondrion display different membrane potentials and are functionally independent. *EMBO J.* **2019**, *38*, e101056. [CrossRef]
9. Zhang, R.; Krigman, J.; Luo, H.; Ozgen, S.; Yang, M.; Sun, N. Mitophagy in cardiovascular homeostasis. *Mech. Ageing Dev.* **2020**, *188*, 111245. [CrossRef]
10. Schlame, M. Protein crowding in the inner mitochondrial membrane. *Biochim. Biophys. Acta Bioenerg.* **2020**, *1862*, 148305. [CrossRef]
11. Gebert, N.; Joshi, A.S.; Kutik, S.; Becker, T.; McKenzie, M.; Guan, X.L.; Mooga, V.P.; Stroud, D.A.; Kulkarni, G.; Wenk, M.R.; et al. Mitochondrial Cardiolipin Involved in Outer-Membrane Protein Biogenesis: Implications for Barth Syndrome. *Curr. Biol.* **2009**, *19*, 2133–2139. [CrossRef] [PubMed]
12. Oemer, G.; Koch, J.; Wohlfarter, Y.; Alam, M.T.; Lackner, K.; Sailer, S.; Neumann, L.; Lindner, H.H.; Watschinger, K.; Haltmeier, M.; et al. Phospholipid Acyl Chain Diversity Controls the Tissue-Specific Assembly of Mitochondrial Cardiolipins. *Cell Rep.* **2020**, *30*, 4281–4291. [CrossRef] [PubMed]
13. Zhang, J.; Guan, Z.; Murphy, A.N.; Wiley, S.E.; Perkins, G.A.; Worby, C.A.; Engel, J.L.; Heacock, P.; Nguyen, O.K.; Wang, J.H.; et al. Mitochondrial Phosphatase PTPMT1 Is Essential for Cardiolipin Biosynthesis. *Cell Metab.* **2011**, *13*, 690–700. [CrossRef] [PubMed]
14. Kasahara, T.; Kubota-Sakashita, M.; Nagatsuka, Y.; Hirabayashi, Y.; Hanasaka, T.; Tohyama, K.; Kato, T. Cardiolipin is essential for early embryonic viability and mitochondrial integrity of neurons in mammals. *FASEB J.* **2019**, *34*, 1465–1480. [CrossRef]
15. Dinca, A.A.; Chien, W.-M.; Chin, M.T. Identification of novel mitochondrial localization signals in human Tafazzin, the cause of the inherited cardiomyopathic disorder Barth syndrome. *J. Mol. Cell. Cardiol.* **2018**, *114*, 83–92. [CrossRef]
16. Hijikata, A.; Yura, K.; Ohara, O.; Go, M. Structural and functional analyses of Barth syndrome-causing mutations and alternative splicing in the tafazzin acyltransferase domain. *Meta Gene* **2015**, *4*, 92–106. [CrossRef]

17. Xu, Y.; Zhang, S.; Malhotra, A.; Edelman-Novemsky, I.; Ma, J.; Kruppa, A.; Cernicica, C.; Blais, S.; Neubert, T.A.; Ren, M.; et al. Characterization of Tafazzin Splice Variants from Humans and Fruit Flies. *J. Biol. Chem.* **2009**, *284*, 29230–29239. [CrossRef]
18. Schlame, M.; Xu, Y.; Ren, M. The Basis for Acyl Specificity in the Tafazzin Reaction. *J. Biol. Chem.* **2017**, *292*, 5499–5506. [CrossRef]
19. Xu, Y.; Anjaneyulu, M.; Donelian, A.; Yu, W.; Greenberg, M.L.; Ren, M.; Owusu-Ansah, E.; Schlame, M. Assembly of the complexes of oxidative phosphorylation triggers the remodeling of cardiolipin. *Proc. Natl. Acad. Sci. USA* **2019**, *116*, 11235–11240. [CrossRef]
20. Taylor, W.A.; Hatch, G.M. Identification of the Human Mitochondrial Linoleoyl-coenzyme A Monolysocardiolipin Acyltransferase (MLCL AT-1). *J. Biol. Chem.* **2009**, *284*, 30360–30371. [CrossRef]
21. Mejia, E.M.; Zegallai, H.; Bouchard, E.D.; Banerji, V.; Ravandi, A.; Hatch, G.M. Expression of human monolysocardiolipin acyltransferase-1 improves mitochondrial function in Barth syndrome lymphoblasts. *J. Biol. Chem.* **2018**, *293*, 7564–7577. [CrossRef] [PubMed]
22. Miklas, J.W.; Clark, E.; Levy, S.; Detraux, D.; Leonard, A.; Beussman, K.; Showalter, M.R.; Smith, A.T.; Hofsteen, P.; Yang, X.; et al. TFPa/HADHA is required for fatty acid beta-oxidation and cardiolipin re-modeling in human cardiomyocytes. *Nat. Commun.* **2019**, *10*, 1–21. [CrossRef] [PubMed]
23. Li, J.; Romestaing, C.; Han, X.; Li, Y.; Hao, X.; Wu, Y.; Sun, C.; Liu, X.; Jefferson, L.S.; Xiong, J.; et al. Cardiolipin Remodeling by ALCAT1 Links Oxidative Stress and Mitochondrial Dysfunction to Obesity. *Cell Metab.* **2010**, *12*, 154–165. [CrossRef] [PubMed]
24. Baandrup, U.; Florio, R.A.; Roters, F.; Olsen, E.G. Electron microscopic investigation of endomyocardial biopsy samples in hypertrophy and cardiomyopathy. A semiquantitative study in 48 patients. *Circulation* **1981**, *63*, 1289–1298. [CrossRef] [PubMed]
25. Sabbah, H.N.; Sharov, V.; Riddle, J.M.; Kono, T.; Lesch, M.; Goldstein, S. Mitochondrial abnormalities in myocardium of dogs with chronic heart failure. *J. Mol. Cell. Cardiol.* **1992**, *24*, 1333–1347. [CrossRef]
26. Schlame, M.; Greenberg, M.L. Biosynthesis, remodeling and turnover of mitochondrial cardiolipin. *Biochim. Biophys. Acta Mol. Cell Biol. Lipids* **2017**, *1862*, 3–7. [CrossRef]
27. Frohman, M.A. Role of mitochondrial lipids in guiding fission and fusion. *J. Mol. Med.* **2015**, *93*, 263–269. [CrossRef]
28. Chen, L.; Gong, Q.; Stice, J.P.; Knowlton, A.A. Mitochondrial OPA1, apoptosis, and heart failure. *Cardiovasc. Res.* **2009**, *84*, 91–99. [CrossRef]
29. Sharov, V.G.; Sabbah, H.N.; Shimoyama, H.; Goussev, A.V.; Lesch, M.; Goldstein, S. Evidence of cardiocyte apoptosis in myocardium of dogs with chronic heart failure. *Am. J. Pathol.* **1996**, *148*, 141–149.
30. DeVay, R.M.; Dominguez-Ramirez, L.; Lackner, L.L.; Hoppins, S.; Stahlberg, H.; Nunnari, J. Coassembly of Mgm1 isoforms requires cardiolipin and mediates mitochondrial inner membrane fusion. *J. Cell Biol.* **2009**, *186*, 793–803. [CrossRef]
31. Meglei, G.; McQuibban, G.A. The Dynamin-Related Protein Mgm1p Assembles into Oligomers and Hydrolyzes GTP To Function in Mitochondrial Membrane Fusion†. *Biochemistry* **2009**, *48*, 1774–1784. [CrossRef] [PubMed]
32. Kiebish, M.A.; Yang, K.; Liu, X.; Mancuso, D.J.; Guan, S.; Zhao, Z.; Sims, H.F.; Cerqua, R.; Cade, W.T.; Han, X.; et al. Dysfunctional cardiac mitochondrial bioenergetic, lipidomic, and signaling in a murine model of Barth syndrome. *J. Lipid Res.* **2013**, *54*, 1312–1325. [CrossRef] [PubMed]
33. Acehan, D.; Vaz, F.; Houtkooper, R.H.; James, J.; Moore, V.; Tokunaga, C.; Kulik, W.; Wansapura, J.; Toth, M.J.; Strauss, A.; et al. Cardiac and Skeletal Muscle Defects in a Mouse Model of Human Barth Syndrome. *J. Biol. Chem.* **2010**, *286*, 899–908. [CrossRef] [PubMed]
34. Gonzalvez, F.; D'Aurelio, M.; Boutant, M.; Moustapha, A.; Puech, J.-P.; Landes, T.; Arnauné-Pelloquin, L.; Vial, G.; Taleux, N.; Slomianny, C.; et al. Barth syndrome: Cellular compensation of mitochondrial dysfunction and apoptosis inhibition due to changes in cardiolipin remodeling linked to tafazzin (TAZ) gene mutation. *Biochim. Biophys. Acta Mol. Basis Dis.* **2013**, *1832*, 1194–1206. [CrossRef]
35. Acehan, D.; Xu, Y.; Stokes, D.L.; Schlame, M. Comparison of lymphoblast mitochondria from normal subjects and patients with Barth syndrome using electron microscopic tomography. *Lab. Investig.* **2007**, *87*, 40–48. [CrossRef]

36. Dudek, J.; Cheng, I.F.; Chowdhury, A.; Wozny, K.; Balleininger, M.; Reinhold, R.; Grunau, S.; Callegari, S.; Toischer, K.; Wanders, R.J.; et al. Cardiac-specific succinate dehydrogenase deficiency in Barth syndrome. *EMBO Mol Med* **2016**, *8*, 139–154. [CrossRef]
37. Dudek, J.; Cheng, I.-F.; Balleininger, M.; Vaz, F.M.; Streckfuss-Bömeke, K.; Hübscher, D.; Vukotic, M.; Wanders, R.J.A.; Rehling, P.; Guan, K. Cardiolipin deficiency affects respiratory chain function and organization in an induced pluripotent stem cell model of Barth syndrome. *Stem Cell Res.* **2013**, *11*, 806–819. [CrossRef]
38. Chatzispyrou, I.A.; Guerrero-Castillo, S.; Held, N.M.; Ruiter, J.P.; Denis, S.W.; Ijlst, L.; Wanders, R.J.; Van Weeghel, M.; Ferdinandusse, S.; Vaz, F.M.; et al. Barth syndrome cells display widespread remodeling of mitochondrial complexes without affecting metabolic flux distribution. *Biochim. Biophys. Acta Mol. Basis Dis.* **2018**, *1864*, 3650–3658. [CrossRef]
39. Wang, S.; Li, Y.; Xu, Y.; Ma, Q.; Lin, Z.; Schlame, M.; Bezzerides, V.J.; Strathdee, D.; Pu, W.T. AAV Gene Therapy Prevents and Reverses Heart Failure in a Murine Knockout Model of Barth Syndrome. *Circ. Res.* **2020**, *126*, 1024–1039. [CrossRef]
40. Gonzalvez, F.; Schug, Z.T.; Houtkooper, R.H.; MacKenzie, E.D.; Brooks, D.G.; Wanders, R.J.A.; Petit, P.X.; Vaz, F.M.; Gottlieb, E. Cardiolipin provides an essential activating platform for caspase-8 on mitochondria. *J. Cell Biol.* **2008**, *183*, 681–696. [CrossRef]
41. Rampelt, H.; Zerbes, R.M.; Van Der Laan, M.; Pfanner, N. Role of the mitochondrial contact site and cristae organizing system in membrane architecture and dynamics. *Biochim. Biophys. Acta Bioenerg.* **2017**, *1864*, 737–746. [CrossRef] [PubMed]
42. Weber, T.A.; Koob, S.; Heide, H.; Wittig, I.; Head, B.; Van Der Bliek, A.; Brandt, U.; Mittelbronn, M.; Reichert, A.S. APOOL Is a Cardiolipin-Binding Constituent of the Mitofilin/MINOS Protein Complex Determining Cristae Morphology in Mammalian Mitochondria. *PLoS ONE* **2013**, *8*, e63683. [CrossRef] [PubMed]
43. Friedman, J.R.; Mourier, A.; Yamada, J.; McCaffery, J.M.; Nunnari, J.; Youle, R.J. MICOS coordinates with respiratory complexes and lipids to establish mitochondrial inner membrane architecture. *eLife* **2015**, *4*, e07739. [CrossRef] [PubMed]
44. Koob, S.; Barrera, M.; Anand, R.; Reichert, A.S. The non-glycosylated isoform of MIC26 is a constituent of the mammalian MICOS complex and promotes formation of crista junctions. *Biochim. Biophys. Acta Bioenerg.* **2015**, *1853*, 1551–1563. [CrossRef]
45. Van Strien, J.; Guerrero-Castillo, S.; Chatzispyrou, I.A.; Houtkooper, R.H.; Brandt, U.; Huynen, M.A. COmplexome Profiling ALignment (COPAL) reveals remodeling of mitochondrial protein complexes in Barth syndrome. *Bioinformatics* **2019**, *35*, 3083–3091. [CrossRef]
46. Sabbah, H.N.; Gupta, R.C.; Singh-Gupta, V.; Zhang, K.; Lanfear, D.E. Abnormalities of Mitochondrial Dynamics in the Failing Heart: Normalization Following Long-Term Therapy with Elamipretide. *Cardiovasc. Drugs Ther.* **2018**, *32*, 319–328. [CrossRef]
47. Anand, R.; Kondadi, A.K.; Meisterknecht, J.; Golombek, M.; Nortmann, O.; Riedel, J.; Peifer-Weiß, L.; Brocke-Ahmadinejad, N.; Schlütermann, D.; Stork, B.; et al. MIC26 and MIC27 cooperate to regulate cardiolipin levels and the landscape of OXPHOS complexes. *Life Sci. Alliance* **2020**, *3*, e202000711. [CrossRef]
48. Kameoka, S.; Adachi, Y.; Okamoto, K.; Iijima, M.; Sesaki, H. Phosphatidic Acid and Cardiolipin Coordinate Mitochondrial Dynamics. *Trends Cell Biol.* **2018**, *28*, 67–76. [CrossRef]
49. Fiedorczuk, K.; Letts, J.A.; Degliesposti, G.; Kaszuba, K.; Skehel, M.; Sazanov, L.A. Atomic structure of the entire mammalian mitochondrial complex I. *Nature* **2016**, *538*, 406–410. [CrossRef]
50. Sedlák, E.; Robinson, N.C. Phospholipase A2 Digestion of Cardiolipin Bound to Bovine CytochromecOxidase Alters Both Activity and Quaternary Structure†. *Biochemistry* **1999**, *38*, 14966–14972. [CrossRef]
51. Sedlák, E.; Panda, M.; Dale, M.P.; Weintraub, S.T.; Robinson, N.C. Photolabeling of Cardiolipin Binding Subunits within Bovine Heart CytochromecOxidase†. *Biochemistry* **2006**, *45*, 746–754. [CrossRef] [PubMed]
52. Shinzawa-Itoh, K.; Seiyama, J.; Terada, H.; Nakatsubo, R.; Naoki, K.; Nakashima, Y.; Yoshikawa, S. Bovine Heart NADH−Ubiquinone Oxidoreductase Contains One Molecule of Ubiquinone with Ten Isoprene Units as One of the Cofactors. *Biochemistry* **2010**, *49*, 487–492. [CrossRef] [PubMed]
53. Sharpley, M.S.; Shannon, R.J.; Draghi, A.F.; Hirst, J. Interactions between Phospholipids and NADH:Ubiquinone Oxidoreductase (Complex I) from Bovine Mitochondria†. *Biochemistry* **2006**, *45*, 241–248. [CrossRef] [PubMed]

54. Robinson, N.C.; Zborowski, J.; Talbert, L.H. Cardiolipin-depleted bovine heart cytochrome c oxidase: Binding stoichiometry and affinity for cardiolipin derivatives. *Biochemistry* **1990**, *29*, 8962–8969. [CrossRef]
55. Spikes, T.E.; Montgomery, M.G.; Walker, J.E. Structure of the dimeric ATP synthase from bovine mitochondria. *Proc. Natl. Acad. Sci. USA* **2020**, *117*, 23519–23526. [CrossRef]
56. Laird, D.M.; Eble, K.S.; Cunningham, C.C. Reconstitution of mitochondrial F0.F1-ATPase with phosphatidylcholine using the nonionic detergent, octylglucoside. *J. Biol. Chem.* **1986**, *261*, 14844–14850.
57. Duncan, A.L.; Robinson, A.J.; Walker, J.E. Cardiolipin binds selectively but transiently to conserved lysine residues in the rotor of metazoan ATP synthases. *Proc. Natl. Acad. Sci. USA* **2016**, *113*, 8687–8692. [CrossRef]
58. Mileykovskaya, E.; Penczek, P.A.; Fang, J.; Mallampalli, V.K.P.S.; Sparagna, G.C.; Dowhan, W. Arrangement of the Respiratory Chain Complexes in Saccharomyces cerevisiae Supercomplex III2IV2 Revealed by Single Particle Cryo-Electron Microscopy. *J. Biol. Chem.* **2012**, *287*, 23095–23103. [CrossRef]
59. Pfeiffer, K.; Gohil, V.; Stuart, R.A.; Hunte, C.; Brandt, U.; Greenberg, M.L.; Schägger, H. Cardiolipin Stabilizes Respiratory Chain Supercomplexes. *J. Biol. Chem.* **2003**, *278*, 52873–52880. [CrossRef]
60. Zhang, M.; Mileykovskaya, E.; Dowhan, W. Cardiolipin Is Essential for Organization of Complexes III and IV into a Supercomplex in Intact Yeast Mitochondria. *J. Biol. Chem.* **2005**, *280*, 29403–29408. [CrossRef]
61. Jiang, F.; Ryan, M.T.; Schlame, M.; Zhao, M.; Gu, Z.; Klingenberg, M.; Pfanner, N.; Greenberg, M.L. Absence of Cardiolipin in thecrd1Null Mutant Results in Decreased Mitochondrial Membrane Potential and Reduced Mitochondrial Function. *J. Biol. Chem.* **2000**, *275*, 22387–22394. [CrossRef] [PubMed]
62. Gómez, L.A.; Hagen, T.M. Age-related decline in mitochondrial bioenergetics: Does supercomplex destabilization determine lower oxidative capacity and higher superoxide production? *Semin. Cell Dev. Biol.* **2012**, *23*, 758–767. [CrossRef] [PubMed]
63. Ostrander, D.B.; Sparagna, G.C.; Amoscato, A.A.; McMillin, J.B.; Dowhan, W. Decreased Cardiolipin Synthesis Corresponds with Cytochromec Release in Palmitate-induced Cardiomyocyte Apoptosis. *J. Biol. Chem.* **2001**, *276*, 38061–38067. [PubMed]
64. Gadicherla, A.K.; Stowe, D.F.; Antholine, W.E.; Yang, M.; Camara, A.K. Damage to mitochondrial complex I during cardiac ischemia reperfusion injury is reduced indirectly by anti-anginal drug ranolazine. *Biochim. Biophys. Acta Bioenerg.* **2012**, *1817*, 419–429. [CrossRef]
65. Saini-Chohan, H.K.; Holmes, M.G.; Chicco, A.J.; Taylor, W.A.; Moore, R.L.; McCune, S.A.; Hickson-Bick, D.L.; Hatch, G.M.; Sparagna, G.C. Cardiolipin biosynthesis and remodeling enzymes are altered during development of heart failure. *J. Lipid Res.* **2008**, *50*, 1600–1608. [CrossRef]
66. Nickel, A.G.; Von Hardenberg, A.; Hohl, M.; Löffler, J.R.; Kohlhaas, M.; Becker, J.; Reil, J.-C.; Kazakov, A.V.; Bonnekoh, J.; Stadelmaier, M.; et al. Reversal of Mitochondrial Transhydrogenase Causes Oxidative Stress in Heart Failure. *Cell Metab.* **2015**, *22*, 472–484. [CrossRef]
67. Chiao, Y.A.; Zhang, H.; Sweetwyne, M.; Whitson, J.; Ting, Y.S.; Basisty, N.; Pino, L.K.; Quarles, E.; Nguyen, N.-H.; Campbell, M.D.; et al. Late-life restoration of mitochondrial function reverses cardiac dysfunction in old mice. *eLife* **2020**, *9*. [CrossRef]
68. Murphy, M.P.; Hartley, R.C. Mitochondria as a therapeutic target for common pathologies. *Nat. Rev. Drug Discov.* **2018**, *17*, 865–886. [CrossRef]
69. Conrad, M.; Jakupoglu, C.; Moreno, S.G.; Lippl, S.; Banjac, A.; Schneider, M.; Beck, H.; Hatzopoulos, A.K.; Just, U.; Sinowatz, F.; et al. Essential Role for Mitochondrial Thioredoxin Reductase in Hematopoiesis, Heart Development, and Heart Function. *Mol. Cell. Biol.* **2004**, *24*, 9414–9423. [CrossRef]
70. Ho, Y.-S.; Magnenat, J.-L.; Gargano, M.; Cao, J. The Nature of Antioxidant Defense Mechanisms: A Lesson from Transgenic Studies. *Environ. Health Perspect.* **1998**, *106*, 1219. [CrossRef]
71. Yen, H.C.; Oberley, T.D.; Vichitbandha, S.; Ho, Y.S.; Clair, D.K.S. The protective role of manganese superoxide dismutase against adriamycin-induced acute cardiac toxicity in transgenic mice. *J. Clin. Investig.* **1996**, *98*, 1253–1260. [CrossRef] [PubMed]
72. Ruprecht, J.J.; King, M.S.; Zögg, T.; Aleksandrova, A.A.; Pardon, E.; Crichton, P.G.; Steyaert, J.; Kunji, E.R. The Molecular Mechanism of Transport by the Mitochondrial ADP/ATP Carrier. *Cell* **2019**, *176*, 435–447. [CrossRef] [PubMed]
73. Beyer, K.; Klingenberg, M. ADP/ATP carrier protein from beef heart mitochondria has high amounts of tightly bound cardiolipin, as revealed by phosphorus-31 nuclear magnetic resonance. *Biochemistry* **1985**, *24*, 3821–3826. [CrossRef] [PubMed]

74. Pebay-Peyroula, E.; Dahout-Gonzalez, C.; Kahn, R.; Trézéguet, V.; Lauquin, G.J.-M.; Brandolin, G. Structure of mitochondrial ADP/ATP carrier in complex with carboxyatractyloside. *Nat. Cell Biol.* **2003**, *426*, 39–44. [CrossRef] [PubMed]
75. Senoo, N.; Kandasamy, S.; Ogunbona, O.B.; Baile, M.G.; Lu, Y.; Claypool, S.M. Cardiolipin, conformation, and respiratory complex-dependent oligomerization of the major mitochondrial ADP/ATP carrier in yeast. *Sci. Adv.* **2020**, *6*, eabb0780. [CrossRef] [PubMed]
76. Palmieri, F.; Bisaccia, F.; Capobianco, L.; Iacobazzi, V.; Indiveri, C.; Zara, V. Structural and functional properties of mitochondrial anion carriers. *Biochim. Biophys. Acta Bioenerg.* **1990**, *1018*, 147–150. [CrossRef]
77. Nałęcz, K.A.; Bolli, R.; Wojtczak, L.; Azzi, A.; Nałcz, K.A. The monocarboxylate carrier from bovine heart mitochondria: Partial purification and its substrate-transporting properties in a reconstituted system. *Biochim. Biophys. Acta Bioenerg.* **1986**, *851*, 29–37. [CrossRef]
78. Indiveri, C.; Tonazzi, A.; Prezioso, G.; Palmieri, F. Kinetic characterization of the reconstituted carnitine carrier from rat liver mitochondria. *Biochim. Biophys. Acta Biomembr.* **1991**, *1065*, 231–238. [CrossRef]
79. Nałecz, K.A.; Kamińska, J.; Nałecz, M.J.; Azzi, A. The activity of pyruvate carrier in a reconstituted system: Substrate specificity and inhibitor sensitivity. *Arch. Biochem. Biophys.* **1992**, *297*, 162–168. [CrossRef]
80. Kaplan, R.S.; A Mayor, J.; Johnston, N.; Oliveira, D.L. Purification and characterization of the reconstitutively active tricarboxylate transporter from rat liver mitochondria. *J. Biol. Chem.* **1990**, *265*, 13379–13385.
81. Vukotic, M.; Nolte, H.; König, T.; Saita, S.; Ananjew, M.; Krüger, M.; Tatsuta, T.; Langer, T. Acylglycerol Kinase Mutated in Sengers Syndrome Is a Subunit of the TIM22 Protein Translocase in Mitochondria. *Mol. Cell* **2017**, *67*, 471–483. [CrossRef] [PubMed]
82. Mårtensson, C.U.; Becker, T. Acylglycerol Kinase: Mitochondrial Protein Transport Meets Lipid Biosynthesis. *Trends Cell Biol.* **2017**, *27*, 700–702. [CrossRef] [PubMed]
83. Müller, M.; Cheneval, D.; Carafoli, E. The Mitochondrial Creatine Phosphokinase is Associated with Inner Membrane Cardiolipin. *Results Probl. Cell Differ.* **1986**, *194*, 151–156. [CrossRef]
84. Müller, M.; Moser, R.; Cheneval, D.; Carafoli, E. Cardiolipin is the membrane receptor for mitochondrial creatine phosphokinase. *J. Biol. Chem.* **1985**, *260*, 3839–3843.
85. Fritz-Wolf, K.; Schnyder, T.; Wallimann, T.; Kabsch, W. Structure of mitochondrial creatine kinase. *Nat. Cell Biol.* **1996**, *381*, 341–345. [CrossRef]
86. Schlattner, U.; Tokarska-Schlattner, M.; Ramirez, S.; Brückner, A.; Kay, L.; Polge, C.; Epand, R.F.; Lee, R.M.; Lacombe, M.-L.; Epand, R.M. Mitochondrial kinases and their molecular interaction with cardiolipin. *Biochim. Biophys. Acta Biomembr.* **2009**, *1788*, 2032–2047. [CrossRef]
87. Li, Y.; Lou, W.; Raja, V.; Denis, S.; Yu, W.; Schmidtke, M.W.; Reynolds, C.A.; Schlame, M.; Houtkooper, R.H.; Greenberg, M.L. Cardiolipin-induced activation of pyruvate dehydrogenase links mitochondrial lipid biosynthesis to TCA cycle function. *J. Biol. Chem.* **2019**, *294*, 11568–11578. [CrossRef]
88. Fatica, E.; Deleonibus, G.A.; House, A.; Kodger, J.V.; Pearce, R.; Shah, R.; Levi, L.; Sandlers, Y. Barth Syndrome: Exploring Cardiac Metabolism with Induced Pluripotent Stem Cell-Derived Cardiomyocytes. *Metabolites* **2019**, *9*, 306. [CrossRef]
89. Li, Y.; Lou, W.; Grevel, A.; Böttinger, L.; Liang, Z.; Ji, J.; Patil, V.A.; Liu, J.; Ye, C.; Hüttemann, M.; et al. Cardiolipin-deficient cells have decreased levels of the iron–sulfur biogenesis protein frataxin. *J. Biol. Chem.* **2020**, *295*, 11928–11937. [CrossRef]
90. Patil, V.A.; Fox, J.L.; Gohil, V.M.; Winge, D.R.; Greenberg, M.L.; Suzuki, K.; Harada, N.; Yamane, S.; Nakamura, Y.; Sasaki, K.; et al. Loss of Cardiolipin Leads to Perturbation of Mitochondrial and Cellular Iron Homeostasis. *J. Biol. Chem.* **2012**, *288*, 1696–1705. [CrossRef]
91. Raja, V.; Salsaa, M.; Joshi, A.S.; Li, Y.; Van Roermund, C.W.; Saadat, N.; Lazcano, P.; Schmidtke, M.; Hüttemann, M.; Gupta, S.V.; et al. Cardiolipin-deficient cells depend on anaplerotic pathways to ameliorate defective TCA cycle function. *Biochim. Biophys. Acta Mol. Cell Biol. Lipids* **2019**, *1864*, 654–661. [CrossRef] [PubMed]
92. Le, C.H.; Benage, L.G.; Specht, K.S.; Puma, L.C.L.; Mulligan, C.M.; Heuberger, A.L.; Prenni, J.E.; Claypool, S.M.; Chatfield, K.C.; Sparagna, G.C.; et al. Tafazzin deficiency impairs CoA-dependent oxidative metabolism in cardiac mitochondria. *J. Biol. Chem.* **2020**, *295*. [CrossRef]
93. Cole, L.K.; Mejia, E.M.; Sparagna, G.C.; Vandel, M.; Xiang, B.; Han, X.; Dedousis, N.; Kaufman, B.A.; Dolinsky, V.W.; Hatch, G.M. Cardiolipin deficiency elevates susceptibility to a lipotoxic hypertrophic cardiomyopathy. *J. Mol. Cell. Cardiol.* **2020**, *144*, 24–34. [CrossRef] [PubMed]

94. Sandlers, Y.; Mercier, K.; Pathmasiri, W.; Carlson, J.; McRitchie, S.; Sumner, S.; Vernon, H.J. Metabolomics Reveals New Mechanisms for Pathogenesis in Barth Syndrome and Introduces Novel Roles for Cardiolipin in Cellular Function. *PLoS ONE* **2016**, *11*, e0151802. [CrossRef] [PubMed]
95. Korman, S.H.; Andresen, B.S.; Zeharia, A.; Gutman, A.; Boneh, A.; Pitt, J. 2-Ethylhydracrylic Aciduria in Short/Branched-Chain Acyl-CoA Dehydrogenase Deficiency: Application to Diagnosis and Implications for the R-Pathway of Isoleucine Oxidation. *Clin. Chem.* **2005**, *51*, 610–617. [CrossRef] [PubMed]
96. Bers, D.M. Altered Cardiac Myocyte Ca Regulation In Heart Failure. *Physiology* **2006**, *21*, 380–387. [CrossRef]
97. Pacher, P.; Beckman, J.S.; Liaudet, L. Nitric Oxide and Peroxynitrite in Health and Disease. *Physiol. Rev.* **2007**, *87*, 315–424. [CrossRef]
98. Lokuta, A.J.; Maertz, N.A.; Meethal, S.V.; Potter, K.T.; Kamp, T.J.; Valdivia, H.H.; Haworth, R.A. Increased Nitration of Sarcoplasmic Reticulum Ca 2+ -ATPase in Human Heart Failure. *Circulation* **2005**, *111*, 988–995. [CrossRef]
99. Periasamy, M.; Janssen, P.M. Molecular Basis of Diastolic Dysfunction. *Hear. Fail. Clin.* **2008**, *4*, 13–21. [CrossRef]
100. Dorn, G.W.; Maack, C. SR and mitochondria: Calcium cross-talk between kissing cousins. *J. Mol. Cell. Cardiol.* **2013**, *55*, 42–49. [CrossRef]
101. Szabadkai, G.; Simoni, A.M.; Chami, M.; Wieckowski, M.R.; Youle, R.J.; Rizzuto, R. Drp-1-Dependent Division of the Mitochondrial Network Blocks Intraorganellar Ca2+ Waves and Protects against Ca2+-Mediated Apoptosis. *Mol. Cell* **2004**, *16*, 59–68. [CrossRef] [PubMed]
102. Kohlhaas, M.; Maack, C. Calcium release microdomains and mitochondria. *Cardiovasc. Res.* **2013**, *98*, 259–268. [CrossRef] [PubMed]
103. Sancak, Y.; Markhard, A.L.; Kitami, T.; Kovács-Bogdán, E.; Kamer, K.J.; Udeshi, N.D.; Carr, S.A.; Chaudhuri, D.; Clapham, D.E.; Li, A.A.; et al. EMRE Is an Essential Component of the Mitochondrial Calcium Uniporter Complex. *Science* **2013**, *342*, 1379–1382. [CrossRef] [PubMed]
104. Perocchi, F.; Gohil, V.M.; Girgis, H.S.; Bao, X.R.; McCombs, J.E.; Palmer, A.E.; Mootha, V.K. MICU1 encodes a mitochondrial EF hand protein required for Ca2+ uptake. *Nat. Cell Biol.* **2010**, *467*, 291–296. [CrossRef] [PubMed]
105. Baughman, J.M.; Perocchi, F.; Girgis, H.S.; Plovanich, M.; Belcher-Timme, C.A.; Sancak, Y.; Bao, X.R.; Strittmatter, L.; A Goldberger, O.; Bogorad, R.L.; et al. Integrative genomics identifies MCU as an essential component of the mitochondrial calcium uniporter. *Nat. Cell Biol.* **2011**, *476*, 341–345. [CrossRef] [PubMed]
106. Raffaello, A.; De Stefani, D.; Sabbadin, D.; Teardo, E.; Merli, G.; Picard, A.; Checchetto, V.; Moro, S.; Szabò, I.; Rizzuto, R. The mitochondrial calcium uniporter is a multimer that can include a dominant-negative pore-forming subunit. *EMBO J.* **2013**, *32*, 2362–2376. [CrossRef] [PubMed]
107. Baradaran, R.; Wang, C.; Siliciano, A.F.; Long, S.B. Cryo-EM structures of fungal and metazoan mitochondrial calcium uniporters. *Nat. Cell Biol.* **2018**, *559*, 580–584. [CrossRef] [PubMed]
108. Ghosh, S.; Ball, W.B.; Madaris, T.R.; Srikantan, S.; Madesh, M.; Mootha, V.K.; Gohil, V.M. An essential role for cardiolipin in the stability and function of the mitochondrial calcium uniporter. *Proc. Natl. Acad. Sci. USA* **2020**, *117*, 16383–16390. [CrossRef]
109. Christodoulou, J.; McInnes, R.R.; Jay, V.; Wilson, G.; Becker, L.E.; Lehotay, D.C.; Platt, B.-A.; Bridge, P.J.; Robinson, B.H.; Clarke, J.T.R. Barth syndrome: Clinical observations and genetic linkage studies. *Am. J. Med. Genet.* **1994**, *50*, 255–264. [CrossRef]
110. Ades, L.C.; Gedeon, A.K.; Wilson, M.J.; Latham, M.; Partington, M.W.; Mulley, J.C.; Nelson, J.; Lui, K.; Sillence, D.O. Barth syndrome: Clinical features and confirmation of gene localisation to distal Xq28. *Am. J. Med. Genet.* **1993**, *45*, 327–334. [CrossRef]
111. Roberts, A.E.; Nixon, C.; Steward, C.G.; Gauvreau, K.; Maisenbacher, M.; Fletcher, M.; Geva, J.; Byrne, B.J.; Spencer, C.T. The Barth Syndrome Registry: Distinguishing disease characteristics and growth data from a longitudinal study. *Am. J. Med. Genet. Part A* **2012**, *158*, 2726–2732. [CrossRef]
112. Spencer, C.T.; Bryant, R.M.; Day, J.; Gonzalez, I.L.; Colan, S.D.; Thompson, W.R.; Berthy, J.; Redfearn, S.P.; Byrne, B.J. Cardiac and Clinical Phenotype in Barth Syndrome. *Pediatrics* **2006**, *118*, e337–e346. [CrossRef] [PubMed]

113. Spencer, C.T.; Byrne, B.J.; Bryant, R.M.; Margossian, R.; Maisenbacher, M.; Breitenger, P.; Benni, P.B.; Redfearn, S.; Marcus, E.; Cade, W.T. Impaired cardiac reserve and severely diminished skeletal muscle O_2 utilization mediate exercise intolerance in Barth syndrome. *Am. J. Physiol. Heart Circ. Physiol.* **2011**, *301*, H2122–H2129. [CrossRef] [PubMed]
114. Wortmann, S.B.; Kluijtmans, L.A.J.; Rodenburg, R.J.; Sass, J.O.; Nouws, J.; Van Kaauwen, E.P.; Kleefstra, T.; Tranebjaerg, L.; De Vries, M.C.; Isohanni, P.; et al. 3-Methylglutaconic aciduria—Lessons from 50 genes and 977 patients. *J. Inherit. Metab. Dis.* **2013**, *36*, 913–921. [CrossRef] [PubMed]
115. Barth, P.; Scholte, H.; Berden, J.; Moorsel, J.V.D.K.-V.; Luyt-Houwen, I.; Veer-Korthof, E.V.; Van Der Harten, J.; Sobotka-Plojhar, M. An X-linked mitochondrial disease affecting cardiac muscle, skeletal muscle and neutrophil leucocytes. *J. Neurol. Sci.* **1983**, *62*, 327–355. [CrossRef]
116. Hornby, B.; McClellan, R.; Buckley, L.; Carson, K.; Gooding, T.; Vernon, H.J. Functional exercise capacity, strength, balance and motion reaction time in Barth syndrome. *Orphanet J. Rare Dis.* **2019**, *14*, 1–12. [CrossRef] [PubMed]
117. Bissler, J.J.; Tsoras, M.; Göring, H.H.H.; Hug, P.; Chuck, G.; Tombragel, E.; A McGraw, C.; Schlotman, J.; A Ralston, M.; Hug, G. Infantile Dilated X-Linked Cardiomyopathy, G4.5 Mutations, Altered Lipids, and Ultrastructural Malformations of Mitochondria in Heart, Liver, and Skeletal Muscle. *Lab. Investig.* **2002**, *82*, 335–344. [CrossRef]
118. Xu, Y.; Phoon, C.K.; Berno, B.; D'Souza, K.; Hoedt, E.; Zhang, G.; A Neubert, E.H.G.Z.T.; Epand, K.D.R.M.; Ren, M.; Schlame, Y.X.M.R.M. Loss of protein association causes cardiolipin degradation in Barth syndrome. *Nat. Chem. Biol.* **2016**, *12*, 641–647. [CrossRef]
119. Kushmerick, M.J.; Moerland, T.S.; Wiseman, R.W. Mammalian skeletal muscle fibers distinguished by contents of phosphocreatine, ATP, and Pi. *Proc. Natl. Acad. Sci. USA* **1992**, *89*, 7521–7525. [CrossRef]
120. Takahashi, H.; Kuno, S.-Y.; Katsuta, S.; Shimojo, H.; Masuda, K.; Yoshioka, H.; Anno, I.; Itai, Y. Relationships between Fiber Composition and NMR Measurements in Human Skeletal Muscle. *NMR Biomed.* **1996**, *9*, 8–12. [CrossRef]
121. Cade, W.T.; Spencer, C.T.; Reeds, D.N.; Waggoner, A.D.; O'Connor, R.; Maisenbacher, M.; Crowley, J.R.; Byrne, B.J.; Peterson, L.R. Substrate metabolism during basal and hyperinsulinemic conditions in adolescents and young-adults with Barth syndrome. *J. Inherit. Metab. Dis.* **2013**, *36*, 91–101. [CrossRef] [PubMed]
122. Sabbah, H.N. Targeting the Mitochondria in Heart Failure: A Translational Perspective. *JACC Basic Transl. Sci.* **2020**, *5*, 88–106. [CrossRef] [PubMed]
123. Bione, S.; D'Adamo, P.; Maestrini, E.; Gedeon, A.K.; Bolhuis, P.A.; Toniolo, D. A novel X-linked gene, G4.5. is responsible for Barth syndrome. *Nat. Genet.* **1996**, *12*, 385–389. [CrossRef] [PubMed]
124. Lu, Y.-W.; Galbraith, L.; Herndon, J.D.; Lu, Y.-L.; Pras-Raves, M.; Vervaart, M.; Van Kampen, A.; Luyf, A.; Koehler, C.M.; McCaffery, J.M.; et al. Defining functional classes of Barth syndrome mutation in humans. *Hum. Mol. Genet.* **2016**, *25*, 1754–1770. [CrossRef]
125. Whited, K.; Baile, M.G.; Currier, P.; Claypool, S.M. Seven functional classes of Barth syndrome mutation. *Hum. Mol. Genet.* **2012**, *22*, 483–492. [CrossRef]
126. Schlame, M.; I Kelley, R.; Feigenbaum, A.; A Towbin, J.; Heerdt, P.M.; Schieble, T.; A Wanders, R.J.; DiMauro, S.; Blanck, T.J.J. Phospholipid abnormalities in children with Barth syndrome. *J. Am. Coll. Cardiol.* **2003**, *42*, 1994–1999. [CrossRef]
127. Houtkooper, R.H.; Rodenburg, R.J.; Thiels, C.; Van Lenthe, H.; Stet, F.; Poll-The, B.T.; Stone, J.E.; Steward, C.G.; Wanders, R.J.; Smeitink, J.A.M.; et al. Cardiolipin and monolysocardiolipin analysis in fibroblasts, lymphocytes, and tissues using high-performance liquid chromatography–mass spectrometry as a diagnostic test for Barth syndrome. *Anal. Biochem.* **2009**, *387*, 230–237. [CrossRef]
128. Xu, Y.; Sutachan, J.J.; Plesken, H.; I Kelley, R.; Schlame, Y.X.M.R.M. Characterization of lymphoblast mitochondria from patients with Barth syndrome. *Lab. Investig.* **2005**, *85*, 823–830. [CrossRef]
129. Cade, W.T.; Laforest, R.; Bohnert, K.L.; Reeds, D.N.; Bittel, A.J.; Fuentes, L.D.L.; Bashir, A.; Woodard, P.K.; Pacak, C.A.; Byrne, B.J.; et al. Myocardial glucose and fatty acid metabolism is altered and associated with lower cardiac function in young adults with Barth syndrome. *J. Nucl. Cardiol.* **2019**, 1–11. [CrossRef]
130. Cade, W.T.; Bohnert, K.L.; Peterson, L.R.; Patterson, B.W.; Bittel, A.J.; Okunade, A.L.; Fuentes, L.D.L.; Steger-May, K.; Bashir, A.; Schweitzer, G.G.; et al. Blunted fat oxidation upon submaximal exercise is partially compensated by enhanced glucose metabolism in children, adolescents, and young adults with Barth syndrome. *J. Inherit. Metab. Dis.* **2019**, *42*, 480–493. [CrossRef]

131. Mayr, J.A.; Haack, T.B.; Graf, E.; Zimmermann, F.A.; Wieland, T.; Haberberger, B.; Superti-Furga, A.; Kirschner, J.; Steinmann, B.; Baumgartner, M.R.; et al. Lack of the Mitochondrial Protein Acylglycerol Kinase Causes Sengers Syndrome. *Am. J. Hum. Genet.* **2012**, *90*, 314–320. [CrossRef] [PubMed]
132. Haghighi, A.; Haack, T.B.; Atiq, M.; Mottaghi, H.; Haghighi-Kakhki, H.; A Bashir, R.; Ahting, U.; Feichtinger, R.G.; A Mayr, J.; Rötig, A.; et al. Sengers syndrome: Six novel AGK mutations in seven new families and review of the phenotypic and mutational spectrum of 29 patients. *Orphanet J. Rare Dis.* **2014**, *9*, 1–12. [CrossRef] [PubMed]
133. Kang, Y.; Stroud, D.A.; Baker, M.J.; De Souza, D.P.; Frazier, A.E.; Liem, M.; Tull, D.; Mathivanan, S.; McConville, M.J.; Thorburn, D.R.; et al. Sengers Syndrome-Associated Mitochondrial Acylglycerol Kinase Is a Subunit of the Human TIM22 Protein Import Complex. *Mol. Cell* **2017**, *67*, 457–470. [CrossRef] [PubMed]
134. Jordens, E.Z.; Palmieri, L.; Huizing, M.; Heuvel, L.V.D.; Sengers, R.C.A.; Dörner, A.; Ruitenbeek, W.; Trijbels, F.J.; Valsson, J.; Sigfusson, G.; et al. Adenine nucleotide translocator 1 deficiency associated with Sengers syndrome. *Ann. Neurol.* **2002**, *52*, 95–99. [CrossRef]
135. Davey, K.M.; Parboosingh, J.S.; McLeod, D.R.; Chan, A.; Casey, R.; Ferreira, P.; Snyder, F.F.; Bridge, P.J.; Bernier, F.P. Mutation of DNAJC19, a human homologue of yeast inner mitochondrial membrane co-chaperones, causes DCMA syndrome, a novel autosomal recessive Barth syndrome-like condition. *J. Med. Genet.* **2005**, *43*, 385–393. [CrossRef]
136. Shen, Z.; Ye, C.; McCain, K.; Greenberg, M.L. The Role of Cardiolipin in Cardiovascular Health. *Biomed. Res. Int.* **2015**, *2015*, 1–12. [CrossRef]
137. Sparkes, R.; Patton, D.; Bernier, F.P. Cardiac features of a novel autosomal recessive dilated cardiomyopathic syndrome due to defective importation of mitochondrial protein. *Cardiol. Young* **2007**, *17*, 215–217. [CrossRef]
138. Janz, A.; Chen, R.; Regensburger, M.; Ueda, Y.; Rost, S.; Klopocki, E.; Günther, K.; Edenhofer, F.; Duff, H.J.; Ergün, S.; et al. Generation of two patient-derived iPSC lines from siblings (LIBUCi001-A and LIBUCi002-A) and a genetically modified iPSC line (JMUi001-A-1) to mimic dilated cardiomyopathy with ataxia (DCMA) caused by a homozygous DNAJC19 mutation. *Stem Cell Res.* **2020**, *46*, 101856. [CrossRef]
139. Rohani, L.; Machiraju, P.; Sabouny, R.; Meng, G.; Liu, S.; Zhao, T.; Iqbal, F.; Wang, X.; Ravandi, A.; Wu, J.C.; et al. Reversible Mitochondrial Fragmentation in iPSC-Derived Cardiomyocytes From Children With DCMA, a Mitochondrial Cardiomyopathy. *Can. J. Cardiol.* **2020**, *36*, 554–563. [CrossRef]
140. Richter-Dennerlein, R.; Korwitz, A.; Haag, M.; Tatsuta, T.; Dargazanli, S.; Baker, M.; Decker, T.; Lamkemeyer, T.; Rugarli, E.I.; Langer, T. DNAJC19, a Mitochondrial Cochaperone Associated with Cardiomyopathy, Forms a Complex with Prohibitins to Regulate Cardiolipin Remodeling. *Cell Metab.* **2014**, *20*, 158–171. [CrossRef]
141. Jefferies, J.L. Barth syndrome. *Am. J. Med Genet. Part C Semin. Med. Genet.* **2013**, *163*, 198–205. [CrossRef] [PubMed]
142. Jefferies, J.L.; Morales, D.L. Mechanical Circulatory Support in Children: Bridge to Transplant Versus Recovery. *Curr. Heart Fail. Rep.* **2012**, *9*, 236–243. [CrossRef] [PubMed]
143. Fraser, C.D.; Jaquiss, R.D.; Rosenthal, D.N.; Humpl, T.; Canter, C.E.; Blackstone, E.H.; Naftel, D.C.; Ichord, R.N.; Bomgaars, L.; Tweddell, J.S.; et al. Prospective Trial of a Pediatric Ventricular Assist Device. *N. Engl. J. Med.* **2012**, *367*, 532–541. [CrossRef] [PubMed]
144. Dedieu, N.; Giardini, A.; Steward, C.G.; Fenton, M.; Karimova, A.; Hsia, T.Y.; Burch, M. Successful mechanical circulatory support for 251 days in a child with intermittent severe neutropenia due to Barth syndrome. *Pediatr. Transpl.* **2012**, *17*, E46–E49. [CrossRef] [PubMed]
145. Coman, D.J.; Yaplito-Lee, J.; Boneh, A. New indications and controversies in arginine therapy. *Clin. Nutr.* **2008**, *27*, 489–496. [CrossRef]
146. Rigaud, C.; Lebre, A.-S.; Touraine, R.; Beaupain, B.; Ottolenghi, C.; Chabli, A.; Ansquer, H.; Ozsahin, H.; Di Filippo, S.; De Lonlay, P.; et al. Natural history of Barth syndrome: A national cohort study of 22 patients. *Orphanet J. Rare Dis.* **2013**, *8*, 70. [CrossRef]
147. Suzuki-Hatano, S.; Saha, M.; Rizzo, S.A.; Witko, R.L.; Gosiker, B.J.; Ramanathan, M.; Soustek, M.S.; Jones, M.D.; Kang, P.B.; Byrne, B.J.; et al. AAV-Mediated TAZ Gene Replacement Restores Mitochondrial and Cardioskeletal Function in Barth Syndrome. *Hum. Gene Ther.* **2019**, *30*, 139–154. [CrossRef]
148. Malhotra, A.; Edelman-Novemsky, I.; Xu, Y.; Plesken, H.; Ma, J.; Schlame, M.; Ren, M. Role of calcium-independent phospholipase A2 in the pathogenesis of Barth syndrome. *Proc. Natl. Acad. Sci. USA* **2009**, *106*, 2337–2341. [CrossRef]

149. Schaloske, R.H.; Dennis, E.A. The phospholipase A2 superfamily and its group numbering system. *Biochim. Biophys. Acta Mol. Cell Biol. Lipids* **2006**, *1761*, 1246–1259. [CrossRef]
150. Huang, Y.; Powers, C.; Moore, V.; Schafer, C.; Ren, M.; Phoon, C.K.L.; James, J.F.; Glukhov, A.V.; Javadov, S.; Vaz, F.M.; et al. The PPAR pan-agonist bezafibrate ameliorates cardiomyopathy in a mouse model of Barth syndrome. *Orphanet J. Rare Dis.* **2017**, *12*, 1–9. [CrossRef]
151. Berger, J.P.; Moller, D.E. The Mechanisms of Action of PPARs. *Annu. Rev. Med.* **2002**, *53*, 409–435. [CrossRef] [PubMed]
152. Lehmann, L.H.; Jebessa, Z.H.; Kreusser, M.M.; Horsch, A.; He, T.; Kronlage, M.; Dewenter, M.; Sramek, V.; Oehl, U.; Krebs-Haupenthal, J.; et al. Faculty Opinions recommendation of A proteolytic fragment of histone deacetylase 4 protects the heart from failure by regulating the hexosamine biosynthetic pathway. *Nat. Med.* **2018**, *24*, 62–72. [CrossRef] [PubMed]
153. Schafer, C.; Moore, V.; Dasgupta, N.; Javadov, S.; James, J.F.; Glukhov, A.I.; Strauss, A.W.; Khuchua, Z. The Effects of PPAR Stimulation on Cardiac Metabolic Pathways in Barth Syndrome Mice. *Front. Pharmacol.* **2018**, *9*, 318. [CrossRef] [PubMed]
154. He, Q.; Harris, N.; Ren, J.; Han, X. Mitochondria-Targeted Antioxidant Prevents Cardiac Dysfunction Induced by Tafazzin Gene Knockdown in Cardiac Myocytes. *Oxid. Med. Cell. Longev.* **2014**, *2014*, 1–12. [CrossRef] [PubMed]
155. Chowdhury, A.; Aich, A.; Jain, G.; Wozny, K.; Lüchtenborg, C.; Hartmann, M.; Bernhard, O.; Balleiniger, M.; Alfar, E.A.; Zieseniss, A.; et al. Defective Mitochondrial Cardiolipin Remodeling Dampens HIF-1α Expression in Hypoxia. *Cell Rep.* **2018**, *25*, 561–570.e6. [CrossRef]
156. Grosberg, A.; Alford, P.W.; McCain, M.L.; Parker, K.K. Ensembles of engineered cardiac tissues for physiological and pharmacological study: Heart on a chip. *Lab Chip* **2011**, *11*, 4165–4173. [CrossRef]
157. Agarwal, A.; Goss, J.A.; Cho, A.; McCain, M.L.; Parker, K.K. Microfluidic heart on a chip for higher throughput pharmacological studies. *Lab Chip* **2013**, *13*, 3599–3608. [CrossRef] [PubMed]
158. Grosberg, A.; Nesmith, A.P.; Goss, J.A.; Brigham, M.D.; McCain, M.L.; Parker, K.K. Muscle on a chip: In vitro contractility assays for smooth and striated muscle. *J. Pharmacol. Toxicol. Methods* **2012**, *65*, 126–135. [CrossRef]
159. Johnson, J.M.; Ferrara, P.J.; Verkerke, A.R.P.; Coleman, C.B.; Wentzler, E.J.; Neufer, P.D.; Kew, K.A.; Brás, L.E.D.C.; Funai, K. Targeted overexpression of catalase to mitochondria does not prevent cardioskeletal myopathy in Barth syndrome. *J. Mol. Cell. Cardiol.* **2018**, *121*, 94–102. [CrossRef]
160. Rajagopal, B.S.; Silkstone, G.G.; Nicholls, P.; Wilson, M.T.; Worrall, J.A.R. An investigation into a cardiolipin acyl chain insertion site in cytochrome c. *Biochim. Biophys. Acta Gen. Subj.* **2012**, *1817*, 780–791. [CrossRef]
161. Birk, A.V.; Liu, S.; Soong, Y.; Mills, W.; Singh, P.; Warren, J.D.; Seshan, S.V.; Pardee, J.D.; Szeto, H.H. The Mitochondrial-Targeted Compound SS-31 Re-Energizes Ischemic Mitochondria by Interacting with Cardiolipin. *J. Am. Soc. Nephrol.* **2013**, *24*, 1250–1261. [CrossRef] [PubMed]
162. Karch, J.; Molkentin, J.D. Identifying the components of the elusive mitochondrial permeability transition pore. *Proc. Natl. Acad. Sci. USA* **2014**, *111*, 10396–10397. [CrossRef] [PubMed]
163. Birk, A.V.; Chao, W.M.; Bracken, C.; Warren, J.D.; Szeto, H.H. Targeting mitochondrial cardiolipin and the cytochromec/cardiolipin complex to promote electron transport and optimize mitochondrial ATP synthesis. *Br. J. Pharmacol.* **2014**, *171*, 2017–2028. [CrossRef] [PubMed]
164. Sabbah, H.N.; Gupta, R.C.; Kohli, S.; Wang, M.; Hachem, S.; Zhang, K. Chronic Therapy With Elamipretide (MTP-131), a Novel Mitochondria-Targeting Peptide, Improves Left Ventricular and Mitochondrial Function in Dogs With Advanced Heart Failure. *Circ. Heart Fail.* **2016**, *9*, e002206. [CrossRef] [PubMed]
165. Szeto, H.H.; Liu, S.; Soong, Y.; Wu, D.; Darrah, S.F.; Cheng, F.-Y.; Zhao, Z.; Ganger, M.; Tow, C.Y.; Seshan, S.V. Mitochondria-Targeted Peptide Accelerates ATP Recovery and Reduces Ischemic Kidney Injury. *J. Am. Soc. Nephrol.* **2011**, *22*, 1041–1052. [CrossRef]
166. Yang, L.; Zhao, K.; Calingasan, N.Y.; Luo, G.; Szeto, H.H.; Beal, M.F. Mitochondria targeted peptides protect against 1-methyl-4-phenyl-1,2,3,6-tetrahydropyridine neurotoxicity. *Antioxid. Redox Signal.* **2009**, *11*, 2095–2104. [CrossRef]
167. Machiraju, P.; Wang, X.; Sabouny, R.; Huang, J.; Zhao, T.; Iqbal, F.; King, M.; Prasher, D.; Lodha, A.; Jimenez-Tellez, N.; et al. SS-31 Peptide Reverses the Mitochondrial Fragmentation Present in Fibroblasts From Patients With DCMA, a Mitochondrial Cardiomyopathy. *Front. Cardiovasc. Med.* **2019**, *6*, 167. [CrossRef]

168. Allen, M.E.; Pennington, E.R.; Perry, J.B.; Dadoo, S.; Makrecka-Kuka, M.; Dambrova, M.; Moukdar, F.; Patel, H.D.; Han, X.; Kidd, G.K.; et al. The cardiolipin-binding peptide elamipretide mitigates fragmentation of cristae networks following cardiac ischemia reperfusion in rats. *Commun. Biol.* **2020**, *3*, 1–12. [CrossRef]
169. Xu, Y.; Condell, M.; Plesken, H.; Edelman-Novemsky, I.; Ma, J.; Ren, M.; Schlame, M. A Drosophila model of Barth syndrome. *Proc. Natl. Acad. Sci. USA* **2006**, *103*, 11584–11588. [CrossRef]
170. Acehan, D.; Khuchua, Z.; Houtkooper, R.H.; Malhotra, A.; Kaufman, J.; Vaz, F.M.; Ren, M.; Rockman, H.A.; Stokes, D.L.; Schlame, M. Distinct effects of tafazzin deletion in differentiated and undifferentiated mitochondria. *Mitochondrion* **2009**, *9*, 86–95. [CrossRef]
171. Sakamoto, T.; Inoue, T.; Otomo, Y.; Yokomori, N.; Ohno, M.; Arai, H.; Nakagawa, Y. Deficiency of Cardiolipin Synthase Causes Abnormal Mitochondrial Function and Morphology in Germ Cells ofCaenorhabditis elegans. *J. Biol. Chem.* **2011**, *287*, 4590–4601. [CrossRef] [PubMed]
172. Serricchio, M.; Bütikofer, P. An essential bacterial-type cardiolipin synthase mediates cardiolipin formation in a eukaryote. *Proc. Natl. Acad. Sci. USA* **2012**, *109*, E954–E961. [CrossRef] [PubMed]
173. Soustek, M.S.; Falk, D.J.; Mah, C.S.; Toth, M.J.; Schlame, M.; Lewin, A.S.; Byrne, B.J. Characterization of a Transgenic Short Hairpin RNA-Induced Murine Model of Tafazzin Deficiency. *Hum. Gene Ther.* **2011**, *22*, 865–871. [CrossRef] [PubMed]
174. Landriscina, C.; Megli, F.M.; Quagliariello, E. Turnover of fatty acids in rat liver cardiolipin: Comparison with other mitochondrial phospholipids. *Lipids* **1976**, *11*, 61–66. [CrossRef] [PubMed]
175. Wahjudi, P.N.; Yee, J.K.; Martinez, S.R.; Zhang, J.; Teitell, M.; Nikolaenko, L.; Swerdloff, R.; Wang, C.; Lee, W.N.P. Turnover of nonessential fatty acids in cardiolipin from the rat heart. *J. Lipid Res.* **2011**, *52*, 2226–2233. [CrossRef]
176. Boynton, T.O.; Shimkets, L.J. MyxococcusCsgA, DrosophilaSniffer, and human HSD10 are cardiolipin phospholipases. *Genes Dev.* **2015**, *29*, 1903–1914. [CrossRef]
177. Shen, Z.; Li, Y.; Gasparski, A.N.; Abeliovich, H.; Greenberg, M.L. Cardiolipin Regulates Mitophagy through the Protein Kinase C Pathway. *J. Biol. Chem.* **2017**, *292*, 2916–2923. [CrossRef]
178. Konno, Y.; Ohno, S.; Akita, Y.; Kawasaki, H.; Suzuki, K. Enzymatic Properties of a Novel Phorbol Ester Receptor/Protein Kinase, nPKC1. *J. Biochem.* **1989**, *106*, 673–678. [CrossRef]

Article

Case Report: Identification of a Novel Variant (m.8909T>C) of Human Mitochondrial *ATP6* Gene and Its Functional Consequences on Yeast ATP Synthase

Qiuju Ding [1,†], Róża Kucharczyk [2,†], Weiwei Zhao [1,†], Alain Dautant [3], Shutian Xu [1], Katarzyna Niedzwiecka [2], Xin Su [3], Marie-France Giraud [3], Kewin Gombeau [3], Mingchao Zhang [1], Honglang Xie [1], Caihong Zeng [1], Marine Bouhier [3], Jean-Paul di Rago [3], Zhihong Liu [1], Déborah Tribouillard-Tanvier [3,4,*] and Huimei Chen [1,*]

1 National Clinical Research Center of Kidney Diseases, Jinling Hospital, Nanjing University School of Medicine, Nanjing 211166, China; dqj3206@163.com (Q.D.); 15205153118@163.com (W.Z.); tinne_xst@163.com (S.X.); zmchj99@163.com (M.Z.); xiehl_doctor@163.com (H.X.); zengch_nj@hotmail.com (C.Z.); liuzhihong@nju.edu.cn (Z.L.)

2 Institute of Biochemistry and Biophysics, Polish Academy of Sciences, 00090 Warsaw, Poland; roza@ibb.waw.pl (R.K.); slimanio@gmail.com (K.N.)

3 Institut de Biochimie et Génétique Cellulaires, Université de Bordeaux, CNRS, UMR 5095, F-33000 Bordeaux, France; a.dautant@ibgc.cnrs.fr (A.D.); xin.su@ibgc.cnrs.fr (X.S.); marie-france.giraud@ibgc.cnrs.fr (M.-F.G.); kewin.gombeau@ibgc.cnrs.fr (K.G.); marine.bouhier@ibgc.cnrs.fr (M.B.); jp.dirago@ibgc.cnrs.fr (J.-P.d.R.)

4 Institut national de la santé et de la recherche médicale, 75000 Paris, France

* Correspondence: deborah.tribouillard-tanvier@ibgc.cnrs.fr (D.T.-T.); chenhuimei@nju.edu.cn (H.C.)

† These authors contributed to this work equally.

Received: 31 July 2020; Accepted: 17 September 2020; Published: 22 September 2020

Abstract: With the advent of next generation sequencing, the list of mitochondrial DNA (mtDNA) mutations identified in patients rapidly and continuously expands. They are frequently found in a limited number of cases, sometimes a single individual (as with the case herein reported) and in heterogeneous genetic backgrounds (heteroplasmy), which makes it difficult to conclude about their pathogenicity and functional consequences. As an organism amenable to mitochondrial DNA manipulation, able to survive by fermentation to loss-of-function mtDNA mutations, and where heteroplasmy is unstable, *Saccharomyces cerevisiae* is an excellent model for investigating novel human mtDNA variants, in isolation and in a controlled genetic context. We herein report the identification of a novel variant in mitochondrial *ATP6* gene, m.8909T>C. It was found in combination with the well-known pathogenic m.3243A>G mutation in mt-tRNA^{Leu}. We show that an equivalent of the m.8909T>C mutation compromises yeast adenosine tri-phosphate (ATP) synthase assembly/stability and reduces the rate of mitochondrial ATP synthesis by 20–30% compared to wild type yeast. Other previously reported *ATP6* mutations with a well-established pathogenicity (like m.8993T>C and m.9176T>C) were shown to have similar effects on yeast ATP synthase. It can be inferred that alone the m.8909T>C variant has the potential to compromise human health.

Keywords: *MT-ATP6*; m.8909T>C; ATP synthase; nephropathy; oxidative phosphorylation; mitochondrial disease

---

1. Introduction

Mitochondria provide aerobic eukaryotes with cellular energy by generating adenosine tri-phosphate (ATP) through the process of oxidative phosphorylation (OXPHOS). As a first step, electrons from nutrients (such as carbohydrates and fatty acids) are transferred by four complexes (CI-CIV) anchored to the mitochondrial inner membrane. This results in a transmembrane electrochemical proton gradient, which drives ATP synthesis from adenosine di-phosphate (ADP) and inorganic phosphate (Pi) by the ATP synthase (CV).

Defects of the OXPHOS system have been implicated in a broad spectrum of human diseases. Typical clinical traits include encephalopathies, cardiomyopathies, myopathies, visual/hearing, liver and renal dysfunctions [1–3]. Many of these diseases are caused by mitochondrial DNA (mtDNA) mutations. This DNA encodes 13 protein subunits of the OXPHOS system and a number of RNAs necessary for the synthesis of these proteins inside the organelle [4]. Pathogenic mtDNA mutations often co-exist with non-mutated mtDNA (heteroplasmy) and are highly recessive (only rare cases of dominancy were reported [5]), which makes it difficult to know how mitochondrial function is affected in patient's cells and tissues. Furthermore, because of the high mutability of the mitochondrial genome due to its exposure to damaging oxygen species (ROS) and the poor activity of mitochondrial DNA repair systems, it can be difficult to establish the pathogenicity of a mtDNA variant [4]. Additionally, the effects of deleterious mtDNA mutations may be aggravated by nucleotide changes in nuclear and mitochondrial DNA that are not pathogenic per se (the so-called modifier genes) [6,7].

Being amenable to mitochondrial genetic transformation [8], and owing to its good fermenting capability that enables survival after the loss of oxidative phosphorylation and inability to stably maintain heteroplasmy [9], *Saccharomyces cerevisiae* has been used as a model to investigate mtDNA mutations identified in patients. We have exploited these attributes to study *MT-ATP6* gene mutations found in patients. This gene encodes the subunit *a* of ATP synthase, which is involved in moving protons through the membrane domain (F_O) of ATP synthase coupled to ATP synthesis [10–15]. These yeast-based studies helped to better define the functional consequences of subunit *a* mutations and provided support for the pathogenicity of rare alleles identified in only a limited number of cases like m.8851T>C [12] and m.8969G>A [16]. Importantly, the extent to which yeast ATP synthase was affected correlated with disease severity, which reflects the strong evolutionary conservation of the regions of subunit *a* where these mutations localize [17–26].

Herein we report the identification of a novel *MT-ATP6* variant, m.8909T>C. It converts into serine a highly conserved phenylalanine residue of subunit *a* ($aF_{128}S$ in humans; $aF_{145}S$ in yeast). We detected it by entirely sequencing the mtDNA of a patient with an extreme clinical presentation including brain, kidney and muscular dysfunctions leading to premature death at the age of 14. This patient also carried the well-known pathogenic m.3243A>G mutation in mt-tRNA$^{Leu(1)}$ (*MT-TL1*). In a yeast model of the m.8909T>C variant, ATP synthase assembly/stability was significantly compromised to an extent comparable to that seen previously in yeast models of other *ATP6* mutations with a well-established pathogenicity. These findings indicate that alone the m.8909T>C variant has the potential to compromise human health.

2. Materials and Methods

2.1. Kidney Analyses

Kidney biopsies were performed under ultrasound guidance by an experienced investigator. Paraffin-embedded sections were routinely stained with periodic acid Schiff and assessed by light microscopy. Fluorescence staining for IgG, IgA, immunoglobulin M (IgM), C3c and C1q was performed on freshly frozen renal tissues and ultra-thin sections stained with uranyl acetate and lead nitrate were examined by electron microscopy as described in [16]. Kidney biopsies were frozen in isopentane chilled with liquid nitrogen. Six-micrometer thick cryostat frozen sections were performed for enzyme histochemical staining for Complex IV (COX), Complex II (SDH) and nicotinamide adenine dinucleotide (NADH) dehydrogenase activities using specific substrates of these enzymes including Cytochrome *c* (C7752, Sigma-Aldrich), β-nicotinamide adenine dinucleotide (N7410, Sigma-Aldrich, St. Louis, MO, USA) and succinic acid (S3674, Sigma-Aldrich), respectively as described [27–29]. Optical density quantification of these activities was assessed from 20 consecutive microscopic fields in each renal section and the adjacent background regions.

2.2. Patient Consent, Ethical Committees, and Adhesion to Biosecurity and Institutional Safety Procedures

The female Chinese patient herein described was hospitalized with recurrent kidney disease and multiple systemic dysfunctions at Jinling Hospital in Nanjing (China). One hundred healthy control adults were randomly recruited from a panel of unaffected, genetically unrelated Han Chinese individuals from the same geographic region. All methods were conducted in accordance with the Ethics Committee of the Jinling Hospital and in the respect of biosafety and public health. Written informed consents were obtained from the patient and her parents as well as the 100 healthy controls. The mother didn't consent for sequencing her mtDNA and medical examination.

2.3. mtDNA Characterization

Whole DNA was prepared from total blood, kidney and urine sediment samples with the DNA extraction kit from Qiagen (Hilden, Germany), and mtDNA was amplified as described in [16]. The revised Cambridge reference sequence (rCRS) of *H. sapiens* mitochondrial DNA (GenBank NC_012920.1) was used to identify variants of this DNA. Variant prioritization and the presence of deletion/depletion was performed as described [16]. The abundance of mutations was assessed using a Pyromark Q24 platform (Qiagen) as described in [16]. After purification using streptavidin Sepharose HP (GE Healthcare), and denaturation with NaOH, the PCR products were annealed and sequenced with the primers listed in Table 1.

Table 1. Primers used for PCR amplification of mitochondrial DNA (mtDNA) and sequencing.

	Forward	Reverse	Product Length (bp)
	For detection of point mutations		
1	CCCACAGTTTATGTAGCTTACC	GTACTATATCTATTGCGCCAGG	1215
2	ACTACCAGACAACCTTAGCC	AACATCGAGGTCGTAAACCC	1293
3	CTTCACCAGTCAAAGCGAAC	AGAAGTAGGGTCTTGGTGAC	1242
4	CGAACTAGTCTCAGGCTTCAAC	TCGTGGTGCTGGAGTTTAAG	1228
5	ACGTAAGCCTTCTCCTCACT	TCGTTACCTAGAAGGTTGCC	1138
6	CCGACCGTTGACTATTCTCT	GATGGCAAATACAGCTCCTA	1160
7	GCAAACTCATCACTAGACATCG	AGCTTTACAGTGGGCTCTAG	1329
8	ACCACAGTTTCATGCCCATC	TGGCCTTGGTATGTGCTTTC	1223
9	CACTTCCACTCCATAACGCT	GTTGAGGGTTATGAGAGTAGC	1308
10	TACCAAATGCCCCTCATTTA	GTAATGAGGATGTAAGCCCG	1272
11	TTCAATCAGCCACATAGCCC	GATGAAACCGATATCGCCGA	1260
12	GAGGGCGTAGGAATTATATCC	GTCAGGTTAGGTCTAGGAGG	1240
13	CATACTCGGATTCTACCCTAG	TGTAATTACTGTGGCCCCTC	1280
14	TCGGCATTATCCTCCTGCTT	GTGCTATGTACGGTAAATGGC	1250
15	TGACTCACCCATCAACAACC	ATAGAAAGGCTAGGACCAAACC	1179
	For detection of mtDNA deletion		
1	GCACCCTATGTCGCAGTATCTGTC TTTG	GGACGAGAAGGGATTTGACTG TAATGTGC	16,255
2	CACTTCCACTCCATAACGCTCC TCATACT	GGGCTATTGGTTGAATGAGTAGG CTGATG	16,250
	For detection of mtDNA copy number		
COX1	TTCGCCGACCGTTGACTATTCTCT	AAGATTATTACAAATGCATGGGC	197
18S	GTCTGTGATGCCCTTAGATG	AGCTTATGACCCGCACTTAC	177
	For pyrosequencing of m.3243A>G		
Amplification	AAGGACAAGAGAAATAAGGC	ATGAGGAGTAGGAGGTTGG	207
Sequencing	TTTTATGCGATTACCG		

2.4. *Media for Growing Yeast*

The following media were used for growing yeast. YPAD: 1% (*w*/*v*) yeast extract, 2% (*w*/*v*) bacto peptone, 40 mg/L adenine and 2% (*v*/*v*) glucose; YPAGly: 1% (*w*/*v*) yeast extract, 2% (*w*/*v*) bacto peptone, 40 mg/L adenine and 2% (*v*/*v*) glycerol; YPGALA: 1% (*w*/*v*) yeast extract, 2% (*w*/*v*) bacto peptone, 40 mg/L adenine and 2% (*v*/*v*) galactose. Solid media contained 2% (*w*/*v*) agar.

2.5. *Construction of S. cerevisiae Strain RKY108 (aF_{145}S)*

We used the QuikChange XL Site-directed Mutagenesis Kit of Stratagene to introduce an equivalent of the m.8909T>C mutation (*a*F_{145}*S*) in the yeast *ATP6* gene cloned in pUC19 [30], utilizing the oligonucleotide 5′-GGTTTATATAAACATGGTTGAGTATTCTTCTCATTATCAGTACCTGCTGGTACACCATTACC-3′. The *atp6*-F_{145}S gene was cloned into the plasmid pJM2 [8]. The resulting plasmid (pRK66) was introduced into mitochondria of the strain DFS160 that totally lacks mitochondrial DNA (ρ^0) as described [8] (see Table 2 for complete genotypes and sources of yeast strains). The resulting mitochondrial transformants (RKY109) were crossed to the *atp6::ARG8m* deletion strain MR10 [30] to yield strain RKY108. This strain has the MR10 nucleus and the *atp6*-F_{145}S gene in a complete (ρ^+) mitochondrial genome. The presence of the *atp6* mutation in these clones was confirmed by DNA sequencing. No other change was detected in the *ATP6* gene. The corresponding wild type strain MR6 (WT) is a derivative of strain W303-1B in which the nuclear *ARG8* gene has been replaced with *HIS3* (*arg8::HIS3*) and with the entirely sequenced mitochondrial genome of strain BY4741 [30].

Table 2. Genotypes and sources of yeast strains.

Strain	Nuclear Genotype	mtDNA	Reference
DFS160	*MATα leu2Δ ura3-52 ade2-101 arg8::URA3 kar1-1*	ρ^o	[31]
NB40-3C	*MATa lys2 leu2-3,112 ura3-52 his3ΔHindIII arg8::hisG*	ρ^+ *cox2-62*	[31]
MR6	*MATa ade2-1 his3-11,15 trp1-1 leu2-3,112 ura3-1 CAN1 arg8::HIS3*	ρ^+	[30]
MR10	*MATa ade2-1 his3-11,15 trp1-1 leu2-3,112 ura3-1 CAN1 arg8::HIS3*	ρ^+ *atp6::ARG8^m*	[30]
RKY109	*MATa leu2Δura3-52 ade2-101 arg8::URA3 kar1-1*	ρ^- *atp6*-F_{145}S	This study
RKY108	*MATa ade2-1 his3-11,15 trp1-1 leu2-3,112 ura3-1 CAN1 arg8::HIS3*	ρ *atp6*-F_{145}S	This study

Yeast genes nomencalure is used in accordance to http://www.yeastgenome.org/help/community/nomenclature-conventions.

2.6. *Yeast-Based Drug Assay*

0.05 OD_{650nm} of exponentially growing cells were homogeneously spread with sterile glass beads on a square Petri dish (12 cm × 12 cm) containing solid YPAGly medium. Sterile filters were deposited on the plate and spotted with oligomycin (purchased from Sigma, St. Louis, MO, USA) dissolved in DMSO.

2.7. Biochemical Investigation of Mitochondria

Oxygen consumption measurements in mitochondria isolated from yeast cells grown in complete galactose medium were performed according to [32], using a Clark electrode (Heito, Paris, France). Freshly prepared mitochondria were added to 1 mL of respiration buffer (10 mM Tris-maleate pH 6.8, 0.65 M sorbitol, 0.3 mM EGTA, and 3 mM potassium phosphate) at 0.15 mg/mL in the reaction chamber maintained at 28 °C. The oxygen consumption was measured after successive additions of 4 mM NADH, 150 μM ADP, 4 μM carbonyl cyanide m-chlorophenylhydrazone (CCCP). The rate of ATP synthesis was measured in the same conditions but with 750 μM ADP. Aliquots were withdrawn every 15 s from the reaction mix and supplemented with 3.5% (*w*/*v*) perchloric acid and 12.5 mM EDTA. After neutralization of the samples to pH 6.5 with KOH 0.3 M/MOPS, ATP was quantified by luciferin/luciferase assay (ATPLite kit from Perkin Elmer, Waltham, MA, USA) on a LKB bioluminometer. Oligomycin (3 μg/mL) was used to determine the amount of ATP produced by ATP synthase. ATPase activity was measured in non-osmotically protected mitochondria at pH 8.4 as described [33]. Blue native polyacrylamide gel electrophoresis (BN-PAGE) was performed as described [34]. For this, 200 μg of mitochondrial proteins were suspended in 50 μL of extraction buffer (30 mM HEPES, 150 mM potassium acetate, 12% glycerol, 2 mM 6-aminocaproic acid, 1 mM EGTA, 2% digitonin (Sigma), one protease inhibitor cocktail tablet (Roche) (pH 7.4) and incubated for 30 min on ice. After centrifugation (14,000 rpm, 4 °C, 30 min) the supernatant containing the solubilized complexes were supplemented with 2.25 μL of loading dye (5% Serva Blue G-250, 750 mM 6-aminocaproic acid) and run into NativePAGE 3–12% Bis-Tris Gel (Invitrogen, Carlsbad, CA, USA). After transfer onto PVDF membrane the yeast ATP synthase complexes were detected with polyclonal antibodies against α-F_1 (Atp1), subunit *c* (Atp9) and subunit *a* (Atp6) used after 1:10,000, 1:5000, and 1:1000 dilutions respectively. The Atp1 antibodies were kindly provided by J. Velours. Anti-Atp9 antibodies were prepared by Eurogentec (Seraing, Belgium) with the synthetic peptide corresponding to the loop connecting the two transmembrane helices of Atp9 as an immunogen. The procedure used to in-gel visualize ATP synthase by its ATPase activity is described in [31].

2.8. Amino-Acid Alignments and Subunit a Topology

Clustal Omega [35] was used to compare amino acid sequences of subunits *a* of various species. The topology of the $aF_{145}S$ mutation was investigated according to the yeast ATP synthase structure described in [36] and drawn using PyMOL [37].

2.9. Statistical Analyses

Chi-square test and SPSS (16.0), Chicago, IL, USA. method were used for evaluating the statistical significance of the data. All of the tests were two tailed, and p values < 0.05 were considered significant.

3. Results

3.1. Case Report

The patient here reported suffered from hemiplegia, epileptic episodes, aphasia, blindness and deafness since she was 10 years old. Cranial CT (Computed Tomography) and MRI (Magnetic Resonance Imaging) revealed cerebral infarction and softening, and inflammatory changes. Moderate to severe abnormalities in electroencephalogram were observed. She also presented severe hearing impairment and hyper lactacidemia. At the age of 14 she developed a nephrotic syndrome with mass proteinuria, hypoproteinemia, and hyperlipidemia (Table 3). Renal biopsy indicated mild mesangial proliferative glomerulonephritis (MPGNs) pattern, with tubule atrophy and interstitial fibrosis (Figure 1A). Diffuse deposition of immunoglobulin M (IgM) was revealed by immunofluorescence microscopy (Figure 1B). Glomerular IgM deposition was observed in the patient's kidney, but also to a lesser extent in healthy controls. IgM deposition can be caused by many reasons and was likely not a crucial factor in disease development. The patient's neuromuscular manifestations partially mitigated with steroid and antiepileptic treatment. A supplemental ATP therapy and additional renal conservative treatment had also some benefits. The initial renal function impairment rapidly aggravated and the patient was at uremia stage when she died after 8 months of follow-up, as revealed by the levels of serum creatinine and urea nitrogen in blood (Table 3).

Table 3. Clinical data.

The Time of First Kidney Biopsy	
Age/years	14
Sex (M/F)	F
Ethnicity	Han Chinese
Course/mouth	8
Urinary protein/g·24 h^{-1}	3.74
BUN/mg·dL^{-1}	20.6
Scr/mg·dL^{-1}	1.23
Alb/g·L^{-1}	27.5
Glo/g·L^{-1}	16.8
TG/mmol·L^{-1}	3.99
Chol/mmol·L^{-1}	8.6
At uremia stage	
BUN/mg·dL^{-1}	58.1
Scr/mg·dL^{-1}	9.27

M: male; F: female; BUN: blood urea nitrogen; Scr: serum creatinine; Alb: albumin; Glo: globulin; TG: triglyceride; Chol: cholesterol.

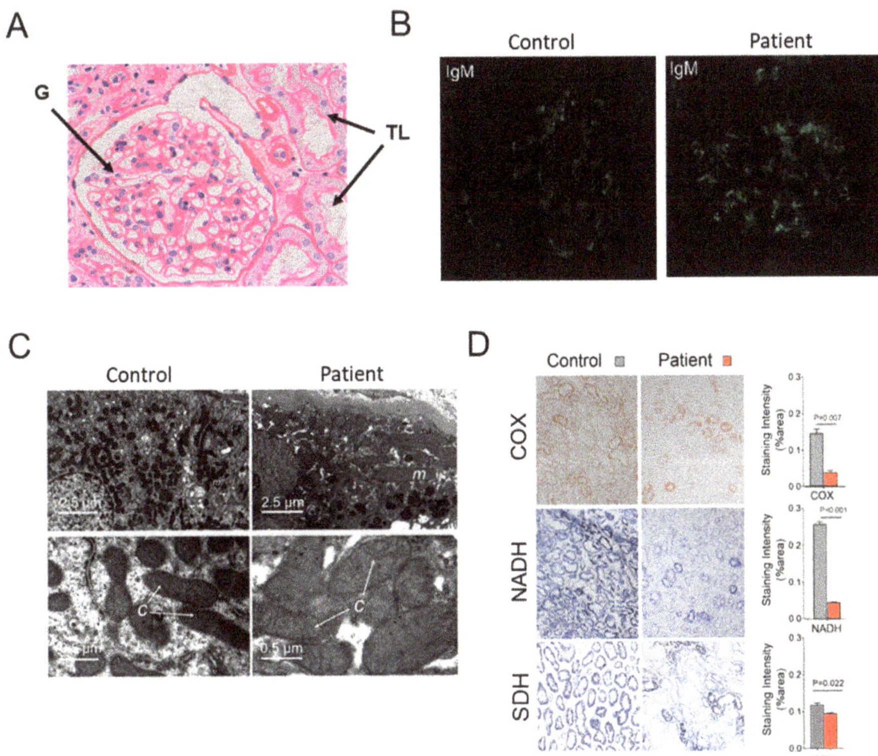

Figure 1. Kidney analyses. (**A**) Light microscopy of kidney samples from the patient show glomeruli (G) mesangial widening, and tubules (TL) atrophy and interstitial fibrosis. (**B**) Fluorescence microscopy reveals immunoglobulin M (IgM) deposits in glomerular cells from the patient. (**C**) Electron micrographs of kidney samples from the healthy control and patient (magnification is ×10,000, and ×80,000 from top to bottom). (**D**) Enzyme histochemical staining of cytochrome *c* oxidase (COX), nicotinamide adenine dinucleotide (NADH) and succinate dehydrogenase (SDH) in freshly frozen kidney biopsies (magnification is ×400).

3.2. Hints for Mitochondrial Dysfunction

Electron microscopy revealed uneven and swelled mitochondria with barely detectable cristae in patient's tubular epithelial cells (Figure 1C). Histochemical staining of cytochrome *c* oxidase (COX or CIV) and NADH dehydrogenase (CI) activities was strongly reduced in renal tissues from the patient in comparison to control kidney samples (Figure 1D), whereas the activity of succinate dehydrogenase (SDH) was much less diminished. No large rearrangement (deletion) and depletion of the patient's mtDNA were detected (not shown). The mitochondrial genome of the patient was entirely sequenced, which revealed a number of nucleotide changes relative to the reference human mitochondrial genome (Table 4), among which m.3243A>G in mt-tRNA^{Leu}, which is the most frequent pathogenic mtDNA allele [38] (Figure 2A). Another point mutation, never reported thus far, was detected in the *MT-ATP6*: m.8909T>C. It is absent in 2704 controls in databases and in 100 age-matched controls from the Nanjing geographic region from which the patient originated (Figure 2B). Pyrosequencing analyses revealed that the *MT-ATP6* variant was homoplasmic in blood, urine sediments (epithelial-like cells detached from tubules) and kidney, whereas the m.3243A>G change was heteroplasmic (50–90%). Defects in mitochondrial translation induced by the m.3243A>G mutation likely explain the poor histochemical staining of CI and CIV, two complexes of mixed genetic origin, and the much better preservation of the entirely nucleus-encoded SDH complex.

Table 4. List of the mtDNA nucleotide changes in the patient relative to the reference sequence of the human mitochondrial genome (http://www.mtdb.igp.uu.se/).

Gene	Nucleotide Changes	Frequency [a]	Amino Acid Change Function Annotation	Type
D-Loop	m.263A>G	1861/1867	non-coding	polymorphic
	m.499G>A *	2106/2144	non-coding	polymorphic
	m.16217T>C *	103/1867	non-coding	polymorphic
	m.16261C>T	111/1867	non-coding	polymorphic
	m.16136T>C *	24/1867	non-coding	polymorphic
12S rRNA	m.750A>G	2682/2704	non-coding	polymorphic
	m.827A>G *	54/2704	non-coding	polymorphic
	m.1438A>G	2620/2704	non-coding	polymorphic
16S rRNA	m.2706A>G	2178/2704	non-coding	polymorphic
TL1	m.3243A>G	0/2704	MELAS	Mutation
ND2	m.4769A>G	2674/2704	$Met_{100}Met$	polymorphic
	m.4820G>A *	45/2704	$Glu_{117}Glu$	polymorphic
	m.5063T>C	3/2704	$Pro_{198}Pro$	polymorphic
COXI	m.6023G>A *	34/2704	$Glu_{40}Glu$	polymorphic
	m.6413T>C *	23/2704	$Asn_{170}Asn$	polymorphic
	m.7028C>T	2199/2704	$Ala_{375}Ala$	polymorphic
ATP6	m.8860A>G	2698/2704	$Thr_{112}Ala$	polymorphic
	m.8909T>C	0/2704	Phe128Ser	Mutation
ND4L	m.10646G>A	13/2704	$Val_{59}Val$	polymorphic
ND4	m.11254T>C	1/2704	$Ile_{165}Ile$	polymorphic
	m.11719G>A	2100/2704	$Gly_{320}Gly$	polymorphic
ND5	m.13590G>A *	110/2704	$Leu_{418}Leu$	polymorphic
	m.13779A>G	0/2704	$Thr_{481}Thr$	polymorphic
CYTB	m.14766T>C	610/2704	$Thr_{7}Ile$	polymorphic
	m.15326A>G	2687/2704	$Thr_{194}Ala$	polymorphic
	m.15535C>T *	48/2704	$Asn_{263}Asn$	polymorphic
	m.15688C>T	0/2704	$Ser_{314}Ser$	polymorphic
	m.15758A>G	29/2704	$Ile_{338}Val$	polymorphic

[a] Frequency refers to the occurrence of the detected nucleotide changes in 2704 control individuals except D-Loop.
* Stands for haplogroup B4b1a nucleotide changes.

The m.8909T>C variant converts a phenylalanine residue into serine at position 128 of human subunit *a* ($aF_{128}S$) of ATP synthase. This residue is highly conserved in a large panel of evolutionary distant species (see below). Therefore, if the m.3243A>G mutation in mt-$tRNA^{Leu}$ likely impacted the two mtDNA encoded subunits (ATP6 and ATP8) of ATP synthase, as was observed in previous studies of cells containing this mutation [39–41], we considered that the m.8909T>C possibly affected also the ATP synthase. It would have been difficult to test this hypothesis from patient's cells and tissues because of their mitochondrial genetic heterogeneity. We, therefore decided, as described below, to investigate the consequences in isolation of an equivalent of this mutation on the yeast ATP synthase.

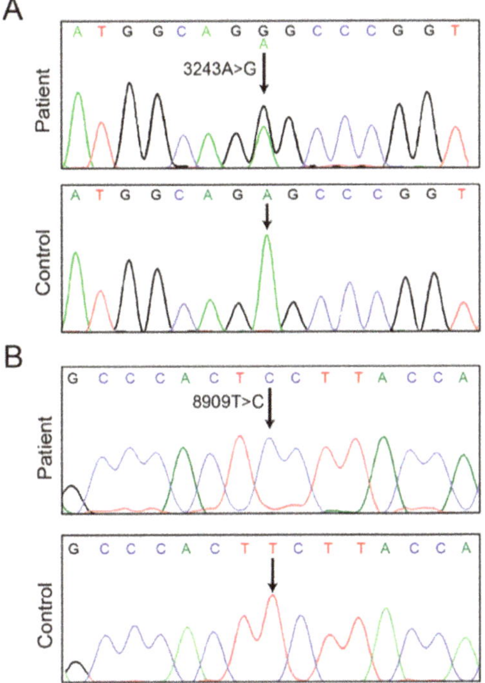

Figure 2. Mitochondrial DNA sequencing. Entire sequencing of the mtDNA of the patient identified the well-known pathogenic mutation m.3243A>G in *MT-TL1* (**A**) and a novel variant in *MT-ATP6*, m.8909T>C (**B**). The former was heteroplasmic whereas the latter was homoplasmic in analyzed cells and tissues (blood, urine and kidney). All the nucleotide changes relative to the reference human mitochondrial genome sequence are listed in Table 2.

3.3. *Consequences of the m.8909T>C Mutation on Yeast ATP Synthase*

Yeast subunit *a* (also called subunit 6 or Atp6) is synthesized as a precursor protein the first ten residues of which are removed during ATP synthase assembly [42]. The phenylalanine residue at position 128 of human subunit *a* that is changed into serine by the m.8909T>C mutation corresponds to the phenylalanine residue at position 145 in the mature yeast protein (155 in the unprocessed form) (see below). A yeast model homoplasmic for the m.8909T>C mutation was created by changing the phenylalanine codon TTC 155 into TCA (see Materials and Methods).

3.3.1. Influence of the $aF_{145}S$ Mutation on Yeast Respiratory Growth

The $a\mathrm{F}_{145}\mathrm{S}$ mutant grew well from fermentable substrates like glucose (Figure 3A), where ATP synthase is not required. Mitochondrial-dependent growth on glycerol was also normal at both 28 °C (the optimal temperature for mitochondrial function in yeast), and at 36 °C (Figure 3A). However, in the presence of increasing concentrations of oligomycin, a chemical inhibitor of ATP synthase [43], respiratory growth of the $a\mathrm{F}_{145}\mathrm{S}$ mutant was less efficient compared to the WT (Figure 3A). An increased sensitivity to oligomycin is usually observed in yeast ATP synthase defective mutants because less of this drug is needed to reach the threshold of ATP synthase activity (20%) below which respiratory growth of yeast becomes obviously compromised [13,44]. The increased sensitivity to oligomycin of the $aF_{145}S$ mutant was further characterized by spreading the cells as a dense layer on glycerol medium and then exposed to a drop of oligomycin deposited on a sterile disk of paper (Figure 3B). Oligomycin diffuses in the growth medium, which results in the establishment of a continuous gradient around the disk. Growth is inhibited until a certain drug concentration. The halos of growth inhibition had a much higher diameter for the mutant vs. wild type yeast (Figure 3B), consistent with the growth tests shown in Figure 3A. These in vivo observations provide a strong indication that the $aF_{145}S$ mutation has detrimental consequences on ATP synthase.

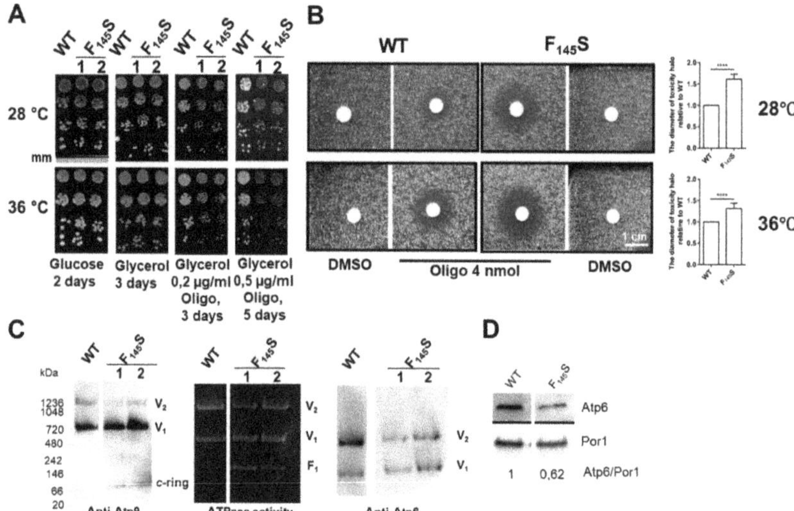

Figure 3. Consequences of an equivalent of the m.8909T>C variant ($aF_{145}S$) in yeast. (**A**) Cells from the WT (MR6) and two genetically independent clones (denoted as 1 and 2) of the $aF_{145}S$ mutant strain (RKY108) grown in glucose were serially diluted and spotted on plates containing glucose, glycerol, and glycerol supplemented with indicated concentrations of oligomycin (Oligo). The plates were scanned after the indicated days of incubation. Representative data from two WT and two independent mutant clones and at least two repeats of each are shown. A mm scale is shown on the upper left image. (**B**) On the left panel, cells from WT and $aF_{145}S$ mutant strains were spread as dense layers onto rich glycerol solid media and then exposed to sterile filters spotted with 4 nmol oligomycin (Oligo) and DMSO as a negative control (solvent). The plates were scanned after 3 days of incubation at 28 °C and 36 °C. The shown scale is 1 cm. The diameters of the halos of growth inhibition (in % of WT) are reported in the shown histograms. **** indicates a p-value < 0.0001. (**C**) Assembly/stability of ATP synthase. Protein extracts were prepared from mitochondria isolated from WT and $aF_{145}S$ strains grown at 36 °C. Samples with a same content in porin were solubilized with 2% digitonin and resolved by Blue native–polyacrylamide gel electrophoresis (BN-PAGE, 200 µg proteins per lane). Dimers (V_2) and monomers (V_1) of F_1F_O complexes and free F_1 were in-gel visualized by their ATPase activity. The protein complexes were transferred onto nitrocellulose and probed with antibodies against subunit c (Atp9) and a (Atp6). On the left margin is a molecular weight ladder. (**D**) Western blot (WB) of mitochondrial proteins resolved by SDS-PAGE with antibodies against subunit a (Atp6) and Porin. The levels of Atp6 are normalized to Porin. The shown plates and gels have been cropped to eliminate samples not linked to this study and that were intercalated between those of interest (WT and $aF_{145}S$). Representative data from at least two repeats are shown.

3.3.2. Influence of the $aF_{145}S$ Mutation on Mitochondrial Respiration and ATP Synthesis

The impact of the $aF_{145}S$ mutation on mitochondrial oxygen consumption and ATP synthesis was investigated in mitochondria extracted from cells grown at 36 °C in a rich galactose medium. Oxygen consumption was measured with NADH as an electron donor, alone (basal or State 4 respiration, which is induced only by the passive permeability to protons of the inner membrane), and after successive additions of ADP (State 3 or phosphorylating conditions, where respiration is normally twice stimulated vs. State 4) and the uncoupler CCCP (thus without any membrane potential, which further stimulates 2-fold the rate of respiration vs. State 3) (Table 5). State 4 respiration was not increased in the $aF_{145}S$ vs. *WT* mitochondria, indicating that the inner membrane had a normal passive permeability to protons and that there were no proton leak through the membrane domain (F_O) of ATP synthase. At State 3 as well as in the presence of CCCP, the rate of oxygen consumption was decreased by 20–30% in the mutant vs. WT mitochondrial samples, and the rate of ATP synthesis (at State 3) was diminished in similar proportions (Table 5). These data indicate that the $aF_{145}S$ mutation slows down the rate of ATP production with no loss in the yield in ATP per electron transferred to oxygen. A decreased capacity to transfer electrons to oxygen is usually observed in yeast ATP synthase defective mutants, except in those with F_O-mediated proton leaks [10–12,30,45], from which it was argued that the proton translocation activity of ATP synthase modulates biogenesis of the respiratory system presumably as a mean to co-regulate mitochondrial electron transfer and ATP synthesis activities [46].

Table 5. Mitochondrial respiration and ATP synthesis.

Strain	Respiration Rate nmoL O min^{-1} mg^{-1}			ATP Synthesis Rate nmoL ATP min^{-1} mg^{-1}	
	NADH	NADH + ADP	NADH + CCCP	−oligo	+oligo
WT	249 ± 21	654 ± 46	1028 ± 21	1393 ± 113	236 +/− 60
$aF_{145}S$	178 ± 2 *	474 ± 5 *	811 ± 9 *	988 ± 101 *	110 +/− 49

Mitochondria were isolated from cell strains grown for 5–6 generations in YPGALA medium (rich galactose) at 36 °C. Reaction mixes contained 0.15 mg/mL of mitochondrial proteins, 4 mM NADH, 150 (for respiration assays) or 750 (for ATP synthesis) μM ADP, 4 μM CCCP, 3 μg/mL oligomycin (*oligo*). Respiratory and ATP synthesis activities were measured using freshly isolated, osmotically protected, mitochondria buffered at pH 6.8. The reported values are averages of two biological replicates and three technical replicates for each assay. Statistical significance of the data was tested using unpaired *t*-test (* indicates a p-value < 0.05).

3.3.3. Influence of the $aF_{145}S$ Mutation on ATP Synthase Assembly/Stability

The ATP synthase organizes into a membrane-extrinsic domain (F_1) and a domain (F_O) largely anchored in the inner membrane [47–49]. The subunit *a* and a ring of identical subunits *c* move protons through the F_O. As a result of this, the *c*-ring rotates and provokes conformational changes in the F_1 that promote ATP synthesis. In current models, the assembly of ATP synthase starts with formation of F_1, followed by its association to the *c*-ring and peripheral stalk subunits that prevent rotation of the catalytic subdomain ($\alpha\beta_3$) of F_1. The process ends with incorporation of subunit *a* [50]. When incorporation of subunit *a* is compromised, the F_1 and the *c*-ring easily dissociate during BN-PAGE analysis of mitochondrial proteins, as was observed in a yeast strain lacking the *ATP6* gene [15,30,51]. In addition to fully assembled monomers and dimers of ATP synthase, free F_1 and *c*-ring were detected in mitochondrial samples from the $aF_{145}S$ mutant whereas these particles were absent in those from *WT* yeast, as revealed by Western blot with subunit *c* (Atp9) antibodies and by the in-gel F_1-mediated ATP hydrolytic activity (Figure 3C). The steady state levels of the mutated subunit *a* were estimated by Western blot (WB) with Atp6 antibodies of mitochondrial proteins resolved in denaturing gels. They were significantly reduced in the mutant vs. *WT* (Figure 3D). Taken together these data show that the $aF_{145}S$ change partially compromises a stable incorporation of subunit *a* within ATP synthase.

3.4. Topology of the Phenylalanine Residue Targeted by the m.8909T>C Mutation

Complete high-resolution structures of mitochondrial ATP synthase were described recently [36,47–49,52–54]. Near the middle of the membrane are located two universally conserved residues, an acidic one in subunit *c* (cE_{59} in yeast) and a positively charged arginine residue in subunit *a*

(aR$_{176}$ in yeast) that are functionally essential [36,55–57] (Figure 4B). Two hydrophilic clefts, one on the *p*-side and *n*-side of the membrane facilitate proton movement from one side of the membrane to the other. Inside the *p*-side channel, *c*E$_{59}$ takes a proton from the intermembrane space (IMS) and releases it into the *n*-side cleft after an almost complete rotation of the *c*-ring. The phenylalanine residue 145 (128 in *H.s.*) that is changed into serine by the m.8909T>C mutation is located at the bottom of the *n*-side cleft within a cluster of hydrophobic residues (aW_{126}, aF_{141}, aF_{142}, aL_{144}, aF_{145}, aY_{166}) beneath the helical domain of subunit *a* (*a*H5) that runs along the subunit *c*-ring (Figure 4C). Replacement of aF_{145} by a polar serine residue may weaken these hydrophobic interactions, and this is possibly responsible for the partial defects in ATP synthase assembly/stability observed in the $aF_{145}S$ mutant (as described above).

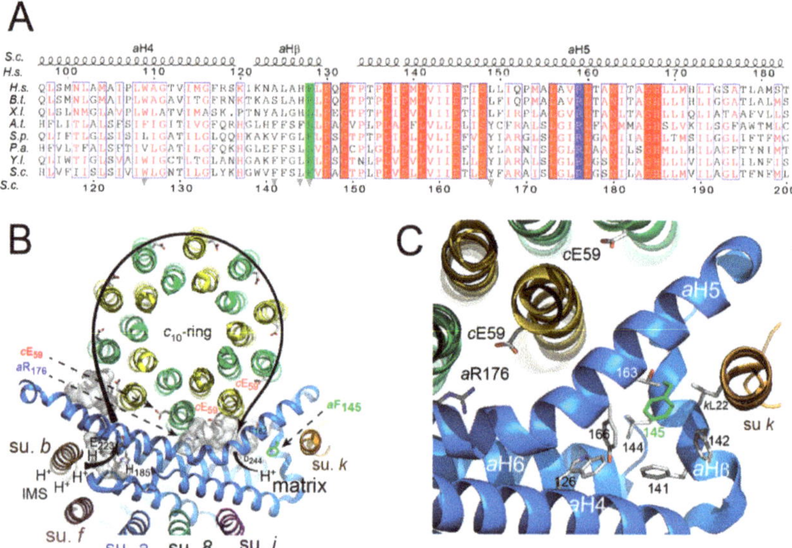

Figure 4. Evolutionary conservation of the phenylalanine residue targeted by the m.8909T>C mutation, and its topology in yeast ATP synthase. (**A**) Amino-acid alignments of subunits *a* from *Homo sapiens* (*H.s.*), *Bos taurus* (*B.t.*), *Xenopus laevis* (*X.l.*), *Arabidopsis thaliana* (*A.t.*), *Schizosaccharomyces pombe* (*S.p.*), *Podospora anserina* (*P.a.*), *Yarrowia lipolytica* (*Y.l.*) and *Saccharomyces cerevisiae* (*S.c.*). At the top and bottom, the residues are numbered according to the *H.s.* protein and mature *S.c.* protein (i.e., without the first 10 N-terminal residues that are cleaved during assembly [42]), respectively. Strictly conserved residues are in white characters on a red background while similar residues are in red on a white background with blue frames. α-helices (*a*H4-5 and *a*Hβ) in the *S.c.* protein marked above the amino-acid alignments are according to [36]. The essential arginine (aR$_{159}$ in humans, aR$_{176}$ in yeast) is on a blue background. The yeast aF_{145} residue corresponding to aF$_{128}$ in *H.s.* targeted by the m.8909T>C mutation is on a green background. The grey arrows mark the residues belonging to the hydrophobic cluster surrounded by *a*H4, *a*Hβ and *a*H5. (**B**) View of the entire *c*-ring and subunits *a, b, 8, I, f and k* from the IMS and the pathway along which protons are transported from the intermembrane space to the mitochondrial matrix. The side chains of the residues that are essential (aR$_{176}$ and cE$_{59}$) and important (aE$_{162}$, aD$_{244}$, aE$_{223}$, H$_{185}$) to this transfer are drawn as stick with their carbon atoms in white. The *p*-side and *n*-side clefts are shown as grey surfaces. (**C**) Enlargement of the region where is located the $aF_{145}S$ mutation. The mutated aF_{145} residue in green belongs to a hydrophobic cluster of six residues (aW_{126}, aF_{141}, aF_{142}, aL_{144}, aF_{145}, aY_{166}), which may stabilize or ease the folding of the bottom of the *n*-side cleft and the proper positioning of *a*H5 along the *c*-ring.

4. Discussion

With the advent of next generation sequencing methods, numerous novel variants of the mitochondrial DNA (mtDNA) are continuously identified in patients suffering from mitochondrial disorders. They are frequently found in only a limited number of cases, sometimes in a single individual (as with the case here reported), which makes it difficult to conclude about their pathogenicity. In a study aiming to probe the possible implication of mtDNA alterations in renal disease in the region of Nanjing, we examined more than 5000 patients with a biopsy-proven kidney dysfunction. Patients retained for entire mtDNA sequencing additionally showed at least two symptoms commonly observed in mitochondrial diseases like diabetes mellitus, deafness, neuromuscular and cardiac manifestations. Another criterion was the presence of abnormal mitochondrial ultrastructure and decreased cytochrome *c* oxidase (COX) and NADH (Complex I) dehydrogenase activities in kidney biopsies. In the patient with all these features here reported, we identified a novel variant in the mitochondrial *ATP6* gene, m8909T>C. It was homoplasmic in blood, urine sediments, and kidney (Figure 2B). It was detected in combination with the well-known pathogenic m.3243A>G mutation in a leucine tRNA gene (*MT-TL1)* [38], at a lesser abundancy (50–90%). There is no doubt that defects in mitochondrial translation induced by the m.3243A>G mutation largely contributed to the disease process and the severe decreases in CI and CIV activities, with only a minimal loss of SDH activity as was observed in other patients carrying this mutation [58].

The m.8909T>C variant leads to replacement of a well conserved phenylalanine residue with serine in ATP synthase subunit *a* ($aF_{128}S$) (Figure 4A), which prompted us to investigate the consequences of an equivalent of this mutation ($aF_{145}S$) in yeast. Several lines of evidence demonstrate a partial impairment of ATP synthase function in the mutant compared to wild type yeast: (i) respiratory growth showed a higher sensitivity to suboptimal concentrations of oligomycin, a chemical that inhibits ATP synthase; (ii) as was observed in many yeast ATP synthase defective mutants, the rate of mitochondrial oxygen consumption was diminished; (iii) ATP was produced less rapidly; and (iv) the presence in BN gels of partial ATP synthase assemblies (free F_1 and *c*-ring particles) attested for a compromised ability of the mutated subunit *a* to be stably incorporated into ATP synthase (Figure 3). According to recently published atomic structures of yeast ATP synthase [36,54], these effects are presumably the consequence of a disorganization of a cluster of hydrophobic residues supposedly important to help subunit *a* to adopt a stable functional conformation around the *c*-ring. Similar ATP synthase defects were previously observed in yeast models of *MT-ATP6* mutations with a well-established pathogenicity, like m.9176T>C [13] and m.8993T>C [11]. It is thus a reasonable hypothesis that the m.8909T>C variant could by itself have detrimental consequences on human health.

In line with this hypothesis, the patient here described showed a very severe clinical presentation in comparison to 35 previously reported patients suffering from kidney problems and presumed to carry only the m.3243A>G mutation (in most of them the mtDNA was not entirely sequenced). The median onset age of disease was 26 years (Table S1). More than 50% of these patients were first diagnosed with a renal disease due to persistent proteinuria. Glomerular lesions were observed in 21 patients, FSGS in 15 patients, and 3 cases showed TIN (Table S2) [59–81]. At the time of diagnosis, two relatively aged patients (41 and 47 years old) developed ESRD. A total of 21 patients were recorded with a median follow-up period of 72 months (range 24–240 months). Kidney function remained stable in two patients after 24 and 72 months, respectively, and one patient died of heart failure 60 months after diagnosis. Eighteen patients evolved toward renal failure or ESRD (Table S1) within 10-years of follow-up after diagnosis (Figure S1). Our patient, carrying both m.8909T>C and m.3243A>G, developed a kidney disease at the age of 14 and showed an extremely rapid progression to ESRD and death after only 8 months after diagnosis of a nephropathic syndrome.

As the m.8909T>C mutation was homoplasmic in the analyzed cells and tissues, it is likely that it is maternally inherited. Unfortunately, the patient's mother did not consent to be sequenced and followed medically. She was apparently healthy at the time her daughter was admitted at the hospital. We had no more contact with her since the death of her child. From her apparent good health, at least at the age she brought her daughter at the hospital, we may conclude that the m.8909T>C mutation has no dramatic consequences on mitochondrial function, which is consistent with the relatively mild effects of an equivalent of this mutation on yeast ATP synthase. However, when combined to a more severe allele like m.3243G>A, it may have the potential to accelerate a disease process. Previous studies already concluded that well-known pathogenic *ATP6* variants can act in synergy with other genetic determinants in patients with a very severe clinical presentation. For instance, while alone it usually provokes mild clinical phenotypes [82], the m.9176T>C mutation was identified in a case of fulminant and fatal Leigh syndrome [23]. As there was no major difference in the amount of mutated mtDNA in relative's probands, it was concluded that additional mitochondrial or nuclear genetic determinants were responsible for the different phenotypic expression. A second study similarly reported a rapid clinical evolution leading to sudden infant death syndrome (SIDS) in families segregating the m.8993T>G mutation [83] or the m.10044A>G mutation of the mt-tRNA^{Gly} gene [84], which suggested that mtDNA abnormalities should be considered as contributing to SIDS. Because of a lack of data from the mother of the patient here described, it would be premature to claim that m.8909T>C is a pathogenic mutation. However, considering the very severe clinical presentation of this patient, and the detrimental consequences of m.8909T>C on yeast ATP synthase, it is a reasonable proposal that this mutation has the potential to impact human health, in particular when combined to other genetic abnormalities that compromise mitochondrial function.

5. Conclusions

In conclusion, we here report a new mtDNA variant in the *ATP6* gene (m.8909T>C) and provide evidence using the yeast model that it has detrimental consequences on ATP synthase similar to those of other *ATP6* mutations with a well-established pathogenicity like m.9176T>C and m.8993T>C. On this basis, it is a reasonable assumption that this variant has the potential to compromise human health. Our work illustrates the power of yeast to help the study of specific human mtDNA variants found in heterogeneous genetic backgrounds and comprehension of human diseases linked to this DNA.

Supplementary Materials: The following are available online at http://www.mdpi.com/2075-1729/10/9/215/s1, Figure S1: Occurrence of renal failure during follow-up of 35 previously reported m.3243G>A patients diagnosed with a nephrotic syndrome, Table S1: Mean clinical characteristics of 35 previously reported m.3243G>A patients diagnosed with a nephrotic syndrome, Table S2. Individual clinical characteristics of 35 previously reported m.3243G>A patients diagnosed with a nephrotic syndrome References [59–81] are cited in the supplementary materials.

Author Contributions: Data analysis and interpretation and manuscript writing, Q.D.; construction and properties of yeast strains, R.K., X.S., M.B. and K.G.; patient recruitment, experiment performance, data analysis and interpretation, drafting the work, W.Z.; experiment performance and data analysis and interpretation, K.N.; patient recruitment, counseling and follow-up, W.Z.; structural modeling and data interpretation and manuscript editing, A.D. and M.-F.G.; patient recruitment, counseling and follow-up, S.X.; laboratory analysis and interpretation, M.Z.; patient selection, H.X.; histological analysis and interpretation, C.Z.; study design, patient selection and clinical analysis and interpretation, Z.L.; study design and coordination, data interpretation and manuscript writing, J.P.d.R., H.C. and D.T.-T. All authors have read and agreed to the published version of the manuscript.

Funding: This study was supported by the National Natural Science Foundation (NSF 81611130131 & 81570714) to H.C., the Association Française contre les Myopathie (AFM 22382) to DTT, the Agence Nationale de la Recherche (ANR-12-BSV8-024) to M.-F.G. and A.D., and by the National Science Center of Poland (UMO-2013/11/B/NZ1/02102) to R.K.

Conflicts of Interest: The authors declare no conflict of interest.

Ethical Approval Number: 2018NL-013. Participants signed a patient consent form for article containing patient details and/or images.

Abbreviations

Alb	Albumin
ADP	Adenosine di phosphate
ATP	Adenosine tri phosphate
BN-PAGE	Blue-native polyacrylamide gel electrophoresis
BUN	Blood urea nitrogen
CCCP	Cyanide m-chlorophenylhydrazone
Chol	Cholesterol
CIV	Cytochrome c oxidase
COX	Complex IV
CI-CIV	Complex I–Complex IV
CV	ATP synthase
DMSO	Dimethylsulfoxide
EGTA	Ethylene glycol-bis(β-aminoethyl ether)-*N,N,N′,N′*-tetraacetic acid
ESRD	End-stage renal disease
F	Female
F_O	Domain FO of ATP synthase
F_1	Domain F1 of ATP synthase
FSGS	FSGS, Focal segmental glomerulosclerosis
Glo	Globulin
HEPES	4-(2-hydroxyethyl)-1-piperazineethanesulfonic acid
IgM	Immunoglobulin M
M	Male
MELAS	Mitochondrial encephalomyopathy, lactic acidosis, and stroke-like episodes
MPGNs	Mesangial proliferative glomerulonephritis
mtDNA	Mitochondrial DNA
NADH	Nicotinamide adenine dinucleotide
Oligo	Oligomycin
OXPHOS	Oxidative phosphorylation
Pi	Inorganic phosphate
PVDF	Polyvinylidene difluoride
ROS	Reactive oxygen species
Scr	Serum creatinine
SDH	Complex II dehydrogenase activities
SDS-PAGE	Sodium dodecyl sulfate–polyacrylamide gel electrophoresis
SIDS	Sudden infant death syndrome
TMPD	*N,N,N,N*,-tetramethyl-p-phenylenediamine
TG	Triglyceride
WB	Western-blot
WT	Wild-type

References

1. DiMauro, S.; Schon, E.A. Mitochondrial respiratory-chain diseases. *N. Engl. J. Med.* **2003**, *348*, 2656–2668. [CrossRef] [PubMed]
2. Vafai, S.B.; Mootha, V.K. Mitochondrial disorders as windows into an ancient organelle. *Nature* **2012**, *491*, 374–383. [CrossRef] [PubMed]
3. Zeviani, M.; Carelli, V. Mitochondrial disorders. *Curr. Opin. Neurol.* **2007**, *20*, 564–571. [CrossRef] [PubMed]
4. Wallace, D.C. Mitochondrial DNA mutations in disease and aging. *Environ. Mol. Mutagen.* **2010**, *51*, 440–450. [CrossRef] [PubMed]
5. Sacconi, S.; Salviati, L.; Nishigaki, Y.; Walker, W.F.; Hernandez-Rosa, E.; Trevisson, E.; Delplace, S.; Desnuelle, C.; Shanske, S.; Hirano, M.; et al. A functionally dominant mitochondrial DNA mutation. *Hum. Mol. Genet.* **2008**, *17*, 1814–1820. [CrossRef] [PubMed]

6. Cai, W.; Fu, Q.; Zhou, X.; Qu, J.; Tong, Y.; Guan, M.X. Mitochondrial variants may influence the phenotypic manifestation of Leber's hereditary optic neuropathy-associated ND4 G11778A mutation. *J. Genet. Genom.* **2008**, *35*, 649–655. [CrossRef]
7. Swalwell, H.; Blakely, E.L.; Sutton, R.; Tonska, K.; Elstner, M.; He, L.; Taivassalo, T.; Burns, D.K.; Turnbull, D.M.; Haller, R.G.; et al. A homoplasmic mtDNA variant can influence the phenotype of the pathogenic m.7472Cins MTTS1 mutation: Are two mutations better than one? *Eur. J. Hum. Genet.* **2008**, *16*, 1265–1274. [CrossRef]
8. Bonnefoy, N.; Fox, T.D. Genetic transformation of Saccharomyces cerevisiae mitochondria. *Methods Cell Biol.* **2001**, *65*, 381–396.
9. Okamoto, K.; Perlman, P.S.; Butow, R.A. The sorting of mitochondrial DNA and mitochondrial proteins in zygotes: Preferential transmission of mitochondrial DNA to the medial bud. *J. Cell Biol.* **1998**, *142*, 613–623. [CrossRef]
10. Rak, M.; Tetaud, E.; Duvezin-Caubet, S.; Ezkurdia, N.; Bietenhader, M.; Rytka, J.; di Rago, J.P. A yeast model of the neurogenic ataxia retinitis pigmentosa (NARP) T8993G mutation in the mitochondrial ATP synthase-6 gene. *J. Biol. Chem.* **2007**, *282*, 34039–34047. [CrossRef]
11. Kucharczyk, R.; Rak, M.; di Rago, J.P. Biochemical consequences in yeast of the human mitochondrial DNA 8993T>C mutation in the ATPase6 gene found in NARP/MILS patients. *Biochim. Biophys. Acta* **2009**, *1793*, 817–824. [CrossRef] [PubMed]
12. Kucharczyk, R.; Giraud, M.F.; Brethes, D.; Wysocka-Kapcinska, M.; Ezkurdia, N.; Salin, B.; Velours, J.; Camougrand, N.; Haraux, F.; di Rago, J.P. Defining the pathogenesis of human mtDNA mutations using a yeast model: The case of T8851C. *Int. J. Biochem. Cell Biol.* **2013**, *45*, 130–140. [CrossRef] [PubMed]
13. Kucharczyk, R.; Ezkurdia, N.; Couplan, E.; Procaccio, V.; Ackerman, S.H.; Blondel, M.; di Rago, J.P. Consequences of the pathogenic T9176C mutation of human mitochondrial DNA on yeast mitochondrial ATP synthase. *Biochim. Biophys. Acta* **2010**, *1797*, 1105–1112. [CrossRef] [PubMed]
14. Kabala, A.M.; Lasserre, J.P.; Ackerman, S.H.; di Rago, J.P.; Kucharczyk, R. Defining the impact on yeast ATP synthase of two pathogenic human mitochondrial DNA mutations, T9185C and T9191C. *Biochimie* **2014**, *100*, 200–206. [CrossRef]
15. Kucharczyk, R.; Salin, B.; di Rago, J.P. Introducing the human Leigh syndrome mutation T9176G into Saccharomyces cerevisiae mitochondrial DNA leads to severe defects in the incorporation of Atp6p into the ATP synthase and in the mitochondrial morphology. *Hum. Mol. Genet.* **2009**, *18*, 2889–2898. [CrossRef]
16. Wen, S.; Niedzwiecka, K.; Zhao, W.; Xu, S.; Liang, S.; Zhu, X.; Xie, H.; Tribouillard-Tanvier, D.; Giraud, M.F.; Zeng, C.; et al. Identification of G8969>A in mitochondrial ATP6 gene that severely compromises ATP synthase function in a patient with IgA nephropathy. *Sci. Rep.* **2016**, *6*, 36313. [CrossRef]
17. Morava, E.; Rodenburg, R.J.; Hol, F.; de Vries, M.; Janssen, A.; van den Heuvel, L.; Nijtmans, L.; Smeitink, J. Clinical and biochemical characteristics in patients with a high mutant load of the mitochondrial T8993G/C mutations. *Am. J. Med. Genet. A* **2006**, *140*, 863–868. [CrossRef]
18. Baracca, A.; Barogi, S.; Carelli, V.; Lenaz, G.; Solaini, G. Catalytic activities of mitochondrial ATP synthase in patients with mitochondrial DNA T8993G mutation in the ATPase 6 gene encoding subunit a. *J. Biol. Chem.* **2000**, *275*, 4177–4182. [CrossRef]
19. Baracca, A.; Sgarbi, G.; Mattiazzi, M.; Casalena, G.; Pagnotta, E.; Valentino, M.L.; Moggio, M.; Lenaz, G.; Carelli, V.; Solaini, G. Biochemical phenotypes associated with the mitochondrial ATP6 gene mutations at nt8993. *Biochim. Biophys. Acta* **2007**, *1767*, 913–919. [CrossRef]
20. Carrozzo, R.; Murray, J.; Santorelli, F.M.; Capaldi, R.A. The T9176G mutation of human mtDNA gives a fully assembled but inactive ATP synthase when modeled in Escherichia coli. *FEBS Lett.* **2000**, *486*, 297–299. [CrossRef]
21. Carrozzo, R.; Rizza, T.; Lucioli, S.; Pierini, R.; Bertini, E.; Santorelli, F.M. A mitochondrial ATPase 6 mutation is associated with Leigh syndrome in a family and affects proton flow and adenosine triphosphate output when modeled in Escherichia coli. *Acta Paediatr. Suppl.* **2004**, *93*, 65–67. [CrossRef]
22. Cortes-Hernandez, P.; Vazquez-Memije, M.E.; Garcia, J.J. ATP6 homoplasmic mutations inhibit and destabilize the human F1F0-ATP synthase without preventing enzyme assembly and oligomerization. *J. Biol. Chem.* **2007**, *282*, 1051–1058. [CrossRef]
23. Dionisi-Vici, C.; Seneca, S.; Zeviani, M.; Fariello, G.; Rimoldi, M.; Bertini, E.; De Meirleir, L. Fulminant Leigh syndrome and sudden unexpected death in a family with the T9176C mutation of the mitochondrial ATPase 6 gene. *J. Inherit. Metab. Dis.* **1998**, *21*, 2–8. [CrossRef]

24. Houstek, J.; Pickova, A.; Vojtiskova, A.; Mracek, T.; Pecina, P.; Jesina, P. Mitochondrial diseases and genetic defects of ATP synthase. *Biochim. Biophys. Acta* **2006**, *1757*, 1400–1405. [CrossRef] [PubMed]
25. Mattiazzi, M.; Vijayvergiya, C.; Gajewski, C.D.; DeVivo, D.C.; Lenaz, G.; Wiedmann, M.; Manfredi, G. The mtDNA T8993G (NARP) mutation results in an impairment of oxidative phosphorylation that can be improved by antioxidants. *Hum. Mol. Genet.* **2004**, *13*, 869–879. [CrossRef] [PubMed]
26. De Meirleir, L.; Seneca, S.; Lissens, W.; Schoentjes, E.; Desprechins, B. Bilateral striatal necrosis with a novel point mutation in the mitochondrial ATPase 6 gene. *Pediatr. Neurol.* **1995**, *13*, 242–246. [CrossRef]
27. Hench, J.; Bratic Hench, I.; Pujol, C.; Ipsen, S.; Brodesser, S.; Mourier, A.; Tolnay, M.; Frank, S.; Trifunovic, A. A tissue-specific approach to the analysis of metabolic changes in Caenorhabditis elegans. *PLoS ONE* **2011**, *6*, e28417. [CrossRef] [PubMed]
28. Whitaker-Menezes, D.; Martinez-Outschoorn, U.E.; Flomenberg, N.; Birbe, R.C.; Witkiewicz, A.K.; Howell, A.; Pavlides, S.; Tsirigos, A.; Ertel, A.; Pestell, R.G.; et al. Hyperactivation of oxidative mitochondrial metabolism in epithelial cancer cells in situ: Visualizing the therapeutic effects of metformin in tumor tissue. *Cell Cycle* **2011**, *10*, 4047–4064. [CrossRef]
29. Kiernan, J.A. Indigogenic substrates for detection and localization of enzymes. *Biotech. Histochem. Off. Publ. Biol. Stain Comm.* **2007**, *82*, 73–103. [CrossRef]
30. Rak, M.; Tetaud, E.; Godard, F.; Sagot, I.; Salin, B.; Duvezin-Caubet, S.; Slonimski, P.P.; Rytka, J.; di Rago, J.P. Yeast cells lacking the mitochondrial gene encoding the ATP synthase subunit 6 exhibit a selective loss of complex IV and unusual mitochondrial morphology. *J. Biol. Chem.* **2007**, *282*, 10853–10864. [CrossRef]
31. Steele, D.F.; Butler, C.A.; Fox, T.D. Expression of a recoded nuclear gene inserted into yeast mitochondrial DNA is limited by mRNA-specific translational activation. *Proc. Natl. Acad. Sci. USA* **1996**, *93*, 5253–5257. [CrossRef] [PubMed]
32. Guerin, B.; Labbe, P.; Somlo, M. Preparation of yeast mitochondria (*Saccharomyces cerevisiae*) with good P/O and respiratory control ratios. *Methods Enzymol.* **1979**, *55*, 149–159. [PubMed]
33. Somlo, M. Induction and repression of mitochondrial ATPase in yeast. *Eur. J. Biochem.* **1968**, *5*, 276–284. [CrossRef] [PubMed]
34. Paumard, P.; Vaillier, J.; Coulary, B.; Schaeffer, J.; Soubannier, V.; Mueller, D.M.; Brethes, D.; di Rago, J.P.; Velours, J. The ATP synthase is involved in generating mitochondrial cristae morphology. *EMBO J.* **2002**, *21*, 221–230. [CrossRef] [PubMed]
35. Sievers, F.; Wilm, A.; Dineen, D.; Gibson, T.J.; Karplus, K.; Li, W.; Lopez, R.; McWilliam, H.; Remmert, M.; Soding, J.; et al. Fast, scalable generation of high-quality protein multiple sequence alignments using Clustal Omega. *Mol. Syst. Biol.* **2011**, *7*, 539. [CrossRef]
36. Guo, H.; Bueler, S.A.; Rubinstein, J.L. Atomic model for the dimeric FO region of mitochondrial ATP synthase. *Science* **2017**, *358*, 936–940. [CrossRef]
37. *The PyMOL Molecular Graphics System*; Version 0.99; Schrödinger, LLC. DeLano Scientific: San Carlos, CA, USA, 2002.
38. Majamaa, K.; Moilanen, J.S.; Uimonen, S.; Remes, A.M.; Salmela, P.I.; Karppa, M.; Majamaa-Voltti, K.A.; Rusanen, H.; Sorri, M.; Peuhkurinen, K.J.; et al. Epidemiology of A3243G, the mutation for mitochondrial encephalomyopathy, lactic acidosis, and strokelike episodes: Prevalence of the mutation in an adult population. *Am. J. Hum. Genet.* **1998**, *63*, 447–454. [CrossRef]
39. Fornuskova, D.; Brantova, O.; Tesarova, M.; Stiburek, L.; Honzik, T.; Wenchich, L.; Tietzeova, E.; Hansikova, H.; Zeman, J. The impact of mitochondrial tRNA mutations on the amount of ATP synthase differs in the brain compared to other tissues. *Biochim. Biophys. Acta* **2008**, *1782*, 317–325. [CrossRef]
40. Sasarman, F.; Antonicka, H.; Shoubridge, E.A. The A3243G tRNALeu(UUR) MELAS mutation causes amino acid misincorporation and a combined respiratory chain assembly defect partially suppressed by overexpression of EFTu and EFG2. *Hum. Mol. Genet.* **2008**, *17*, 3697–3707. [CrossRef]
41. McMillan, R.P.; Stewart, S.; Budnick, J.A.; Caswell, C.C.; Hulver, M.W.; Mukherjee, K.; Srivastava, S. Quantitative Variation in m.3243A>G Mutation Produce Discrete Changes in Energy Metabolism. *Sci. Rep.* **2019**, *9*, 5752. [CrossRef]
42. Michon, T.; Galante, M.; Velours, J. NH2-terminal sequence of the isolated yeast ATP synthase subunit 6 reveals post-translational cleavage. *Eur. J. Biochem.* **1988**, *172*, 621–625. [CrossRef] [PubMed]
43. Symersky, J.; Osowski, D.; Walters, D.E.; Mueller, D.M. Oligomycin frames a common drug-binding site in the ATP synthase. *Proc. Natl. Acad. Sci. USA* **2012**, *109*, 13961–13965. [CrossRef] [PubMed]

44. Mukhopadhyay, A.; Uh, M.; Mueller, D.M. Level of ATP synthase activity required for yeast Saccharomyces cerevisiae to grow on glycerol media. *FEBS Lett.* **1994**, *343*, 160–164. [CrossRef]
45. Lefebvre-Legendre, L.; Vaillier, J.; Benabdelhak, H.; Velours, J.; Slonimski, P.P.; di Rago, J.P. Identification of a nuclear gene (FMC1) required for the assembly/stability of yeast mitochondrial F(1)-ATPase in heat stress conditions. *J. Biol. Chem.* **2001**, *276*, 6789–6796. [CrossRef] [PubMed]
46. Su, X.; Rak, M.; Tetaud, E.; Godard, F.; Sardin, E.; Bouhier, M.; Gombeau, K.; Caetano-Anollés, D.; Salin, B.; Chen, H.; et al. Deregulating mitochondrial metabolite and ion transport has beneficial effects in yeast and human cellular models for NARP syndrome. *Hum. Mol. Genet.* **2019**. [CrossRef]
47. Morales-Rios, E.; Montgomery, M.G.; Leslie, A.G.; Walker, J.E. Structure of ATP synthase from Paracoccus denitrificans determined by X-ray crystallography at 4.0 A resolution. *Proc. Natl. Acad. Sci. USA* **2015**, *112*, 13231–13236. [CrossRef]
48. Zhou, A.; Rohou, A.; Schep, D.G.; Bason, J.V.; Montgomery, M.G.; Walker, J.E.; Grigorieff, N.; Rubinstein, J.L. Structure and conformational states of the bovine mitochondrial ATP synthase by cryo-EM. *Elife* **2015**, *4*. [CrossRef]
49. Allegretti, M.; Klusch, N.; Mills, D.J.; Vonck, J.; Kuhlbrandt, W.; Davies, K.M. Horizontal membrane-intrinsic alpha-helices in the stator a-subunit of an F-type ATP synthase. *Nature* **2015**, *521*, 237–240. [CrossRef]
50. Tzagoloff, A.; Barrientos, A.; Neupert, W.; Herrmann, J.M. Atp10p assists assembly of Atp6p into the F0 unit of the yeast mitochondrial ATPase. *J. Biol. Chem.* **2004**, *279*, 19775–19780. [CrossRef]
51. Zeng, X.; Kucharczyk, R.; di Rago, J.P.; Tzagoloff, A. The leader peptide of yeast Atp6p is required for efficient interaction with the Atp9p ring of the mitochondrial ATPase. *J. Biol. Chem.* **2007**, *282*, 36167–36176. [CrossRef]
52. Hahn, A.; Parey, K.; Bublitz, M.; Mills, D.J.; Zickermann, V.; Vonck, J.; Kuhlbrandt, W.; Meier, T. Structure of a Complete ATP Synthase Dimer Reveals the Molecular Basis of Inner Mitochondrial Membrane Morphology. *Mol. Cell* **2016**, *63*, 445–456. [CrossRef] [PubMed]
53. Hahn, A.; Vonck, J.; Mills, D.J.; Meier, T.; Kuhlbrandt, W. Structure, mechanism, and regulation of the chloroplast ATP synthase. *Science* **2018**, *360*. [CrossRef] [PubMed]
54. Srivastava, A.P.; Luo, M.; Zhou, W.; Symersky, J.; Bai, D.; Chambers, M.G.; Faraldo-Gomez, J.D.; Liao, M.; Mueller, D.M. High-resolution cryo-EM analysis of the yeast ATP synthase in a lipid membrane. *Science* **2018**. [CrossRef] [PubMed]
55. Vik, S.B.; Antonio, B.J. A mechanism of proton translocation by F1F0 ATP synthases suggested by double mutants of the a subunit. *J. Biol. Chem.* **1994**, *269*, 30364–30369. [PubMed]
56. Junge, W.; Lill, H.; Engelbrecht, S. ATP synthase: An electrochemical transducer with rotatory mechanics. *Trends Biochem. Sci.* **1997**, *22*, 420–423. [CrossRef]
57. Pogoryelov, D.; Krah, A.; Langer, J.D.; Yildiz, O.; Faraldo-Gomez, J.D.; Meier, T. Microscopic rotary mechanism of ion translocation in the F(o) complex of ATP synthases. *Nat. Chem. Biol.* **2010**, *6*, 891–899. [CrossRef]
58. Davidson, M.M.; Walker, W.F.; Hernandez-Rosa, E. The m.3243A>G mtDNA mutation is pathogenic in an in vitro model of the human blood brain barrier. *Mitochondrion* **2009**, *9*, 463–470. [CrossRef]
59. Seidowsky, A.; Hoffmann, M.; Glowacki, F.; Dhaenens, C.M.; Devaux, J.P.; de Sainte Foy, C.L.; Provot, F.; Gheerbrant, J.D.; Hummel, A.; Hazzan, M.; et al. Renal involvement in MELAS syndrome–a series of 5 cases and review of the literature. *Clin. Nephrol.* **2013**, *80*, 456–463. [CrossRef]
60. Kurogouchi, F.; Oguchi, T.; Mawatari, E.; Yamaura, S.; Hora, K.; Takei, M.; Sekijima, Y.; Ikeda, S.; Kiyosawa, K. A case of mitochondrial cytopathy with a typical point mutation for MELAS, presenting with severe focal-segmental glomerulosclerosis as main clinical manifestation. *Am. J. Nephrol.* **1998**, *18*, 551–556. [CrossRef]
61. Nakamura, S.; Yoshinari, M.; Doi, Y.; Yoshizumi, H.; Katafuchi, R.; Yokomizo, Y.; Nishiyama, K.; Wakisaka, M.; Fujishima, M. Renal complications in patients with diabetes mellitus associated with an A to G mutation of mitochondrial DNA at the 3243 position of leucine tRNA. *Diabetes Res. Clin. Pract.* **1999**, *44*, 183–189. [CrossRef]
62. Löwik, M.M.; Hol, F.A.; Steenbergen, E.J.; Wetzels, J.F.; van den Heuvel, L.P. Mitochondrial tRNALeu(UUR) mutation in a patient with steroid-resistant nephrotic syndrome and focal segmental glomerulosclerosis. *Nephrol. Dial. Transplant.* **2005**, *20*, 336–341. [CrossRef] [PubMed]

63. Guéry, B.; Choukroun, G.; Noël, L.H.; Clavel, P.; Rötig, A.; Lebon, S.; Rustin, P.; Bellané-Chantelot, C.; Mougenot, B.; Grünfeld, J.P.; et al. The spectrum of systemic involvement in adults presenting with renal lesion and mitochondrial tRNA (Leu) gene mutation. *J. Am. Soc. Nephrol.* **2003**, *14*, 2099–2108. [CrossRef] [PubMed]
64. Fujii, H.; Mori, Y.; Kayamori, K.; Igari, T.; Ito, E.; Akashi, T.; Noguchi, Y.; Kitamura, K.; Okado, T.; Terada, Y.; et al. A familial case of mitochondrial disease resembling Alport syndrome. *Clin. Exp. Nephrol.* **2008**, *12*, 159–163. [CrossRef]
65. Suzuki, T.; Fujino, T.; Sugiyama, M.; Ishida, M. A case of mitochondrial encephalomyopathy (MELAS). *Nihon Jinzo Gakkai Shi* **1996**, *38*, 109–114.
66. Mima, A.; Shiota, F.; Matsubara, T.; Iehara, N.; Akagi, T.; Abe, H.; Nagai, K.; Matsuura, M.; Murakami, T.; Kishi, S.; et al. An autopsy case of mitochondrial myopathy, encephalopathy, lactic acidosis, and stroke-like episodes (MELAS) with intestinal bleeding in chronic renal failure. *Ren. Fail.* **2011**, *33*, 622–625. [CrossRef] [PubMed]
67. Takahashi, N.; Shimada, T.; Ishibashi, Y.; Yoshitomi, H.; Oyake, N.; Murakami, Y.; Nishino, I.; Nonaka, I.; Goto, Y.; Kitamura, J. Marked left ventricular hypertrophy in a patient with mitochondrial myopathy, encephalopathy, lactic acidosis, and stroke-like episodes. *Int. J. Cardiol.* **2008**, *129*, e77–e80. [CrossRef] [PubMed]
68. Li, J.Y.; Hsieh, R.H.; Peng, N.J.; Lai, P.H.; Lee, C.F.; Lo, Y.K.; Wei, Y.H. A follow-up study in a Taiwanese family with mitochondrial myopathy, encephalopathy, lactic acidosis and stroke-like episodes syndrome. *J. Formos. Med. Assoc.* **2007**, *106*, 528–536. [CrossRef]
69. Jansen, J.J.; Maassen, J.A.; van der Woude, F.J.; Lemmink, H.A.; van den Ouweland, J.M.; t' Hart, L.M.; Smeets, H.J.; Bruijn, J.A.; Lemkes, H.H. Mutation in mitochondrial tRNA(Leu(UUR)) gene associated with progressive kidney disease. *J. Am. Soc. Nephrol.* **1997**, *8*, 1118–1124.
70. Ireland, J.; Rossetti, S.; Haugen, E.; Ireland, J.; Michels, V.; Harris, P. Mitochondrial causes of renal insufficiency and hearing loss. *Kidney Int.* **2004**, *65*, 2444–2445. [CrossRef]
71. Yamagata, K.; Tomida, C.; Umeyama, K.; Urakami, K.; Ishizu, T.; Hirayama, K.; Gotoh, M.; Iitsuka, T.; Takemura, K.; Kikuchi, H.; et al. Prevalence of Japanese dialysis patients with an A-to-G mutation at nucleotide 3243 of the mitochondrial tRNA(Leu(UUR)) gene. *Nephrol. Dial. Transplant.* **2000**, *15*, 385–388. [CrossRef]
72. Azevedo, O.; Vilarinho, L.; Almeida, F.; Ferreira, F.; Guardado, J.; Ferreira, M.; Lourenço, A.; Medeiros, R.; Almeida, J. Cardiomyopathy and kidney disease in a patient with maternally inherited diabetes and deafness caused by the 3243A>G mutation of mitochondrial DNA. *Cardiology* **2010**, *115*, 71–74. [CrossRef]
73. Manouvrier, S.; Rötig, A.; Hannebique, G.; Gheerbrandt, J.D.; Royer-Legrain, G.; Munnich, A.; Parent, M.; Grünfeld, J.P.; Largilliere, C.; Lombes, A.; et al. Point mutation of the mitochondrial tRNA(Leu) gene (A 3243 G) in maternally inherited hypertrophic cardiomyopathy, diabetes mellitus, renal failure, and sensorineural deafness. *J. Med. Genet.* **1995**, *32*, 654–656. [CrossRef] [PubMed]
74. Tsujita, Y.; Kunitomo, T.; Fujii, M.; Furukawa, S.; Otsuki, H.; Fujino, K.; Hamamoto, T.; Tabata, T.; Matsumura, K.; Sasaki, T.; et al. A surviving case of mitochondrial cardiomyopathy diagnosed from the symptoms of multiple organ dysfunction syndrome. *Int. J. Cardiol.* **2008**, *128*, e43–e45. [CrossRef] [PubMed]
75. Bergamin, C.S.; Rolim, L.C.; Dib, S.A.; Moisés, R.S. Unusual occurrence of intestinal pseudo obstruction in a patient with maternally inherited diabetes and deafness (MIDD) and favorable outcome with coenzyme Q10. *Arq. Bras. Endocrinol. Metabol.* **2008**, *52*, 1345–1349. [CrossRef] [PubMed]
76. Ihara, M.; Tanaka, H.; Yashiro, M.; Nishimura, Y. Mitochondrial myopathy, encephalopathy, lactic acidosis, and stroke-like episodes (MELAS) with chronic renal failure: Report of mother-child cases. *Rinsho Shinkeigaku.* **1996**, *36*, 1069–1073. [PubMed]
77. Ueda, Y.; Ando, A.; Nagata, T.; Yanagida, H.; Yagi, K.; Sugimoto, K.; Okada, M.; Takemura, T. A boy with mitochondrial disease: asymptomatic proteinuria without neuromyopathy. *Pediatr. Nephrol.* **2004**, *19*, 107–110. [CrossRef]
78. Cheong, H.I.; Chae, J.H.; Kim, J.S.; Park, H.W.; Ha, I.S.; Hwang, Y.S.; Lee, H.S.; Choi, Y. Hereditary glomerulopathy associated with a mitochondrial tRNA(Leu) gene mutation. *Pediatr. Nephrol.* **1999**, *13*, 477–480. [CrossRef]

79. Yorifuji, T.; Kawai, M.; Momoi, T.; Sasaki, H.; Furusho, K.; Muroi, J.; Shimizu, K.; Takahashi, Y.; Matsumura, M.; Nambu, M.; et al. Nephropathy and growth hormone deficiency in a patient with mitochondrial tRNA(Leu(UUR)) mutation. *J. Med. Genet.* **1996**, *33*, 621–622. [CrossRef]
80. Hotta, O.; Inoue, C.N.; Miyabayashi, S.; Furuta, T.; Takeuchi, A.; Taguma, Y. Clinical and pathologic features of focal segmental glomerulosclerosis with mitochondrial tRNALeu(UUR) gene mutation. *Kidney Int.* **2001**, *59*, 1236–1243. [CrossRef]
81. Cao, X.Y.; Wei, R.B.; Wang, Y.D.; Zhang, X.G.; Tang, L.; Chen, X.M. Focal segmental glomerulosclerosis associated with maternally inherited diabetes and deafness: Clinical pathological analysis. *Indian J. Pathol. Microbiol.* **2013**, *56*, 272–275. [CrossRef]
82. Thyagarajan, D.; Shanske, S.; Vazquez-Memije, M.; De Vivo, D.; DiMauro, S. A novel mitochondrial ATPase 6 point mutation in familial bilateral striatal necrosis. *Ann. Neurol.* **1995**, *38*, 468–472. [CrossRef] [PubMed]
83. Pastores, G.M.; Santorelli, F.M.; Shanske, S.; Gelb, B.D.; Fyfe, B.; Wolfe, D.; Willner, J.P. Leigh syndrome and hypertrophic cardiomyopathy in an infant with a mitochondrial DNA point mutation (T8993G). *Am. J. Med. Genet.* **1994**, *50*, 265–271. [CrossRef] [PubMed]
84. Santorelli, F.M.; Schlessel, J.S.; Slonim, A.E.; DiMauro, S. Novel mutation in the mitochondrial DNA tRNA glycine gene associated with sudden unexpected death. *Pediatr. Neurol.* **1996**, *15*, 145–149. [CrossRef]

Article

Palmitate but Not Oleate Exerts a Negative Effect on Oxygen Utilization in Myoblasts of Patients with the m.3243A>G Mutation: A Pilot Study

Leila Motlagh Scholle [1,*], Helena Schieffers [1], Samiya Al-Robaiy [2], Annemarie Thaele [1], Diana Lehmann Urban [3] and Stephan Zierz [1]

1. Department of Neurology, Martin-Luther-University Halle-Wittenberg, 06120 Halle (Saale), Germany; helena@schieffers.de (H.S.); annemarie.thaele@uk-halle.de (A.T.); stephan.zierz@medizin.uni-halle.de (S.Z.)
2. Center for Basic Medical Research, Martin-Luther-University Halle-Wittenberg, 06120 Halle (Saale), Germany; samiya.al-robaiy@uk-halle.de
3. Department of Neurology, Ulm University, 89070 Ulm, Germany; diana.lehmann@rku.de

* Correspondence: Leila.scholle@medizin.uni-halle.de; Tel.: +49-345-557-3628; Fax: +49-345-557-3505

Received: 5 August 2020; Accepted: 15 September 2020; Published: 16 September 2020

Abstract: It is known that exposure to excess saturated fatty acids, especially palmitate, can trigger cellular stress responses interpreted as lipotoxicity. The effect of excessive free fatty acids on oxidative phosphorylation capacity in myoblasts of patients with the m.3243A>G mutation was evaluated with the mitochondrial (Mito) stress test using a Seahorse XF96 analyzer. ß-oxidation, measured with the Seahorse XF96 analyzer, was similar in patients and controls, and reduced in both patients and controls at 40 °C compared to 37 °C. Mito stress test in the absence of fatty acids showed lower values in patients compared to controls. The mitochondrial activity and ATP production rates were significantly reduced in presence of palmitate, but not of oleate in patients, showing a negative effect of excessive palmitate on mitochondrial function in patients. Diabetes mellitus is a frequent symptom in patients with m.3243A>G mutation. It can be speculated that the negative effect of palmitate on mitochondrial function might be related to diacylglycerols (DAG) and ceramides (CER) mediated insulin resistance. This might contribute to the elevated risk for diabetes mellitus in m.3243A>G patients.

Keywords: fatty acid oxidation; palmitate; oleate; m.3243A>G mutation

---

1. Introduction

Fatty acid oxidation (FAO) provides the major source of energy supply for skeletal muscles in situations requiring simultaneous glucose sparing, and major energy supply, such as prolonged fasting or exercise. ß-oxidation is the main pathway of fatty acid catabolism in mitochondria [1]. The accumulation of oversupplied fatty acids (FAs) in different tissues, such as the pancreas, liver, and skeletal muscle, might lead to a cellular dysfunction in these tissues, and an apoptotic cell death, commonly referred to as "lipotoxicity" [2,3]. The toxic effects of FAs seem to be dependent on their chain length and degree of saturation. The two-common long-chain saturated FAs (SFA), palmitate (C16:0) and stearate (C18:0), are known to be the most lipotoxic ones. In contrast, monounsaturated FAs such as oleate (C18:1), have been reported to protect against the above-mentioned SFA-induced toxicity [2,4]. Moreover, saturated FAs have worsened the insulin-resistance, while monounsaturated and polyunsaturated ones improved it [5].

The effect of temperature on the functional properties of skeletal rat muscle mitochondria was reported in one study [6]. Increasing the assay temperature in above-mentioned study within the range of 25–42 °C increased the mitochondrial respiratory chain activities, resulting in an elevated phosphorylation rate. They reported a temperature-induced decrease in oxidative phosphorylation (OXPHOS) efficiency as well.

It has been shown that the increase of temperature from 35 °C to 40 °C induced the uncoupling of substrate oxidation from adenosine diphosphate (ADP) phosphorylation, and decreased the efficiency of mitochondria to produce adenosine triphosphate (ATP) in skeletal myofibers. This uncoupling effect was more pronounced for FAs than for carbohydrates as substrate [7]. It is already known that a reduction in substrate oxidation, caused by mitochondrial dysfunctions, leads to a lipid accumulation, such as the deposition of lipid mediators, such as diacylglycerols (DAG) and ceramides (CER). Both, DAG and CER are reported to inhibit insulin signaling [8–10].

The m.3243A>G point mutation in the *MT-TL1* gene (encoding mt-tRNA$^{Leu(UUR)}$) can be found in approximately 80% of patients with mitochondrial encephalopathy, lactic acidosis, and stroke like episodes (MELAS)-syndrome. In these patients, the ATP production rate was significantly less than in controls [11] and the mitochondrial respiration is known to be impaired [12]. Diabetes mellitus frequently accompanies the m.3243A>G mutation [13]. Thus, the evaluation of effects of saturated free fatty acids (FFAs), which could lead to an insulin resistance, might be relevant for these patients.

The aim of the present study was to evaluate the oxidative phosphorylation and ß-oxidation pathway in patients harboring the m.3243A>G mutation. It was of interest to examine whether palmitate and oleate show a toxic or protective effect on OXPHOS in myoblasts of patients with the m.3243A>G mutation. Moreover, the effect of a fever-stimulating temperature on ß-oxidation was simulated in 40 °C cultured cells in both, m.3243A>G patients and controls.

2. Materials and Methods

2.1. Human Myoblasts

Muscle primary cells from five patients harboring the genetically confirmed m.3243A>G mutation and 5 controls were provided by the Muscle Tissue Culture Collection (MTCC) at the Friedrich-Baur-Institute (Department of Neurology, Ludwig Maximilian University, Munich, Germany; part of the German network on muscular dystrophies, MD-NET, partner of EuroBioBank) [14].

Cells were collected and processed by MTCC in compliance with all applicable laws, rules, regulations, and other requirements of any applicable governmental authority. The cells were grown in skeletal muscle cell growth medium (PromoCell, Heidelberg, Germany) supplemented with 10% fetal bovine serum (FBS). Further details are given in Table 1 and as described earlier [12]. As controls served 5 patients (2 males, 3 females), who had muscle biopsy for diagnosis of a suspected neuromuscular disorder. They were deemed 'normal controls' if they were ultimately found to have no muscle disease by combined clinical and histological criteria. Age of controls ranged from 35 to 53 years.

Table 1. Sex and age of five patients with genetically confirmed m.3243A>G mutation and five healthy controls as previously described [12], F: female, M: male.

	Gender	Age at Biopsy
Patients		
P 1	M	43
P 2	M	42
P 3	F	70
P 4	M	34
P 5	F	40
Controls		
C 1	F	50
C 2	M	53
C 3	F	40
C 4	M	35
C 5	F	49

2.2. Metabolic Function Measurements with the Seahorse XF96 Cell Analyzer

2.2.1. Mito Stress Test

To evaluate mitochondrial function, the Mito Stress test was performed using a Seahorse XF96 Cell Analyzer according to the manufacturer's recommendations. Briefly, myoblasts from patients and controls were seeded to Seahorse XF96 cell culture microplates (2.5×10^4 cells per well) in skeletal muscle cell growth medium supplemented with 10% fetal bovine serum (FBS). After a 24-h incubation at 37 °C, the cells were washed twice with a pre-warmed assay medium (XF base medium supplemented with 10 mM glucose, 2 mM glutamine, and 1 mM sodium pyruvate; pH 7.4).

Oxygen-consumption rate (OCR) values were simultaneously measured following sequential injections of inhibitors of mitochondrial oxidative phosphorylation: (I) oligomycin (2 µM), an ATP synthase inhibitor; (II) carbonyl cyanide p-(trifluoromethoxy) phenylhydrazone (FCCP, 2 µM), an uncoupler; (III) antimycin A (0.5 µM) (inhibitor of complex III) and (IV) rotenone (0.5 µM) (inhibitor of complex I), inhibiting uncoupled respiration. Key parameters of mitochondrial function such as basal respiration (BR), ATP production rate (ATP-R) and spare respiratory capacity (SRC) were analyzed using the above-described measurements. The data was normalized to cell numbers by measurement of Hoechst dye staining of nuclei with excitation and emission wavelengths 355 nm and 465 nm using a Tecan Infinite TM M1000 and plotted as OCR (pmol/min/U fluorescence/well ± SD), accordingly.

The Effect of FFAs on the Mitochondrial Respiration

To test the effect of FFAs on the mitochondrial respiration, patient and control myoblasts were plated (as described in Section 2.2.1) 16 h prior to Mito stress test measurements, Bovine serum albumin (BSA)-conjugated palmitic acid or oleic acid (ratio FFA: BSA = 6:1) was added to the media. The final concentration of oleate or palmitate was 300 µM. Comparable amounts of BSA were added in the untreated group.

2.2.2. Fatty Acid Oxidation

To assess the ability of myoblasts to oxidize exogenous fatty acids, the oxygen-consumption rate (OCR) was analyzed using a XF96 Cell Analyzer according to the manufacturer's protocol (Seahorse Bioscience). For this purpose, the myoblast of patients and controls were plated (as described in Section 2.1) 16-h prior to measurements, growth medium was replaced by substrate-limited medium (Dulbecco's modified Eagle's medium (DMEM) with 0.5 mM glucose, 1.0 mM glutamine, 0.5 mM carnitine, and 1% FBS) and incubated at 37 °C or 40 °C. Moreover, 45 min before the beginning of OCR measurement, the medium was changed into pre-warmed FAO Assay Medium (111 mM NaCl, 4.7 mM KCl, 2.0 mM $MgSO_4$, 1.2 mM Na_2HPO_4, 2.5 mM glucose, 0.5 mM carnitine, and 5 mM HEPES). Just before measurements, palmitic acid or oleic acid were added to a final concentration of 167 µM total FA (bound in a 6:1 molar ratio to BSA) as substrate. The data were normalized to cell numbers.

All Seahorse experiments were repeated at least twice. All data shown are the means ± SD of at least 3 different wells per group.

All chemicals used for these assays were obtained from Sigma Aldrich (St Louis, MO, USA).

2.3. Statistical Analysis

Statistical analysis, calculation, and visualization were performed using Prism 8 (GraphPad, San Diego, CA, USA). Analysis of correlation was carried out using a two-way analysis of variance (ANOVA) followed by Tukey's post hoc test. Significance was set at $p = 0.05$. The statistical tests chosen were predetermined by the size of the study group and the numerical range of values.

2.4. Ethical Statement

The study was conducted in accordance with the Declaration of Helsinki and was approved by the local Ethical Committee of the Medical Faculty of Martin Luther University of Halle-Wittenberg (Project identification codes 215/20.01.10/3 and 2020-019).

3. Results

3.1. Metabolic Function Measurements with the Seahorse XF96 Cell Analyzer

3.1.1. Mito Stress Test

Without any additional FAs, almost all important parameters of mitochondrial function, such as basal respiration (BR), maximal respiration (MR) and spare respiratory capacity (SRC) were significantly lower in patients compared to controls. Only the ATP linked respiration rate (ATP-R) was not significantly different in patients and controls (Figure 1, Table 2).

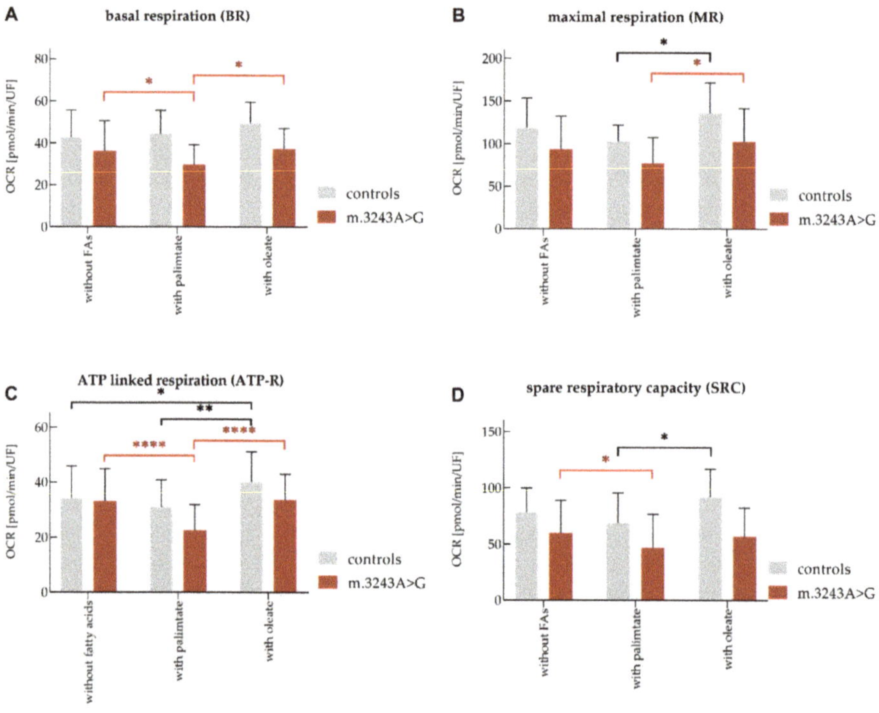

Figure 1. Evaluation of mitochondrial function using a Seahorse XF96 Cell Analyzer in myoblasts from patients ($n = 5$) and controls ($n = 5$). Key parameters of mitochondrial function were analyzed as described in methods. (**A**) basal respiration (BR), (**B**) maximal respiration (MR), (**C**) ATP linked respiration (ATP-R), and (**D**) spare respiratory capacity (SRC). In some experiments, FAs oleate or palmitate were added as excessive nutrients. The significant differences between patients and controls are presented in Table 2. The significance between values without FAs or after treatment with them is only shown if there are significant differences: * = p-values ≤ 0.05, ** = p-values ≤ 0.01, **** = p-values ≤ 0.0001. OCR: oxygen consumption rate, FAs: Fatty acids, UF: unit fluorescence/well.

Table 2. Comparison of the mean values of key parameters for mitochondrial function (basal, MR, ATP production, and SRC) in myoblasts of patients (n = 5) and controls (n = 5) with or without 300 µM palmitate or oleate. Significant differences between patients and controls are shown if $p \leq 0.05$.

	Mito Stress Test								
				+Palmitate			+Oleate		
	Controls (Mean)	Patients (Mean)	*p*-Value	Controls (Mean)	Patients (Mean)	*p*-Value	Controls (Mean)	Patients (Mean)	*p*-Value
Basal	42.7	36.3	0.004	44.5	29.7	<0.0001	49.6	37.3	0.0006
MR	118.9	93.6	<0.0001	102.6	76.7		136	102.7	0.002
SRC	78.6	59.9	0.0003	68.8	46.8	0.006	92.3	56.9	<0.0001
ATP	34.2	33.2		30.9	22.6	0.009	40.2	33.7	0.04

3.1.2. The Effect of FFAs on Mitochondrial Respiration

Differences between Patients and the Controls

After treatment of myoblasts with palmitate or oleate, the key respiratory parameters BR and SRC were still significantly higher in controls compared to patients. MR was significantly higher in controls than in patients only upon addition of oleate (Table 2).

The Effect of Palmitate

The treatment of patient's myoblasts with palmitate for 16 h prior to measurement led to a significant reduction of BR, SRC, and ATP-R values compared to the values without FAs. In contrast, controls showed similar BR, MR, SRC, and ATP-R values to those without FAs after treatment with palmitate (Figure 1, Table 3).

Table 3. The effect of treatment of patients (n = 5) and controls (n = 5) myoblasts with 300 µM palmitate or oleate conjugated with BSA on key parameters for mitochondrial function (basal, MR, ATP production and SRC). Significant differences between values obtained with or without FAs are presented if $p \leq 0.05$. *p*.: *p*-values, –FAs: without Fatty acids, Pa: palmitate. O: oleate.

	Controls							
	–FAs (Mean)	Pa (Mean)	*p*. –FAs/Pa (Mean)	O (Mean)	*p*. –FAs/O (Mean)	Pa (Mean)	O (Mean)	*p*. Pa/O (Mean)
Basal	42.7	44.5		49.6		44.5	49.6	
MR	118.9	102.6		136		102.6	136	0.01
SRC	78.6	68.9		92.3		68.9	92.3	0.01
ATP	34.2	30.9		40.2	0.02	30.9	40.2	0.003
	Patients							
	–FAs (Mean)	Pa (Mean)	*p*. –FAs/Pa (Mean)	O (Mean)	*p*. –FAs/O (Mean)	Pa (Mean)	O (Mean)	*p*. Pa/O (Mean)
Basal	36.3	29.7	0.03	37.3		29.7	37.3	0.04
MR	93.6	76.7		102.7		76.7	102.7	0.02
SRC	59.9	46.8	0.04	56.9		46.8	56.9	
ATP	33.2	22.6	<0.0001	33.7		22.6	33.7	<0.0001

The Effect of Oleate

The treatment of cells with oleate only led to a significant increase of ATP-R in controls compared to the values without FAs. However, BR, MR, and SRC remained unchanged (Figure 1, Table 3).

BR was significantly higher in patients in presence of oleate compared to those in presence of palmitate. MR and ATP-R were higher in both patients and controls after treatment of the cells with oleate, compared to values in presence of palmitate (Table 3).

3.1.3. Fatty Acid Oxidation (FAO)

The Difference between Patients and the Controls

Independent of the applied substrate (oleate or palmitate), MR, ATP-R, and SRC were similar in patients and controls at 37 °C and 40 °C. BR was higher in controls than in patients, however, only at 37 °C and in presence of palmitate (Figure 2, Table 4). The increase of temperature from 37 °C to 40 °C led to a significant decrease in all key parameters (BR, MR, ATP-R, and SRC) in both, patients and controls (Table 5).

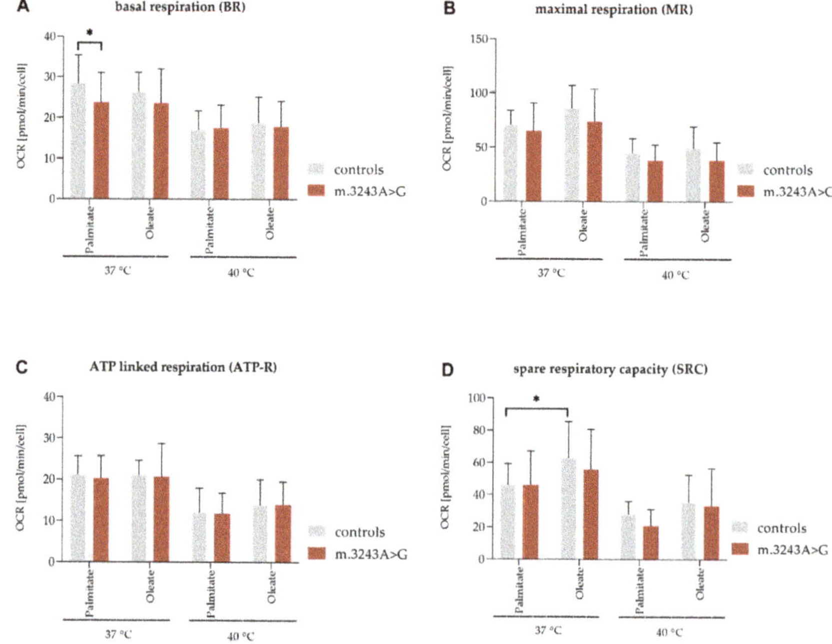

Figure 2. Evaluation of ß-oxidation using a Seahorse XF96 Cell Analyzer in myoblasts from patients (n = 5) and controls (n = 5). The measurements were performed at 37 °C or 40 °C as described in methods. (**A**) basal respiration (BR), (**B**) maximal respiration (MR), (**C**) ATP linked respiration (ATP-R), and (**D**) spare respiratory capacity (SRC) by use of palmitate or oleate as substrate. BR, MR, ATP-R, and SRC (key parameters) were significantly reduced upon increased temperature from 37 °C to 40 °C in both patients and controls (significance shown in Table 5) * = p-values ≤ 0.05. OCR: oxygen consumption rate, UF: unit fluorescence/well.

Table 4. Measurement of ß-oxidation using a Seahorse XF96 Cell Analyzer in myoblasts of patients (n = 5) and controls (n = 5) with palmitate or oleate as substrate at 37 °C and 40 °C. Significance was set at p = 0.05. FAO: fatty acid oxidation.

	FAO 37 °C					
	Palmitate as Substrate			**Oleate as Substrate**		
	Controls (Mean)	**Patients (Mean)**	***p*-Value**	**Controls (Mean)**	**Patients (Mean)**	***p*-Value**
Basal	28.4	23.7	0.04	26.3	23.5	
MR	70.7	64.9		85.9	73.7	
SRC	46	45.7		62.8	55.5	
ATP	21.1	20.2		21.1	20.6	

Table 4. *Cont.*

FAO 40 °C						
	Palmitate as Substrate			Oleate as Substrate		
	Controls (Mean)	Patients (Mean)	*p*-Value	Controls (Mean)	Patients (Mean)	*p*-Value
Basal	17.2	17.5		18.8	17.9	
MR	45.1	38		49.5	44.3	
SRC	28	20.8		35.5	33.3	
ATP	12	11.7		13.9	13.9	

Table 5. Evaluation of ß-oxidation using a Seahorse XF96 Cell Analyzer in myoblasts of patients ($n = 5$) and controls ($n = 5$) with palmitate or oleate as substrate. The difference between values obtained at 37 °C and 40 °C is shown if the results were significant (*p*-values ≤ 0.05).

Controls						
	Palmitate as Substrate			Oleate as Substrate		
	37 °C (Mean)	40 °C (Mean)	*p*-Value	37 °C (Mean)	40 °C (Mean)	*p*-Value
Basal	28.4	17.2	<0.0001	26.3	18.8	0.005
MR	70.7	45.1	0.006	85.9	49.5	<0.0001
SRC	46	28	0.05	62.8	35.5	0.0002
ATP	21.1	12	<0.0001	21.1	13.9	0.001
Patients						
	Palmitate as Substrate			Oleate as Substrate		
	37 °C (Mean)	40 °C (Mean)	*p*-Value	37 °C (Mean)	40 °C (Mean)	*p*-Value
Basal	23.7	17.5	0.02	23.5	17.9	0.04
MR	64.9	38	0.0006	73.7	44.3	0.0003
SRC	45.7	20.8	0.0006	55.5	33.3	0.003
ATP	20.2	11.7	<0.0001	20.6	13.9	0.004

The Difference between Palmitate and Oleate

The important parameters BR, MR, ATP-R, and SRC were mostly similar with palmitate or oleate as substrate. Only SRC was higher in controls in the presence of oleate as substrate compared to the value resulted from palmitate as substrate at 37 °C (Table 6).

Table 6. Evaluation of ß-oxidation using a Seahorse XF96 Cell Analyzer at 37 °C and 40 °C in myoblasts of patients ($n = 5$) and controls ($n = 5$) with palmitate or oleate as substrate. The difference between values obtained using palmitate or oleate as substrate is shown if the results were significant (*p*-values ≤ 0.05), Pa: palmitate. O: oleate.

FAO 37 °C						
	Controls			Patients		
	Pa (Mean)	O (Mean)	*p*-Value	Pa (Mean)	O (Mean)	*p*-Value
Basal	28.4	26.3		23.7	23.5	
MR	70.7	85.9		64.9	73.7	
SRC	46	62.8	0.04	45.7	55.5	
ATP	21.1	21.1		20.2	20.6	
FAO 40 °C						
	Controls			Patients		
	Pa (Mean)	O (Mean)	*p*-Value	Pa (Mean)	O (Mean)	*p*-Value
Basal	17.2	18.8		17.5	17.9	
MR	45.1	49.5		38	44.3	
SRC	28	35.5		20.8	33.3	
ATP	12	13.9		11.7	13.9	

4. Discussion

Mutations of the mitochondrial DNA (mtDNA) involve different organs. These genetic changes can affect the translation of mtDNA-encoded proteins, including respiratory chain complexes [15,16]. The m.3243A>G mutation has been described to be of strong diabetogenic nature [17]. Elevated FFAs are discussed to be a major cause of insulin resistance in skeletal muscle and liver [18].

It is generally accepted, that saturated FAs like palmitate induce insulin resistance, whereas the monounsaturated ones like oleate increase insulin sensitivity in diabetic patients and healthy individuals [4]. It has been shown, that oleate, but not palmitate, increased the expression of genes related to the FAO pathway in a sirtuin (Sirt)1-peroxisome proliferator-activated receptor gamma coactivator (PGC)-1α dependent manner. This, in turn, led to an increase in complete FA oxidation in mice skeletal muscle [19]. There are other reports about the preventive effect of oleate on saturated-fatty-acid-induced endoplasmic reticulum (ER) stress, inflammation, and insulin resistance through adenosine monophosphate-activated protein kinase (AMPK). In the above-mentioned study, oleate, in contrast to palmitate, did not increase the levels of ER stress markers, which is involved in the link between lipid-induced inflammation and insulin resistance [20].

In the present study, taking palmitate as standard substrate for FAO measurements and oleate as a comparison, the data showed generally similar ß-oxidation rates in m.3243A>G patients and controls. Both, patients and controls showed decreased parameters in this pathway at 40 °C compared to 37 °C. ATP-R was reduced to about 35–45% at 40 °C compared to that at 37 °C for both, patients and controls.

Using glucose as substrate for the Mito stress test, all key parameters of mitochondrial respiration were significantly lower in patients than in controls.

The long-chain FFAs are poorly soluble in aqueous solutions. Moreover, in vitro exposure to high levels of FFAs might lead to lipotoxicity and cellular dysfunction [21,22]. The FAs used in the present study were conjugated with BSA in all experiments (ratio FFA: BSA = 6:1).

The treatment of fibroblasts from healthy controls with palmitic acid for 16 h prior to OCR measurement has been reported to significantly increase the MR and SRC for about 20% and 45%, accordingly [23]. However, oleic acid displayed no effect on OCR in the above-mentioned study. In the present study, the extra addition of FAs oleate and palmitate affected the OCR key parameters in Mito stress test differently. While oleate generally showed positive effects on OXPHOS values, palmitate had an impaired them. Considering the values obtained in presence of FAs excess, oleate seemed to show a slightly positive effect on respiratory factors in patients. After addition of oleate, ATP-R was about 30% higher in controls compared to the values with palmitate. In contrast, the negative effect of palmitate excess was significant in patients. MR and ATP-R were reduced about 17% and 30% in patients after addition of palmitate, accordingly.

The different effects in myoblasts in presence of oleate and palmitate could be explained as follows: before performance of FAO measurements, the cells are cultured for 16 h in a nutrient restricted medium (Section 2.2.2). Thus, the cells are forced to take the FAs as the only present substrate to survive. On the other hand, by Mito stress test, the cells have sufficient nutrients and the added FAs (oleate and palmitate) count as excessive substrates. FAs are the major energy source for skeletal muscle. However, the balance between energy demand, uptake, and β-oxidation of FAs should be regulated. An imbalance between fatty acid uptake and β-oxidation might lead to an insulin resistance [24]. The elevated lipid levels, which exceed the cell's capacity to store or utilize FAs, can, as well, lead to a lipotoxic response to activate stress pathways and apoptosis. For ß-oxidation in mitochondria, both, saturated and unsaturated long-chain FAs are used as substrate. However, only long-chain saturated fatty acyl CoAs serve as substrates for de novo ceramide synthesis, which is involved in initiation of apoptosis [25].

A reduction in palmitic acid, but not in oleic acid oxidation, has been reported in myotubes of patients with diabetes type II compared to controls [26]. It has been shown that monounsaturated fatty acids, such as oleic acid, are metabolized and then accumulated in the form of low-toxic triacylglycerol (TAG). However, a large amount of palmitate inhibits the TAG synthesis at the DAG stage, which leads to the accumulation of DAG in the cell [27,28]. Based on these findings, a substitution of palmitate and other saturated or unsaturated FFAs has been recommended in patients with diabetes type II to reduce the accumulation of DAG and TAG to not promote insulin resistance [26].

The data confirmed the reported negative effect of palmitate on respiratory function of cells. This negative effect was more pronounced in m.3243A>G patients, which led to approximately 30% reduction of ATP production. The controls seemed to benefit slightly more than patients from the positive effect of oleate clearing the excess of saturated FAs via increasing the FA Oxidation. This counteracts with inflammation and insulin resistance in skeletal muscle [19]. Nevertheless, this positive effect should be noticed for keeping the balance of the body weight in over nutrition or pathological states, such as mitochondrial disorders.

An induced mitochondrial reactive oxygen species (ROS) production, mitochondrial dysfunction, and insulin resistance in skeletal muscle cells have been reported upon supplementation of palmitate, but not of oleate. Oleate should even have a protective benefit against palmitate-induced insulin resistance and might enhance the mitochondrial function, protecting against apoptosis, and increasing insulin sensitivity [29]. The analysis of potential apoptosis and ROS production might be evaluated in upcoming projects.

5. Conclusions

The data in the present study showed different effects of palmitate and oleate on oxygen utilization in myoblasts of both m.3243A>G patients and controls. The myoblasts, indeed, showed similar ß-oxidation values using palmitate or oleate as substrate. However, the presence of excessive palmitate showed a negative effect on respiratory rates of patients. This could confirm the already reported negative effect of excessive palmitate on mitochondrial function in other studies. Since saturated fatty acids increase the insulin resistance, these results might reveal why the patients harboring the m.3243A>G mutation have a higher risk for developing of diabetes type II.

Limitations: the study was performed on a small study population. In particular, in the case of the 70-year-old patient listed in the cohort, the aging might have an effect on mitochondrial dysfunction. However, it has to be noticed that mitochondrial diseases are rare diseases with a huge variety of mutations and the present study should be considered as a pilot one; it is not always responsible to perform a tissue/muscle biopsy. It is known that patients harboring the m.3243A>G mutation have a higher risk of developing diabetes mellitus. However, in the present study, only one patient is diagnosed with diabetes mellitus (Patient P1).

Author Contributions: Conceptualization, L.M.S., H.S., S.Z.; methodology, L.M.S., H.S., S.A.-R., A.T.; software, L.M.S., D.L.U., H.S., S.A.-R.; data curation, L.M.S., D.L.U., H.S., S.A.-R.; writing—original draft preparation, L.M.S., D.L.U.; writing—review and editing, L.M.S., H.S., S.A.-R., A.T.; D.L.U.; S.Z.; project administration, L.M.S., S.Z. All authors have read and agreed to the published version of the manuscript.

Funding: D.L.U. is funded by the Hertha-Nathorff-Programm (HNP) of the Medical University of Ulm, Germany.

Acknowledgments: We thank the Muscle Tissue Culture Collection (MTCC) at the University Hospital of Munich Muenchen for providing the samples, and Julia Emmerich for excellent technical assistance. The Muscle Tissue Culture Collection is part of the German network on muscular dystrophies (MD-NET) and the German network for mitochondrial disorders (mito-NET, 01GM1113A) funded by the German Ministry of Education and Research (BMBF, Bonn, Germany). The Muscle Tissue Culture Collection is a partner of EuroBioBank (www.eurobiobank.org) and TREAT-NMD (www.treat-nmd.eu). L.M.S., D.L.U., and S.Z. are members of the German mitoNET funded by the German Ministry of Education and Research. We thank the Center of Basic Medical Research (ZMG) of the medical school at the University of Halle-Wittenberg for providing the Seahorse device. We acknowledge the financial support within the funding program Open Access Publishing by the German Research Foundation (DFG).

Conflicts of Interest: The authors declare no conflict of interest.

References

1. Bonnefont, J.P.; Demaugre, F.; Prip-Buus, C.; Saudubray, J.M.; Brivet, M.; Abadi, N.; Thuillier, L. Carnitine palmitoyltransferase deficiencies. *Mol. Genet. Metab.* **1999**, *68*, 424–440. [CrossRef] [PubMed]

2. Henique, C.; Mansouri, A.; Fumey, G.; Lenoir, V.; Girard, J.; Bouillaud, F.; Prip-Buus, C.; Cohen, I. Increased mitochondrial fatty acid oxidation is sufficient to protect skeletal muscle cells from palmitate-induced apoptosis. *J. Biol. Chem.* **2010**, *285*, 36818–36827. [CrossRef] [PubMed]
3. Turpin, S.M.; Lancaster, G.I.; Darby, I.; Febbraio, M.A.; Watt, M.J. Apoptosis in skeletal muscle myotubes is induced by ceramides and is positively related to insulin resistance. *Am. J. Physiol. Endocrinol. Metab.* **2006**, *291*, E1341–E1350. [CrossRef] [PubMed]
4. Coll, T.; Eyre, E.; Rodríguez-Calvo, R.; Palomer, X.; Sánchez, R.M.; Merlos, M.; Laguna, J.C.; Vázquez-Carrera, M. Oleate reverses palmitate-induced insulin resistance and inflammation in skeletal muscle cells. *J. Biol. Chem.* **2008**, *283*, 11107–11116. [CrossRef]
5. Riccardi, G.; Giacco, R.; Rivellese, A. Dietary fat, insulin sensitivity and the metabolic syndrome. *Clin. Nutr.* **2004**, *23*, 447–456. [CrossRef]
6. Jarmuszkiewicz, W.; Woyda-Ploszczyca, A.; Koziel, A.; Majerczak, J.; Zoladz, J.A. Temperature controls oxidative phosphorylation and reactive oxygen species production through uncoupling in rat skeletal muscle mitochondria. *Free Radic. Biol. Med.* **2015**, *83*, 12–20. [CrossRef]
7. Tardo-Dino, P.-E.; Touron, J.; Baugé, S.; Bourdon, S.; Koulmann, N.; Malgoyre, A. The effect of a physiological increase in temperature on mitochondrial fatty acid oxidation in rat myofibers. *J. Appl. Physiol.* **2019**, *127*, 312–319. [CrossRef]
8. Montgomery, M.K.; Turner, N. Mitochondrial dysfunction and insulin resistance: An update. *Endocr. Connect.* **2015**, *4*, R1–R15. [CrossRef]
9. Bruce, C.R.; Risis, S.; Babb, J.R.; Yang, C.; Kowalski, G.M.; Selathurai, A.; Lee-Young, R.S.; Weir, J.M.; Yoshioka, K.; Takuwa, Y. Overexpression of sphingosine kinase 1 prevents ceramide accumulation and ameliorates muscle insulin resistance in high-fat diet–fed mice. *Diabetes* **2012**, *61*, 3148–3155. [CrossRef]
10. Samuel, V.T.; Petersen, K.F.; Shulman, G.I. Lipid-induced insulin resistance: Unravelling the mechanism. *Lancet* **2010**, *375*, 2267–2277. [CrossRef]
11. Lin, D.-S.; Kao, S.-H.; Ho, C.-S.; Wei, Y.-H.; Hung, P.-L.; Hsu, M.-H.; Wu, T.-Y.; Wang, T.-J.; Jian, Y.-R.; Lee, T.-H. Inflexibility of AMPK-mediated metabolic reprogramming in mitochondrial disease. *Oncotarget* **2017**, *8*, 73627. [CrossRef] [PubMed]
12. Motlagh Scholle, L.; Schieffers, H.; Al-Robaiy, S.; Thaele, A.; Dehghani, F.; Lehmann Urban, D.; Zierz, S. The Effect of Resveratrol on Mitochondrial Function in Myoblasts of Patients with the Common m. 3243A>G Mutation. *Biomolecules* **2020**, *10*, 1103. [CrossRef] [PubMed]
13. Ohkubo, K.; Yamano, A.; Nagashima, M.; Mori, Y.; Anzai, K.; Akehi, Y.; Nomiyama, R.; Asano, T.; Urae, A.; Ono, J. Mitochondrial gene mutations in the tRNALeu (UUR) region and diabetes: Prevalence and clinical phenotypes in Japan. *Clin. Chem.* **2001**, *47*, 1641–1648. [CrossRef] [PubMed]
14. Mora, M.; Angelini, C.; Bignami, F.; Bodin, A.-M.; Crimi, M.; Di Donato, J.-H.; Felice, A.; Jaeger, C.; Karcagi, V.; LeCam, Y. The EuroBioBank Network: 10 years of hands-on experience of collaborative, transnational biobanking for rare diseases. *Eur. J. Hum. Genet.* **2015**, *23*, 1116–1123. [CrossRef] [PubMed]
15. Tan, T.M.; Caputo, C.; Medici, F.; Pambakian, A.L.; Dornhorst, A.; Meeran, K.; Williams, G.R.; Khoo, B. MELAS syndrome, diabetes and thyroid disease: The role of mitochondrial oxidative stress. *Clin. Endocrinol.* **2009**, *70*, 340–341. [CrossRef] [PubMed]
16. Janssen, G.M.; Hensbergen, P.J.; van Bussel, F.J.; Balog, C.I.; Maassen, J.A.; Deelder, A.M.; Raap, A.K. The A3243G tRNALeu (UUR) mutation induces mitochondrial dysfunction and variable disease expression without dominant negative acting translational defects in complex IV subunits at UUR codons. *Hum. Mol. Genet.* **2007**, *16*, 2472–2481. [CrossRef]
17. Maassen, J.A.; M't Hart, L.; van Essen, E.; Heine, R.J.; Nijpels, G.; Tafrechi, R.S.J.; Raap, A.K.; Janssen, G.M.; Lemkes, H.H. Mitochondrial diabetes: Molecular mechanisms and clinical presentation. *Diabetes* **2004**, *53*, S103–S109. [CrossRef]
18. Boden, G. Effects of free fatty acids (FFA) on glucose metabolism: Significance for insulin resistance and type 2 diabetes. *Exp. Clin. Endocrinol. Diabetes* **2003**, *111*, 121–124. [CrossRef]
19. Lim, J.-H.; Gerhart-Hines, Z.; Dominy, J.E.; Lee, Y.; Kim, S.; Tabata, M.; Xiang, Y.K.; Puigserver, P. Oleic acid stimulates complete oxidation of fatty acids through protein kinase A-dependent activation of SIRT1-PGC1α complex. *J. Biol. Chem.* **2013**, *288*, 7117–7126. [CrossRef]

20. Salvado, L.; Coll, T.; Gomez-Foix, A.; Salmeron, E.; Barroso, E.; Palomer, X.; Vazquez-Carrera, M. Oleate prevents saturated-fatty-acid-induced ER stress, inflammation and insulin resistance in skeletal muscle cells through an AMPK-dependent mechanism. *Diabetologia* **2013**, *56*, 1372–1382. [CrossRef]
21. Oliveira, A.F.; Cunha, D.A.; Ladriere, L.; Igoillo-Esteve, M.; Bugliani, M.; Marchetti, P.; Cnop, M. In vitro use of free fatty acids bound to albumin: A comparison of protocols. *Biotechniques* **2015**, *58*, 228–233. [CrossRef] [PubMed]
22. Alsabeeh, N.; Chausse, B.; Kakimoto, P.A.; Kowaltowski, A.J.; Shirihai, O. Cell culture models of fatty acid overload: Problems and solutions. *Biochim. Biophys. Acta Mol. Cell Biol. Lipids* **2018**, *1863*, 143–151. [CrossRef] [PubMed]
23. Theunissen, T.E.J.; Gerards, M.; Hellebrekers, D.; van Tienen, F.H.; Kamps, R.; Sallevelt, S.; Hartog, E.; Scholte, H.R.; Verdijk, R.M.; Schoonderwoerd, K.; et al. Selection and characterization of palmitic acid responsive patients with an oxphos complex I defect. *Front. Mol. Neurosci.* **2017**, *10*, 336. [CrossRef]
24. Zhang, L.; Keung, W.; Samokhvalov, V.; Wang, W.; Lopaschuk, G.D. Role of fatty acid uptake and fatty acid β-oxidation in mediating insulin resistance in heart and skeletal muscle. *Biochim. Biophys. Acta Mol. Cell Biol. Lipids* **2010**, *1801*, 1–22. [CrossRef] [PubMed]
25. Brookheart, R.T.; Michel, C.I.; Schaffer, J.E. As a matter of fat. *Cell Metab.* **2009**, *10*, 9–12. [CrossRef]
26. Gaster, M.; Rustan, A.C.; Beck-Nielsen, H. Differential utilization of saturated palmitate and unsaturated oleate: Evidence from cultured myotubes. *Diabetes* **2005**, *54*, 648–656. [CrossRef] [PubMed]
27. Montell, E.; Turini, M.; Marotta, M.; Roberts, M.; Noé, V.; Ciudad, C.J.; Macé, K.; Gómez-Foix, A.M. DAG accumulation from saturated fatty acids desensitizes insulin stimulation of glucose uptake in muscle cells. *Am. J. Physiol. Endocrinol. Metab.* **2001**, *280*, E229–E237. [CrossRef]
28. Korbecki, J.; Bajdak-Rusinek, K. The effect of palmitic acid on inflammatory response in macrophages: An overview of molecular mechanisms. *Inflamm. Res.* **2019**, *68*, 915–932. [CrossRef]
29. Yuzefovych, L.; Wilson, G.; Rachek, L. Different effects of oleate vs. palmitate on mitochondrial function, apoptosis, and insulin signaling in L6 skeletal muscle cells: Role of oxidative stress. *Am. J. Physiol. Endocrinol. Metab.* **2010**, *299*, E1096–E1105. [CrossRef]

Article

Aging Promotes Mitochondria-Mediated Apoptosis in Rat Hearts

Mi-Hyun No [1], Youngju Choi [2], Jinkyung Cho [2], Jun-Won Heo [1,2], Eun-Jeong Cho [1,2], Dong-Ho Park [1,2], Ju-Hee Kang [2,3], Chang-Ju Kim [4], Dae Yun Seo [5], Jin Han [5] and Hyo-Bum Kwak [1,2,*]

1 Department of Biomedical Science, Program in Biomedical Science & Engineering, Department of Kinesiology, Inha University, Incheon 22212, Korea; 77nodaji@hanmail.net (M.-H.N.); gjwnsdnjs03@naver.com (J.-W.H.); cejeong97@naver.com (E.-J.C.); dparkosu@inha.ac.kr (D.-H.P.)
2 Institute of Sports and Arts Convergence, Inha University, Incheon 22212, Korea; choiyoungju0323@gmail.com (Y.C.); lovebuffalo@gmail.com (J.C.); johykang@inha.ac.kr (J.-H.K.)
3 Department of Pharmacology, College of Medicine, Inha University, Incheon 22212, Korea
4 Department of Physiology, College of Medicine, Kyung Hee University, Seoul 02447, Korea; changju@khu.ac.kr
5 National Research Laboratory for Mitochondrial Signaling, Department of Physiology, College of Medicine, Cardiovascular and Metabolic Disease Center, Inje University, Busan 47392, Korea; sdy925@gmail.com (D.Y.S.); phyhanj@gmail.com (J.H.)
* Correspondence: kwakhb@inha.ac.kr; Tel.: +82-32-860-8183; Fax: +82-32-860-8188

Received: 20 July 2020; Accepted: 4 September 2020; Published: 5 September 2020

Abstract: Aging represents a major risk for developing cardiac disease, including heart failure. The gradual deterioration of cell quality control with aging leads to cell death, a phenomenon associated with mitochondrial dysfunction in the heart. Apoptosis is an important quality control process and a necessary phenomenon for maintaining homeostasis and normal function of the heart. However, the mechanism of mitochondria-mediated apoptosis in aged hearts remains poorly understood. Here, we used male Fischer 344 rats of various ages, representing very young (1 month), young (4 months), middle-aged (12 months), and old (20 months) rats, to determine whether mitochondria-mediated apoptotic signals and apoptosis in the left ventricle of the heart are altered notably with aging. As the rats aged, the extramyocyte space and myocyte cross-sectional area in their left ventricle muscle increased, while the number of myocytes decreased. Additionally, mitochondrion-mediated apoptotic signals and apoptosis increased remarkably during aging. Therefore, our results demonstrate that aging promotes remarkable morphological changes and increases the degree of mitochondrion-mediated apoptosis in the left ventricle of rat hearts.

Keywords: aging heart; Bcl-2 family; mitochondria; programmed cell death

---

1. Introduction

Aging is commonly characterized by a gradual deterioration of tissue and organ function, and has been identified as the main cause of cardiac dysfunction. Several factors, including mitochondrial dysfunction, genomic instability, defective proteostasis, and epigenetic changes, have been considered important contributors to aging [1,2]. In particular, aging causes significant alterations in the structure and function of the heart, especially in the left ventricle, in addition to increasing cardiac oxidative stress and inflammation [1,3]. Impaired cardiac structure and function with aging result in enhanced susceptibility to cardiovascular diseases (such as heart failure), which can contribute to morbidity and mortality.

Mitochondria are the energy power plants of cells, controlling their fate. They are an important player in various processes and have the ability to maintain metabolic homeostasis and manage aging-related mechanisms [2]. Several studies suggested that mitochondria have evolved to regulate other cellular functions, including those contributing to biological aging, such as the generation of reactive oxygen species (ROS), inflammation, senescence, and resultant necrotic and apoptotic cell death [4]. Our studies have revealed that aging mainly results in damage to the electron transport chain during mitochondrial respiration, due to a decrease in electron transport by complex II-IV, which can promote electron leakage and ROS production [5].

Apoptosis, which is known as programmed cell death, is an essential process for cellular and tissue homeostasis, maintaining cell growth and differentiation, and controlled tissue repair [4,6]. However, excessive apoptosis in the heart induces cardiac dysfunction [7]. In particular, it was reported that reduced heart muscle fiber mass due to apoptosis directly results in the development of myocardial contraction abnormalities, heart failure, and other cardiac conditions [8,9]. The mitochondrion-driven apoptotic pathway, regulated by members of the Bcl-2 family, plays a pivotal regulatory role in aging [10]. An imbalance between the pro-apoptotic proteins Bax and Bid and the anti-apoptotic proteins Bcl-2 and Bcl-xl provokes the opening of the mitochondrial permeability transition pore (mPTP), which, in turn, triggers the translocation of cytochrome c from the mitochondrial intermembrane to the cytoplasm [11]. The released cytochrome c binds to adenosine triphosphate, apoptotic protease-activating factor 1, and pro-caspase 9, which subsequently activates caspase-3, thereby causing DNA fragmentation [12–14]. Data from our group and other studies have shown that aging alters the mitochondrial apoptotic pathway in the skeletal and cardiac muscles of rats, resulting in increased Bax/Bcl-2 ratio, cleaved caspase-3 levels, and DNA fragmentation [14–16]. However, Nitahara et al. [17] reported that aging has no effect on mitochondria-dependent apoptosis, in particular, the expression of Bax and Bcl-2, in the rat heart. Nevertheless, aging induced cardiomyocyte apoptosis. Despite these findings, the effects of aging on the regulation of mitochondria-mediated apoptosis in the rat heart remain unclear.

In the present study, we aimed to determine whether mitochondria-mediated apoptosis on cardiac muscle would induce greater changes by aging. To distinguish the alterations that occur during biological aging (i.e., growth, development, and aging phases), we investigated the changes in the myocardial structure and mitochondria-mediated apoptotic signaling in the cardiac muscles of male Fischer 344 rats during the four stages of life, based on the lifespan characteristics of this rat strain, i.e., very young (VYG; 1 month old), young (YG; 4 months old), middle-aged (MG; 10 months old), and old (OG; 20 months old). We hypothesized that aging would induce more drastic changes in myocardial structure and apoptosis mediated through the mitochondrial pathway—including alterations in the levels of Bcl-2 family proteins, mPTP opening, cleaved caspase-3, and DNA fragmentation—that correspond to the growth and development phases of the rat heart.

2. Materials and Methods

2.1. Animal Experiments and Ethical Approval

Forty male Fischer 344 rats were randomly divided into four groups according to age: very young (VYG, 1 month old; n = 10), young (YG, 4 months old; n = 10), middle-aged (MG, 10 months old; n = 10), and old (OG, 20 months old; n = 10). All rats were housed (two per cage) under standardized conditions (20 ± 2 °C; 12 h light:dark cycle). Standard laboratory chow (LabDiet 5L79, Orient Bio, Gyeonggi-Do, Korea) and water were provided ad libitum. The experimental protocol was approved by the Institutional Animal Care and Use Committee of the Kyung Hee University (approval number: KHUASP [SE]-17-089), and was performed in accordance with the animal care guidelines of the ethics committee.

2.2. Tissue Preparation

Frozen left ventricle (LV) tissues were homogenized in CETi lysis buffer (pH 7.6; TransLab, Korea) containing a non-ionic detergent, protease inhibitors (4 mM AEBSF, 1 μg/mL benzamidine, 1 μg/mL leupeptin, 1 μg/mL pepstatin, 1 mM EDTA, and 1 mM EGTA), and phosphatase inhibitors (1 mM sodium fluoride, 1 mM sodium orthovanadate, 1 mM beta-glycerophosphate, and 2.5 mM sodium pyrophosphate). The LV tissues were prepared with the OMNI TH115 Tissue Homogenizer (OMNI International, Kennesaw, GA, USA). The samples were centrifuged at 10,000× *g* twice (20 and 10 min, respectively) at 4 °C, and the supernatants were collected for protein analysis.

2.3. Hematoxylin and Eosin Staining

LV tissues were first subjected to paraffin-embedding; then, the paraffin-embedded tissue blocks were cut into 5-μm-thick cross-sections, and the sections were placed on glass slides. Next, the sections were deparaffinized by xylene treatment and hydrated in running distilled water for at least 2 min. The samples were stained with hematoxylin for 5 min, rinsed with running tap water, and stained with eosin solution for 2 min. The slides were then rinsed with absolute alcohol. They were then mounted using synthetic resin and dried overnight at room temperature. Images of the section were captured on a microscope of Axioplan 2 (Carl Zeiss, Jena, Germany) and the images were quantified using the ImageJ analysis program (NIH, Bethesda, MD, USA). Two sections of LV in 4 rats per group were analyzed for percentage of extramyocyte space, number of myocytes, and cross-sectional area (CSA). Each section was targeted to the endocardial region. The average myocyte count of the sections of the group was calculated per 100,000 μm^2 area, and the mean myocyte CSA for the left ventricular histological section is square micrometers.

2.4. Western Immunoblotting

The levels of proteins involved in apoptotic signaling—Bax, Bcl-2, and cleaved caspase-3—were determined via Western blotting. Total protein (20 μg) from homogenized LV tissues was denatured at 95 °C for 5 min. The samples were loaded onto 10–12% SDS–polyacrylamide gels and resolved by electrophoresis at 110 V for 2 h. The proteins were electro-transferred onto a nitrocellulose membrane (Pall Corporation, Port Washington, NY, USA) for 1 h at 260 mA. Thereafter, the membranes were stained with Ponceau S (Sigma-Aldrich, St. Louis, MO, USA) to ensure equal loading and protein transfer of all samples. The membranes were blocked with 5% skim milk in TBS buffer with 0.1% Tween-20 (TBST) for 2 h at room temperature, washed thrice for 10 min with TBST, and incubated at 4 °C for 12 h, with the following primary antibodies (diluted in TBS): anti-Bax (1:1000; Santa Cruz Biotechnology, Dallas, TX, USA), anti-Bcl-2 (1:1000; Santa Cruz Biotechnology), and anti-cleaved caspase-3 antibodies (1:500; Cell Signaling Technology, Beverly, MA, USA). The membranes were then washed thrice for 10 min with TBST. Thereafter, the membranes were slowly incubated on a shaker for 1 h with horseradish peroxidase-conjugated anti-mouse or anti-rabbit secondary antibodies (1:3000 in TBST; Santa Cruz) at room temperature. After washing with TBST, protein bands were developed using an enhanced chemiluminescence detection kit (Thermo Scientific, Waltham, MA, USA). Relative protein quantification was performed by densitometry analysis using the Image-Pro Plus software (Media Cybernetics, Rockville, MD, USA).

2.5. Mitochondrial Permeability Transition Pore (mPTP) Opening Sensitivity

The mitochondrial Ca^{2+}-retention capacity is commonly used to evaluate the susceptibility of mPTP opening sensitivity, as previously reported [5]. The mPTP opening was determined through the Ca^{2+}-retention capacity graph. The value of the Ca^{2+}-retention capacity of the control group (VYG) was calculated indirectly by dividing the other groups, and comparisons were expressed as percentage values.

2.6. Immunohistochemistry

LV tissues were first embedded in paraffin; these paraffin blocks were cut into 5-μm-thick sections, which were placed onto slides and washed thrice for 3 min with phosphate buffered saline (PBS) solution and incubated in 3% hydrogen peroxide for 30 min. Afterwards, the slides were washed thrice with PBS and incubated in PBS with 10% goat serum and 1% bovine serum albumin (BSA) at room temperature for 2 h. The sections were incubated overnight at 4 °C with anti-cleaved caspase-3 antibodies (1:600; Cell Signaling Technology) diluted in PBS with 1% BSA. The slides were gently washed thrice with PBS and incubated in anti-rabbit secondary antibodies (1:1000; Santa Cruz Biotechnology) at room temperature for 1 h, and washed thrice with PBS. The ABC reagent kit (1:100; Vector Laboratories, Burlingame, CA, USA) was used to amplify the protein signals, and the sections were visualized by staining with 0.03% diaminobenzidine (a chromogenic substrate) at room temperature. The slides were stained and dehydrated via a hematoxylin control method, and then cover-slipped. The cleaved caspase-3–positive cells were quantified in the LV tissue sections using the ImageJ software (version 1.52a; NIH, Bethesda, MD, USA).

2.7. Terminal Deoxynucleotidyl Transferase-Mediated dUTP Nick-End Labeling (TUNEL Assay)

To identify DNA fragmentation, we performed TUNEL staining using an ApopTag Plus Peroxidase In Situ Apoptosis Detection Kit (Intergen Company, Purchase, NY, USA), according to the manufacturer's instructions. The slides were soaked in an ethanol-acetic acid solution (2:1 v/v dilution), placed in a freezer for 5 min, and washed thrice with PBS. Sections were sequentially incubated with 0.5% Triton X-100, Protease K (100 mg/mL), 3% hydrogen peroxide, and TUNEL reaction mixture, which was used to rinse the samples after each step. The staining was performed using horseradish peroxidase-tagged antibodies and 0.03% diaminobenzidine, with counterstaining using the Nissl dye. TUNEL-positive myonuclei were counted in two sections of LV in 4 rats per group, and the averages were calculated as percentages of the total number of labeled myonuclei.

2.8. Statistical Analysis

Data were expressed as the means ± standard errors of means (SEMs). The differences among the four study groups were analyzed by one-way analysis of variance (ANOVA) with post-hoc Tukey's test. Statistical significance was considered at $p < 0.05$.

3. Results

3.1. Effects of Aging on Cardiac Muscle Morphology

LV cross-sections were stained with hematoxylin and eosin to assess their morphology. Cardiac muscle remodeling was determined by measuring extramyocyte space, extent of apparent fibrosis, myocyte cross-sectional area (CSA), and the number of myocytes per 100,000 μm^2 (Figure 1A). We found that the samples from the OG group showed more extensive cardiac muscle remodeling compared with those from the other groups. No difference in the percentage of extramyocyte space in the samples from the VYG and YG groups was observed, while the samples from the OG group showed significantly more extramyocyte space than those from the VYG, YG, and MG groups; furthermore, the samples from the MG group had more extramyocyte space than those from the VYG and YG groups (mean ± SEM; VYG: 4.48 ± 0.47%; YG: 7.13 ± 0.83%; MG: 13.51 ± 0.97%; OG: 20.85 ± 1.23%)) ($p < 0.05$ in all comparisons, Figure 1B). The number of myocytes per 100,000 μm^2 in the LV tissues was significantly lower in case of the samples from the MG and OG groups compared with those from the VYG and YG groups (mean ± SEM; VYG: 113.70 ± 6.79; YG: 115.38 ± 5.10; MG: 72.17 ± 5.83; OG: 50.86 ± 3.83)) ($p < 0.05$ in all comparisons, Figure 1C). In contrast, the mean myocyte CSA in the combined myocardial and endocardial regions was significantly higher in the OG group than in the VYG, YG, and MG groups (mean ± SEM; 377.66 ± 28.38 μm^2 vs. 120.94 ± 9.34, 134.04 ± 10.79, and 264.22 ± 18.90 μm^2, respectively) ($p < 0.05$ in all comparisons, Figure 1D), and higher in the

MG rats than in the VYG and YG animals (mean ± SEM; 264.22 ± 18.90 μm^2 vs. 120.94 ± 9.34 and 134.04 ± 10.79 μm^2, respectively) ($p < 0.05$ in all comparisons, Figure 1D).

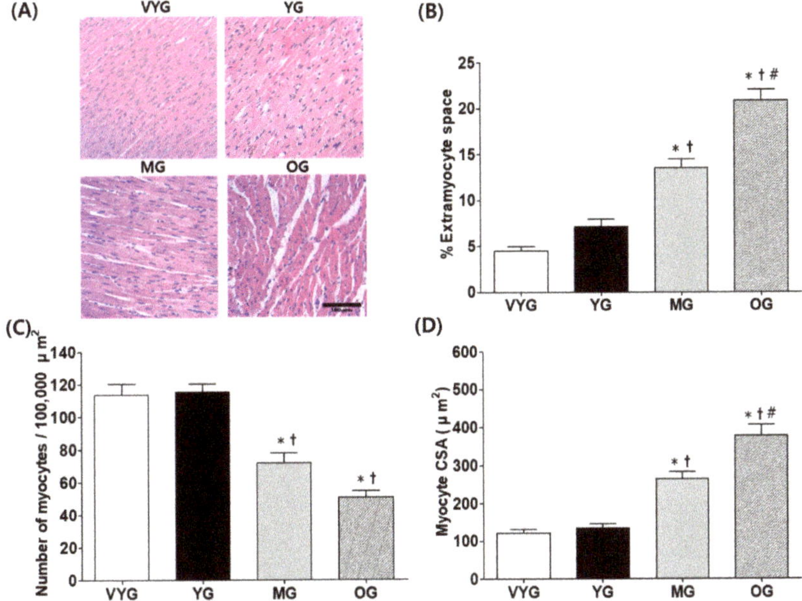

Figure 1. (**A**) Representative histological cross-sections of the left ventricle (LV) tissues of rats from the very young (VYG), young (YG), middle-aged (MG), and old (OG) groups stained with hematoxylin and eosin (amplification: 40×). Unstained areas indicate extramyocyte space. (**B**) Quantification of the percentage of extramyocyte space. (**C**) Number of myocytes per 100,000 μm^2 in the LV tissues. (**D**) Myocyte cross-sectional area (CSA) of LV tissue histological sections, in square micrometers. The scale bar indicates 100 μm. Data presented as the means ± SEMs. * $p < 0.05$ vs. VYG. † $p < 0.05$ vs. YG. # $p < 0.05$ vs. MG.

3.2. *Effects of Aging on Mitochondria-Mediated Apoptotic Signaling in Cardiac Muscles*

The expression of the pro-apoptotic Bax protein increased by 1960%, 78%, and 69% in the samples from the OG group (4.12 ± 0.35), compared with that in samples from the VYG (0.20 ± 0.04), YG (2.32 ± 0.34), and MG groups (2.44 ± 0.31), respectively ($p < 0.05$ in all comparisons, Figure 2A). The samples from the YG and MG groups showed similar Bax levels ($p > 0.05$). In contrast, the levels of the anti-apoptotic Bcl-2 protein were reduced by 80%, 82%, and 70% in the samples from the OG group (0.29 ± 0.04), compared with those in the samples from the VYG (1.46 ± 0.17), YG (1.61 ± 0.22), and MG (0.98 ± 0.08) groups, respectively ($p < 0.05$ in all comparisons, Figure 2B), while the samples from the VYG and YG groups showed similar Bcl-2 levels ($p > 0.05$). The Bax/Bcl-2 ratio, which is important during the early stage of mitochondria-mediated apoptosis, increased more significantly in the samples from the OG group than in those from the VYG, YG, and MG groups ($p < 0.05$, Figure 2C). Analysis of mPTP opening sensitivity in the VYS group was set to 100.00; the mPTP opening sensitivity was increased by 154%, 303%, and 587% in the samples from the OG group (687.19 ± 60.83), compared with that in the samples from the VYG (100.00), YG (254.16 ± 23.76), and MG groups (403.55 ± 68.45), respectively ($p < 0.05$ in all comparisons, Figure 2D). Furthermore, the mPTP opening sensitivity was higher in the samples from the MG group than in those from the VYG group ($p < 0.05$, Figure 2D), while no significant difference was seen between the mPTP opening sensitivity in the samples from the VYG and YG groups (100.00 vs. 254.16 ± 23.76; Figure 2D). Analysis of the expression of cleaved caspase-3, which is present downstream to the Bcl-2 family in the mitochondria-mediated apoptotic pathway, revealed that its levels were significantly higher in OG rats (1.05 ± 0.08) than in the VYG (0.20 ± 0.03), YG (0.54 ± 0.08), and MG (0.76 ± 0.05) rats ($p < 0.05$ in all comparisons, Figure 2E). Furthermore, the levels of cleaved caspase-3 were higher in the samples from the YG and MG groups than in those from the VYG group ($p < 0.05$ in both comparisons, Figure 2E).

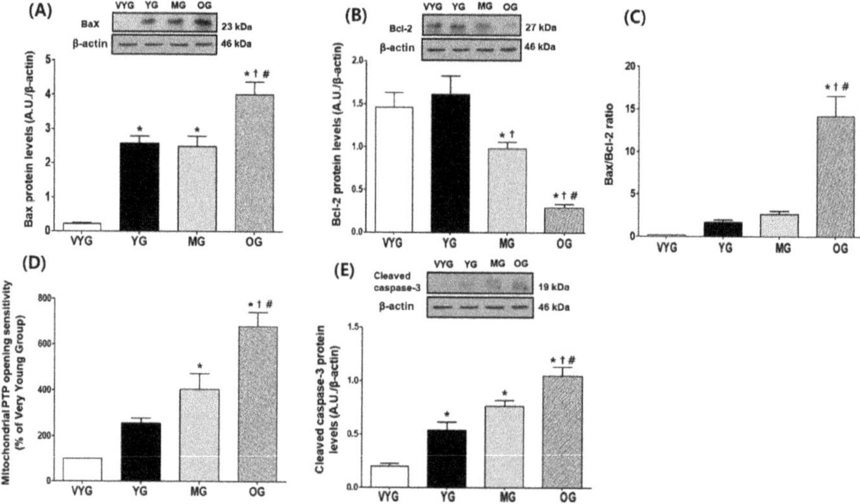

Figure 2. Mitochondria-mediated apoptotic signaling in the left ventricle (LV) tissues of rats from the very young (VYG), young (YG), middle-aged (MG), and old (OG) groups. (**A**) Bax levels detected by Western blotting with actin as the normalization control. (**B**) Representative Western blot results and relative quantification of Bcl-2 levels in LV tissues. (**C**) Representative Bax/Bcl-2 ratios in LV tissues. (**D**) Representative mPTP opening sensitivity with that of the VYG samples as the normalization control. (**E**) Cleaved caspase-3 levels detected by Western blotting, with actin as the normalization control. Data are presented as the means ± SEMs. * $p < 0.05$ vs. VYG. † $p < 0.05$ vs. YG. # $p < 0.05$ vs. MG.

3.3. *Effects of Aging on Cleaved Caspase-3-Positive Cells and TUNEL-Positive Myonuclei in Cardiac Muscles*

The number of cleaved caspase-3-positive cells increased by 250% and 185% in the samples from the OG group (20.64 ± 1.96), compared with that in samples from the VYG (5.90 ± 0.85) and YG groups (7.24 ± 1.00), respectively ($p < 0.05$ in both comparisons; Figure 3A,C). Additionally, the number of TUNEL-positive myonuclei in the cardiac muscles of OG rats was significantly higher than that in the cardiac muscles of the VYG (by 2196%), YG (by 810%), and MG (by 107%) rats (mean ± SEM; 21.35 ± 0.42 vs. 0.93 ± 0.05, 1.92 ± 0.08, and 8.46 ± 0.35, respectively) ($p < 0.05$ in all comparisons; Figure 3B,D). The number of TUNEL-positive myonuclei was higher in the samples from the MG group than in those from the VYG and YG groups ($p < 0.05$ in both comparisons, Figure 3B,D). However, the samples from the VYG and YG groups did not show significant differences in the numbers of cleaved caspase-3-positive cells and TUNEL-positive myonuclei ($p > 0.05$).

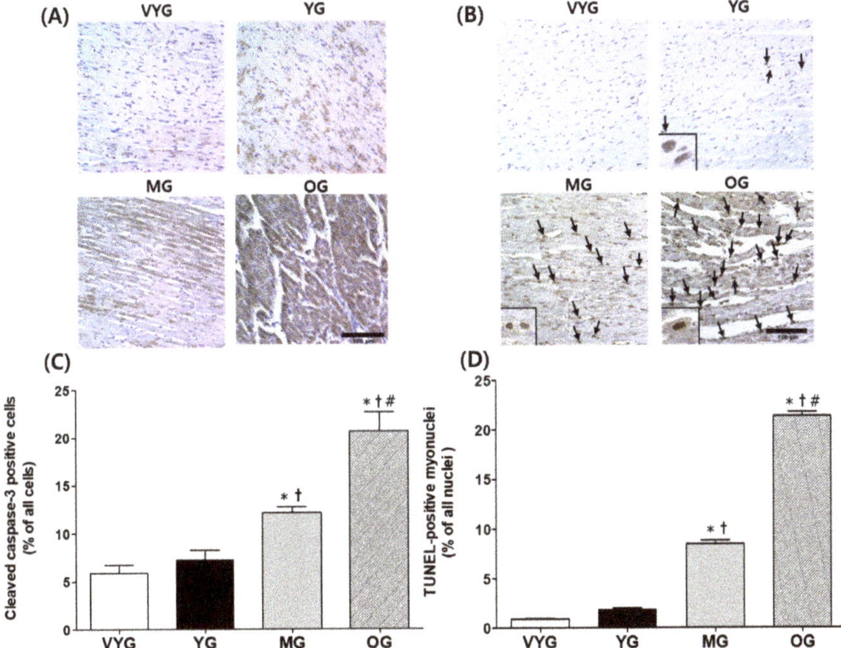

Figure 3. Cleaved caspase-3–positive cells and TUNEL-positive myonuclei in the left ventricle (LV) tissues of rats from the very young (VYG), young (YG), middle-aged (MG), and old (OG) groups (scale bar: 100 μm and magnification: 40×). (**A**) Representative photographs of LV tissue sections stained with anti-cleaved caspase-3 antibodies (**B**) TUNEL staining images in which brown-stained regions represent TUNEL-positive myonuclei. Apoptotic myonuclei are indicated by arrows which are typically enlarged in the lower left corner. (**C**) Quantification of cleaved caspase-3–positive cells via immunohistochemical staining. (**D**) Quantification of TUNEL-positive myonuclei. Data are presented as the means ± SEMs. * $p < 0.05$ vs. VYG. † $p < 0.05$ vs. YG. # $p < 0.05$ vs. MG.

4. Discussion

The main findings of this study were as follows: (i) cardiac muscle remodeling, assessed based on morphological changes in tissues, increased with advancing age; (ii) mitochondria-dependent apoptotic signaling (including Bax/Bcl-2 ratio, mPTP opening sensitivity, and cleaved caspase-3 protein levels) remarkably increased with advancing age; and (iii) apoptosis (including numbers of cleaved caspase-3-positive cells and TUNEL-positive myonuclei) also increased with advancing age. These results reveal that aging induces significant alterations in the myocardial structure and mitochondria-mediated apoptotic signaling in the rat heart, and that these changes are more drastic during the old-age phase and not in the developing-age phase.

To our knowledge, this is the first report regarding the changes in the myocardial structure and mitochondria-mediated apoptotic signaling, which occurs via Bcl-2 family proteins, in rat cardiac muscles throughout their lifespan—including each phase of growth (very young vs. young), development (young vs. middle-aged), and aging (middle-aged vs. old). As age-related changes in mitochondria-mediated apoptotic signaling are debatable, determination of such signals, as well as the myocardial structure at different ages will likely provide valuable information on the cellular and molecular mechanisms underlying aging in the heart.

Several studies using models of young and old animals have shown that cardiac remodeling and function deteriorate with age [18–20]. Consistent with previous findings, we also observed significant cardiac morphological changes, such as increased extramyocyte space and myocyte CSA, in aged rats, compared with younger animals during the growth and development phases. However, these changes seemed gradual between the development and aging phases. Aging is generally perceived as the most drastic morphological change in the cardiac muscle, with more subtle and progressive changes in the growth and development phases. Progressive aging of the heart is caused by the excessive deposition of extracellular matrix (ECM) elements, such as collagen and fibronectin, triggered by the uncontrolled activation of the fibrosis pathway and suppression of anti-fibrosis signals, which can lead to cardiac fibrosis [21]. Hence, aging may cause an excessive accumulation of ECM components in cardiomyocytes, leading to heart failure caused by pathological mechanisms, including diastolic decline and cardiac hypertrophy [22,23]. Our results suggest that aging is the main cause of increased extramyocyte space and myocyte CSA, and the reduced number of cardiomyocytes.

Mitochondrial dysfunction leads to an imbalance of Bax and Bcl-2 levels, which will, in turn, activate caspase-3, a pivotal protein involved in mitochondria-mediated apoptosis [14,24]. Oxidative stress is also known to cause the release of Bax into the cytoplasm, which will promote the mPTP opening sensitivity and the activation of caspase-9 and caspase-3. These signals eventually result in DNA fragmentation and programmed cell death [12,16,25]. We recently found increased mitochondrial hydrogen peroxide production with aging in rat cardiac muscles [5]. Interestingly, the current study revealed that the Bax/Bcl-2 ratio was markedly higher in the cardiac muscles of older rats, indicating that the aging phase was related to early-stage mitochondria-mediated apoptosis. Moreover, progressively increased mPTP opening sensitivity and the cleavage of caspase-3 were seen during the aging process, and they occurred more prominently in the aging phase. We believe that apoptosis is primarily driven by mitochondrial ROS accumulation and mPTP opening. Therefore, it is possible that aging could enhance mitochondria-mediated apoptotic signaling [26,27], which will contribute further towards cardiac muscle apoptosis.

Additionally, we found that the number of TUNEL-positive myonuclei, which is an apoptosis marker, was affected by aging, especially during the development and aging phases. Fannin et al. [28] investigated the mitochondria-mediated apoptotic pathway in aging female F344xBN rats and suggested that aging was associated with increases in the Bax/Bcl-2 ratio, caspase-3 activation, and the number of TUNEL-positive myonuclei. Apoptosis has an important homeostatic role in normal, healthy hearts; however, excessive apoptosis leads to pathological, life-threatening heart dysfunction during the aging process [29]. We did not measure the cardiac function in addition to it being relative to aging; however, apoptosis might have affected cardiac function [30]. Further studies are necessary to better understand the role of mitochondria-mediated apoptosis in aging-induced cardiac dysfunction.

The present study has some limitations. First, we did not include a histochemistry analysis of mitochondrial respiratory chain complexes and mitochondrial ultrastructure examination in heart tissues that can provide important information to prove the relationship between mitochondrial dysfunction and aging. Second, since female hormones (e.g., estrogen) regulate mitochondrial function [31], we used only male rats, including very young (1 months), young (4 months), middle-aged (10 months), and old (20 months) rats as animal models in this study.

5. Conclusions

This study showed that aging induces cardiac muscle remodeling in rats, promoting increases in extramyocyte space and the CSA and reducing the number of myocytes, which are key features involved in cardiac fibrosis. Additionally, aging induces mitochondria-mediated apoptotic signals (Bax expression, increased Bax/Bcl-2 ratio and mPTP opening sensitivity, and the activation of caspase-3) and apoptosis (TUNEL-positive myonuclei) in cardiac muscles. These results provide strong evidence supporting our hypothesis that aging would cause drastic changes in the myocardial structure, mitochondria-mediated apoptotic signaling, and DNA fragmentation in rat hearts. Nevertheless, morphological and molecular changes in cardiac muscles are substantially less prominent during the growth and development phases of the lifespan of rats.

Author Contributions: M.-H.N. and H.-B.K. contributed to the conception and design of the study. M.-H.N., J.-W.H. and E.-J.C. collected the data. M.-H.N., Y.C., J.C., D.-H.P., J.-H.K., C.-J.K., D.Y.S., J.H. and H.-B.K. conducted the critical discussion. M.-H.N., Y.C. and H.-B.K. wrote the manuscript. All authors have read and agreed to the published version of the manuscript.

Funding: This study was supported by the Ministry of Education of the Republic of Korea and the National Research Foundation of Korea (NRF-2016S1A5A8018954).

Conflicts of Interest: The authors declare no conflict of interest.

References

1. Zhang, Y.; Herman, B. Apoptosis and successful aging. *Mech. Ageing Dev.* **2002**, *123*, 563–565. [CrossRef]
2. Nilsson, M.I.; Tarnopolsky, M.A. Mitochondria and aging-the role of exercise as a countermeasure. *Biology* **2019**, *8*, 40. [CrossRef] [PubMed]
3. Higashi, Y.; Sukhanov, S.; Anwar, A.; Shai, S.Y.; Delafontaine, P. Aging, atherosclerosis, and IGF-1. *J. Gerontol. A Biol. Sci. Med. Sci.* **2012**, *67*, 626–639. [CrossRef] [PubMed]
4. Venditti, P.; Di Meo, S. The role of reactive oxygen species in the life cycle of the mitochondrion. *Int. J. Mol. Sci.* **2020**, *21*, 2173. [CrossRef] [PubMed]
5. No, M.H.; Heo, J.W.; Yoo, S.Z.; Jo, H.S.; Park, D.H.; Kang, J.H.; Seo, D.Y.; Han, J.; Kwak, H.B. Effects of aging on mitochondrial hydrogen peroxide emission and calcium retention capacity in rat heart. *J. Exerc. Rehabil.* **2018**, *14*, 920–926. [CrossRef] [PubMed]
6. Kwak, H.B. Effects of aging and exercise training on apoptosis in the heart. *J. Exerc. Rehabil.* **2013**, *9*, 212–219. [CrossRef]
7. Orogo, A.M.; Gustafsson, A.B. Cell death in the myocardium: My heart won't go on. *IUBMB Life* **2013**, *65*, 651–656. [CrossRef]
8. Kumar, D.; Jugdutt, B.I. Apoptosis and oxidants in the heart. *J. Lab. Clin. Med.* **2003**, *142*, 288–297. [CrossRef]
9. Jose Corbalan, J.; Vatner, D.E.; Vatner, S.F. Myocardial apoptosis in heart disease: Does the emperor have clothes? *Basic Res. Cardiol.* **2016**, *111*, 31. [CrossRef]
10. Martin-Fernandez, B.; Gredilla, R. Mitochondria and oxidative stress in heart aging. *Age (Dordr)* **2016**, *38*, 225–238. [CrossRef]
11. Yoo, S.Z.; No, M.H.; Heo, J.W.; Chang, E.; Park, D.H.; Kang, J.H.; Seo, D.Y.; Han, J.; Jung, S.J.; Hwangbo, K.; et al. Effects of a single bout of exercise on mitochondria-mediated apoptotic signaling in rat cardiac and skeletal muscles. *J. Exerc. Rehabil.* **2019**, *15*, 512–517. [CrossRef] [PubMed]
12. Pollack, M.; Leeuwenburgh, C. Apoptosis and aging: Role of the mitochondria. *J. Gerontol. A Biol. Sci. Med. Sci.* **2001**, *56*, B475–B482. [CrossRef] [PubMed]
13. Takemura, G.; Kanoh, M.; Minatoguchi, S.; Fujiwara, H. Cardiomyocyte apoptosis in the failing heart–a critical review from definition and classification of cell death. *Int. J. Cardiol.* **2013**, *167*, 2373–2386. [CrossRef]
14. Kwak, H.B.; Song, W.; Lawler, J.M. Exercise training attenuates age-induced elevation in Bax/Bcl-2 ratio, apoptosis, and remodeling in the rat heart. *FASEB J.* **2006**, *20*, 791–793. [CrossRef]
15. Song, W.; Kwak, H.B.; Lawler, J.M. Exercise training attenuates age-induced changes in apoptotic signaling in rat skeletal muscle. *Antioxid. Redox Signal.* **2006**, *8*, 517–528. [CrossRef] [PubMed]
16. Liu, L.; Azhar, G.; Gao, W.; Zhang, X.; Wei, J.Y. Bcl-2 and Bax expression in adult rat hearts after coronary occlusion: Age-associated differences. *Am. J. Physiol.* **1998**, *275*, R315–R322. [CrossRef] [PubMed]
17. Nitahara, J.A.; Cheng, W.; Liu, Y.; Li, B.; Leri, A.; Li, P.; Mogul, D.; Gambert, S.R.; Kajstura, J.; Anversa, P. Intracellular calcium, DNase activity and myocyte apoptosis in aging Fischer 344 rats. *J. Mol. Cell. Cardiol.* **1998**, *30*, 519–535. [CrossRef]
18. Horn, M.A.; Graham, H.K.; Richards, M.A.; Clarke, J.D.; Greensmith, D.J.; Briston, S.J.; Hall, M.C.; Dibb, K.M.; Trafford, A.W. Age-related divergent remodeling of the cardiac extracellular matrix in heart failure: Collagen accumulation in the young and loss in the aged. *J. Mol. Cell. Cardiol.* **2012**, *53*, 82–90. [CrossRef]
19. Jugdutt, B.I.; Jelani, A.; Palaniyappan, A.; Idikio, H.; Uweira, R.E.; Menon, V.; Jugdutt, C.E. Aging-related early changes in markers of ventricular and matrix remodeling after reperfused ST-segment elevation myocardial infarction in the canine model: Effect of early therapy with an angiotensin II type 1 receptor blocker. *Circulation* **2010**, *122*, 341–351. [CrossRef]

20. Bujak, M.; Kweon, H.J.; Chatila, K.; Li, N.; Taffet, G.; Frangogiannis, N.G. Aging-related defects are associated with adverse cardiac remodeling in a mouse model of reperfused myocardial infarction. *J. Am. Coll. Cardiol.* **2008**, *51*, 1384–1392. [CrossRef]
21. Bonnans, C.; Chou, J.; Werb, Z. Remodelling the extracellular matrix in development and disease. *Nat. Rev. Mol. Cell Biol.* **2014**, *15*, 786–801. [CrossRef]
22. Piek, A.; de Boer, R.A.; Sillje, H.H. The fibrosis-cell death axis in heart failure. *Heart Fail. Rev.* **2016**, *21*, 199–211. [CrossRef]
23. Dai, D.F.; Chen, T.; Johnson, S.C.; Szeto, H.; Rabinovitch, P.S. Cardiac aging: From molecular mechanisms to significance in human health and disease. *Antioxid. Redox Signal.* **2012**, *16*, 1492–1526. [CrossRef] [PubMed]
24. Xiong, S.; Mu, T.; Wang, G.; Jiang, X. Mitochondria-mediated apoptosis in mammals. *Protein Cell* **2014**, *5*, 737–749. [CrossRef] [PubMed]
25. Antonsson, B. Mitochondria and the Bcl-2 family proteins in apoptosis signaling pathways. *Mol. Cell. Biochem.* **2004**, *256–257*, 141–155. [CrossRef] [PubMed]
26. Redza-Dutordoir, M.; Averill-Bates, D.A. Activation of apoptosis signalling pathways by reactive oxygen species. *Biochim. Biophys. Acta* **2016**, *1863*, 2977–2992. [CrossRef]
27. Smith, M.A.; Schnellmann, R.G. Calpains, mitochondria, and apoptosis. *Cardiovasc. Res.* **2012**, *96*, 32–37. [CrossRef]
28. Fannin, J.; Rice, K.M.; Thulluri, S.; Arvapalli, R.K.; Wehner, P.; Blough, E.R. The effects of aging on indices of oxidative stress and apoptosis in the female fischer 344/Nnia X Brown Norway/BiNia rat heart. *Open Cardiovasc. Med. J.* **2013**, *7*, 113–121. [CrossRef]
29. Roh, J.; Rhee, J.; Chaudhari, V.; Rosenzweig, A. The role of exercise in cardiac aging: From physiology to molecular mechanisms. *Circ. Res.* **2016**, *118*, 279–295. [CrossRef]
30. Lesnefsky, E.J.; Chen, Q.; Hoppel, C.L. Mitochondrial metabolism in aging heart. *Circ. Res.* **2016**, *118*, 1593–1611. [CrossRef]
31. Klinge, C.M. Estrogenic control of mitochondrial function. *Redox Biol.* **2020**, *31*, 101435. [CrossRef] [PubMed]

MDPI
St. Alban-Anlage 66
4052 Basel
Switzerland
Tel. +41 61 683 77 34
Fax +41 61 302 89 18
www.mdpi.com

Life Editorial Office
E-mail: life@mdpi.com
www.mdpi.com/journal/life

www.ingramcontent.com/pod-product-compliance
Lightning Source LLC
LaVergne TN
LVHW071622170726
843515LV00009B/2455

* 9 7 8 3 0 3 6 5 4 6 4 8 3 *